Arithmetic of *L*-functions

Ω IAS/PARK CITY MATHEMATICS SERIES

Volume 18

Arithmetic of L-functions

Cristian Popescu
Karl Rubin
Alice Silverberg
Editors

American Mathematical Society
Institute for Advanced Study

IAS/Park City Mathematics Institute runs mathematics education programs that bring together high school mathematics teachers, researchers in mathematics and mathematics education, undergraduate mathematics faculty, graduate students, and undergraduates to participate in distinct but overlapping programs of research and education. This volume contains the lecture notes from the Graduate Summer School program.

2010 *Mathematics Subject Classification.* Primary 11G40, 11R42, 14G10.

Library of Congress Cataloging-in-Publication Data

Arithmetic of L-functions / Cristian Popescu, Karl Rubin, Alice Silverberg, editors.
p. cm. — (IAS/Park City mathematics series ; v. 18)
Includes bibliographical references.
ISBN 978-0-8218-5320-7 (alk. paper)
1. L-functions. 2. Number theory. I. Popescu, Cristian, 1964– II. Rubin, Karl III. Silverberg, Alice.

QA246.A75 2011
512.7′3—dc23

2011020548

♾ The paper used in this book is acid-free and falls within the guidelines
established to ensure permanence and durability.
Visit the AMS home page at `http://www.ams.org/`

10 9 8 7 6 5 4 3 2 1 16 15 14 13 12 11

Contents

Karl Rubin
Euler Systems and Kolyvagin Systems 449

Preface

The IAS/Park City Mathematics Institute (PCMI) was founded in 1991 as part of the "Regional Geometry Institute" initiative of the National Science Foundation. In mid 1993 the program found an institutional home at the Institute for Advanced Study (IAS) in Princeton, New Jersey.

The IAS/Park City Mathematics Institute encourages both research and education in mathematics and fosters interaction between the two. The three-week summer institute offers programs for researchers and postdoctoral scholars, graduate students, undergraduate students, high school teachers, undergraduate faculty, and researchers in mathematics education. One of PCMI's main goals is to make all of the participants aware of the total spectrum of activities that occur in mathematics education and research: we wish to involve professional mathematicians in education and to bring modern concepts in mathematics to the attention of educators. To that end the summer institute features general sessions designed to encourage interaction among the various groups. In-year activities at the sites around the country form an integral part of the High School Teachers Program.

Each summer a different topic is chosen as the focus of the Research Program and Graduate Summer School. Activities in the Undergraduate Summer School deal with this topic as well. Lecture notes from the Graduate Summer School are being published each year in this series. The first eighteen volumes are:

- Volume 1: *Geometry and Quantum Field Theory* (1991)
- Volume 2: *Nonlinear Partial Differential Equations in Differential Geometry* (1992)
- Volume 3: *Complex Algebraic Geometry* (1993)
- Volume 4: *Gauge Theory and the Topology of Four-Manifolds* (1994)
- Volume 5: *Hyperbolic Equations and Frequency Interactions* (1995)
- Volume 6: *Probability Theory and Applications* (1996)
- Volume 7: *Symplectic Geometry and Topology* (1997)
- Volume 8: *Representation Theory of Lie Groups* (1998)
- Volume 9: *Arithmetic Algebraic Geometry* (1999)
- Volume 10: *Computational Complexity Theory* (2000)
- Volume 11: *Quantum Field Theory, Supersymmetry, and Enumerative Geometry* (2001)
- Volume 12: *Automorphic Forms and their Applications* (2002)
- Volume 13: *Geometric Combinatorics* (2004)
- Volume 14: *Mathematical Biology* (2005)
- Volume 15: *Low Dimensional Topology* (2006)
- Volume 16: *Statistical Mechanics* (2007)
- Volume 17: *Analytic and Algebraic Geometry: Common Problems, Different Methods* (2008)
- Volume 18: *Arithmetic of L-functions*

Volumes are in preparation for subsequent years.

Some material from the Undergraduate Summer School is published as part of the Student Mathematical Library series of the American Mathematical Society. We hope to publish material from other parts of the IAS/PCMI in the future. This will include material from the High School Teachers Program and publications documenting the interactive activities which are a primary focus of the PCMI. At the summer institute late afternoons are devoted to seminars of common interest to all participants. Many deal with current issues in education: others treat mathematical topics at a level which encourages broad participation. The PCMI has also spawned interactions between universities and high schools at a local level. We hope to share these activities with a wider audience in future volumes.

John C. Polking
Series Editor
July 2011

Introduction

Cristian Popescu, Karl Rubin, and Alice Silverberg

IAS/Park City Mathematics Series
Volume 18, 2009

Introduction

Cristian Popescu, Karl Rubin, and Alice Silverberg

The 2009 Graduate Summer School of the Institute for Advanced Study/Park City Mathematics Institute was held in Park City, Utah from June 29 to July 17. The topic was "Arithmetic of L-functions". This volume contains the written versions of the graduate lecture courses.

The theory of special values of L-functions and their connection with arithmetic began with the classical class number formula, which expresses the leading term in the Taylor expansion at $s = 0$ of the zeta function $\zeta_K(s)$ of a number field K in terms of arithmetic invariants of K. Due essentially to Dirichlet for quadratic fields and to Dedekind for arbitrary number fields, this formula says

$$\lim_{s\to 0} \frac{\zeta_K(s)}{s^{r_1+r_2-1}} = -\frac{h_K R_K}{w_K}, \tag{1}$$

where r_1 is the number of real embeddings of K, r_2 is the number of conjugate pairs of complex embeddings of K, w_K is the number of roots of unity in K, h_K is the class number of K, and R_K is the regulator of K, the determinant of an $(r_1 + r_2 - 1) \times (r_1 + r_2 - 1)$ matrix whose entries are logarithms of fundamental units of K.

The general theme of the lectures in this volume is various generalizations of the class number formula (1), with an emphasis on the Birch and Swinnerton-Dyer Conjecture and on Stark's Conjecture.

Beginning in the late 1950's, Birch and Swinnerton-Dyer developed a conjectural analogue of (1) for the L-function of an elliptic curve. For the statement see Conjecture 2.10 of the lectures of Gross in this volume.

Stark formulated the initial version of his conjecture in the 1970's. If K/k is a finite Galois extension of number fields, then the zeta function $\zeta_K(s)$ factors into a product of Artin L-functions

$$\zeta_K(s) = \prod_{\rho} L(s,\rho)^{\dim(\rho)} \tag{2}$$

where ρ runs through irreducible complex representations of $\mathrm{Gal}(K/k)$. Stark's Conjecture can be viewed as a factorization of the right hand side of (1), corresponding to the factorization (2) of the left hand side. For the statement see §2.4 of the lectures of Tate in this volume.

Department of Mathematics, University of California, San Diego, La Jolla, CA 92093-0112, USA
E-mail address: cpopescu@math.ucsd.edu
Department of Mathematics, University of California, Irvine, CA 92697-3875, USA
E-mail address: krubin@math.uci.edu
Department of Mathematics, University of California, Irvine, CA 92697-3875, USA
E-mail address: asilverb@math.uci.edu

It is tempting to ask for a single general conjecture that will encompass all imaginable generalizations of (1). Currently the best approach in this direction is the Equivariant Tamagawa Number Conjecture (ETNC) of Burns and Flach, which refines and extends an earlier conjecture of Bloch and Kato and deals with L–functions of motives with coefficients. In this volume, we do not go into detail concerning this conjecture in full generality. However, the lectures of Burns and Kings discuss how the ETNC specializes to Stark's Conjecture and to the Birch and Swinnerton-Dyer Conjecture when the motive in question is an Artin motive or an elliptic curve defined over a global field, respectively.

Here is the organization of this volume in more detail. The lectures of Tate introduce Stark's basic conjecture, and Stark's lecture describes the origin of his conjecture. Popescu's lectures discuss integral and p-adic refinements of the conjecture in the case of abelian extensions, and the lectures of Kolster describe a conjecture due to Coates and Sinnott for values of abelian L-functions at negative integers. The lectures of Burns describe the connection between the ETNC and Stark's Conjecture.

Silverberg's lecture gives an introduction to elliptic curves, and the lectures of Gross give an introduction to the Birch and Swinnerton-Dyer Conjecture for elliptic curves over global fields. The lectures of Ulmer discuss the Birch and Swinnerton-Dyer Conjecture for elliptic curves over function fields. The lectures of Birch and Vatsal describe the application of the theory of complex multiplication to the construction of Heegner points (not covered in this volume is Kolyvagin's use of Heegner points to attack the Birch and Swinnerton-Dyer Conjecture). The lectures of Kings describe the connection between the ETNC and the Birch and Swinnerton-Dyer Conjecture.

Rohrlich's lectures discuss root numbers of L-functions. Thanks to the functional equation, in certain cases these root numbers determine the parity of the order of vanishing of an L-function, and thereby have arithmetic implications. Rubin's lectures describe a general theory of Euler systems and Kolyvagin systems, and their application to proving results relating L-functions and arithmetic.

Finally, we would like to express our thanks to all the lecturers for their lectures and their written contributions to this volume. Making this material accessible to a graduate student audience is not easy, and we appreciate their efforts. Without their hard work, this program would not have been possible. We would also like to thank the teaching assistants and the participants, all of whom contributed in essential ways to the success of the program.

Cristian D. Popescu
Karl Rubin
Alice Silverberg

Part I

Stark's Conjecture

Stark's Basic Conjecture

John Tate

IAS/Park City Mathematics Series
Volume 18, 2009

Stark's Basic Conjecture

John Tate

These notes are an expanded version of my three lectures of introduction to Stark's conjectures at the 2009 Park City Math Institute. They are basically a rehash, in English, of some early parts of my old book [**15**].

I would like to thank Daniel Vallières heartily for his assistance in the preparation of these notes. He was a great help to me by writing a first draft of section 2, and putting the whole into LaTeX.

Some notation.
In these notes, k denotes a number field, i.e., a finite extension of the field $\mathbb{Q}$ of rationals, and O_k is its ring of integers. If I is an ideal in O_k, we denote its number of residue classes $[O_k : I]$ by $\mathbf{N}I$. This number is also the positive generator of the ideal $N_{K/\mathbb{Q}}I \subset \mathbb{Z}$, and is a multiplicative function of ideals. By a place v of k we mean an equivalence class of non-trivial absolute values on k. These are of three types, finite, real and complex, corresponding, respectively, to a non zero prime ideal P in O_k, to an embedding of k into $\mathbb{R}$, and to a pair of distinct complex conjugate embeddings of k into $\mathbb{C}$. We use the usual normalization for the absolute values: If v is real, corresponding to the embedding σ, then $|x|_v = |\sigma(x)|$; if v is complex corresponding to a pair of complex conjugate embeddings $\{\sigma, \bar{\sigma}\}$, then $|x|_v = |\sigma(x)|^2 = \sigma(x)\bar{\sigma}(x)$, and if v is finite, corresponding to the prime ideal P, then $|x|_v = (\mathbf{N}P)^{-\operatorname{ord}_P(x)}$. With these normalizations, the product formula holds:

$$\prod_v |x|_v = 1, \text{ for all } x \in k^\times. \tag{0.0.1}$$

We denote the number of real places by $r_1 = r_1(k)$ and the number of complex ones by $r_2 = r_2(k)$. For each finite place v we denote the corresponding prime ideal by P_v and the residue field by $\mathbb{F}_v$. We sometimes denote the cardinality of $\mathbb{F}_v$ by $q_v = \mathbf{N}P_v$.

1. L-functions.

The L-functions associated with number fields are of two fundamentally different types. We will first discuss the very classical abelian ones, made with characters of generalized ideal class groups, which go back to Dirichlet and are the simplest sort of "automorphic" L-functions. Then we will discuss the "nonabelian" L-functions introduced by E. Artin, which are made with representations of Galois groups,

Department of Mathematics, The University of Texas at Austin, 1 University Station – C1200, Austin, TX 78712-0257
E-mail address: tate@math.utexas.edu

and were the first examples of explicitly defined "motivic" L-functions. Artin L-functions made with one dimensional representations of a Galois group are equal to classical abelian L-functions, thanks to Artin's reciprocity law.

1.1. Ideal class characters.

Define a "level" to be a pair $\mathfrak{m} = (\mathfrak{m}_f, \mathfrak{m}_\infty)$ consisting of an ideal $\mathfrak{m}_f$ in O_k and a set $\mathfrak{m}_\infty$ of real places of k. For $a, b \in O_k$, write $a \equiv b \bmod^\times \mathfrak{m}$ if a and b are prime to $\mathfrak{m}$, $a \equiv b \bmod \mathfrak{m}_f$ and $a/b > 0$ at each place in $\mathfrak{m}_\infty$. A (generalized) ideal class character mod $\mathfrak{m}$ is a multiplicative complex valued function χ of ideals I prime to $\mathfrak{m}$ such that $\chi(I) = 1$ if I is a principal ideal generated by a number $a \equiv 1 \bmod^\times \mathfrak{m}$. For example, if $k = \mathbb{Q}$, then a character mod $(m\mathbb{Z}, \{\infty\})$ is the same as a Dirichlet character mod m, if we identify ideals prime to m with their positive integer generators.

Exercise 1.1.1.

(1) Let $\mathcal{I}_\mathfrak{m}$ denote the group of fractional ideals prime to $\mathfrak{m}$, i.e., generated by prime ideals not dividing $\mathfrak{m}_f$. Let $\mathcal{P}_\mathfrak{m}$ denote the subgroup of the principal ideals which are generated by elements $c \in k^\times$ of the form $c = a/b$ with $a \equiv b \bmod^\times \mathfrak{m}$. Show that the group $\mathcal{C}_\mathfrak{m} := \mathcal{I}_\mathfrak{m}/\mathcal{P}_\mathfrak{m}$ is finite (see (2) below) and that a character mod $\mathfrak{m}$ in the above sense is the same as the restriction to integral ideals of a homomorphism $\chi : \mathcal{I}_\mathfrak{m} \to \mathbb{C}^*$ which is trivial on $\mathcal{P}_\mathfrak{m}$, hence is the same as a character of the group $\mathcal{C}_\mathfrak{m}$.

(2) Show that there is an exact sequence

$$O_k^\times \to (O_k/\mathfrak{m}_f)^\times \times \{\pm 1\}^{\mathfrak{m}_\infty} \to \mathcal{C}_\mathfrak{m} \to \mathcal{C} \to 0,$$

where $\mathcal{C} = \mathcal{C}_{(O_k, \emptyset)}$ is the usual ideal class group and $\{\pm 1\}^{\mathfrak{m}_\infty}$ denotes the product of $|\mathfrak{m}_\infty|$ groups of order two, representing the possible choices of signs at the real places $v \in \mathfrak{m}_\infty$. It follows that the groups $\mathcal{C}_\mathfrak{m}$ are finite. Hence the values of an ideal class character χ are complex roots of 1, and in particular, have absolute value 1.

(3) The set of levels is partially ordered under the relation

$$\mathfrak{m}_1 \leq \mathfrak{m}_2 \Leftrightarrow (\mathfrak{m}_1)_f | (\mathfrak{m}_2)_f \text{ and } (\mathfrak{m}_1)_\infty \subset (\mathfrak{m}_2)_\infty.$$

If $\mathfrak{m}_1 \leq \mathfrak{m}_2$, the identity map on ideals induces a homomorphism $\mathcal{C}_{\mathfrak{m}_2} \to \mathcal{C}_{\mathfrak{m}_1}$. Show this homomorphism is surjective (even if there are primes dividing $\mathfrak{m}_2$ which do not divide $\mathfrak{m}_1$). This allows us to identify characters mod $\mathfrak{m}_1$ with certain characters mod $\mathfrak{m}_2$ and then to view the group of all ideal class characters as the Pontrjagin dual of the profinite group $\varprojlim \mathcal{C}_\mathfrak{m}$. Show for each character χ that there is a smallest level $\mathfrak{m}$ such that χ is a character mod $\mathfrak{m}$. This level is called the *conductor* of χ and denoted by $\mathfrak{m}_\chi$. One says χ is *primitive* mod $\mathfrak{m}_\chi$, and *imprimitive* mod $\mathfrak{m}$ for all $\mathfrak{m} > \mathfrak{m}_\chi$.

1.2. Classical abelian L-functions.

Associated with each ideal class character χ is a function $L(s, \chi)$ of a complex variable s, defined in the right half-plane $\Re(s) > 1$ by

$$L(s, \chi) = \prod_P \frac{1}{1 - \chi(P)\mathbf{N}P^{-s}} = \sum_I \chi(I)\mathbf{N}I^{-s}, \tag{1.2.1}$$

where the product is over the prime ideals P not dividing $\mathfrak{m}_\chi$ and the sum over all integral ideals I prime to $\mathfrak{m}_\chi$. The formal equality of the product and sum follows from unique factorization and the fact that the function of ideals $X(I) := \chi(I)\mathbf{N}I^{-s}$ is multiplicative , so that for $x > 0$

$$\begin{aligned}\prod_{\mathbf{N}P\leq x}(1-X(P))^{-1} &= \prod_{\mathbf{N}P\leq x}\sum_{n_P=0}^{\infty} X(P)^{n_P}\\ &= \sum_{(\ldots,n_P,\ldots)}\prod_{\mathbf{N}P\leq x} X(P)^{n_P}\\ &= \sum_{(\ldots,n_P,\ldots)} X\left(\prod_{\mathbf{N}P\leq x} P^{n_P}\right)\\ &= \sum_I X(I),\end{aligned}$$

where the last sum is over all integral ideals I prime to $\mathfrak{m}_\chi$ whose prime factorization involves only prime ideals with norm $\leq x$. The absolute and uniform convergence in $\Re(s) \geq \sigma > 1$ as $x \to \infty$ follows from

$$\sum_P |X(P)| \leq \sum_P \mathbf{N}P^{-\sigma} \leq [k:\mathbb{Q}]\sum_p p^{-\sigma} \leq \frac{[k:\mathbb{Q}]}{\sigma-1}.$$

(The last sum is estimated for $\sigma > 1$ by comparison with the integral $\int_1^\infty x^{-\sigma}dx$). Hecke proved that $L(s,\chi)$ has an analytic continuation to the whole s-plane if $\chi \neq 1$. For $\chi = 1$, the zeta function $\zeta_k(s) := L(s,1)$ is analytic in the whole plane except for a simple pole at $s = 1$. The proof of analytic continuation gives also a functional equation relating $L(1-s,\chi)$ and $L(s,\bar{\chi})$. One way to express this functional equation is the following. For each infinite place v of k, define $\gamma_v(s,\chi)$ to be $\Gamma(\frac{s}{2})$ if v is real and not contained in the conductor of χ, to be $\Gamma(\frac{s+1}{2})$ if v is real and in the conductor of χ, and $\Gamma(\frac{s}{2})\Gamma(\frac{s+1}{2}) = 2^{1-s}\pi^{1/2}\Gamma(s)$ if v is complex. Then there exist constants $B_\chi > 0$ and $C_\chi \in \mathbb{C}^*$ such that the functions

$$\Lambda(s,\chi) := \prod_{v\,|\,\infty}\gamma_v(s,\chi)L(s,\chi) \tag{1.2.2}$$

satisfy

$$\Lambda(1-s,\chi) = C_\chi B_\chi^s \Lambda(s,\bar{\chi}). \tag{1.2.3}$$

Since the gamma function appears in the functional equation of L-functions, we remind the reader of some of its properties.

(1) It is a meromorphic function on the complex plane.
(2) It satisfies $\Gamma(s+1) = s\Gamma(s)$, and $\Gamma(1) = 1$; hence $\Gamma(n+1) = n!$ for $n = 0,1,2,\ldots$.
(3) It is nowhere zero and has simple poles at $s = -n$, where $n = 0,1,2,\ldots$, with residue $(-1)^n/n!$. These are the only poles of $\Gamma(z)$.
(4) **Legendre duplication formula**: $\Gamma(\frac{s}{2})\Gamma(\frac{s+1}{2}) = 2^{1-s}\pi^{1/2}\Gamma(s)$.
(5) $\Gamma(1/2) = \sqrt{\pi}$. (Take $s = 1$ in (4).)

One can find a proof of these facts in [**9**] for example.

1.3. Representations of finite groups.

Let G be a finite group. By a *representation* of G we mean a finite dimensional left $\mathbb{C}[G]$-module V. To give such representation is the same as to give a finite dimensional $\mathbb{C}$-vector space V, together with a homomorphism $\rho : G \to GL(V)$ giving the action of G on V. (We use the symbol $GL(V)$ to denote $\operatorname{Aut}(V)$ because a basis for V gives an isomorphism $\operatorname{Aut}(V) \xrightarrow{\sim} GL_n(\mathbb{C})$ for $n = \dim(V)$.) But we take the module point of view in which the symbol V includes the action of G, and write simply σx instead of $\rho(\sigma)x$ for $x \in V$ and $\sigma \in G$. Two representations V and W are said to be isomorphic if they are isomorphic as $\mathbb{C}[G]$-modules.

Examples:

(1) The *regular* representation is $V = \mathbb{C}[G]$ with the action of G given by left multiplication.
(2) $V = \mathbb{C}$, with $\sigma x = x$ for all $\sigma \in G, x \in \mathbb{C}$ is the *trivial* representation.
(3) If V and W are representations, then $V \otimes_{\mathbb{C}} W$ is also, with $\sigma(x \otimes x') := \sigma x \otimes \sigma x'$.
(4) The space $\operatorname{Hom}_{\mathbb{C}}(V, W)$ of all linear maps $f : V \to W$ is a representation, with $(\sigma f)(x) := \sigma(f(\sigma^{-1}x))$. (Note: We often write f^σ instead of σf, so $f^\sigma(x) = \sigma(f(\sigma^{-1}x))$. But remember: $(f^\sigma)^\tau = f^{(\tau\sigma)}$. Our action is always a left action.)
(5) The *dual* representation to V is $V^* := \operatorname{Hom}_{\mathbb{C}}(V, \mathbb{C})$. We have $(V^*)^* = V$, and $\operatorname{Hom}_{\mathbb{C}}(V^*, W^*) \xrightarrow{\sim} \operatorname{Hom}_{\mathbb{C}}(W, V)$, canonically. Also, $\operatorname{Hom}_{\mathbb{C}}(V, W) \xrightarrow{\sim} V^* \otimes_{\mathbb{C}} W$, canonically.

We assume the reader knows the basic theory and simply list the results we need and some notation and terminology we will use.

(i) Semisimplicity. $\mathbb{C}[G]$ is semisimple, i.e., every representation is a direct sum of irreducible representations. (V is *irreducible* if it is a simple $\mathbb{C}[G]$-module, i.e., has exactly two submodules, V and $\{0\}$.)

(ii) Characters. The *character* of a representation V is the function $\chi_V : G \to \mathbb{C}$ defined by $\chi_V(\sigma) = \operatorname{Tr}(\sigma \mid V)$, the trace of the map $x \mapsto \sigma x$ of V into V. This χ_V is a *central function* on G, meaning that it is constant on each conjugacy class, or, equivalently, that the element $\sum_{\sigma \in G} \chi(\sigma)\sigma$ is in the center of $\mathbb{C}[G]$. The *degree* of the character χ of a representation V is $\chi(1) = \dim V$.

Examples:

(1)
$$\chi_{\mathbb{C}}(\sigma) = 1, \quad \text{for all } \sigma \in G.$$

(2)
$$\chi_{\mathbb{C}[G]}(1) = |G|, \quad \text{and} \quad \chi_{\mathbb{C}[G]}(\sigma) = 0 \quad \text{for} \quad \sigma \neq 1.$$

(3)
$$\chi_{V \otimes_{\mathbb{C}} W}(\sigma) = \chi_V(\sigma) \cdot \chi_W(\sigma).$$

(4)
$$\chi_{V^*}(\sigma) = \chi_V(\sigma^{-1}) = \bar{\chi}_V(\sigma).$$

($\chi_V(\sigma)$ is a sum of complex m-th roots of 1, if $\sigma^m = 1$.)

$$(5) \qquad \chi_{\mathrm{Hom}(V,W)} = \chi_{V^*\otimes W} = \bar{\chi}_V(\sigma)\cdot\chi_W(\sigma).$$

(iii) Orthogonality. One makes the space C of central functions on G into a Hilbert space by putting $\langle f,g\rangle_G = \frac{1}{|G|}\sum_{\sigma\in G} f(\sigma)\bar{g}(\sigma)$. The equality

$$(1.3.1) \qquad \langle\chi_V,\chi_W\rangle_G = \dim(\mathrm{Hom}_G(V,W))$$

is easy to prove. For each $\mathbb{C}[G]$-module V, the operator $|G|^{-1}\sum_{\sigma\in G}\sigma$ is a projection of V onto V^G (see (iv) below). The trace of a projection operator is the dimension of its image. Hence

$$\dim(V^G) = \mathrm{Tr}(|G|^{-1}\sum_{\sigma\in G}\sigma|V) = \frac{1}{|G|}\cdot\sum_{\sigma\in G}\chi_V(\sigma).$$

It follows that

$$\dim(\mathrm{Hom}_G(V,W)) = \dim(\mathrm{Hom}(V,W)^G) = \frac{1}{|G|}\cdot\sum_{\sigma\in G}\bar{\chi}_V(\sigma)\chi_W(\sigma),$$

and since this quantity is an integer, hence real, (1.3.1) follows.

Let $\{V_i\}$ be a complete family of non-isomorphic irreducible representations of G, and let χ_i be the character of V_i. By (1.3.1) and Schur's lemma, the χ_i form an orthonormal system in C. In fact, the χ_i span C, showing that the number of isomorphism classes of irreducible representations V_i is equal to $\dim C$, that is, to the number of conjugacy classes in G. To show that the χ_i span C it suffices to show that a central function f which is orthogonal to every χ_i, is 0. Let $x = \sum_{\sigma\in G} f(\sigma)\sigma$. This element x is in the center of $\mathbb{C}[G]$, hence acts G-linearly on every V, and by Schur's Lemma acts as a constant, say c_i, on V_i^*. We have

$$< f,\chi_i >= \frac{1}{|G|}\cdot\sum_{\sigma\in G} f(\sigma)\bar{\chi}_i(\sigma) = \frac{1}{|G|}\cdot\mathrm{Tr}(x|V_i^*) = c_i\dim(V_i).$$

Thus, $< f,\chi_i >= 0 \Leftrightarrow c_i = 0 \Leftrightarrow xV_i^* = \{0\}$. It follows that if f is orthogonal to every irreducible character then x annihilates every irreducible representation, hence every representation by semisimplicity. In particular, $x\,\mathbb{C}[G] = \{0\}$, so $x = 0$ as claimed.

(iv) Isotypical components; projections. Let V be a representation. Then by semisimplicity,

$$(1.3.2) \qquad V = \oplus_i V[i],$$

where $V[i]$ is isomorphic to a direct sum of copies of V_i. The decomposition (1.3.2) is unique. If $x = \sum_i x_i$ with $x_i \in V[i]$, then

$$(1.3.3) \qquad x_i = \frac{\dim V_i}{|G|}\sum_{\sigma\in G}\bar{\chi}_i(\sigma)\sigma x.$$

In other words, $e_{\chi_i} := \frac{\dim V_i}{|G|}\sum_{\sigma\in G}\bar{\chi}_i(\sigma)\sigma \in \mathbb{C}[G]$ is the projection of V onto $V[i]$ which kills all the other $V[j]$'s. To see this, note that e_{χ_i} is in the center of $\mathbb{C}[G]$, hence its action on V_j is multiplication by a scalar a_{ij}. Then compute a_{ij} using $a_{ij} = \frac{1}{|V_j|}\mathrm{Tr}(e_{\chi_i}\mid V_j)$.

The $V[i]$'s are called the *isotypical* components of V. The decomposition of $V[i]$ into a direct sum of copies of V_i corresponds to expressing $\mathrm{Hom}_G(V_i,V)$ as a direct sum of 1-dimensional subspaces and is not at all unique, but the number of copies of V_i in such an expression, the *multiplicity* of V_i in V, is unique and equals $\langle \chi_i, \chi_V \rangle_G = \dim \mathrm{Hom}_G(V_i, V)$.

One denotes the isotypical component of V corresponding to the trivial representation $\mathbb{C}$ by $V^G = \{x \in V \mid \sigma x = x \text{ for all } \sigma \in G\}$. The corresponding projection, $\frac{1}{|G|}\sum_\sigma \sigma$ is especially useful, as in the proof of (1.3.1) above.

(v) Induced representation; Frobenius reciprocity. Let Rep_G denote the category of representations of G. Let $f : H \to G$ be a homomorphism of groups. In this paragraph we discuss adjoint functors $f^* : \mathrm{Rep}_G \to \mathrm{Rep}_H$ and $f_* : \mathrm{Rep}_H \to \mathrm{Rep}_G$. The classic case is that in which f is an inclusion $H \subset G$. In that case f^*V is called the restriction of V to H and denoted by $\mathrm{res}\, V$ or $\mathrm{res}_H^G V$, and f_*W is called the representation of G *induced by* the representation W of the subgroup H, and is denoted by $\mathrm{Ind}\, W$ or $\mathrm{Ind}_H^G W$. In general, if V is a representation of G, then f^*V is the representation of H with the same underlying vector space as V, with H acting through f, that is, $\tau x = f(\tau)x$ for $\tau \in H$ and $x \in V$. The functor f_* in the other direction comes with canonical elements $\iota_W \in \mathrm{Hom}_H(W, f^*f_*W)$ and is characterized by the fact that for every pair of representations W of H and V of G, the map $\mathrm{Hom}_G(f_*W, V) \to \mathrm{Hom}_H(W, f^*V)$ given by $\phi \mapsto (f^*\phi) \circ \iota_W$ is bijective. Since every group homomorphism f is a passage to quotient followed by an isomorphism followed by an inclusion, it suffices, in order to describe f_* explicitly, to treat the two cases $H \to H/N$ and $H \subset G$. In the first of these, $f_*W = W^N$, and ι_W is the projection of W onto W^N discussed in **(iv)** just above. In the second, classical, case, ι_W is an inclusion $W \subset f_*W$, and $\mathrm{Ind}_H^G = f_*W = \oplus_j \sigma_j W$, where the σ_j are representatives of the left cosets $\sigma_j H$ of H in G. The action of G is forced by this: $\sigma \sum_i \sigma_i y_i = \sum_i \sigma_{\sigma(i)} \tau_{\sigma,i} y_i$ for $y_i \in W$, where the subscript $\sigma(i)$ and $\tau_{\sigma,i} \in H$ are uniquely determined by $\sigma\sigma_i = \sigma_{\sigma(i)}\tau_{\sigma,i}$.

If χ is the character of V, we denote by $f^*\chi$ the character of f^*V. Clearly, $f^*\chi(\sigma) = \chi(f(\sigma))$. If ψ is the character of W, we denote by $f_*\psi$ the character of f_*W. It is not hard to see that

$$(1.3.4) \qquad (f_*\psi)(\sigma) = \frac{1}{|H|} \sum_{\rho \in G} \sum_{\tau \in H; f(\tau) = \rho\sigma\rho^{-1}} \psi(\tau).$$

Equating the dimensions of the spaces on each side of the isomorphism

$$\mathrm{Hom}_G(f_*W, V) \xrightarrow{\sim} \mathrm{Hom}_H(W, f^*V)$$

characterizing f_* gives the relation

$$(1.3.5) \qquad \langle f_*\psi, \chi \rangle_G = \langle \psi, f^*\chi \rangle_H.$$

This is known as *Frobenius reciprocity*.

(vi) Abelian characters; duality for finite abelian groups. The characters χ of 1-dimensional representations are the same thing as the group homomorphisms $G \to \mathbb{C}^\times$. We will call them *abelian characters*. They form an abelian group $\mathrm{Hom}(G, \mathbb{C}^\times)$ under multiplication. If G is abelian, they are the only irreducible characters. The group they form is called the *character group* of G and is denoted in these notes by $\hat{G}$. We have a perfect duality: the natural maps $G \to \hat{\hat{G}}$, and $\mathrm{Hom}(\hat{G}', \hat{G}) \to \mathrm{Hom}(G, G')$ are isomorphisms.

1.4. Decomposition, inertia, Frobenius.

Let K/k be a finite Galois extension and $G = G_{K/k}$ its Galois group. Recall that for each place v of k, G acts transitively on the set of places w of K which lie above (i.e., extend) v. If w is one of these places, its stabilizer is a subgroup of G called the *decomposition group* of w and denoted by G_w. It can be identified with the Galois group of the corresponding extension of the completions K_w/k_v.

For finite v, an element τ of G_w induces an automorphism $\tilde{\tau}$ of the residue field $\mathbb{F}_w$ of w. The map $\tau \mapsto \tilde{\tau}$ is a homomorphism of G_w onto the Galois group of the w/v residue field extension $\mathbb{F}_w/\mathbb{F}_v$. The kernel is called the *inertia group* of w and is denoted by I_w. The order of I_w is the ramification index $e_{w/v}$ and is 1 for all but a finite number of places v, those dividing the relative discriminant ideal $d_{K/k}$. As Galois group of the finite field extension $\mathbb{F}_w/\mathbb{F}_v$, the quotient group G_w/I_w is cyclic with a canonical generator, the Frobenius automorphism $\tilde{\sigma}_w$ which raises elements of $\mathbb{F}_w$ to the q_v-th power, where $q_v = \mathbf{N}P_v$ is the number of elements in $\mathbb{F}_v$.

1.5. Artin L-functions.

Let K/k be a finite Galois extension and $G = G_{K/k}$ its group. Let V be a representation of G. For a finite place v of k let $F_v(T,V) = \det(1 - \sigma_w T \mid V^{I_w})$ be the characteristic polynomial of the action on V^{I_w} of a Frobenius automorphism σ_w attached to a place w of K above v. Although σ_w is determined only up to multiplication by an element of I_w, its action on V^{I_w} is independent of which element of a coset $\sigma_w I_w$ we chose. The polynomial $F_v(T,V)$ obviously depends only on the isomorphism class of V, and it also depends only on v, not on the choice of w above v, because the places w above v are all conjugate. Note that for v unramified in K, hence for all but a finite number of v we have $I_w = \{1\}$, hence $V^{I_w} = V$, and $F_v(T,V)$ is of degree $\dim V$. Note also that $F_v(T,V)$ depends only on V as G_w-module, not as G-module. The L-function $L(s,V)$ defined by Artin is an "Euler product" over the finite places v of k of "local L-functions" $L_v(s,V) := F_v(\mathbf{N}P_v^{-s}, V)^{-1}$.

$$L(s,V) = \prod_v L_v(s,V) = \prod_v \frac{1}{\det(1 - \sigma_w q_v^{-s} \mid V^{I_w})}. \tag{1.5.1}$$

Exercise 1.5.2. Show that this product converges absolutely and defines an analytic function in the half plane $\Re(s) > 1$.

In fact, it has a meromorphic continuation to the whole plane, and a deep conjecture of Artin is that it is an entire function if the trivial representation does not occur in V — more about this soon. It is really enough to consider irreducible representations, because

$$L(s, V \oplus W) = L(s,V)L(s,W). \tag{1.5.3}$$

Exercise 1.5.4. Since $L(s,V)$ depends only on the isomorphism class of the representation V, we can also denote it by $L(s,\chi_V)$, where $\chi_V : G \to \mathbb{C}$ is the character of V. Show for each finite place v that for $\chi := \chi_V$,

$$\log L_v(s,\chi) = \sum_{n=1}^{\infty} \frac{\operatorname{Tr}(\sigma_w^n \mid V^{I_w})}{n q_v^{ns}} = \sum_{n=1}^{\infty} \frac{\chi(\sigma_w^n)}{n q_v^{ns}},$$

where $\chi(\sigma_w^n) = \frac{1}{|I_w|}\sum_{\tau \in \sigma_w^n} \chi(\tau)$ is the average value of χ on the coset $\sigma_w^n I_w$ of I_w.

Another property of these L-functions concerns the situation in which K is contained in a larger Galois extension K' of k, so $k \subset K \subset K'$. Then $G = G_{K/k}$ is a quotient group of $G' := G_{K'/k}$. Let V' denote the G'-module with the same underlying vector space as V, with G' acting through G (this representation is also denoted by $\mathrm{Infl}(V)$, inflation of V). Then

$$L(s, V') = L(s, V) \tag{1.5.5}$$

This is true because if we let w' be a place of K' above w, then $I_w = I'_{w'} G_{K'/K}$, hence $(V')^{I_{w'}} = V^I$, and $\sigma_w = \sigma_{w'} G_{K'/K}$. Property (1.5.5) shows that $L(s, V)$ really depends only on V viewed as module $\bar{V}$ for "the" absolute Galois group $G_k = \mathrm{Gal}(\bar{k}/k)$ of k. Note that the isomorphism class of $\bar{V}$ as G_k-module is independent of how we view K/k as subextension of $\bar{k}/k$. The representations of the form $\bar{V}$ for some K/k and V are, up to isomorphism, simply the $\mathbb{C}[G_k]$–modules X of finite dimension over $\mathbb{C}$ for which the action map $G_k \times X \to X$ is continuous for the profinite (Krull) topology in G_k and the discrete topology in X (or the usual complex vector space topology in X – it's the same, because $GL_n(\mathbb{C})$ has "no small subgroups").

The really key property of Artin's L-functions relates L-functions of different fields. Suppose k' is an intermediate field, $k \subset k' \subset K$, the fixed field of a subgroup H of G. Let W be a representation of H and let $V = \mathrm{Ind}_H^G W$ be the representation of G induced by W. Then

$$L(s, W) = L(s, V) \qquad (V = \mathrm{Ind}_{k'}^k W). \tag{1.5.6}$$

Example. If $k' = K$, and $W = \mathbb{C}$ is the trivial representation of the identity subgroup of G, then $V = \mathbb{C}[G]$ is the regular representation of G, and since $\mathbb{C}[G] \xrightarrow{\sim} \oplus_i V_i^{\dim V_i}$, where the V_i are (representatives of the isomorphism classes of) the irreducible representations of G, we have, by (1.5.3) and (1.5.6)

$$\zeta_K(s) = \prod_i L(s, V_i)^{\dim V_i} = \zeta_k(s) \prod_{i \neq 1} L(s, V_i)^{\dim V_i}, \tag{1.5.7}$$

if we number the V_i so that V_1 is the trivial representation of G, for by (1.5.1) we have then $L(s, V_1) = \zeta_k(s)$. More generally, if k' is any intermediate field, and W the trivial representation of H, then $V = \mathbb{C}[G/H]$ is the permutation representation of G acting on the set G/H of cosets of H, and by Frobenius reciprocity, $V \xrightarrow{\sim} \oplus_i V_i^{m_i}$, where $m_i = \dim\ V_i^H$. In particular, $m_1 = 1$ and we have

$$\zeta_{k'}(s) = \zeta_k(s) \cdot \prod_{i \neq 1} L(s, V_i)^{m_i}. \tag{1.5.8}$$

Thus, if the $L(s, V_i)$ are entire functions for $i \neq 1$ as Artin conjectured, then the zeta function of k divides the zeta function of every extension field k'/k, and the Riemann hypothesis for all zeta functions would imply it for all L-functions. It was Artin's investigation of the interrelations among the zeta functions of different fields, in particular the intermediate fields of a non-abelian Galois extension K/k which led him to define his new kind of L-functions.

Exercise 1.5.9. Fill in the details of the following proof of (1.5.6). Let v be a finite place of k and w be a place of K above v. Let $G = \coprod_{i=1}^r G_w \rho_i H$ be the expression of G as disjoint union of double cosets of G_w and H. Let $w_i = \rho_i^{-1} w$

and let v_i' be the place of k' below w_i. Then $v_1', v_2', ..., v_r'$ are the places of k' above v, so it suffices to show that

$$L_v(s,V) = \prod_{i=1}^{r} L_{v_i'}(s,W). \tag{1.5.10}$$

For each i, let $G_w = \coprod_{j=1}^{m_i} \tau_{ij}(G_w \cap H^{\rho_i})$. (Notation: For $\rho \in G$ and $X \subset G$ we write $X^\rho := \rho X \rho^{-1}$.) Then

$$G = \coprod_{i=1}^{r} \coprod_{j=1}^{m_i} \tau_{ij}\rho_i H.$$

Hence, by definition of induced representation, V contains W as an H-submodule, and

$$V = \oplus_{i=1}^{r} \oplus_{j=1}^{m_i} \tau_{ij}\rho_i W = \oplus_{i=1}^{r} V_i, \tag{1.5.11}$$

where $V_i = \oplus_{j=1}^{m_i} \tau_{ij}\rho_i W$ is a G_w-module for each i, and is in fact isomorphic to the G_w-module induced from the $G_w \cap H^{\rho_i}$-module $\rho_i W$. Applying the automorphism ρ_i to our situation, ($H \mapsto H^{\rho_i}$, $k' \mapsto \rho_i k'$, $W \mapsto \rho_i W$, etc.), we have by transport of structure, $L_{v'}(s, \rho_i W) = L_{v_i'}(s, \rho_i W)$, where v' is the place of $\rho_i k$ below w. Comparing with (1.5.10) and (1.5.11) and using (1.5.3) shows that we are now reduced to the local case, in which $G = G_w$.

Next we reduce to the local unramified case by showing that $V^I \xrightarrow{\sim} \mathrm{Ind}_{H/J}^{G/I}(W^J)$, where $I = I_w$, and $J = H \cap I$ are the inertia subgroups of G and H, and where we view H/I as subgroup of G/J by identifying H/J with HI/I in the obvious way. One way to do this is to factor the homomorphism $f : H \to G/I$ in two ways

$$H \hookrightarrow G \to G/I \ \text{ and } \ H \to H/J \hookrightarrow G/I.$$

Calculating f_*W in two ways accordingly we find

$$f_*W \xrightarrow{\sim} (\mathrm{Ind}_H^G W)^I \xrightarrow{\sim} V^I \ \text{ and } \ f_*W \xrightarrow{\sim} \mathrm{Ind}_{H/J}^{G/I}(W^J).$$

Finally, in the local unramified case, G is cyclic, generated by the v-Frobenius σ_v and H the subgroup generated by the v'-Frobenius $\sigma_{v'} = \sigma_v^f$, where $f = (G:H)$. Then $V = \oplus_{i=0}^{f-1} \sigma_v^i W$ and we can assume W is 1-dimensional, with basis x. Let $\sigma_v' x = \eta x$. For each solution ζ to $\zeta^f = \eta$ the element $\sum_{i=0}^{f-1} \zeta^{-i}\sigma^i x \in V$ is an eigenvector for σ_v with eigenvalue ζ. Hence

$$L_v(s,V) = \det(1 - \sigma_v q_v^{-s} \,|\, V) = \prod_{\zeta^f = \eta} (1 - \zeta q^{-s}) = (1 - \eta q^{-fs})$$

$$= \det(1 - \eta q_v^{-s} \,|\, W) = L_{v'}(s,W).$$

1.6. Reciprocity and the relation between the two kinds of L-functions. Artin had quickly mastered[1] Takagi's recent work proving that "class fields" are just abelian extensions. Takagi had shown that for each level $\mathfrak{m}$ (see §1.1) there is an abelian extension $K_\mathfrak{m}$ of k with group $G_{K_\mathfrak{m}/k}$ having the same invariants as, hence isomorphic to, the generalized ideal class group $\mathcal{C}_\mathfrak{m}$ and such that the way in

[1]He told the story that when he was a post-doc in Göttingen, he asked Siegel if he could borrow Siegel's reprint of Takagi's paper. When Siegel said yes, he could have it for 24 hours, Artin took the 90-page paper and copied by hand as much of the essential part as he could in that time.

which a prime ideal P of k decomposes in K is determined by the class of P in $\mathcal{C}_{\mathfrak{m}}$. From Takagi's decomposition law it follows that

$$\zeta_K(s) = \prod_{\chi} L(s, \chi) = \zeta_k(s) \cdot \prod_{\chi \neq 1} L(s.\chi), \tag{1.6.1}$$

where the first product is over all characters χ of $\mathcal{C}_{\mathfrak{m}}$ and the $L(s, \chi)$ are the classical abelian L-functions of §1.2. Artin realized that in the abelian case $K_{\mathfrak{m}}/k$ the simplest explanation for the existence of the two factorizations (1.5.7) and (1.6.1) of $\zeta_k(s)$ was that they be the same, and the simplest explanation for that would be the existence of a **canonical** isomorphism $\mathcal{C}_{\mathfrak{m}} \overset{\sim}{\to} G_{K_{\mathfrak{m}}/k}$ which, for each finite place v of k unramified in K, associated to the class of the prime P_v the Frobenius substitution σ_v. To prove this he had to show, for every element $a \in O_k$ such that $a \equiv 1 \bmod^{\times} \mathfrak{m}$, that

$$(a) = \prod_v P_v^{m_v} \quad \Rightarrow \quad \prod_v \sigma_v^{m_v} = 1. \tag{1.6.2}$$

It took him 4 years. Artin called (1.6.2) *the* reciprocity law, because every known explicit reciprocity law could be interpreted as (1.6.2) for some special Kummer extension K/k. For example, if q is a prime $\equiv 1 \bmod 4$, then for the extension $\mathbb{Q}(\sqrt{q})/\mathbb{Q}$, (1.6.2) implies that there is a non-trivial character χ of order 2 mod q such that $\chi(p) = (\frac{q}{p})$ for primes $p \neq q$. The only such character is $p \mapsto (\frac{p}{q})$. Hence, $(\frac{q}{p}) = (\frac{p}{q})$.

1.7. Brauer's theorem, meromorphicity.

A representation V of G and its character χ_V are called *monomial* if V is induced from a 1-dimensional representation of some subgroup of G, or in other words, V is a direct sum of 1-dimensional subspaces which are permuted transitively by G.

Theorem 1.7.1 (R. Brauer). *Every character χ of a finite group G is a linear combination with integral coefficients of monomial characters.*

For a proof of the theorem, see e.g. [12], Ch. 10, Th. 20. One of Brauer's motivations for proving it was the fact that if the ψ_i are 1-dimensional characters of some subgroups H_i of a Galois group G such that

$$\chi = \sum_i n_i \operatorname{Ind}_{H_i}^G \psi_i, \tag{1.7.2}$$

then

$$L(s, \chi) = \prod_i L(s, \psi_i)^{n_i}. \tag{1.7.3}$$

By the reciprocity law, the $L(s, \psi_i)$ are classical abelian L-series, hence meromorphic on $\mathbb{C}$, and even entire for $\psi_i \neq 1$ (See §1.2). Hence Brauer's theorem implies

Corollary 1.7.4. *Every Artin L-function is meromorphic in the whole complex plane.*

Exercise 1.7.5. Show that if the trivial representation does not occur in χ, then one can cancel on the right side of (1.7.3) all of the terms in which ψ_i is the trivial character of H_i. (Write the hypothesis in the form $\langle \chi, 1_G \rangle_G = 0$ and use Frobenius reciprocity.)

If in (1.7.3) $n_i \geq 0$ for all i, and $\psi_i \neq 1$ for all i, then $L(s,\chi)$ is entire. However for most χ such an expression does not exist. Artin's conjecture that $L(s,\chi)$ is holomorphic for irreducible $\chi \neq 1$ is much deeper, implying a miraculous cancellation of the zeros of the $L(s,\psi_i)$ with $n_i < 0$ by those of those with $n_i > 0$ in (1.7.3).

1.8. Functional equation.

To get a neater functional equation, Artin defined local L-functions also for infinite places v of k. For such a v the decomposition group G_w of a place w of K above v is of order 1 or 2, hence has a unique generator, which we denote by σ_w, and we put

$$(1.8.1)\quad L_v(s,\chi) = \left(\Gamma\left(\frac{s}{2}\right)\Gamma\left(\frac{s+1}{2}\right)\right)^{\chi(1)} = (2^{1-s}\sqrt{\pi}\Gamma(s))^{\chi(1)}, \text{ for } v \text{ complex,}$$

and

$$(1.8.2)\qquad L_v(s,\chi) = \Gamma\left(\frac{s}{2}\right)^{\frac{\chi(1)+\chi(\sigma_w)}{2}} \cdot \Gamma\left(\frac{s+1}{2}\right)^{\frac{\chi(1)-\chi(\sigma_w)}{2}}, \text{ for } v \text{ real.}$$

Note that these L_v's depend only on the action of σ_w on V, i.e., only on V as G_w-module, just as was the case for finite v. Remark also that

$$\frac{\chi(1)+\chi(\sigma_w)}{2} = \dim V^{G_w}, \text{ and } \frac{\chi(1)-\chi(\sigma_w)}{2} = \operatorname{codim} V^{G_w},$$

since $\dim V^{G_w} = \langle \chi|_{G_w}, 1_{G_w}\rangle_{G_w}$.

Exercise 1.8.3. Show that (1.5.10) holds for infinite v as well.

Let

$$(1.8.4)\qquad \Lambda(s,V) := \prod_{\text{all } v} L_v(s,V) = \Gamma\left(\frac{s}{2}\right)^a \cdot \Gamma\left(\frac{s+1}{2}\right)^b L(s,V),$$

where $a = a_1 + r_2 \dim V$, and $b = a_2 + r_2 \dim V$, with

$$a_1 = \sum_{v \text{ real}} \dim V^{G_w},$$

$$a_2 = \sum_{v \text{ real}} \operatorname{codim} V^{G_w}.$$

The three basic properties (1.5.3), (1.5.5) and (1.5.6) of $L(s,V)$ as function of V hold for $\Lambda(s,V)$ as well. This is trivial, given Exercise (1.8.3). For an abelian character ψ, our definition of $\Lambda(s,\psi)$ is consistent with the definition in §1.2. Writing an arbitrary $L(s,\chi)$ in the form (1.7.3) and recalling the functional equation (1.2.3) for abelian L-functions, one sees that there are constants $B_\chi = \prod_i B_{\psi_i}^{n_i} > 0$ and $C_\chi = \prod_i C_{\psi_i}^{n_i}$, such that

$$(1.8.5)\qquad \Lambda(1-s,\chi) = C_\chi B_\chi^s \Lambda(s,\bar{\chi}).$$

Note that the B_χ and C_χ are uniquely determined by this equation, independent of the choice of expression (1.7.2), because $CB^s = 1$ for all s implies $B = C = 1$.

Exercise 1.8.6. Assuming (1.2.3) show that there exist unique constants $A_\chi > 0$ and $W_\chi \in \mathbb{C}^*$ such that $A_\chi = A_{\bar\chi}$ and such that the functions

$$\xi(s,\chi) := A_\chi^{s/2}\Lambda(s,\chi) = A_\chi^{s/2}\Gamma\left(\frac{s}{2}\right)^a \cdot \Gamma\left(\frac{s+1}{2}\right)^b L(s,\chi), \tag{1.8.7}$$

satisfy

$$\xi(1-s,\chi) = W_\chi \cdot \xi(s,\bar\chi). \tag{1.8.8}$$

Show also that $|W_\chi| = 1$, and $W_{\bar\chi} = \overline{W}_\chi$. (Suggestion: Note that $L(\bar s,\bar\chi) = \overline{L(s,\chi)}$ and consider the functional equation on the vertical line $\Re(s) = 1/2$.) The constant W_χ is called an "Artin root number". Artin showed that

$$A_\chi = \frac{|d_k|^{\chi(1)}\mathbf{N}\mathfrak{f}(\chi)}{\pi^{[k:\mathbb{Q}]\chi(1)}}, \tag{1.8.9}$$

where $\mathfrak{f}(\chi)$ is an integral ideal of $\mathcal{O}_k$ involving only primes ramified in K called the *conductor* of χ.

Exercise 1.8.10. Specializing to the case $\chi = 1$, and assuming that $W_1 = 1$, show that the Dedekind zeta function $\zeta_k(s)$ satisfies the following functional equation. Let

$$\xi(s) = \left(\frac{|d_k|}{\pi^{[k:\mathbb{Q}]}}\right)^{s/2} \Gamma\left(\frac{s}{2}\right)^{r_1+r_2} \Gamma\left(\frac{s+1}{2}\right)^{r_2} \zeta_k(s),$$

then

$$\xi(1-s) = \xi(s).$$

2. Basic Stark Conjecture.

2.1. Class number formula.

The following famous theorem is due to Dedekind who generalized results of Dirichlet on quadratic forms.

Theorem 2.1.1. *The Dedekind zeta function $\zeta_k(s)$ has a simple pole at $s = 1$ with residue*

$$\frac{2^{r_1}(2\pi)^{r_2}}{\sqrt{|d_k|}}\frac{h_k R_k}{w_k},$$

where h_k is the class number of k, R_k the regulator of k, and w_k the number of roots of unity contained in k.

By means of the functional equation of ζ_k, this last result can be translated into one at $s = 0$. Using (1.8.10) and properties of the gamma function listed in §1.2, one gets

Corollary 2.1.2. *The Taylor expansion of $\zeta_k(s)$ at $s = 0$ starts as follows*

$$\zeta_k(s) = -\frac{h_k R_k}{w_k} s^{r_1+r_2-1} + \dots$$

Thus the first non-vanishing Taylor coefficient of ζ_k at $s = 0$ is the product of a rational number $-h_k/w_k$ with the regulator R_k, which is the determinant of a matrix whose entries involve logarithms of absolute values of units and whose size $r_1 + r_2 - 1$ is the order of vanishing of $\zeta_k(s)$ at $s = 0$. The basic Stark conjecture is a similar statement for Artin L-functions.

2.2. S-imprimitive L-functions.

Let S be a finite set of places of k containing S_∞, the set of infinite places of k, and let V be a representation of the Galois group $G = G_{K/k}$ of a Galois extension K/k. The S-imprimitive Artin L-function is defined for $\Re(s) > 1$ as

$$L_S(s,V) = \prod_{v \notin S} L_v(s,V).$$

These imprimitive L-functions satisfy the same properties (1.5.3), (1.5.5) and (1.5.6) as the primitive ones, since these properties hold place by place for the "local L-functions" $L_v(s,V)$. For the trivial character $\chi = 1$, we get the S-Dedekind zeta function of the field k

$$\zeta_{k,S}(s) = \prod_{P \notin S} (1 - \mathbf{N}P^{-s})^{-1},$$

which can be viewed as the zeta function associated with the Dedekind domain $O_{k,S}$ consisting of the S-integers of k. We denote the S-class number by $h_{k,S}$ and the S-regulator by $R_{k,S}$.

Exercise 2.2.1. Let v be a place of k which is not contained in S and let $S' = S \cup \{v\}$. Let also m be the order of v in the S-class group of k. Show that

(1) $h_S = m h_{S'}$,
(2) $R_{S'} = m(\log \mathbf{N}v) R_S$, where h_S is the class number of the ring of S-integers, and R_S its regulator.

From this there follows easily a generalization of (2.1.2).

Exercise 2.2.2. The Taylor expansion of $\zeta_{k,S}(s)$ at $s = 0$ starts as follows

$$\zeta_{k,S} = -\frac{h_S R_S}{w_k} s^{|S|-1} + \ldots,$$

Given an Artin L-function, we write

$$L_S(s,\chi) = c_S(\chi) s^{r_S(\chi)} + \ldots$$

The order of vanishing $r(\chi) = r_S(\chi)$ at $s = 0$ is well known. In order to compute it, we introduce some notation. Let S_K be the set of places of K lying above places of S. We let $Y = Y_{S_K}$ be the free abelian group on S_K and let $s : Y \longrightarrow \mathbb{Z}$ be the homomorphism such that $s(w) = 1$ for all $w \in S_K$. The kernel of this map is denoted by $X = X_{S_K}$ and we have a short exact sequence of G-modules

$$0 \longrightarrow X \longrightarrow Y \xrightarrow{s} \mathbb{Z} \longrightarrow 0. \tag{2.2.3}$$

Exercise 2.2.4. Let w_0 be an arbitrary place in S_K. Show that the elements $w - w_0$ where w runs over all places of S_K different from w_0 form a $\mathbb{Z}$-basis of X.

Given an abelian group B and a subring A of $\mathbb{C}$, we shall denote $A \otimes_{\mathbb{Z}} B$ simply by AB. We denote the character of the representation $\mathbb{C}X$ by χ_X and the one of $\mathbb{C}Y$ by χ_Y. For each $v \in S$, let us choose a place w of K lying above v. We then have the formula

$$r_S(\chi) = \left(\sum_{v \in S} \dim V^{G_w}\right) - \dim V^G = \langle \chi, \chi_X \rangle_G = \dim \operatorname{Hom}_G(V^*, \mathbb{C}X). \tag{2.2.5}$$

PROOF OF 2.2.5. Since $\mathbb{C}Y$ is a permutation representation, χ_Y is integer valued. By (2.2.3) we have $\chi_Y = \chi_X + 1$. Hence χ_X is integral valued as well. The right hand equality in (2.2.5) follows from (1.3.1) and the fact that χ_X is real valued.

The middle equality follows from Frobenius reciprocity and

$$\mathbb{C}X \oplus \mathbb{C} = \mathbb{C}Y = \bigoplus_{v \in S} \mathrm{Ind}_{G_w}^{G}(\mathbb{C}). \tag{2.2.6}$$

The left hand inequality in (2.2.5) follows from the functional equation and the two facts:

(i) The order at $s = 1$ of $L(s, V)$ is $-\dim V^G$.

(ii) For each v, real, complex or finite, the Euler factor $L_v(s, V)$ has order 0 at $s = 1$ and order $-\dim V^{G_w}$ at $s = 0$, where as usual w is a place above v.

Both claims are additive in V, and if true for V they are true for representations induced by V. Hence, by Brauer's Theorem it suffices to prove these for abelian characters, for which (i) is standard, and (ii) is easily checked from the definitions of L_v, in the three cases (1.5.1), (1.8.1), (1.8.2). □

Corollary 2.2.7. *If V is a one dimensional representation, then*

$$r_S(\chi) = \begin{cases} |\{v \in S \,|\, G_w \subseteq \mathrm{Ker}(\chi)\}|, & \text{if } \chi \neq 1; \\ |S| - 1, & \text{if } \chi = 1. \end{cases}$$

2.3. Stark regulator.

Given a character χ, we will now introduce a regulator attached to χ which will appear in the formulation of the conjecture. We denote the group of S_K-units of K by $U = U_{K,S_K}$ and we consider the logarithm map $\lambda : U \longrightarrow \mathbb{R}Y$ defined by

$$u \mapsto \sum_{w \in S_K} \log |u|_w \cdot w.$$

Its image is in $\mathbb{R}X$ because of the product formula (0.0.1). This λ is the map used in the proof of the S-unit theorem. Its kernel consists of μ_K, the group of roots of unity in K, and its image is a lattice of full rank in $\mathbb{R}X$, see [**16**]. Tensoring with either $\mathbb{R}$ or $\mathbb{C}$, λ induces isomorphisms (which we denote by the same symbol λ)

$$\mathbb{R}U \xrightarrow{\simeq} \mathbb{R}X, \qquad \mathbb{C}U \xrightarrow{\simeq} \mathbb{C}X,$$

which are G-equivariant as is easily checked. This implies that $\mathbb{Q}X$ and $\mathbb{Q}U$ are isomorphic as $\mathbb{Q}[G]$-modules, but not canonically so (see [**12**], remark in small print after the proof of Prop. 33).

Originally, Stark used a so-called Artin system of units. This is a set of units ε_w, one for each place $w \in S_K$, such that $\varepsilon_w^\sigma = \varepsilon_{w^\sigma}$ for all $\sigma \in G$ and such that the only relation among them is $\prod_w \varepsilon_w = 1$. In other words, an Artin system of units gives a $\mathbb{Z}[G]$-module morphism $Y \longrightarrow U$ which induces an injective $\mathbb{Z}[G]$-morphism $X \hookrightarrow U$. After tensoring with $\mathbb{Q}$, we get a $\mathbb{Q}[G]$-isomorphism since $\mathbb{Q}X$ and $\mathbb{Q}U$ have the same dimension over $\mathbb{Q}$. When the base field is $\mathbb{Q}$ and $S = \{\infty\}$ any of the units ε_w is called a Minkowski unit. For a really concrete proof of the existence of an Artin system of units, see [**2**], or [**3**], Ch. V,§ 3.

Let $f : \mathbb{Q}X \longrightarrow \mathbb{Q}U$ be a $\mathbb{Q}[G]$-isomorphism. It induces an isomorphism of $\mathbb{C}[G]$-modules $\mathbb{C}X \xrightarrow{\simeq} \mathbb{C}U$ which we denote by the same symbol. Composing with

λ, we get a $\mathbb{C}[G]$-automorphism of $\mathbb{C}X$, and given any representation V, it induces an automorphism

$$(\lambda \circ f)_V : \mathrm{Hom}_G(V^*, \mathbb{C}X) \longrightarrow \mathrm{Hom}_G(V^*, \mathbb{C}X)$$

$$\varphi \mapsto \lambda \circ f \circ \varphi.$$

Recall that the dimension of $\mathrm{Hom}_G(V^*, \mathbb{C}X)$ is $r(\chi)$ by (2.2.5). The *Stark regulator* is then defined to be

$$R(V, f) = R_S(V, f) = \det((\lambda \circ f)_V).$$

Remark that this last definition does not depend on the choice of the realization V of χ and thus we shall often write $R(\chi, f)$ instead of $R(V, f)$.

2.4. Stark's Basic Conjecture.

Given a representation V with character χ of a finite group G, we define $\mathbb{Q}(\chi) = \mathbb{Q}(\{\chi(\sigma) \mid \sigma \in G\})$; it is an abelian extension of $\mathbb{Q}$ since it is contained in a cyclotomic field. We denote its Galois group over $\mathbb{Q}$ by Δ_χ. From time to time we will use without comment the following well known lemma relating the Galois theory over $\mathbb{Q}$ to the group $\mathrm{Aut}(\mathbb{C})$) of all automorphisms of the field $\mathbb{C}$ of complex numbers.

Lemma 2.4.1. *Let K be a Galois extension of $\mathbb{Q}$ contained in $\mathbb{C}$. The map $\alpha \mapsto \alpha_{|K}$ is a surjective homomorphism of* $\mathrm{Aut}(\mathbb{C})$ *onto* $\mathrm{Gal}(K/\mathbb{Q})$, *and K is the fixed field of its kernel.*

For $\alpha \in \mathrm{Aut}(\mathbb{C})$ we can define a functor $V \mapsto V^\alpha$ from the category of representations of G to itself by viewing the map $\alpha : \mathbb{C} \to \mathbb{C}$ as a base extension and putting $V^\alpha := \mathbb{C} \otimes_\alpha V$. The map $v \mapsto v^\alpha := 1 \otimes v$ is an isomorphism $V \tilde{\to} V^\alpha$ of $\mathbb{Q}[G]$-modules, but is an "α-linear" map of $\mathbb{C}[G]$-modules, that is, $(cv)^\alpha = c^\alpha v^\alpha$ for $c \in \mathbb{C}$ and $v \in V$. If V has character χ then V^α has character χ^α. Of course the character χ^α, and hence the isomorphism class of the representation V^α, depend only on the image $\alpha_{|\mathbb{Q}(\chi)}$ of α in Δ_χ.

If G is the Galois group of a Galois extension of number fields K/k and V is a representation of G with character χ, it thus makes sense to consider $L_S(s, \chi^\alpha)$ as well as $c_S(\chi^\alpha)$ and $R_S(\chi^\alpha, f)$, for $\alpha \in \Delta$ or $\mathrm{Aut}\,\mathbb{C}$. By (2.2.5) we have

$$r(\chi) = \langle \chi, \chi_X \rangle = \frac{1}{|G|} \sum_{\sigma \in G} \chi(\sigma) \chi_X(\sigma),$$

since, as we remarked before, χ_X is integer valued. We thus have

$$r(\chi^\alpha) = r(\chi),$$

for all $\alpha \in \Delta_\chi$.

Define now $A(\chi, f) = A_S(\chi, f) := R_S(\chi, f)/c_S(\chi)$. Stark basic's conjecture can be stated as follows.

Conjecture 2.4.2. *Given any representation V of $G_{K/k}$ with character χ, we have*

- $A(\chi, f) \in \mathbb{Q}(\chi)$,
- $A(\chi, f)^\alpha = A(\chi^\alpha, f)$, *for all* $\alpha \in \Delta_\chi$.

We will soon see that the truth of this conjecture is independent of the choice of G-isomorphism $f : \mathbb{Q}X \to \mathbb{Q}U$. Lemma 2.4.1 above essentially solves

Exercise 2.4.3. Show that Stark's basic conjecture is equivalent to the statement $A(\chi, f)^\alpha = A(\chi^\alpha, f)$ for all $\alpha \in \mathrm{Aut}(\mathbb{C})$.

For $\alpha \in \mathrm{Aut}(\mathbb{C})$, a homomorphism of $\mathbb{C}[G]$-modules $\theta : V \to W$ induces the homomorphism $\theta^\alpha : V^\alpha \to W^\alpha$ defined by $\theta^\alpha(v^\alpha) = (\theta(v))^\alpha$.

Let $\theta : \mathbb{C}X \to \mathbb{C}X$ be a $\mathbb{C}[G]$-endomorphism of $\mathbb{C}X$. For each representation V of G, define $\delta(V, \theta) := \det(\theta_V)$, where θ_V denotes the $\mathbb{C}$-linear map $\phi \mapsto \theta \circ \phi$ of $\mathrm{Hom}_{\mathbb{C}[G]}(V^*, \mathbb{C}X)$ into itself. Clearly, $\delta(V, \theta)$ depends only on the isomorphism class of V, and we sometimes write $\delta(\chi, \theta)$ instead of $\delta(V, \theta)$ if χ is the character of V.

These determinants δ satisfy

(1) $\delta(\chi, \theta \circ \theta') = \delta(\chi, \theta)\delta(\chi, \theta')$
(2) $\delta(\chi^\alpha, \theta^\alpha) = \delta(\chi, \theta)^\alpha$ for $\alpha \in \mathrm{Aut}(\mathbb{C})$
(3) $\delta(\chi_1 + \chi_2, \theta) = \delta(\chi_1, \theta)\delta(\chi_2, \theta)$
(4) If $\chi = \mathrm{Ind}_H^G \psi$, then $\delta(\chi, \theta) = \delta(\psi, \theta)$.
(5) If $\chi = \mathrm{Infl}_{G/H}^G \xi$ then $\delta(\chi, \theta) = \delta(\xi, \theta \circ \iota_{K^H}^K)$, .

Property (1) is obvious. Property (2) comes by transport of structure and the fact that $\mathbb{C}X$ and $(\mathbb{C}X)^\alpha$ are isomorphic representations, since χ_X is rational valued. Property (3) is trivial. Property (4) comes from the fact that for a representation W of a subgroup H of G there is a canonical isomorphism $\mathrm{Hom}_G(\mathrm{Ind}_H^G W, \mathbb{C}X) \xrightarrow{\sim} \mathrm{Hom}_H(W, \mathbb{C}X)$. The key to understanding and proving (5) is that for a subfield $K' \subset K$, we use the imbedding $i_{K'}^K : Y_{S_{K'}} \to Y_{S_K}$ defined by $w' \mapsto \sum_{w|w'} [w : w']w$, where $[w : w']$ denotes the local degree $[K_w : K'_{w'}]$ The point is that $i_{K'}^K$ is the imbedding such that for each $S_{K'}$-unit $u' \in K'$ we have $i_{K'}^K \lambda_{K'}(u') = \lambda_K(u')$, where on the right we view u' as an S_K-unit in K. We leave the details to the reader.

Exercise 2.4.4. Using properties (1) and (2) above, show that Stark's basic conjecture (2.4.2), (2.4.3) is independent of the choice of f. In fact, show that if we extend the definition of $A(\chi, f)$ to arbitrary $\mathbb{C}[G]$-isomorphisms $f : \mathbb{C}X \to \mathbb{C}U$, by defining $A(\chi, f) = \delta(\chi, \lambda f)/c(\chi)$, then if the equation $A(\chi^\alpha, f^\alpha) = A(\chi, f)^\alpha$ is true for one f, it is true for all f.

Exercise 2.4.5. Using properties (3), (4), (5) of δ, show that for fixed f the function $\chi \mapsto A(\chi, f)$ obeys the same formalism as Artin L-functions in connection with $\chi_1 + \chi_2, \mathrm{Ind}_H^G \chi$, and $\mathrm{infl}_{G/H}^G \chi$. Hence conclude

Theorem 2.4.6.

- *If Basic Stark is true for all abelian characters of every subgroup H of G, then it is true for all characters of G.*
- *If it is true for all abelian characters of all Galois groups of number fields, it is true for all characters of all Galois groups of number fields.*
- *If it is true for all characters of a Galois group G, then it is true for all characters of each subgroup H of G.*
- *If it is true for all characters of Galois extensions of $\mathbb{Q}$, then it is true for all characters of all Galois extensions of number fields.*

Theorem 2.4.7. *The truth of Stark's Basic Conjecture for given K/k and χ is independent of S.*

PROOF. Let $S' = S \cup \{v\}$, where v is a place of k not in S. Denote by $X', U', \ldots$ things associated to S'. Let f, f' be as in our first statement of the basic Stark conjecture and define

$$B(\chi) := \frac{A'(\chi, f')}{A(\chi, f)} = \frac{R'(\chi, f')}{R(\chi, f)} / \frac{c(\chi')}{c(\chi)} \tag{2.4.8}$$

We must show that $B(\chi^\alpha) = B(\chi)^\alpha$ for all $\alpha \in \operatorname{Aut} \mathbb{C}$. Since the exact sequences

$$\{0\} \to \mathbb{Q}X \to \mathbb{Q}X' \to \mathbb{Q}X'/\mathbb{Q}X \to \{0\}$$

and

$$\{0\} \to \mathbb{Q}U \to \mathbb{Q}U' \to \mathbb{Q}U'/\mathbb{Q}U \to \{0\}$$

split, we can assume that $f'_{|\mathbb{C}X} = f$. Then

$$\frac{R(\chi, f')}{R(\chi, f)} = \det(\lambda' f' | \operatorname{Hom}_{\mathbb{C}[G]}(V^*, \frac{\mathbb{C}X'}{\mathbb{C}X}), \tag{2.4.9}$$

the determinant of the automorphism of

$$\operatorname{Hom}(V^*, \frac{\mathbb{C}X'}{\mathbb{C}X}) = \frac{\operatorname{Hom}(V^*, \mathbb{C}X')}{\operatorname{Hom}(V^*, \mathbb{C}X)} \tag{2.4.10}$$

induced by $\lambda' f'$.

On the other hand,

$$\frac{c'(\chi)}{c(\chi)} = \text{lead. coeff. of the expansion of} \det(1 - \chi(\sigma_v) NP_v^{-s} | V^{I_w}) \text{ at } s = 0. \tag{2.4.11}$$

By what we have seen, it is enough to treat the case in which V is 1-dimensional and its character $\chi : G \to \mathbb{C}$ is an injective homomorphism. Let w be a place of K above v.

First case, $\chi(G_w) \neq 1$: Then $V^{G_w} = \{0\}, r'(\chi) = r(\chi)$, and $\frac{R(\chi, f')}{R(\chi, f)} = 1$. On the other hand, $c'(\chi)/c(\chi)$ is equal to 1 if χ is non trivial on inertia, because then $V^{I_w} = \{0\}$ and is equal to $(1 - \chi(\sigma_v)\mathbf{N}P_v^{-s})\,|_{s=0} = 1 - \chi(\sigma_v) \neq 0$, if χ is trivial on inertia. Therefore

$$B(\chi) = 1 \text{ or } \frac{1}{1 - \chi(\sigma_v)}$$

in this case.

Second case, $\chi(G_w) = 1$: Then $G_w = \{1\}$, v splits completely in K, we have $P_v O_K = \coprod_{\sigma \in G} P_{\sigma w}$, and $\mathbf{N}P_w = \mathbf{N}P_v$. The $\mathbb{Q}[G]$-module $\mathbb{Q}X'/\mathbb{Q}X$ is isomorphic to $\mathbb{Q}[G]$. Hence, the $\bar{\chi}$-component of $\mathbb{C}X'/\mathbb{C}X$ is $\mathbb{C}e_{\bar{\chi}} w \bmod \mathbb{C}X$, of dimension 1, and $R'(\chi, f')/R(\chi, f)$ is the number $a \in \mathbb{C}$ such that

$$\lambda f(e_{\bar{\chi}} w) = a e_{\bar{\chi}} w \mod \mathbb{C}X$$

(In this formula and the following, we view w as a representative of an element $w - w_0$ in X' mod X, where w_0 is a fixed archimedean place, and we write simply λ and f instead of λ' and f'.)

Let $u = f(w - w_0)$. We have

$$\lambda f(e_{\bar{\chi}} w) = e_{\bar{\chi}} \lambda f(w) = e_{\bar{\chi}} \lambda(u) = e_{\bar{\chi}} \sum_{\sigma \in G} \log |u|_{\sigma w}) \sigma w$$

$$= \log \mathbf{N}P_v (\sum_{\sigma \in G} (-\operatorname{ord}_{\sigma w}(u) \bar{\chi}(\sigma)) e_{\bar{\chi}} w,$$

since P_v splits completely in K and $e_{\bar{\chi}}\sigma = \bar{\chi}(\sigma)e_{\bar{\chi}}$. On the other hand,

$$\frac{c'(\chi)}{c(\chi)} = \frac{d}{ds}(1 - (\mathbf{N}P_v)^{-s})\mid_{s=0} = \log \mathbf{N}P_v.$$

Conclusion:

$$B(\chi) = \sum_{\sigma \in G}(-\mathrm{ord}_{\sigma w} u)\chi(\sigma^{-1}).$$

Since in both cases $B(\chi)$ is a rational function of the quantities $\chi(\sigma)$, $\sigma \in G$, with rational coefficients, we have $B(\chi^\alpha) = (B(\chi))^\alpha$, as was to be shown. □

We finish this section by discussing the situation for representations whose character is rational valued. Stark's basic conjecture is true for these, but not trivial. We begin by showing that Stark's basic conjecture is true for the trivial representation $\mathbb{C}$. In this case the conjecture is reduced to proving that $A(\chi, f) \in \mathbb{Q}$ since $\mathbb{Q}(\chi) = \mathbb{Q}$. Since the truth of the conjecture is unchanged by inflation, we can assume that $K = k$, $G = 1$, and let $\chi = 1$ be the trivial character. Since the truth of the conjecture is independent of the choice of f, we can also suppose that f arises from an injection of $\mathbb{Z}$-modules $X \hookrightarrow U$. Under these assumptions, we shall compute $R(\chi, f)$. Remark first that $\mathrm{Hom}_G(\mathbb{C}^*, \mathbb{C}X) \simeq \mathbb{C}X$ in this case, and under this identification the automorphism $(\lambda \circ f)_{\mathbb{C}}$ of $\mathrm{Hom}_G(\mathbb{C}^*, \mathbb{C}X)$ is nothing else than the automorphism $\lambda \circ f$ of $\mathbb{C}X$. Therefore, we have $R(\chi, f) = \det(\lambda \circ f)$, and we can compute this last determinant explicitly as follows. Choose a place $v_0 \in S$. A $\mathbb{Z}$-basis for X is given by the $v - v_0$ where v runs over all places of S different from v_0. Let $\varepsilon_v = f(v - v_0)$. Since f is injective, we see that the ε_v are independent units and we have

$$\begin{aligned} \lambda \circ f(v - v_0) &= \lambda(\varepsilon_v) \\ &= \sum_{w \in S} \log |\varepsilon_v|_w \cdot w \\ &= \sum_{w \in S \smallsetminus \{v_0\}} \log |\varepsilon_v|_w (w - v_0), \end{aligned}$$

where this last equality is true by the product formula (0.0.1) and the fact that the ε_v are S-units. Hence, we have $\det(\lambda \circ f) = \det(\log |\varepsilon_v|_w)_{v,w \in S \smallsetminus \{v_0\}}$, and therefore

$$R(\chi, f) = \det(\lambda \circ f) = \pm \frac{[U : f(X)] \cdot R_S}{w_k}.$$

It follows that

$$A(\chi, f) = \frac{R(\chi, f)}{c(\chi)} = \pm \frac{[U : f(X)] \cdot R_S}{w_k} \cdot \frac{w_k}{h_S R_S} = \pm \frac{[U : f(X)]}{h_S} \in \mathbb{Q},$$

because $c(\chi) = -h_S R_S / w_k$, as we have seen in §2.2.

Note that it is possible to choose f such that $f(X)$ generates U mod torsion and $\det(\lambda \circ f)$ is positive. For such an f the ε_v's are a fundamental system of units, the Stark regulator $R(\chi, f)$ is the S-regulator R_S, and $A(\chi, f) = -w_k/h_S$.

Being true for the trivial representation, Stark's basic conjecture is true also for every representation which is induced from a trivial one, i.e., for any transitive permutation representation, hence for any permutation representation. By a theorem of Artin, if V is a representation whose character χ is rational valued, some positive integral multiple $m\chi$ of χ is the character of a permutation representation. Hence $A(\chi, f)^m \in \mathbb{Q}$ for some positive integer m. But one can in fact prove that $A(\chi, f) \in \mathbb{Q}$. Ted Chinburg pointed the way to a proof by conjecturing that for any irreducible χ, the character $\psi = \mathrm{Tr}_{\mathbb{Q}(\chi)/\mathbb{Q}}\chi = \sum_{\alpha\in\Delta}\chi^\alpha$ has the property that not only is $A(\psi, f) \in \mathbb{Q}$, but in fact $\pm A(\psi, f)$ is the norm from $\mathbb{Q}(\chi)$ to $\mathbb{Q}$ of a fractional ideal in $\mathbb{Q}(\chi)$. (See [**15**], Ch.II, §7 for a proof of this stronger statement.)

2.5. The case $r(\chi) = 0$.
If $r_S(\chi) = 0$, there is a unique f, the identity map of a $\{0\}$ vector space, and $R(\chi, f) = 1$. Thus $A(\chi, f)^{-1} = c_S(\chi) = L_S(0, \chi)$. In order to prove Stark's basic conjecture (2.4.2) is true in this case, we must show that $L_S(0, \chi^\alpha) = L_S(0, \chi)^\alpha$ for all $\alpha \in \mathrm{Aut}(\mathbb{C})$. Siegel,[**16**], has proved this for one dimensional χ. (Siegel proves, for abelian G, that the partial zeta functions $\zeta(\sigma, s)$, for which $L(s, \chi) = \sum_{\sigma\in G}\chi(\sigma)\zeta(\sigma, s)$, have rational values $\zeta(\sigma, 1-k) \in \mathbb{Q}$, for $k = 1, 2, \dots.$ Hence $L(1-k, \chi^\alpha) = L(1-k, \chi)^\alpha$, and in particular, $L(0, \chi^\alpha) = L(0, \chi)^\alpha$.) Thus the idea is to use Brauer's Theorem and write

$$\chi = \sum_i n_i \operatorname{Ind}_{H_i}^G \psi_i,$$

where the ψ_i are one dimensional characters of some subgroups H_i. Write S_i for $S_{K^{H_i}}$. The Artin L-function can then be decomposed as a product

$$L_S(s, \chi) = \prod_i L_{S_i}(s, \psi_i)^{n_i}.$$

One would like to say

$$L_S(0, \chi) = \prod_i L_{S_i}(0, \psi_i)^{n_i},$$

and then use Siegel's Theorem to conclude. Unfortunately, it is not that simple since some L-functions $L_{S_i}(s, \psi_i)$ might vanish at $s = 0$ and some of the n_i might be negative integers. In order to avoid this problem, we use a refinement of Brauer's Theorem due to Serre, for which a proof can be found in [**15**], Ch.III, p.72.

Lemma 2.5.1. *Let G be a finite group with center C, and let χ be an irreducible character of G. The restriction of χ to C is a multiple of a character ψ of degree one of C and we can write*

$$\chi = \sum_i n_i \operatorname{Ind}_{H_i}^G \psi_i,$$

where, for each i, H_i is a subgroup of G containing C, ψ_i is a character of degree one of H_i whose restriction to C is ψ, and the n_i are integers.

Now remark that, by (2.2.5), $r_S(\chi)$ is additive in χ, is ≥ 0 for all χ and is an increasing function of S for each χ, Therefore we can suppose that the representation V is irreducible and $S = S_\infty$. Since the truth of the conjecture is invariant under inflation, we can assume that V is faithful (that is, the associated homomorphism $G \longrightarrow GL(V)$ is injective). We have seen previously that if $\chi = 1$, the trivial character, then Stark's basic conjecture is true. We can thus assume as well that $\chi \neq 1$. Under these assumptions $V^G = \{0\}$, hence (2.2.5) shows that $V^{G_w} = 0$ for every infinite place w of K. This implies that $G_w \neq \{1\}$, hence $G_w = \{1, \tau_w\}$ for some τ_w of order 2. Moreover τ_w acts on V as -1, since its eigenvalues are ± 1. But it has no eigenvalue 1, since $V^{G_w} = \{0\}$. The representation V being faithful, we conclude that all the τ_w are equal to the same element τ in G, and that τ acts as the scalar -1 on V. Hence $\chi(\tau) = -\dim(V) = -\chi(1)$ and, since it acts as a scalar, τ is in the center C of G.

We can then apply Lemma 2.5.1 in order to conclude that the restriction of χ to C is equal to $n \cdot \psi$ for some $n \in \mathbb{Z}$ and some character ψ of degree 1 of C and

$$\chi = \sum_i n_i \operatorname{Ind}_{H_i}^G \psi_i,$$

where, for each i, H_i is a subgroup of G containing C, ψ_i is a character of degree one of H_i whose restriction to C is ψ, and the n_i are integers.

We thus have

$$\psi_i(\tau) = \psi(\tau) = \frac{\psi(\tau)}{\psi(1)} = \frac{n\psi(\tau)}{n\psi(1)} = \frac{\chi(\tau)}{\chi(1)} = -1, \tag{2.5.2}$$

Therefore each ψ_i is non-trivial on the decomposition groups of the places $v \in S$ which are all equal to $\{1, \tau\}$. It follows (2.2.5) that $r(\psi_i) = 0$, hence $L(0, \psi_i) \neq 0$ for all i.

2.6. The case $r(\chi) = 1$ and Stark units.

Let χ be an irreducible character of $G_{K/k}$ for which $r_S(\chi) = 1$ for some S. This means that χ has multiplicity 1 in $\mathbb{C}X$, and since $\mathbb{C}X$ comes from a representation $\mathbb{Q}X$ defined over $\mathbb{Q}$, it follows that $r_S(\chi^\alpha) = 1$ for all $\alpha \in \Delta = \mathrm{Gal}(\mathbb{Q}(\chi)/\mathbb{Q}$. For each $\alpha \in \Delta$, let V^α denote the unique irreducible constituent of $\mathbb{C}X$ with character χ^α, and let $\bar{V}$ denote the one of these with character $\bar{\chi}$, which is isomorphic to V^*. The central idempotent

$$e_{\chi^\alpha} = \frac{\chi(1)}{|G|} \sum_{\sigma \in G} \chi(\sigma^{-1})^\alpha \, \sigma = e_\chi^\alpha \in \mathbb{C}[G]$$

projects $\mathbb{C}X$ onto V^α, and kills all other irreducible constituents of $\mathbb{C}X$. Let π be the sum of these idempotents.

$$\pi := \sum_{\alpha \in \Delta} e_\chi^\alpha. \tag{2.6.1}$$

Clearly, π lies in $\mathbb{Q}[G]$ and therefore projects $\mathbb{Q}X$ onto a $\mathbb{Q}[G]$-submodule W of $\mathbb{Q}X$ such that $\mathbb{C}W = \sum_{\alpha \in \Delta} V^\alpha$ and has character $\sum_{\alpha \in \Delta} \chi^\alpha$. For each $\mathbb{C}$-valued function $h : \Delta \to \mathbb{C}$, the endomorphism of $\mathbb{C}W$ which acts on V^α as scalar multiplication by $h(\alpha)$ is induced by the action of the element

$$\theta_h := \sum_{\alpha \in \Delta} h(\alpha) e_{\chi^\alpha}. \tag{2.6.2}$$

These operators θ_h form a commutative ring isomorphic to $\mathbb{C}^{|\Delta|}$ (in fact, canonically isomorphic to $\mathbb{C} \otimes_{\mathbb{Q}} \mathbb{Q}(\chi)$ via the map which, for $a \in \mathbb{C}$ and $z \in \mathbb{Q}(\chi)$, associates the tensor $a \otimes z$ with the operator $a\,\theta_{h_z}$, where

$$h_z(\alpha) := z^\alpha. \tag{2.6.3}$$

Lemma 2.6.4. *The following are equivalent conditions on h.*

(1) $h(1) \in \mathbb{Q}(\chi)$ *and* $h(\alpha) = h(1)^\alpha$ *for all* $\alpha \in \Delta$.
(2) $\theta_h \in \mathbb{Q}[G]$.
(3) $\theta_h W \subset W$.

PROOF. Let H_i denote the set of h's satisfying condition (i).

$H_1 \subset H_2$: For $h \in H_1$ and $\sigma \in G$, the coefficient of σ in θ_h is the trace from $\mathbb{Q}(\chi)$ to $\mathbb{Q}$ of the element $h(1)\chi^{-1}(\sigma)$ and is therefore in $\mathbb{Q}$.

$H_2 \subset H_3$: This is evident since $\sigma W \subset W$ for all $\sigma \in G$.

$H_3 \subset H_1$: We know $H_1 \subset H_3$ and must show that this inclusion is an equality. The map $h \mapsto h(1)$ of H_1 into $\mathbb{Q}(\chi)$ is bijective (the inverse map is $z \mapsto h_z$, (2.6.3)). Hence it will suffice to show $\dim_{\mathbb{Q}} H_3 \leq [\mathbb{Q}(\chi) : \mathbb{Q}] = |\Delta|$. This is true because the obvious maps $H_3 \to \mathrm{End}_{\mathbb{Q}}(W)$ and $\mathbb{C} \otimes_{\mathbb{Q}} \mathrm{End}_{\mathbb{Q}[G]}(W) \to \mathrm{End}_{\mathbb{C}[G]}(\mathbb{C}W)$ are injective, hence

$$\begin{aligned}\dim_Q H_3 &\leq \dim_{\mathbb{Q}} \mathrm{End}_{\mathbb{Q}[G]}(W) = \dim_{\mathbb{C}}(\mathbb{C} \otimes_{\mathbb{Q}} \mathrm{End}_{\mathbb{Q}[G]}(W)) \\ &\leq \dim_{\mathbb{C}} \mathrm{End}_{\mathbb{C}[G]}(\mathbb{C}W) = \dim_{\mathbb{C}} \prod_{\alpha \in \Delta} \mathrm{End}_{\mathbb{C}[G]}(V^{\alpha}) = |\Delta|.\end{aligned}$$

□

Now let $f : \mathbb{Q}X \to \mathbb{Q}U$ be a $\mathbb{Q}[G]$-homomorphism and put:
- $A = \sum_{\alpha \subset \Delta_\chi} A(\chi^\alpha, f) e_{\bar{\chi}^\alpha}$
- $c = \sum_{\alpha \in \Delta_\chi} c(\chi^\alpha) e_{\bar{\chi}^\alpha} = \sum_{\alpha \subset \Delta_\chi} L'(0, \chi^\alpha) e_{\bar{\chi}^\alpha}$
- $R = \sum_{\alpha \in \Delta_\chi} R(\chi^\alpha, f) e_{\bar{\chi}^\alpha}$

Theorem 2.6.5. *Let χ be an irreducible character of G such that $r(\chi) = 1$. Let $W_U = f(W) = \pi\mathbb{Q}U$. The following assertions are equivalent.*

(1) *Stark's Basic Conjecture is true for χ.*
(2) $cW = \lambda(W_U)$
(3) $h_z c\mathbb{Q}X \subset \lambda\mathbb{Q}U$, *for all $z \in \mathbb{Q}(\chi)^*$*

PROOF. (1) $\Leftrightarrow$ (2): Lemma 2.6.4 with $h(\alpha) = A(\chi^\alpha, f)$ shows that Stark's basic conjecture holds for χ if and only if $AW \subset W$. Each of the three operators A, c and R act injectively on $\mathbb{C}W$. Thus, we have

$$AW \subset W \;\Leftrightarrow\; W = AW \;\Leftrightarrow\; cW = cAW.$$

We have $cA = R$ by the definition of $A(\chi, f)$, and to conclude we have only to show that $RW = \lambda W_U$. For each α, the operator R acts on $\bar{V}^\alpha$ by the scalar $R(\chi^\alpha, f)$. By definition, the scalar $R(\chi, f)$ is the determinant of the endomorphism $(\lambda f)_V$ of the 1-dimensional vector space $\mathrm{Hom}_{\mathbb{C}[G]}(\bar{V}^\alpha, \mathbb{C}X) = \mathrm{Hom}_{\mathbb{C}[G]}(\bar{V}^\alpha, \bar{V}^\alpha)$, which is the same as the scalar by which λf acts on $\bar{V}^\alpha$. Hence $RW = \lambda f W = \lambda W_U$ as claimed

(2) $\Leftrightarrow$ (3) : Condition (3) holds for $z = 0$. If $z \neq 0$ then $h_z\mathbb{Q}X = \mathbb{Q}W$, so the left hand sides of (2) and (3) are the same, and (2) $\Rightarrow$ (3) because $W = \pi U \subset \mathbb{Q}U$. On the other hand, applying π to (3), say with $z = 1$, gives $cW \subset \lambda\mathbb{Q}W_U$, and this inclusion is an equality because the two spaces have the same $\mathbb{Q}$-dimension. □

We can now explain the notion of Stark units. Let Ψ be a set of irreducible characters of G such that

(1) The trivial character is not contained in Ψ,
(2) For all $\chi \in \Psi$, $\chi^\alpha \in \Psi$, for all $\alpha \in \mathrm{Aut}(\mathbb{C})$.

Let $(z_\chi)_{\chi\in\Psi}$ be a family of complex numbers satisfying $z_{\chi^\alpha} = z_\chi^\alpha$ for all $\chi \in \Psi$ and for all $\alpha \in \mathrm{Aut}(\mathbb{C})$. Assuming Stark's Basic Conjecture, Theorem 2.6.5 implies that

$$\left(\sum_{\chi\in\Psi} z_\chi L'_S(0,\chi)e_{\bar\chi}\right)\cdot X \subset \lambda(\mathbb{Q}U), \tag{2.6.6}$$

because the left hand side of (2.6.6) is a sum, over non-conjugate $\chi \in \Psi$, of left hand sides of assertion (3) of the theorem. Since the trivial character is not in ψ, we can replace X by Y in (2.6.6) and we get the following result: If w is a place in S_K, then there exists a positive integer m and a S-unit ε of K such that

$$\lambda(\varepsilon) = m\sum_{\chi\in\Psi} z_\chi L'_S(0,\chi)e_{\bar\chi}\cdot w.$$

Suppose for simplicity that there is a $v \in S$ which splits completely, and fix a place $w \in K$ lying above v. Then recalling that $e_{\bar\chi} = \frac{\chi(1)}{|G|}\sum_{\sigma\in G}\chi(\sigma)\sigma$, we can rewrite this last equation as

$$\begin{cases} |\varepsilon|_{w'} = 1, & \text{if } w' \nmid v,\\ \log|\varepsilon|_{\sigma w} = \frac{m}{|G|}\sum_{\chi\in\Psi}\chi(1)z_\chi L'_S(0,\chi)\chi(\sigma), & \text{for all } \sigma\in G.\end{cases}$$

Given m, these equations determine the "Stark unit" ε, if it exists, up to a root of unity in K.

For example, suppose v is a real place which splits completely in K. Then w is given by an imbedding $K \subset \mathbb{R}$, and $\varepsilon = \pm e^{\log|\varepsilon|_w}$. Hence, using $|\varepsilon|_{\sigma w} = |\varepsilon^{\sigma^{-1}}|_w$ we find

$$\varepsilon^{\sigma^{-1}} = \pm e^{\frac{m}{|G|}\sum_{\chi\in\Psi}\chi(1)z_\chi L'_S(0,\chi)\chi(\sigma)}, \quad \text{for all } \sigma\in G. \tag{2.6.7}$$

(Note that the exponent is real, because, complex conjugation being continuous, we have $L(\bar s,\bar\chi) = \overline{L(s,\chi)}$.) Equation (2.6.7) is a marvelous conjectural expression of an algebraic number and its conjugates over k in terms of special values of analytic functions, of a sort different from what Hilbert expected in his 12th problem. But what can we take for m? If K/k is abelian and S contains the places v ramified in K, Stark has a stronger conjecture for which there is overwhelming evidence, both numerical and structural, namely, that $m = |\mu(K)|$ will do for the case $z_\chi = 1$ for all χ. The non-abelian situation is unclear, but Chinburg [**5**] has an interesting discussion of the case $k = \mathbb{Q}$ and $\dim V = 2$. In general one supposes m is small, at least in case K is totally real.

If we guess a valid m and double it, then all the signs in (2.6.7) are $+$, and we can calculate real numbers close to the conjugates of ε at the real place w. We can then compute approximations to the coefficients of the polynomial $f(X) = \prod_\sigma(X - \varepsilon^\sigma)$ and try to find the small integers in k which are those coefficients, and check that the roots of the resulting polynomial $f(X) \in k[X]$ generate the extension K/k which we started with. This experiment has been carried out successfully hundreds of times in the past 35 years, with extensions having many different small groups G, giving overwhelming experimental evidence for Stark's basic conjecture. There are also special cases in which the conjecture is a classical result, as in the case in which $k = \mathbb{Q}$, and K is the real subfield of the group of nth roots of 1, in which case the Stark unit is the cyclotomic unit $2 - 2\cos(\frac{2\pi}{n})$. But to my knowledge we are no closer to a proof of Stark's basic conjecture for real v in the general case, or even in case of abelian χ with $r(\chi) - 1$, than we were when Stark first announced it 35 years ago.

Bibliography

[1] E. Artin, *Zur Theorie der L-Reihen mit allgemeinen Gruppencharakteren*, Hamb. Abh. **8** (1930), 292–306.

[2] E. Artin, *Über Einheiten relativ galoisscher Zahlkörper*, J. reine angew. Math., (1932), 153–156.

[3] E. Artin and J. Tate, "Class Field Theory," AMS Chelsea, 2008.

[4] J.W.S. Cassels and A. Fröhlich, "Algebraic number theory," Academic Press, New York, 1967, 2d edition, LMS, 2010.

[5] T. Chinburg, *Stark's conjecture for L-functions with first order zeroes at* $s = 0$, Advances in Math. **48** (1983), 82–113.

[6] T. Chinburg, *On the Galois structure of algebraic integers and S-units*, Inventiones Math. **74** (1983), no. 3, 321–349.

[7] P. Deligne, *Les constantes des équations fonctionnelles des fonctions L*, in "Modular forms of one variable II," Springer Lecture Notes n° **359** (1972), 501–597.

[8] P. Deligne and J.-P. Serre, *Formes modulaires de poids 1*, Ann. Sci. E.N.S., 4e sér., t7 (1974), 507–530.

[9] K. Knopp, "Theory of Functions, Part II", Dover Publications, 1947.

[10] J. Martinet, *Character theory and Artin L-functions*, in "Algebraic number fields," ed. A. Fröhlich, Academic Press, New York, 1977.

[11] J.-P. Serre, "Corps Locaux," Hermann, Paris, 1962.

[12] J.-P. Serre, "Linear Representations of Finite Groups," Springer, New York, 1971. (Translation of "Représentations linéaires des groupes finis," Hermann, Paris, 1967.)

[13] C.L. Siegel, *Berechnung von Zetafunktionen an ganzzahligen Stellen*, Nachr. Akad. Wiss. Göttingen Math.-Phys. Kl. II (1969) 87–102.

[14] H.M. Stark, *L-functions at s=1, I,II,III,IV*, Advances in Math., 7 (1971), p. 301-343; 17 (1975), p. 60-92; 22 (1976), p 64-84; 35 (1980), p. 197-235.

[15] J. Tate, "Les Conjectures de Stark sur les Fonctions L d'Artin en $s = 0$," Progress in Mathematics, Vol.47, Birkhäuser, Boston-Basel-Stuttgart, 1984.

[16] A. Weil, "Basic Number Theory," 3rd Edition, Springer-Verlag, 1974.

The Origin of the "Stark Conjectures"

Harold Stark

IAS/Park City Mathematics Series
Volume 18, 2009

The Origin of the "Stark Conjectures"

Harold Stark

Introduction

The aim of this paper is to explain how my first order zero conjecture on values of L-functions [**23**] came to be. The fact that [**23**] is Part IV of a series of papers indicates that the conjecture did not spring into existence overnight. But in fact, it goes further back than one might think. This paper is more a personal history than a mathematics paper, although Appendix 2 should be of interest mathematically.

1. 1966: Class-number one complex quadratic fields

In this paper, k_d will denote the quadratic field of a discriminant d and χ_d will denote the corresponding Dirichlet character of conductor $|d|$ given explicitly for $n > 0$ by the Kronecker symbol

$$\chi_d(n) = \left(\frac{d}{n}\right).$$

The class number of k_d will be denoted by $h(d)$. The origins of the conjectures go back all the way to my 1966 proof that there are precisely nine complex quadratic fields of class-number one. This appeared at the start of 1967 [**14**]. This work made use of functions used by Heilbronn in 1934 who proved the Gauss conjecture (in the language of quadratic forms) that $h(d) \to \infty$ as $d \to -\infty$. The proof was ineffective and did not allow a finite search to determine all fields of a given class-number. In a followup paper, Heilbronn and Linfoot [**7**] showed that there were at most ten complex quadratic fields of class number one (nine are well known) and the tenth, if it existed, would have violated GRH.

To study $h(d)$, Heilbronn's papers made use of two analytic functions. The first of these is the Epstein zeta function for binary positive definite quadratic forms,

$$\zeta(s, Q) = \frac{1}{2}\sum_{m,n}{}' Q(m,n)^{-s}$$

where $\sum{}'$ means the sum is over all $m, n \neq 0, 0$ and

$$Q(x, y) = ax^2 + bxy + cy^2$$

is a positive definite quadratic form of discriminant

$$d = b^2 - 4ac < -4.$$

In this paper, d is restricted to be a "fundamental discriminant" other than -3 and -4. In other words, d is the discriminant of a complex quadratic field other

Department of Mathematics, UCSD, La Jolla CA 92093 USA
E-mail address: stark@math.ucsd.edu

than k_{-3} and k_{-4}. We are skipping -3 and -4 because the number of units in all k_d with $d < 0$ is 2 except for these two fields and I am too lazy to account for the changes in these two instances.

This particular case of the Epstein zeta function goes back to Dirichlet who found in essence that $\zeta(s, Q)$ has a first order pole at $s = 1$ of residue

$$\text{res}_{s=1}\Big(\zeta(s, Q)\Big) = \frac{\pi}{\sqrt{|d|}}.$$

When summed over a complete set of inequivalent quadratic forms Q, we get the zeta function of k_d,

$$\sum_Q \zeta(s, Q) = \zeta(s, k_d) = \zeta(s)L(s, \chi_d), \tag{1}$$

as Dirichlet knew. Obviously, Dirichlet did not call the character a Kronecker symbol! Nor did he put in the middle entry $\zeta(s, k_d)$ in Equation (1), as he was dealing with quadratic forms. But the total number of representations of a positive integer n by a complete set of inequivalent quadratic forms of discriminant d was known and that suffices to give the outer ends of Equation (1).

The Epstein zeta functions were individually analytically continuable to the entire complex s-plane except for the above mentioned first order pole at $s = 1$, but while I was in graduate school, I didn't even know that. The expansions for $\zeta(s, Q)$ and later for $L(s, \chi_f, Q)$ that Heilbronn gave were valid for $\sigma = Re(s) > 1/2$. But for class-number one, I did know that $\zeta(s)$ and more generally, the Dirichlet L-functions were analytically continuable.

Heilbronn also introduced characters into the mix. Suppose that f is also the discriminant of a quadratic field and for simplicity let

$$(f, d) = 1.$$

Define for the same quadratic forms as above,

$$L(s, \chi_f, Q) = \frac{1}{2}\sum_{m,n}{}'\chi_f\Big(Q(m, n)\Big)\cdot Q(m, n)^{-s}.$$

The analogue of Equation (1) is then

$$\sum_Q L(s, \chi_f, Q) = L(s, \chi_f)L(s, \chi_{fd}). \tag{2}$$

While Heilbronn and Linfoot were interested in potential real zeros of $\zeta(s, Q)$ and $L(s, \chi_f, Q)$, in **[14]** I was interested in the value of $L(1, \chi_f, Q)$. When $h(d) = 1$, we know the value of $L(1, \chi_f, Q)$ from Equation (2) and Dirichlet,

$$L(1, \chi_f, Q) = \frac{2\pi h(fd)h(f)\log(\varepsilon_f)}{f\sqrt{|d|}} \text{ when } f > 0 \tag{3}$$

and a similar formula holds when $f < 0$. In **[14]**, I made use of the value when $f > 0$.

I expanded $L(1, \chi_f, Q)$ via the Poisson summation formula. These expansions were very rapidly convergent. That allowed me to find a relation between the two values of $L(1, \chi_f, Q)$ in Equation (3) with $f = 8$ and $f = 12$. After many simplifications, there resulted a Diophantine equation with rational integer values having precisely six solutions, corresponding to the six already known fields with

$d = -3$ (not covered in our discussion here), -11, -19, $-43, -67, -163$. A tenth field would have provided a larger solution to the equation and so could not exist.

My proof that there were exactly nine $h(d) = 1$ complex quadratic fields was the first accepted complete proof of this theorem. At the same time, there was an effective proof given by Baker which appeared at the end of 1966 [**1**]. Baker made a major advance in the transcendence theory of linear forms of logarithms of algebraic numbers which allowed one for the first time to specify by transcendence methods and upper bound on the absolute value of the hypothetical "tenth discriminant" and so it became effectively decidable whether or not the tenth discriminant would exist.

Interestingly, the final Diophantine equations that I found had already been encountered in a 1952 paper of Heegner [**6**]! For the moment, suffice it to say Heegner's proof was (to those who still knew of its existence) universally deemed incorrect. I will say more about this in Appendix 1.

2. 1970: The beginnings of the conjectures

I learned of the general analytic continuation and functional equations of Epstein zeta functions in 1965. But unlike the situation with no characters, there was no analytic continuation to the left of $\sigma = 1/2$ for the individual $L(s, \chi_f, Q)$ that I was aware of in 1966. And needless to say, there was no functional equation or value of $L(1, \chi_f, Q)$ outside the class-number one setting that I knew of. But motivated by the fact that individual Epstein zeta functions could be treated without a class-number assumption, I followed the classical methods and in 1968 found an analytic continuation and functional equation for the $L(s, \chi_f, Q)$ [**15**]. I then turned to the question of evaluation $L(1, \chi_f, Q)$.

I was also becoming more comfortable with the number field version of everything and in fact it turned out that all three of the problems at the start of the Section had already had answers of a sort. If $\mathfrak{a}$ is an integral ideal in a quadratic field and χ is a Dirichlet character defined on Z with conductor f, then the function

$$\widetilde{\chi}(\mathfrak{a}) = \chi\Big(\mathbb{N}(\mathfrak{a})\Big) \tag{4}$$

is a ray class character of k_d whose conductor divides f, but can occasionally be smaller. As a result, Hecke's general analytic continuations and functional equations [**5**] which he proved for the partial L-functions (the contributions from just one ideal class) would have instantly given the analytic continuation and functional equations for the $L(s, \chi_f, Q)$.

The Kronecker second limit formula would also have given the value of $L(1, \chi_f, Q)$, although there are, very loosely speaking, around f^2 ray classes in a single ideal class, so it was not an appealing prospect from the point of view of doing any numerical work. It was actually even worse than this, because at $s = 1$, the second limit formula is not just one ray class out of $L(1, \chi_f, Q)$. However, in 1968, Siegel published his version of a class-number one proof [**13**]. It was in this paper where I learned that for $\chi = \chi_f$, the character $\widetilde{\chi}_f = \widetilde{\chi}$ given by Equation (4) is not only a ray class character of k_d but also a ring class character whose conductor is precisely f when $(f, d) = 1$. This is nice because there are very loosely speaking, around f ring classes per ideal class. Besides the lowering from f^2 to f, the first Kronecker limit formula applies, and as Siegel showed, it does so one ring class at a time.

This led me to an evaluation of $L(1, \chi_f, Q)$. It did not have the optimum denominator, but it had an explicit coefficient times the log of a number in the ring class field of k_d of conductor f. This number was given by the theory of complex multiplication as products of Δ-quotients evaluated at points of k_d. (This is also why the denominator was not optimal. But it was easier to deal with delta-quotients than eta-quotients.) This result appeared as the first part of my four part series on values of L-functions at $s = 1$ [**18**]. I had not yet turned to $s = 0$. I had this result already in early 1970, which is the time when I spent several months in Göttingen visiting Siegel.

I saw Siegel practically ever weekday during the time I was there. I was also beginning to wonder how general my evaluation of $L(1, \chi_f, Q)$ could be; that is, for how general an L-function would there be an interesting value at $s = 1$? This was motivated in part by my earlier experiences with $\zeta(s, Q)$ and $L(s, \chi_f, Q)$ in the class-number one setting, but which ultimately behaved nicely far more generally. Naturally I suggested to Siegel the possibility that a general Artin L-function could be evaluated at $s = 1$. I rather liked the corollary that regulators could be factored into pieces if such a possibility were true, but I didn't get the chance to mention the corollary.

Siegel had precisely the opposite reaction! This is absurd, he said (there are no quotation marks, because I don't remember his exact words), because it would imply that a regulator could be factored into pieces. That same evening, I factored a few S_3 type regulators of the sixth degree fields by hand and playing with the determinants. The method was to allow extra fudge factors and just find nice units, preferably from subfields when possible, which generate a subgroup of the unit group of finite index. With such units, there is a good chance that row and column operations can simplify the determinant and split it into pieces.

I presented my factorizations to Siegel the next day and this reminded him of a result of Brauer on class-number relations from 1951 [**3**]. There are many multiplicative identities of products of zeta functions of number fields. This is equivalent to finding linear dependencies among induced trivial characters of subgroups of a group. One looks at the value at $s = 1$ of the corresponding zeta product (the poles all cancel each other out) and gets on one side a product and quotient combination of class-numbers and on the other side, a product and quotient combination of regulators. (At $s = 1$, one also has to deal with the pi and discriminant terms; group theory gets rid of these. In retrospect, all this should have been done at $s = 0$).

Thus the regulator combination is a rational number (the class-number combination) and so it is no shock that there should be a factorization theory for regulators. Brauer showed via representation theory that the regulator combination could be bounded above and below, thereby providing a class-number relation up to a bounded factor.

Siegel also told me about group determinants. It seems that as a young man, Siegel had a group theory course from Frobenius. Frobenius founded the theory of representations of finite non-abelian groups and group determinants played a central role in developing his theory. Nowadays, factorizations of group determinants are a consequence of representation theory rather than the other way around. So I started to learn a little about representations of finite groups in 1970. I published my first very vague version of the conjectures in my 1970 ICM talk [**17**].

3. 1975: The first calculations

In 1975, I learned that it was possible at MIT to program a remote computer on an interactive basis via an electronic typewriter. This was a huge advance over punching in a program on IBM cards. (I had resisted using punch cards beginning graduate school; probably my advisor was not happy with that. My thesis result cited in Appendix 1, was produced on a mechanical desk calculator!) And so, I decided to test my rather vague conjecture on real quadratic fields. Siegel and Ramachandra had proved my vague conjecture for complex quadratic fields via complex multiplication and the Kronecker limit formulas. The fudge factors were horrendous, but explicit. But nothing at all was known for real quadratic ground fields, and seemingly, nothing had been suspected either.

The first order zero conjecture seemed a reasonable starting point since I have never seen a way to extract a matrix entry from the value of a 2×2 determinant. What I was already after was a way to find the class field via the values of the L-functions. This presented me with a problem. There were two potential calculations that could be made: one at $s = 1$ and one at $s = 0$. I had discovered that the zeta function of a field had a much nicer expression at $s = 0$ than at $s = 1$. It was also the case that Dirichlet L-functions were nicer at $s = 0$ than at $s = 1$. All the complications that Dirichlet faced disappeared when the derivation was switched over to $s = 0$. See Appendix 2 to see an example of how much simpler the classic Dirichlet evaluations become at $s = 0$.

I had seen in Part II [**19**] that relative quadratic class-fields could be dealt with theoretically and so I settled upon trying to find a cyclic quartic class-field of k_5, four of whose conjugates over $\mathbb{Q}$ are real and the other four conjugates over $\mathbb{Q}$ are complex. This would give me an L-function $L(s)$ with a first order zero at $s = 0$ for which I had no theoretical interpretation of $L'(0)$. When I factored up the regulator using a set of Artin units (a generalization of Minkowski units when $\mathbb{Q}$ is the ground field), it gave me a hint as to the shape of my desired $L'(0)$ value. Half of the algebraic conjugates of my units were to be real and, I hoped, would show up in $L'(0)$, the other half of the algebraic conjugates were to be complex and my Artin unit factorization suggested the possibility that the complex conjugates could have absolute value one. This was the origin of the absolute value part of the first order zero conjecture. I was hoping that a fudge factor here would be at least rational; if not, I was prepared to look at $L(1)$, but everything where a result was known suggested that results would be simpler at $s = 0$.

And so, I decided to program up a specific example over k_5 in double precision Fortran (fifteen significant digits!!). I also had to program in the arithmetic of the field. The calculated values turned out to give exactly, or at least (more or less) 15 digits of, what I'd hoped for and I had calculated my first cyclic fourth degree class-field over a real quadratic field. The conjectured evaluation is still unproved in the case of this example. This example and one other example over k_{229}, a class-number three field, were presented at the Durham conference in 1975 [**20**].

Over the next few months, I ran many more examples of class fields over real quadratic fields with class-numbers out to 5 or 6 and the class fields having degrees into the 20's. Fifteen significant digits were a serious restriction on how much further I could go and still convincingly find the defining equations of the class-fields. The examples included many instances of non-prime conductors and the data clearly showed that what was being matched were the imprimitive versions of those

L-functions whose characters had lower conductors. All this was formulated into an initial conjecture in Part III [**21**]. I still did not have the fudge factor completely nailed down in its ultimate form, but it was close. In 1977 I also programed up a single totally real cubic ground field example [**22**].

4. The full "over $\mathbb{Z}$" conjecture

Siegel [**12**] for Hilbert class fields and Ramachandra [**10**] for ray class fields, had provided evaluations of L-functions over complex quadratic fields at $s = 1$, but their fudge factors were really awful, resulting in units with amazingly high indices in the full unit group of the class field. A natural goal would be an index equal to the class-number of the class field. Lower than that would imply the class-number of the class field would be larger than one (although I do have a few examples of precisely that). The fudge factors I was getting over real quadratic fields were much better than what Siegel and Ramachandra had obtained, so if I was going to cover all first order zero situations, it seemed to me that I had to improve the known results for complex quadratic ground fields.

One feature of the Kronecker limit formula results was that the first and second limit formulas produce units (or sometimes a unit outside a single $\mathfrak{p}$ in a larger ray class field than one is seeking. A high power of this unit then turns out to be in the desired class field. This was responsible for a part of the indices produced by Siegel and Ramachandra. To attack this, I developed in Part IV [**23**] a version of Shimura's reciprocity law [**11**] (see also Lang [**8**]) that I was capable of using.

Shimura's reciprocity law is the ultimate in a series of special cases going back to Hasse's papers [**4**] and is extremely useful for determining in which class field a modular function at a CM point lands. I found a way of using the reciprocity law to lower the power to get into the class field and the $-1/W$ factor in the conjecture was born. Here W is the number of roots of unity in the class field. The minus sign is there because with it we get algebraic integers; in complex quadratic ground fields when there are prime power conductors, we do not get units and without the minus sign, we would not get integers. The same thing happens over $\mathbb{Q}$.

We are almost to the "over $\mathbb{Z}$" part of the conjecture. The conjecture had now become that each (imprimitive version of) $L'(0, \chi)$ would be $-1/W$ times a character combination of logs of normalized absolute values of the conjugates of one S-unit in our desired class-field and this S-unit would be independent of the choice of χ. But when we multiply all these $L'(0, \chi)$ values out to get (a good chunk of) the lead term at $s = 0$ of the zeta function of the class field (which is $-HR/W$ with the same W; H and R are the class-number and regulator of the class field), we get W to a power in the denominator and are thus doomed to have at least a power of W times H as the index of our units in the unit group of the class field.

This problem can be overcome when we are over complex quadratic ground fields, thanks to the fact that the Wth root of our S-unit, ε, is also in a larger class field. This means a conjugate of ε divided by an appropriate power of ε itself is a Wth power in our desired class-field and this allows us to remove all the W's from the index. The easiest way to make this happen for the general conjecture is to demand that $\varepsilon^{1/W}$ is in a nice extension of our ground field and our full conjecture is now accounted for. In fairness to others braver than me, I did not demand my sufficiently nice extension to be abelian over the ground field, but just that it was a central extension. This sufficed to get rid of the excess W terms. All examples to date (and there were already many in 1980) have turned out to be abelian and "abelian" is now the common phrasing.

Appendix 1 Heegner.

In 1952 Heegner [**6**] purported to prove that the "tenth discriminant" does not exist via the theory of complex multiplication. He derived a Diophantine equation whose rational integer solutions would be in 1-1 correspondence with the class-number one complex quadratic fields of discriminants other than -4, -7, -8. Heegner found that there were six solutions corresponding to the remaining known class-number one discriminants, -3, -11, -19, -43, -67, -163.

Sadly, the mathematical community at the time dismissed Heegner's proof as incorrect. The commonly stated reason was that it depended on an unproved conjecture of Weber [**24**]. (The referee notes that Weber presented observations but did not state conjectures; however, "conjecture" is how most people refer to the problem and so I will continue to use the word here.)

This conjecture stated that the classical modular function $f(z)$ has$f(\sqrt{d})$ in the ring class field modulo 2 of $Q(\sqrt{d})$ for certain discriminants including the class-number one discriminants from -11 onwards. Heegner was not anointed as a professional mathematician (see [**9**] for a one paragraph biography). This may have been part of the problem. (The referee of this article noted that Heegner certainly behaved as a mathematician.) The other part of the problem is the presentation in his paper. The paper is devoted to solving Diophantine equations with the assistance of the theory of complex multiplication. At some points of the paper, Heegner indeed uses Weber's conjecture. But weaving in and out of the paper is the material needed for the class-number one theorem, and Weber's conjecture is not used in these parts of the paper. What Heegner does need, and what Weber purports to prove, is that $f(\sqrt{d})^2$ generates the desired ring class-field.

As noted, Weber's conjecture does not appear as a conjecture in [**24**]; in fact it appears as a "Theorem" with capital T! Of course, one expects nothing less than a capital T from a German writer. The "Theorem" in question is item 8 of Section 127, between equations (23) and (24). Weber makes it clear that this "Theorem" has been arrived at by what we would call now incomplete induction and not by rigorous proof. His "Theorem" is that $f(\sqrt{d})^3$ is in the desired ring class field; combined with $f(\sqrt{d})^2$, we deduce that $f(\sqrt{d})$ is in (and ultimately generates) the same class field.

By the time I was in graduate school, Heegner's paper was no longer known. Even the reviewer of his paper for Math Reviews (under whom I originally got interested in the class-number one problem as an undergraduate) never mentioned the paper to me. And many published papers since Heegner's continued to discuss the "unsolved" problem with no references to Heegner. I believe I am the modern rediscoverer of Heegner's paper. I came across it in Math Reviews somewhere in academic 1962-63 while working on my PhD thesis. The last sentence of the review is a thesis killer!

My thesis research, well underway, was going to push the bound on the tenth discriminant out much, much further than then known. (from Lehmer's 1930's $5 \cdot 10^9$ out to $10^{10000000}$, if you care. When the constants in Baker's theorem were later provided, the upper bound on the absolute value of the tenth discriminant came in under my thesis bound and so completed the effective transcendence proof of the class-number one problem.) Fortunately for me and my thesis, my thesis advisor, DH Lehmer, vaguely remembered that Heegner was incorrect, but that he was going to attend a conference in Boulder in the summer of 1963 and verify with the experts that was the case. He returned and confirmed once again that "the experts" said Heegner's proof was incorrect. Lehmer did not tell me who "the experts" were and I never found out. However, two of my three main suspects in the whole affair were at the Boulder meeting.

It is commonly said that my proof and Heegner's proof are the same. They are not! The only thing common to the two papers is the final Diophantine equation. There is no complex multiplication in my paper, and no modular functions appear. Even Kronecker's limit formulas do not appear. All this was replaced by Dirichlet's class-number formula.

My 1967 class-number one paper was the first place to state what Heegner's proof entailed. I don't think that anyone ever tried to read Heegner's paper before me! I was totally ignorant of complex multiplication and still figured out the outline of the proof, and the lack of Weber's conjecture therein. I spent part of the summer of 1967 learning the classical theory of complex multiplication from Volume III of Weber's *Algebra* and by the end of that summer, I had provided patches of those parts of Weber's book used in Heegner's paper that I thought needed patching. If asked, I would say Heegner's proof was fixably incomplete.

All this appeared in Volume 1, Number 1 of the Journal of Number Theory which unfortunately didn't come out until 1969 [**16**]. My patches were based on Weber's methods which were more of an art form than methods. I did not manage to find a Weber style proof of the conjecture that Heegner never used in the class-number one part of his paper. However, Birch [**2**] proved the full Weber conjecture using the best pre-Shimura version of the reciprocity law and a little bit of trickery. He announced his result in Bombay in 1967 and also presented it at a 1968 Oberwolfach meeting I attended. With the Shimura law, Weber's conjecture becomes a homework problem.

Appendix 2 An evaluation at $s = 0$.

I have stated above that evaluations are simpler at $s = 0$ than they are at $s = 1$. Let me give a quick example for the partial zeta functions related to the real subfield of the cyclotomic field of the fth roots of unity. The partial zeta function for the sum of the two congruence classes of a and f-a (mod f) where $0 < a < f$ then takes the form

$$g(s, a; f) = \sum_{n=0}^{\infty} (nf + a)^{-s} + \sum_{n=1}^{\infty} (nf - a)^{-s}$$

What we have here is the sum of two rescaled classical Hurwitz zeta functions at a and $f - a$. The reason for writing it this way will become apparent shortly.

The series for $g(s, a; f)$ converges absolutely for $\sigma > 1$. For $\sigma > 1$, we rewrite $g(s, a; f)$ as

$$(5) \qquad g(s, a; f) = a^{-s} + \sum_{n=1}^{\infty} \left[(nf + a)^{-s} + (nf - a)^{-s} - 2(nf)^{-s} \right] + 2f^{-s}\zeta(s).$$

Via the binomial theorem, the quantity in [ ] in Equation (5) is $O(n^{-\sigma-2})$ with the constant in the O uniform on compact sets in the s-plane. Thus the series in Equation (5) converges absolutely and uniformly on compact sets in the right half plane given by $\sigma > -1$ and therefore gives an analytic continuation of $g(s, a; f)$ back to $\sigma > -1$.

As a warmup, we evaluate

$$g(0, a; f) = 1 + \sum_{n=1}^{\infty} \left[1 + 1 - 2\right] + 2\zeta(0) = 1 + 2\zeta(0). \tag{6}$$

If you don't know that $\zeta(0) = -1/2$, set $f = 2$, $a = 1$ and look at

$$g(s, 1; 2) = 2\sum_{n=0}^{\infty}(2n+1)^{-s} = 2\zeta(s)\left(1 - 2^{-s}\right)$$

so that $g(0, 1; 2) = 0$. We plug this into Equation (6) and get

$$\zeta(0) = -1/2,$$

and

$$g(0, a; f) = 0,$$

for all f and a.

Our interest turns to $g'(0, a; f)$ and this is actually the goal of this Appendix. Since the series converges absolutely and uniformly on compact sets in $\sigma > -1$, we may differentiate the expression in Equation (5) term by term and then set $s = 0$. We get

$$\begin{aligned} g'(0, a; f) &= -\left\{ \log(a) + \sum_{n=1}^{\infty} \left[\log(nf + a) + \log(nf - a) - 2\log(nf)\right] \right\} \\ &\qquad - 2\zeta(0)\log(f) + 2\zeta'(0) \\ &= -\log\left\{ a \prod_{n=1}^{\infty} \left[\frac{(nf+a)(nf-a)}{nf \cdot nf}\right] \right\} + \log(f) + 2\zeta'(0) \\ &= -\log\left\{ \frac{2\pi a}{f} \prod_{n=1}^{\infty} \left(1 - \frac{(a/f)^2}{n^2}\right) \right\} + \log(2\pi) + 2\zeta'(0) \\ &= -\log\left[2\sin\left(\frac{\pi a}{f}\right)\right] + \log(2\pi) + 2\zeta'(0). \end{aligned} \tag{7}$$

To evaluate the extra terms most easily, let $f = 6$ and $a = 1$. We then have

$$g(s, 1; 6) = \sum_{\substack{n=1 \\ (n,6)=1}}^{\infty} n^{-s} = \zeta(s)(1 - 2^{-s})(1 - 3^{-s})$$

which has a double zero at $s = 0$. Therefore $g'(0, 1; 6) = 0$. Inserting this into Equation (7) shows that

$$\zeta'(0) = -\frac{1}{2}\log(2\pi)$$

and we get our final evaluation

$$g'(0, a; f) = -\log\left[2\sin\left(\frac{\pi a}{f}\right)\right].$$

The number in [ ] is an algebraic integer and is a unit if f has two distinct prime factors. (In general, the conjectures allow us to derive the relative norm; the S-unit claim is a consequence of the algebraic integer conjecture.) The square of this number is always in the real subfield of the cyclotomic field generated by the fth roots of unity.

Bibliography

[1] Baker, A. Linear forms in logarithms of algebraic numbers. I, Mathematika, 13 (1966), 204-216.

[2] Birch, B.J. Weber's class invariants. Mathematika 16 (1969), 283-294.

[3] Brauer, R. Beziehungen zwischen Klassenzahlen von Teilkörpen eines galoisisschen Körpers. Math. Nachr. 4 (1951), 158-174.

[4] Hasse, H. Neue Begründung der komplexen Multiplikation. II. Aufbau ohne Benutzung der allgemeinen Klassenkörpertheorie, J. Reine Angew. Math. 165(1931), 64-88.

[5] Hecke, E. Über die L-Funktionen und den Dirichletschen Primzahlsatz für einen beliebigen Zahlkörper. Nach. K. Ges. Göttingen, Math-phy Kl. (1917), 299-318. (Werke, #9)

[6] Heegner, K. Diophantische Analysis und Modulfunktionen. Math. Z. 56 (1952), 227-253.

[7] Heilbronn, H. and Linfoot, E.H. On the imaginary quadratic corpora of class-number one. Quart. J. Math. Oxford Ser 5 (1934) 293-301.

[8] Lang, S. Elliptic Functions. Addison Wesley, Reading 1973, 326pp.

[9] Meyer, C. Bemerkungen zum Satz von Heeger-Stark über die imaginär-quadratischen Zahlkörper mit der Klassenzahl Eins. J. Reine Angew. Math. 242 (1970), 179-214.

[10] Ramachandra, K. Some applications of Kronecker's limit formulas. Ann. of Math. (2) 80 (1964), 104-148.

[11] Shimura, G. Introduction to the arithmetic theory of automorphic functions. Publications of the Mathematical Soceity of Japan 11, Iwanami Shoten, Publishers, Tokyo; Princeto University Press, Princeton 1971, 267pp.

[12] Siegel, C.L. Lectures on advanced analytic number theory. Tata Institute for Fundamental Research Lectures on Mathematics, 23, Tata Institute of Fundamental Research, Bombay 1965, 331pp.

[13] Siegel, C.L. Zum Beweise des Starkschen Satzes. Invent. Math. 5 (1968), 180-191.

[14] Stark, H.M. A complete determination of the complex quadratic fields of class-number one. Michigan Math. J. (1967), 1-27.

[15] Stark, H.M. L-functions and characters sums for quadratic forms. II, Acta Arith. 15 (1968/69) 307-317.

[16] Stark, H.M. On the "gap" in a theorem of Heegner. J. Number Theory 1 (1969), 16-27.

[17] Stark, H.M. Class-number problems in quadratic fields. Actes du Congrès International des Mathématiciens (Nice), Gauthier Villars, Paris 1971, Tome 1, 511-518.

[18] Stark, H.M. Values of L-functions at $s = 1$. I. L-functions for quadratic forms. Advances in Math. 7 (1971), 301-343.

[19] Stark, H.M. Values of L-functions at $s = 1$. II. Artin L-functions with rational characters. Advances in Math. 17 (1975), 60-92.

[20] Stark, H.M. Class fields for real quadratic fields and L-series at $s = 1$. Algebraic number fields: L-functions and Galois properties (Proc. Sympos., Univ. Durham, Durham 1975) Academic Press, London 1977, 355-375.

[21] Stark, H.M. Values of L-functions at $s = 1$. III. Totally real fields and Hilbert's twelfth problem. Advances in Math. 22 (1976) 64-84.

[22] Stark, H.M. Hilbert's twelfth problem and L-series. Bull. Amer. Math. Soc. 83 (1977) 1072-1074.

[23] Stark, H.M. Values of L-functions at $s = 1$. IV. First derivatives at $s = 0$. Advances in Math. 35 (1980), 197-235.

[24] Weber, H. Lehrbuch der Algebra, Vol. 3. Chelsea, New York, 1961.

Integral and p-adic Refinements of the Abelian Stark Conjecture

Cristian D. Popescu

IAS/Park City Mathematics Series
Volume 18, 2009

Integral and p-adic Refinements of the Abelian Stark Conjecture

Cristian D. Popescu

In loving memory of my father, Dumitru D. Popescu (1937–2005) and my mentor, Nicolae Popescu (1937–2010).

Abstract. We give a formulation of the abelian case of Stark's Main Conjecture in terms of determinants of projective modules and briefly show how this formulation leads naturally to its Equivariant Tamagawa Number Conjecture (ETNC) – type integral refinements. We discuss the Rubin-Stark integral refinement of an idempotent piece of Stark's Abelian Main Conjecture. In the process, we give a new formulation of its particular case, the Brumer-Stark conjecture, in terms of annihilators of generalized Arakelov class-groups (first Chow groups.) In this context, we discuss somewhat in detail our recent theorems with Greither, settling refinements of the Brumer-Stark conjecture, under certain hypotheses. We formulate a general Gross–type refinement of the Rubin–Stark conjecture, which has emerged fairly recently from work of Gross, Tate, Tan, Burns, Greither and the author, and interpret it in terms of special values of p–adic L–functions. Finally, we discuss the recent overwhelming evidence in support of the combined Gross-Rubin-Stark conjecture, which is a consequence of independent work of Burns and that of Greither and the author on the ETNC and various Equivariant Main Conjectures in Iwasawa theory, respectively.

Introduction

In the 1970s and early 1980s, Stark (see [**34**], [**35**], [**36**], [**37**]) developed a remarkable Galois–equivariant conjectural link between the values at $s = 0$ of the first non–vanishing derivatives of the S–imprimitive Artin L–functions $L_{K/k,S}(\rho, s)$

Department of Mathematics, University of California, San Diego, La Jolla, CA 92093-0112, USA
E-mail address: cpopescu@math.ucsd.edu

Key words and phrases.
Global and p–adic L-functions; Regulators

2000 *Mathematics Subject Classification.* 11M38, 11G20, 11G25, 11G45, 14F30

The author was partially supported by NSF Grants DMS-0600905 and DMS-0901447.

associated to a Galois extension K/k of number fields and a certain $\mathbb{Q}[\mathrm{Gal}(K/k)]$–module invariant of the group U_S of S–units in K. Stark's Main Conjecture should be viewed as a vast Galois–equivariant generalization of the unrefined, *rational version* of Dirichlet's class–number formula

$$\lim_{s\to 0}\frac{1}{s^r}\zeta_k(s)\in\mathbb{Q}^\times\cdot R_k\,,$$

in which the zeta function $\zeta_k(s)$ is replaced by a Galois–equivariant L–function

$$\Theta_{K/k,S}(s)=\sum_{\rho\in\widehat{G}}L_{K/k,S}(\rho,s)\cdot e_{\rho^\vee}\,,$$

with values in the center of the group–ring $\mathcal{Z}(\mathbb{C}[\mathrm{Gal}(K/k)])$, the regulator R_k and the rank r of the group of units in k are replaced by a Galois–equivariant regulator with values in $\mathcal{Z}(\mathbb{C}[\mathrm{Gal}(K/k)])$ and the local rank function of the projective $\mathbb{Q}[\mathrm{Gal}(K/k)]$–module $\mathbb{Q}U_S$, respectively. In [**18**], Gross formulated a p–adic version of Stark's Main Conjecture, bringing into the picture the p–adic L–functions (p–adic interpolations of Artin's L–functions), which had just been constructed in [**7**] and [**11**], independently.

In the 1980s, work of Stark, loc.cit., Tate [**42**], and Chinburg [**8**], [**9**], among others, revealed not only the depth and importance of Stark's Main Conjecture for number theory, but also the fact that an *integral refinement* of this statement, in the spirit of the full strength, integral Dirichlet class–number formula

$$\lim_{s\to 0}\frac{1}{s^r}\zeta_k(s)=-\frac{h_k}{w_k}\cdot R_k\,,$$

would have very far reaching applications to major unsolved problems in the field (e.g. Hilbert's 12–th problem, asking for an explicit class-field theory over arbitrary number fields.)

In [**37**], Stark himself formulated such an integral refinement for the particular case of *abelian extensions* K/k and their associated imprimitive L–functions $L_{K/k,S}(\chi,s)$ of *order of vanishing at most* 1 *at* $s=0$, under additional assumptions on S, and proved it in the cases where k is either $\mathbb{Q}$ or imaginary quadratic. The proofs revealed deep connections between the (however limited in scope) integral conjecture and the theory of Gauss sums, cyclotomic units, and elliptic units, respectively. In [**18**] and [**19**], Gross developed a p–adic refinement of Stark's integral conjecture in this context and proved it for imaginary, abelian extensions of $\mathbb{Q}$, as a direct consequence of his joint work with Koblitz [**21**].

Then, Tate's influential book [**42**] was published, casting the conjectures developed so far and their potential applications into a completely new light and extending them to function fields (characteristic p global fields.) An ℓ–adic étale cohomological proof of the integral Stark conjecture for function fields, due to Deligne and Tate, was given in Chpt. V of [**42**]. Almost immediately after, Hayes (see [**22**] and [**23**]) gave explicit proofs of Stark's integral conjecture and its refinement by Gross, in the case of function fields, establishing beautiful connections with the theory of torsion points of rank 1, sign normalized Drinfeld modules.

In the late 1980s and early 1990s, after the extremely successful emergence of Kolyvagin's theory of Euler systems, Rubin (see [**31**]) realized that, if true, Stark's integral conjecture is a rich source of Euler systems. The Euler systems of Gauss sums, cyclotomic units, elliptic units, and torsion points of rank 1 Drinfeld modules are very particular cases of this very general fact. Driven by this realization, Rubin (see [**32**]) generalized Stark's integral conjecture to the case of abelian extensions of number fields K/k and their associated S–imprimitive L–functions of minimal order of vanishing $r \geq 0$, under certain additional assumptions on S. This is the Rubin–Stark conjecture, one of the central themes of this paper. The Rubin-Stark conjecture was extended to function fields, and proved in fairly general cases in that context in [**27**] by the author. As a consequence, a general Gras-type conjecture for function fields was proved in [**26**]. (The cyclotomic Gras conjecture in number fields had been proved earlier by Mazur–Wiles and Kolyvagin.) Recently, a Gross–type refinement of the Rubin–Stark conjecture has emerged out of work of Tate, Gross, Tan, Burns, Greither and the author. This has led to a combined Gross-Rubin-Stark Conjecture, which encapsulates the Rubin-Stark and Gross conjectures and predicts, among other things, a subtle link between special values of derivatives of global and p–adic L–functions.

In [**1**], Bloch and Kato formulated a very general conjecture on special values of L–functions associated to motives with abelian coefficients (in particular, Dirichlet motives, whose L–functions are the object of study of Stark's Abelian Main Conjecture.) Now, this statement is known as the Equivariant Tamagawa Number Conjecture (ETNC.) Burns and Flach (see [**3**] and [**4**]) extended this statement to motives with non-abelian coefficients. They also showed that, in the case of Artin motives, the ETNC can be viewed as an integral refinement of Stark's Main Conjecture in its most general setting. Recently, Burns has shown in [**2**] that the ETNC for Dirichlet motives implies (refined versions of) the combined Gross–Rubin–Stark Conjecture. In [**5**], Burns and Greither (with later contributions by Flach on the 2–primary side) proved the ETNC for Dirichlet motives with base field $\mathbb{Q}$. Recently, Burns (see [**2**] and the references therein) has proved the ETNC for Dirichlet motives over function fields as well. Therefore, the Gross–Rubin–Stark conjecture was proved in those cases.

In recent joint work with Greither (see [**17**], [**16**], [**15**]), the author has proved an Equivariant Main Conjecture in Iwasawa theory for both function fields and number fields (away from the 2–primary part and under a certain μ–invariant hypothesis, in the number field case.) It turns out that this conjecture implies the Gross–Rubin–Stark Conjecture (in full generality, in function fields and in an imprimitive form and away from 2, for CM abelian extensions of totally real number fields.) In particular, it implies refined versions of an imprimitive Brumer–Stark Conjecture and of the full étale Coates–Sinnott Conjecture (discussed in Manfred Kolster's contribution to this volume.) Ultimately, our Equivariant Main Conjecture implies the full ETNC for function fields and its "minus" piece under the action of complex conjugation, for abelian CM extensions of totally real number fields, providing new evidence in support of the ETNC and the combined Gross–Rubin–Stark Conjecture.

This paper is organized as follows. After a brief §1 on various algebraic tools needed throughout, in §2, we give a formulation of the abelian case of Stark's Main Conjecture in terms of determinants of projective modules. In §3, we briefly show how this formulation leads naturally to its ETNC–type integral refinements and discuss in detail the Rubin-Stark integral refinement of an idempotent piece of Stark's Abelian Main Conjecture. In §4, we restrict ourselves to the order of vanishing 1 case of the Rubin-Stark conjecture. In particular, in §4.3, we give a new formulation of the Brumer-Stark conjecture (a special case of the order of vanishing 1 Rubin-Stark conjecture) in terms of annihilators of generalized Arakelov class-groups (first Chow groups.) In this context, we discuss somewhat in detail our recent theorems with Greither, settling refinements of the Brumer-Stark conjecture, under certain hypotheses. In §5, we formulate a Gross–type refinement of the Rubin–Stark conjecture (the so-called combined Gross–Rubin–Stark Conjecture) and interpret it in terms of special values of p–adic L–functions. In §5.5, we discuss the recent overwhelming evidence in support of the combined Gross-Rubin-Stark conjecture, which has emerged from work of Burns and that of Greither and the author on the ETNC and various Equivariant Main Conjectures in Iwasawa theory, respectively.

We conclude this introduction by extending our apologies to several people who have made significant contributions to this area over the years and whose names and specific results will not be mentioned in what follows, due to our lack of time, space and expertise. These are: Noboru Aoki, Samit Dasgupta, Henri Darmon, David Dummit, Caleb Emmons, Anthony Hayward, Radan Kučera, Joongul Lee, Jonathan Sands, Barry Smith, David Solomon, Brett Tangedal.

1. Algebraic preliminaries

In what follows, R will denote a commutative, reduced, Noetherian ring. Typically, R will be a group algebra associated to a finite, commutative group G, with coefficients in $\mathbb{Z}$, $\mathbb{Q}$, $\mathbb{R}$, $\mathbb{C}$ or the ring of ℓ–adic integers $\mathbb{Z}_\ell$, for an arbitrary prime number ℓ. If M is an R–module and R' is a commutative R–algebra, then $R'M := R'\otimes_R M$.

1.1. Duals

If M is an R–module, we let $M^* := \mathrm{Hom}_R(M, R)$ denote the R–dual of M. We endow M^* with the usual R–module structure $(r\cdot\phi)(m) = r\phi(m)$, for all $r \in R$, $\phi \in M^*$ and $m \in M$.

Remark 1.1.1. *If M is a $\mathbb{Z}[G]$–module, where G is a finite, commutative group, then there is a canonical $\mathbb{Z}[G]$–module isomorphism*

$$\mathrm{Hom}_{\mathbb{Z}}(M,\mathbb{Z}) \xrightarrow{\sim} \mathrm{Hom}_{\mathbb{Z}[G]}(M,\mathbb{Z}[G]), \qquad \phi \longrightarrow \psi\,,$$

where $\psi(m) := \sum_{\sigma\in G}\phi(\sigma^{-1}\cdot m)\cdot\sigma$, for all $\phi \in \mathrm{Hom}_{\mathbb{Z}}(M,\mathbb{Z})$ and all $m \in M$. Here, $\mathrm{Hom}_{\mathbb{Z}}(M,\mathbb{Z})$ is endowed with the usual $\mathbb{Z}[G]$–module structure given by $(\sigma\cdot\phi)(m) = \phi(\sigma\cdot m)$, for all $\sigma \in G$, $\phi \in \mathrm{Hom}_{\mathbb{Z}}(M,\mathbb{Z})$, and $m \in M$.

Any $\phi \in M^*$ induces an R–linear morphism

$$\phi^{(r)} : \wedge_R^r M \longrightarrow \wedge_R^{r-1} M, \qquad x_1\wedge\cdots\wedge x_r \longrightarrow \sum_{i=1}^{r}(-1)^{i-1}\phi(x_i)\cdot x_1\wedge\cdots\wedge\widehat{x_i}\wedge\cdots\wedge x_r\,,$$

for all $r \in \mathbb{Z}_{\geq 1}$. If the context is clear and r is fixed, then we will sometimes abusively write ϕ instead of $\phi^{(r)}$. The construction above leads to the more general R–linear morphism

$$\begin{array}{rcl} \wedge_R^i M^* \otimes_R \wedge_R^r M & \longrightarrow & \wedge_R^{r-i} M \\ (\phi_1\wedge\cdots\wedge\phi_i)\otimes(x_1\wedge\cdots\wedge x_r) & \longrightarrow & \phi_i^{(r-i+1)}\circ\cdots\circ\phi_1^{(r)}(x_1\wedge\cdots\wedge x_r), \end{array}$$

for all $r, i \in \mathbb{Z}_{\geq 1}$ with $i \leq r$ and all $\phi_1, \dots, \phi_i \in M^*$ and $x_1, \dots, x_r \subset M$,

Remark 1.1.2. *With notations as above, note that if $r = i$, then we have*

$$\phi_r^{(1)} \circ \cdots \circ \phi_1^{(r)}(x_1 \wedge \cdots \wedge x_r) = \det(\phi_i(x_j);\ 1 \leq i, j \leq r),$$

for all $\phi_1, \dots, \phi_r \in M^*$ and all $x_1, \dots, x_r \in M$. This establishes an R–module morphism

$$\wedge_R^r M^* \longrightarrow (\wedge_R^r M)^*, \qquad \phi_1 \wedge \cdots \wedge \phi_r \longrightarrow \phi_r^{(1)} \circ \cdots \circ \phi_1^{(r)}.$$

It is very easy to check that if M is a free R–module (or, more generally, a projective R–module), then the morphism above is an isomorphism.

Let $\mathcal{K} := Q(R)$ be the total ring of fractions of R (i.e. the ring of fractions obtained from R by inverting all its non–zero–divisors.) Note that R is a subring of $\mathcal{K}$. Since R is Noetherian and reduced, $\mathcal{K}$ is a finite direct sum of fields. Indeed, if $\{\mathfrak{p}_1, \dots, \mathfrak{p}_n\}$ are the minimal prime ideals of R, then it is easy to show that we have a ring isomorphism $\mathcal{K} \simeq \oplus_{i=1}^n Q(R/\mathfrak{p}_i)$, where $Q(R/\mathfrak{p}_i)$ is the field of fractions of the integral domain $R/\mathfrak{p}_i$, for all $i = 1, \dots, n$. Now, let M be a finitely generated R–module. We have a canonical R–module morphism

$$\text{(1)} \qquad \operatorname{Hom}_R(M, R) \longrightarrow \operatorname{Hom}_{\mathcal{K}}(\mathcal{K}M, \mathcal{K}), \qquad \phi \longrightarrow \psi,$$

where $\psi(x/s) = \phi(x)/s$, for all $x \in M$ and all non–zero–divisors $s \in R$. This map happens to be injective in this case and it induces a canonical isomorphism of $\mathcal{K}$–modules

$$\operatorname{Hom}_R(M, R) \otimes_R \mathcal{K} \simeq \operatorname{Hom}_{\mathcal{K}}(\mathcal{K}M, \mathcal{K}).$$

For every $r \in \mathbb{Z}_{\geq 0}$, we have an R–module morphism

$$\begin{aligned} E_r : \wedge_R^r M^* \otimes_R \wedge_{\mathcal{K}}^r \mathcal{K}M &\longrightarrow \mathcal{K} \\ (\phi_1 \wedge \cdots \wedge \phi_r) \otimes (x_1 \wedge \cdots \wedge x_r) &\longrightarrow \psi_r^{(1)} \circ \cdots \circ \psi_1^{(r)}(x_1 \wedge \cdots \wedge x_r) = \det(\psi_i(x_j)) \end{aligned}$$

where $\phi_i \in M^*$, $x_i \in \mathcal{K}M$ and $\phi_i \longrightarrow \psi_i$ via the map (1) above, for all $i = 1, \dots, r$.

Definition 1.1.3. *For R, M and r as above, let $\mathcal{L}_R(M, r)$ be the R–submodule of $\wedge_{\mathcal{K}}^r \mathcal{K}M$ given by*

$$\mathcal{L}_R(M, r) := \{\epsilon \in \wedge_{\mathcal{K}}^r \mathcal{K}M \mid E_r((\phi_1 \wedge \cdots \wedge \phi_r) \otimes \epsilon) \in R, \quad \forall\, (\phi_1 \wedge \cdots \wedge \phi_r) \in \wedge_R^r M^*\}.$$

Remark 1.1.4. *Note the following.*

(1) $\mathcal{L}_R(M, 0) = R$, *for all R and M as above.*

(2) *The canonical $\mathcal{K}$–module isomorphism $\mathcal{K} \otimes_R \wedge_R^r M \simeq \wedge_{\mathcal{K}}^r \mathcal{K}M$ induces an obvious R–linear map $\wedge_R^r M \longrightarrow \wedge_{\mathcal{K}}^r \mathcal{K}M$. This map is not injective, in general. Its kernel consists of those elements in $\wedge_R^r M$ which are annihilated by non–zero–divisors in R. We denote by $\widetilde{\wedge_R^r M}$ its image. We have an obvious double-inclusion of R–modules*

$$\widetilde{\wedge_R^r M} \subseteq \mathcal{L}_R(M, r) \subseteq \wedge_{\mathcal{K}}^r \mathcal{K}M.$$

(3) *If M is a projective R–module, then we leave it as an exercise for the reader to check that*

$$\mathcal{L}_R(M, r) = \widetilde{\wedge_R^r M} \simeq \wedge_R^r M.$$

Note that since M is R–projective, so is $\wedge_R^r M$. Consequently, the R–linear morphism $\wedge_R^r M \longrightarrow \wedge_{\mathcal{K}}^r \mathcal{K}M$ is injective in this case, which leads to the isomorphism above. Standard properties of exterior powers of direct sums of modules reduce the verification of the equality above to the case where M is R–free, which is straightforward.

(4) *In particular, if $R = \mathbb{Z}[1/|G|][G]$, where G is a finite, abelian group, then R is easily seen to be isomorphic to a finite direct sum of Dedekind domains (rings of integers in cyclotomic fields.) Consequently, any finitely generated $\mathbb{Z}[1/|G|][G]$–module M which has no $\mathbb{Z}$–torsion is $\mathbb{Z}[1/|G|][G]$–projective. Consequently, (3) above applies, for all r.*

Example 1.1.5. Let G be a finite, abelian group and $r \in \mathbb{Z}_{\geq 1}$. Let $R := \mathbb{Z}[G]$ and $M := \mathbb{Z}^r$, viewed as a $\mathbb{Z}[G]$–module with trivial G–action. In this case, it is easily seen that $\mathcal{K} := Q(\mathbb{Z}[G]) = \mathbb{Q}[G]$. Consequently, $\mathcal{K}M = \mathbb{Q}^r$, with trivial G–action. If $\{e_1, \dots, e_r\}$ is a $\mathbb{Z}$–basis for M, then the reader can check that the following equalities and isomorphisms of $\mathbb{Z}[G]$–modules hold

$$\wedge_{\mathcal{K}}^r \mathcal{K}M = \mathbb{Q}e_1 \wedge \cdots \wedge e_r \simeq \mathbb{Q}, \quad \wedge_R^r M \simeq \widetilde{\wedge_R^r M} = \mathbb{Z}e_1 \wedge \cdots \wedge e_r \simeq \mathbb{Z},$$
$$\mathcal{L}_R(M, r) = \tfrac{1}{|G|^{r-1}}\mathbb{Z}e_1 \wedge \cdots \wedge e_r,$$

where the G–actions are trivial everywhere. If $r \geq 2$ and $|G| > 1$, the inclusion $\widetilde{\wedge_R^r M} \subseteq \mathcal{L}_R(M, r)$ is strict. If $r = 1$, the inclusion in question is an equality. Obviously, the module M is not R–projective for any $r \geq 1$ and non–trivial group G (more on this in the next section.)

Example 1.1.6. If G is a finite, abelian group, $R := \mathbb{Z}[G]$, and M is an R–module, then we have

$$\mathcal{L}_R(M, 1) = \widetilde{M},$$

as it easily follows from Remark 1.1.1 above.

1.2. Evaluation maps

Let M denote a finitely generated $\mathbb{Z}[G]$–module, where G is a finite, abelian group. As usual, we let $M^* := \mathrm{Hom}_{\mathbb{Z}[G]}(M, \mathbb{Z}[G])$. In this section, we take a closer look at the elements of the lattice $\mathcal{L}_{\mathbb{Z}[G]}(M, r)$, for a fixed $r \in \mathbb{Z}_{\geq 0}$. The properties discussed below will not be used until the very last section of this article.

Let us fix $\varepsilon \in \mathcal{L}_{\mathbb{Z}[G]}(M, r)$. Then, from the definition of $\mathcal{L}_{\mathbb{Z}[G]}(M, r)$, the element ε gives rise to a $\mathbb{Z}$–linear evaluation map

$$\mathrm{ev}_{\varepsilon, \mathbb{Z}} : \wedge_{\mathbb{Z}}^r M^* \longrightarrow \mathbb{Z}[G], \qquad \phi_1 \wedge \cdots \wedge \phi_r \to (\phi_r^{(1)} \circ \cdots \circ \phi_1^{(r)})(\varepsilon),$$

for all $\phi_1, \dots, \phi_r \in M^*$. It is easy to see that the map $\mathrm{ev}_{\varepsilon, \mathbb{Z}}$ factors through a $\mathbb{Z}[G]$–linear map

$$\widetilde{\mathrm{ev}_{\varepsilon, \mathbb{Z}}} : \wedge_{\mathbb{Z}[G]}^r M^* \longrightarrow \mathbb{Z}[G].$$

The main goal of this section is to generalize the evaluation map $\mathrm{ev}_{\varepsilon, \mathbb{Z}}$ to some extent. For that purpose, we let $\mathcal{R}$ denote an arbitrary commutative ring with 1. We view $\mathcal{R}[G]$ as a $\mathbb{Z}[G]$–algebra in the usual way and will make use of the ring isomorphism

$$i_{\mathcal{R}} : \mathbb{Z}[G] \otimes_{\mathbb{Z}} \mathcal{R} \simeq \mathcal{R}[G], \qquad (x \otimes a) \to xa.$$

We let $M^*_{\mathcal{R}} := \operatorname{Hom}_{\mathbb{Z}[G]}(M, \mathcal{R}[G])$, endowed with the usual $\mathcal{R}[G]$–module structure. For the sake of consistency, note that $M^* = M^*_{\mathbb{Z}}$. It is easy to see that the isomorphism in Remark 1.1.1 leads to an isomorphism of $\mathcal{R}[G]$–modules

$$j_{M,\mathcal{R}} : M^* \otimes_{\mathbb{Z}} \mathcal{R} \simeq M^*_{\mathcal{R}}\,, \qquad \phi \otimes a \to \psi\,,$$

where $\psi(x) = \phi(x)a$, for all $\phi \in M^*$, $x \in M$ and $a \in \mathcal{R}$. Consequently, for any such ring $\mathcal{R}$, the element ε gives rise to the more general $\mathcal{R}[G]$–linear evaluation map

$$\begin{array}{rl} \mathrm{ev}_{\varepsilon,\mathcal{R}} : \wedge^r_{\mathcal{R}} M^*_{\mathcal{R}} \xrightarrow[\sim]{\wedge^r j^{-1}_{M,\mathcal{R}}} \wedge^r_{\mathcal{R}}(M^* \otimes_{\mathbb{Z}} \mathcal{R}) \xrightarrow[\sim]{} & (\wedge^r_{\mathbb{Z}} M^*) \otimes_{\mathbb{Z}} \mathcal{R} \\ & \Big\downarrow {\scriptstyle \mathrm{ev}_{\varepsilon,\mathbb{Z}} \otimes \mathbf{1}_{\mathcal{R}}} \\ & \mathbb{Z}[G] \otimes_{\mathbb{Z}} \mathcal{R} \xrightarrow[\sim]{i_{\mathcal{R}}} \mathcal{R}[G]\,. \end{array}$$

Above, the only non-labeled isomorphism is a consequence of the canonical $\mathcal{R}$–linear isomorphism $\wedge^r_{\mathcal{R}}(N \otimes_{\mathbb{Z}} \mathcal{R}) \simeq (\wedge^r_{\mathbb{Z}} N) \otimes_{\mathbb{Z}} \mathcal{R}$, for all $\mathbb{Z}$–modules N. Just as in the case $\mathcal{R} = \mathbb{Z}$, it is easy to see that the map $\mathrm{ev}_{\varepsilon,\mathcal{R}}$ factors through an $\mathcal{R}[G]$–linear map

$$\widetilde{\mathrm{ev}_{\varepsilon,\mathcal{R}}} : \wedge^r_{\mathcal{R}[G]} M^*_{\mathcal{R}} \longrightarrow \mathcal{R}[G]\,.$$

Remark 1.2.1. *In order to get a feel for the map* $\mathrm{ev}_{\varepsilon,\mathcal{R}}$, *note that, if* $\psi_1, \dots, \psi_r$ *are elements in* $M^*_{\mathcal{R}}$, *such that* $\psi_i = j_{M,\mathcal{R}}(\phi_i \otimes a_i)$, *with* $\phi_i \in M^*$ *and* $a_i \in \mathcal{R}$, *for all* $i = 1, \dots, r$, *then we have*

$$\mathrm{ev}_{\varepsilon,\mathcal{R}}(\psi_1 \wedge \dots \wedge \psi_r) = \mathrm{ev}_{\varepsilon,\mathbb{Z}}(\phi_1 \wedge \dots \wedge \phi_r) \cdot \prod_{i=1}^{r} a_i\,. \tag{2}$$

This equality permits us to make the following useful observation. For that purpose, let us assume further that $\mathcal{R}$ *is a* $\mathbb{N}$*–graded ring*

$$\mathcal{R} = \mathcal{R}^{(0)} \oplus \mathcal{R}^{(1)} \oplus \dots \oplus \mathcal{R}^{(i)} \oplus \cdots,$$

where $\mathcal{R}^{(i)}$ *denotes the homogeneous component of degree* i *of* $\mathcal{R}$ *in the given grading, for all* $i \in \mathbb{N}$. *For a* $\mathbb{Z}[G]$*–module* M *as above, let*

$$M^*_{\mathcal{R}^{(i)}} := \operatorname{Hom}_{\mathbb{Z}[G]}(M, \mathcal{R}^{(i)} \otimes_{\mathbb{Z}} \mathbb{Z}[G]),$$

for all $i \in \mathbb{N}$. *Then,* $M^*_{\mathcal{R}}$ *is an* $\mathbb{N}$*–graded* $\mathcal{R}$*–module, with the grading given by*

$$M^*_{\mathcal{R}} = \bigoplus_{i \in \mathbb{N}} M^*_{\mathcal{R}^{(i)}}\,,$$

where $M^*_{\mathcal{R}^{(i)}}$ *is viewed as a* $\mathbb{Z}[G]$*–submodule of* $M^*_{\mathcal{R}}$ *via the injection*

$$\mathcal{R}^{(i)} \otimes_{\mathbb{Z}} \mathbb{Z}[G] \hookrightarrow \mathcal{R} \otimes_{\mathbb{Z}} \mathbb{Z}[G] \simeq \mathcal{R}[G],$$

for all $i \in \mathbb{N}$. *Equality (2), combined with the* $\mathcal{R}$*–linearity of* $\mathrm{ev}_{\varepsilon,\mathcal{R}}$, *leads to*

$$\mathrm{ev}_{\varepsilon,\mathcal{R}}(\psi_1 \wedge \dots \wedge \psi_r) \in \mathcal{R}^{(r)}[G] := i_{\mathcal{R}}(\mathcal{R}^{(r)} \otimes_{\mathbb{Z}} \mathbb{Z}[G]),$$

for all $\psi_1, \dots, \psi_r \in M^*_{\mathcal{R}^{(1)}}$. *This observation will be very useful in the last section of this article.*

1.3. Projective modules

Recall that a finitely generated R–module P is called projective if there exists an R–module Q, an $n \in \mathbb{Z}_{\geq 0}$ and an R–module isomorphism $P \oplus Q \simeq R^n$. Recall that P is projective if and only if it is locally free. This means that for all $\mathfrak{p} \in \mathrm{Spec}(R)$, there exists a (unique) $r_{\mathfrak{p}} \in \mathbb{Z}_{\geq 0}$ and an isomorphism of $R_{\mathfrak{p}}$–modules $P_{\mathfrak{p}} \simeq R_{\mathfrak{p}}^{r_{\mathfrak{p}}}$, where $P_{\mathfrak{p}}$ and $R_{\mathfrak{p}}$ denote the usual localizations at the prime ideal $\mathfrak{p}$.

Consequently, any finitely generated projective R–module P gives rise to a rank–function

$$\mathrm{rk}_P : \mathrm{Spec}(R) \longrightarrow \mathbb{Z}, \qquad \mathrm{rk}_P(\mathfrak{p}) := \mathrm{rank}_{R_{\mathfrak{p}}}(P_{\mathfrak{p}}) = r_{\mathfrak{p}}\,.$$

It is easy to see that this function is locally constant (i.e. continuous with respect to the Zariski topology on $\mathrm{Spec}(R)$ and the discrete topology on $\mathbb{Z}$.) Indeed, for any prime $\mathfrak{p} \in \mathrm{Spec}(R)$, this function is constant on $\mathrm{Spec}(R/\mathfrak{p}) \subseteq \mathrm{Spec}(R)$. Consequently, if $\mathfrak{p}_1, \dots, \mathfrak{p}_n$ are the minimal prime ideals of R, then rk_P is constant on the irreducible components $\mathrm{Spec}(R/\mathfrak{p}_i)$ of $\mathrm{Spec}(R)$ and therefore locally constant on $\mathrm{Spec}(R)$.

Let $R = \oplus_{i=1}^s R_i$ be a decomposition of R as a direct sum of rings R_i with connected spectra. Of course, this corresponds to a decomposition $1_R = e_1 + \cdots + e_s$ of the identity element 1_R into a sum of indecomposable idempotents e_i and also to the disjoint union decomposition $\mathrm{Spec}(R) = \cup_{i=1}^s \mathrm{Spec}(R_i)$ of $\mathrm{Spec}(R)$ into its connected components. Let P be a finitely generated projective R–module and let r_i be the (constant) value of rk_P restricted to $\mathrm{Spec}(R_i)$, for all $i = 1, \dots, s$. Obviously, we have a direct sum decomposition $P = \oplus_{i=1}^s P_i$, where $P_i := e_i \cdot P \simeq P \otimes_R R_i$ is a projective R_i–module of constant rank r_i, for all $i = 1, \dots, s$.

Definition 1.3.1. *1) With notations as above, the determinant–module associated to P (or, simply, the determinant of P) is the projective R–module of constant rank 1 given by*

$$\det_R(P) := \bigoplus_{i=1}^{s} \wedge_{R_i}^{r_i} P_i\,.$$

2) Further, if Q is a projective R–module with the same rank function as P (i.e. $\mathrm{rk}_Q = \mathrm{rk}_P$) and $f \in \mathrm{Hom}_R(P, Q)$, then the determinant–morphism $\tilde{f}$ associated to f is the R–module morphism

$$\tilde{f} : \det_R(P) \longrightarrow \det_R(Q), \qquad \tilde{f} := \oplus_{i=1}^s \wedge^{r_i} f_i\,,$$

where f_i is the restriction of f to P_i viewed in $\mathrm{Hom}_{R_i}(P_i, Q_i)$, for all $i = 1, \dots, s$.

3) If $f \in \mathrm{Hom}_R(P, P)$, then the determinant of f is the unique element $\det_R(f) \in R$ satisfying the property that for all primes $\mathfrak{p} \in \mathrm{Spec}(R)$ its image via the canonical map $R \longrightarrow R_{\mathfrak{p}}$ is the usual determinant of the endomorphism $f_{\mathfrak{p}} := 1_{R_{\mathfrak{p}}} \otimes f$ of the free $R_{\mathfrak{p}}$–module $P_{\mathfrak{p}} \simeq R_{\mathfrak{p}} \otimes_R P$.

Remark 1.3.2. *The uniqueness of $\det_R(f)$ in part 3) of the above definition is clear since the diagonal map $R \longrightarrow \prod_{\mathfrak{p}} R_{\mathfrak{p}}$ is injective. Its existence is proved by picking an R–module Q, such that $P \oplus Q \simeq R^n$, for some $n \in \mathbb{Z}_{\geq 1}$, and then defining $\det_R(f)$ as the usual determinant of the endomorphism $f \oplus 1_Q$ of the free R-module $P \oplus Q$.*

Recall that if Q is a projective R–module of constant rank 1, there is an R–module isomorphism

$$\mathrm{ev}_Q : Q \otimes_R Q^* \xrightarrow{\sim} R, \qquad x \otimes f \longrightarrow f(x),$$

called the evaluation map associated to Q. More generally, if P is a finitely generated projective R–module P, then we denote by ev_P the evaluation map associated to $\det_R(P)$

$$\mathrm{ev}_P : \det_R(P) \otimes_R \det_R(P)^* \xrightarrow{\sim} R.$$

Example 1.3.3. Let us consider the ring $R := F[G]$, where G is a finite abelian group and F a field of characteristic 0. We denote by $\widehat{G}(F)$ the set of characters of all irreducible linear representations of G defined over F. If F is algebraically closed, then $\widehat{G}(F)$ is the group consisting of the characters of all the 1–dimensional representations of G defined over F. In general, there is a one–to–one correspondence

$$\widehat{G}(\overline{F})/\sim_F \longleftrightarrow \widehat{G}(F), \qquad [\chi] \longrightarrow \psi := \sum_{\tau \in \mathrm{Gal}(F(\chi)/F)} \chi^\tau, \tag{3}$$

where $\overline{F}$ is the algebraic closure of F, $G_F := \mathrm{Gal}(\overline{F}/F)$, $\sim_F$ is the equivalence relation given by the natural G_F–action on $\widehat{G}(\overline{F})$ and $[\chi]$ is the equivalence class of a character $\chi \in \widehat{G}(\overline{F})$. There is an F–algebra isomorphism

$$F[G] \xrightarrow{\sim} \bigoplus_{[\chi]} F(\chi), \qquad \sum_{\sigma \in G} a_\sigma \cdot \sigma \longrightarrow (\sum_{\sigma \in G} a_\sigma \cdot \chi(\sigma))_{[\chi]}, \tag{4}$$

where $[\chi]$ runs through $\widehat{G}(\overline{F})/\sim_F$ and $F(\chi)$ is the minimal field-extension of F which contains the values of χ. Consequently, there is a one-to-one correspondence

(5)

$$\widehat{G}(F) \longleftrightarrow \mathrm{Spec}(F[G]), \quad \psi \longrightarrow \mathfrak{p}_\psi := \ker(F[G] \xrightarrow{\chi} F(\chi)), \text{ if } [\chi] \longrightarrow \psi \text{ via (3)}.$$

Here, $\chi : F[G] \longrightarrow F(\chi)$ is the morphism of F–algebras sending σ to $\chi(\sigma)$, for all $\sigma \in G$. Note that the prime ideal $\mathfrak{p}_\psi$ only depends on the conjugacy class $[\chi]$ of χ. In what follows, e_ψ denotes the usual idempotent associated to $\psi \in \widehat{G}(F)$ in $F[G]$, defined by

$$e_\psi := \frac{1}{|G|} \sum_{\sigma \in G} \psi(\sigma) \cdot \sigma^{-1}.$$

Obviously, if $[\chi] \longrightarrow \psi$ via (3), then we have $e_\psi = \sum_\tau e_{\chi^\tau}$ in $\overline{F}[G]$. Consequently, we have $\mathfrak{p}_\psi = (1 - e_\psi)F[G]$ and $e_\psi F[G] \xrightarrow{\sim} F(\chi)$ via isomorphism (4), for all $\psi \in \widehat{G}(F)$. The decomposition of $R := F[G]$ into direct sum of rings with connected spectra is therefore given by

$$F[G] = \bigoplus_{\psi \in \widehat{G}(F)} e_\psi F[G],$$

where $e_\psi F[G]$ is a field (finite extension of F), for all ψ.

Now, if M is a finitely generated $F[G]$–module, then $M = \oplus_\psi M^\psi$, where

$$M^\psi := e_\psi M$$

is a finite dimensional vector space over the field $e_\psi F[G]$. Consequently, M is always $F[G]$–projective and its rank function is given by

$$\mathrm{rk}_M : \mathrm{Spec}(F[G]) \longrightarrow \mathbb{Z}, \qquad \mathrm{rk}_M(\mathfrak{p}_\psi) = \dim_{e_\psi F[G]}(M^\psi), \quad \forall \psi \in \widehat{G}(F).$$

Note that M is $F[G]$–free if and only if rk_M is a constant function. If $h : M \longrightarrow M$ is an $F[G]$–module endomorphism, then $h = \oplus_\psi h^\psi$, where $h^\psi : M^\psi \longrightarrow M^\psi$ is the restriction of h to M^ψ. Consequently, we have the following equality in $F[G]$.

$$\det_{F[G]}(h) = \sum_{\psi \in \widehat{G}(F)} \det_{e_\psi F[G]}(h^\psi)\,.$$

Example 1.3.4. Now, let us consider the ring $R := \mathbb{Z}[G]$. It is well–known that $\mathrm{Spec}(\mathbb{Z}[G])$ is connected. Indeed, otherwise, there would exist a non-trivial idempotent $\alpha = \sum_{\sigma \in G} \alpha_\sigma \cdot \sigma$ in $\mathbb{Z}[G]$. Under the embedding $\mathbb{Z}[G] \hookrightarrow \mathbb{C}[G]$, α is a non–trivial idempotent in $\mathbb{C}[G]$ and therefore $\alpha = \sum_{\chi \in \mathcal{S}} e_\chi$, where χ runs through a non-empty, strict subset $\mathcal{S}$ of $\widehat{G}(\mathbb{C})$. This means that

$$\frac{1}{|G|} \sum_{\chi \in \mathcal{S}} \chi(\sigma) = \alpha_\sigma \in \mathbb{Z}\,, \quad \forall \sigma \in G\,.$$

However, since $|\mathcal{S}| < |G| = |\widehat{G}(\mathbb{C})|$ and $|\chi(\sigma)| = 1$, for all χ and σ, we have $|\alpha_\sigma| < 1$, therefore $\alpha_\sigma = 0$, for all σ. Consequently, $\alpha = 0$ and therefore $\mathrm{Spec}(\mathbb{Z}[G])$ is connected.

A well–known criterion of Nakayama (see [**42**], Chpt. II, §4) states that a finitely generated $\mathbb{Z}[G]$–module is projective if and only if it is $\mathbb{Z}$–free and G–cohomologically trivial (i.e. the Tate cohomology groups $\widehat{\mathrm{H}}^i(\mathfrak{g}, M)$ vanish, for all $i \in \mathbb{Z}$ and all subgroups $\mathfrak{g}$ of G.) Assuming that we have such a projective module P, its rank function rk_P is constant, say equal to $r \in \mathbb{Z}_{\geq 0}$, on $\mathrm{Spec}(\mathbb{Z}[G])$. Consequently, the projective $\mathbb{Q}[G]$–module $\mathbb{Q} \otimes_{\mathbb{Z}} P$ has constant rank r and it is therefore free of rank r over $\mathbb{Q}[G]$ (see Example 1.3.3 above for $F := \mathbb{Q}$.) The $\mathbb{Z}[G]$–module P may not be free, of course. However, a well–known theorem of Swan (see [**42**], Chpt. II, §4) states that there exists an ideal $\mathfrak{a}$ of $\mathbb{Z}[G]$ whose index in $\mathbb{Z}[G]$ is finite and coprime to $|G|$ and an isomorphism of $\mathbb{Z}[G]$–modules

$$P \xrightarrow{\sim} \mathfrak{a} \oplus \mathbb{Z}[G]^{r-1}.$$

Note that any such ideal $\mathfrak{a}$ is $\mathbb{Z}[G]$–projective of rank 1. (To prove projectivity, apply Nakayama's criterion stated above !)

1.4. Extension of scalars.

Next, we discuss briefly the behavior of some of the constructions in the previous section under extension of scalars. For that purpose, let $\rho : R \longrightarrow R'$ be a morphism of commutative, reduced, Noetherian rings and let P be a finitely generated projective R–module. We let $P' := R'P := R' \otimes_R P$, which is obviously a finitely generated projective R'–module. There are canonical (and obvious) isomorphisms of R'–modules

$$R' \otimes_R \det_R(P) \simeq \det_{R'}(P'), \qquad R' \otimes_R \det_R(P)^* \simeq \det_{R'}(P')^*\,. \tag{6}$$

The isomorphisms above induce canonical R–module morphisms

$$\iota : \det_R(P) \longrightarrow \det_{R'}(P'), \qquad \iota^* : \det_R(P)^* \longrightarrow \det_{R'}(P')^*\,.$$

It is easy to check that the evaluation maps commute with extension of scalars, which is to say that the following diagram is commutative.

$$\begin{array}{ccc} \det_R(P)\otimes_R \det_R(P)^* & \xrightarrow{\iota\otimes\iota^*} & \det_{R'}(P')\otimes_{R'}\det_{R'}(P')^* \\ {\scriptstyle \mathrm{ev}_P}\downarrow\wr & & {\scriptstyle \mathrm{ev}_{P'}}\downarrow\wr \\ R & \xrightarrow{\quad\rho\quad} & R' \end{array}$$

An example of extension of scalars which will be of particular interest to us is the case where $R' := eR$, where e is an idempotent in R and $\rho : R \longrightarrow eR$ sends x to ex, for all $x \in R$. In this case, isomorphisms (6) above can be rewritten as

$$(7)\qquad e\det_R(P) \simeq \det_{eR}(eP) \qquad e\det_R(P)^* \simeq \det_{eR}(eP)^*\,.$$

One last property we would like to point out is the following. Consider an R'–linear endomorphism $h : P' \longrightarrow P'$. Now, we take the composition of the following R–linear maps

$$\begin{array}{ccc} \det_R(P)\otimes_R \det_R(P)^* & \xrightarrow{\iota\otimes\iota^*} & \det_{R'}(P')\otimes_{R'}\det_{R'}(P')^* \\ & & \downarrow{\scriptstyle \tilde h\otimes \mathrm{id}} \\ & & \det_{R'}(P')\otimes_{R'}\det_{R'}(P')^* \xrightarrow[\sim]{\mathrm{ev}_{P'}} R'. \end{array}$$

It is easy to check (locally !) that we have an equality of R–submodules of R'.

$$(8)\qquad \mathrm{Im}(\mathrm{ev}_{P'}\circ(\tilde h\otimes \mathrm{id})\circ(\iota\otimes\iota^*)) = \rho(R)\cdot \det_{R'}(h)\,.$$

In particular, if $R = R'$, $h \in \mathrm{End}_R(P)$, and $\rho = \mathrm{id}_R$, then we have an equality of R–ideals

$$\mathrm{Im}(\mathrm{ev}_P\circ(\tilde h\otimes \mathrm{id})) = R\cdot \det_R(h)\,.$$

2. Stark's Main Conjecture in the abelian case

2.1. Notations

In this section, we develop a formulation of Stark's Main Conjecture (see [**34**], [**35**], [**36**], [**37**]), in the particular case of Artin L functions associated to 1 dimensional complex representations (i.e. abelian L–functions.) Although closer in spirit to the formulation of Conjecture 5.1 in §5 of [**42**], our formulation will be made in terms of determinants of projective modules. The determinant–approach will make the transition to the Equivariant Tamagawa Number Conjecture–type integral refinements of Stark's Main Conjecture seem much more intuitive and natural.

Throughout this section, K/k will denote an abelian extension of global fields of Galois group $G := \mathrm{Gal}(K/k)$. We will fix a finite, non-empty set S of primes in k, containing the set S_{ram} of primes which ramify in K/k as well as the set S_∞ of infinite primes. (If k is a function field, then S_∞ is empty.) We let S_K denote the set of primes in K which sit above primes in S. The group of S_K–units in K is denoted by U_S and its torsion subgroup is denoted by μ_K. As usual, we let $w_K := |\mu_K|$ denote the cardinality of μ_K. The group of S_K–divisors in K is denoted by Y_S. This consists of formal finite integral linear combinations of primes in S_K.

The subgroup of Y_S consisting of divisors of (formal) degree 0 is denoted by X_S. We have an exact sequence of $\mathbb{Z}[G]$–modules

$$0 \longrightarrow X_S \longrightarrow Y_S \xrightarrow{\text{aug}} \mathbb{Z} \longrightarrow 0\,, \tag{9}$$

where (the augmentation map) "aug" is the $\mathbb{Z}$–linear map sending every $w \in S_K$ to 1 and G acts on $\mathbb{Z}$ trivially.

For every prime v in k we denote by G_v the decomposition group of a prime v (relative to the abelian extension K/k). If v is unramified in K/k, then σ_v denotes the Frobenius automorphism associated to v in G. If we pick a prime $w(v) \in S_K$ sitting above v, for every $v \in S$, then we have an isomorphism of $\mathbb{Z}[G]$–modules

$$\bigoplus_{v\in S} \mathbb{Z}[G/G_v] \xrightarrow{\sim} Y_S\,, \qquad (x_v)_{v\in S} \longrightarrow \sum_{v\in S} x_v \cdot w(v), \tag{10}$$

where $(x_v)_{v\in S}$ is an element in the direct sum on the left.

2.2. The L–functions

For every finite prime v of k, we denote $\mathbf{N}v$ the cardinality of its residue field. If v is unramified in K/k, we let σ_v denote its Frobenius automorphism in G. As usual, the S–imprimitive L–function $L_S(\chi, s)$ associated to a character $\chi \in \widehat{G}(\mathbb{C})$ is defined as the meromorphic extension to the whole complex plane of the holomorphic function given by the infinite product

$$\prod_{v\notin S}(1-\chi(\sigma_v)\cdot \mathbf{N}v^{-s})^{-1}\,, \qquad \Re(s) > 1\,.$$

This product is known to be uniformly and absolutely convergent on compact subsets of the half-plane $\Re(s) > 1$. If $\chi \neq \mathbf{1}_G$, then the L–function $L_S(\chi, s)$ obtained this way is holomorphic everywhere on the complex plane. If $\chi = \mathbf{1}_G$, then $\zeta_{k,S}(s) := L_S(\mathbf{1}_G, s)$ is holomorphic at all $s \neq 1$ and it has a pole of order 1 at $s = 1$. For every χ as above, we let $r_{\chi,S}$ and $L_S^*(\chi, 0)$ denote the order of vanishing and leading term in the Taylor expansion of $L_S(\chi, s)$ at $s = 0$, respectively. As a consequence of the functional equation of $L_S(\chi, s)$, one can easily show that

$$\begin{aligned} r_{S,\chi} := \text{ord}_{s=0} L_S(\chi, s) &= \dim_{\mathbb{C}}(\mathbb{C}X_S)^{\chi} \\ &= \begin{cases} \text{card}\,\{v \in S \mid \chi(G_v) = \{1\}\}, & \text{if } \chi \neq \mathbf{1}_G; \\ \text{card}\, S - 1, & \text{if } \chi = \mathbf{1}_G. \end{cases} \end{aligned} \tag{11}$$

In the present context, it is more convenient to work with the so-called G–equivariant L–function

$$\Theta_S : \mathbb{C} \longrightarrow \mathbb{C}[G], \qquad \Theta_S(s) := \sum_{\chi\in\widehat{G}(\mathbb{C})} L_S(\chi, s)\cdot e_{\chi^{-1}}.$$

An equivalent and, at times, more convenient way of writing $\Theta_S(s)$ is

$$\Theta_S(s) = \sum_{\sigma\in G} \zeta_S(\sigma, s)\cdot \sigma^{-1}\,, \tag{12}$$

where $\zeta_S(\sigma, s)$ is the S–incomplete partial zeta function associated to σ, for all $\sigma \in G$. The leading term in the Taylor expansion of $\Theta_S(s)$ at $s = 0$ is the following:

$$\Theta_S^*(0) := \sum_{\chi\in\widehat{G}(\mathbb{C})} L_S^*(\chi, 0)\cdot e_{\chi^{-1}} \in \mathbb{C}[G]^{\times}\,.$$

Of course, we could talk about the order of vanishing r_S of $\Theta_S(s)$ at $s=0$. However, r_S is not a simple integer any longer, but an integral–valued function

$$r_S : \operatorname{Spec}\mathbb{C}[G] \longrightarrow \mathbb{Z}, \qquad r_S(\mathfrak{p}_\chi) = r_{S,\chi}, \quad \forall\, \chi \in \widehat{G}(\mathbb{C}).$$

Remark 2.2.1. *Equalities (11) show that r_S is in fact the rank function $\mathrm{rk}_{\mathbb{C}X_S}$ associated to the finitely generated projective $\mathbb{C}[G]$–module $\mathbb{C}X_S$. In the next section we will see that this coincides with the rank function $\mathrm{rk}_{\mathbb{C}U_S}$ of $\mathbb{C}U_S$.*

2.3. Stark's Main Conjecture (the abelian case)

In what follows, if w is a (finite or infinite) prime in K, we denote by $|\cdot|_w$ the canonically normalized absolute value associated to w. A classical theorem of Dirichlet gives a canonical $\mathbb{C}[G]$–module isomorphism

$$\lambda_S : \mathbb{C}U_S \xrightarrow{\sim} \mathbb{C}X_S, \qquad \tilde{u} \longrightarrow -\sum_{w\in S_K} \log|u|_w \cdot w\,,$$

where u is an arbitrary element in U_S and $\tilde{u}$ is its image via the canonical morphism $U_S \longrightarrow \mathbb{C}U_S$. Note that although a priori the image of λ_S sits inside $\mathbb{C}Y_S$, the classical product formula shows that this image is in fact contained in $\mathbb{C}X_S$. However, since the two complex representations of G involved in the isomorphism above are in fact obtained from the rational representations $\mathbb{Q}X_S$ and $\mathbb{Q}U_S$ by extension of scalars, the original rational representations have to be isomorphic as well. Let us fix a $\mathbb{Q}[G]$–module isomorphism

$$f : \mathbb{Q}X_S \xrightarrow{\sim} \mathbb{Q}U_S\,.$$

This leads to a $\mathbb{C}[G]$–linear automorphism of the projective $\mathbb{C}[G]$–module $\mathbb{C}X_S$

$$\lambda_S \circ (f \otimes \mathbf{1}_{\mathbb{C}}) : \mathbb{C}X_S \xrightarrow{\sim} \mathbb{C}X_S\,.$$

Definition 2.3.1. *The element $\det_{\mathbb{C}[G]}\lambda_S \circ (f\otimes \mathbf{1}_C) \in \mathbb{C}[G]^\times$ is called the Stark regulator associated to the set of data $(K/k, S, f)$.*

With notations as in Example 1.3.3, the following statement is easily seen to be equivalent to Stark's Main Conjecture (as stated in [**42**], Chpt. I, §5), restricted to the case of abelian extensions of global fields.

Conjecture 2.3.2. $A_{\mathbb{Q}}(K/k, S, f)$. *For every f as above, the following equivalent statements hold.*

(1) $\frac{\det_{\mathbb{C}[G]}(\lambda_S\circ(f\otimes\mathbf{1}_C))}{\Theta_S^*(0)} \in \mathbb{Q}[G]^\times$.

(2) *If $A_S(\chi, f) := \frac{\det_{\mathbb{C}}(\lambda_S\circ(f\otimes\mathbf{1}_C))^\chi}{L_S^*(\chi^{-1},0)}$, then $A_S(\chi,f)^\tau = A_S(\chi^\tau, f)$, for all characters $\chi \in \widehat{G}(\mathbb{C})$ and all $\tau \in \mathrm{Aut}(\mathbb{C})$.*

(3) *There is an equality of rank 1, free $\mathbb{Q}[G]$–submodules of $\mathbb{C}[G]$*

$$\Theta_S^*(0)\mathbb{Q}[G] = \det_{\mathbb{C}[G]}(\lambda_S \circ (f\otimes \mathbf{1}_{\mathbb{C}}))\mathbb{Q}[G]\,.$$

Remark 2.3.3. *It is not difficult to show that the validity of the conjecture above is independent of S, as long as S contains all the K/k–ramified and all the infinite primes of k. (See [**42**], Proposition 7.3, Chpt. I, §7 for a quick proof of this fact.)*

Our next goal is to arrive at a formulation of the conjecture above which is independent of the map f. The first step consists of applying equality (8) above for $P := \mathbb{Q}X_S$, $R := \mathbb{Q}[G]$, $R' := \mathbb{C}[G]$, the natural inclusion $\rho : \mathbb{Q}[G] \longrightarrow \mathbb{C}[G]$ and $h := \lambda_S \circ (f \otimes \mathbf{1}_C)$. We obtain the following equality of $\mathbb{Q}[G]$–submodules of $\mathbb{C}[G]$.

$$\det_{\mathbb{C}[G]}(\lambda_S \circ (f \otimes \mathbf{1}_{\mathbb{C}}))\mathbb{Q}[G] = \mathrm{Im}(\mathrm{ev}_{\mathbb{C}X_S} \circ (\widetilde{\lambda_S \circ (f \otimes \mathbf{1}_{\mathbb{C}})} \otimes \mathrm{id}) \circ (\iota \otimes \iota^*)).$$

However, since the maps $\widetilde{\lambda}_S$, $\widetilde{f}$, ι and ι^* are injective, the equality above can be rewritten as

$$\det_{\mathbb{C}[G]}(\lambda_S \circ (f \otimes \mathbf{1}_{\mathbb{C}}))\mathbb{Q}[G] = \mathrm{ev}_{\mathbb{C}X_S}(\widetilde{\lambda_S}(\tilde{f}(\det_{\mathbb{Q}[G]}(\mathbb{Q}X_S))) \otimes_{\mathbb{Q}[G]} \det_{\mathbb{Q}[G]}(\mathbb{Q}X_S)^*).$$

Now, since f is an isomorphism, we have $\tilde{f}(\det_{\mathbb{Q}[G]}(\mathbb{Q}X_S)) = \det_{\mathbb{Q}[G]}(\mathbb{Q}U_S)$. Therefore, we arrive at the following equality of $\mathbb{Q}[G]$–modules:

$$\det_{\mathbb{C}[G]}(\lambda_S \circ (f \otimes \mathbf{1}_{\mathbb{C}}))\mathbb{Q}[G] = \mathrm{ev}_{\mathbb{C}X_S}(\widetilde{\lambda_S}(\det_{\mathbb{Q}[G]}(\mathbb{Q}U_S)) \otimes_{\mathbb{Q}[G]} \det_{\mathbb{Q}[G]}(\mathbb{Q}X_S)^*).$$

Consequently, the abelian Main Conjecture of Stark can be restated independently of f as follows.

Conjecture 2.3.4. $A_{\mathbb{Q}}(K/k, S)$. *For K/k and S as above, the following equivalent statements hold.*

(1) *We have an equality of $\mathbb{Q}[G]$–submodules of $\mathbb{C}[G]$:*

$$\Theta_S^*(0)\mathbb{Q}[G] = \mathrm{ev}_{\mathbb{C}X_S}(\widetilde{\lambda_S}(\det_{\mathbb{Q}[G]}(\mathbb{Q}U_S)) \otimes_{\mathbb{Q}[G]} \det_{\mathbb{Q}[G]}(\mathbb{Q}X_S)^*).$$

(2) *If $\mathfrak{X}_S^*$ is a generator for the free, rank one $\mathbb{Q}[G]$–module $\det_{\mathbb{Q}[G]}(\mathbb{Q}X_S)^*$, then there exists a unique $\mathfrak{U}_S \in \det_{\mathbb{Q}[G]}(\mathbb{Q}U_S)$ (necessarily a generator of $\det_{\mathbb{Q}[G]}(\mathbb{Q}U_S)$), such that*

$$\Theta_S^*(0) = \mathrm{ev}_{\mathbb{C}X_S}(\widetilde{\lambda_S}(\mathfrak{U}_S) \otimes \mathfrak{X}_S^*).$$

Remark 2.3.5. *Since λ_S is an isomorphism, we have*

$$\widetilde{\lambda_S}(\det_{\mathbb{Q}[G]}(\mathbb{Q}U_S)) = \det_{\mathbb{Q}[G]}(\lambda_S(\mathbb{Q}U_S)).$$

So, one can replace the equality in statement (1) *of the conjecture above with*

$$\Theta_S^*(0)\mathbb{Q}[G] = \mathrm{ev}_{\mathbb{C}X_S}(\det_{\mathbb{Q}[G]}(\lambda_S(\mathbb{Q}U_S)) \otimes_{\mathbb{Q}[G]} \det_{\mathbb{Q}[G]}(\mathbb{Q}X_S)^*).$$

The right-hand side of the equality above is an invertible $\mathbb{Q}[G]$–submodule of $\mathbb{C}[G]$ which measures the "relative position" of the full rank projective $\mathbb{Q}[G]$–submodules $\lambda_S(\mathbb{Q}U_S)$ and $\mathbb{Q}X_S$ of $\mathbb{C}X_S$. This "relative position" is measured by the $\mathbb{Q}[G]$–submodule generated by the elements of $\mathbb{C}[G]$ obtained by evaluating every element in the determinant of the first module against every element in the dual of the determinant of the second. The conjecture above simply states that the leading term $\Theta_S^*(0)$ in the Taylor expansion at $s = 0$ of the equivariant L–function Θ_S captures this "relative position".

$$\begin{array}{ccc}
\det_{\mathbb{C}[G]}(\mathbb{C}X_S) \otimes_{\mathbb{C}[G]} \det_{\mathbb{C}[G]}(\mathbb{C}X_S)^* & \xrightarrow[\sim]{\mathrm{ev}_{\mathbb{C}X_S}} & \mathbb{C}[G] \\
\uparrow & & \uparrow \\
\det_{\mathbb{Q}[G]}(\lambda_S(\mathbb{Q}U_S)) \otimes_{\mathbb{Q}[G]} \det_{\mathbb{Q}[G]}(\mathbb{Q}X_S)^* & \dashrightarrow & \mathbb{Q}[G]\Theta_S^*(0)
\end{array}$$

Example 2.3.6. Let us consider the case $K = k$. Assume that $|S| = r + 1$, for some $r \in \mathbb{Z}_{\geq 1}$. Let $S := \{v_0, v_1, \ldots, v_r\}$ and denote by $\{u_1, u_2, \ldots, u_r\}$ a set of fundamental S–units. This means that the images $\widetilde{u_i}$ of u_i via the natural morphism $j : U_S \longrightarrow \mathbb{Q}U_S$ form a $\mathbb{Z}$–basis for

$$\widetilde{U_S} := j(U_S) \simeq U_S/\mu_k,$$

where μ_k denotes the group of roots of unity in k. In this case, we have $\Theta_S(s) = \zeta_{k,S}(s)$, where $\zeta_{k,S}(s)$ is the S–incomplete Dedekind zeta-function associated to the field k. The class number formula gives us the equalities

$$\Theta_S^*(0) = \zeta_{k,S}^*(0) = -\frac{R_{k,S} \cdot h_{k,S}}{w_k}.$$

Here, $h_{k,S}$ is the cardinality of the ideal–class group of the ring $O_{k,S}$ of S–integers in k, w_k is the cardinality of μ_k and $R_{k,S}$ is the Dirichlet S–regulator, given by

$$R_{k,S} := |\det(\log |u_i|_{v_j})| = \mathrm{vol}(\mathbb{R}X_S/\lambda_S(\widetilde{U_S})),$$

where $i, j \in \{1, 2, \ldots, r\}$ and "vol" denotes the (Lebesgue) volume of the quotient of $\mathbb{R}X_S \simeq \mathbb{R}^{|S|-1}$ by its full rank generating sublattice $\lambda_S(\widetilde{U_S})$. Now, since $\mathbb{Q}U_S$ and $\mathbb{Q}X_S$ are free $\mathbb{Q}$–modules of rank r with bases

$$\{\bar{u}_1, \ldots, \bar{u}_r\}, \quad \{(v_1 - v_0), \ldots, (v_r - v_0)\},$$

respectively, $\det_{\mathbb{Q}}(\mathbb{Q}U_S) = \wedge_{\mathbb{Q}}^r \mathbb{Q}U_S$ and $\det_{\mathbb{Q}}(\mathbb{Q}X_S) = \wedge_{\mathbb{Q}}^r \mathbb{Q}X_S$ are free $\mathbb{Q}$–modules of rank 1 with bases

$$\{\widetilde{u_1} \wedge \cdots \wedge \widetilde{u_r}\}, \quad \{(v_1 - v_0) \wedge \cdots \wedge (v_r - v_0)\},$$

respectively, where the exterior products are viewed over $\mathbb{Q}$. We let

$$\{(v_1 - v_0)^*, \ldots, (v_r - v_0)^*\}$$

denote the basis of $(\mathbb{Q}X_S)^*$ which is dual to $\{(v_1 - v_0), \ldots, (v_r - v_0)\}$. Remark 1.1.2 above establishes an explicit isomorphism of $\mathbb{Q}$–modules

$$\begin{array}{rcl} \wedge_{\mathbb{Q}}^r(\mathbb{Q}X_S)^* & \xrightarrow{\sim} & (\wedge_{\mathbb{Q}}^r \mathbb{Q}X_S)^*, \\ (v_1 - v_0)^* \wedge \cdots \wedge (v_r - v_0)^* & \longrightarrow & (v_r - v_0)^{*(1)} \circ \cdots \circ (v_1 - v_0)^{*(r)}. \end{array}$$

Consequently, we obtain the following equalities of $\mathbb{Q}$–submodules of $\mathbb{C}$.

$$\begin{aligned} \mathrm{ev}_{\mathbb{C}}(\det_{\mathbb{Q}}(\lambda_S(\mathbb{Q}U_S)) \otimes_{\mathbb{Q}} \det_{\mathbb{Q}}(\mathbb{Q}X_S)^*) &= \mathbb{Q} \cdot \det((v_i - v_0)^*(\lambda_S(u_j))) \\ &= \mathbb{Q} \cdot \det(-\log |u_j|_{v_i}) \\ &= \mathbb{Q} \cdot R_{k,S} \\ &= \mathbb{Q} \cdot \Theta_S^*(0). \end{aligned}$$

This verifies Conjecture $A_{\mathbb{Q}}(K/k, S)$ in the case $K = k$, assuming that $|S| \geq 2$. The interested reader is invited to check that in this case, if we take

$$\mathfrak{X}_S^* := (v_r - v_0)^{*(1)} \circ \cdots \circ (v_1 - v_0)^{*(r)},$$

then we have

$$\mathfrak{U}_S := \pm \frac{h_{k,S}}{w_k} \cdot \widetilde{u_1} \wedge \cdots \wedge \widetilde{u_r},$$

where the notations are as in formulation (2) of Conjecture 2.3.4. We leave the case $K = k$ and $|S| = 1$ as an exercise.

2.4. Idempotents

Let us consider a splitting

$$1 = \sum_{e \in E} e ,$$

of $1 \in \mathbb{Q}[G]$ into a finite sum of idempotents $e \in \mathbb{Q}[G]$. Equation (7) above shows that Conjecture $A_{\mathbb{Q}}(K/k, S)$ holds true if and only if

$$e \cdot \Theta_S^*(0)\mathbb{Q}[G] = \mathrm{ev}_{\mathbb{C}X_S}(\widetilde{\lambda_S}(\det_{e\mathbb{Q}[G]}(e\mathbb{Q}U_S)) \otimes_{e\mathbb{Q}[G]} \det_{e\mathbb{Q}[G]}(e\mathbb{Q}X_S)^*), \quad \forall\, e \in E.$$

We will denote the equality above by $eA_{\mathbb{Q}}(K/k, S)$ and refer to it as the "e–component of Stark's Main conjecture $A_{\mathbb{Q}}(K/k, S)$", for all idempotents $e \in \mathbb{Q}[G]$.

A useful decomposition of 1 as a sum of idempotents in $\mathbb{Q}[G]$ is the following. For all $r \in \mathbb{Z}_{\geq 0}$, let

$$e_{S,r} := \sum_{\chi \in \widehat{G}_{S,r}} e_\chi ,$$

where $\widehat{G}_{S,r} := \{\chi \in \widehat{G} \mid r_{S,\chi} = r\}$. Then $e_{S,r}$ is an idempotent in $\mathbb{C}[G]$. However, equalities (11) show that $r_{S,\chi} = r_{S,\chi'}$ whenever $\chi \sim_{\mathbb{Q}} \chi'$. Consequently, $e_{S,r} \in \mathbb{Q}[G]$, for all $r \in \mathbb{Z}_{\geq 0}$. This means that every set S as above leads to a canonical decomposition

$$1 = \sum_{r \in \mathbb{Z}_{\geq 0}} e_{S,r}$$

of 1 as a sum of orthogonal idempotents in $\mathbb{Q}[G]$. The advantage of using this decomposition over any other stems from the fact that for all $r \in \mathbb{Z}_{\geq 0}$ the projective $e_{S,r} \cdot \mathbb{Q}[G]$–modules $e_{S,r}\mathbb{Q}U_S$ and $e_{S,r}\mathbb{Q}X_S$ are free of rank r (see Remark 2.2.1.) Consequently, we have equalities

$$\det_{e_{S,r}\mathbb{Q}[G]}(e_{S,r}\mathbb{Q}U_S) = e_{S,r}\wedge^r_{\mathbb{Q}[G]}\mathbb{Q}U_S, \qquad \det_{e_{S,r}\mathbb{Q}[G]}(e_{S,r}\mathbb{Q}X_S) = e_{S,r}\wedge^r_{\mathbb{Q}[G]}\mathbb{Q}X_S .$$

At the analytic (L–function) level, we have the following equalities, for all $r \in \mathbb{Z}_{\geq 0}$.

$$e_{S,r}\Theta_S^*(0) = e_{S,r} \cdot \frac{1}{r!}\lim_{s \to 0}\frac{1}{s^r}\Theta_S(s).$$

Now, let $\overline{u} := \{u_1, \dots, u_r\}$ and $\overline{x} := \{x_1, \dots, x_r\}$ be bases of the free $e_{S,r} \cdot \mathbb{Q}[G]$–modules $e_{S,r}\mathbb{Q}U_S$ and $e_{S,r}\mathbb{Q}X_S$, respectively and let $\overline{x}^* := \{x_1^*, \dots, x_r^*\}$ be the basis of $e_{S,r}(\mathbb{Q}X_S)^* = (e_{S,r}\mathbb{Q}X_S)^*$ (the first dual is over $\mathbb{Q}[G]$ and the second over $e_{S,r}\mathbb{Q}[G]$) which is dual to $\overline{x}$. Then, it is obvious that for every $r \in \mathbb{Z}_{\geq 0}$, the $e_{S,r}$–component of Stark's Main Conjecture is in fact the following.

Conjecture 2.4.1. $e_{S,r}A_{\mathbb{Q}}(K/k, S)$. *The following equivalent statements hold.*

(1)

$$\frac{e_{S,r} \cdot \frac{1}{r!}\lim_{s \to 0}\frac{1}{s^r}\Theta_S(s)}{\det(x_i^*(\lambda_S(u_j)))} \in (e_{S,r}\mathbb{Q}[G])^\times .$$

(2) *If* $\mathfrak{X}_{S,r}^* := x_1^* \wedge \cdots \wedge x_r^*$, *there exists a unique* $\alpha \in (e_{S,r}\mathbb{Q}[G])^\times$, *such that*

$$e_{S,r} \cdot \frac{1}{r!}\lim_{s \to 0}\frac{1}{s^r}\Theta_S(s) = \mathrm{ev}_{\mathbb{C}X_S}(\widetilde{\lambda_S}(\mathfrak{U}_{S,r}) \otimes \mathfrak{X}_{S,r}^*),$$

where $\mathfrak{U}_{S,r} := \alpha \cdot u_1 \wedge \cdots \wedge u_r \in e_{S,r}\wedge^r_{\mathbb{Q}[G]}\mathbb{Q}U_S$.

Above, we have identified x_i^* with an element in $(e_{S,r}\mathbb{C}X_S)^*$. This is the scalar extension $(x_i^* \otimes_{\mathbb{Q}} \mathbf{1}_{\mathbb{C}})$ of x_i^* to $e_{S,r}\mathbb{C}X_S$. Also, we have made the identification

$$\wedge^r_{(e_{S,r}\mathbb{C}G)}(e_{S,r}\mathbb{C}X_S)^* = (\wedge^r_{(e_{S,r}\mathbb{C}G)}e_{S,r}\mathbb{C}X_S)^*,$$

as in Remark 1.1.2. Of course, if $r = 0$, the determinant $\det(x_i^*(\lambda_S(u_j)))$ is equal to 1, by definition. Also, note that the uniqueness in the statement above is an immediate consequence of the fact that the map $\mathrm{ev}_{\mathbb{C}X_S}$ is an isomorphism.

Example 2.4.2. In this example, we examine the $e_{S,0}$–component $e_{S,0}A_{\mathbb{Q}}(K/k, S)$ of Conjecture $A_{\mathbb{Q}}(K/k, S)$. According to the above considerations and equality (12), we have the equivalences

$$e_{S,0}A_{\mathbb{Q}}(K/k, S) \quad \Leftrightarrow \quad \Theta_S(0) \in \mathbb{Q}[G] \quad \Leftrightarrow \quad \zeta_S(\sigma, 0) \in \mathbb{Q}\,, \text{ for all } \sigma \in G\,.$$

In the number field case, the third statement above is a deep, well-known theorem of Klingen and Siegel (see [**33**].) In the function field case, a much sharper form of this statement is true, as we will see in the next example. So, conjecture $e_{S,0}A_{\mathbb{Q}}(K/k, S)$ is true in general.

In fact, both the Klingen-Siegel theorem mentioned above and its function field counterpart state that $\Theta_S(1-n) \in \mathbb{Q}[G]$, for all $n \in \mathbb{Z}_{\geq 1}$ (see Remark 3.2.4 below for a refinement of these results.)

Example 2.4.3. In this example, we examine conjecture $A_{\mathbb{Q}}(K/k, S)$ in the case of abelian extensions K/k of global fields of characteristic p (i.e. function fields), where p is an arbitrary prime number. Let us assume that the exact field of constants of k is $\mathbb{F}_q$, where q is a power of p. For a prime v in either k or K, we let d_v be the degree of v over $\mathbb{F}_q$. This means that the residue field $\kappa(v)$ of v is $\mathbb{F}_{q^{d_v}}$. Consequently, we have $\mathbf{N}v = q^{d_v}$, for all v. Therefore, the Euler product expansion of $L_S(\chi, s)$ has the form

$$L_S(\chi, s) = \prod_{v \notin S}(1 - \chi(\sigma_v) \cdot (q^{-s})^{d_v})^{-1}, \qquad \Re(s) > 1, \quad \chi \in \widehat{G}.$$

This leads naturally to considering the power series of variable u

$$P_S(u) := \prod_{v \notin S}(1 - \sigma_v^{-1} \cdot u^{d_v})^{-1} \in \mathbb{Q}[G][[u]]\,.$$

Note that the infinite product converges in the u–adic topology of $\mathbb{Q}[G][[u]]$ because for every $n \in \mathbb{N}$ there are only finitely many primes v such that $d_v \leq n$. In fact, as shown in [**42**], Chpt. IV, it turns out that $P_S(u)$ is a rational function in the variable u with coefficients in $\mathbb{Q}[G]$,

$$P_S(u) \in \mathbb{Q}[G](u)\,,$$

with a single pole at $u = q^{-1}$ and that we have an equality

$$\Theta_S(s) = P_S(q^{-s}), \qquad \text{for all } s \in \mathbb{C} \setminus \{1\}\,.$$

The last equality implies that we have

(13)

$$e_{S,r}\Theta_S^*(0) = e_{S,r} \cdot \frac{1}{r!}\frac{\mathrm{d}^r P}{\mathrm{d}u^r}(1) \cdot (-\log q)^r \in (e_{S,r}\mathbb{Q}[G])^{\times} \cdot (\log q)^r\,, \quad \text{for all } r \in \mathbb{Z}_{\geq 0}\,.$$

For $r = 0$, this simply says that $\Theta_S(0) \in \mathbb{Q}[G]$, which settles the $e_{S,0}$–component of conjecture $A_{\mathbb{Q}}(K/k, S)$, as promised at the end of the last example. Next, let us consider an arbitrary $r \in \mathbb{Z}_{\geq 1}$ and analyze the determinant in Conjecture 2.4.1(1) for two bases $\overline{x}$ and $\overline{u}$. For all $u \in U_S$, we have

$$\lambda_S(\tilde{u}) = -\sum_{w \in S_K} \log |u|_w \cdot w = -(\log q) \cdot (\sum_{w \in S_K} d_w \cdot \mathrm{ord}_w(u) \cdot w) \in (\log q) \cdot X_S \,,$$

where $\mathrm{ord}_w(u) \in \mathbb{Z}$ is the evaluation at u of the canonically normalized valuation associated to w. However, this means that $\lambda_S(\mathbb{Q}U_S) \subseteq (\log q) \cdot \mathbb{Q}X_S$. Therefore, we have

$$x_i^*(\lambda_S(u_j)) \in (\log q) \cdot e_{S,r}\mathbb{Q}[G], \text{ for all } i, j = 1, \ldots, r.$$

Consequently, we have $\det(x_i^*(\lambda_S(u_j)) \in (\log q)^r \cdot (e_{S,r}\mathbb{Q}[G])^\times$. If we combine this statement with (13) above, we obtain

$$\frac{e_{S,r}\Theta_S^*(0)}{\det(x_i^*(\lambda_S(u_j))} \in (e_{S,r}\mathbb{Q}[G])^\times \,.$$

This verifies conjecture $e_{S,r}A_{\mathbb{Q}}(K/k, S)$, for all $r \in \mathbb{Z}_{\geq 0}$. Therefore, conjecture $A_{\mathbb{Q}}(K/k, S)$ holds true for characteristic p global fields.

3. Integral refinements of Stark's Main Conjecture

The main goal of this section is to try to put an integral structure on conjecture $A_{\mathbb{Q}}(K/k, S)$. The meaning of the phrase "integral structure" will become clear as the section progresses. For the moment, let us just note that in the case $K = k$ (see Example 2.3.6), the left-hand side of the equality in conjecture $A_{\mathbb{Q}}(K/k, S)$ is

$$\mathbb{Q}\zeta_{k,S}^*(0) = \mathbb{Q}R_{k,S} \,,$$

which only determines the regulator $R_{k,S}$ up to a rational factor, missing completely the important integral invariants $h_{k,S}$ and w_k captured by the class number formula for the global field k. So, roughly speaking, we would like to have a conjectural equality which captures the invertible $\mathbb{Z}[G]$–submodule $\mathbb{Z}[G]\Theta_S^*(0)$ of $\mathbb{C}[G]$, rather than the $\mathbb{Q}[G]$–submodule $\mathbb{Q}[G]\Theta_S^*(0)$. The main difficulty in achieving this goal lies in that, at first sight, there is no canonical invertible $\mathbb{Z}[G]$–submodule $\mathcal{M}_S$,

$$\mathcal{M}_S \subseteq \widetilde{\lambda_S}(\det_{\mathbb{Q}[G]}(\mathbb{Q}U_S)) \otimes_{\mathbb{Q}[G]} \det_{\mathbb{Q}[G]}(\mathbb{Q}X_S)^* \,.$$

If we had a canonical $\mathcal{M}_S$, then a natural conjectural equality would simply state

$$\mathrm{ev}_{\mathbb{C}X_S}(\mathcal{M}_S) = \mathbb{Z}[G]\Theta_S^*(0) \,.$$

The difficulty in finding such a module $\mathcal{M}_S$ stems from the non-triviality of the $\mathbb{Z}[G]$–module structure of U_S and X_S. For example, if these $\mathbb{Z}[G]$–modules were *perfect* (i.e. of finite projective dimension, i.e. G–cohomologically trivial and therefore of projective dimension at most 1), the construction of $\mathcal{M}_S$ would be more or less immediate, via an extension of the "determinant" functor from the category of projective modules to that of modules of finite projective dimension, due to Grothendieck and Mumford-Knudsen (see [**24**].) Although X_S and U_S are rarely perfect (see Example 3.1.3 below), it turns out that, under mild hypotheses on S, the difference between their $\mathbb{Z}[G]$–module structures is controlled by a length 2, *perfect complex* of $\mathbb{Z}[G]$–modules (i.e. a complex which is quasi-isomorphic to a complex of perfect $\mathbb{Z}[G]$–modules.)

The main idea behind the most general integral refinement of $A_{\mathbb{Q}}(K/k, S)$ to date is to set the module $\mathcal{M}_S$ above equal to a canonical isomorphic image of the

Grothendieck-Mumford-Knudsen determinant of this perfect complex. Although it is not our goal to focus on this general integral refinement (deferring that for Burns's contribution to this volume), we will describe it very briefly in the next section. This will permit the reader to see how the integral refinements which will be covered in detail in our lectures fit in the big picture.

3.1. The Equivariant Tamagawa Number Conjecture – a brief introduction

For simplicity, in this section we will work under the hypothesis that S is sufficiently large, so that the ideal class–group of the ring of S–integers $O_{K,S}$ is trivial (i.e. $h_{K,S} = 1$, in the notations of Example 2.3.6.) Note that, since the validity of $A_{\mathbb{Q}}(K/k, S)$ does not depend on S (see Remark 2.3.3), this assumption is technically inconsequential in the context of the "rational" conjecture $A_{\mathbb{Q}}(K/k, S)$. Also, note that Tchebotarev's density theorem allows us to enlarge S by adjoining a finite set of primes which split completely in K/k in order achieve $h_{K,S} = 1$, which turn out to be relevant in the context of the Rubin-Stark integral refinement discussed later in this paper.

Theorem 3.1.1 (Tate, [**41**] and [**42**]). *Under the above hypotheses, there exists a canonical class $c_2 \in \mathrm{Ext}^2_{\mathbb{Z}[G]}(X_S, U_S)$ which has a representative*

$$0 \longrightarrow U_S \longrightarrow P_{-1} \longrightarrow P_0 \longrightarrow X_S \longrightarrow 0\,,$$

where P_1 and P_0 are finitely generated $\mathbb{Z}[G]$–modules of projective dimension ≤ 1 (i.e. G–cohomologically trivial.)

The element $c_2 := c_2(K/k, S)$ is called the Tate class and it is constructed via a clever gluing process out of local and global fundamental classes (see Chpt. II, §5 of [**42**] and also [**9**].) The four term exact sequence in the theorem is called a Tate sequence. An immediate consequence of the above theorem is the following.

Corollary 3.1.2. *For all $i \in \mathbb{Z}$, we have group isomorphisms*

$$\widehat{\mathrm{H}}^i(G, X_S) \xrightarrow[\sim]{\cup\, c_2} \widehat{\mathrm{H}}^{i+2}(G, U_S)\,,$$

given by the cup product with $c_2 \in \mathrm{Ext}^2_{\mathbb{Z}[G]}(X_S, U_S) \simeq \widehat{\mathrm{H}}^2(G, \mathrm{Hom}_{\mathbb{Z}}(X_S, U_S))$.

Example 3.1.3. The Corollary above allows us to compute the cohomology of the "mysterious" $\mathbb{Z}[G]$–module U_S in terms of that of the more down-to-earth module X_S. This is illustrated in the following example. Let p be an odd prime number. Let $K := \mathbb{Q}(\zeta_p)$, $k := \mathbb{Q}$ and $S := \{p, v_\infty, v_1, \ldots, v_r\}$, where v_∞ is the infinite prime of $\mathbb{Q}$ and $v_1, \ldots, v_r$ are distinct primes in k, which split completely in K/k and are chosen so that $h_{K,S} = 1$. Exact sequence (9) and isomorphism (10) imply immediately that we have an isomorphism of $\mathbb{Z}[G]$–modules

$$X_S \simeq \mathbb{Z}[G]/(1-j) \bigoplus \mathbb{Z}[G]^r,$$

where $j \in G$ is the complex conjugation automorphism of K, obviously satisfying $G_{v_\infty} = \{1, j\}$. For all $i \in \mathbb{Z}$, we have the following group isomorphisms.

$$\widehat{\mathrm{H}}^{2i}(G_{v_\infty}, X_S) \simeq \widehat{\mathrm{H}}^0(G_{v_\infty}, X_S) \simeq \widehat{\mathrm{H}}^0(G_{v_\infty}, \mathbb{Z}[G/G_{v_\infty}]) \simeq (\mathbb{Z}/2\mathbb{Z})^{|G|/2} \neq 0,$$
$$\widehat{\mathrm{H}}^{2i+1}(G_{v_\infty}, X_S) \simeq \widehat{\mathrm{H}}^1(G_{v_\infty}, X_S) \simeq \widehat{\mathrm{H}}^1(G_{v_\infty}, \mathbb{Z}[G/G_{v_\infty}]) \simeq 0.$$

In the two sequences of isomorphisms above, the first isomorphism is a consequence of the cyclicity of G_{v_∞}, the second is due to the fact that $\mathbb{Z}[G]^r$ is G–cohomologically trivial, and the third is based on the definitions of $\widehat{\mathrm{H}}^0$ and $\widehat{\mathrm{H}}^1$, respectively, and the fact that G_{v_∞} acts trivially on $\mathbb{Z}[G/G_{v_\infty}]$. The isomorphisms above combined with the last Corollary, lead to

$$\widehat{\mathrm{H}}^{2i}(G_{v_\infty}, U_S) \simeq (\mathbb{Z}/2\mathbb{Z})^{|G|/2} \neq 0, \qquad \widehat{\mathrm{H}}^{2i+1}(G_{v_\infty}, U_S) = 0,$$

for all $i \in \mathbb{Z}$. In particular, these calculations show that the $\mathbb{Z}[G]$–modules X_S and U_S are not of finite projective dimension in this case. We leave it as an exercise for the reader to compute the cohomology groups $\widehat{\mathrm{H}}^i(G, U_S)$, for all $i \in \mathbb{Z}$ in this case. Also, the reader might be interested in repeating this exercise for the case $k = \mathbb{Q}$, $K = \mathbb{Q}(\zeta_p)^+$. He will discover that X_S and U_S are of finite projective dimension in that case.

Roughly speaking, the Equivariant Tamagawa Number Conjecture (ETNC) of Bloch-Kato (see [**1**]) states that the elusive module $\mathcal{M}_S$ in the introduction to this section is a specific isomorphic image of the Grothendieck-Mumford-Knudsen determinant $\mathbf{det}(P^\bullet)$ of the perfect complex of $\mathbb{Z}[G]$–modules

$$P^\bullet : \cdots \longrightarrow 0 \longrightarrow P_{-1} \longrightarrow P_0 \longrightarrow \cdots$$

(extended by 0 at levels other than -1 and 0) arising from the Tate sequence above (see [**24**] for the definition of $\mathbf{det}(P^\bullet)$.) More precisely, $\mathcal{M}_S$ is obtained as follows. One considers the following injective morphism of $\mathbb{Z}[G]$–modules:

$$\begin{aligned} \rho_S \ : \ \mathbf{det}(P^\bullet) \overset{\iota}{\hookrightarrow} \mathbb{Q}\mathbf{det}(P^\bullet) \ &\overset{e}{\simeq} \ \mathbf{det}(\mathbb{Q}P^\bullet) \\ &\overset{c}{\simeq} \ \det_{\mathbb{Q}[G]}(\mathbb{Q}U_S) \otimes_{\mathbb{Q}[G]} \det_{\mathbb{Q}[G]}(\mathbb{Q}X_S)^* \\ &\simeq \ \widetilde{\lambda_S}(\det_{\mathbb{Q}[G]}(\mathbb{Q}U_S)) \otimes_{\mathbb{Q}[G]} \det_{\mathbb{Q}[G]}(\mathbb{Q}X_S)^*. \end{aligned}$$

Above, the morphism ι sends x to $1_{\mathbb{Q}} \otimes x$, for all $x \in \mathbf{det}(P^\bullet)$. The isomorphism e is the analogue of (6) for the Grothendieck-Mumford-Knudson determinant functor at the level of perfect complexes. It turns out that $\mathbf{det}$ also commutes with base-change. The isomorphism c arises as follows: the perfect complex of $\mathbb{Q}[G]$–modules $\mathbb{Q}P^\bullet$ is in fact cohomologically perfect. This means that its cohomology groups $\mathrm{H}^{-1}(\mathbb{Q}P^\bullet) = \mathbb{Q}U_S$ and $\mathrm{H}^0(\mathbb{Q}P^\bullet) = \mathbb{Q}X_S$ are perfect $\mathbb{Q}[G]$–modules (i.e. of finite projective dimension over $\mathbb{Q}[G]$.) Obviously, these cohomology groups are in fact projective $\mathbb{Q}[G]$–modules. If R is a commutative ring and $\mathcal{C}^\bullet$ is a perfect complex of R–modules which is cohomologically perfect, there is a canonical isomorphism of R–modules

$$c_{\mathcal{C}^\bullet} : \mathbf{det}(\mathcal{C}^\bullet) \simeq \left(\bigotimes_{i\in\mathbb{Z}} \mathbf{det}(\mathrm{H}^{2i+1}(\mathcal{C}^\bullet))\right) \otimes_R \left(\bigotimes_{i\in\mathbb{Z}} \mathbf{det}(\mathrm{H}^{2i}(\mathcal{C}^\bullet))^*\right).$$

The isomorphism c above is the canonical $c_{\mathbb{Q}P^\bullet}$. Also, note that since $\mathbb{Q}U_S$ and $\mathbb{Q}X_S$ are projective $\mathbb{Q}[G]$–modules, their Grothendieck-Mumford-Knudson determinants $\mathbf{det}$ coincide with their classical determinants $\det_{\mathbb{Q}[G]}$ defined earlier in these lectures. Finally, the last isomorphism above is $\widetilde{\lambda_S} \otimes \mathrm{id}$. Now, we define

$$\mathcal{M}_S := \mathrm{Im}(\rho_S) = \rho_S(\mathbf{det}(P^\bullet)).$$

This is an invertible $\mathbb{Z}[G]$–submodule of $\det_{\mathbb{C}[G]}(\mathbb{C}U_S)\otimes_{\mathbb{C}[G]}\det_{\mathbb{C}[G]}(\mathbb{C}X_S)^*$ which turns out to be independent on the Tate sequence representing the Tate canonical class c_2.

Conjecture 3.1.4 (ETNC, Bloch-Kato [**1**]). *Under the above hypotheses, we have the following equality of invertible $\mathbb{Z}[G]$–submodules of $\mathbb{C}[G]$.*

$$\mathrm{ev}_{\mathbb{C}X_S}(\mathcal{M}_S)=\mathbb{Z}[G]\Theta_S^*(0)\,.$$

As mentioned above, we will not insist upon commenting further on this conjecture in these lectures. For further details, we refer the reader to Burns's contribution to this volume. At this point, it suffices to say that all refinements (both integral and p–adic) of $A_{\mathbb{Q}}(K/k,S)$ which will be discussed in detail in these lectures are consequences of the ETNC, as shown in [**2**]. (See also §5.5, below.)

3.2. The integral conjecture of Rubin-Stark

In this section, we will discuss a conjecture stated by Rubin in [**32**], which generalizes an earlier integral conjecture of Stark [**37**]. It would be misleading to say that Rubin's conjecture is an integral refinement of $A_{\mathbb{Q}}(K/k,S)$. As it will become very clear by the end of this section, Rubin's conjecture is in fact an integral refinement of an idempotent piece $e_{S,r}A_{\mathbb{Q}}(K/k,S)$ of Stark's Main Conjecture $A_{\mathbb{Q}}(K/k,S)$, under some additional hypotheses on S and for an appropriately chosen integer r. (See the notations and definitions of section 2.4.)

3.2.1. **Additional hypotheses on S.** Let us fix an integer $r\geq 0$. Let us suppose that the set S contains r distinct primes $v_1,\dots,v_r$ which split completely in K/k and that $\mathrm{card}(S)\geq r+1$. Let us pick a prime $v_0\in S\setminus\{v_1,\dots,v_r\}$. Equalities (11) show that $e_{S,r'}=0$, for all $r'\leq r$. Rubin's conjecture aims for an integral refinement of $e_{S,r}A_{\mathbb{Q}}(K/k,S)$ (i.e. the first possibly non–trivial idempotent piece of $A_{\mathbb{Q}}(K/k,S)$), under the above additional hypotheses on the set S.

Our first goal is to give the rationale behind the new hypotheses on S as well as the choice of the particular idempotent $e_{S,r}\in\mathbb{Q}[G]$. For that purpose, we use the notations introduced in §2.4. Let $S_0:=S\setminus\{v_1,\dots,v_r\}$. Let $W=(w_0,w_1,\dots,w_r)$ be an ordered r–tuple of primes in K, such that w_i sits above v_i, for all i. Note that since v_i splits completely in K/k, we have $\mathrm{Ann}_{\mathbb{Z}[G]}(w_i-w_0)=\mathrm{Ann}_{\mathbb{Z}[G]}(w_i)=0$, for all $i=1,\dots,r$ (see isomorphism (10).) Consequently, X_S has a natural direct sum decomposition

$$X_S=X_{S_0}\oplus(\oplus_{i=1}^r\mathbb{Z}[G](w_i-w_0))\,,$$

where the second direct summand is a free $\mathbb{Z}[G]$–module of rank r and basis $\{w_1-w_0,\dots,w_r-w_0\}$. The above decomposition is a direct consequence of the exact sequence of $\mathbb{Z}[G]$–modules

$$0\longrightarrow X_{S_0}\longrightarrow X_S\xrightarrow{\ \pi\ }\oplus_{i=1}^r\mathbb{Z}[G]w_i\longrightarrow 0\,,$$

where π is the forgetful map sending $\sum_{w\in S_K}a_w\cdot w$ to $\sum_{w\in S_K\setminus S_{0,K}}a_w\cdot w$. The map π has an obvious section sending w_i to (w_i-w_0), for all $i=1,\dots,r$, which splits the exact sequence and leads to the above decomposition. Since $e_{S,r}\mathbb{Q}X_S$ is a free $e_{S,r}\mathbb{Q}[G]$–module of rank r, we have

$$e_{S,r}\mathbb{Q}X_S=\oplus_{i=1}^r e_{S,r}\mathbb{Q}[G](w_i-w_0)\,,$$

and a natural $e_{S,r}\mathbb{Q}[G]$–basis $\overline{x} = \{x_1, \dots, x_r\}$ for $e_{S,r}\mathbb{Q}X_S$, given by

$$x_i := e_{S,r}(w_i - w_0), \qquad \text{for all } i = 1, \dots, r.$$

The free (full rank r) $e_{S,r}\mathbb{Z}[G]$–submodule of $e_{S,r}\mathbb{Q}X_S$ of basis $\overline{x}$ puts a natural integral structure on $e_{S,r}\mathbb{Q}X_S$. This leads to a natural integral structure on $\det_{e_{S,r}\mathbb{Q}[G]}(e_{S,r}\mathbb{Q}X_S)^*$ given by its free (full rank, i.e. rank one) submodule $e_{S,r}\mathbb{Z}[G](x_1^* \wedge \cdots \wedge x_r^*)$, where the exterior powers are viewed over $e_{S,r}\mathbb{Z}[G]$. (See the notations and identifications made in §2.4.)

The conclusion is that the additional conditions imposed upon S and the choice of the particular idempotent $e_{S,r}$ give rise to a canonical generator

$$\mathfrak{X}_{S,r}^* := x_1^* \wedge \cdots \wedge x_r^*$$

of the free, rank one $e_{S,r}\mathbb{Q}[G]$–module $\det_{e_{S,r}\mathbb{Q}[G]}(e_{S,r}\mathbb{Q}X_S)^*$. Also, it is important to note that evaluating against $\mathfrak{X}_{S,r}^*$ is particularly simple. Indeed, if we let

$$R_W : e_{S,r} \wedge_{\mathbb{C}[G]}^r \mathbb{C}U_S \xrightarrow{\sim} e_{S,r}\mathbb{C}[G], \qquad R_W(\mathfrak{U}) := \mathrm{ev}_{\mathbb{C}X_S}(\widetilde{\lambda_S}(\mathfrak{U}) \otimes \mathfrak{X}_{S,r}^*),$$

then for $\mathfrak{U} = e_{S,r}\widetilde{u}_1 \wedge \cdots \wedge \widetilde{u_r}$, with $u_1, \dots, u_r \in U_S$, we have

$$R_W(\mathfrak{U}) = e_{S,r} \cdot \det(-\sum_{\sigma \in G} \log|u_i^\sigma|_{w_j} \cdot \sigma^{-1}).$$

This is implied by the equality $x_j^*(\lambda_S(u_i)) = (-\sum_{\sigma\in G} \log|u_i^\sigma|_{w_j} \cdot \sigma^{-1}) \cdot e_{S,r}$, for all $i, j = 1, \dots, r$, which follows from the definition of λ_S and from the fact that v_j is completely split in K/k.

Definition 3.2.1. *The map R_W defined above is called the Rubin-Stark regulator.*

Remark 3.2.2. *It is easy to see that for two different choices of ordered r–tuples W and W' corresponding to possibly different orderings of the set $\{v_1, \dots, v_r\}$, we have $\mathcal{R}_W = \pm\sigma \cdot \mathcal{R}_{W'}$, where $\sigma \in G$. Since $\pm\sigma \in \mathbb{Z}[G]^\times$, the truth of the conjectures below is independent of these choices.*

Note that, under the current hypotheses, we have

$$e_{S,r}\Theta_S^*(0) = \frac{1}{r!}\lim_{s\to 0}\frac{1}{s^r}\Theta_S(s).$$

Following Rubin, we will denote this special value by $\Theta_S^{(r)}(0)$. Under the current hypotheses, conjecture $e_{S,r}A_{\mathbb{Q}}(K/k, S)$ can be restated as follows.

Conjecture 3.2.3. $A_{\mathbb{Q}}(K/k, S, r)$. *For all K/k, S, and r satisfying the above hypotheses, there exists a unique element $\varepsilon_S \in e_{S,r} \wedge_{\mathbb{Q}[G]}^r \mathbb{Q}U_S$, such that*

$$R_W(\varepsilon_S) = \Theta_S^{(r)}(0).$$

Note that the uniqueness in the conjecture holds true and is an immediate consequence of the fact that $\mathrm{ev}_{\mathbb{C}X_S}$ is an isomorphism which renders the Rubin-Stark regulator R_W bijective. The equality in the conjecture above can be rewritten as

$$\mathrm{ev}_{\mathbb{C}X_S}(\widetilde{\lambda_S}(\mathbb{Z}[G]\varepsilon_S) \otimes_{\mathbb{Z}[G]} \mathbb{Z}[G]\mathfrak{X}_S^*) = \mathbb{Z}[G]\Theta_S^{(r)}(0).$$

This may lure the reader into believing that we have arrived this way at an integral refinement of $A_{\mathbb{Q}}(K/k, S, r)$. However, this is deceiving: although the $\mathbb{Z}[G]$–module $\mathbb{Z}[G]\mathfrak{X}_S^*$ is indeed canonical, that is not the case with $\mathbb{Z}[G]\varepsilon_S$, as we have no integral information on the conjectural element ε_S. The only information we have on ε_S is rational in nature. Namely, if ε_S indeed exists, then we have an equality

$$\mathbb{Q}[G]\varepsilon_S = e_{S,r} \cdot \wedge^r_{\mathbb{Q}[G]}\mathbb{Q}U_S.$$

As we will see in the next few subsections, Rubin's Conjecture does not quite identify the module $\mathbb{Z}[G]\varepsilon_S$ with a canonical $\mathbb{Z}[G]$–submodule of $e_{S,r} \cdot \wedge^r_{\mathbb{Q}[G]}\mathbb{Q}U_S$. Instead, it specifies a canonical $\mathbb{Z}[G]$–submodule of $e_{S,r}\cdot\wedge^r_{\mathbb{Q}[G]}\mathbb{Q}U_S$ in which a certain "multiple" of $\mathbb{Z}[G]\varepsilon_S$ sits with a finite index.

3.2.2. **The additional set of primes T.** To the set of data $(K/k, S, r)$ satisfying the hypotheses in the previous section, we add a new finite, non-empty set of primes T of k, which is disjoint from S and satisfies the following condition:

$$\{\zeta \in \mu_K \mid \zeta \equiv 1 \mod w, \quad \forall w \in T_K\} = \{1\}. \tag{14}$$

Note that, if $\mathrm{char}(k) = p > 0$, then $T \neq \emptyset$ implies the condition above. If $\mathrm{char}(k) = 0$, then the condition above is satisfied if, for example, T contains two primes of distinct residual characteristics or a prime of residual characteristic larger than $w_K = |\mu_K|$. We let

$$U_{S,T} := \{u \in U_S \mid u \equiv 1 \mod w, \quad \forall w \in T_K\}.$$

The new group of units is a $\mathbb{Z}[G]$ submodule of U_S with no $\mathbb{Z}$ torsion (see (14)). There is an obvious exact sequence of $\mathbb{Z}[G]$ modules (see [**32**]):

(15)

$$1 \longrightarrow U_{S,T} \longrightarrow U_S \longrightarrow \Delta_T := \bigoplus\nolimits_{w\in T_K} \kappa(w)^\times \longrightarrow A_{S,T} \longrightarrow A_S \longrightarrow 1,$$

where A_S is the ideal class-group of $O_{K,S}$ and $A_{S,T}$ is a T–modified version of A_S, defined by

$$A_{S,T} := \frac{\{\text{fractional ideals of } O_{K,S}, \text{ coprime to } T\}}{\{fO_{K,S} \mid f \in K^\times, \quad f \equiv 1 \mod w, \quad \forall w \in T_K\}}.$$

The Artin reciprocity map associated to K establishes a group isomorphism between $A_{S,T}$ and the Galois group of the maximal abelian extension of K which is completely split at primes in S, is unramified away from T, and at most tamely ramified at all primes in T. We define the T–modified S–imprimitive ζ–function associated to K by

$$\zeta_{K,S,T}(s) := \prod_{w\in T_K}(1 - \mathrm{N}w^{1-s}) \cdot \zeta_{K,S}(s), \qquad \forall s \in \mathbb{C}.$$

This way, we obtain a holomorphic function. Note the simple pole at $s = 1$ of $\zeta_{K,S}$ is annihilated by the zero at $s = 1$ of the product

$$\delta_{K,T}(s) := \prod_{w\in T_K}(1 - \mathrm{N}w^{1-s}).$$

If we define the (S, T)–regulator of K by $R_{K,S,T} := \mathrm{vol}(\mathbb{R}X_S/\lambda_S(U_{S,T}))$ and let $h_{K,S} := |A_S|$ and $h_{K,S,T} := |A_{S,T}|$, then the exact sequence (15) gives us an equality

$$R_{K,S,T} = R_{K,S} \cdot [\widetilde{U_S} : U_{S,T}] = |\Delta_T| \cdot w_K^{-1} \cdot h_{K,S,T}^{-1} \cdot h_{K,S}.$$

If we combine this equality with the S–class number formula (see Example 2.3.6) and the observation that $\delta_{K,T}(0) = \pm|\Delta_T|$, we obtain the (S, T)–class number formula

$$\zeta^*_{K,S,T}(0) = \delta_{K,T}(0) \cdot \zeta^*_{K,S}(0) = \pm h_{K,S,T} \cdot R_{K,S,T}, \tag{16}$$

where "+" arises if and only if $\delta_{K,T}(0) < 0$. These constructions can be extended to the G–equivariant L–function. We define the T–modified version of Θ_S by setting

$$\Theta_{S,T}(s) := \prod_{v \in T} (1 - \mathrm{N}v^{1-s} \cdot \sigma_v^{-1}) \cdot \Theta_S(s)\,.$$

This is a $\mathbb{C}[G]$–valued, holomorphic function. The simple pole at $s = 1$ of $\Theta_S(s)$, supported at the trivial character idempotent $e_{\mathbf{1}_G}$, is annihilated by the zero at $s = 1$ of the $\mathbb{C}[G]$–valued, holomorphic

$$\delta_{K/k,T}(s) := \prod_{v \in T} (1 - \mathrm{N}v^{1-s} \cdot \sigma_v^{-1})\,,$$

supported at the same idempotent.

Remark 3.2.4. *Above, we noticed that* $\delta_{K,T}(0) = \pm|\Delta_T|$. *A better way of writing this equality is* $\mathrm{Fit}_{\mathbb{Z}}(\Delta_T) = \delta_{K,T}(0)\mathbb{Z}$. *It turns out that the more general* $\delta_{K/k,T}(0)$ *has a similar interpretation. Indeed, we have isomorphisms of* $\mathbb{Z}[G]$*–modules*

$$\begin{aligned}\Delta_T \simeq \bigoplus_{v \in T} \kappa(w(v))^\times \otimes_{\mathbb{Z}[G_v]} \mathbb{Z}[G] \quad &\simeq \quad \bigoplus_{v \in T} \mathbb{Z}[G_v]/(1 - \mathrm{N}v \cdot \sigma_v^{-1}) \otimes_{\mathbb{Z}[G_v]} \mathbb{Z}[G] \\ &\simeq \quad \bigoplus_{v \in T} \mathbb{Z}[G]/(1 - \mathrm{N}v \cdot \sigma_v^{-1})\,,\end{aligned}$$

where $w(v)$ *is a fixed prime in* K *sitting above* v, *for each* $v \in T$ *(see* [**28**].*) Consequently, we have*

$$\delta_{K/k,T}(0) \cdot \mathbb{Z}[G] = \prod_{v \in T} (1 - \mathrm{N}v \cdot \sigma_v^{-1}) \cdot \mathbb{Z}[G] = \mathrm{Fit}_{\mathbb{Z}[G]}(\Delta_T).$$

In particular, since $\mu_K \subseteq \Delta_T$ *(see property (14)), we have* $\delta_{K/k,T}(0) \in \mathrm{Ann}_{\mathbb{Z}[G]}(\mu_K)$. *In fact, the following equality of* $\mathbb{Z}[G]$*–ideals holds*

$$\mathrm{Ann}_{\mathbb{Z}[G]}(\mu_K) = \langle \delta_{K/k,T}(0) \mid T \rangle_{\mathbb{Z}[G]}\,,$$

where the ideal on the right is generated by $\delta_{K/k,T}(0)$, *for all sets* T *satisfying the above properties, provided that* S *is fixed (see* [**42**], *Lemme 1.1, Chpt. IV.) Next, we observe that* $\mu_K = (\mathbb{Q}/\mathbb{Z})'(1)^{G_K}$, *where "(1)" denotes the Tate twist with respect to the first power of the cyclotomic character* $c_k : G_k \longrightarrow \mathrm{Aut}(\mu_{\overline{k}}) \subseteq \widehat{\mathbb{Z}}^\times$, G_L *denotes the absolute Galois group of the global field* L *and*

$$(\mathbb{Q}/\mathbb{Z})' = \begin{cases} \mathbb{Q}/\mathbb{Z}, & \text{if } char(k) = 0; \\ \mathbb{Q}/\mathbb{Z}[1/p], & \text{if } char(k) = p > 0. \end{cases}$$

We remind the reader that if $n \in \mathbb{Z}$ *and* $\mathcal{M}$ *is a torsion* $\mathbb{Z}$*–module endowed with a left* G_k*–action* $m \longrightarrow {}^\sigma m$, *for all* $\sigma \in G_k$ *and* $m \in \mathcal{M}$, *then the* n*–th Tate twist* $\mathcal{M}(n)$ *of* $\mathcal{M}$ *is, by definition, the module* $\mathcal{M}$ *with a new* G_k*–action given by*

$$\sigma * m := c_k(\sigma)^n \cdot {}^\sigma m.$$

It turns out that if we assume in addition that the sets T *contain at least two primes of distinct residual characteristic (if* $\mathrm{char}(k) > 0$ *this assumption is unnecessary), then we have an equality*

$$\mathrm{Ann}_{\mathbb{Z}[G]}\left((\mathbb{Q}/\mathbb{Z})'(n)^{G_K}\right) = \langle \delta_{K/k,T}(1-n) \mid T \rangle_{\mathbb{Z}[G]}\,,$$

*for all $n \in \mathbb{Z}_{\geq 1}$ (see Lemma 2.3 in [**10**].) The reader familiar with étale cohomology of arithmetic schemes, may want to rethink the above equality in terms of the well known $\mathbb{Z}[G]$–module isomorphisms*

$$\mathrm{H}^1_{\mathrm{et}}(O_{K,S}, \mathbb{Z}(n))_{\mathrm{tors}} \simeq (\mathbb{Q}/\mathbb{Z})'(n)^{G_K}, \qquad \textit{for all } n \in \mathbb{Z}_{\geq 1}.$$

*An even more remarkable property of the elements $\delta_{K/k,T}(1-n)$ considered above is the following theorem due to Deligne-Ribet [**11**], if $\mathrm{char}(k) = 0$ and to Deligne-Tate ([**42**], Chpt. V), if $\mathrm{char}(k) > 0$. This gives an integral refinement of the Klingen-Siegel Theorem and its function field counterpart mentioned in Example 2.4.2 above.*

Theorem 3.2.5. *For all $n \in \mathbb{Z}_{\geq 1}$ and all sets T as above, we have*

$$\Theta_{S,T}(1-n) = \delta_{K/k,T}(1-n) \cdot \Theta_S(1-n) \in \mathbb{Z}[G]\,,$$

provided that T contains at least two primes of different residual chars., if $n > 1$ and $\mathrm{char}(k) = 0$.

Now, we reformulate conjecture $A_{\mathbb{Q}}(K/k, S, r)$ in terms of the T–modified equivariant L–function $\Theta_{S,T}$. This is a simple task for two reasons. First, since $\delta_{K/k,T}(0) \in \mathbb{Q}[G]^{\times}$, we have

$$\Theta^{(r)}_{S,T}(0) = \delta_{K/k,T}(0) \cdot \Theta^{(r)}_S(0)\,.$$

Second, since $U_S/U_{S,T}$ is finite (see exact sequence (15)), we have $\mathbb{Q}U_S = \mathbb{Q}U_{S,T}$. Consequently, conjecture $A_{\mathbb{Q}}(K/k, S, r)$ is equivalent to the following.

Conjecture 3.2.6. $A_{\mathbb{Q}}(K/k, S, T, r)$. *For all K/k, S, T and r satisfying the above hypotheses, there exists a unique element $\varepsilon_{S,T} \in e_{S,r} \wedge^r_{\mathbb{Q}[G]} \mathbb{Q}U_{S,T}$, such that*

$$R_W(\varepsilon_{S,T}) = \Theta^{(r)}_{S,T}(0)\,.$$

Of course, the conjectural ε_S and $\varepsilon_{S,T}$ are linked by $\varepsilon_{S,T} = \delta_{K/k,T}(0) \cdot \varepsilon_S$. Since $\delta_{K/k,T}(0) \in \mathbb{Q}[G]^{\times}$, we have an equality of $\mathbb{Q}[G]$–modules $\mathbb{Q}[G]\varepsilon_S = \mathbb{Q}[G]\varepsilon_{S,T}$. However, we only have an inclusion

$$\mathbb{Z}[G]\varepsilon_{S,T} \subseteq \mathbb{Z}[G]\varepsilon_S.$$

The index $[\,\mathbb{Z}[G]\varepsilon_S : \mathbb{Z}[G]\varepsilon_{S,T}\,]$ is finite and supported at primes dividing μ_K (a direct consequence of Remark 3.2.4.) As we will see in the next section, Rubin's Conjecture is an integral refinement of the conjecture above, in the sense that it specifies a canonical $\mathbb{Z}[G]$–submodule of $e_{S,r} \cdot \wedge^r_{\mathbb{Q}[G]}\mathbb{Q}U_{S,T}$ which contains $\mathbb{Z}[G]\varepsilon_{S,T}$ with a finite index.

3.2.3. **The Rubin-Stark lattice.** In this section, we assume that the set of data $(K/k, S, T, r)$ satisfies the hypotheses listed in the previous section. In the following, we use the notations of Definition 1.1.3 in the "Algebraic Preliminaries" section.

Definition 3.2.7. *The Rubin-Stark lattice $\Lambda_{S,T}$ is given by*

$$\Lambda_{S,T} := \mathcal{L}_{\mathbb{Z}[G]}(U_{S,T},\, r) \cap e_{S,r} \cdot \wedge^r_{\mathbb{Q}[G]}\mathbb{Q}U_{S,T}\,.$$

More explicitly, $\Lambda_{S,T}$ consists of all $\epsilon \in e_{S,r} \cdot \wedge^r_{\mathbb{Q}[G]}\mathbb{Q}U_{S,T}$, such that

$$(\phi_r^{(1)} \circ \cdots \circ \phi_1^{(r)})(\epsilon) \in \mathbb{Z}[G]\,,$$

for all $\phi_1, \ldots, \phi_r \in U^*_{S,T}$, where $U^*_{S,T} := \mathrm{Hom}_{\mathbb{Z}[G]}(U_{S,T}, \mathbb{Z}[G])$.

Remark 3.2.8. *If M is a $\mathbb{Z}[G]$–module, then we let*

$$M_{S,r} := \widetilde{M} \cap e_{S,r}\mathbb{Q}M = \{x \in \widetilde{M} \mid (1 - e_{S,r})x = 0 \text{ in } \mathbb{Q}M\}.$$

The following are direct consequences of Remark 1.1.4 and Example 1.1.6.

(1) *If $r = 0$, then $\Lambda_{S,T} = \mathbb{Z}[G]_{S,0}$*
(2) *If $r = 1$, then $\Lambda_{S,T} = (U_{S,T})_{S,1}$. Note that since $U_{S,T}$ has no non-trivial $\mathbb{Z}$–torsion, we have an equality $U_{S,T} = \widetilde{U_{S,T}}$.*
(3) *For $r \geq 1$, we have a double inclusion*

$$(\wedge^r_{\mathbb{Z}[G]} U_{S,T})_{S,r} \subseteq \Lambda_{S,T} \subseteq \mathbb{Z}[1/|G|](\wedge^r_{\mathbb{Z}[G]} U_{S,T})_{S,r}.$$

3.2.4. **The Rubin-Stark conjecture.** The integral refinement proposed by Rubin in [**32**] for conjecture $A_{\mathbb{Q}}(K/k, S, T, r)$ is the following.

Conjecture 3.2.9. *$A_{\mathbb{Z}}(K/k, S, T, r)$ (Rubin-Stark). For all K/k, S, T and r satisfying the above hypotheses, there exists a unique element $\varepsilon_{S,T} \in \Lambda_{S,T}$, such that*

$$R_W(\varepsilon_{S,T}) = \Theta^{(r)}_{S,T}(0)\,.$$

In the case $r = 1$, the above statement is easily seen to be equivalent to the integral conjecture formulated by Stark himself in [**37**]. The key ingredients in establishing this equivalence are Proposition 1.2 in [**42**], Chpt. IV and Remark 3.2.4 above regarding the various properties of $\delta_{K/k,T}(0)$. From now on, we will refer to the unique element $\varepsilon_{S,T} \in \mathbb{C}(\wedge^r_{\mathbb{Z}[G]} U_{S,T})_{S,r}$ satisfying the regulator equality in the conjecture above as *the Rubin-Stark element* for the data $(K/k, S, T, r)$.

Example 3.2.10. Let us assume that $r = 0$. Then, $\Lambda_{S,T} = \mathbb{Z}[G]_{S,0}$ and the Rubin-Stark regulator is the inclusion $e_{S,0}\mathbb{Q}[G] \subseteq e_{S,0}\mathbb{C}[G]$. Consequently, the Rubin-Stark conjecture states that

$$\Theta_{S,T}(0) \in \mathbb{Z}[G]_{S,0}\,.$$

Theorem 3.2.5 above combined with the obvious equality $(1 - e_{S,0})\Theta_{S,T}(0) = 0$ implies that the conjecture holds true in this case.

Example 3.2.11. (compare to Example 2.3.6.) Let us consider the case where S contains at least $r+1$ distinct primes $v_0, v_1, \ldots, v_r$ which split completely in K/k. Unless $S = \{v_0, \ldots, v_r\}$, we have $e_{S,r} = 0$ and the conjecture is trivially true in that case. So, let us assume that $S = \{v_0, \ldots, v_r\}$. Then, $e_{S,r} = e_{\mathbf{1_G}}$ and therefore we have

$$\Theta^{(r)}_{S,T}(0) = \zeta^{(r)}_{k,S,T}(0) \cdot e_{\mathbf{1_G}} = \zeta^{*}_{k,S,T}(0) \cdot e_{\mathbf{1_G}}\,.$$

Let $u_1, \ldots, u_r$ be a $\mathbb{Z}$–basis for the group of (S,T)–units $U_{k,S,T}$ of k. Since

$$\sum_{\sigma \in G} \log |u_i^{\sigma}|_{w_j} \cdot \sigma^{-1} = \log |u_i|_{v_j} \cdot |G| e_{\mathbf{1_G}},$$

for all $i, j = 1, \ldots, n$, the (S,T)–class number formula (16) implies that in this case the Rubin-Stark element $\varepsilon_{S,T}$ is given by

$$\varepsilon_{S,T} = \pm \frac{h_{k,S,T}}{|G|^r} u_1 \wedge \cdots \wedge u_r\,,$$

where the exterior products are viewed over $\mathbb{Z}[G]$. The question that needs to be answered is whether $\varepsilon_{S,T}$ belongs to the Rubin-Stark lattice $\Lambda_{S,T}$ or not. Let $\phi_1, \ldots, \phi_r \in U_{S,T}^*$. Since u_i belongs to $U_{k,S,T} \subseteq (U_{S,T})^G$, we have

$$\phi_j(u_i) \in \mathbb{Z}[G]^G = \mathbb{Z} \cdot |G| e_{\mathbf{1_G}},$$

for all i, j as above. Consequently, we have

$$(\phi_r^{(1)} \circ \cdots \circ \phi_1^{(r)})(\varepsilon_{S,T}) \in \mathbb{Z} h_{k,S,T} e_{\mathbf{1_G}} \subseteq \frac{h_{k,S,T}}{|G|} \mathbb{Z}[G].$$

Now, since K/k is unramified everywhere and completely split above S, the class-field theoretical interpretation of the (S,T)–class group $A_{k,S,T}$ (see §3.2.2 above) implies that $|G| \mid h_{k,S,T}$. This shows that, indeed, $\varepsilon_{S,T} \in \Lambda_{S,T}$ in this case. Note that, in particular, this settles the Rubin-Stark conjecture in the case $K = k$.

Example 3.2.12. A question which arises naturally in the context of the Rubin-Stark conjecture is what prevents the Rubin-Stark element $\varepsilon_{S,T}$ from lying in the somewhat more natural guess for a Rubin-Stark lattice given by $(\wedge^r_{\mathbb{Z}[G]} U_{S,T})_{S,r}$. Of course, if $r = 1$, then $\Lambda_{S,T} = (U_{S,T})_{S,1}$. For r large, Rubin gave number field examples where $\varepsilon_{S,T} \notin (\wedge^r_{\mathbb{Z}[G]} U_{S,T})_{S,r}$ (see [**32**].) In [**29**] and [**30**], such examples were constructed in the case $r = 2$, for both function fields and number fields. All these examples arise in situations where $U_{S,T}$ is not $\mathbb{Z}[G]$–projective (i.e. not G–cohomologically trivial.) When the $\mathbb{Z}[G]$–module $U_{S,T}$ is projective, then $\mathcal{L}_{\mathbb{Z}[G]}(U_{S,T}, r) = \widetilde{\wedge^r_{\mathbb{Z}[G]} U_{S,T}}$ (see Remark 1.1.4(3)) and therefore

$$\Lambda_{S,T} = (\wedge^r_{\mathbb{Z}[G]} U_{S,T})_{S,r}.$$

Next, we describe a somewhat systematic way of producing examples where $U_{S,T}$ is a projective $\mathbb{Z}[G]$–module.

Let k be a number field with non-trivial class-number h_k and let $S_0 := S_\infty$ be the set of infinite primes in k. Let p be a prime number which divides h_k and T be a finite set of non-archimedean primes in k, containing at least two primes of distinct residual characteristics. Let K/k be an abelian extension of degree p which is unramified everywhere. (The existence of K/k is a consequence of the assumption $p \mid h_k$.) Let $G := \mathrm{Gal}(K/k)$. Tchebotarev's density theorem permits us to construct a finite set of primes S, which contains S_0 and is disjoint from T, such that all but one prime v_0 in $S \setminus S_0$ splits completely in K/k, the prime v_0 is inert in K/k, and $h_{K,S,T} = 1$. We examine conjecture $A_{\mathbb{Z}}(K/k, S, T, r)$, where $r = |S| - 1$. Note that every prime in $S \setminus \{v_0\}$ splits completely in K/k. Since v_0 is inert in K/k, we have $e_{S,r} = 1$ in this case. Since $h_{K,S,T} = 1$, we have an exact sequence of $\mathbb{Z}[G]$–modules (see (15)):

$$1 \longrightarrow U_{S,T} \longrightarrow U_S \longrightarrow \Delta_T \longrightarrow 1.$$

Further, since $\delta_{T,K/k}(0) := \prod_{v \in T}(1 - \mathrm{N}v \cdot \sigma_v^{-1})$ is a not a zero-divisor in $\mathbb{Z}[G]$ (check this !), Remark 3.2.4 leads to an exact sequence of $\mathbb{Z}[G]$–modules

$$0 \longrightarrow \bigoplus_{v \in T} \mathbb{Z}[G] \xrightarrow{\bigoplus_{v \in T} j_v} \bigoplus_{v \in T} \mathbb{Z}[G] \longrightarrow \Delta_T \longrightarrow 0,$$

where j_v is the multiplication by $(1 - \mathrm{N}v \cdot \sigma_v^{-1})$ map. This implies that the $\mathbb{Z}[G]$–module Δ_T is G–cohomologically trivial. Consequently, U_S and $U_{S,T}$ have isomorphic G–cohomology groups. On the other hand, since $h_{K,S,T} = 1$, we have $h_{K,S} = 1$

(note that $h_{K,S} \mid h_{K,S,T}$.) Consequently, Corollary 3.1.2 combined with the fact that G is cyclic (and therefore its Tate cohomology is periodic of period 2) implies that U_S and X_S have isomorphic G–cohomology groups. However, it is easy to check that in this case we have isomorphisms of $\mathbb{Z}[G]$–modules

$$Y_S \simeq \mathbb{Z} \oplus \mathbb{Z}[G]^r, \qquad X_S \simeq \mathbb{Z}[G]^r,$$

where G acts trivially in $\mathbb{Z}$. Consequently, X_S is G–cohomologically trivial. Therefore, $U_{S,T}$ is G–cohomologically trivial and therefore $\mathbb{Z}[G]$–projective. As discussed above, this implies that

$$\Lambda_{S,T} = \widetilde{\wedge^r_{\mathbb{Z}[G]} U_{S,T}} = \wedge^r_{\mathbb{Z}[G]} U_{S,T}.$$

The $\mathbb{Z}[G]$–projectivity of $U_{S,T}$ is not necessary in order to have an equality $\Lambda_{S,T} = (\wedge^r_{\mathbb{Z}[G]} U_{S,T})_{S,r}$. In order to see this, let us examine further the example above. However, instead of assuming that v_0 is inert, assume that v_0 splits completely. Therefore, S consists of $(r+1)$ distinct primes which split completely, as in Example 3.2.11. Obviously, in order to show that $\Lambda_{S,T} = (\wedge^r_{\mathbb{Z}[G]} U_{S,T})_{S,r}$, it suffices to show that $\mathbb{Z}_{(p)}\Lambda_{S,T} = \mathbb{Z}_{(p)}(\wedge^r_{\mathbb{Z}[G]} U_{S,T})_{S,r}$, where $\mathbb{Z}_{(p)}$ denotes the usual localization of $\mathbb{Z}$ at its prime ideal $p\mathbb{Z}$ (see part (3) of Remark 3.2.8.) We have an exact sequence of $\mathbb{Z}[G]$–modules

$$0 \longrightarrow X_S \longrightarrow Y_S \simeq \mathbb{Z}[G]^{r+1} \longrightarrow \mathbb{Z} \longrightarrow 0,$$

which implies that

$$\widehat{H}^0(G, X_S) \simeq \widehat{H}^1(G, \mathbb{Z}) = 0$$
$$\widehat{H}^1(G, X_S) \simeq \widehat{H}^0(G, \mathbb{Z}) \simeq \mathbb{Z}/p\mathbb{Z}.$$

Of course, the same isomorphisms hold true for the G–cohomology groups of $U_{S,T}$ (see above.) Since $U_{S,T}$ is not G–cohomologically trivial, $U_{S,T}$ is not $\mathbb{Z}[G]$–projective in this case. Now, since G is of prime order p and $U_{S,T}$ is $\mathbb{Z}$-free, Theorem 4.19 in [**39**] implies that we have an isomorphism

$$\mathbb{Z}_{(p)}U_{S,T} \simeq \mathbb{Z}_{(p)}I_G \oplus \mathbb{Z}_{(p)}[G]^r$$

in the category of $\mathbb{Z}_{(p)}[G]$–modules, where I_G is the augmentation ideal of $\mathbb{Z}[G]$, generated by $(\sigma - 1)$, for any generator σ of G. (In applying Theorem 4.19 loc. cit., note that $\mathbb{Z}_{(p)}[G]$ is a local ring of maximal ideal generated by p and $(\sigma - 1)$.) Pick units $U_0, U_1, \dots, U_r$ in $U_{S,T}$, such that

$$\mathbb{Z}_{(p)}U_{S,T} \simeq \oplus_{i=0}^r \mathbb{Z}_{(p)}[G]U_i, \qquad \mathbb{Z}_{(p)}[G]U_0 \simeq \mathbb{Z}_{(p)}I_G, \quad \mathbb{Z}_{(p)}[G]U_i \simeq \mathbb{Z}_{(p)}[G],$$

for all $i = 1, \dots, r$. Since $e_{S,r} = e_{\mathbf{1_G}}$ and $e_{\mathbf{1_G}}I_G = 0$, we have an equality

$$e_{S,r} \cdot (\mathbb{Q} \wedge^r_{\mathbb{Z}[G]} U_{S,T}) = e_{\mathbf{1_G}} \cdot \mathbb{Q}(U_1 \wedge \cdots \wedge U_r)\,,$$

where the exterior products are viewed over $\mathbb{Z}[G]$. Now, it is very easy to see that

$$\mathbb{Z}_{(p)}\Lambda_{S,T} = \mathbb{Z}_{(p)}(\wedge^r_{\mathbb{Z}[G]} U_{S,T})_{S,r} = |G| e_{\mathbf{1_G}} \cdot \mathbb{Z}_{(p)}(U_1 \wedge \cdots \wedge U_r)\,.$$

(To prove this, consider the particular elements $\phi_i \in \mathbb{Z}_{(p)}U^*_{S,T}$, satisfying $\phi_i(U_j) = \delta_{i,j}$, for all values $i = 1, \dots, n$ and $j = 0, 1, \dots, n$.) In order to reconcile the above description of $\mathbb{Z}_{(p)}\Lambda_{S,T}$ with the concrete formula for $\varepsilon_{S,T}$ obtained in Example 3.2.11, observe that since $\widehat{H}^0(G, U_{S,T}) = 0$, we have

$$U_{k,S,T} = (U_{K,S,T})^G = N_G \cdot U_{K,S,T}, \quad \mathbb{Z}_{(p)}U_{k,S,T} = \oplus_{i=1}^r \mathbb{Z}_{(p)}(N_G U_i)\,,$$

where $N_G := |G|e_{\mathbf{1_G}}$ is the usual norm element in $\mathbb{Z}[G]$. Note that $N_G U_0 = 1$. Consequently, the $\mathbb{Z}$–basis $\{u_1, \ldots, u_r\}$ for $U_{k,S,T}$ considered in Example 3.2.11 can be taken to satisfy $u_i^{\alpha_i} = N_G U_i^{\beta_i}$, for some $\alpha_i, \beta_i \in \mathbb{Z} \setminus p\mathbb{Z}$, for all $i = 1, \ldots, r$. Let $\alpha := \prod_i \alpha_i$ and $\beta := \prod_i \beta_i$. Since $N_G^r = |G|^{r-1} N_G$, we have the following

$$\begin{aligned}\varepsilon_{S,T} = \pm\frac{\beta}{\alpha} \cdot \frac{h_{k,S,T}}{|G|^r}(N_G U_1 \wedge \cdots \wedge N_G U_r) &= \pm\tfrac{\beta}{\alpha} \cdot \tfrac{h_{k,S,T}}{|G|} N_G \cdot (U_1 \wedge \cdots \wedge U_r) \\ &\in |G|e_{\mathbf{1_G}} \cdot \mathbb{Z}_{(p)}(U_1 \wedge \cdots \wedge U_r)\,.\end{aligned}$$

Indeed, this reconciles the two calculations.

3.2.5. **Functorial behavior.** In this section, we will describe very briefly several functorial properties of conjecture $A_{\mathbb{Z}}(K/k, S, T, r)$: dependence on the top field K; dependence on S (and implicitly on r); dependence on T. All these properties are consequences of the existence and uniqueness of the Rubin-Stark element $\varepsilon_{S,T}$ in the space $e_{S,r}(\mathbb{C} \wedge^r_{\mathbb{Z}[G]} U_{S,T})$. Below, we assume that the set of data $(K/k, S, T, r)$ satisfies the hypotheses in the Rubin-Stark conjecture.

First, let K' be an intermediate field, $k \subseteq K' \subseteq K$. Then, it is obvious that $(K'/k, S, T, r)$ also satisfies the hypotheses in the Rubin-Stark conjecture.

Proposition 3.2.13. *If conjecture $A_{\mathbb{Z}}(K/k, S, T, r)$ holds, then so does conjecture $A_{\mathbb{Z}}(K'/k, S, T, r)$.*

PROOF. (Sketch; see [**32**] for details.) Let $G' := G(K'/k)$ and $H := G(K/K')$. We let $\mathcal{R}_{W'}$ be the Rubin-Stark regulator for $(K'/k, S, T, r)$ and the r–tuple $W' := (w_1', \ldots, w_r')$, where w_i' is the prime in K' sitting below w_i, for all $i = 1, \ldots, r$. We let $\varepsilon'_{S,T}$ and $\Lambda'_{S,T}$ denote the corresponding Rubin-Stark element and lattice, respectively. If $U'_{S,T} := U_{K',S,T} = (U_{S,T})^H$, then $N_H : U_{S,T} \to U'_{S,T}$ denotes the usual norm map. It is easy to check that the $\mathbb{C}[G]$–module morphism

$$\wedge^r N_H : e_{S,r} \wedge^r_{\mathbb{C}[G]} \mathbb{C}U_{S,T} \twoheadrightarrow e_{S,r} \wedge^r_{\mathbb{C}[G']} \mathbb{C}U'_{S,T}\,,$$

obtained by first extending N_H by $\mathbb{C}$–linearity to $\mathbb{C}U_S$ and then taking exterior powers, maps $\varepsilon_{S,T}$ to $\varepsilon'_{S,T}$ and also maps $\Lambda_{S,T}$ into $\Lambda'_{S,T}$. □

Next, let us assume that $S' = S \cup \{v_{r+1}, \ldots, v_{r'}\}$, where $r' > r$ and $v_{r+1}, \ldots, v_{r'}$ are distinct primes in k, which do not belong to $S \cup T$ and split completely in K/k. Then, it is obvious that $(K/k, S', T, r')$ also satisfy the hypotheses of the Rubin-Stark conjecture.

Proposition 3.2.14. *If conjecture $A_{\mathbb{Z}}(K/k, S', T, r')$ holds, then so does conjecture $A_{\mathbb{Z}}(K/k, S, T, r)$.*

PROOF. (Sketch; see [**32**] for details.) First, it is obvious that $e_{S,r} = e_{S',r'}$. We extend the ordered r-tuple $W = (w_1, \cdots, w_r)$ to an ordered r'–tuple $W' = (w_1, \ldots, w_r, w_{r+1}, \ldots, w_{r'})$ of primes w_i in K sitting above v_i and denote by $\varepsilon_{S',T}$ the Rubin-Stark element in $e_{S,r} \wedge^{r'}_{\mathbb{C}[G]} \mathbb{C}U_{S',T}$ corresponding to the regulator $R_{W'}$ associated to W'. For all $i = r+1, \ldots, r'$, we define $w_i^* \in (U_{S',T})^*$ by

$$w_i^*(u) = \sum_{\sigma \in G} \mathrm{ord}_{w_i}(u^\sigma) \cdot \sigma^{-1}, \qquad \text{for all } u \in U_{S',T}\,,$$

and extend these by $\mathbb{C}$–linearity to elements in $\mathrm{Hom}_{\mathbb{C}[G]}(\mathbb{C}U_{S',T}, \mathbb{C}[G])$. The extensions will be abusively denoted by w_i^* as well. It is not difficult to see that the $\mathbb{C}[G]$–module morphism

$$(w_{r'}^*)^{(r+1)} \circ \cdots \circ (w_{r+1}^*)^{(r')} : e_{S,r} \wedge_{\mathbb{C}[G]}^{r'} \mathbb{C}U_{S',T} \longrightarrow e_{S,r} \wedge_{\mathbb{C}[G]}^{r} \mathbb{C}U_{S',T}$$

is an isomorphism onto $e_{S,r} \wedge_{\mathbb{C}[G]}^{r} \mathbb{C}U_{S,T}$, which maps $\varepsilon_{S',T}$ to $\varepsilon_{S,T}$ and also maps the lattice $\Lambda_{S',T}$ into the lattice $\Lambda_{S,T}$. Note that since $\mathbb{C}U_{S,T}$ is a direct summand of $\mathbb{C}U_{S',T}$ in the category of $\mathbb{C}[G]$–modules, $e_{S,r} \wedge_{\mathbb{C}[G]}^{r} \mathbb{C}U_{S,T}$ is a direct summand of $e_{S,r} \wedge_{\mathbb{C}[G]}^{r} \mathbb{C}U_{S',T}$. □

Remark 3.2.15. *As remarked before, Burns has shown in* [**2**] *that the ETNC (Conjecture 3.1.4 above) implies the Rubin-Stark Conjecture. In fact, one starts with data $(K/k, S, T, r)$ satisfying the hypotheses in conjecture $A_{\mathbb{Z}}(K/k, S, T, r)$, then one uses Tchebotarev's density theorem to construct a set S' as in Proposition 3.2.14 above and satisfying the additional hypothesis that the S'–ideal class group of K is trivial (i.e. $h_{K,S'} = 1$) and then one proves that the ETNC for data $(K/k, S')$ implies conjecture $\mathbb{A}_{\mathbb{Z}}(K/k, S', T, r')$. Finally, one uses Proposition 3.2.14 above to derive conjecture $A_{\mathbb{Z}}(K/k, S, T, r)$.*

Next, let S' be any finite set of primes in k, such that $S \subseteq S'$ and $S' \cap T = \emptyset$. Then, the set of data $(K/k, S', T, r)$ satisfies the hypotheses in the Rubin-Stark conjecture as well.

Proposition 3.2.16. *If conjecture $A_{\mathbb{Z}}(K/k, S, T, r)$ holds, then so does conjecture $A_{\mathbb{Z}}(K/k, S', T, r)$.*

Proof. Since $\Theta_{S',T}^{(r)}(0) = \prod_{v \in S' \setminus S}(1 - \sigma_v^{-1}) \cdot \Theta_{S,T}^{(r)}(0)$, we have

$$e_{S',r}\mathbb{C}[G] = \prod_{v \in S' \setminus S} (1 - \sigma_v^{-1}) \cdot e_{S,r}\mathbb{C}[G]$$

(check this equality !) and

$$\varepsilon_{S',T} = (\prod_{v \in S' \setminus S} (1 - \sigma_v^{-1})) \cdot \varepsilon_{S,T}\,.$$

Since $U_{S',T}/U_{S,T}$ has no $\mathbb{Z}$–torsion (check this !), Remark 1.1.1 implies that the inclusion of $\mathbb{Z}[G]$–modules $U_{S,T} \hookrightarrow U_{S',T}$ induces a surjective $\mathbb{Z}[G]$–module morphism $(U_{S',T})^* \twoheadrightarrow (U_{S',T})^*$. This implies right away that we have an inclusion

$$(\prod_{v \in S' \setminus S} (1 - \sigma_v^{-1})) \cdot \Lambda_{S,T} \subseteq \Lambda_{S',T}.$$

Combined with the last displayed equality, this concludes the proof. Note that, tacitly, we have used the fact that since $\mathbb{C}U_{S,T}$ is a direct summand of $\mathbb{C}U_{S',T}$, $e_{S,r} \wedge_{\mathbb{C}[G]}^{r} \mathbb{C}U_{S,T}$ is a direct summand of $e_{S,r} \wedge_{\mathbb{C}[G]}^{r} \mathbb{C}U_{S',T}$ as well (all in the category of $\mathbb{C}[G]$–modules.) □

Finally, let us assume that T' is a finite set of primes in k, such that $T \subseteq T'$ and $S \cap T' = \emptyset$. Then, the set of data $(K/k, S, T', r)$ satisfies the hypotheses in the Rubin-Stark conjecture as well.

Proposition 3.2.17. *If conjecture $A_{\mathbb{Z}}(K/k, S, T, r)$ holds, then so does conjecture $A_{\mathbb{Z}}(K/k, S, T', r)$.*

Proof. See Proposition 5.3.1 of [**28**]. □

4. The Rubin-Stark conjecture in the case $r = 1$

4.1. General considerations

In this section, we take a closer look at the predictions of conjecture $A_{\mathbb{Z}}(K/k, S, T, r)$ in the important particular case when $r = 1$. As we mentioned earlier, in this case, the conjecture had already been formulated by Stark himself for number fields in [**37**] and extended by Tate to function fields in [**42**]. Although Stark's and Tate's formulations only depend on the set S and do not involve the set T, it is not difficult to prove that for a given S as above the Stark-Tate conjecture for the data $(K/k, S, 1)$ is true if and only if the conjectures $A_{\mathbb{Z}}(K/k, S, T, 1)$ are true, for all T, such that $(K/k, S, T, 1)$ satisfy the above hypotheses (see Remark 4.1.4 below.) The case $r = 1$ of the Rubin-Stark conjecture reads as follows.

Conjecture 4.1.1. $A_{\mathbb{Z}}(K/k, S, T, 1)$ *(Stark, Tate) If the set of data* $(K/k, S, T, 1)$ *satisfies the hypotheses above and* w_1 *is a prime in* K *sitting above the distinguished prime* v_1 *in* S *which splits completely in* K/k, *then there exists a unique* $\varepsilon_{S,T} \in (U_{S,T})_{S,1}$, *such that*

$$\Theta'_{S,T}(0) = -\sum_{\sigma \in G} \log |\varepsilon_{S,T}^{\sigma^{-1}}|_{w_1} \cdot \sigma .$$

Let $v_0 \in S$, with $v_0 \neq v_1$ and let us fix a prime w_0 in K sitting above v_0. Note that the $e_{S,1}\mathbb{R}[G]$–module $e_{S,1}(\mathbb{R}X_S)$ is free of rank one of basis $e_{S,1}(w_1 - w_0)$. This observation, combined with the obvious equality $e_{S,1}\mathbb{R}[G] = \Theta'_{S,T}(0)\mathbb{R}[G]$ (by the definition of $e_{S,1}$) and the injectivity of the $\mathbb{Z}[G]$–module morphism $\lambda_S : U_{S,T} \hookrightarrow \mathbb{R}X_S \subseteq \mathbb{R}Y_S$ leads right away to the following alternative description of $(U_{S,T})_{S,1}$.

$$(U_{S,T})_{S,1} = \{u \in U_{S,T} \mid \lambda_S(u) \in \Theta'_{S,T}(0)\mathbb{R}[G](w_1 - w_0)\}. \tag{17}$$

In view of the above remarks, the following is an equivalent formulation of conjecture $A_{\mathbb{Z}}(K/k, S, T, 1)$.

Conjecture 4.1.2 (Stark, Tate). *Under the previous assumptions and notations, the following equivalent statements hold.*

(1) *There exists* $\varepsilon_{S,T} \in U_{S,T}$, *such that*

$$\Theta'_{S,T}(0)(w_1 - w_0) = \lambda_S(\varepsilon_{S,T}) .$$

(2) *There exists* $\varepsilon_{S,T} \in K_T^{\times}$, *such that*

$$\Theta'_{S,T}(0)(w_1 - w_0) = -\sum_{w} \log |\varepsilon_{S,T}|_w \cdot w ,$$

where the sum is taken over all primes w *in* K *and*

$$K_T^{\times} := \{x \in K^{\times} \mid x = 1 \mod v, \ \forall v \in T\} .$$

Remark 4.1.3. *Let* $U_{S,T}^{(v_1)} := \{u \in U_{S,T} \mid -\sum_{\sigma \in G} \log |\varepsilon_{S,T}^{\sigma^{-1}}|_{w_1} \cdot \sigma \in \mathbb{R}[G]\Theta'_{S,T}(0)\}$. *Since* $e_{S,1}w_0 = 0$, *if* $S| > 2$ *and* $e_{S,1}w_0 = e_{\mathbf{1_G}}w_0$, *if* $|S| = 2$ *(by the definition of* $e_{S,1}$ *and the fact that* $e_{S,1}w_1 \neq 0$*), equality (17) leads to the following additional description of* $(U_{S,T})_1$.

$$(U_{S,T})_{S,1} = U_{S,T}^{(v_1)} \cap \mathcal{A}_{S,T}^{(v_1)} \tag{18}$$

where $\mathcal{A}_{S,T}^{(v_1)}$ is a group of so-called anti-units defined by

$$\mathcal{A}_{S,T}^{(v_1)} := \begin{cases} \{u \in U_{S,T} \mid |u|_w = 1, \ \text{for all } w \in S_K, w \nmid v_1\}, & \text{if } |S| > 2; \\ \{u \in U_{S,T} \mid |u|_w = |u|_{w'}, \ \text{for all } w, w' \in S_K, w, w' \nmid v_1\}, & \text{if } |S| = 2. \end{cases}$$

Remark 4.1.4. *The unique (S,T)–unit $\varepsilon_{S,T}$ which would render the above conjecture true is called the T–modified Stark unit associated to $(K/k, S, T, 1)$. Let us fix S and vary T in a finite family $\mathcal{T}$, such that $(K/k, S, T, 1)$ satisfies the above hypotheses for all $T \in \mathcal{T}$ and the set $\{\delta_{K/k,T}(0) \mid T \in \mathcal{T}\}$ generates the $\mathbb{Z}[G]$–ideal $\mathrm{Ann}_{\mathbb{Z}[G]}(\mu_K)$ (see Remark 3.2.4). Let us pick elements $\alpha_T \in \mathbb{Z}[G]$, such that*

$$w_K = \sum_{T \in \mathcal{T}} \alpha_T \cdot \delta_{K/k,T}(0)\,.$$

Then, since $\Theta'_{S,T}(0) = \delta_{K/k,T}(0)\Theta'_S(0)$, for all $T \in \mathcal{T}$, we have

$$\Theta'_S(0) = -\frac{1}{w_K} \sum_{\sigma \in G} \log |\varepsilon_S^{\sigma^{-1}}|_{w_1} \cdot \sigma\,, \qquad \text{for } \varepsilon_S := \prod_{T \in \mathcal{T}} \varepsilon_{S,T}^{\alpha_T}\,,$$

and ε_S is the unique element in $(\widetilde{U_S})_{S,1}$ satisfying the first equality above. The (conjectural) element ε_S is called the Stark unit associated to the data $(K/k, S, 1)$. Note that ε_S denotes in fact a class of S–units in $\widetilde{U_S} = U_S/\mu_K$. Also, note that the uniqueness of ε_S and $\varepsilon_{S,T}$ (which are consequences of the injectivity of the Rubin-Stark regulator) give us equalities in U_S

$$\varepsilon_S^{\delta_{K/k,T}(0)} = \varepsilon_{S,T}^{w_K}, \qquad \text{for all } T \in \mathcal{T}\,.$$

Note that the left-hand side of the above equalities does not depend on the representative we choose for ε_S in U_S. The above equalities combined with Proposition 1.2, Chpt. IV of **[42]** *show that the field extension $K(\varepsilon_S^{1/w_K})/k$ generated by any w_K–root of any representative in U_S of the conjectural Stark unit ε_S is an abelian extension. This abelian condition is built into the original ("T-less") Stark–Tate formulation of the above conjecture.*

At this point, it is natural to discuss the conjecture above separately in the case where the split prime v_1 is an archimedean (infinite) prime (which will force k to be a number field) and a non-archimedean prime (in which case k can be either a number field or a function field), respectively.

4.2. Archimedean v_1

Under the hypothesis that the distinguished prime v_1 is archimedean, the conjecture is known to hold in two major particular cases, both due to Stark: the case where $k = \mathbb{Q}$, which we will treat somewhat in detail below, and the case where k is a quadratic imaginary field for which we send the reader to **[42]**, Chpt. IV, Proposition 3.9 or **[37]**.

Let us assume that $k = \mathbb{Q}$. Since the (unique) infinite prime of k is assumed to split completely in K/k, the top field K has to be a real, abelian extension of $\mathbb{Q}$. Class-field theory forces K to be a subfield of some real cyclotomic field $\mathbb{Q}(\zeta_m)^+ := \mathbb{Q}(\zeta_m + \zeta_m^{-1})$, where ζ_m is root of unity of order m and m can be chosen to be any multiple of the conductor of K/k. Propositions 3.2.13, 3.2.16 and 3.2.17 above permit us to reduce the proof of conjecture $\mathbb{A}_{\mathbb{Z}}(K/\mathbb{Q}, S, T, 1)$ to the situation where $K = \mathbb{Q}(\zeta_m)^+$, $S = S_m := \{v_1 = \infty\} \cup \{p\mathbb{Z} \mid p \text{ prime in } \mathbb{Z},\ p \mid m\}$ and

$T = T_q := \{q\mathbb{Z}\}$, where q is a prime in $\mathbb{Z}$ which does not divide $2m$. (Note that $\mu_K = \{\pm 1\}$ and $w_K = 2$ in this case. Also, note that for such prime q, there are no non-trivial roots of unity in the full cyclotomic field $\mathbb{Q}(\zeta_m)$ which are congruent to 1 modulo q.) Let us recall the classical formulas

$$L'_S(\chi, s) = -\frac{1}{2}\sum_{\sigma\in G} \log|\epsilon_m^\sigma|_{w_1} \cdot \chi(\sigma)\,, \qquad \text{for all } \chi \in \mathrm{Gal}(K/\mathbb{Q})\,,$$

where $\epsilon_m := (1-\zeta_m)(1-\zeta_m^{-1})$ is obviously an element of U_S and w_1 is the prime sitting over v_1 in K corresponding to the embedding of $\mathbb{Q}(\zeta_m)^+$ into $\mathbb{C}$ sending ζ_m to $\exp(2\pi i/m)$. (For a quick proof, combine Theorem 4.9 in [**43**] with the functional equation for $L_S(\chi, s)$.) We combine the above character-dependent equalities to obtain the following (equivalent) G–equivariant equality.

$$\Theta'_S(0) = -\frac{1}{2}\sum_{\sigma\in G} \log|\epsilon_m^{\sigma^{-1}}|_{w_1} \cdot \sigma.$$

Now, since the prime q is not ramified in the full cyclotomic extension $\mathbb{Q}(\zeta_m)/\mathbb{Q}$, we can afford to abuse the notation a bit and denote by σ_q its Frobenius automorphism in $\mathrm{Gal}(\mathbb{Q}(\zeta_m)/\mathbb{Q})$. We leave it as an (easy) exercise for the reader to check that

$$(1-\zeta_m)^{(1-q\sigma_q^{-1})} \in \mathbb{Q}(\zeta_m)^+\,, \qquad \{(1-\zeta_m)^{(1-q\sigma_q^{-1})}\}^2 = \epsilon_m^{(1-q\sigma_q^{-1})}.$$

Now, if we let $\varepsilon_{S,T} := (1-\zeta_m)^{(1-q\sigma_q^{-1})}$, the last two displayed equalities combined give:

$$\Theta'_{S,T}(0) = (1-q\sigma_q^{-1}) \cdot \Theta'_S(0) = -\sum_{\sigma\in G} \log|\varepsilon_{S,T}^{\sigma^{-1}}|_{w_1} \cdot \sigma.$$

Consequently, if we show that $\varepsilon_{S,T} \in (U_{S,T})_{S,1}$, the equality above verifies the Rubin-Stark conjecture $\mathbb{A}_{\mathbb{Z}}(\mathbb{Q}(\zeta_m)^+/\mathbb{Q}, S = S_m, T = T_q, 1)$. Now, since $(1-\zeta_m)$ is an S–unit in $\mathbb{Q}(\zeta_m)$,

$$\varepsilon_{S,T} := (1-\zeta_m)^{(1-q\sigma_q^{-1})}$$

is an (S,T)–unit in $\mathbb{Q}(\zeta_m)$ (use the exact sequence (15) and Remark 3.2.4 for top field $\mathbb{Q}(\zeta_m)$ and sets S_m and T_q). Therefore, $\varepsilon_{S,T}$ is an (S,T)–unit in $\mathbb{Q}(\zeta_m)^+$. In order to show that $\varepsilon_{S,T}$ sits in the subspace $(U_{S,T})_{S,1}$ of $U_{S,T}$, one uses the equality (18) combined with the well-known fact that $(1-\zeta_m)$ is a unit in $\mathbb{Q}(\zeta_m)$, if m is not a power of a prime, and $(1-\zeta_m)$ is a generator of the unique and totally ramified prime $\mathfrak{P}$ sitting above p in $\mathbb{Q}(\zeta_m)$, if m is a power of the prime p. The conclusion of the above considerations is the following.

Theorem 4.2.1 (Stark). *Conjecture $A_{\mathbb{Z}}(K/k, S, T, 1)$ is true if $k = \mathbb{Q}$ and K is any real abelian extension of $\mathbb{Q}$.*

Remark 4.2.2. *Assume that the set of data $(K/\mathbb{Q}, S, T, 1)$ satisfies the usual hypotheses and that K is a any real abelian extension of $\mathbb{Q}$. With notations as above, we can always find a natural number m divisible by the conductor of $K/\mathbb{Q}$, such that $S = S_m$. Then, the calculations preceding the Theorem above combined with Remark 4.1.4 and Propositions 3.2.13, 3.2.16 and 3.2.17 show that the Rubin-Stark unit associated to the data $(K/\mathbb{Q}, S, T, 1)$ is given by*

$$\varepsilon_{S,T} := \mathbf{N}_{\mathbb{Q}(\zeta_m)^+/K}((1-\zeta_m)^{\delta_{\mathbb{Q}(\zeta_m)/\mathbb{Q},T}(0)})\,,$$

whereas the Stark unit associated to the data $(K/\mathbb{Q}, S, 1)$ is given by

$$\varepsilon_S := \mathbf{N}_{\mathbb{Q}(\zeta_m)^+/K}((1-\zeta_m)(1-\zeta_m^{-1})) = \mathbf{N}_{\mathbb{Q}(\zeta_m)/K}(1-\zeta_m).$$

It should be noted that the last equality shows that the Stark units corresponding to real abelian extensions of $\mathbb{Q}$ are norms (down from full cyclotomic fields) of m–torsion points $(1-\zeta_m)$ of the multiplicative formal group $\widehat{G_m}(X,Y) = (X+1)(Y+1)-1$. Saying that these m–torsion points, for varying m, generate the maximal abelian extension $\mathbb{Q}^{ab}$ of $\mathbb{Q}$ (a restatement of the celebrated Kronecker-Weber Theorem) is equivalent to saying that the Stark units $\epsilon_m = (1-\zeta_m)(1-\zeta_m^{-1})$ for the full real cyclotomic extensions $\mathbb{Q}(\zeta_m)^+/\mathbb{Q}$ generate the maximal real abelian extension $\mathbb{Q}^{ab,+}$ of $\mathbb{Q}$. This is not an accident, as Stark himself noticed in [**36**]*: in general, there is a known and surprising link between Hilbert's 12–th problem for a totally real number field k (explicit generation of abelian extensions of k) and the $r=1$ case of the Rubin-Stark conjecture for abelian extensions K/k in which only one of the real primes of k splits completely in K (see* [**42**]*, Chpt. IV, Proposition 3.8 and the comments which follow.) Similar, but more general consequences can be derived by assuming the Rubin-Stark conjecture for arbitrary r. We will not pursue these ideas here.*

As mentioned above, Stark also proved the following (see [**37**] and [**42**], Chpt. IV, Prop. 3.9.)

Theorem 4.2.3 (Stark). *If k is an imaginary quadratic field and $(K/k, S, T, 1)$ satisfies the usual hypotheses, then conjecture $A_{\mathbb{Z}}(K/k, S, T, 1)$ holds true.*

We will not give the proof of the theorem above. However, it is worth mentioning that in this case the Stark unit corresponding to $(K/k, S, 1)$ is the norm down to K of an elliptic unit in the ray–class field of k of conductor equal to that of K/k. So, the role of the cyclotomic units $\epsilon_m = (1-\zeta_m)(1-\zeta_m^{-1})$ is played by elliptic units in this case. Moreover, the elliptic units in question are constructed out of torsion points of an elliptic curve with CM by O_k. The role of the formal group $\widehat{G_m}$ (or, more precisely that of the multiplicative group G_m) is taken by these CM elliptic curves. The analogue of the Kronecker-Weber Theorem in this case is the Main Theorem of Complex Multiplication which gives a precise recipe for the construction of the maximal abelian extension k^{ab} of k out of the coordinates of the torsion points and the j–invariant of any such elliptic curve with CM by O_k.

4.3. Non-archimedean v_1 – the Brumer-Stark conjecture

In this section, we take a closer look at conjecture $A_{\mathbb{Z}}(K/k, S, T, 1)$ in the situation where the distinguished split prime $v_1 \in S$ is a finite prime. This is the classical Brumer-Stark conjecture (formulated by Stark and strengthening an older conjecture of Brumer in the case of number fields, and extended by Tate to the case of function fields.) Our goal in this section is twofold: *First,* we give a new interpretation of the Brumer-Stark conjecture for arbitrary global fields in terms of the annihilation by the special L–value $\Theta_{S_0,T}(0)$ of certain Arakelov class-groups (Chow groups) associated to the top field K, where $S_0 := S \setminus \{v_1\}$; *Second,* we discuss the overwhelming evidence in support of the Brumer-Stark conjecture in terms of this interpretation, some very recent, some going back to the early 1980s and before.

We start by fixing two finite, disjoint sets of primes $\mathcal{S}$ and $\mathcal{T}$ in K, such that $\mathcal{S}$ contains the archimedean (infinite) primes of K. Note that $\mathcal{T}$ may be empty and, if K is a function field, $\mathcal{S}$ may be empty as well. We will define an Arakelov class-group $CH^1_{\mathcal{S}}(K)^0_{\mathcal{T}}$, generalizing the classical Arakelov class-group $CH^1(K)^0$ of a number field K (see [**25**]) and also generalizing the Picard group of degree zero divisors $\mathrm{Pic}^0(K)$ of a function field K. First, we define the group of Arakelov $(\mathcal{S},\mathcal{T})$–divisors of K by

$$\mathrm{Div}_{\mathcal{S}}(K)_{\mathcal{T}} := \left(\bigoplus_{w\notin\mathcal{S}\cup\mathcal{T}} \mathbb{Z}w\right) \oplus \left(\bigoplus_{w\in\mathcal{S}} \mathbb{R}w\right),$$

where the direct sums are taken with respect to all the primes w of K and the first direct summand is the free $\mathbb{Z}$–module generated by all the primes in K which are outside of $\mathcal{S}\cup\mathcal{T}$ and the second direct summand is the $\mathbb{R}$–vector space of basis $\mathcal{S}$. We define the degree map

$$\deg_{K,\mathcal{S}} : \mathrm{Div}_{\mathcal{S}}(K)_{\mathcal{T}} \longrightarrow \mathbb{R}$$

as the unique map which is $\mathbb{Z}$–linear on the first summand and $\mathbb{R}$–linear on the second and satisfies

$$\deg_{K,\mathcal{S}}(w) = \begin{cases} \log|Nw|, & \text{if } w\notin\mathcal{S}\cup\mathcal{T}\,; \\ 1, & \text{if } w\in\mathcal{S}. \end{cases}$$

We let $\mathrm{Div}_{\mathcal{S}}(K)^0_{\mathcal{T}}$ denote the kernel of $\deg_{K,\mathcal{S}}$ (i.e. the group of degree 0 Arakelov $(\mathcal{S},\mathcal{T})$–divisors.) Further, we let $K^\times_{\mathcal{T}}$ denote the subgroup of $K^\times$ consisting of all $x\in K^\times$, such that $x\equiv 1 \mod w$ (meaning $\mathrm{ord}_w(x-1)>0$), for all $w\in\mathcal{T}$. The product formula leads to a divisor morphism

$$\mathrm{div}_{K,\mathcal{S}} : K^\times_{\mathcal{T}} \longrightarrow \mathrm{Div}_{\mathcal{S}}(K)^0_{\mathcal{T}}, \qquad x \to \sum_{w\notin\mathcal{S}\cup\mathcal{T}} \mathrm{ord}_w(x)\cdot w + \sum_{w\in\mathcal{S}} (-\log|x|_w)\cdot w\,,$$

where $\mathrm{ord}_w(\cdot)$ and $|\cdot|_w$ denote the canonically normalized valuation and metric associated to w, respectively. Note that, for all $x\in K^\times_{\mathcal{T}}$, we have an equality

$$-\sum_{w}\log|x|_w\cdot w = \sum_{w\notin\mathcal{S}\cup\mathcal{T}} \deg_{K,\mathcal{S}}(w)\mathrm{ord}_w(x)\cdot w + \sum_{w\in\mathcal{S}} \deg_{K,\mathcal{S}}(w)(-\log|x|_w)\cdot w, \tag{19}$$

where the left sum is taken over all primes w of K. The Arakelov $(\mathcal{S},\mathcal{T})$–class group is given by

$$CH^1_{\mathcal{S}}(K)^0_{\mathcal{T}} := \frac{\mathrm{Div}_{\mathcal{S}}(K)^0_{\mathcal{T}}}{\mathrm{div}_{K,\mathcal{S}}(K^\times_{\mathcal{T}})}.$$

Note that, if K is a number field and $\mathcal{S}=S_\infty$ and $\mathcal{T}=\emptyset$, then $CH^1_{\mathcal{S}}(K)^0_{\mathcal{T}}$ is the classical Arakelov class-group $CH^1(K)^0$ defined in [**25**], for example. Also, if K is a function field and $\mathcal{S}=\mathcal{T}=\emptyset$, then $CH^1_{\mathcal{S}}(K)^0_{\mathcal{T}}$ is the Picard group $\mathrm{Pic}^0(K)$ of divisors of K of degree 0 (which is the function field analogue of $CH^1(K)^0$.) There is an obvious natural exact sequence of groups

$$0 \longrightarrow \frac{\mathbb{R}X_{\mathcal{S}}}{\mathrm{div}_{K,\mathcal{S}}(U_{\mathcal{S},\mathcal{T}})} \xrightarrow{\ \iota\ } CH^1_{\mathcal{S}}(K)^0_{\mathcal{T}} \xrightarrow{\ \pi\ } Cl(O_{K,\mathcal{S}})_{\mathcal{T}} \longrightarrow 0\,, \tag{20}$$

where $Cl(O_{K,\mathcal{S}})_{\mathcal{T}}$ is the $(\mathcal{S},\mathcal{T})$–ideal class group $A_{K,\mathcal{S},\mathcal{T}}$ defined just below (15), if $\mathcal{S}\neq\emptyset$ and

$$Cl(O_{K,\mathcal{S}})_{\mathcal{T}} = CH^1_{\emptyset}(K)^0_{\mathcal{T}} =: \mathrm{Pic}^0(K)_{\mathcal{T}},$$

if $\mathcal{S} = \emptyset$ (and therefore K is a function field.) The map ι is induced by the natural inclusion $\mathbb{R}X_{\mathcal{S}} \subseteq \mathrm{Div}_{\mathcal{S}}(K)^0_{\mathcal{T}}$ and the map π sends the class of a divisor $D = \sum_{w \notin \mathcal{T}} a_w \cdot w$ in $\mathrm{Div}_{\mathcal{S}}(K)^0_{\mathcal{T}}$ to the class of $D' := \sum_{w \notin \mathcal{T} \cup \mathcal{S}} a_w \cdot w$ in $Cl(O_{K,\mathcal{S}})_{\mathcal{T}}$.

Let us note that the map $\mathrm{div}_{K,\mathcal{S}}$ restricted to the group of $(\mathcal{S}, \mathcal{T})$–units $U_{\mathcal{S},\mathcal{T}}$ of K coincides with the map $\lambda_{\mathcal{S}}$ of the previous sections. Consequently, if $\mathcal{S}, \mathcal{T} \neq \emptyset$, then $CH^1_{\mathcal{S}}(K)^0_{\mathcal{T}}$ is a compact group (in the obvious topology) whose volume (with respect to the Lebesgue and counting measures on the infinite and finite sides, respectively) satisfies the remarkable equality

$$\mathrm{vol}(CH^1_{\mathcal{S}}(K)^0_{\mathcal{T}}) = \mathrm{vol}\left(\frac{\mathbb{R}X_{\mathcal{S}}}{\lambda_{\mathcal{S}}(U_{\mathcal{S},\mathcal{T}})}\right) \cdot \mid Cl(O_{K,\mathcal{S}})_{\mathcal{T}} \mid = R_{K,\mathcal{S},\mathcal{T}} \cdot h_{K,\mathcal{S},\mathcal{T}} = \pm\zeta^*_{K,\mathcal{S},\mathcal{T}}(0)\,.$$

Obviously, if K/k is a Galois extension of Galois group G and the sets $\mathcal{S}$ and $\mathcal{T}$ are G–invariant, then all the groups defined above are naturally endowed with $\mathbb{Z}[G]$–module structures and all the above morphisms and exact sequences can be viewed in the category of $\mathbb{Z}[G]$–modules.

Now, we return to conjecture $A_{\mathbb{Z}}(K/k, S, T, 1)$. As promised, we will assume that the distinguished prime $v_1 \in S$ which splits completely in K/k is a finite prime and let $S_0 := S \setminus \{v_1\}$. As before, for simplicity, we will also use S_0 and T to denote the sets of primes in K sitting above primes in S_0 and T, respectively. All the calculations that follow will happen at the level of K, so no confusion will ensue. Now, from the definitions, since $w_1 \notin S_0 \cup T$, we have

$$\Theta'_{S,T}(0) = \log|Nw_1| \cdot \Theta_{S_0,T}(0) = \deg_{K,S_0}(w_1) \cdot \Theta_{S_0,T}(0)\,.$$

The last equality combined with (19) leads to the following equivalent formulation of Conjecture 4.1.2, for fixed sets S_0 and T and arbitrary finite split prime $v \notin S_0 \cup T$.

Conjecture 4.3.1. *$BrSt(K/k, S_0, T)$. Let K/k be an abelian extension of global fields. Let S_0 be a finite, non-empty set of primes in k, containing all the infinite primes and all the primes which ramify in K/k. Let T be a finite set of primes in k, disjoint from S_0, such that $K^\times_T$ has no torsion. Then, for every prime v of k which is not in $S_0 \cup T$ and splits completely in K/k and every prime w in K sitting above v, there exists a unique $\varepsilon_w \in K^\times_T$, such that*

$$\Theta_{S_0,T}(0) \cdot (w - \deg_{K,S_0}(w) \cdot w_0) = \mathrm{div}_{K,S_0}(\varepsilon_w)\,,$$

for any (some) prime w_0 in K sitting above any (some) prime $v_0 \in S_0$.

The statement above is the Brumer-Stark conjecture for data $(K/k, S_0, T)$ and will be denoted $BrSt(K/k, S_0, T)$ in what follows. Although the original formulation due to Stark and Tate did not depend on a set T, Remark 4.1.4 above makes the equivalence between the two formulations clear.

Proposition 4.3.2. *The conjecture $BrSt(K/k, S_0, T)$ is equivalent to*

$$\Theta_{S_0,T}(0) \in \mathrm{Ann}_{\mathbb{Z}[G]} CH^1_{S_0}(K)^0_T\,.$$

PROOF. Let v, w, v_0 and w_0 be as in the statement of $BrSt(K/k, S_0, T)$. Then,

$$(w - \deg_{K,S_0}(w) \cdot w_0) \in \mathrm{Div}_{S_0}(K)^0_T.$$

So, if $\Theta_{S_0,T}(0) \in \mathrm{Ann}_{\mathbb{Z}[G]} CH^1_{S_0}(K)^0_T$, then we have

$$\Theta_{S_0,T}(0)(w - \deg_{K,S_0}(w) \cdot w_0) = \mathrm{div}_{K,S_0}(\varepsilon_w)\,,$$

for some $\varepsilon_w \in K_T^\times$, by the definition of $CH^1_{S_0}(K)^0_T$. This shows that conjecture $BrSt(K/k, S_0, T)$ holds.

Now, assume that $BrSt(K/k, S_0, T)$ holds. Let us fix v_0 and w_0 as in its statement. Let $D \in \mathrm{Div}_{S_0}(K)^0_T$. The class-field theoretical interpretation of $Cl(O_{K,S_0})_T$ (see the paragraphs below (15)) combined with Tchebotarev's density theorem and exact sequence (20) for $\mathcal{S} = S_0$ and $\mathcal{T} = T$ show that there exists a prime w in K sitting above a prime v in k, such that $v \notin S_0 \cup T$, v splits completely in K/k and

$$D = (w - \deg_{K,S_0}(w) \cdot w_0) + \mathfrak{X}_0 ,$$

for some $\mathfrak{X}_0 \in \mathbb{R}X_{S_0}$. Note that in the exact sequence (20) the image via π of the class of the degree 0 divisor $(w - \deg_{K,S_0}(w) \cdot w_0)$ in $CH^1_{S_0}(K)^0_T$ is the class of the prime w in $Cl(O_{K,S_0})_T$. Obviously, we have $\Theta_{S_0,T}(0) \cdot \mathbb{R}X_{S_0} = 0$. This is so because for every $\chi \in \widehat{G}(\mathbb{C})$, if $\chi(\Theta_{S_0,T}(0)) \neq 0$, then

$$\dim_{\mathbb{C}}(\mathbb{C}X_{S_0})^\chi = \mathrm{ord}_{s=0} L_{S_0,T}(\chi, s) = 0$$

(see (11) above.) Consequently, we have

$$\Theta_{S_0,T}(0) \cdot D = \Theta_{S_0,T}(0) \cdot (w - \deg_{K,S_0}(w) \cdot w_0) = \deg_{K,S_0}(\varepsilon_w) ,$$

for some $\varepsilon_w \in K_T^\times$. Therefore, $\Theta_{S_0,T}(0) \in \mathrm{Ann}_{\mathbb{Z}[G]} CH^1_{S_0}(K)^0_T$, as required. □

Corollary 4.3.3 (top change). *Assume that $(K/k, S_0, T)$ satisfy the above hypotheses. Let K' be a global field, such that $K \subseteq K'$ and K'/k is abelian of Galois group G'. Assume that $(K'/k, S_0, T)$ satisfy the above hypotheses as well. Then*

$$BrSt(K'/k, S_0, T) \Rightarrow BrSt(K/k, S_0, T) .$$

PROOF. Assume that $BrSt(K'/k, S_0, T)$ holds. Let $\widetilde{\Theta}_{S_0,T}$ be the equivariant L–function associated to $(K'/k, S_0, T)$. The inflation property of Artin L–functions implies that $\widetilde{\Theta}_{S_0,T}(0)$ maps to $\Theta_{S_0,T}(0)$ under the natural projection $\mathbb{Z}[G'] \twoheadrightarrow \mathbb{Z}[G]$. Let $D \in \mathrm{Div}_{S_0}(K)^0_T$. Let j be the natural injective $\mathbb{Z}[G]$–linear morphism $j : \mathrm{Div}_{S_0}(K)^0_T \to \mathrm{Div}_{S_0}(K')^0_T$. Under our working hypotheses, we have

$$j(\Theta_{S_0,T}(0) \cdot D) = \widetilde{\Theta}_{S_0,T}(0) \cdot j(D) = \mathrm{div}_{K',S_0}(\varepsilon') ,$$

for some $\varepsilon' \in K'^\times_T$. However, since $j(\Theta_{S_0,T}(0) \cdot D) \in (\mathrm{Div}_{S_0}(K')^0_T)^{G(K'/K)}$, this implies that for every $\sigma \in G(K'/K)$, we have $\mathrm{div}_{K',S_0}(\varepsilon'^\sigma/\varepsilon') = 0$. Since div_{K',S_0} is injective on $K'^\times_T$ (recall that $K'^\times_T$ has no $\mathbb{Z}$–torsion), this implies that $\varepsilon' \in K_T^\times$. Since j is injective, this implies that

$$\Theta_{S_0,T}(0) \cdot D = \mathrm{div}_{K,S_0}(\varepsilon') ,$$

which concludes the proof of the Corollary. □

Remark 4.3.4. *The proof of the above proposition can be used to show that if K'/K is a Galois extension of global fields and S_0 and T are finite, disjoint, sets of primes in K, such that $S_\infty \subseteq S_0$ and $K'^\times_T$ has no torsion, then the natural map*

$$CH^1_{S_0}(K)^0_T \longrightarrow CH^1_{S_0}(K')^0_T$$

is an injective group morphism. In addition, if $S_{ram}(K'/K) \subseteq S_0$, then have

$$CH^1_{S_0}(K)^0_T \simeq (CH^1_{S_0}(K')^0_T)^{G(K'/K)}$$

via the above injective group morphism. This last isomorphism is a consequence of the equality $\widehat{H}^1(G(K'/K), K'^\times_T) = 0$, which can be checked without difficulty.

Next, we focus on **the number field case** of the Brumer-Stark conjecture. Note that in this case the conjecture is trivially true unless the base field k is totally real and the top field K is totally complex. Indeed, if this condition is not satisfied, then there is an infinite prime in S_0 which splits completely in K/k. Consequently, the conjecture is a particular case of Example 3.2.11 above.

Proposition 4.3.5. *Assume that k is a number field.*

(1) *Then the Brumer-Stark conjecture $BrSt(K/k, S_0, T)$ is equivalent to*

$$\Theta_{S_0,T}(0) \in \mathrm{Ann}_{\mathbb{Z}[G]} CH^1(K)^0_T,$$

where $CH^1(K)^0_T := CH^1_{S_\infty}(K)^0_T$.

(2) *In particular, conjecture $BrSt(K/k, S_0, T)$ implies that*

$$\Theta_{S_0,T}(0) \in \mathrm{Ann}_{\mathbb{Z}[G]} Cl(O_K)_T, \quad \Theta_{S_0,T}(0) \in \mathrm{Ann}_{\mathbb{Z}[G]} Cl(O_K),$$

where $Cl(O_K)_T := Cl(O_{K,S_\infty})_T = A_{K,S_\infty,T}$.

PROOF. In order to show (1), in the proof of the previous proposition pick v_0 to be an infinite prime (which is allowed, since $S_\infty \subseteq S_0$) and use the exact sequence (20) for $\mathcal{S} := S_\infty$ and $\mathcal{T} = T$.

Part (2) is a direct consequence of (1) combined with the surjective $\mathbb{Z}[G]$–module morphisms

$$CH^1(K)^0_T \twoheadrightarrow Cl(O_K)_T \twoheadrightarrow Cl(O_K).$$

The first surjection above is the map π in exact sequence (20) for $\mathcal{S} := S_\infty$ and $\mathcal{T} := T$ and the second comes from exact sequence (15) with $S = S_\infty$. □

Now, let us assume that K is a CM field and k is a totally real field. In this case, the decomposition groups in G of all the infinite primes of k are equal and of order 2. Let us denote by j the (common) generator of these groups. (j is the so-called complex conjugation k–algebra endomorphism of K.) For any $\mathbb{Z}[G]$–module M, one can split the $\mathbb{Z}[1/2][G]$–module $M[1/2] := M \otimes_{\mathbb{Z}} \mathbb{Z}[1/2]$ into a direct sum $\pm$–eigenspaces with respect to the action of j in the usual way:

$$M[1/2] := M^- \oplus M^+, \quad M^- := \frac{1}{2}(1-j) \cdot M[1/2], \quad M^+ := \frac{1}{2}(1+j) \cdot M[1/2].$$

Proposition 4.3.6. *Assume that K is CM and k is totally real. Then, the following equivalence holds.*

$$\Theta_{S_0,T}(0) \in \mathrm{Ann}_{\mathbb{Z}[G]}(CH^1(K)^0_T[1/2]) \Leftrightarrow \Theta_{S_0,T}(0) \in \mathrm{Ann}_{\mathbb{Z}[G]} Cl(O_K)^-_T.$$

PROOF. First, under our working hypotheses, S_∞ (the set of infinite primes in k) has to contain more than one element. Indeed, if that were not the case, then $k = \mathbb{Q}$ and K/k is unramified away from S_∞, which means that $K = \mathbb{Q}$ and therefore K would not be CM. This observation, combined with (11), implies that $\Theta_{S_0,T}(0) \in \mathbb{Z}[G]^-$. Consequently, we have an equality

$$\Theta_{S_0,T}(0) \cdot (CH^1(K)^0_T[1/2]) = \Theta_{S_0,T}(0) \cdot (CH^1(K)^0_T)^-.$$

Now, if we take "$-$"–eigenspaces (which is an exact functor) in exact sequence (20) with $\mathcal{S} = S_\infty$ and $\mathcal{T} = T$ and observe that $(\mathbb{R}X_{S_\infty})^- = 0$, we obtain an equality

$$(CH^1(K)^0_T)^- = Cl(O_K)^-_T,$$

which concludes the proof of the Proposition. □

Proposition 4.3.7. *Assume that K is totally complex, k is totally real and let $\widetilde{K}$ denote the maximal CM subfield of K. Then, we have an equivalence*

$$\Theta_{S_0,T}(0) \in \mathrm{Ann}_{\mathbb{Z}[G]}(CH^1(K)^0_T[1/2]) \Leftrightarrow \widetilde{\Theta}_{S_0,T}(0) \in \mathrm{Ann}_{\mathbb{Z}[\widetilde{G}]}Cl(O_{\widetilde{K}})^-_T,$$

where $\widetilde{G} := G(\widetilde{K}/k)$ and $\widetilde{\Theta}_{S_0,T}$ is the equivariant L-function associated to the data $(\widetilde{K}/k, S_0, T)$.

PROOF. Obviously, $\widetilde{K} = K^{\mathfrak{g}}$ (the maximal subfield of K which is fixed by $\mathfrak{g}$), where $\mathfrak{g}$ is the subgroup of G generated by the set $\{\sigma_v \cdot \sigma_{v'} \mid v, v' \in S_\infty\}$. Here, σ_v denotes the Frobenius (complex conjugation) morphism associated to v in G, for all $v \in S_\infty$. Since $\mathfrak{g}$ is a finite (elementary) 2-group, it is easy to see that the natural morphisms $CH^1(\widetilde{K})^0_T \to CH^1(K)^0_T$ and $Cl(O_{\widetilde{K}})_T \to Cl(O_K)_T$ induce isomorphisms of $\mathbb{Z}[\widetilde{G}]$-modules

$$CH^1(\widetilde{K})^0_T[1/2] \simeq CH^1(K)^0_T[1/2]^{\,\mathfrak{g}}, \qquad Cl(O_{\widetilde{K}})_T[1/2] \simeq Cl(O_K)_T[1/2]^{\,\mathfrak{g}}. \tag{21}$$

(The second isomorphism above is well known, and it is a consequence of the surjectivity of the norm map $N_{\mathfrak{g}} : Cl(O_K)_T[1/2] \to Cl(O_{\widetilde{K}})_T[1/2]$, which can be obtained from the class-field theoretical interpretation of $Cl(O_K)_T[1/2]$ and $Cl(O_{\widetilde{K}})_T[1/2]$ and the fact that the order of $\mathfrak{g}$ is a power of 2. The first isomorphism is a consequence of the second, exact sequence (20) for $\mathcal{S} = S_\infty$ and $\mathcal{T} = T$ and the obvious equalities $\mathbb{R}X_{S_\infty}[1/2]^{\,\mathfrak{g}} = \mathbb{R}\widetilde{X}_{S_\infty}[1/2]$, $U^{\mathfrak{g}}_{S_\infty,T} = \widetilde{U}_{S_\infty,T}$ and $H^1(\mathfrak{g}, \mathbb{R}X_{S_\infty}/\mathrm{div}_{K,S_\infty}(U_{S_\infty,T})[1/2]) = 0$. Here, $\widetilde{*}$ refers to the usual objects at the $\widetilde{K}$-level.)

Next, we observe that $CH^1(K)^0_T[1/2]^{\,\mathfrak{g}} = e_{\mathbf{1}_{\mathfrak{g}}} \cdot CH^1(K)^0_T[1/2]$. Consequently,

$$CH^1(K)^0_T[1/2] = CH^1(K)^0_T[1/2]^{\,\mathfrak{g}} \oplus (1 - e_{\mathbf{1}_{\mathfrak{g}}}) \cdot CH^1(K)^0_T[1/2], \tag{22}$$

where $e_{\mathbf{1}_{\mathfrak{g}}} := 1/|\mathfrak{g}| \sum_{\sigma \in \mathfrak{g}} \sigma$ is the usual idempotent associated to the trivial character of $\mathfrak{g}$. Also, let us note that if $\chi \in \widehat{G}(\mathbb{C})$ is non-trivial when restricted to $\mathfrak{g}$, then $\chi(\sigma_v) = 1$, for some $v \in S_\infty$. (Otherwise $\chi(\sigma_v) = -1$, for all $v \in S_\infty$ and therefore χ is trivial on the generators $\sigma_v \cdot \sigma_{v'}$ of $\mathfrak{g}$.) Consequently, we have $\chi(\Theta_{S_0,T}) = 0$, for any such character χ (see (11).) Consequently, we have

$$(1 - e_{\mathbf{1}_{\mathfrak{g}}})\Theta_{S_0,T}(0) = 0\,.$$

Now, combine the last equality with (22) and the first isomorphism in (21) to get

$$\Theta_{S_0,T}(0) \cdot CH^1(K)^0_T[1/2] = \Theta_{S_0,T}(0) \cdot CH^1(K)^0_T[1/2]^{\,\mathfrak{g}} = \widetilde{\Theta}_{S_0,T}(0) \cdot CH^1(\widetilde{K})^0_T[1/2]\,.$$

(Note that the natural surjection $\mathbb{Z}[G] \twoheadrightarrow \mathbb{Z}[\widetilde{G}]$ maps $\Theta_{S_0,T}(0)$ to $\widetilde{\Theta}_{S_0,T}(0)$ by the inflation property of Artin L-functions.) Now, we combine the last equality with Proposition 4.3.6 applied to $\widetilde{K}/k$ to conclude the proof. □

Remark 4.3.8. *If M is a $\mathbb{Z}[G]$-module, we let*

$$M[1/2] := M \otimes_{\mathbb{Z}} \mathbb{Z}[1/2], \qquad M_{(\ell)} := M \otimes_{\mathbb{Z}} \mathbb{Z}_{(\ell)},$$

where $\mathbb{Z}_{(\ell)}$ denotes the localization of $\mathbb{Z}$ at its maximal ideal $\ell\mathbb{Z}$, for any prime number ℓ. We have obvious injective (diagonal) morphisms of $\mathbb{Z}[G]$-modules

$$M \hookrightarrow M[1/2] \bigoplus M_{(2)}, \qquad M[1/2] \hookrightarrow \bigoplus_{\ell \neq 2} M_{(\ell)}\,.$$

This leads to the following equalities of $\mathbb{Z}[G]$–ideals

$$\mathrm{Ann}_{\mathbb{Z}[G]}(M) = \mathrm{Ann}_{\mathbb{Z}[G]}(M[1/2]) \bigcap \mathrm{Ann}_{\mathbb{Z}[G]}(M_{(2)}),$$
$$\mathrm{Ann}_{\mathbb{Z}[G]}(M[1/2]) = \bigcap_{\ell \neq 2} \mathrm{Ann}_{\mathbb{Z}[G]}(M_{(\ell)}).$$

We apply this observation to the $\mathbb{Z}[G]$–module $M := CH^1(K)_T$ and combine it with Propositions 4.3.5 and 4.3.7 to conclude that if K and k are number fields (necessarily totally complex and totally real, respectively), then the Brumer-Stark conjecture $BrSt(K/k, S_0, T)$ is equivalent to

$$\Theta_{S_0,T}(0) \in \mathrm{Ann}_{\mathbb{Z}[G]}(CH^1(K)_T)_{(2)} \text{ and } \widetilde{\Theta}_{S_0,T}(0) \in \mathrm{Ann}_{\mathbb{Z}[\widetilde{G}]}(Cl(O_{\widetilde{K}})_T^-)_{(\ell)}, \ \forall \ell \neq 2,$$

where the notations are as in Proposition 4.3.7. For a given prime ℓ (which can be 2), we refer to the statement above which corresponds to ℓ as "the ℓ–primary component of $BrSt(K/k, S_0, T)$."

The following theorem settles a refinement of the ℓ–primary component of the Brumer-Stark conjecture for number fields, for any prime $\ell \neq 2$, provided that the Iwasawa μ–invariant associated to ℓ and K vanishes and that the set S_0 contains the set S_ℓ of primes in k which divide ℓ.

Theorem 4.3.9 (Greither-Popescu [**16**]). *Let K/k be an abelian extension of number fields of Galois group G, with K CM and k totally real. Let S_0 be a finite set of primes in k, such that $S_\infty \cup S_{\mathrm{ram}}(K/k) \subseteq S_0$. Let $\ell \neq 2$ be a prime number. Assume that the Iwasawa μ–invariant $\mu_{K,\ell}$ associated to K and ℓ vanishes. Then, we have*

$$\Theta_{S_0 \cup S_\ell, T}(0) \in \mathrm{Fitt}_{\mathbb{Z}_{(\ell)}[G]}\left((Cl(O_K)_T^-)_{(\ell)}^\vee\right) \subseteq \mathrm{Ann}_{\mathbb{Z}_{(\ell)}[G]}(Cl(O_K)_T^-)_{(\ell)}.$$

for any finite set T of primes in k, disjoint from $S_0 \cup S_\ell$ and such that $K(\mu_\ell)_T^\times$ has no $\mathbb{Z}$–torsion.

We need to explain some notations in the theorem above. If M is a finite $\mathbb{Z}[G]$–module, $M_{(\ell)}^\vee := \mathrm{Hom}_{\mathbb{Z}_{(\ell)}}(M_{(\ell)}, \mathbb{Q}_\ell/\mathbb{Z}_\ell)$ is the usual Pontrjagin dual of $M_{(\ell)}$, viewed as a $\mathbb{Z}_{(\ell)}[G]$–module with the co-variant G–action given by ${}^\sigma f(x) := f(\sigma \cdot x)$, for all $f \in M_{(\ell)}^\vee$, $x \in M_{(\ell)}$ and $\sigma \in G$. As usual, Fitt stands for the first Fitting ideal, which is a smaller, refined version of the annihilator of a finitely generated module over an arbitrary commutative ring. The inclusion of $\mathbb{Z}_{(\ell)}[G]$–ideals in the statement above is clear from the definitions and it is strict inclusion, in general.

The theorem above is a consequence of an equivariant Main Conjecture in Iwasawa theory proved in [**16**] and refining the Main Conjecture proved in [**44**]. The arguments involve the Iwasawa cyclotomic tower $K(\mu_{\ell^\infty})/K$, which is ramified at all primes above ℓ, explaining the presence of S_ℓ in the statement of the theorem. The assumption $\mu_{K,\ell} = 0$ (needed for technical reasons in our proof of the equivariant Main Conjecture) is a classical conjecture of Iwasawa, known to hold, for example, for all fields K which are abelian over $\mathbb{Q}$ and all primes ℓ, due to Ferrero-Washington and Sinnott. Of course, for abelian extensions of $\mathbb{Q}$, the following classical theorem of Stickelberger, dated 1890 (see [**38**]), settles the full Brumer-Stark conjecture in an explicit (constructive) way.

Theorem 4.3.10 (Stickelberger [**38**]). *Let $K := \mathbb{Q}(\mu_m)$, where $m \geq 3$ and m is either odd or divisible by 4. Let $S_0 := \{p \mid p \text{ prime}, p|m\} \cup \{\infty\}$. Let $\mathfrak{P}$ be a prime in K, which sits above a rational prime $p \notin S_0$. Let $\chi_{\mathfrak{P}} : (O_K/\mathfrak{P})^\times \to \mu_m$ be* *the unique morphism satisfying* $\chi_{\mathfrak{P}}(x) \equiv x^{(N\mathfrak{P}-1)/m} \mod \mathfrak{P}$, *for all* $x \in O_K$. *Let* $\psi : O_K/\mathfrak{P} \to \mu_p$ *be any nontrivial, additive character. Let*

$$\mathcal{G}(\chi_{\mathfrak{P}}, \psi) := \sum_{x \in O_K/\mathfrak{P}} \chi_{\mathfrak{P}}(x)\psi(x) \in \mathbb{Q}(\mu_{pm}), \quad g(\chi_{\mathfrak{P}}, \psi) := \frac{\mathcal{G}(\chi_{\mathfrak{P}}, \psi)}{\sqrt{\pm N\mathfrak{P}}},$$

with the sign chosen so that $g(\chi_{\mathfrak{P}}, \psi) \in \mathbb{Q}(\mu_{4pm})$. *Then, the following are satisfied.*

(1) *The element* $g(\mathfrak{P}) := g(\chi_{\mathfrak{P}}, \psi)^{w_K}$ *belongs to* K *and is independent of the choice of* ψ. *Here,* w_K *denotes the number of roots of unity in* K.
(2) *We have* $|g(\mathfrak{P})|_v = 1$, *for all infinite primes* v *of* K.
(3) *The* O_K*-ideal* $\mathfrak{P}^{w_K \Theta_{S_0}(0)}$ *is principal, generated by* $g(\mathfrak{P})$.

The elements $\mathcal{G}(\chi_{\mathfrak{P}}, \psi)$ and $g(\mathfrak{P})$ defined above are particular cases of Gauss and Jacobi sums, respectively. Statements (1) and (2) are well known properties of Gauss sums. Statement (3) is a consequence of the prime power decomposition of $\mathcal{G}(\chi_{\mathfrak{P}}, \psi)^m$, which is due to Stickelberger. The reader can consult §3 of [**10**] for very lucid proofs of all these facts.

Corollary 4.3.11. *Let* $K/\mathbb{Q}$ *be an abelian extension and assume that* $(K/\mathbb{Q}, S_0, T)$ *satisfy the hypotheses of the Brumer-Stark conjecture. Then,* $BrSt(K/\mathbb{Q}, S_0, T)$ *holds true.*

Proof. First of all, Proposition 3.2.16, combined with Corollary 4.3.3 and the Kronecker-Weber Theorem permit us to reduce the proof to the case where K and S_0 are as in Stickelberger's Theorem. Let $p \notin S_0 \cup T$ be a prime which splits completely in $K/\mathbb{Q}$ and let $\mathfrak{P}$ be a prime in K dividing p. Since $g(\mathfrak{P})^{1/w_K} = g(\chi_{\mathfrak{P}}, \psi)$ belongs to $\mathbb{Q}(\mu_{4pm})$, which is an abelian extension of $\mathbb{Q}$, Proposition 1.2 in Chpt. IV of [**42**] combined with our Remark 3.2.4 implies that

$$g(\mathfrak{P})^{\delta_T(0)} \in K_T^{\times}.$$

Now, (2) and (3) in Stickelberger's Theorem imply that

$$\mathrm{div}_{K,S_0}(g(\mathfrak{P})^{\delta_T(0)}) = \Theta_{S_0,T}(0) \cdot \mathfrak{P}.$$

However, since $|S_0| > 1$, we have $\Theta_{S_0,T}(0) \cdot \mathfrak{P} = \Theta_{S_0,T}(0)(\mathfrak{P} - \deg_{K,S_0}(\mathfrak{P}) \cdot \infty)$, where ∞ denotes (abusively) a prime in K which sits above the unique infinite prime in $\mathbb{Q}$ (also denoted ∞.) Consequently, the last displayed equality settles conjecture $BrSt(K/\mathbb{Q}, S_0, T)$. (Set $w := \mathfrak{P}$, $v_0 := \infty$ and $\varepsilon_w := g(\mathfrak{P})^{\delta_T(0)}$ in the statement of Conjecture 4.3.1.) □

Finally, we discuss **the function field case** of the Brumer-Stark conjecture. As mentioned before, the conjecture was formulated by Tate in this case (see [**42**], Chpt. V.) and proved independently and with very different methods by Deligne-Tate in loc. cit. and Hayes in [**22**].

Theorem 4.3.12 (Deligne-Tate [**42**], Hayes [**22**]). *The Brumer-Stark conjecture* $BrSt(K/k, S_0, T)$ *holds true if* k *is a function field.*

Hayes's proof [**22**] is based on a wonderful explicit construction of the Brumer-Stark (Rubin-Stark) element $\varepsilon_{S,T}$ out of an $\mathfrak{f}$–torsion point of a sign-normalized, rank one Drinfeld module ρ_{v_1} with "CM" by the ring $O_{k,\{v_1\}}$ (the subring of k consisting of elements which are integral away from v_1.) Here, v_1 is an arbitrary prime in k which is not in S_0 and splits completely in K/k, $S := S_0 \cup \{v_1\}$ and $\mathfrak{f}$ is the conductor of K/k (viewed as an ideal of $O_{k,\{v_1\}}$.) Hayes's proof is the perfect function field analogue of the proof of Theorem 4.2.1 outlined above. In Hayes's proof, the role of the formal multiplicative group $\widehat{G_m}$ (which has "CM" by $\mathbb{Z}$, the ring of elements in $\mathbb{Q}$ which are integral away from the split prime $v_1 := \infty$) is taken by the Drinfeld module ρ_{v_1} (with "CM" by $O_{k,\{v_1\}}$.) In fact, the analogy runs much deeper: just the way the $\mathbb{Z}$–torsion points of $\widehat{G_m}$ generate the maximal abelian extension $\mathbb{Q}^{\mathrm{ab},+}$ of $\mathbb{Q}$ in which ∞ splits completely, the $O_{k,\{v_1\}}$–torsion points of ρ_{v_1} generate the maximal abelian extension of k in which v_1 splits completely, provided that the Dedekind domain $O_{k,\{v_1\}}$ has class-number 1. If the class-number of $O_{k,\{v_1\}}$ is not 1, then the situation is a bit more complicated (but still fully understood), as one has to generate the Hilbert-class field of $O_{k,\{v_1\}}$ first. This is in fact a rough description of explicit (abelian) class-field theory over an arbitrary function field k, due independently to Drinfeld and Hayes.

In light of our interpretation of the Brumer-Stark conjecture in terms of Arakelov class-groups, the Deligne-Tate proof can be described roughly as follows. *First,* one constructs a Picard 1–motive $\mathcal{M}_{S_0,T}$ defined over $\mathbb{F}_q$ and depending on the data $(K/k, S_0, T, \mathbb{F}_q)$, where $\mathbb{F}_q$ is the exact field of constants of k. *Second,* one uses the Lefschetz trace formula for the ℓ–adic étale cohomology of smooth, projective curves defined over $\mathbb{F}_q$, for ℓ prime, $\ell \neq p := \mathrm{char}(\mathbb{F}_q)$, to show that

$$\Theta_{S_0,T}(s) = P(q^{-s}), \quad P(t) := \det_{\mathbb{Q}_\ell[G]}(1 - \gamma_q \cdot t \mid T_\ell(\mathcal{M}_{S_0,T}) \otimes_{\mathbb{Z}_\ell} \mathbb{Q}_\ell).$$

Note that $P(t)$ is a slight variation on the G–equivariant characteristic polynomial of the q–power arithmetic Frobenius morphism γ_q acting on the ℓ–adic Tate module $T_\ell(\mathcal{M}_{S_0,T})$ of $\mathcal{M}_{S_0,T}$, for all primes ℓ, $\ell \neq p$. The polynomial P is independent on the prime $\ell \neq p$ and it turns out that $P \in \mathbb{Z}[G][t]$. For a brief description of the *third step,* we make the simplifying assumption that $\mathbb{F}_q$ is the exact field of constants of K as well. For every $n \in \mathbb{Z}_{\geq 1}$, let $K_n := K \cdot \mathbb{F}_{q^n}$, where the field compositum is viewed inside a fixed separable closure $\overline{k}$ of k. Note that $(K_n/k, S_0, T)$ satisfy the hypotheses of the Brumer-Stark conjecture, for all such n. (Note that K_n/K is unramified and $T \neq \emptyset$.) We let $K_\infty := K \cdot \overline{\mathbb{F}_q} = \cup_n K_n$ and define

$$CH^1_\emptyset(K_\infty)^0_T = \varinjlim_n CH^1_\emptyset(K_n)^0_T\,, \tag{23}$$

where the injective limit is taken with respect to the natural injections

$$CH^1_\emptyset(K_n)^0_T \hookrightarrow CH^1_\emptyset(K_m)^0_T,$$

for all n, m, with $n \mid m$ (see Remark 4.3.4.) Now, it turns out that there exists a semiabelian variety J_T defined over $\mathbb{F}_q$, whose group of $\overline{\mathbb{F}_q}$-rational points satisfies

$$J_T(\overline{\mathbb{F}_q}) = CH^1_\emptyset(K_\infty)^0_T, \qquad J_T(\overline{\mathbb{F}_q})^{\gamma_q^n=1} = CH^1_\emptyset(K_n)^0_T, \text{ for all } n\,.$$

In fact, J_T is an extension of the Jacobian J of a smooth, projective model of K by a torus which depends on T. There are $\mathbb{Z}_\ell[G]$–linear injective morphisms

$$T_\ell(J_T) \hookrightarrow T_\ell(\mathcal{M}_{S_0,T})\,,$$

preserving the γ_q–action, for all primes ℓ. The second step and the above inclusion imply that

$$P(\gamma_q^{-1}) \cdot T_\ell(\mathcal{M}_{S_0,T}) = 0, \quad P(\gamma_q^{-1}) \cdot T_\ell(J_T) = 0, \qquad \text{for all } \ell \neq p\,,$$

where $P(\gamma_q)$ is viewed inside $\mathbb{Z}[G][\gamma_q^{-1}]$, which maps naturally to the rings $\mathrm{End}(J_T)$ and $\mathrm{End}(\mathcal{M}_{S_0,T})$. Now, it is very easy to show that if an endomorphism of a semiabelian variety annihilates its ℓ-adic Tate module, for some prime ℓ different from the characteristic of its field of definition, then the endomorphism in question is trivial. Consequently, we have

$$P(\gamma_q^{-1}) \cdot CH^1_\emptyset(K_\infty)^0_T = 0\,, \quad P(\gamma_q^{-1}) \cdot CH^1_\emptyset(K_n)^0_T = 0, \text{ for all } n\,. \tag{24}$$

Now, we note that γ_q is the canonical topological generator of $\Gamma := \mathrm{Gal}(K_\infty/K)$, which is isomorphic (as a topological group) to the profinite completion $\widehat{\mathbb{Z}}$ of $\mathbb{Z}$. Under the assumption that $\mathbb{F}_q$ is the exact field of constants of both K and k, we have an obvious isomorphism $G(K_\infty/k) \simeq G \times \Gamma$. Consequently, we can view $P(\gamma_q^{-1})$ as an element of $\mathbb{Z}[G(K_\infty/k)]$. It turns out that, under the natural projections $\mathbb{Z}[G(K_\infty/k)] \twoheadrightarrow \mathbb{Z}[G(K_n/k)]$, the element $P(\gamma_q^{-1})$ maps to $\Theta^{(n)}_{S_0,T}(0)$, where $\Theta^{(n)}_{S_0,T}(s)$ is the equivariant L–function associated to $(K_n/k, S_0, T)$, for all $n \in \mathbb{Z}_{\geq 1}$. We combine this fact with (23) and (24) to conclude that

$$\Theta^{(n)}_{S_0,T}(0) \in \mathrm{Ann}_{\mathbb{Z}[G(K_n/k)]} CH^1_\emptyset(K_n)^0_T\,, \qquad \text{for all } n \in \mathbb{Z}_{\geq 1}\,.$$

Now, the proof of $BrSt(K_n/k, S_0, T)$, for all $n \in \mathbb{Z}_{\geq 1}$, is a consequence of the following, which is the function field analogue of Proposition 4.3.5.

Proposition 4.3.13. *The following statements are equivalent.*

(1) *$BrSt(K_n/k, S_0, T)$ holds true, for all $n \in \mathbb{Z}_{\geq 1}$.*
(2) *$\Theta^{(n)}_{S_0,T}(0) \in \mathrm{Ann}_{\mathbb{Z}[G(K_n/k)]} CH^1_\emptyset(K_n)^0_T$, for all $n \in \mathbb{Z}_{\geq 1}$.*

PROOF. The implication (1) $\Rightarrow$ (2) is an immediate consequence of Proposition 4.3.2 and the canonical injective morphisms of $\mathbb{Z}[G(K_n/k)]$–modules

$$j_n : CH^1_\emptyset(K_n)^0_T \hookrightarrow CH^1_{S_0}(K_n)^0_T\,,$$

sending the class in $CH^1_\emptyset(K_n)^0_T$ of the divisor $D := (\sum_{w \notin T} x_w \cdot w) \in \mathrm{Div}_\emptyset(K_n)^0_T$ to the class in $CH^1_{S_0}(K_n)^0_T$ of the divisor

$$D' := (\sum_{w \notin T \cup S_0} x_w \cdot w + \sum_{w \in S_0} (x_w \cdot \log |Nw|) \cdot w) \in \mathrm{Div}_{S_0}(K_n)^0_T.$$

Here, the sums are taken with respect to all the primes w in K_n and, as usual, we abuse notation and let S_0 and T denote the sets of primes in K_n sitting above prime in S_0 and T, respectively. A more appropriate notation would be $S_0(K_n)$ and $T(K_n)$, respectively (see below.)

The implication (2) $\Rightarrow$ (1) is proved as follows. First, we leave it to the reader to check that the cokernel of the maps $\pi_n \circ j_n$ in the diagrams

$$\begin{array}{ccccccccc} & & & & CH^1_\emptyset(K_n)^0_T & & & & \\ & & & & \downarrow{\scriptstyle j_n} & \searrow^{\pi_n \circ j_n} & & & \\ 0 & \longrightarrow & \dfrac{\mathbb{R}X_{S_0}}{\mathrm{div}_{K_n,S_0}(U_{S_0,T})} & \xrightarrow{\iota_n} & CH^1_{S_0}(K_n)^0_T & \xrightarrow{\pi_n} & Cl(O_{K_n,S_0})_T & \longrightarrow & 0\,, \end{array}$$

satisfies $\mathrm{coker}(\pi_n \circ j_n) \simeq \mathbb{Z}/d_{S_0,n}\mathbb{Z}$, for all n, where

$$d_{S_0,n} := \gcd\left([\kappa(w) : \mathbb{F}_{q^n}] \mid w \in S_0(K_n)\right).$$

Here, $\kappa(w)$ denotes the residue field of the prime w. This observation implies right away that for n "sufficiently large" (meaning n divisible by $[\kappa(w) : \mathbb{F}_q]$, for all $w \in S_0(K)$), we have $d_{S_0,n} = 1$ and therefore $(\pi_n \circ j_n)$ is surjective. Assume that (2) in the Proposition holds and let us fix an $m \in \mathbb{Z}_{\geq 1}$. We will show that $BrSt(K_m/k, S_0, T)$ holds. Let n be sufficiently large in the above sense and such that $m|n$. Since $(\pi_n \circ j_n)$ is surjective, we have an equality

$$\mathrm{Im}(\iota_n) + \mathrm{Im}(j_n) = CH^1_{S_0}(K_n)^0_T\,.$$

Part (2) of the Proposition combined with $\Theta^{(n)}_{S_0,T}(0) \cdot \mathbb{R}X_{S_0} = 0$ implies that

$$\Theta^{(n)}_{S_0,T}(0) \in \mathrm{Ann}_{\mathbb{Z}[G(K_n/k)]} CH^1_{S_0}(K_n)^0_T\,.$$

Now, Proposition 4.3.2 combined with Proposition 4.3.3 imply that conjecture $BrSt(K_m/k, S_0, T)$ holds true. □

We conclude this section by stating the function field analogue of Theorem 4.3.9 above, which refines the above results of Deligne and Tate and, implicitly, leads to a refinement of the Brumer-Stark conjecture for function fields.

Theorem 4.3.14 (Greither-Popescu [**17**]). *The following hold.*

(1) *The $\mathbb{Z}_\ell[[G(K_\infty/k)]]$–module $T_\ell(\mathcal{M}_{S_0,T})$ has projective dimension* 1 *and*

$$\mathrm{Fitt}_{\mathbb{Z}_\ell[[G(K_\infty/k)]]}(T_\ell(\mathcal{M}_{S_0,T})) = (P(\gamma_q^{-1})),$$

for all prime numbers ℓ.

(2) *Let $T_\ell(J_T)^* := \mathrm{Hom}_{\mathbb{Z}_\ell}(T_\ell(J_T), \mathbb{Z}_\ell)$ endowed with the co-variant $\mathbb{G}(K_\infty/k)$–action. Then, for all prime numbers ℓ we have*

$$P(\gamma_q^{-1}) \in \mathrm{Fitt}_{\mathbb{Z}_\ell[[G(K_\infty/k)]]}(T_\ell(J_T)^*) \subseteq \mathrm{Ann}_{\mathbb{Z}_\ell[[G(K_\infty/k)]]}(T_\ell(J_T)^*).$$

(3) *For all $n \in \mathbb{Z}_{\geq 1}$, we have*

$$\Theta^{(n)}_{S_0,T}(0) \in \mathrm{Fitt}_{\mathbb{Z}[G(K_n/k)]}(CH^1_\emptyset(K_n)^{0\ \vee}_T) \subseteq \mathrm{Ann}_{\mathbb{Z}[G(K_n/k)]} CH^1_\emptyset(K_n)^0_T\,,$$

where, for any finite $\mathbb{Z}[G(K_n/k)]$–module M, we let $M^\vee := \mathrm{Hom}_{\mathbb{Z}}(M, \mathbb{Q}/\mathbb{Z})$ and endow it with the co-variant $G(K_n/k)$–action.

Remark 4.3.15. (1) *Part (1) in Theorem 4.3.14 is the function field analogue of the Equivariant Main Conjecture mentioned (but not stated, for technical reasons) in the paragraphs after Theorem 4.3.9.*

(2) *Parts (2) and (3) of Theorem 4.3.14 are more or less immediate consequences of part (1).*

(3) *Note that, in* [**17**], *$CH^1_\emptyset(K)^0_T$ is denoted by $\mathrm{Pic}^0(K)_T$. This notation is justified by the fact that, indeed, $CH^1_\emptyset(K)^0_T$ is a T–modified version of the classical Picard group $\mathrm{Pic}^0(K)$. In fact, there is a canonical short exact sequence of groups linking the two:*

$$0 \longrightarrow \frac{\bigoplus_{w\in T(K)} \kappa(w)^\times}{\mathbb{F}_q^\times} \longrightarrow CH^1_\emptyset(K)^0_T \longrightarrow \mathrm{Pic}^0(K) \longrightarrow 0.$$

(4) *In recent work* [**15**], *we have shown that part (1) of Theorem 4.3.14 implies the Equivariant Tamagawa Number Conjecture (Conjecture 3.1.4) for all abelian extensions K/k of function fields. (In this case, the ETNC was also proved by Burns with different methods.)*

(5) *In fact, the Deligne-Tate 1–motive constructions and the techniques involved in the proof of Theorem 4.3.14 were the main inspiration for our proof of the Equivariant Main Conjecture for number fields and, implicitly, the proof of Theorem 4.3.9 above. In the number field case, we construct Iwasawa modules which behave like the ℓ–adic Tate modules $T_\ell(\mathcal{M}_{S_0,T})$ of the Picard 1–motive $\mathcal{M}_{S_0,T}$. Then, we prove the analogue of Theorem 4.3.14(1) for these Iwasawa modules , with $P(\gamma_q^{-1})$ replaced by an appropriate G–equivariant ℓ–adic L–function. The ℓ–adic L–function in question is described somewhat in detail in the next section.*

5. Gross–type (p–adic) refinements of the Rubin-Stark conjecture

With notations as in the previous sections, assume that the data $(K/k, S, T, r)$ satisfies the hypotheses in the Rubin-Stark conjecture $A_{\mathbb{Z}}(K/k, S, T, r)$. In [**18**] and [**19**], Gross proposed a refinement of the Rubin–Stark conjecture in the case where the order of vanishing is $r = 1$. In the number field case, if the distinguished split prime $v_1 \in S$ is finite (call it $\mathfrak{p}$ and assume it sits over the rational prime p), then Gross's conjecture predicts the $\mathfrak{P}$–adic expansion of the Rubin-Stark S–unit $\varepsilon_{S,T} \in U_{S,T}$ at a prime $\mathfrak{P}$ in $S(K)$ sitting above the split prime $\mathfrak{p}$ in terms of values of derivatives of p–adic L–functions. In the function field case, Gross's conjecture was proved by Hayes in [**23**]. In the case where $k = \mathbb{Q}$ and K is imaginary, the conjecture was proved by Gross in [**18**] (see also [**21**] and [**12**].)

In the early 1990s, Gross and Tate (see [**20**]) expressed interest in formulating and hinted at a Gross–type refinement for the Rubin–Stark conjecture for arbitrary orders of vanishing r. Tan formulated and partly proved such a refinement for function fields (see [**40**].) In this section, we describe a general Gross–type refinement of the Rubin-Stark Conjecture which has emerged recently out of work of Burns, the present author, and their collaborators. In the number field case, we interpret a particular case of this conjecture in terms of special values of derivatives of p–adic L–functions. Finally, we discuss some recent results supporting this conjecture.

5.1. The set-up

As above, let us assume that the data $(K/k, S, T, r)$ satisfies the hypotheses in the Rubin-Stark conjecture $A_{\mathbb{Z}}(K/k, S, T, r)$. We let $\varepsilon := \varepsilon_{S,T}$ denote the corresponding Rubin-Stark element. Further, assume that L/k is an abelian extension, *not necessarily finite,* such that $k \subseteq K \subseteq L$, $S_{\mathrm{ram}}(L/k) \subseteq S$ and $L_T^\times$ has no $\mathbb{Z}$–torsion. (Recall that $L_T^\times$ denotes the subgroup of $L^\times$ consisting of all elements congruent to 1 module every prime in T.) We let $\Gamma := \mathrm{Gal}(L/K)$ and $\mathcal{G} := \mathrm{Gal}(L/k)$ and $G := \mathrm{Gal}(K/k)$.

The main idea behind the conjecture we are about to describe is as follows. We assume that conjecture $A_{\mathbb{Z}}(K/k, S, T, r)$ holds, meaning that $\varepsilon \in \Lambda_{S,T} \subseteq \mathcal{L}_{\mathbb{Z}[G]}(U_{S,T}, r)$, where $\Lambda_{S,T}$ is Rubin's lattice. Since $\varepsilon \in \mathcal{L}_{\mathbb{Z}[G]}(U_{S,T}, r)$, the constructions in §1.2 permit us to associate evaluation maps $\mathrm{ev}_{\varepsilon,\mathcal{R}}$ with values in $\mathcal{R}[G]$, to arbitrary unital commutative rings $\mathcal{R}$. The conjecture aims at interpreting the value $\Theta_{L/k,S,T}(0)$ of the equivariant L–function associated to $(L/k, S, T)$ in terms of a value of the evaluation map $\mathrm{ev}_{\varepsilon,\mathcal{R}(\Gamma)}$ associated to a certain graded ring $\mathcal{R}(\Gamma)$, which depends on Γ and will be described below.

5.2. The relevant graded rings $\mathcal{R}(\Gamma)$ and $\mathcal{R}_\Gamma$

With notations as above, we let $\mathbb{Z}[[\mathcal{G}]]$ and $\mathbb{Z}[[\Gamma]]$ denote the integral profinite group rings associated to the profinite (possibly infinite) groups $\mathcal{G}$ and Γ, respectively. If Γ and $\mathcal{G}$ are finite, then these are the usual group rings $\mathbb{Z}[\mathcal{G}]$ and $\mathbb{Z}[\Gamma]$. Otherwise,

$$\mathbb{Z}[[\Gamma]] := \varprojlim_{L'} \mathbb{Z}[G(L'/K)],$$

where L' runs over the finite extensions L'/K, with $K \subseteq L' \subseteq L$ and the projective limit is taken with respect to the usual group-ring projections induced by Galois restriction at the level of Galois groups. We have a similar description of $\mathbb{Z}[[\mathcal{G}]]$.

Definition 5.2.1.

(1) *We let $I(\Gamma)$ denote the augmentation ideal of $\mathbb{Z}[[\Gamma]]$, given by*

$$I(\Gamma) := \varprojlim_{L'} I(G(L'/K)),$$

where $I(G(L'/K))$ is the augmentation ideal of the group ring $\mathbb{Z}[G(L'/K)]$, i.e. the kernel of the $\mathbb{Z}$–linear map $\mathrm{aug}_{\mathrm{L}'} : \mathbb{Z}[G(L'/K)] \longrightarrow \mathbb{Z}$ satisfying the property that $\mathrm{aug}_{L'}(\sigma) = 1$, for all $\sigma \in G(L'/K)$ and all L' as above.

(2) *We let I_Γ denote the Γ–relative augmentation ideal of $\mathbb{Z}[[\mathcal{G}]]$, given by*

$$I_\Gamma := \varprojlim_{L'} I_{G(L'/K)},$$

where $I_{G(L'/K)}$ is the kernel of the natural projection $\mathbb{Z}[G(L'/k)] \twoheadrightarrow \mathbb{Z}[G]$, for all L'.

Now, we are ready to define the relevant graded rings $\mathcal{R}(\Gamma)$ and $\mathcal{R}_\Gamma$.

Definition 5.2.2. *We define $\mathcal{R}(\Gamma)$ and $\mathcal{R}_\Gamma$ to be the graded rings*

$$\mathcal{R}(\Gamma) := \bigoplus_{n\geq 0} I(\Gamma)^n/I(\Gamma)^{n+1}, \qquad \mathcal{R}_\Gamma := \bigoplus_{n\geq 0} I_\Gamma^n/I_\Gamma^{n+1},$$

with the obvious addition and multiplication.

Lemma 5.2.3. *The following hold.*

(1) *For all $n \in \mathbb{Z}_{\geq 0}$, we have a direct sum decomposition*

$$I_\Gamma^n = \bigoplus_{\sigma\in G} \widetilde{\sigma} \cdot I(\Gamma)^n,$$

where $\widetilde{\sigma}$ is a fixed element of $\mathcal{G}$ which maps to σ via the natural surjection $\mathcal{G} \twoheadrightarrow G$.

(2) *For all $n \in \mathbb{Z}_{\geq 0}$, we have an isomorphism of $\mathbb{Z}[G]$–modules*

$$I(\Gamma)^n/I(\Gamma)^{n+1} \otimes_{\mathbb{Z}} \mathbb{Z}[G] \simeq I_\Gamma^n/I_\Gamma^{n+1}, \qquad \widehat{\iota} \otimes \sigma \to \widehat{\widetilde{\sigma}\iota},$$

where $\widehat{\iota}$ is the class of $\iota \in I(\Gamma_n)^n$ in the quotient $I(\Gamma)^n/I(\Gamma)^{n+1}$, $\widetilde{\sigma}$ is defined as in (1), for all $\sigma \in G$ and $\widehat{\widetilde{\sigma}\iota}$ is the class of $\widetilde{\sigma}\iota \in I_\Gamma^n$ in the quotient $I_\Gamma^n/I_\Gamma^{n+1}$.

(3) *We have isomorphisms of graded $\mathbb{Z}[G]$–algebras*

$$i : \mathcal{R}(\Gamma)[G] \simeq \mathcal{R}(\Gamma) \otimes_{\mathbb{Z}} \mathbb{Z}[G] \simeq \mathcal{R}_\Gamma$$

where the first isomorphism is $i_{\mathcal{R}(\Gamma)}^{-1}$ of §1.2 and the second is the direct sum over $n \in \mathbb{Z}_{\geq 0}$ of the isomorphisms in (2).

PROOF. (1) Obviously, it suffices to prove the analogue of (1) for $I_{G(L'/K)}$ and $I(G(L'/K))$, for all L' as above and then take a projective limit. We leave this as an exercise for the reader. Part (2) is an immediate consequence of (1). Of course, one needs to note that the element $\widehat{\widetilde{\sigma}\iota}$ defined above does not depend on the lift $\widetilde{\sigma} \in \mathcal{G}$ of $\sigma \in G$ and that $I_\Gamma^n/I_\Gamma^{n+1}$ has a natural $\mathbb{Z}[G]$–module structure given by $\sigma \cdot \widehat{\iota} := \widehat{\widetilde{\sigma}\iota}$, for all $\sigma \in G$ and $\iota \in I_\Gamma^n$. Part (3) is a direct consequence of part (2). □

Remark 5.2.4. *In what follows, we will denote by $\mathcal{R}(\Gamma)^{(n)} := I(\Gamma)^n/I(\Gamma)^{n+1}$ and $\mathcal{R}_\Gamma^{(n)} := I_\Gamma^n/I_\Gamma^{n+1}$ the degree n homogeneous components of $\mathcal{R}(\Gamma)$ and $\mathcal{R}_\Gamma$, respectively. By abuse of notation, we let*

$$\mathcal{R}(\Gamma)^{(n)}[G] := i_{\mathcal{R}(\Gamma)}(\mathcal{R}(\Gamma)^{(n)} \otimes_{\mathbb{Z}} \mathbb{Z}[G]).$$

Obviously, the map i in part (3) of the Lemma above establishes an isomorphism of $\mathbb{Z}[G]$–modules $\mathcal{R}(\Gamma)^{(n)}[G] \simeq \mathcal{R}_\Gamma^{(n)}$, for all n.

5.3. The conjecture

We assume that the data $(L/K/k, S, T, r)$ satisfies the above hypotheses. As in the set-up for the Rubin-Stark conjecture $A_{\mathbb{Z}}(K/k, S, T, r)$, we let $(v_1, \dots, v_r)$ be an ordered r–tuple of distinct primes in S which split completely in K/k. For each $i = 1, \dots, r$, we fix a prime w_i in K sitting above v_i and consider the ordered r–tuple $W := (w_1, \dots, w_r)$.

Let $v := v_i$ and $w := w_i$, for some $i \in \{1, \dots, r\}$. We denote by $G_w(L/K)$ the decomposition group associated to w in the abelian extension L/K. Note that, because v splits completely in K/k, we have $G_w(L/K) = G_v(L/k)$. We denote by K_w the completion of K in the metric $|\cdot|_w$. In what follows, we identify $G_w(L/K)$ with the Galois group $G(L \cdot K_w/K_w)$ in the usual way. Here, $L \cdot K_w$ denotes a field compositum of K_w and L. Furthermore, we let

$$\rho_w : K_w^\times \longrightarrow G_w(L/K) \subseteq \Gamma$$

denote the composition of the local Artin reciprocity map associated to the abelian extension $L \cdot K_w/K_w$ of the local field K_w with the inclusion $G_w(L/K) \subseteq \Gamma$.

Definition 5.3.1. *Following Gross, we define $\mathbb{Z}[G]$–linear morphisms ϕ_w and ψ_w, for all w as above:*

$$\psi_w : U_{S,T} \xrightarrow{\phi_w} \mathcal{R}(\Gamma)[G] \xrightarrow[\sim]{i} \mathcal{R}_\Gamma, \qquad \phi_w(u) := \sum_{\sigma \in G} (\widehat{\rho_w(u^\sigma)} - 1) \cdot \sigma^{-1},$$

for all $u \in U_{S,T}$, where $(\widehat{\rho_w(u^\sigma) - 1})$ is the class of $(\rho_w(u^\sigma) - 1)$ in

$$I(\Gamma)/I(\Gamma)^2 = \mathcal{R}(\Gamma)^{(1)} \subseteq \mathcal{R}(\Gamma).$$

Now, let us assume that the Rubin-Stark conjecture $A_{\mathbb{Z}}(K/k, S, T, r)$ holds, i.e. the Rubin-Stark element $\varepsilon := \varepsilon_{S,T}$ lies in $\Lambda_{S,T} \subseteq \mathcal{L}_{\mathbb{Z}[G]}(U_{S,T}, r)$. Since the maps $\phi_{w_1}, \dots, \phi_{w_r}$ defined above take values in $\mathcal{R}(\Gamma)^{(1)}[G]$, Remark 2 in §1.2 implies that

$$\mathrm{ev}_{\varepsilon, \mathcal{R}(\Gamma)}(\phi_1 \wedge \dots \wedge \phi_r) \in \mathcal{R}(\Gamma)^{(r)}[G]\,.$$

This remark, combined with the grading-preserving ring isomorphism i of Lemma 5.2.3, implies that

$$\mathrm{ev}_{\varepsilon, \mathcal{R}_\Gamma}(\psi_{w_1} \wedge \dots \wedge \psi_{w_r}) := i(\mathrm{ev}_{\varepsilon, \mathcal{R}(\Gamma)}(\phi_{w_1} \wedge \dots \wedge \phi_{w_r})) \in \mathcal{R}_\Gamma^{(r)} = I_\Gamma^r/I_\Gamma^{r+1}\,.$$

Definition 5.3.2. *If L/k is finite, then $\Theta_{L/k,S,T}(0)$ denotes the value of the usual $G(L/k)$–equivariant L–function associated to the data $(L/k, S, T)$. If L/k is infinite, then $\Theta_{L/k,S,T}(0)$ is the unique element in $\mathbb{Z}[[\mathcal{G}]]$ which maps to $\Theta_{L'/k,S,T}(0)$ via the projections*

$$\pi_{L/L'} : \mathbb{Z}[[\mathcal{G}]] \twoheadrightarrow \mathbb{Z}[G(L'/k)]\,,$$

for all finite field extensions L'/k, with $k \subseteq L' \subseteq L$. (Note that the inflation property of Artin L–functions ensures that such an element exists and is unique.)

Now, we are finally fully prepared to state the Gross–type refinement of the Rubin-Stark conjecture announced in the introduction.

Conjecture 5.3.3. $\widetilde{A}_{\mathbb{Z}}(L/K/k, S, T, r)$. *Under the above hypotheses, the following hold true.*

(1) $\Theta_{L/k,S,T}(0) \in I_\Gamma^r$.
(2) *Assume that the Rubin-Stark Conjecture $A_{\mathbb{Z}}(K/k, S, T, r)$ holds. Let $\varepsilon := \varepsilon_{S,T}$ denote the corresponding Rubin–Stark element. Then, we have an equality in $I_\Gamma^r/I_\Gamma^{r+1}$:*

$$\Theta_{L/k,S,T}(0) \bmod I_\Gamma^{r+1} = \mathrm{ev}_{\varepsilon,\mathcal{R}_\Gamma}(\psi_{w_1} \wedge \cdots \wedge \psi_{w_r}).$$

A more enlightening formulation, incorporating the Rubin-Stark conjecture, can be achieved as follows. Given $(L/K/k, S, T, r)$ and the ordered r–tuple W as above, we can define a new $\mathbb{Z}[G]$–linear regulator $R_{W,\Gamma}$ on the Rubin-Stark lattice $\Lambda_{S,T}$, by setting

$$R_{W,\Gamma} : \Lambda_{S,T} \longrightarrow I_\Gamma^r/I_\Gamma^{r+1}, \quad R_{W,\Gamma}(\epsilon) := \mathrm{ev}_{\epsilon,\mathcal{R}_\Gamma}(\psi_{w_1} \wedge \cdots \wedge \psi_{w_r}), \qquad \text{for all } \epsilon \in \Lambda_{S,T}.$$

Also, note that if we view $\Theta_{K/k,S,T}(s)$ as an element of the power series ring $\mathbb{C}[G][[s]]$, we have

$$\mathrm{ord}_{s=0}\Theta_{K/k,S,T}(s) \geq r \Leftrightarrow \Theta_{K/k,S,T}(s) \in s^r\mathbb{C}[G][[s]]\,.$$

Further, since the above equivalent statements hold (under our current hypotheses), we have

$$\Theta_{K/k,S,T}(s) \mod s^{r+1}\mathbb{C}[G][[s]] = \frac{1}{r!}\frac{\mathrm{d}^r}{\mathrm{d}s^r}\Theta_{K/k,S,T}(0)\,,$$

under the isomorphism of $\mathbb{C}[G]$–modules $s^r\mathbb{C}[G][[s]]/s^{r+1}\mathbb{C}[G][[s]] \simeq \mathbb{C}[G]$ which sends $\widehat{s^r \cdot f}$ to $f(0)$, for all $f \in \mathbb{C}[G][[s]]$. Consequently, the following statement holds if and only if $A_{\mathbb{Z}}(K/k, S, T, r)$ and $\widetilde{A}_{\mathbb{Z}}(L/K/k, S, T, r)$ hold simultaneously.

Conjecture 5.3.4. $A_{\mathbb{Z}}(L/K/k, S, T, r)$. *Under the above hypotheses, there exists a unique $\varepsilon := \varepsilon_{S,T}$ in the Rubin-Stark lattice $\Lambda_{S,T}$, such that the following hold.*

(1) *We have $\Theta_{K/k,S,T}(s) \in s^r\mathbb{C}[G][[s]]$ and an equality*

$$R_W(\varepsilon) = \Theta_{K/k,S,T}(s) \mod s^{r+1}\mathbb{C}[G][[s]]$$

in $\mathbb{C}[G] \simeq s^r\mathbb{C}[G][[s]]/s^{r+1}\mathbb{C}[G][[s]]$.

(2) *We have $\Theta_{L/k,S,T}(0) \in I_\Gamma^r$ and an equality*

$$R_{W,\Gamma}(\varepsilon) = \Theta_{L/k,S,T}(0) \mod I_\Gamma^{r+1}$$

in $I_\Gamma^r/I_\Gamma^{r+1}$.

Remark 5.3.5. *Note that the Rubin-Stark conjecture $A_{\mathbb{Z}}(K/k, S, T, r)$ is the particular case $L = K$ of conjecture $A_{\mathbb{Z}}(L/K/k, S, T, r)$.*

The following proposition makes the link between Conjecture $\widetilde{A}_{\mathbb{Z}}(L/K/k, S, T, r)$ above and the classical conjectures of Gross (Conjectures 4.1 and 7.6 in [**19**] and Conjecture 3.13 in [**18**].)

Proposition 5.3.6. *Assume that the data $(L/K/k, S, T, r)$ satisfies the above hypotheses and* $\mathrm{char}(k) = 0$. *Then, we have the following.*

(1) *If $K = k$, then $\widetilde{A}_{\mathbb{Z}}(L/K/k, S, T, r)$ is equivalent to Gross's Global Conjecture 4.1 in* [**19**].
(2) *If $r = 1$, then $\widetilde{A}_{\mathbb{Z}}(L/K/k, S, T, r)$ is equivalent to Gross's Global Conjecture 7.1 in* [**19**].
(3) *Assume that K is CM, k is totally real and $L := K_{p^\infty}$ is the cyclotomic $\mathbb{Z}_p$–extension of K, for some prime number p. Then $\widetilde{A}_{\mathbb{Z}}(L/K/k, S, T, 1)$ is equivalent to Gross's p–adic Conjecture 3.13 of* [**18**].

PROOF. The reader is advised to look up the statements of Gross's conjectures in the references cited above. He will realize that (2) is immediate and (1) is a consequence of the explicit formula for the Rubin-Stark element given in Example 3.2.11. Note that, if $K = k$, then S contains $r + 1$ primes which split completely in K/k, as required in Example 3.2.11.

The proof of (3) is more subtle and will be given in detail in the next section. □

5.4. Linking values of derivatives of p–adic and global L–functions

One of the main features of Conjecture $A_{\mathbb{Z}}(L/K/k, S, T, r)$ stated above is that, under the appropriate hypotheses, it establishes a deep connection between the values of the r–th derivatives of (equivariant) global and p–adic L–functions at $s = 0$. We discuss this connection below. Once this connection is made, the proof of part (3) of Proposition 5.3.6 will be clear.

Let us assume that $\mathrm{char}(k) = 0$, p is a prime number and $L := K_{p^\infty}$ is the cyclotomic $\mathbb{Z}_p$–extension of K. Note that, under these assumptions, we have $S_p \subseteq S$, where S_p denotes the set of all primes in k sitting above p. Let us fix an isomorphism of (profinite) topological groups $\Gamma \simeq \mathbb{Z}_p$ and denote by γ the topological generator of Γ which is sent to $1 \in \mathbb{Z}_p$ via this isomorphism. Since Γ is (topologically) cyclic, we have group isomorphisms

$$(\mathbb{Z}_p \simeq)\ \Gamma \simeq I(\Gamma)^n/I(\Gamma)^{n+1}, \qquad \sigma \to \widehat{(\sigma - 1)^n}\,, \tag{25}$$

for all $\sigma \in \Gamma$ and all $n \geq 1$. (Check this for finite quotients of Γ and pass to a projective limit.) the isomorphisms above, combined with Lemma 5.2.3 lead to $\mathbb{Z}_p[G]$–linear isomorphisms

$$I_\Gamma^n/I_\Gamma^{n+1} \simeq I(\Gamma)^n/I(\Gamma)^{n+1} \otimes_{\mathbb{Z}} \mathbb{Z}[G] \simeq \mathbb{Z}_p[G]\,, \qquad \text{for all } n \geq 1\,. \tag{26}$$

Since the algebras $\mathbb{Z}_p[[\Gamma]]$ and $\mathbb{Z}_p[[\mathcal{G}]]$ have well understood structures, it would be more convenient to work with $I(\Gamma)_p$ and $I_{\Gamma,p}$ (the analogues of $I(\Gamma)$ and I_Γ in $\mathbb{Z}_p[[\Gamma]]$ and $\mathbb{Z}_p[[\mathcal{G}]]$) instead of $I(\Gamma)$ and I_Γ, respectively. Fortunately, we have $I(\Gamma)_p^n \cap \mathbb{Z}[[\Gamma]] = I(\Gamma)^n$, $I_{\Gamma,p}^n \cap \mathbb{Z}[[\mathcal{G}]] = I_{\Gamma,p}^n$ and isomorphisms

$$I(\Gamma)^n/I(\Gamma)^{n+1} \simeq I(\Gamma)_p^n/I(\Gamma)_p^{n+1}\,, \qquad I_\Gamma^n/I_\Gamma^{n+1} \simeq I_{\Gamma,p}^n/I_{\Gamma,p}^{n+1}\,,$$

for all $n \geq 1$, as it can be easily checked by using (25) and (26) above. So, from now on, we will work with the algebras $\mathbb{Z}_p[[\Gamma]]$ and $\mathbb{Z}_p[[\mathcal{G}]]$ and their corresponding augmentation ideals and view the equivariant L–value $\Theta_{L/k,S,T}(0)$ inside $\mathbb{Z}_p[[\mathcal{G}]]$.

For simplicity, we will assume from this point on that k is totally real, $e^{2\pi i/p} \in K$ (and therefore K is totally complex) and k_{p^∞} and K are linearly disjoint over k. Consequently, Galois restriction leads to the following group and ring isomorphism, respectively.

$$\mathcal{G} \simeq \Gamma \times G\,, \qquad \mathbb{Z}_p[[\mathcal{G}]] \simeq \mathbb{Z}_p[G][[\Gamma]]\,.$$

Now, we have well–known (p–adically continuous) ring isomorphisms

$$\mathbb{Z}_p[[\Gamma]] \simeq \mathbb{Z}_p[[t]]\,, \qquad \mathbb{Z}_p[G][[\Gamma]] \simeq \mathbb{Z}_p[G][[t]]\,,$$

which are $\mathbb{Z}_p$–linear and $\mathbb{Z}_p[G]$–linear, respectively, and map $(\gamma - 1)$ to t. Here, t is a variable. These lead to isomorphisms at the level of augmentation ideals

$$I(\Gamma)_p^n \simeq t^n\mathbb{Z}_p[[t]]\,, \qquad I_{\Gamma,p}^n \simeq t^n\mathbb{Z}_p[G][[t]]\,,$$

for all n. Via the above ring isomorphism, the equivariant L–value $\Theta_{L/k,S,T}(0)$ can be viewed as a power series $\Theta^{(p^\infty)}_{K/k,S,T}(t)$ in the variable t, with coefficients in the group ring $\mathbb{Z}_p[G]$. Consequently,

$$\Theta_{L/k,S,T}(0) \in I_{\Gamma,p}^r \;\Leftrightarrow\; \mathrm{ord}_{t=0}\Theta^{(p^\infty)}_{K/k,S,T}(t) \geq r\,.$$

If the above equivalent statements hold true, then the above established isomorphism $I_{\Gamma,p}^r/I_{\Gamma,p}^{r+1} \simeq \mathbb{Z}_p[G]$ maps

$$\Theta_{L/k,S,T}(0) \mod I_{\Gamma,p}^{r+1} \;\to\; \frac{1}{r!}\cdot\frac{\mathrm{d}^r}{\mathrm{d}t^r}(\Theta^{(p^\infty)}_{K/k,S,T}(t))\mid_{t=0}\,.$$

The regulator $R_{W,\Gamma}$ defined in the previous section gives rise to the p–adic regulator

$$R_{W,p} : \Lambda_{S,T} \xrightarrow{R_{W,\Gamma}} I_{\Gamma,p}^r/I_{\Gamma,p}^{r+1} \simeq t^r\mathbb{Z}_p[G][[t]]/t^{r+1}\mathbb{Z}_p[G][[t]] \simeq \mathbb{Z}_p[G]\,.$$

In light of these considerations, we can rewrite conjecture $A_{\mathbb{Z}}(K_{p^\infty}/K/k,S,T,r)$ as follows.

Conjecture 5.4.1. $A_{\mathbb{Z},p}(K/k,S,T,r)$. *Assume that p is a prime number and $(K_{p^\infty}/K/k,S,T,r)$ satisfy the above hypotheses. Then, there exists a unique $\varepsilon := \varepsilon_{S,T}$ in the Rubin-Stark lattice $\Lambda_{S,T}$, such that the following hold.*

(1) *We have $\Theta_{K/k,S,T}(s) \in s^r\mathbb{C}[G][[s]]$ and an equality in $\mathbb{C}[G]$:*

$$R_W(\varepsilon) = \frac{1}{r!}\cdot\frac{\mathrm{d}^r}{\mathrm{d}s^r}(\Theta_{K/k,S,T}(s))\mid_{s=0}$$

(2) *We have $\Theta^{(p^\infty)}_{K/k,S,T}(t) \in t^r\mathbb{Z}_p[G][[t]]$ and an equality in $\mathbb{Z}_p[G]$:*

$$R_{W,p}(\varepsilon) = \frac{1}{r!}\cdot\frac{\mathrm{d}^r}{\mathrm{d}t^r}(\Theta^{(p^\infty)}_{K/k,S,T}(t))\mid_{t=0}\,.$$

Now, the main point is that the power series $\Theta^{(p^\infty)}_{K/k,S,T}(t)$ can be viewed in fact as a G–equivariant p–adic L–function associated to the data $(K/k,S,T)$. We clarify this connection (essentially due to Iwasawa, Deligne-Ribet, and Pi. Cassou-Noguès) below. For that purpose, we start by fixing an embedding $\mathbb{C} \hookrightarrow \mathbb{C}_p$ (where

$\mathbb{C}_p$ denotes the completion of an algebraic closure $\overline{\mathbb{Q}_p}$ of $\mathbb{Q}_p$.) Via this embedding, we view the characters $\chi \in \widehat{G}(\mathbb{C})$ as taking values in $\mathbb{C}_p$. Let

$$\omega_p, c_p : \mathrm{Gal}(K_{p^\infty}/k) \longrightarrow \mathbb{Z}_p^\times$$

denote the p–adic Teichmüller and cyclotomic characters of $\mathrm{Gal}(K_{p^\infty}/k)$, respectively. Note that, under the assumption $e^{2\pi i/p} \in K$, the character ω_p factors through $G := \mathrm{Gal}(K/k)$. Let

$$u := c_p\omega_p^{-1}(\gamma) \in 1 + p\mathbb{Z}_p.$$

Now, let us split $\Theta^{(p^\infty)}_{K/k,S,T}(t)$ into its character components

$$\Theta^{(p^\infty)}_{K/k,S,T}(t) = \sum_{\chi\in\widehat{G}(\mathbb{C})} f_\chi(t)\cdot e_\chi\,,$$

where the $f_\chi(t) := \chi(\Theta^{(p^\infty)}_{K/k,S,T}(t))$ are power series in $\mathbb{Z}_p(\chi)[[t]]$ which are uniquely determined by and uniquely determine $\Theta^{(p^\infty)}_{K/k,S,T}(t)$. The following theorem, proved by Iwasawa in the case $k = \mathbb{Q}$, gives a remarkable p–adic interpolation property of the power series f_χ.

Theorem 5.4.2 (Deligne-Ribet [**11**], Cassou-Noguès [**7**]). *Under the above assumptions, we have the following equalities*

$$f_\chi(u^{1-n} - 1) = L_{S,T}(\chi^{-1}\omega_p^{1-n}, 1-n)\,,$$

for all $\chi \in \widehat{G}(\mathbb{C})$ *and all* $n \in \mathbb{Z}_{\geq 1}$.

Definition 5.4.3. *For a character* $\chi\in\widehat{G}(\mathbb{C})$*, the* (S,T)*–modified* p*–adic* L*–function* $L_{S,T,p}(\chi, s)$ *of variable* $s \in \mathbb{Z}_p$ *associated to* χ *is defined by*

$$L_{S,T,p}(\chi, s) := f_{\chi^{-1}\omega_p}(u^s - 1)\,.$$

Equivalently (apply the Theorem above), $L_{S,T,p}(\chi, s)$ *is the unique p-adically analytic function defined on* $\mathbb{Z}_p$ *with values in* $\mathbb{C}_p$*, such that*

$$L_{S,T,p}(\chi, 1-n) = L_{S,T}(\chi\omega_p^{-n}, 1-n)\,, \qquad \textit{for all } n \in \mathbb{Z}_{\geq 1}.$$

Next, we note that if $\log_p : 1 + p\mathbb{Z}_p \to \mathbb{Z}_p$ denotes the p–adic logarithm, then

$$\frac{1}{r!}\cdot\frac{\mathrm{d^r}}{\mathrm{dt^r}}(f_\chi(t))\mid_{t=0} = \frac{1}{(\log_p(u))^r}\frac{1}{r!}\cdot\frac{\mathrm{d^r}}{\mathrm{ds^r}}(f_\chi(u^s - 1))\mid_{s=0}\,,$$

for all $\chi \in \widehat{G}(\mathbb{C})$. Consequently, an equivalent, character-by-character formulation of Conjecture $A_{\mathbb{Z},p}(K/k, S, T, r)$ is the following.

Conjecture 5.4.4. $A_{\mathbb{Z},p}(K/k, S, T, r)$. *Assume that* p *is a prime number and* $(K_{p^\infty}/K/k, S, T, r)$ *satisfy the above hypotheses. Then, there exists a unique* $\varepsilon := \varepsilon_{S,T}$ *in the Rubin-Stark lattice* $\Lambda_{S,T}$*, such that the following hold, for all the characters* $\chi \in \widehat{G}(\mathbb{C})$.

(1) *We have* $\mathrm{ord}_{s=0}L_{S,T}(\chi, s) \geq r$ *and an equality in* $\mathbb{C}$*:*

$$\chi^{-1}(R_W(\varepsilon)) = \frac{1}{r!}\cdot\frac{\mathrm{d^r}}{\mathrm{ds^r}}(L_{S,T}(s))\mid_{s=0}$$

(2) *We have* $\mathrm{ord}_{s=0} L_{S,T,p}(\chi, s) \geq r$ *and an equality in* $\mathbb{C}_p$*:*

$$\chi^{-1}\omega_p(R_{W,p}(\varepsilon)) = \frac{1}{(\log_p(u))^r}\frac{1}{r!} \cdot \frac{\mathrm{d^r}}{\mathrm{ds^r}}(L_{S,T,p}(\chi, s)) \mid_{s=0} .$$

Remark 5.4.5. *We conclude this section by noting that the statement of conjecture* $A_{\mathbb{Z},p}(K/k, S, T, r)$ *predicts the existence of a unique* **global**, *arithmetically meaningful element* ε, *whose evaluation against the global regulator* R_W *essentially equals the* r*–th derivative of the* G*–equivariant global* L*–function at* $s = 0$ *and whose evaluation against the* p*–adic regulator* $R_{W,p}$ *essentially equals the* r*–th derivative of the* G*–equivariant* p*–adic* L*–function at* $s = 0$. *Another remarkable prediction is that if the* G*–equivariant global* L*–function has order of vanishing at least* r *at* $s = 0$ *due to the presence of* r *distinct primes in* S *which split completely in* K/k, *then the* G*–equivariant* p*–adic* L*-function has the same property.*

5.5. Sample evidence

In this final section, we discuss some evidence in support of $A_{\mathbb{Z}}(L/K/k, S, T, r)$ and, implicitly, its particular cases $A_{\mathbb{Z}}(K/k, S, T, r)$ (the Rubin-Stark conjecture, which is the case $L = K$) and $A_{\mathbb{Z},p}(K/k, S, T)$ (which is the case $L = K_{p^\infty}$, for some prime p.) We start with classical evidence, in the case $r = 1$, and then move on to very recent evidence, for arbitrary values of r.

The following theorem motivated Gross to formulate his p–adic conjecture (Conjecture 3.13 in [**18**].)

Theorem 5.5.1 (Gross-Koblitz [**21**]). *Assume that* $(K/\mathbb{Q}, S, T, 1)$ *satisfy the above hypotheses,* K *is imaginary and* p *is the distinguished splitting prime in* $K/\mathbb{Q}$. *Then, conjecture* $A_{\mathbb{Z},p}(K/\mathbb{Q}, S, T, 1)$ *holds true.*

First, it turns out that one can reduce the problem to the case $K = \mathbb{Q}(\zeta_m)$, where m is the conductor of $K/\mathbb{Q}$. In this case, the Rubin-Stark (Brumer-Stark) element $\varepsilon_{S,T}$ is more or less a Gauss sum $g(\mathfrak{P})_T$, where $\mathfrak{P}$ is the chosen prime in $\mathbb{Q}(\zeta_m)$ which sits above p (see Theorem 4.3.10 and its Corollary.) In our current notations, Gross and Koblitz [**21**] linked $g(\mathfrak{P})_T$ to a special value of Morita's p–adic Γ–function. In [**12**], Ferrero and Greenberg had already related the values of our regulators $(\chi^{-1}\omega_p) \circ R_{W,p}$ at this special p–adic Γ-value to the first derivatives at $s = 0$ of the p–adic L–functions $L_{S,T,p}(\chi, s)$, for all $\chi \in \widehat{G}(\mathbb{C})$. The proof of the Theorem above is the combination of these two results.

Building upon his explicit, Drinfeld-module theoretic proof of the Brumer-Stark conjecture in function fields (see Theorem 4.3.12 above), Hayes proved the following.

Theorem 5.5.2 (Hayes [**23**]). *If* k *is a function field, then* $A_{\mathbb{Z}}(L/K/k, S, T, 1)$ *holds true.*

More recent techniques, mostly stemming from efforts to prove the Equivariant Tamagawa Number Conjecture (ETNC) for Dirichlet motives and various Equivariant Main Conjectures in Iwasawa theory, have led to the following results.

Theorem 5.5.3 (Burns [**2**]). *Conjecture* $A_{\mathbb{Z}}(L/K/\mathbb{Q}, S, T, r)$ *holds true.*

In [**2**], Burns shows that the ETNC for abelian extensions of a global field k implies conjecture $A_{\mathbb{Z}}(L/K/k, S, T, r)$, as long as the data $(L/K/k, S, T, r)$ satisfies the above hypotheses. (See Burns's contribution to this volume as well.) In [**5**], Burns and Greither proved the ETNC for abelian extensions of $\mathbb{Q}$, away from its 2–primary part. Its 2–primary part was proved by Flach later (see [**14**]; [**13**] in [**6**] contains an incomplete proof.) As a consequence, the theorem above follows.

Theorem 5.5.4 (Greither-Popescu [**15**]). *If k and K are a totally real and a CM number field, respectively, p is an odd prime and the Iwasawa μ–invariant $\mu_{K,p}$ associated to K and p vanishes, then conjecture $A_{\mathbb{Z},p}(K/k, S, T, r)$ holds true.*

This result is a consequence of our proof of an Equivariant Main Conjecture in Iwasawa theory (mentioned in the paragraph below Theorem 4.3.9 as well.) Actually, we prove the p–primary part of the more general conjecture $A_{\mathbb{Z}}(L/K/k, S, T, r)$, for all odd primes p, under the additional hypotheses that $S_p \subseteq S$. This hypothesis is automatically satisfied if $L = K_{p^\infty}$, which leads to the Theorem above.

Theorem 5.5.5 (Burns [**2**], Greither-Popescu [**15**]). *Assume that k is a function field. Then, conjecture $A_{\mathbb{Z}}(L/K/k, S, T, r)$ holds true.*

Burns's proof is based on the fact that the ETNC for abelian extensions of k implies conjecture $A_{\mathbb{Z}}(L/K/k, S, T, r)$ and on his proof of the ETNC for abelian extensions of function fields (see [**2**] and the references therein, as well as Burns's contribution to this volume.)

The Greither-Popescu proof is a consequence of Theorem 4.3.14(1) above (the Equivariant Main Conjecture in function fields.) Of course, as mentioned before, the ETNC for function fields is a consequence of our Theorem 4.3.14(1) as well, but one can derive the Theorem above directly, without invoking the ETNC.

Bibliography

[1] Spencer Bloch and Kazuya Kato. L-functions and Tamagawa numbers of motives. In *The Grothendieck Festschrift, Vol. I*, volume 86 of *Progr. Math.*, pages 333–400. Birkhäuser Boston, Boston, MA, 1990.

[2] David Burns. Congruences between derivatives of abelian L-functions at $s = 0$. *Invent. Math.*, 169(3):451–499, 2007.

[3] David Burns and Matthias Flach. Tamagawa numbers for motives with (non-commutative) coefficients. *Doc. Math.*, 6:501–570 (electronic), 2001.

[4] David Burns and Matthias Flach. Tamagawa numbers for motives with (noncommutative) coefficients. II. *Amer. J. Math.*, 125(3):475–512, 2003.

[5] David Burns and Cornelius Greither. On the equivariant Tamagawa number conjecture for Tate motives. *Invent. Math.*, 153(2):303–359, 2003.

[6] David Burns, Cristian D. Popescu, Jonathan Sands, and David Solomon, editors. *Stark's conjectures: recent work and new directions*, volume 358 of *Contemporary Mathematics*. American Mathematical Society, Providence, RI, 2004. Papers from the International Conference on Stark's Conjectures and Related Topics held at Johns Hopkins University, Baltimore, MD, August 5–9, 2002.

[7] Pierrette Cassou-Noguès. Valeurs aux entiers négatifs des fonctions zêta et fonctions zêta p-adiques. *Invent. Math.*, 51(1):29–59, 1979.

[8] Ted Chinburg. On the Galois structure of algebraic integers and S–units. *Invent. Math.*, 74(321–349), 1983.

[9] Ted Chinburg. Exact sequences and Galois module structure. *Ann. of Math.*, 121(351–376), 1985.

[10] John Coates. p-adic L-functions and Iwasawa's theory. In *Algebraic number fields: L-functions and Galois properties (Proc. Sympos., Univ. Durham, Durham, 1975)*, pages 269–353. Academic Press, London, 1977.

[11] Pierre Deligne and Kenneth A. Ribet. Values of abelian L-functions at negative integers over totally real fields. *Invent. Math.*, 59(3):227–286, 1980.

[12] Bruce Ferrero and Ralph Greenberg. On the behavior of p-adic L-functions at $s = 0$. *Invent. Math.*, 50(1):91–102, 1978/79.

[13] Matthias Flach. The equivariant Tamagawa number conjecture: a survey. In *Stark's conjectures: recent work and new directions*, volume 358 of *Contemp. Math.*, pages 79–125. Amer. Math. Soc., Providence, RI, 2004. With an appendix by C. Greither.

[14] Matthias Flach. On the cyclotomic main conjecture for the prime 2. *to appear in J. Reine Angew. Math.*, 2011.

[15] Cornelius Greither and Cristian D. Popescu. Explicit ℓ–adic Tate sequences and applications. *in progress*, 2010.

[16] Cornelius Greither and Cristian D. Popescu. An equivariant Main Conjecture in Iwasawa theory and applications. *submitted for publication*, 2011.

[17] Cornelius Greither and Cristian D. Popescu. The Galois module structure of ℓ–adic realizations of Picard 1–motives and applications. *to appear in Intl. Math. Res. Notices*, 2011.

[18] Benedict H. Gross. p-adic L-series at $s = 0$. *J. Fac. Sci. Univ. Tokyo Sect. IA Math.*, 28(3):979–994 (1982), 1981.

[19] Benedict H. Gross. On the values of abelian L-functions at $s = 0$. *J. Fac. Sci. Univ. Tokyo Sect. IA Math.*, 35(1):177–197, 1988.

[20] Benedict H. Gross. Letter to John Tate. 1993.

[21] Benedict H. Gross and Neal Koblitz. Gauss sums and the p-adic Γ-function. *Ann. of Math. (2)*, 109(3):569–581, 1979.

[22] David R. Hayes. Stickelberger elements in function fields. *Compositio Math.*, 55(2):209–239, 1985.

[23] David R. Hayes. The refined $\mathfrak{p}$-adic abelian Stark conjecture in function fields. *Invent. Math.*, 94(3):505–527, 1988.

[24] Finn Faye Knudsen and David Mumford. The projectivity of the moduli space of stable curves. I. Preliminaries on "det" and "Div". *Math. Scand.*, 39(1):19–55, 1976.

[25] Jürgen Neukirch. *Algebraic number theory*, volume 322 of *Grundlehren der Mathematischen Wissenschaften [Fundamental Principles of Mathematical Sciences]*. Springer-Verlag, Berlin, 1999. Translated from the 1992 German original and with a note by Norbert Schappacher, With a foreword by G. Harder.

[26] Cristian D. Popescu. Gras–type conjectures for function fields. *Comp. Math.*, 118(3):263–290, 1999.

[27] Cristian D. Popescu. On a refined Stark conjecture for function fields. *Comp. Math.*, 116:321–367, 1999.

[28] Cristian D. Popescu. Base change for stark–type conjectures "over $\mathbb{Z}$". *J. Reine Angew. Math.*, 542:85–111, 2002.

[29] Cristian D. Popescu. Stark's Question and a strong form of Brumer's Conjecture. *Compos. Math.*, 140:631–646, 2004.

[30] Cristian D. Popescu. Stark's Question on special values of L–functions. *Math. Res. Letters*, 14(3):531–545, 2007.

[31] Karl Rubin. Stark units and Kolyvagin's "Euler systems". *J. Reine Angew. Math.*, 425:141–154, 1992.

[32] Karl Rubin. A Stark conjecture "over $\mathbf{Z}$" for abelian L-functions with multiple zeros. *Ann. Inst. Fourier (Grenoble)*, 46(1):33–62, 1996.

[33] Carl Ludwig Siegel. Über die Fourierschen Koeffizienten von Modulformen. *Nachr. Akad. Wiss. Göttingen Math.-Phys. Kl. II*, 1970:15–56, 1970.

[34] Harold M. Stark. Values of L-functions at $s = 1$. I. L-functions for quadratic forms. *Advances in Math.*, 7:301–343 (1971), 1971.

[35] Harold M. Stark. L-functions at $s = 1$. II. Artin L-functions with rational characters. *Advances in Math.*, 17(1):60–92, 1975.

[36] Harold M. Stark. L-functions at $s = 1$. III. Totally real fields and Hilbert's twelfth problem. *Advances in Math.*, 22(1):64–84, 1976.

[37] Harold M. Stark. L functions at $s = 1$. IV. First derivatives at $s = 0$. *Adv. in Math.*, 35(3):197–235, 1980.

[38] Ludwig Stickelberger. Über eine Verallgemeinerung der Kreistheilung. *Math. Annalen*, 37:321–367, 1890.

[39] Richard G. Swan. *K-theory of finite groups and orders*. Lecture Notes in Mathematics, Vol. 149. Springer-Verlag, Berlin, 1970.

[40] Ki-Seng Tan. A note on Stickelberger elements for cyclic p-extensions over global function fields of characteristic p. *Math. Res. Lett.*, 11(2-3):273–278, 2004.

[41] John Tate. The cohomology groups of tori in finite Galois extensions of number fields. *Nagoya Math. J.*, 27:709–719, 1966.

[42] John Tate. *Les conjectures de Stark sur les fonctions L d'Artin en $s = 0$*, volume 47 of *Progress in Mathematics*. Birkhäuser Boston Inc., Boston, MA, 1984. Lecture notes written up by Dominique Bernardi and Norbert Schappacher.

[43] Lawrence C. Washington. *Introduction to cyclotomic fields*, volume 83 of *Graduate Texts in Mathematics*. Springer-Verlag, New York, second edition, 1997.

[44] Andrew Wiles. The Iwasawa conjecture for totally real fields. *Ann. of Math. (2)*, 131(3):493–540, 1990.

Special Values of L-functions at Negative Integers

Manfred Kolster

IAS/Park City Mathematics Series
Volume 18, 2009

Special Values of L-functions at Negative Integers

Manfred Kolster

Introduction

Let χ be a 1-dimensional Artin character over a number field F. The values of the Artin L-function $L(\chi, s)$ at negative integers $1-n$ for $n \geq 2$ are trivial unless F is totally real and χ has parity $(-1)^n$, i.e. the field $F_\chi := \bar{F}^{ker\,\chi}$ is totally real for even n and a CM-field for odd n. In these cases the values are non-zero algebraic numbers contained in $\mathbb{Q}(\chi)$, the field obtained by adjoining to $\mathbb{Q}$ the values of χ.

In these lectures we discuss the arithmetic meaning of these values. The approach is via Iwasawa theory, p-adic L-functions and the Main Conjecture in Iwasawa theory (proved by Wiles), which provides a p-adic interpretation of the values for each prime p. In the case of the trivial character we will describe the relation to the Birch-Tate Conjecture (the case where F is real and $n = 2$) as well as to the more general Lichtenbaum Conjectures. We will mostly ignore the prime 2, which causes technical problems, and for which the results are less complete. However, the 2-primary information is important for the formulation of some of the conjectures.

The arithmetic interpretations for a fixed prime p are in terms of étale cohomology groups attached to the ring $o'_F = o_F[1/p]$ of p-integers of F. We will discuss two "global" interpretations in terms of algebraic K-groups and in terms of motivic cohomology groups, which in some cases differ by non-trivial powers of 2. The known results for $p = 2$ suggest that in general motivic cohomology contains the "correct" number-theoretic information.

In the final section we will discuss a conjecture of Coates-Sinnott – the analog of Stickelberger's Theorem – about annihilation of higher algebraic K-theory groups in relative abelian extensions. The conjecture can be approached prime by prime, and we sketch the proof of the cohomological version of the p-part of this conjecture for those odd primes p, which do not divide the order of the Galois group of the relative abelian extension (the "semi-simple case").

Department of Mathematics and Statistics, McMaster University, Hamilton ON L8S 4K1, Canada
E-mail address: kolster@mcmaster.ca

LECTURE 1

Iwasawa theory and cohomology

1.1. The classical Main Conjecture

Let F be a number field and let p be a prime number. A Galois extension F_∞/F is called a $\mathbb{Z}_p$-*extension*, if $\Gamma := Gal(F_\infty/F) \cong \mathbb{Z}_p$. Since the closed subgroups of $\mathbb{Z}_p$ are of the form 0 or $p^n\mathbb{Z}_p$, we have for each $n \geq 0$ a unique subfield F_n of degree p^n over F and $Gal(F_n/F) \cong \mathbb{Z}/p^n\mathbb{Z}$. Hence we obtain a tower

$$F = F_0 \subset F_1 \subset F_2 \subset \cdots \subset F_\infty,$$

such that $[F_n : F] = p^n$ and $F_\infty = \cup_{n\geq 0} F_n$.

A typical example of a $\mathbb{Z}_p$-extension F_∞/F is the so-called *cyclotomic* $\mathbb{Z}_p$-*extension*, which is constructed as follows: Let $L_\infty = F(\mu_{p^\infty})$ be obtained from F by adjoining the group μ_{p^∞} of all p-power roots of unity. Then $Gal(L_\infty/F) \cong \mathbb{Z}_p \times \Delta$, where Δ is finite. Now take $F_\infty = L_\infty^\Delta$.

Let γ denote a topological generator of Γ, and let $\Gamma_n = Gal(F_n/F)$. Passing to the inverse limit over the group rings $\mathbb{Z}_p[\Gamma_n]$ we obtain the *Iwasawa-algebra* $\mathbb{Z}_p[[\Gamma]] := \varprojlim \mathbb{Z}_p[\Gamma_n]$. The group rings $\mathbb{Z}_p[\Gamma_n]$ are generally quite complicated, but the Iwasawa-algebra has a rather simple structure. Namely, it is isomorphic to the power series ring $\Lambda := \mathbb{Z}_p[[T]]$, the isomorphism being induced by $\gamma \mapsto 1 + T$.

In the following we have to allow slightly more general coefficients: Let $\mathcal{O}$ denote a finite extension of $\mathbb{Z}_p$, let π be a uniformizer for $\mathcal{O}$, let v denote the discrete valuation on $\mathcal{O}$, normalized so that $v(\pi) = 1$, and let $|\ |_v$ denote the corresponding absolute value with $|a|_v = p^{-f\cdot v(a)}$, where f denotes the residue degree.

We now consider $\Lambda := \mathcal{O}[[T]] \cong \mathcal{O}[[\Gamma]]$. This is a two-dimensional Noetherian local Krull domain, and the structure of finitely generated Λ-modules is known up to pseudo-isomorphism (cf. [**6**], Chapter VII,4, Theorem 5). If M and N are finitely generated Λ-modules, then we write $M \sim N$ if there exists a pseudo-isomorphism $f : M \to N$, *i.e.*, a module homomorphism with finite kernel and cokernel. Now, the structure theorem for finitely generated Λ-modules says that for every finitely generated Λ-module M there is a pseudo-isomorphism

$$M \sim \Lambda^r \oplus \bigoplus_{i=1}^m \Lambda/\mathfrak{p}_i^{n_i}.$$

Here $\mathfrak{p}_i$ are height 1 prime ideals of Λ, hence they are either equal to (π) or to $(F(T))$, where $F(T)$ is an irreducible distinguished polynomial, *i.e.*, , of the form

$$F(T) = T^n + b_{n-1}T^{n-1} + \cdots + b_0$$

with $\pi | b_i$ for all i. The prime ideals $\mathfrak{p}_i$ and the integers $r \geq 0, m \geq 0$ and $n_i \geq 1$ are uniquely determined by M. The ideal $\prod_{i=1}^m \mathfrak{p}_i^{n_i}$ is the *characteristic ideal* of M, which has a unique generator of the form

$$f(T) = \pi^{\mu} \cdot f^*(T)$$

where $f^*(T)$ is a distinguished polynomial. $f^*(T)$ is the *characteristic polynomial* of M. The exponent μ is the *μ-invariant of M* and $\lambda := deg\, f^*(T)$ is called the *λ-invariant of M.*

The characteristic polynomial is in fact a characteristic polynomial in the sense of linear algebra: Let $\overline{\mathbb{Q}_p}$ denote an algebraic closure of $\mathbb{Q}_p$, and let $V = M \otimes_{\mathcal{O}} \overline{\mathbb{Q}_p}$. This is a $\overline{\mathbb{Q}_p}$-vectorspace of rank λ and $f^*(T)$ is the characteristic polynomial of the endomorphism $\gamma - 1$ acting on V.

The following result is extremely useful: Assume that M is a finitely generated Λ-torsion module with characteristic polynomial $f^*(T)$. Let μ denote the μ-invariant of M and let $f(T) = \pi^{\mu} \cdot f^*(T)$. We denote by M^{Γ} the invariants of M under Γ and by $M_{\Gamma} = M/(\gamma - 1)M$ the coinvariants of M.

Lemma 1.1.1 (cf. [**27**]). *The following statements are equivalent :*

(a) M^{Γ} *is finite*
(b) M_{Γ} *is finite*
(c) $f(0) \neq 0$.

If these conditions are satisfied, then

$$\frac{|M^{\Gamma}|}{|M_{\Gamma}|} = |f(0)|_v.$$

Let us assume now that F_∞/F is the cyclotomic $\mathbb{Z}_p$-extension. Let $G_\infty = Gal(F(\mu_{p^\infty})/F) \cong \Gamma \times \Delta$, where $\Delta \cong Gal(F(\zeta_{2p})/F)$. G_∞ acts on μ_{p^∞} and this action gives rise to the *cyclotomic character*

$$\rho : G_\infty \to \mathbb{Z}_p^*$$

defined by

$$\zeta^\sigma = \zeta^{\rho(\sigma)}$$

for all $\sigma \in G_\infty$ and all $\zeta \in \mu_{p^\infty}$. We denote by κ the restriction of ρ to Γ and by ω the restriction of ρ to Δ. ω is the *Teichmüller character.*

Let M be a $\mathbb{Z}_p$-module with a G_∞-action, denoted by $m \mapsto m^\sigma$. For $n \in \mathbb{Z}$ the n-th Tate twist $M(n)$ of M is defined as the $\mathbb{Z}_p$-module M with the new G_∞-action

$$m \mapsto \rho(\sigma)^n \cdot m^\sigma.$$

In particular, $\mathbb{Z}_p(1) \cong \varprojlim \mu_{p^n} =: \mathcal{T}$, which is the so-called *Tate-module*, and $\mathbb{Q}_p/\mathbb{Z}_p(1) \cong \mu_{p^\infty}$. In general: $M(n) \cong M \otimes_{\mathbb{Z}_p} \mathbb{Z}_p(n)$. If M and N are two $\mathbb{Z}_p$-modules with a G_∞-action, then we turn $Hom_{\mathbb{Z}_p}(M, N)$ into a G_∞-module in the following way: For $f \in Hom_{\mathbb{Z}_p}(M, N)$ and $\sigma \in G_\infty$ we define f^σ via

$$f^\sigma(m) = (f(m^{\sigma^{-1}}))^\sigma.$$

It is easy to see that with this definition of the G_∞-action on Hom-groups we obtain canonical isomorphisms for all $n \in \mathbb{Z}$:

$$Hom_{\mathbb{Z}_p}(M(n), \mathbb{Q}_p/\mathbb{Z}_p) \cong Hom_{\mathbb{Z}_p}(M, \mathbb{Q}_p/\mathbb{Z}_p(-n)) \cong Hom_{\mathbb{Z}_p}(M, \mathbb{Q}_p/\mathbb{Z}_p)(-n).$$

We note the following:

Lemma 1.1.2 (cf. [**27**]). *Assume that M is a Λ-torsion module with characteristic polynomial $f^*(T)$. Then the characteristic polynomial of $M(n)$ is given by*

$$f^*(\kappa(\gamma)^{-n}(1+T)-1).$$

The most interesting Λ-modules arise as Galois groups of certain abelian pro-p extensions of F_∞, where F_∞/F is an arbitrary $\mathbb{Z}_p$-extension of a number field F. Assume then that K_∞ is an abelian pro-p extension of F_∞, let $X = Gal(K_\infty/F_\infty)$, and assume that K_∞/F is again a Galois extension (although not necessarily abelian). Let $G = Gal(K_\infty/F)$. We obtain an extension of $\mathbb{Z}_p$-modules

$$0 \to X \to G \to \Gamma \to 0.$$

Since X is abelian, Γ acts on X by inner automorphisms, and this action turns X into a compact Λ-module. As examples we can take for K_∞ the maximal abelian unramified pro-p extension of F_∞, usually denoted by L_∞, or the maximal subextension of L_∞, in which all p-adic primes of F_∞ split completely, usually denoted by L'_∞. The corresponding Galois groups $X_\infty := Gal(L_\infty/F_\infty)$ and $X'_\infty := Gal(L'_\infty/F_\infty)$ are examples of finitely generated Λ - torsion modules.

The main example in the current framework is the following: Let S be a finite set of primes in F containing the primes above p and the infinite primes. S_p will denote the minimal such set, i.e. the set consisting exactly of the primes above p and the infinite primes. Let M_∞^S denote the maximal abelian pro-p-extension of F_∞, which is unramified outside primes in S, and let $\mathfrak{X}^S = Gal\,(M_\infty^S/F_\infty)$. This is a finitely generated Λ-module, which we will call the *standard Iwasawa module* over F_∞ for the set S. Let us again specialize to the case of the cyclotomic $\mathbb{Z}_p$-extension. Iwasawa([**22**]) has shown that in this case $\mathfrak{X}^S$ has no non-trivial finite Λ-submodules and that the Λ-rank of $\mathfrak{X}^S$ is equal to the number r_2 of different pairs of complex conjugate embeddings of F. In particular, $\mathfrak{X}^S$ is a Λ-torsion module if and only if F is totally real.

From now on F will be a fixed totally real number field and F_∞ will denote the cyclotomic $\mathbb{Z}_p$-extension with p being an odd prime number. We note that under Leopoldt's Conjecture the cyclotomic $\mathbb{Z}_p$-extension is the only $\mathbb{Z}_p$-extension of a totally real number field.

We consider now a 1-dimensional p-adic valued Artin character ψ over F of finite order:

$$\psi : Gal(\overline{F}/F) \to \overline{\mathbb{Q}_p}^\times,$$

and we denote by F_ψ the fixed field of the kernel of ψ, so that ψ is a faithful character on $Gal(F_\psi/F)$. We assume that ψ is even, i.e. that F_ψ is again a totally real number field. We recall Greenberg's terminology (cf. [**17**]) about the different *types* of the characters ψ: ψ is of *type S*, if

$$F_\psi \cap F_\infty = F,$$

and ψ is of *type W*, if

$$F_\psi \subset F_\infty.$$

We note that the trivial character is the only character, which is both of type S and of type W.

We fix an embedding $\overline{\mathbb{Q}_p} \to \mathbb{C}$, which allows us to view ψ as a complex-valued character as well.

Deligne-Ribet ([**14**]) have shown that there exists a p-adic L-function $L_p(s,\psi)$, which interpolates the special values of certain Artin L-functions in the following way: For all integers $n \geq 1$

$$L_p(1-n,\psi) = L(1-n,\psi\omega^{-n}) \cdot \prod_{\mathfrak{p}|p}(1-\psi\omega^{-n}(\mathfrak{p})N(\mathfrak{p})^{n-1}).$$

These values uniquely determine the p-adic L-function.

Now let S be again a finite set of primes in F containing S_p. One defines an "imprimitive" p-adic L-function $L_p^S(s,\psi)$ by specifying the values at $s = 1-n$ for all $n \geq 1$ as above via

$$L_p^S(1-n,\psi) = L(1-n,\psi\omega^{-n}) \cdot \prod_{\mathfrak{p}\in S}(1-\psi\omega^{-n}(\mathfrak{p})N(\mathfrak{p})^{n-1}).$$

Removing the Euler factors for primes in S from the Artin L-functions we obtain the L-functions $L^S(s,\psi\omega^{-n})$ and the relation between the imprimitive L-functions at $1-n$ is then simply expressed as (cf. [**16**], section 3):

$$L_p^S(1-n,\psi) = L^S(1-n,\psi\omega^{-n})$$

for all $n \geq 1$.

We define

$$H_\psi(T) = \begin{cases} \psi(\gamma)(1+T)-1 & \text{if } \psi \text{ is of type W} \\ 1 & \text{otherwise} \end{cases},$$

and we denote the extension of $\mathbb{Z}_p$ obtained by adjoining the values $\psi(g)$, $g \in Gal(F_\psi/F)$ by $\mathcal{O}_\psi = \mathbb{Z}_p[\psi]$. It was shown by Deligne-Ribet ([**14**]) that there exists a power series $G_{\psi,S}(T) \in \mathcal{O}_\psi[[T]]$, so that

$$L_p^S(1-s,\psi) = \frac{G_{\psi,S}(\kappa(\gamma)^s-1)}{H_\psi(\kappa(\gamma)^s-1)}.$$

The power series $G_{\psi,S}(T)$ can be written uniquely as

$$G_{\psi,S}(T) = \pi^{\mu(G_{\psi,S})} \cdot g_{\psi,S}^*(T) \cdot u_{\psi,S}(T),$$

where $g_{\psi,S}^*(T)$ is a distinguished polynomial, $u_{\psi,S}(T)$ is a unit power series and π is a uniformizer in $\mathcal{O}_\psi$.

The classical Main Conjecture in Iwasawa Theory, proven by Wiles in [**38**] for odd p (and also for $p=2$, if $F=\mathbb{Q}$) relates the polynomial $g_{\psi,S}^*(T)$ to the following characteristic polynomial: Let $F_{\psi,\infty}$ denote the cyclotomic $\mathbb{Z}_p$-extension of F_ψ, and let $\mathfrak{X}^S$ denote the standard Iwasawa module over $F_{\psi,\infty}$ for the set S. The Galois group $G = Gal(F_\psi/F)$ acts on the finite-dimensional vectorspace

$$V = \mathfrak{X}^S \otimes_{\mathbb{Z}_p} \overline{\mathbb{Q}}_p,$$

and we denote by V^ψ the eigenspace of V corresponding to the action of G via ψ. Now let $f_{\psi,S}^*(T)$ denote the characteristic polynomial of $\gamma-1$ acting on V^ψ.

Iwasawa's Main Conjecture 1.1.3 (Wiles). *Let F be a totally real number field, let p be an odd prime, and let ψ be a 1-dimensional p-adic Artin character over F of type S. Then for any finite set of primes S of F containing S_p:*

$$g^*_{\psi,S}(T) = f^*_{\psi,S}(T).$$

It is important to note that the characteristic polynomial $f^*_{\psi,S}(T)$ does not change if we take instead of F_ψ any finite abelian extension E of F containing F_ψ with $E \cap F_\infty = F$ and then consider the standard Iwasawa module over E_∞. (cf. [**17**], Proposition 1).

1.2. Cohomology

We are going to use the Main Conjecture to relate special values of Artin L-functions at negative integers to orders of étale cohomology groups. For our purposes it suffices to use a description of étale cohomology in terms of Galois cohomology: Fix an arbitrary prime p and an arbitrary number field F. Let $\Omega_F^{(p)}$ denote the maximal algebraic extension of F, which is unramified outside primes above p and infinite primes, and let $G_F^{(p)} = Gal(\Omega_F^{(p)}/F)$. The étale cohomology groups $H^*_{\text{ét}}(\operatorname{Spec} o_F[\frac{1}{p}], \mu_{p^m}^{\otimes n})$ of the scheme $\operatorname{Spec} o_F[\frac{1}{p}]$ with values in the étale sheaf $\mu_{p^m}^{\otimes n}$ as defined by Grothendieck (cf. e.g. [**29**]) can be identified with the Galois cohomology groups $H^*(G_F^{(p)}, \mu_{p^m}^{\otimes n})$. To simplify notations we will write $H^*_{\text{ét}}(o'_F, \mathbb{Z}/p^m(n))$, where $o'_F = o_F[\frac{1}{p}]$. Similarly, if S is a finite set of primes of F containing S_p, then we obtain the étale cohomology groups $H^*_{\text{ét}}(o_F^S, \mathbb{Z}/p^m(n))$ as Galois cohomology groups, where we replace the extension $\Omega_F^{(p)}$ by the maximal algebraic S-ramified extension Ω_F^S of F.

Passing to the inverse or the direct limit with respect to m of the cohomology groups $H^*_{\text{ét}}(o'_F, \mu_{p^m}^{\otimes n})$ we obtain the p-adic cohomology groups

$$H^*_{\text{ét}}(o'_F, \mathbb{Z}_p(n)) := \varprojlim H^*_{\text{ét}}(o'_F, \mu_{p^m}^{\otimes n}),$$

and the groups

$$H^*_{\text{ét}}(o'_F, \mathbb{Q}_p/\mathbb{Z}_p(n)) = \varinjlim H^*_{\text{ét}}(o'_F, \mu_{p^m}^{\otimes n}).$$

.

We note the following: For each $n \in \mathbb{Z}$ the exact sequence

$$0 \to \mathbb{Z}_p(n) \to \mathbb{Q}_p(n) \to \mathbb{Q}_p(n)/\mathbb{Z}_p(n) \to 0$$

gives rise to a long exact sequence in étale cohomology and the kernels and cokernels of the boundary maps

$$\delta_i : H^{i-1}_{\text{ét}}(o'_F, \mathbb{Q}_p/\mathbb{Z}_p(n)) \to H^i_{\text{ét}}(o'_F, \mathbb{Z}_p(n)) \ \ (i \geq 1)$$

can be described as follows (cf. [**35**]):

The kernel of δ_i is the maximal divisible subgroup of $H^{i-1}_{\text{ét}}(o'_F, \mathbb{Q}_p/\mathbb{Z}_p(n))$ and the image of δ_i is the torsion subgroup of $H^i_{\text{ét}}(o'_F, \mathbb{Z}_p(n))$. In particular this implies that the torsion subgroup of $H^1_{\text{ét}}(o'_F, \mathbb{Z}_p(n))$ is isomorphic to $H^0_{\text{ét}}(o'_F, \mathbb{Q}_p/\mathbb{Z}_p(n)) = H^0(F, \mathbb{Q}_p/\mathbb{Z}_p(n))$:

$$H^1_{\text{ét}}(o'_F, \mathbb{Z}_p(n))_{\text{tors}} \cong H^0_{\text{ét}}(o'_F, \mathbb{Q}_p/\mathbb{Z}_p(n)).$$

In the following proposition we summarize some known results about the finitely generated p-adic étale cohomology groups for rings of integers. We only list the results for odd primes p and integers $n \geq 2$.

Proposition 1.2.1. *Let p be an odd prime and let $n \geq 2$. Then*

1. $H^0_{ét}(o'_F, \mathbb{Z}_p(n)) = 0$.
2. $H^k_{ét}(o'_F, \mathbb{Z}_p(n)) = 0$ *for* $k \geq 3$.
3. *There are isomorphisms*
$$H^1_{ét}(o'_F, \mathbb{Z}_p(n)) \cong H^1_{ét}(F, \mathbb{Z}_p(n)).$$
4. *The groups $H^2_{ét}(o'_F, \mathbb{Z}_p(n))$ are finite and trivial for almost all primes p.*
5. *The groups $H^1_{ét}(o'_F, \mathbb{Z}_p(n))$ are finitely-generated $\mathbb{Z}_p$-modules and*
$$rk_{\mathbb{Z}_p} H^1_{ét}(F, \mathbb{Z}_p(n)) = \begin{cases} r_1 + r_2 & \text{if } n \text{ is odd} \\ r_2 & \text{if } n \text{ is even.} \end{cases}$$

Except for property 2 the statements in Proposition 1.2.1 are also true for $p = 2$. If F has real places, then property 2 is false for $p = 2$, which causes technical problems in this case.

Property 5 implies that for $n \geq 2$ the étale cohomology groups $H^1_{ét}(F, \mathbb{Z}_p(n))$ are finite precisely when F is totally real and n is even. If this is the case, then the boundary map δ_2 is an isomorphism:

$$H^1_{ét}(o'_F, \mathbb{Q}_p/\mathbb{Z}_p(n)) \cong H^2_{ét}(o'_F, \mathbb{Z}_p(n)).$$

Let us assume now that $n > 1$ is odd and that E is a CM-field with maximal real subfield E^+. Then by property 5 $H^1_{ét}(E, \mathbb{Z}_p(n))$ has the same $\mathbb{Z}_p$-rank as $H^1_{ét}(E^+, \mathbb{Z}_p(n))$. Since p is odd $H^1_{ét}(E, \mathbb{Z}_p(n))$ splits into eigenspaces

$$H^1_{ét}(E, \mathbb{Z}_p(n)) = H^1_{ét}(E, \mathbb{Z}_p(n))^+ \oplus H^1_{ét}(E, \mathbb{Z}_p(n))^-$$

under complex conjugation with $H^1_{ét}(E, \mathbb{Z}_p(n))^+ \cong H^1_{ét}(E^+, \mathbb{Z}_p(n))$. Therefore for odd $n > 1$ $H^1_{ét}(E, \mathbb{Z}_p(n))^-$ is finite and hence

$$H^1_{ét}(o'_E, \mathbb{Q}_p/\mathbb{Z}_p(n))^- \cong H^2_{ét}(o'_E, \mathbb{Z}_p(n))^-.$$

We obtain:

Corollary 1.2.2. *a) If F is totally real and $n \geq 2$ is even, then*

$$H^1_{ét}(o'_F, \mathbb{Q}_p/\mathbb{Z}_p(n)) \cong H^2_{ét}(o'_F, \mathbb{Z}_p(n))$$

b) If E is CM, and $n > 1$ is odd, then

$$H^1_{ét}(o'_E, \mathbb{Q}_p/\mathbb{Z}_p(n))^- \cong H^2_{ét}(o'_E, \mathbb{Z}_p(n))^-.$$

There are two global "cohomology theories", closely related to étale cohomology: Algebraic K-theory and motivic cohomology. The precise relationship depends on the validity of the Bloch-Kato Conjecture, which appears to have been proven by Rost and Voevodsky – at least all the details are now either published or submitted for publication. If we assume the Bloch-Kato Conjecture, then the picture is the following – the 2-primary information here is unconditional. We refer to [**24**] for more details and further references:

For $i = 1, 2$ and all integers $n \geq 2$ there are isomorphisms

$$K_{2n-i}(o_F) \cong H^i_{\mathcal{M}}(o_F, \mathbb{Z}(n))$$

up to (known) 2-torsion, and for all p there are isomorphisms

$$H^i_{\mathcal{M}}(o_F, \mathbb{Z}(n)) \otimes \mathbb{Z}_p \cong H^i_{\text{ét}}(o'_F, \mathbb{Z}_p(n)).$$

Here the K-groups are Quillen's K-groups, and the motivic cohomology groups can e.g. be defined as Bloch's higher Chow groups:

$$H^i_{\mathcal{M}}(o_F, \mathbb{Z}(n)) := CH^n(o_F, \mathbb{Z}(2n - i)).$$

If we do not want to assume the Bloch-Kato Conjecture, then we can still find global cohomological "models" $H^i(o_F, \mathbb{Z}(n))$, $i = 1, 2$, for the étale cohomology groups. For $i = 2$ this is easy. We simply define

$$H^2(o_F, \mathbb{Z}(n)) = \prod_p H^2_{\text{ét}}(o_F[\frac{1}{p}], \mathbb{Z}_p(n)),$$

which are finite groups and play in many situations a role similar to that of the class group.

For $i = 1$ the construction is more involved (cf. [**9**]). The resulting group $H^1(o_F, \mathbb{Z}(n))$ is an analog of the group of units. It is a finitely generated abelian group of rank

$$\begin{array}{ll} r_2 & \text{if } n \geq 3 \text{ is odd} \\ r_1 + r_2 & \text{if } n \geq 2 \text{ is even,} \end{array}$$

which satisfies

$$H^i(o_F, \mathbb{Z}(n)) \otimes \mathbb{Z}_p \cong H^i_{\text{ét}}(o'_F, \mathbb{Z}_p(n))$$

for all primes p.

In any case it is important to note that for certain indices there is a known difference between the 2-primary parts of the K-groups and the cohomology groups, which has an impact on the formulation of the conjectures we are going to discuss.

Finally in this section we want to discuss Galois actions on étale cohomology groups. Let E/F is a finite Galois extension of number fields with Galois group G. Let p be an odd prime and let S denote a finite set of primes of F containing all primes above p as well as all primes which ramify in F, so that the extension E/F is S-ramified. Using the properties of the étale cohomology groups the following results about Galois descent and co-descent follow easily from the Hochschild-Serre and the Tate spectral sequences:

Proposition 1.2.3. *Let E/F be a Galois extension of number fields with Galois group G. Let p be an odd prime and let S be a finite set of primes of F containing S_p and all primes ramified in E. Then*

1. $H^1_{ét}(E, \mathbb{Z}_p(n))^G \cong H^1_{ét}(F, \mathbb{Z}_p(n))$.
2. $H^2_{ét}(o_E^S, \mathbb{Z}_p(n))_G \cong H^2_{ét}(o_F^S, \mathbb{Z}_p(n))$.
3. *For all $k \geq 0$ there are isomorphisms*

$$\hat{H}^k(G, H^2_{ét}(o_E^S, \mathbb{Z}_p(n))) \cong \hat{H}^{k+2}(G, H^1_{ét}(E, \mathbb{Z}_p(n))).$$

LECTURE 2

Conjectures and results

2.1. The Lichtenbaum Conjecture

We are now ready to apply the Main Conjecture in the case where ψ is of order prime to p (p odd). The character ψ is then automatically of type S. We now choose a finite abelian extension E of F containing F_ψ and μ_p (e.g. $E = F_\psi(\mu_p)$), so that the Galois group G of E/F is of order prime to p. Then E_∞ contains all p-power roots of unity. For any finite set S of primes containing S_p the standard Iwasawa module $\mathfrak{X}^S$ over E_∞ is a $\mathbb{Z}_p[G][[T]]$-module. The following arguments are valid for an arbitrary finite set S of primes containing S_p, and so we simply drop the index S from the notations.

Since the order of G is prime to p, the idempotents of the group algebra $\mathbb{Q}_p[G]$ are contained in $\mathbb{Z}_p[G]$ and $\mathbb{Z}_p[G]$ is a maximal order in $\mathbb{Q}_p[G]$, isomorphic to a finite product of discrete valuation rings $\mathcal{O}_\rho$ for certain (absolutely irreducible) characters ρ of G. Given a finitely generated $\mathbb{Z}_p[G][[T]]-$ module M and a character ρ, the ρ-th component M^ρ of M is defined as

$$M^\rho = e_\rho(M \otimes_{\mathbb{Z}_p} \mathcal{O}_\rho).$$

This is a finitely generated $\mathcal{O}_\rho[[T]]-$ module.

We now take $M = \mathfrak{X}$ and $\rho = \psi$, and let $\Lambda = \mathcal{O}_\psi[[T]]$. Since ψ is even, the ψ-component $\mathfrak{X}^\psi$ is a finitely generated Λ-torsion module. We also note that $\mathcal{O}_\psi$ is unramified over $\mathbb{Z}_p$, and so we can take $\pi = p$ as the uniformizer. We denote the characteristic polynomial of $\mathfrak{X}^\psi$ by $f_\psi^*(T)$, and we let

$$f(T) = p^\mu \cdot f_\psi^*(T),$$

which is a generator of the characteristic ideal of $\mathfrak{X}^\psi$.

Wiles has shown ([**38**],Theorem 1.4) that the μ-invariant μ of $\mathfrak{X}^\psi$ coincides with $\mu(G_\psi)$, and therefore by the Main Conjecture the characteristic ideal of $\mathfrak{X}^\psi$ is generated by $G_\psi(T)$.

Let us fix now an integer $n \geq 2$ and consider the Λ-module $\mathfrak{X}^\psi(-n)$ and its Pontryagin dual $Hom_{\mathbb{Z}_p}(\mathfrak{X}^\psi(-n), \mathbb{Q}_p/\mathbb{Z}_p)$. We let $\chi = \psi\omega^{-n}$, and note that

$$\mathfrak{X}^\psi(-n) = \mathfrak{X}(-n)^\chi.$$

Taking duals we obtain

$$\begin{aligned} Hom_{\mathbb{Z}_p}(\mathfrak{X}^\psi(-n), \mathbb{Q}_p/\mathbb{Z}_p) &= Hom_{\mathbb{Z}_p}(\mathfrak{X}(-n)^\chi, \mathbb{Q}_p/\mathbb{Z}_p) = \\ &= Hom_{\mathbb{Z}_p}(\mathfrak{X}(-n), \mathbb{Q}_p/\mathbb{Z}_p)^{\chi^{-1}} = Hom_{\mathbb{Z}_p}(\mathfrak{X}, \mathbb{Q}_p/\mathbb{Z}_p(n))^{\chi^{-1}}. \end{aligned}$$

Because E_∞ contains all p-power roots of unity, the Galois group $Gal(\Omega_E^{(p)}/E_\infty)$ acts trivially on the abelian group $\mathbb{Q}_p/\mathbb{Z}_p(n)$, and therefore

$$Hom_{\mathbb{Z}_p}(\mathfrak{X}, \mathbb{Q}_p/\mathbb{Z}_p(n)) = H^1_{\text{ét}}(o'_{E_\infty}, \mathbb{Q}_p/\mathbb{Z}_p(n)),$$

where

$$H^1_{\text{ét}}(o'_{E_\infty}, \mathbb{Q}_p/\mathbb{Z}_p(n)) = \varinjlim H^1_{\text{ét}}(o'_{E_m}, \mathbb{Q}_p/\mathbb{Z}_p(n)).$$

Galois descent implies that

$$H^1_{\text{ét}}(o'_{E_\infty}, \mathbb{Q}_p/\mathbb{Z}_p(n))^\Gamma = H^1_{\text{ét}}(o'_E, \mathbb{Q}_p/\mathbb{Z}_p(n)).$$

The parity of χ is equal to $(-1)^n$ and therefore by Corollary 1.2.2 the χ^{-1}-eigenspace of $H^1_{\text{ét}}(o'_E, \mathbb{Q}_p/\mathbb{Z}_p(n))$ is finite and isomorphic to $H^2_{\text{ét}}(o'_E, \mathbb{Z}_p(n))^{\chi^{-1}}$. We have shown:

Proposition 2.1.1. *The Pontryagin dual of $\mathfrak{X}^\psi(-n)_\Gamma$ is isomorphic to the finite group $H^2_{\text{ét}}(o'_E, \mathbb{Z}_p(n))^{\chi^{-1}}$.*

Proposition 2.1.1 shows that $\mathfrak{X}^\psi(-n)_\Gamma$ is finite. By Lemma 1.1.1 the same is then true for the Γ-invariants $\mathfrak{X}^\psi(-n)^\Gamma$. However, $\mathfrak{X}^\psi(-n)$ has no non-trivial finite Λ-submodules, hence the Γ-invariants of $\mathfrak{X}^\psi(-n)$ are trivial, and therefore we can compute the order of $\mathfrak{X}^\psi(-n)_\Gamma$ in terms of the valuation of the characteristic polynomial at 0. By Lemma 1.1.2 the characteristic polynomial of $\mathfrak{X}^\psi(-n)$ is given by $f^*(\kappa(\gamma)^n(1+T)-1)$. Hence by the Main Conjecture:

Proposition 2.1.2.

$$|H^2_{\text{ét}}(o'_E, \mathbb{Z}_p(n))^{\chi^{-1}}| = |\mathfrak{X}^\psi(-n)_\Gamma| = |f(\kappa(\gamma)^n - 1)|_v{}^{-1} = |L_p(1-n, \psi)|_v{}^{-1} = |L(1-n,\chi)|_v{}^{-1},$$

provided that $\psi \neq 1$.

We note that the last equality holds, since the Euler factors relating the values of $L_p(1-n,\psi)$ and $L(1-n,\chi)$ are v-adic units for all primes v above p and integers $n \geq 2$.

We can slightly reformulate the result: Let us write $a \sim_p b$ if the two rational numbers a, b have the same p-adic valuation. Let d_χ denote the degree of $\mathcal{O}_\chi$ over $\mathbb{Z}_p$. Then

$$L(1-n,\chi)^{d_\chi} \sim_p |H^2_{\text{ét}}(o'_E, \mathbb{Z}_p(n))^{\chi^{-1}}|,$$

provided that $\chi \neq \omega^n$.

If $\psi = 1$, then we have to consider the polynomial $H_1(T) = T$ as well. A similar, but much easier calculation for the Iwasawa module $X = \mathbb{Z}_p$ over E_∞, whose characteristic polynomial equals T yields

$$|X(-n)_\Gamma|^{-1} = \kappa(\gamma)^n - 1.$$

The dual of $X(-n)$ is equal to $\mathbb{Q}_p/\mathbb{Z}_p(n)^{\omega^n}$, and hence

$$|H^0(E, \mathbb{Q}_p/\mathbb{Z}_p(n))^{\omega^n}| = \kappa(\gamma)^n - 1.$$

Since $H^0(E, \mathbb{Q}_p/\mathbb{Z}_p(n)$ is cyclic, there is only one non -trivial eigenspace, hence we have $H^0(E, \mathbb{Q}_p/\mathbb{Z}_p(n)^\chi \neq 1 \leftrightarrow \chi = \omega^n$. This finally leads to the main result, which we formulate for a general finite set S of primes:

Theorem 2.1.3. *Let χ be a 1-dimensional Artin character of order prime to p over a real field F. Then for any finite set S of primes of F containing S_p, and any $n \geq 2$, so that $\chi(-1) = (-1)^n$, we have*

$$L^S(1-n,\chi)^{d_\chi} \sim_p \frac{|H^2_{\acute{e}t}(o_E^S, \mathbb{Z}_p(n))^{\chi^{-1}}|}{|H^0(E, \mathbb{Q}_p/\mathbb{Z}_p(n)^{\chi^{-1}}|},$$

where E is any finite abelian extension of F of degree prime to p, containing F_χ.

Let us consider the special case that $\chi = 1$ and $n \geq 2$ is even. Recall that

$$H^2(o_F, \mathbb{Z}(n)) \cong \prod_p H^2_{\acute{e}t}(o_F', \mathbb{Z}_p(n))$$

is a finite group, and we denote its order simply by $h_n(F)$ indicating the similarity with the class number of F. Furthermore we denote the order of $H^0(F, \mathbb{Q}/\mathbb{Z}(n))$ simply by $w_n(F)$. With these notations we obtain the following

Corollary 2.1.4. *Let F be a totally real number field and let $n \geq 2$ be an even integer. Then*

$$\zeta_F(1-n) = \pm \frac{h_n(F)}{w_n(F)}.$$

up to powers of 2.

We remark that the 2-primary part of the Corollary is also true if F is abelian over $\mathbb{Q}$. (cf. [**24**]).

In the special case $n = 2$ the group $H^2(o_F, \mathbb{Z}(2))$ is isomorphic to $K_2(o_F)$ ([**35**]) and the statement of Corollary 2.1.4 in this case is the Birch-Tate Conjecture:

Birch-Tate Conjecture 2.1.5. *Let F be a totally real number field. Then*

$$\zeta_F(-1) = \pm \frac{|K_2(o_F)|}{w_2(F)}$$

(up to possible powers of 2 if F is not abelian over $\mathbb{Q}$).

Corollary 2.1.4 is a special case of the (cohomological version of the) Lichtenbaum Conjecture ([**28**]): Let F be an arbitrary number field, and let $n \geq 2$. The order of vanishing of the zeta-function of F at $1-n$ is equal to

$$\begin{array}{ll} r_2 & \text{if } n \text{ is even} \\ r_1 + r_2 & \text{if } n \text{ is odd.} \end{array}$$

These numbers are equal to the ranks of $K_{2n-1}(o_F)$ by a result of Borel's ([**4**]). Let $\zeta_F^*(1-n)$, the special value of ζ_F at $1-n$, denote the first non-vanishing coefficient in a Taylor expansion of the zeta-function $\zeta_F(s)$ around $1-n$.

Lichtenbaum Conjecture 2.1.6. *Up to powers of 2:*

$$\zeta_F^*(1-n) = \pm \frac{|K_{2n-2}(o_F)|}{|K_{2n-1}(o_F)_{tors}|} \cdot R_n^B(F).$$

Here $R_n^B(F)$ denotes the Borel regulator (cf. [**4**]). If the Bloch-Kato Conjecture is true, then the Lichtenbaum Conjecture is true for abelian number fields (cf. [**25, 26, 1, 21, 7**]).

If we want to include the 2-primary parts into this conjecture, then we should replace the K-groups by motivic cohomology groups, i.e. we are led to the motivic reformulation:

Motivic Lichtenbaum Conjecture 2.1.7.

$$\zeta_F^*(1-n) = \pm \frac{|H^2_{\mathcal{M}}(o_F, \mathbb{Z}(n))|}{|H^1_{\mathcal{M}}(o_F, \mathbb{Z}(n))_{tors}|} \cdot R_n^{\mathcal{M}}(F).$$

Here $R_n^{\mathcal{M}}(F)$ differs from the Borel regulator by a power of 2. This conjecture is known to be true (assuming Bloch-Kato) if F is totally real abelian and $n \geq 2$ is even (cf. Corollary 2.1.4) and in a few other cases.

2.2. The Coates-Sinnott Conjecture

We now consider an arbitrary abelian extension E/F of number fields with Galois group G, and let S be a finite set of primes in F containing the primes ramified in E and the infinite primes. It is well-known that there exists a function $\theta^S_{E/F}(s)$ with values in the complex group ring $\mathbb{C}[G]$, such that

$$\chi(\theta^S_{E/F}(s)) = L^S(\chi^{-1}, s)$$

for all characters χ of G. We simply define

$$\theta^S_{E/F}(s) = \sum_{\chi} L^S(\chi^{-1}, s) e_\chi \in \mathbb{C}[G],$$

where, as before, the sum extends over all absolutely irreducible characters of G, and $e_\chi = \frac{1}{|G|}\sum_{g\in G} \chi-(g)g^{-1}$ denotes the idempotent belonging to χ. By a result of Klingen-Siegel $\theta^S_{E/F}(1-n)$ is contained in $\mathbb{Q}[G]$ for all $n \geq 1$, and it was shown by Deligne-Ribet that suitable multiples of $\theta^S_{E/F}(1-n)$ are actually contained in the integral group ring $\mathbb{Z}[G]$. More precisely

$$Ann_{\mathbb{Z}[G]}(H^0(E, \mathbb{Q}/\mathbb{Z}(n))) \cdot \theta^S_{E/F}(1-n) \subset \mathbb{Z}[G].$$

The ideal $Ann_{\mathbb{Z}[G]}(H^0(E, \mathbb{Q}/\mathbb{Z}(n))) \cdot \theta^S_{E/F}(1-n)$ is called the *n-th higher Stickelberger ideal* and denoted by $\mathrm{Stick}^S_{E/F}(n)$. The classical Stickelberger Theorem states that

$$\mathrm{Stick}^S_{E/\mathbb{Q}}(1) \subset Ann_{\mathbb{Z}[G]}(Cl(o_E)),$$

and Brumer conjectured that the same result holds for arbitrary abelian extensions E/F. For $n \geq 2$ another generalization of Stickelberger's theorem, involving higher Quillen K-groups, was suggested by Coates-Sinnott in the concluding remarks of [**12**]:

Coates-Sinnott Conjecture 2.2.1. *Let E/F be an abelian Galois extension of number fields with Galois group G, and let $n \geq 2$. Then*

$$Stick^S_{E/F}(n) \subset Ann_{\mathbb{Z}[G]}(K_{2n-2}(o_E)).$$

We note that at negative integers $1-n$, $n \geq 2$, the Artin L-function $L^S(\chi, s)$ vanishes unless F is totally real and $\chi(-1) = (-1)^n$. Therefore one usually restricts attention to F totally real, and either E totally real and n even or E CM and n odd.

Conjecture 2.2.1 was proven by Coates and Sinnott in [**12**] for $n = 2$ and E abelian over $\mathbb{Q}$ up to powers of 2.

As before, the 2-primary information about this conjecture suggests that the K-groups should be replaced by motivic cohomology groups, i.e. the correct version should read

Motivic Coates-Sinnott Conjecture 2.2.2. *Let E/F be an abelian extension of number fields with Galois group G, and let $n \geq 2$. Then*

$$Stick^S_{E/F}(n) \subset Ann_{\mathbb{Z}[G]}(H^2_{\mathcal{M}}(o_E, \mathbb{Z}(n))).$$

To approach the conjecture one considers each prime p separately, and shows that

$$Ann_{\mathbb{Z}_p[G]}(H^0(E, \mathbb{Q}_p/\mathbb{Z}_p(n))) \cdot \theta^S_{E/F}(1-n) \subset Ann_{\mathbb{Z}_pG]}(H^2_{\text{ét}}(o^S_E, \mathbb{Z}_p(n)).$$

This gives the p-part of the cohomological version of the conjecture.

We want to show now that the Classical Main Conjecture for and odd prime p implies the p-part of the conjecture in the semi-simple case, i.e. we are considering an odd prime p, which does not divide the order of G. In the setting of the previous section (we are enlarging E to contain μ_p) we fix $n \geq 2$ and a character χ of G with parity $(-1)^n$, so that the character $\psi := \chi\omega^n$ is real.

We first recall the definition and some of the properties of Fitting ideals.

The (first)*Fitting ideal* $Fitt_R(M)$ of a finitely generated R-module M over a commutative ring R is defined as follows: Choose a free resolution

$$R^m \xrightarrow{\beta} R^n \to M \to 0$$

of M. The Fitting ideal $Fitt_R(M)$ of M is the R-ideal generated by all $n \times n$-minors of the $n \times m$-matrix representing β. This definition is independent of the choice of the free resolution. One of the properties of the Fitting ideal is that it is contained in the annihilator of M:

$$Fitt_R(M) \subset Ann_R(M),$$

and the two ideals are equal if M is a cyclic R-module.

It is now rather straightforward to compute the Fitting ideal of the Iwasawa-module

$$\mathfrak{X}^S(-n)^\chi = \mathfrak{X}^{S,\psi}(-n).$$

We note that this is a finitely generated torsion $\mathcal{O}_\chi[[T]]$- module without non-trivial finite submodules, and therefore by a result of Greither ([**18**], Theorem 2.2, [**31**], Lemma 2.3) has projective dimension ≤ 1. We emphasize that the proof does not need the module to be finitely generated as a $\mathbb{Z}_p$-module, hence one does not have to assume that the μ-invariant of $\mathfrak{X}^{S,\psi}(-n)$ is trivial.

Now, if M is a f.g. torsion R-module of projective dimension ≤ 1, then there is a resolution of M of the form

$$0 \to R^n \xrightarrow{\beta} R^n \to M \to 0,$$

hence

$$Fitt_R(M) = (det\,\beta)$$

is a principal ideal generated by the determinant of β.

In our case we have an injection

$$0 \to \mathfrak{X}^{S,\psi}(-n) \to \Lambda/(f_{\psi,S}(\kappa(\gamma)^n(1+T)-1))$$

with finite cokernel, where as in section 3 $f_{\psi,S}(T)$ is a generating polynomial of the characteristic ideal of $\mathfrak{X}^{S,\psi}$. At all height 1 primes of Λ the two principal ideals $Fitt_\Lambda(\mathfrak{X}^S(-n)^\chi)$ and $(f_{\psi,S}(\kappa(\gamma)^n(1+T)-1))$ coincide, and it is then well known (cf. [**20**],Proposition 3.2.1) that this implies the equality of the two ideals. We obtain:

Proposition 2.2.3.

$$Fitt_\Lambda(\mathfrak{X}^S(-n)^\chi) = (f_{\psi,S}(\kappa(\gamma)^n(1+T)-1)).$$

As an immediate consequence we obtain a reformulation of the Classical Main Conjecture:

Corollary 2.2.4. *The Main Conjecture for ψ is equivalent to*

$$Fitt_\Lambda(\mathfrak{X}^{S,\psi}(-n)) = (G_{\psi,S}(\kappa(\gamma)^n(1+T)-1))$$

for all $n \geq 2$

We now descend to $\mathfrak{X}^{S,\psi}(-n)_\Gamma$. Its Fitting ideal over $\mathcal{O}_\psi$ is the image of $Fitt_\Lambda(\mathfrak{X}^{S,\psi}(-n)$ under the map $T \mapsto 0$, hence

Corollary 2.2.5.

$$Fitt_{\mathcal{O}_\psi}(\mathfrak{X}^{S,\psi}(-n)_\Gamma) = (L_p^S(1-n,\psi)),$$

if $\psi \neq 1$.

To treat the case $\psi = 1$ we consider the Iwasawa module $\mathbb{Z}_p(-n)$ and obtain with a similar argument:

$$Fitt_{\mathbb{Z}_p}(\mathbb{Z}_p(-n)_\Gamma) = (\kappa(\gamma)^n - 1),$$

and therefore in this case

Corollary 2.2.6.

$$Fitt_{\mathbb{Z}_p}(\mathfrak{X}^S(-n)_\Gamma) = Fitt_{\mathbb{Z}_p}(\mathbb{Z}_p(-n)_\Gamma)\cdot(L_p^S(1-n,1)).$$

For a finite module M the Fitting ideals over $\mathbb{Z}_p$ of M and its dual M^* are the same, and so we can dualize and finally take the sum over all eigenspaces to obtain

Theorem 2.2.7.

$$Fitt_{\mathbb{Z}_p[G]}(H^2_{\acute{e}t}(o_E^S,\mathbb{Z}_p(n))) = Fitt_{\mathbb{Z}_p[G]}(H^0(E,\mathbb{Q}_p/\mathbb{Z}_p(n)))\cdot\theta^S_{E/F}(1-n).$$

We note that this implies the p-part of the cohomological version of the Coates-Sinnott Conjecture, because the right-hand side equals $Ann_{\mathbb{Z}_p[G]}(H^0(E, \mathbb{Q}_p/\mathbb{Z}_p(n) \cdot \theta^S_{E/F}(1-n)$, which is then contained in $Ann_{\mathbb{Z}_p[G]}(H^2_{\text{ét}}(o^S_E, \mathbb{Z}_p(n)))$. Since

$$H^2_{\text{ét}}(o'_E, \mathbb{Z}_p(n)) \subset H^2_{\text{ét}}(o^S_E, \mathbb{Z}_p(n))$$

we obtain

Corollary 2.2.8. *If $p \nmid |G|$, p an odd prime, then the p-part of the cohomological version of the Coates-Sinnott Conjecture holds:*

$$Ann_{\mathbb{Z}_p[G]}(H^0(E, \mathbb{Q}_p/\mathbb{Z}_p(n))) \cdot \theta^S_{E/F}(1-n) \subset Ann_{\mathbb{Z}_pG]}(H^2_{ét}(o^S_E, \mathbb{Z}_p(n)).$$

To prove the cohomological version of the Coates-Sinnott Conjecture in general, one has to replace the Classical Main Conjecture by an Equivariant Main Conjecture, because the decomposition into eigenspaces is no longer available. A version of the Equivariant Main Conjecture has been formulated and proven by Ritter-Weiss ([**32**]) under the hypothesis that the μ-invariant of the Iwasawa-module $\mathfrak{X}$ is trivial and that p is odd. As a consequence Nguyen Quang Do proved the cohomological version of the Coates-Sinnott Conjecture (cp. [**30**]). Independently, this was also proven by Burns-Greither ([**8**]) under the same assumptions (and some additional restrictions on the primes p, if $F \neq \mathbb{Q}$) as a consequence of the Equivariant Tamagawa Number Conjecture. Most recently, Greither and Popescu gave a new and much more general approach to an Equivariant Main Conjecture, again under the assumption that the μ-invariant vanishes and that p is odd, which seems to have many more applications, e.g. it covers the function field case as well. The situation for the prime 2 remains almost completely open.

Bibliography

[1] D. Benois and T. Nguyen Quang Do, *Les nombres de Tamagawa locaux et la conjecture de Bloch et Kato pour les motifs* $\mathbb{Q}(m)$ *sur un corps abelien*, Ann. Sci. École Norm. Sup. (4) **35** (2000), 641–672.

[2] S. Bloch, *Algebraic cycles and higher K-theory*, Adv. in Math. **61** (1086), 267–304.

[3] S. Bloch and K. Kato, *L-functions and Tamagawa numbers of motives* The Grothendieck Fest-schrift, vol. 1, Progr. Math 86, Birkhäuser, Boston, 1990, pp. 333–400.

[4] A. Borel, *Stable real cohomology of arithmetic groups*, Ann. Sci. École Norm. Sup. **7** (1977), 613–636.

[5] A. Borel, *Cohomologie de* SL_n *et valeurs de fonction zêta*, Ann. Sci. Scuola Norm. Sup. Pisa **4** (1974), 235–272.

[6] N.Bourbaki, *Commutative Algebra, Chapters 1-7*, Springer, Berlin, Heidelberg, New York, London, Paris, Tokyo, 1989

[7] D. Burns and C. Greither, *On the equivariant Tamagawa number conjecture for Tate motives*, Invent. math. **153** (2003), 303–359

[8] D. Burns and C. Greither, *Equivariant Weierstrass preparation and values of L-funtions at negative integers*, Doc. Math (2003) Extra volume in honor of Kazuya Kato, 157–185

[9] T. Chinburg, M. Kolster, G. Pappas and V. Snaith, *Galois structure of K-groups of rings of integers*, K-Theory **14** (1998), 319–369.

[10] J. Coates, *p-adic L-functions and Iwasawa theory*, Algebraic Number Theory (A. Fröhlich, ed.), Academic Press, London, 1977, pp. 269–353.

[11] J. Coates, *On* K_2 *and some classical conjectures in algebraic number theory*, Ann. of Math. **95** (1972), 99–116.

[12] J. Coates and W. Sinnott, *An analogue of Stickelberger's theorem for the higher K-groups*, Inv. Math. **24** (1974), 149–161

[13] P. Cornacchia and P.-A. Østvær, *On the Coates-Sinnott Conjecture*, K-Theory **19** (2000), 195–209

[14] P. Deligne and K. Ribet, *Values of Abelian L-functions at negative integers over totally real fields*, Inv. Math **59** 1980, 227–286

[15] R.Greenberg, *On p-adic L-functions and cyclotomic fields*, Nagoya Math. J. **56** (1974), 61–77

[16] R.Greenberg, *On p-adic L-functions and cyclotomic fields. II*, Nagoya Math. J. **67** (1977), 139–158

[17] R.Greenberg, *On p-adic Artin L-functions*, Nagoya Math. J. **89** (1983), 77–87

[18] C. Greither, *The structure of some minus class groups and Chinburg's third conjecture for abelian fields*, Math. Z. **229** (1998), 107–136

[19] C. Greither, *Some cases of Brumer's conjecture for abelian CM extensions of totally real fields*, Math. Z. **233** (2000), 515–534

[20] C. Greither, *Computing Fitting ideals of Iwasawa modules*, Math. Z. **246** (2004), 733–767

[21] A. Huber and G. Kings, *Bloch-Kato conjecture and main conjecture of Iwasawa theory for Dirichlet characters*, Duke Math. J. **119** (2003), no. 3, 393–464

[22] K. Iwasawa, *On* $\mathbb{Z}_l$*-extensions of algebraic number fields*, Ann. of Math. **98** (1973), 246–326.

[23] M. Kolster, *Higher relative class number formulae*, Math. Annalen **323** (2002), 667–692.

[24] M. Kolster, *K-theory and arithmetic*, Contemporary developments in algebraic K-theory (Dedicated to H. Bass on the occasion of is 70th birthday), (M. Karoubi, A.O. Kuku, C. Pedrini, eds.) ICTP Lecture Notes **15** 2003, 191–258

[25] M. Kolster, T. Nguyen Quang Do and V. Fleckinger, *Twisted S-units, p-adic class number formulas, and the Lichtenbaum conjectures*, Duke Math. J. **84** (1996), 679–717.

[26] M. Kolster, T. Nguyen Quang Do and V. Fleckinger, *Correction to "Twisted S-units, p-adic class number formulas, and the Lichtenbaum conjectures"*, Duke Math. J. **90** (1997), 641–643.

[27] S. Lichtenbaum, *On the values of zeta and L-functions, I*, Annals of Math. **96** (1972), 338–360.

[28] S. Lichtenbaum, *Values of zeta functions, étale cohomology, and algebraic K-theory*, Lecture Notes in Math., vol. 342, Springer Verlag, 1973, pp. 489–501.

[29] J.S. Milne, *Étale Cohomology*, Princeton University Press, 1980.

[30] T. Nguyen Quang Do, *Conjeture principale équivariante, idéaux de Fitting et annulateurs en theéorie d'Iwasawa*, J. de Théorie. des Nombres de Bordeaux **3** (2005), 643–668.

[31] C. Popescu, *On the Coates-Sinnott Conjecture*, Math. Nachrichten **282** (2009), no. 10, 1370–1390.

[32] J. Ritter and A. Weiss *The Lifted Root Number Conjecture and Iwasawa Theory*, Memoirs of the AMS **748**, AMS, Providence, Rhode Island 2002

[33] C. Soulé, *K-théorie des anneaux d'entiers de corps de nombres et cohomologie étale*, Invent. Math. **55** (1979), 251–295.

[34] A. A. Suslin and V. Voevodsky, *The Bloch–Kato conjecture and motivic cohomology with finite coefficients*, preprint (1995); revised version (1999).

[35] J. Tate, *Relations between K_2 and Galois cohomology*, Invent. Math. **36** (1976), 257–274.

[36] V. Voevodsky, A. A. Suslin and E. Friedlander, *Cycles, Transfers, and Motivic Homology Theories*, Annals of Math. Studies, vol. 143, Princeton University Press, New JA. A. Suslin-ersey, 2000.

[37] V. Voevodsky *Motivic Cohomology with $\mathbb{Z}/2$-coefficients*, Publ. Math. de l'IHÉS, vol. 98 #1 (2003), 59–104.

[38] A. Wiles, *The Iwasawa Conjecture for totally real fields*, Annals of Math. **131** (1990), 493–540.

An Introduction to the Equivariant Tamagawa Number Conjecture: the Relation to Stark's Conjecture

David Burns

IAS/Park City Mathematics Series
Volume 18, 2009

An Introduction to the Equivariant Tamagawa Number Conjecture: the Relation to Stark's Conjecture

David Burns

Introduction

The 'Equivariant Tamagawa Number Conjecture' (or 'ETNC' for short) was formulated by Flach and the present author in [**11**] as a natural refinement of the 'Tamagawa Number Conjecture' of Bloch and Kato. The latter conjecture was originally formulated in [**4**] and then extended and refined by both Kato [**26, 27**] and Fontaine and Perrin-Riou [**21**]. This earlier seminal work of Bloch, Kato, Fontaine and Perrin-Riou was restricted to motives with commutative coefficients and also left unresolved an important 'sign ambiguity' problem. By combining Deligne's notion of virtual objects with the methods of relative algebraic K-theory the ETNC simultaneously extended the formalism of Tamagawa number conjectures to motives with non-commutative coefficients and also resolved the sign ambiguity problem (in the commutative case).

The formalism of the ETNC now underlies much of the important recent work in non-commutative Iwasawa theory (see, for example, Fukaya and Kato [**22**] or Venjakob's survey article [**43**]). At the same time it has also provided a universal framework which, upon appropriate specialisation, has refined a very wide variety of well known and rather explicit conjectures and results concerning the leading terms at integer values of motivic L-functions ranging, for example, from refinements of Stark's Conjecture such as the Rubin-Stark Conjecture to the conjectures of Chinburg in Galois module theory and the refinement of the Birch and Swinnerton-Dyer Conjecture formulated by Mazur and Tate. By now there is also a considerable amount of evidence, both theoretical and numerical and due to many authors, in support of important special cases of the ETNC (see, for example, Flach's survey article [**19**] or [**7**]).

The reader can find a discussion of the general motivic setting of the ETNC in the accompanying article of Guido Kings [**29**]. In this article however our aim is to discuss in as explicit a fashion as possible the special case of the ETNC that is relevant to the leading term at zero of Dirichlet L-functions and hence also to the abelian case of Stark's Conjecture. We shall also explain how this case of the

---

King's College London, Department of Mathematics, Strand, London WC2R 2LS, United Kingdom

E-mail address: david.burns@kcl.ac.uk

I am very grateful to the Cristian Popescu, Karl Rubin and Alice Silverberg for inviting me to give a series of lectures at the Summer School that was held at Park City in 2009. I am also grateful to Henri Johnston for several useful comments about these notes.

ETNC leads to the formulation (and in some important cases proof) of natural refinements of Stark's Conjecture. Our intention throughout is to stress the main ideas which underlie this approach and so we prefer to suppress some (occasionally important) technical details. However, any interested reader can of course find all missing details in the indicated references. For details of the analogous results and conjectures for non-abelian Galois extensions the reader can consult [**7, 9**].

The main contents of this lecture series is as follows. In Lecture 1 we quickly review the necessary aspects of the theory of determinant functors and then explain how this theory allows one to associate canonical invertible modules to certain types of Yoneda two-extensions. In Lecture 2 we use the theory reviewed in Lecture 1 to give an explicit statement of the relevant special case of the ETNC (this is Conjecture 2.5). We then prove that Conjecture 2.5 implies the abelian case of Stark's Conjecture. In Lecture 3 we discuss links between Conjecture 2.5 and two sorts of explicit refinement of Stark's Conjecture. We first sketch how one derives from Conjecture 2.5 the type of explicit congruences for the values of L-series that occur in the 'refined class number formulas' conjectured by Gross and others. Then we discuss connections between Conjecture 2.5 and a possible new refinement of Stark's Conjecture that would give a natural analogue of Brumer's Conjecture for the values of higher derivatives of L-series.

At various points in the lectures we have included exercises. We hope that these exercises (which should not be difficult) will help an interested reader develop a better understanding of the algebraic methods being used.

LECTURE 1

Determinant modules

In this lecture we quickly review the necessary aspects of the theory of determinant functors and then explain how this theory allows one to associate canonical invertible modules to certain types of Yoneda two-extensions.

1. Free modules

Let R be a commutative unital noetherian ring with identity element 1_R. For each finitely generated free R-module M of rank $\mathrm{rk}_R(M)$ we set

$$[M]_R := \wedge_R^{\mathrm{rk}_R(M)} M. \tag{1.1}$$

This is a free R-module of rank one and if one fixes an R-basis $\{m_i : 1 \le i \le \mathrm{rk}_R(M)\}$ of M, then one obtains a (non-canonical) isomorphism of R-modules $[M]_R \xrightarrow{\sim} R$ by sending $\wedge_{i=1}^{i=\mathrm{rk}_R(M)} m_i$ to 1_R. If $\phi : M \to N$ is a homomorphism of finitely generated free R-modules of the same rank, then we write $[\phi]_R$ for the induced homomorphism of R-modules $\wedge_R^{\mathrm{rk}_R(M)}\phi : [M]_R \to [N]_R$.

When R is clear from context we will often abbreviate $[M]_R$ to $[M]$. If N is another finitely generated free R-module, then we will abbreviate the tensor product $[M]_R \otimes_R [N]_R$ to $[M]_R[N]_R$, or even occasionally just $[M][N]$.

We will use the following basic properties of this construction.

P1) $[0]_R = R$.

P2) If $\mathcal{E} : 0 \to M \to N \to P \to 0$ is a short exact sequence of finitely generated free R-modules, then there is a canonical isomorphism of R-modules of the form

$$\iota(\mathcal{E}) : [N]_R \cong [M]_R[P]_R.$$

(For an explicit description of this isomorphism see Remark 1.1 below.)

P3) For each finitely generated free R-module M we set

$$[M]_R^{-1} := \mathrm{Hom}_R([M]_R, R),$$

regarded as an R-module via $r(\theta)(\delta) = \theta(r\delta)$ for each $r \in R, \theta \in [M]_R^{-1}$ and $\delta \in [M]_R$. Then there are canonical isomorphisms of R-modules of the form

$$\mathrm{ev}_{M,1} : [M]_R[M]_R^{-1} \xrightarrow{\sim} R, \quad \delta \otimes \theta \mapsto \theta(\delta)$$
$$\mathrm{ev}_{M,2} : [M]_R^{-1}[M]_R \xrightarrow{\sim} R, \quad \theta \otimes \delta \mapsto \theta(\delta).$$

P4) Each isomorphism of finitely generated free R-modules $\phi : M \to N$ gives rise to canonical isomorphisms of R-modules of the form

$$t(\phi)_1 : [M]_R[N]_R^{-1} \xrightarrow{[\phi]_R \otimes \mathrm{id}} [N]_R[N]_R^{-1} \xrightarrow{\mathrm{ev}_{N,1}} R$$

$$t(\phi)_2 : [M]_R^{-1}[N]_R \xrightarrow{[\phi]_R^{-1} \otimes \mathrm{id}} [N]_R^{-1}[N]_R \xrightarrow{\mathrm{ev}_{N,2}} R$$

where $[\phi]_R^{-1}$ denotes the isomorphism $\mathrm{Hom}_R([M]_R, R) \to \mathrm{Hom}_R([N]_R, R)$ which sends each θ to $\theta \circ [\phi^{-1}]_R$.

Remark 1.1. Fix a short exact sequence $\mathcal{E}$ as in property P2) and choose R-bases $\{m_i : 1 \leq i \leq \mathrm{rk}_R(M)\}$ and $\{p_j : 1 \leq j \leq \mathrm{rk}_R(P)\}$ of M and P respectively. For each index j fix a pre-image $\hat{p}_j$ of p_j under the given surjection $N \to P$. Then $\{m_i : 1 \leq i \leq \mathrm{rk}_R(M)\} \cup \{\hat{p}_j : 1 \leq j \leq \mathrm{rk}_R(P)\}$ is an R-basis of N and, whilst the isomorphism $\iota(\mathcal{E})$ is independent of any choice of bases, it is characterised by the equality

$$\iota(\mathcal{E})((\wedge_{i=1}^{i=\mathrm{rk}_R(M)} m_i) \wedge_R (\wedge_{j=1}^{j=\mathrm{rk}_R(P)} \hat{p}_j)) = (\wedge_{i=1}^{i=\mathrm{rk}_R(M)} m_i) \otimes_R (\wedge_{j=1}^{j=\mathrm{rk}_R(P)} p_j).$$

Now, since $\mathcal{E}$ splits (because P is free), there is also an associated short exact sequence of R-modules $\mathcal{E}' : 0 \to M \to N \to P \to 0$ and hence an isomorphism $\iota(\mathcal{E}') : [N]_R \cong [P]_R[M]_R$. However, writing τ for the canonical isomorphism of R-modules $[M]_R[P]_R \cong [P]_R[M]_R$, the diagram

$$\begin{array}{ccc} [N]_R & \xrightarrow{\iota(\mathcal{E})} & [M]_R[P]_R \\ \downarrow{\scriptstyle \mathrm{id}} & & \downarrow{\scriptstyle \tau} \\ [N]_R & \xrightarrow{\iota(\mathcal{E}')} & [P]_R[M]_R \end{array}$$

commutes in general only 'up to sign'. This is actually an important and delicate technical problem which must be resolved before one can formulate precise special value conjectures in an equivariant setting (see, for example, Exercise 2.7 and Remark 2.11). The resolution of this problem is due to Grothendieck, Knudsen and Mumford (cf. [**30**]) and involves defining the determinant functor $[-]_R$ to take values in the category of *graded* R-modules. However, since a thorough treatment of this issue would obscure the present exposition, we have chosen to make no further reference to it. For the same reason we shall always lazily abbreviate both $\mathrm{ev}_{M,1}$ and $\mathrm{ev}_{M,2}$ to ev_M and both $t(\phi)_1$ and $t(\phi)_2$ to $t(\phi)$.

Remark 1.2. If $\mathcal{E}$ is an exact sequence as in property P2), then the isomorphisms $\mathrm{ev}_M, \mathrm{ev}_N$ and ev_P allow one to regard $\iota(\mathcal{E})$ as an isomorphism of the form $[M]_R^{-1}[N]_R \xrightarrow{\sim} [P]_R$ or $[N]_R^{-1} \xrightarrow{\sim} [M]_R^{-1}[P]_R^{-1}$ or even $[M]_R[N]_R^{-1}[P]_R \xrightarrow{\sim} R$. When applying isomorphisms of the form $\iota(\mathcal{E})$ we will usually not specify explicitly which one of these (or several other) possible interpretations it is that we have in mind, believing that it will always be clear from context.

The above theory extends directly to the setting of modules over semisimple algebras. To explain this we let A be any finite dimensional semisimple commutative algebra. Then A decomposes as a finite product of fields $\prod_{i \in I} A_i$ and there is a corresponding decomposition of any finitely generated A-module M as a direct sum

$\bigoplus_{i\in I} M_i$ where each M_i is a finite dimensional A_i-space. Since each M_i is a free A_i-module we may therefore use the notation introduced above to define an A-module by setting

$$[M]_A := \bigoplus_{i\in I}[M_i]_{A_i}.$$

If M is a free A-module, then this definition of $[M]_A$ agrees with that given above. In general, there is a (non-canonical) isomorphism of A-modules $[M]_A \cong A$ and the above properties P1)-P4) (with R equal to each field A_i) combine to give analogous properties and isomorphisms $\iota(\mathcal{E}), \mathrm{ev}_M$ (where we set $[M]_A^{-1} := \mathrm{Hom}_A([M]_A, A)$) and $t(\phi)$ in this new setting.

Exercise 1.3. In the context of property P4) set $d := \mathrm{rk}_R(M) = \mathrm{rk}_R(N)$ and fix R-bases $\{m_i : 1 \le i \le d\}$ and $\{n_i : 1 \le i \le d\}$ of M and N respectively. Set $\underline{m} := \wedge_{i=1}^{i=d} m_i \in [M]_R$ and define $\underline{n}^*$ in $[N]_R^{-1}$ by the equality $\underline{n}^*(\wedge_{i=1}^{i=d} n_i) = 1_R$. Let β_M and β_N denote the R-module isomorphisms $[M]_R \xrightarrow{\sim} R$ and $[N]_R^{-1} \xrightarrow{\sim} R$ which send $\underline{m}$ and $\underline{n}^*$ to 1_R respectively. Let Φ denote the matrix of the isomorphism $\phi : M \to N$ with respect to the bases $\{m_i : 1 \le i \le d\}$ and $\{n_i : 1 \le i \le d\}$. Prove that the following diagram commutes

$$\begin{array}{ccc}
[M]_R \otimes_R [N]_R^{-1} & \xrightarrow{t(\phi)} & R \\
\downarrow{\scriptstyle \beta_M \otimes_R \beta_N} & & \uparrow{\scriptstyle \times\det(\Phi)} \\
R \otimes_R R & \xrightarrow{\mathrm{id}} & R.
\end{array}$$

2. Modules of finite projective dimension

For simplicity we now focus on the rings that are of most relevance for us. So we fix a finite abelian group G and consider the group ring $\mathbb{Q}[G]$ (this is a finite dimensional semisimple commutative $\mathbb{Q}$-algebra) and the finitely generated subring $\mathbb{Z}[G]$ of $\mathbb{Q}[G]$. For each prime p we write $\mathbb{Z}_{(p)}$ for the localisation of $\mathbb{Z}$ at p. For each module M we write $M_{(p)}$ for the localised module $\mathbb{Z}_{(p)} \otimes_{\mathbb{Z}} M$.

2.1. Projective modules

If M is a finitely generated projective G-module, then a well-known theorem of Swan (see, for example, [**17**, (32.11)]) implies that $M_{\mathbb{Q}} := \mathbb{Q} \otimes_{\mathbb{Z}} M$ is a (finitely generated) free $\mathbb{Q}[G]$-module and further that for each prime p the localisation $M_{(p)}$ is a free $\mathbb{Z}_{(p)}[G]$-module of rank equal to $\mathrm{rk}_{\mathbb{Q}[G]}(M_{\mathbb{Q}})$. This fact implies that there is an equality of non-zero G-submodules $\wedge_{\mathbb{Z}[G]}^{\mathrm{rk}_{\mathbb{Q}[G]}(M_{\mathbb{Q}})} M = \bigcap_p [M_{(p)}]_{\mathbb{Z}_{(p)}[G]}$ of $[M_{\mathbb{Q}}]_{\mathbb{Q}[G]}$, where in the intersection p runs over all prime numbers, and so motivates the following definition

$$[M]_{\mathbb{Z}[G]} := \bigcap_p [M_{(p)}]_{\mathbb{Z}_{(p)}[G]}. \tag{1.2}$$

This construction has the following basic properties:

- If M is a finitely generated free G-module, then the definition of $[M]_{\mathbb{Z}[G]}$ in (1.2) coincides with that given in (1.1).
- It is in general no longer true that $[M]_{\mathbb{Z}[G]}$ is a free $\mathbb{Z}[G]$-module (of rank one). However, for every prime p the localisation $([M]_{\mathbb{Z}[G]})_{(p)}$ is equal to $[M_{(p)}]_{\mathbb{Z}_{(p)}[G]}$ and hence is isomorphic (non-canonically) to $\mathbb{Z}_{(p)}[G]$.

This means that $[M]_{\mathbb{Z}[G]}$ is an 'invertible $\mathbb{Z}[G]$-module', or equivalently a 'locally-free $\mathbb{Z}[G]$-module of rank one'.

- There are natural analogues of the properties P1)-P4) in §1 and of the isomorphisms $\iota(\mathcal{E}), \mathrm{ev}_M$ (with $[M]_{\mathbb{Z}[G]}^{-1} := \mathrm{Hom}_{\mathbb{Z}[G]}([M]_{\mathbb{Z}[G]}, \mathbb{Z}[G])$) and $t(\phi)$ in this new setting.

2.2. Modules of finite projective dimension

It is well known that a finitely generated G-module M has finite projective dimension if and only if it is cohomologically-trivial (that is, for all subgroups J of G and all degrees i the Tate cohomology group $\hat{H}^i(J, M)$ vanishes) and that this condition is automatically satisfied if, for example, M is a finite module of order prime to the order of G (cf. [**2**, §9, Th. 9]). Further, if M is any such module then, since the Krull dimension of $\mathbb{Z}[G]$ is equal to 1, there exists an exact sequence of $\mathbb{Z}[G]$-modules of the form

$$\mathcal{E}: 0 \to P \to N \to M \to 0$$

in which P and N are both finitely generated and projective.

Definition 1.4. If M is a finitely generated G-module that has finite projective dimension, then one chooses a resolution $\mathcal{E}$ of M as above and defines $[M]_{\mathbb{Z}[G]}$ to be the image under the isomorphism $\iota(\mathbb{Q}\otimes_{\mathbb{Z}}\mathcal{E}) : [P_{\mathbb{Q}}]_{\mathbb{Q}[G]}^{-1}[N_{\mathbb{Q}}]_{\mathbb{Q}[G]} \cong [M_{\mathbb{Q}}]_{\mathbb{Q}[G]}$ of the submodule $[P]_{\mathbb{Z}[G]}^{-1}[N]_{\mathbb{Z}[G]}$ of $[P_{\mathbb{Q}}]_{\mathbb{Q}[G]}^{-1}[N_{\mathbb{Q}}]_{\mathbb{Q}[G]}$.

This construction has the following properties:

- $[M]_{\mathbb{Z}[G]}$ is independent of the choice of resolution $\mathcal{E}$.
- If M is a finitely generated projective G-module, then (M has finite projective dimension and) $[M]_{\mathbb{Z}[G]}$ coincides with the module defined in (1.2).
- There are natural analogues of the properties P1)-P4) in §1 and of the isomorphisms $\iota(\mathcal{E}), \mathrm{ev}_M$ (with $[M]_{\mathbb{Z}[G]}^{-1} := \mathrm{Hom}_{\mathbb{Z}[G]}([M]_{\mathbb{Z}[G]}, \mathbb{Z}[G])$) and $t(\phi)$ in this new setting.

Exercise 1.5. Prove that $[M]_{\mathbb{Z}[G]}$ is independent of the choice of $\mathcal{E}$.

Exercise 1.6. Let M be a finite module. Show that M has finite projective dimension as a $\mathbb{Z}$-module and that $[M]_{\mathbb{Z}}$ is equal to the submodule $|M|^{-1}\cdot\mathbb{Z}$ of $\mathbb{Q} = [M_{\mathbb{Q}}]_{\mathbb{Q}}$.

Exercise 1.7. As a natural generalisation of Exercise 1.6 prove the following result: if M is any finite G-module that has finite projective dimension, then $[M]_{\mathbb{Z}[G]}$ is equal to the submodule $\mathrm{Fit}_{\mathbb{Z}[G]}(M)^{-1}$ of $\mathbb{Q}[G] = [M_{\mathbb{Q}}]_{\mathbb{Q}[G]}$, where $\mathrm{Fit}_{\mathbb{Z}[G]}(M)$ denotes the (initial) Fitting ideal of the G-module M.

3. Yoneda two-extensions

We now fix a finite abelian group G and a finitely generated subring Λ of $\mathbb{Q}$ and set $\mathfrak{A} := \Lambda[G]$. We also assume to be given $\mathfrak{A}$-modules M and N.

3.1. Perfect two-extensions

A 'two-extension of M by N' is an exact sequence of $\mathfrak{A}$-modules of the form

$$\mathcal{T}: 0 \to N \to E_0 \xrightarrow{d} E_1 \to M \to 0.$$

If $\mathcal{T}'$ is another such exact sequence $0 \to N \to E_0' \xrightarrow{d'} E_1' \to M \to 0$, then we write $\mathcal{T} \rightsquigarrow \mathcal{T}'$ if there exists a commutative diagram of $\mathfrak{A}$-modules of the form

$$\begin{array}{ccccccccc} 0 & \longrightarrow & N & \longrightarrow & E_0 & \xrightarrow{d} & E_1 & \longrightarrow & M & \longrightarrow & 0 \\ & & \downarrow{\scriptstyle \mathrm{id}_N} & & \downarrow{\scriptstyle \phi^0} & & \downarrow{\scriptstyle \phi^1} & & \downarrow{\scriptstyle \mathrm{id}_M} & & \\ 0 & \longrightarrow & N & \longrightarrow & E_0' & \xrightarrow{d'} & E_1' & \longrightarrow & M & \longrightarrow & 0. \end{array} \tag{1.3}$$

This relation $\rightsquigarrow$ fails to be an equivalence relation on the set of such two-extensions (it is not in general symmetric) but nevertheless generates an equivalence relation. The associated set $\mathrm{Ext}^2_{\mathfrak{A}}(M,N)$ of equivalence classes is finite, of cardinality dividing a power of $|G|$, and has a natural abelian group structure. (For more details see, for example, [**25**, Chap. IV, §9].)

Definition 1.8. We say that an element ϵ of $\mathrm{Ext}^2_{\mathfrak{A}}(M,N)$ is *perfect* if it can be represented by an extension $0 \to N \to E_0 \to E_1 \to M \to 0$ in which the $\mathfrak{A}$-modules E_0 and E_1 are both finitely generated and of finite projective dimension.

Example 1.9. Even if M and N are finitely generated $\mathfrak{A}$-modules, the existence of a perfect element in $\mathrm{Ext}^2_{\mathfrak{A}}(M,N)$ is a strong restriction on the triple $(\mathfrak{A}, M, N)$ and is far from guaranteed. To describe an explicit example, we investigate the existence of perfect extensions in the case that $\mathfrak{A} = \mathbb{Z}[G]$ and $M = N = \mathbb{Z}$. We note first that any exact sequence of G-modules $0 \to \mathbb{Z} \to E_0 \xrightarrow{d} E_1 \to \mathbb{Z} \to 0$ splits into two short exact sequences $0 \to \mathbb{Z} \to E_0 \to \mathrm{im}(d) \to 0$ and $0 \to \mathrm{im}(d) \to E_1 \to \mathbb{Z} \to 0$. Thus, if E_0 and E_1 are cohomologically-trivial G-modules, then the connecting homomorphisms in the long exact sequences of Tate cohomology that are associated to these short exact sequences induce isomorphisms $\hat{H}^0(G,\mathbb{Z}) \cong \hat{H}^{-1}(G,\mathrm{im}(d)) \cong \hat{H}^{-2}(G,\mathbb{Z})$. Since $\hat{H}^0(G,\mathbb{Z}) = \mathbb{Z}/|G|$ whilst $\hat{H}^{-2}(G,\mathbb{Z}) = H_1(G,\mathbb{Z})$ is isomorphic to (the abelianisation of) G by [**2**, §3, Prop. 1], it follows that the existence of a perfect element in $\mathrm{Ext}^2_G(\mathbb{Z},\mathbb{Z})$ implies that G is cyclic. Conversely, if G is cyclic, with generating element g, and we set $T_G := \sum_{h\in G} h \in \mathbb{Z}[G]$ then for any natural number a that is coprime to $|G|$ the exact sequence

$$0 \to \mathbb{Z}[G] \xrightarrow{y \mapsto (0,y)} (\mathbb{Z} \oplus \mathbb{Z}[G])/\mathbb{Z}(a, -T_G) \xrightarrow{(x,y)\mapsto x(\mathrm{mod}\, a)} \mathbb{Z}/a \to 0$$

implies that the G-module $(\mathbb{Z} \oplus \mathbb{Z}[G])/\mathbb{Z}(a, -T_G)$ is cohomologically-trivial and hence that the exact sequence

$$\begin{aligned} 0 \to \mathbb{Z} \xrightarrow{m \mapsto (m,0)} & (\mathbb{Z} \oplus \mathbb{Z}[G])/\mathbb{Z}(a, -T_G) \\ & \xrightarrow{(x,y)\mapsto (g-1)y} \mathbb{Z}[G] \xrightarrow{\sum_{h\in G} n_h h \mapsto \sum_{h\in G} n_h} \mathbb{Z} \to 0 \end{aligned} \tag{1.4}$$

represents a perfect element in $\mathrm{Ext}^2_G(\mathbb{Z},\mathbb{Z})$.

3.2. Determinant lattices

We set $N := \mathbb{R}\otimes_\Lambda N$ and $M := \mathbb{R}\otimes_\Lambda M$ and now assume to be given an isomorphism of $\mathbb{R}[G]$-modules of the form

$$\lambda : N_{\mathbb{R}} \xrightarrow{\sim} M_{\mathbb{R}}.$$

We show that if ϵ is any perfect element of $\mathrm{Ext}^2_{\mathfrak{A}}(M, N)$, then one can associate to the pair (ϵ, λ) a canonical (invertible) $\mathfrak{A}$-submodule of $\mathbb{R}[G]$.

To do this we first choose a representative

$$\mathcal{T} : 0 \to N \to E_0 \xrightarrow{d} E_1 \to M \to 0$$

of ϵ. This exact sequence gives rise to short exact sequences of $\mathbb{Q}[G]$-modules

$$\mathcal{E}_1 : 0 \to N_{\mathbb{Q}} \to E_{0,\mathbb{Q}} \to \mathrm{im}(d)_{\mathbb{Q}} \to 0$$
$$\mathcal{E}_2 : 0 \to \mathrm{im}(d)_{\mathbb{Q}} \to E_{1,\mathbb{Q}} \to M_{\mathbb{Q}} \to 0$$

and hence to a composite isomorphism of $\mathbb{Q}[G]$-modules

$$\iota(\mathcal{T}) : [E_{0,\mathbb{Q}}][E_{1,\mathbb{Q}}]^{-1} \xrightarrow{\iota(\mathcal{E}_1)\iota(\mathcal{E}_2)} ([N_{\mathbb{Q}}][\mathrm{im}(d)_{\mathbb{Q}}])([\mathrm{im}(d)_{\mathbb{Q}}]^{-1}[M_{\mathbb{Q}}]^{-1})$$
$$\xrightarrow{\mathrm{ev}_{\mathrm{im}(d)_{\mathbb{Q}}}} [N_{\mathbb{Q}}][M_{\mathbb{Q}}]^{-1}.$$

We then write $\iota(\mathcal{T}, \lambda)$ for the composite isomorphism of $\mathbb{R}[G]$-modules

$$\iota(\mathcal{T}, \lambda) : [E_{0,\mathbb{R}}][E_{1,\mathbb{R}}]^{-1} \xrightarrow{\mathbb{R}\otimes_{\mathbb{Q}}\iota(\mathcal{T})} [N_{\mathbb{R}}][M_{\mathbb{R}}]^{-1} \xrightarrow{t(\lambda)} \mathbb{R}[G].$$

Definition 1.10. We define the *determinant lattice* of ϵ and λ by setting

$$\Xi(\epsilon, \lambda) := \iota(\mathcal{T}, \lambda)([E_0]_{\mathfrak{A}}[E_1]_{\mathfrak{A}}^{-1}),$$

where we regard $[E_0]_{\mathfrak{A}}[E_1]_{\mathfrak{A}}^{-1}$ as a submodule of $[E_{0,\mathbb{R}}]_{\mathbb{R}[G]}[E_{1,\mathbb{R}}]_{\mathbb{R}[G]}^{-1}$ in the natural way.

The key result in respect of the above construction is the following.

Proposition 1.11. *The lattice $\Xi(\epsilon, \lambda)$ depends only upon ϵ and λ.*

Proof. We only explain the main point (but see Remark 1.12 below). If ϵ is perfect then for any two representatives $\mathcal{T}$ and $\mathcal{T}'$ of ϵ one knows that $\mathcal{T} \rightsquigarrow \mathcal{T}'$ (note that this is a special property of perfect extensions and is not true in general). In particular, there exists a commutative diagram of the form (1.3). We consider the following related commutative diagram

$$\begin{array}{ccccc} & & E_0' & \xrightarrow{d'} & E_1' \\ & & \downarrow{\scriptstyle (0,\mathrm{id})} & & \downarrow{\scriptstyle \mathrm{id}} \\ E_0 & \xrightarrow{(d,\phi^0)} & E_1 \oplus E_0' & \xrightarrow{(-\phi^1,d')} & E_1' \\ \downarrow{\scriptstyle \mathrm{id}} & & \downarrow{\scriptstyle (\mathrm{id},0)} & & \\ E_0 & \xrightarrow{d} & E_1. & & \end{array} \tag{1.5}$$

The central column of this diagram is clearly a short exact sequence, call it $\mathcal{E}$, and it is also straightforward to show that the central row is a short exact sequence, call it $\mathcal{E}'$. The diagram therefore induces a composite isomorphism of $\mathfrak{A}$-modules

$$\iota(\mathcal{T}',\mathcal{T}) : [E_0'][E_1']^{-1} \xrightarrow{\iota(\mathcal{E})} ([E_1 \oplus E_0'][E_1]^{-1})[E_1']^{-1}$$
$$\cong ([E_1 \oplus E_0'][E_1']^{-1})[E_1]^{-1} \xrightarrow{\iota(\mathcal{E}')} [E_0][E_1]^{-1}$$

where the central isomorphism is the obvious one. It can be shown that $\iota(\mathcal{T}',\lambda) = \iota(\mathcal{T},\lambda) \circ (\mathbb{R}[G] \otimes_{\mathfrak{A}} \iota(\mathcal{T}',\mathcal{T}))$ and hence that

$$\iota(\mathcal{T}',\lambda)([E_0'][E_1']^{-1}) = \iota(\mathcal{T},\lambda)(\iota(\mathcal{T}',\mathcal{T})([E_0'][E_1']^{-1})) = \iota(\mathcal{T},\lambda)([E_0][E_1]^{-1}),$$

as required. □

Remark 1.12. The proof of Proposition 1.11 can be understood conceptually as follows. The central square of (1.3) is a quasi-isomorphism between the two-term complexes with differentials d and d' respectively. The central row of (1.5) is the mapping cone of this morphism (and is thus exact since the mapping cone of any quasi-isomorphism is acyclic) and so (1.5) corresponds to the canonical short exact sequence of complexes associated to the mapping cone. The computation of lattices given in the proof of Proposition 1.11 then results from two general facts: determinant lattices are multiplicative on short exact sequences of (compatibly trivialised) perfect complexes and the determinant lattice of an acyclic complex is trivial.

It is now convenient to introduce some more notation. Since G is a finite abelian group the algebra $\mathbb{R}[G]$ decomposes as a finite product of fields $\prod_{i\in I} F_i$. For any finitely generated $\mathbb{R}[G]$-module V there is a corresponding direct sum decomposition $V = \bigoplus_{i\in I} V_i$ where each V_i is a finite dimensional F_i-space. Any endomorphism ϕ of the $\mathbb{R}[G]$-module V therefore decomposes as a sum $\phi = \oplus_{i\in I}\phi_i$ with $\phi_i \in \mathrm{End}_{F_i}(V_i)$ for each $i \in I$ and so one can define an $\mathbb{R}[G]$-valued determinant of ϕ by setting $\det_{\mathbb{R}[G]}(\phi) := \prod_{i\in I} \det_{F_i}(\phi_i)$.

Exercise 1.13. Let ϵ be a perfect element of $\mathrm{Ext}^2_G(M,N)$ and λ an isomorphism of $\mathbb{R}[G]$-modules $N_\mathbb{R} \to M_\mathbb{R}$. Prove that if λ' is any other isomorphism of $\mathbb{R}[G]$-modules $N_\mathbb{R} \to M_\mathbb{R}$, then in $\mathbb{R}[G]$ one has $\Xi(\epsilon,\lambda') = \det_{\mathbb{R}[G]}(\lambda^{-1}\circ\lambda')\cdot\Xi(\epsilon,\lambda)$.

Exercise 1.14. Let G be the trivial group. Then the group $\mathrm{Ext}^2_G(M,N)$ vanishes and its unique element $0_{M,N}$ is perfect if and only if the modules M and N are both finitely generated. Assuming that $0_{M,N}$ is perfect, for any isomorphism λ of $\mathbb{R}$-modules $N_\mathbb{R} \to M_\mathbb{R}$ compute explicitly the submodule $\Xi(0_{M,N},\lambda)$ of $\mathbb{R}$. (Recall Exercises 1.3 and 1.6.)

Example 1.15. In some cases the lattice $\Xi(\epsilon,\lambda)$ actually determines the extension class ϵ. To describe an explicit example we return to the setting, and notation, of Example 1.9. We also now assume that G is a cyclic group of prime order p. In this case Tate cohomology is periodic (by [**2**, §8, Th.5]) and so the group $\mathrm{Ext}^2_G(\mathbb{Z},\mathbb{Z})$ is isomorphic to $H^2(G,\mathbb{Z}) \cong \hat{H}^0(G,\mathbb{Z}) = \mathbb{Z}/|G|$ and so has order p. Further, if for each integer a with $0 < a < p$ we write ϵ_a for the element of $\mathrm{Ext}^2_G(\mathbb{Z},\mathbb{Z})$ that is represented by the two-extension (1.4) for this value of a, then one can check that $\mathrm{Ext}^2_G(\mathbb{Z},\mathbb{Z}) = \{0\} \cup \{\epsilon_a : 0 < a < p\}$ and so every non-zero element of $\mathrm{Ext}^2_G(\mathbb{Z},\mathbb{Z})$ is perfect. Now if λ is any isomorphism of (trivial) $\mathbb{R}[G]$-modules $\mathbb{R} \cong \mathbb{R}$, then an explicit computation using the representative (1.4) of ϵ_a shows that the $\mathbb{Z}$-lattice $T_G(\Xi(\epsilon_a, \lambda))$ is generated by the element $\frac{\lambda(1)}{pa}T_G$ of $\mathbb{R}[G]$ and hence that $\Xi(\epsilon_a, \lambda) \neq \Xi(\epsilon_{a'}, \lambda)$ if $a \neq a'$.

LECTURE 2

A particular case of the ETNC

In this lecture we shall give (as Conjecture 2.5) an explicit statement of the relevant special case of the ETNC and also explain its relationship to the abelian case of Stark's Conjecture.

To do this we fix a finite abelian extension of global fields K/k and set $G := \mathrm{Gal}(K/k)$. We also fix a finite non-empty set of places S of k which contains all archimedean places (if any) and all places which ramify in K/k.

We write $S(K)$ for the set of places of K which lie above those in S. We also write $\mathcal{O}_S$ for the ring of $S(K)$-integers in K and set $U_S := \mathcal{O}_S^\times$, $J_S := \prod_{w\in S(K)} K_w^\times$ and $C_S := J_S/\Delta(U_S)$, where Δ is the natural diagonal embedding $U_S \to J_S$. We write Y_S for the free abelian group on the set $S(K)$ and X_S for the kernel of the homomorphism $\epsilon : Y_S \to \mathbb{Z}$ that sends each element of $S(K)$ to 1. We note that all of these groups are naturally G-modules.

1. Canonical classes

In this section we quickly recall the definition of certain canonical extension classes coming from class field theory.

1.1

We first assume S is large enough so that the S-ideal class group $\mathrm{Cl}(\mathcal{O}_S)$ vanishes and consider exact commutative diagrams of G-modules of the following form

$$
\begin{array}{ccccccccccc}
 & & 0 & & 0 & & 0 & & 0 & & \\
 & & \downarrow & & \downarrow & & \downarrow & & \downarrow & & \\
0 & \longrightarrow & U_S & \longrightarrow & A & \longrightarrow & B & \longrightarrow & X_S & \longrightarrow & 0 \\
 & & \downarrow{\scriptstyle\Delta} & & \downarrow & & \downarrow & & \downarrow{\scriptstyle\subset} & & \\
0 & \longrightarrow & J_S & \longrightarrow & A' & \longrightarrow & B' & \longrightarrow & Y_S & \longrightarrow & 0 \\
 & & \downarrow & & \downarrow & & \downarrow & & \downarrow{\scriptstyle\epsilon} & & \\
0 & \longrightarrow & C_S & \longrightarrow & A'' & \longrightarrow & B'' & \longrightarrow & \mathbb{Z} & \longrightarrow & 0 \\
 & & \downarrow & & \downarrow & & \downarrow & & \downarrow & & \\
 & & 0 & & 0 & & 0 & & 0. & &
\end{array}
\tag{2.1}
$$

The proof of the following result is due essentially to Tate.

Lemma 2.1. *Assume that* $\mathrm{Cl}(\mathcal{O}_S)$ *vanishes. Then the extension classes of the second and third rows of the diagram (2.1) determine the extension class of the* *first row. Further, if all of the G-modules A', B', A'' and B'' have finite projective dimension, then the extension class of the first row of (2.1) is perfect.*

PROOF. The key point in proving the first assertion is that if $\mathrm{Cl}(\mathcal{O}_S)$ vanishes, then C_S is the module of a class formation and so the group $H^1(G_v, C_S)$ vanishes for each place v in S, where G_v denotes the decomposition group in G of any place of K above v (for more details see the proof of [**10**, Lem. 4]).

The key point regarding the second assertion is that if A', B', A'' and B'' all have finite projective dimension, then the exactness of the two central columns in (2.1) implies that the modules A and B also have finite projective dimension. Given this, and the fact that the modules U_S and X_S are both finitely generated, a standard argument of homological algebra shows that the extension class of the first row is perfect. □

Using the earlier fundamental work of Tate in [**38**] Chinburg proved in [**16**] the existence of a diagram (2.1) in which the second, resp. third, row is a representative of the semi-local, resp. global, canonical class of class field theory for the extension K/k and set of places S (for more details of these canonical classes see, for example, either [**38**] or [**39**]). Further, it was shown that in any such diagram the modules A', B', A'', B'' can be chosen to be of finite projective dimension. Following Lemma 2.1 we may therefore make the following definition.

Definition 2.2. We let τ_S denote the perfect element of $\mathrm{Ext}^2_G(X_S, U_S)$ which is represented by the extension class of the first row of any diagram (2.1) in which the second, resp. third, row is a representative of the semi-local, resp. global, canonical class for the extension K/k and set of places S.

1.2

The following result extends the above definition of τ_S to the general case.

Lemma 2.3. *Let S be any finite non-empty set of places S of k which contains all archimedean places (if any) and all places which ramify in K/k. Then there exists a (finitely generated) G-module $\tilde{X}_S$ which is unique up to unique isomorphism and has the following two properties:*

(i) *There is an exact sequence of G-modules $0 \to \mathrm{Cl}(\mathcal{O}_S) \to \tilde{X}_S \to X_S \to 0$.*
(ii) *There exists a canonical perfect extension class τ_S in $\mathrm{Ext}^2_G(\tilde{X}_S, U_S)$ such that if $\mathrm{Cl}(\mathcal{O}_S)$ vanishes (so $\tilde{X}_S = X_S$), then τ_S coincides with the extension class in Definition 2.2.*

PROOF. This follows by the same argument as used to prove [**10**, Prop. 3.1]. (The argument of loc. cit. is phrased solely in terms of number fields, and assumes that $\mathrm{Cl}(\mathcal{O}_S)$ vanishes, but it extends immediately to the case of global function fields and to the case of general $\mathrm{Cl}(\mathcal{O}_S)$.) □

Remark 2.4. Whilst the descriptions given in Lemma 2.3 are sufficient for our present purposes they obviously give no idea of the nature of the module $\tilde{X}_S$ and extension class τ_S. A more natural way to define both $\tilde{X}_S$ and τ_S is in terms of the compactly supported étale cohomology of the constant sheaf $\mathbb{Z}$ on $\mathrm{Spec}(\mathcal{O}_S)$ and uses the language of complexes rather than Yoneda extensions. For details of this approach (which is used in the proof of Lemma 2.3 indicated above) see [**10**]. See also [**6**] for a discussion of the relation of this construction to Lichtenbaum's conjectural theory of Weil-étale cohomology.

2. Statement of the conjecture

We set $G^\wedge := \mathrm{Hom}(G, \mathbb{C}^\times)$ and for each $\chi \in G^\wedge$ we write e_χ for the idempotent $\frac{1}{|G|}\sum_{g\in G}\chi(g^{-1})g$ of $\mathbb{C}[G]$ and $L_S(\chi, s)$ for the S-truncated Dirichlet L-function that is associated to χ. We then obtain a $\mathbb{C}[G]$-valued meromorphic function of a complex variable z by setting

$$\theta_S(z) := \sum_{\chi\in G^\wedge} L_S(\overline{\chi}, z)e_\chi. \tag{2.2}$$

where $\overline{\chi}$ denotes the contragredient of χ. It is straightforward to show (by using the algebra decomposition (2.4) below and the fact that the action of complex conjugation is continuous) that the leading term $\theta_S^*(0)$ of $\theta_S(z)$ at $z = 0$ belongs to $\mathbb{R}[G]^\times \subset \mathbb{C}[G]$ The relevant case of the ETNC then predicts an explicit formula for the full sublattice $\mathbb{Z}[G]\cdot\theta_S^*(0)$ of $\mathbb{R}[G]$. To state this conjecture we write

$$\lambda_S : U_{S,\mathbb{R}} \xrightarrow{\sim} \tilde{X}_{S,\mathbb{R}} = X_{S,\mathbb{R}}$$

for the isomorphism of $\mathbb{R}[G]$-modules that is induced by (the exact sequence in Lemma 2.3(i) and) the negative of the Dirichlet regulator map. We also recall Definition 1.10.

Conjecture 2.5. $\mathbb{Z}[G]\cdot\theta_S^*(0) = \Xi(\tau_S, \lambda_S)$.

Exercise 2.6. Show that if $K = k$ (so $\mathbb{Z}[G] = \mathbb{Z}$), then Conjecture 2.5 recovers the analytic class number formula up to a sign. (Recall Exercise 1.14.)

Exercise 2.7. Using the result of Exercise 1.13 show that, in general, Conjecture 2.5 is sensitive to a change of sign of either τ_S or λ_S.

We now end this subsection with some general remarks about Conjecture 2.5.

Remark 2.8. We will prove in Proposition 2.12 below that Conjecture 2.5 is a refinement of the abelian case of Stark's Conjecture. However, even if (the abelian case of) Stark's Conjecture is known to be valid, it is still in general very difficult to verify Conjecture 2.5. The main reason for this is that obtaining explicit representatives of a given extension class (and, conversely, computing explicitly the class of a given extension) is a difficult algebraic problem and constitutes a considerable obstacle to computing the lattice $\Xi(\tau_S, \lambda_S)$ and hence to verifying Conjecture 2.5.

Remark 2.9. The validity of Conjecture 2.5, and more generally of the ETNC for any motive that is base changed through a Galois extension K/k, is known to be independent of the choice of S and to behave functorially under change of extension (more precisely, if Conjecture 2.5 is valid for K/k then for each intermediate field E it is valid for both K/E and E/k). Despite the difficulties indicated in Remark 2.8, Conjecture 2.5 is also known to be valid unconditionally in each of the following special cases.

- K is a finite abelian extension of $\mathbb{Q}$. (The validity of Conjecture 2.5 in this case follows from results of Greither and the present author [**12**, Th. 8.1, Rem. 8.1] and later results of Flach dealing with the 2-primary part of the conjecture [**20**].)
- K/k is quadratic. (This case is proved by Kim in [**28**, §2.4, Rem. i)].)
- K is a global function field. (This case is proved in [**8**].)

- If K is an abelian extension of an imaginary quadratic field F and k any intermediate field of K/F, then the theory of elliptic units implies that Stark's Conjecture is valid for every complex character of G and so Conjecture 2.5 decomposes naturally into 'p-primary components'. In this case Bley has shown that the p-primary component of Conjecture 2.5 is valid for every odd prime p which splits in $F/\mathbb{Q}$ and is coprime to the class number of F (cf. [**3**, Th. 4.2]).

Remark 2.10. The formulation of Conjecture 2.5 given above actually looks nothing like the conjectures formulated in [**11**]. The main reason for this is that the case we are dealing with is very special. Indeed, in this case one can formulate a natural leading term conjecture by using a single perfect two-extension of G-modules (namely τ_S) unlike the general case in which one has to work with perfect complexes (rather than Yoneda extensions) over $\mathbb{Z}_p[G]$ for each prime p and then piece things together in the style of the definition (1.2). The proof that Conjecture 2.5 is indeed a special case of the ETNC formulated in [**11**, Conj. 4] is in fact rather technical and is given in [**10**].

Remark 2.11. If, as already alluded to in Remark 2.4, one reinterprets Conjecture 2.5 in terms of étale cohomology (by using the results of [**10**]), then connections with other (commutative) conjectures become apparent. In particular, one finds that in the context of global function fields, resp. number fields, Conjecture 2.5 is an equivariant refinement of the relevant case of Lichtenbaum's conjecture [**31**, Conj. 8.1e)] relating special values of Zeta functions of curves over finite fields to Weil-étale cohomology, resp. is a version without 'sign ambiguities' of the relevant case of Kato's earlier 'generalized Iwasawa main conjecture' [**27**, Conj. 3.2.2]. (These sign ambiguities are of the same type as those discussed in Remark 1.1 and mean that the relevant case of [**27**, Conj. 3.2.2] is itself insufficient to imply either type of the refinements of Stark's Conjecture that we will discuss in Lecture 3.)

3. The relation to Stark's Conjecture

In this section we prove the following result.

Proposition 2.12. *Conjecture 2.5 implies the abelian case of Stark's (Main) Conjecture.*

PROOF. For each homomorphism of $\mathbb{Q}[G]$-modules $\phi : U_{S,\mathbb{Q}} \to X_{S,\mathbb{Q}}$, the composite $\lambda_S^{-1} \circ \phi_{\mathbb{R}}$ is an endomorphism of the (finitely generated) $\mathbb{R}[G]$-module $U_{S,\mathbb{R}}$ and so we obtain an element of $\mathbb{R}[G]$ by setting

$$R(\phi) := \det_{\mathbb{R}[G]}(\lambda_S^{-1} \circ \phi_{\mathbb{R}}) \in \mathbb{R}[G] \tag{2.3}$$

(where we use the notation introduced just prior to Exercise 1.13).

We now fix an *iso*morphism of $\mathbb{Q}[G]$-modules $\phi : U_{S,\mathbb{Q}} \to X_{S,\mathbb{Q}}$ (so that $\phi^{-1}(X_{S,\mathbb{Q}}) = U_{S,\mathbb{Q}}$ and $R(\phi) \in \mathbb{R}[G]^\times$). We also fix a representative of τ_S of the form

$$\mathcal{T}_S : 0 \to U_S \to E_0 \xrightarrow{d} E_1 \to \tilde{X}_S \to 0$$

where the G-modules E_0 and E_1 are both finitely generated and of finite projective dimension. Then, upon unwinding the definition of the homomorphism $\iota(\mathcal{T}_S, \lambda_S) =$

$t(\lambda_S) \circ (\mathbb{R} \otimes_{\mathbb{Q}} \iota(\mathcal{T}_S))$ which occurs in Definition 1.10, one finds that the validity of Conjecture 2.5 implies equalities

$$\begin{aligned}
\mathbb{Q}[G] \cdot \theta_S^*(0) &= \Xi(\tau_S, \lambda_S)_{\mathbb{Q}} \\
&= \iota(\mathcal{T}_S, \lambda_S)([E_{0,\mathbb{Q}}][E_{1,\mathbb{Q}}]^{-1}) \\
&= t(\lambda_S)([U_{S,\mathbb{Q}}][X_{S,\mathbb{Q}}]^{-1}) \\
&= \mathrm{ev}_{X_{S,\mathbb{Q}}}([\lambda_S(U_{S,\mathbb{Q}})][X_{S,\mathbb{Q}}]^{-1}) \\
&= \mathrm{ev}_{X_{S,\mathbb{Q}}}([\lambda_S \circ \phi^{-1}(X_{S,\mathbb{Q}})][X_{S,\mathbb{Q}}]^{-1}) \\
&= \mathrm{ev}_{X_{S,\mathbb{Q}}}([X_{S,\mathbb{Q}}][X_{S,\mathbb{Q}}]^{-1}) \cdot \det_{\mathbb{R}[G]}(\lambda_S \circ \phi_{\mathbb{R}}^{-1}) \\
&= \mathrm{ev}_{X_{S,\mathbb{Q}}}([X_{S,\mathbb{Q}}][X_{S,\mathbb{Q}}]^{-1}) \cdot R(\phi)^{-1} \\
&= \mathbb{Q}[G] \cdot R(\phi)^{-1}.
\end{aligned}$$

Thus, if Conjecture 2.5 is valid, then $\theta_S^*(0)R(\phi)$ belongs to $\mathbb{Q}[G]$.

To explicate the containment $\theta_S^*(0)R(\phi) \in \mathbb{Q}[G]$ we use the following notation: for each character χ in $G^\wedge$ we write $\tilde{\chi} : \mathbb{C}[G] \to \mathbb{C}$ for the $\mathbb{C}$-linear ring homomorphism which sends each element g of G to $\chi(g)$. Then there is a natural identification of $\mathbb{C}$-algebras

$$\mathbb{C}[G] \xrightarrow{\sim} \prod_{\chi \in G^\wedge} \mathbb{C}, \; x \mapsto (\tilde{\chi}(x))_{\chi \in G^\wedge} \tag{2.4}$$

and, with respect to this identification, one has

$$\mathbb{Q}[G] = \{(x_\chi)_{\chi \in G^\wedge} : (x_\chi)^\omega = x_{\chi^\omega} \text{ for all } \chi \in G^\wedge \text{ and all } \omega \in \mathrm{Aut}(\mathbb{C})\}. \tag{2.5}$$

In particular, the containment $\theta_S^*(0)R(\phi) \in \mathbb{Q}[G]$ shown above implies that for all $\chi \in G^\wedge$ and all $\omega \in \mathrm{Aut}(\mathbb{C})$ one has

$$(\tilde{\chi}(\theta_S^*(0)R(\phi)))^\omega = \widetilde{\chi^\omega}(\theta_S^*(0)R(\phi)).$$

To deduce the abelian case of Stark's Main Conjecture (in the formulation given by Tate in [**40**, Chap. I, Conj. 5.1]) from this family of equalities we now only need to note that for each $\chi \in G^\wedge$ one has

$$\tilde{\chi}(\theta_S^*(0)R(\phi)) = \frac{L_S^*(\overline{\chi}, 0)}{\det_{\mathbb{C}}(\lambda_{S,\mathbb{C}} \circ \phi_{\mathbb{C}}^{-1} \mid e_\chi(X_{S,\mathbb{C}}))}. \tag{2.6}$$

□

Exercise 2.13. Prove the equalities (2.5) and (2.6).

LECTURE 3

Explicit refinements of Stark's Conjecture

It is by now known that Conjecture 2.5 either recovers or implies refinements of a wide variety of more explicit (and often better known) refinements of Stark's Conjecture, including all of the following:

- The 'Rubin-Stark Conjecture' of Rubin [**36**].
- Popescu's Conjecture [**35**].
- The 'refined class number formula' of Gross [**23**].
- The 'refined class number formula' of Tate [**41, 42**].
- The 'refined class number formula' of Aoki, Lee and Tan [**1**].
- The 'guess' formulated by Gross in [**23**, top of p. 195].
- The 'refined $\mathfrak{p}$-adic abelian Stark Conjecture' of Gross [**23**].
- Brumer's Conjecture.
- The Brumer-Stark Conjecture.
- The '$\Omega(3)$-Conjecture' of Chinburg [**15, 16**].
- The 'Strong Stark Conjecture' of Chinburg [**15**].
- The 'Lifted Root Number Conjecture' of Gruenberg, Ritter and Weiss [**24**].

In [**7**] the reader can find a discussion of the methods needed to deduce the conjectures listed above from Conjecture 2.5. To give an example of the sort of techniques that are used we will now sketch in §1 how one derives from Conjecture 2.5 the kind of explicit congruences for the value of $\theta_S(z)$ at $z = 0$ that are formulated by Gross in [**23**]. Then in §2 we shall discuss a possible new refinement of Stark's Conjecture that would provide an analogue of Brumer's Conjecture for the values of higher derivatives of $\theta_S(z)$ at $z = 0$.

Remark 3.1. The list given above indicates not just that Conjecture 2.5 is in effect a 'universal' refinement of the abelian case of Stark's Conjecture, but also that a wide array of more explicit, and already much studied, conjectures in the literature can be used to test the validity of Conjecture 2.5, and hence also of the very general ETNC formalism. As an explicit example of this we note that Dasgupta has recently conjectured in [**18**, Conj. 3.21] a precise formula for the unit which occurs in Gross's refined $\mathfrak{p}$-adic abelian Stark Conjecture and has also shown that, under suitable conditions, this precise formula is implied, up to a root of unity, by Gross's conjecture. Explicit computational evidence for Dasgupta's conjectural formula (such as that described in [**18**, §4]) can thus be regarded as evidence for Conjecture 2.5. I am very grateful to Samit Dasgupta for pointing this out.

1. Congruences for values of L-series

We now explain how Conjecture 2.5 predicts certain explicit integral congruences for the value

$$\theta_S(0) := \sum_{\chi \in G^\wedge} L_S(\overline{\chi}, 0) e_\chi$$

of the function $\theta_S(z)$ at $z = 0$.

In order not to hide the main ideas we prefer to avoid certain technical difficulties that are caused by the presence of roots of unity in K. To do this we write w_K for the number of roots of unity in K, set $\mathbb{Z}' := \mathbb{Z}[\frac{1}{w_k}]$ and only consider the image of the equality of Conjecture 2.5 after applying the functor $\mathbb{Z}' \otimes_{\mathbb{Z}} -$. We set $\mathfrak{A} := \mathbb{Z}[G]$ and $\mathfrak{A}' := \mathbb{Z}'[G]$ and for any $\mathfrak{A}$-module M we write M' for the $\mathfrak{A}'$-module $\mathbb{Z}' \otimes_{\mathbb{Z}} M$. For any $\mathbb{Z}'$-module M we set $M_{\mathbb{R}} := \mathbb{R} \otimes_{\mathbb{Z}'} M$.

1.1

We write τ_S' for the (perfect) element of $\mathrm{Ext}^2_{\mathfrak{A}'}(U_S', \tilde{X}_S')$ that is obtained as the image of $1_{\mathbb{Z}'} \otimes_{\mathbb{Z}} \tau_S$ under the natural isomorphism of finite groups

$$\mathbb{Z}' \otimes_{\mathbb{Z}} \mathrm{Ext}^2_{\mathfrak{A}}(U_S, \tilde{X}_S) \cong \mathrm{Ext}^2_{\mathfrak{A}'}(U_S', \tilde{X}_S').$$

In the following preliminary result we set $n := |S| - 1$.

Lemma 3.2. *Assume that Conjecture 2.5 is valid for K/k.*

(i) *Then, regarding λ_S as an isomorphism of $\mathbb{R}[G]$-modules of the form $U'_{S,\mathbb{R}} = U_{S,\mathbb{R}} \to \tilde{X}_{S,\mathbb{R}} = \tilde{X}'_{S,\mathbb{R}}$, one has $\mathfrak{A}' \cdot \theta_S^*(0) = \Xi(\tau_S', \lambda_S)$.*

(ii) *There exists a representative of τ_S' of the form*

$$0 \to U_S' \to F \xrightarrow{\phi} F \xrightarrow{\pi} \tilde{X}_S' \to 0 \tag{3.1}$$

where F is a finitely generated free $\mathfrak{A}'$-module with a basis $\{b_i : 1 \le i \le d\}$ such that each of the following is true.

(a) *$d \ge n$.*

(b) *The module $F_1 := \mathfrak{A}' \cdot \{b_i : 1 \le i \le n\}$ satisfies $F_1^G = \ker(\phi^G)$.*

(c) *The module $F_2 := \mathfrak{A}' \cdot \{b_i : n < i \le d\}$ satisfies $\phi(F_2^G) \subseteq F_2^G$.*

Proof. We only sketch the key points of the proof.

Since $\Xi(\tau_S', \lambda_S) = \Xi(\tau_S, \lambda_S)'$ the equality of claim (i) follows immediately from that of Conjecture 2.5.

To prove claim (ii) we first choose a representative of the extension class τ_S' of the form

$$0 \to U_S' \xrightarrow{\alpha} M \xrightarrow{\beta} F \xrightarrow{\gamma} \tilde{X}_S' \to 0 \tag{3.2}$$

in which F is a finitely generated free $\mathfrak{A}'$-module of rank $d \ge n$ and M is a finitely generated $\mathfrak{A}'$-module which has finite projective dimension (such a representative exists because τ_S' is perfect). Then the explicit definition of $\Xi(\tau_S', \lambda_S)$ (in Definition 1.10) implies that there is an isomorphism of $\mathfrak{A}'$-modules of the form

$$[M]_{\mathfrak{A}'}[F]_{\mathfrak{A}'}^{-1} \cong \Xi(\tau_S', \lambda_S). \tag{3.3}$$

We claim that this isomorphism implies that M is isomorphic to F. To see this we note first that, since M has finite projective dimension and is also $\mathbb{Z}'$-torsion-free (since both U_S' and F are $\mathbb{Z}'$-torsion-free) it is a projective $\mathfrak{A}'$-module (cf. [**2**, Th. 8]), hence a locally-free $\mathfrak{A}'$-module by Swan's Theorem (cf. [**17**, (32.11)]) and so is isomorphic to a module of the form $(\mathfrak{A}')^{\mathrm{rk}_{\mathfrak{A}'}(M)-1} \oplus I$ for a locally-free $\mathfrak{A}'$-module I of rank one (by [**17**, (49.3)]). This implies in particular that I is isomorphic to $\wedge_{\mathfrak{A}'}^{\mathrm{rk}_{\mathfrak{A}'}(M)} M = [M]_{\mathfrak{A}'}$. Now, since F is a free $\mathfrak{A}'$-module, $[F]_{\mathfrak{A}'}$ is isomorphic to $\mathfrak{A}'$. In addition, since $\theta_S^*(0)$ is a unit of $\mathbb{R}[G]$, the equality of claim (i) implies that the map $x \mapsto x\theta_S^*(0)^{-1}$ gives an isomorphism of $\mathfrak{A}'$-modules $\Xi(\tau_S', \lambda_S) \cong \mathfrak{A}'$. From the isomorphism (3.3) it therefore follows that the module $[M]_{\mathfrak{A}'}$, and hence also I, is isomorphic to $\mathfrak{A}'$. This implies that $M \cong (\mathfrak{A}')^{\mathrm{rk}_{\mathfrak{A}'}(M)-1} \oplus I$ is a free $\mathfrak{A}'$-module and, by considering ranks of the modules in the sequence obtained by applying $\mathbb{Q} \otimes_{\mathbb{Z}'} -$ to (3.2), one finds that M must be isomorphic to F, as claimed.

After making suitable changes to α and β one can therefore replace M by F in (3.2) without changing the extension class τ_S' that the sequence represents. It is then a straightforward exercise to show that, if necessary after replacing β by $\phi := \iota \circ \beta$ and γ by $\pi := \gamma \circ \iota^{-1}$ for a suitable automorphism ι of F, one obtains an exact sequence (3.1) which represents τ_S' and also satisfies all of the conditions in claim (ii). $\square$

Exercise 3.3. With ϕ and π as in (3.1), choose sections ι_1 and ι_2 to the surjective $\mathbb{R}[G]$-module homomorphisms $F_{\mathbb{R}} \xrightarrow{\phi_{\mathbb{R}}} \mathrm{im}(\phi)_{\mathbb{R}}$ and $F_{\mathbb{R}} \xrightarrow{\pi_{\mathbb{R}}} \tilde{X}'_{S,\mathbb{R}}$ respectively (such sections exist because the algebra $\mathbb{R}[G]$ is semisimple). Write $\langle \phi, \lambda_S \rangle$ for the unique element of $\mathrm{Aut}_{\mathbb{R}[G]}(F_{\mathbb{R}})$ that is equal to $\iota_2 \circ \lambda_S$ on $U'_{S,\mathbb{R}}$ and to $\phi_{\mathbb{R}}$ on $\iota_1(\mathrm{im}(\phi)_{\mathbb{R}})$. Prove that $\Xi(\tau_S', \lambda_S) = \mathfrak{A}' \cdot \det_{\mathbb{R}[G]}(\langle \phi, \lambda_S \rangle)$.

1.2

We now consider the element $\theta_S(0)$. To do this we first define an idempotent in $\mathbb{C}[G]$ by setting

$$e_0 := \sum_{\chi \in G_0^\wedge} e_\chi$$

where $G_0^\wedge$ is the set of characters χ in $G^\wedge$ with the property that $L_S(\overline{\chi}, 0) \neq 0$.

If for each v in S we write G_v for the decomposition group in G of any place of K above v, then the algebraic order $r_S(\chi)$ of the function $L_S(\overline{\chi}, z)$ at $z = 0$ is equal to

$$r_S(\chi) = \dim_{\mathbb{C}}(e_\chi(X_{S,\mathbb{C}})) = \begin{cases} \sum_{v \in S} \dim_{\mathbb{C}}(e_\chi(\mathbb{C}[G/G_v])), & \text{if } \chi \neq 1_G, \\ |S| - 1, & \text{if } \chi = 1_G \end{cases} \tag{3.4}$$

where 1_G is the trivial character of G (cf. [**40**, Chap. I, Prop. 3.4]). As an immediate consequence of this formula (and the existence of the isomorphism λ_S) one finds that

$$G_0^\wedge = \{\chi \in G^\wedge : e_\chi(X_{S,\mathbb{C}}) = 0\} = \{\chi \in G^\wedge : e_\chi(U_{S,\mathbb{C}}) = 0\}. \tag{3.5}$$

We claim that if Conjecture 2.5 is valid, then there exists a unit u of $\mathfrak{A}'$ such that

$$\begin{aligned} \theta_S(0) &= \theta_S(0)e_0 + \theta_S(0)(1 - e_0) \\ &= \theta_S^*(0)e_0 \\ &= u \cdot \det_{\mathbb{R}[G]}(\langle \phi, \lambda_S \rangle)e_0 \\ &= u \cdot \det_{\mathbb{R}[G]}(\phi_{\mathbb{R}})e_0 \\ &= u \cdot \det_{\mathfrak{A}'}(\phi). \end{aligned} \tag{3.6}$$

Indeed the second equality here follows because the definition of e_0 implies that $\theta_S(0) = \theta_S(0)e_0 = \theta_S^*(0)e_0$; the third equality follows (for a suitable unit u) by combining the (conjectural) equality of Lemma 3.2(i) with the explicit formula of Exercise 3.3; the fourth equality is because the description of (3.5) implies that $e_0(U_{S,\mathbb{Q}}) = 0$ and hence that $\langle \phi, \lambda_S \rangle$ and $\phi_{\mathbb{R}}$ restrict to give the same endomorphism of $e_0 F_{\mathbb{R}}$; the last equality follows from Exercise 3.4 below.

We write I_G for the kernel of the natural augmentation map $\epsilon : \mathfrak{A}' \to \mathbb{Z}'$. Then Lemma 3.2(ii)(b) and (c) combine to imply that the matrix of ϕ with respect to the given basis $\{b_i : 1 \le i \le d\}$ is a block matrix of the form

$$\left(\begin{array}{c|c} A & B \\ \hline C & D \end{array}\right) \tag{3.7}$$

where A belongs to $\mathrm{M}_n(I_G)$, D belongs to $\mathrm{M}_{d-n}(\mathfrak{A}')$ and all entries of the matrices B and C belong to I_G. It follows that $\det(A)$ belongs to I_G^n and hence that the equality (3.6) implies

$$\begin{aligned} \theta_S(0) &\equiv u\det(D)\det(A) \pmod{I_G^{n+1}} \\ &\equiv \epsilon(u)\epsilon(\det(D))\det(A) \pmod{I_G^{n+1}}. \end{aligned} \tag{3.8}$$

The key observation now is that the right hand side of this congruence is equal to the discriminant of the algebraic height pairing defined in the next subsection.

Exercise 3.4. Using the description (3.5) prove that $e_\chi(\ker(\phi)_{\mathbb{C}}) \neq 0$ for all characters $\chi \notin G_0^\wedge$. Deduce that $\det_{\mathfrak{A}'}(\phi) = \det_{\mathfrak{A}'}(\phi)e_0$.

1.3

For any G-module M we write M^G and M_G for the maximal submodule and quotient module upon which G acts trivially. We use similar notation for homomorphisms of G-modules.

To review the construction of the relevant height pairing (which is motivated by the general approach developed by Nekovář in [**34**, §11]) we write $F^\bullet$, $F_G^\bullet$ and $F^{\bullet,G}$ for the complexes $F \xrightarrow{\phi} F$, $F_G \xrightarrow{\phi_G} F_G$ and $F^G \xrightarrow{\phi^G} F^G$ where in each case the first term is placed in degree 0 and the homomorphism ϕ is as in (3.1). We note that the action of the element $\mathrm{Tr}_G := \sum_{g\in G} g$ of $\mathfrak{A}'$ on F induces an isomorphism $F_G \cong F^G$ and hence also a quasi-isomorphism from $F_G^\bullet$ to $F^{\bullet,G}$. In particular, by tensoring (over $\mathfrak{A}'$) the tautological short exact sequence $0 \to I_G \to \mathfrak{A}' \to \mathbb{Z}' \to 0$ with $F^\bullet$ and using the given quasi-isomorphism $\mathbb{Z}' \otimes_{\mathfrak{A}'} F^\bullet = F_G^\bullet \cong F^{\bullet,G}$ one obtains a short exact sequence of complexes of the form

$$0 \to I_G \otimes_{\mathfrak{A}'} F^\bullet \to F^\bullet \to F^{\bullet,G} \to 0.$$

From the long exact cohomology sequence of this exact sequence one obtains a connecting homomorphism

$$\ker(\phi)^G = H^0(F^{\bullet,G}) \to H^1(I_G \otimes_{\mathfrak{A}'} F^\bullet) = I_G \otimes_{\mathfrak{A}'} H^1(F^\bullet) = I_G \otimes_{\mathfrak{A}'} \mathrm{cok}(\phi)$$

and hence also, by passing to G-coinvariants, a homomorphism

$$\beta_\phi : \ker(\phi)^G \to I_G/I_G^2 \otimes_{\mathbb{Z}'} \mathrm{cok}(\phi)_G.$$

We write

$$\rho_\phi : \ker(\phi)^G \times \mathrm{Hom}_{\mathbb{Z}'}(\mathrm{cok}(\phi)_G, \mathbb{Z}') \to I_G/I_G^2$$

for the pairing that is induced by β_ϕ.

Lemma 3.5. *There are ordered $\mathbb{Z}'$-bases $\{x_i : 1 \leq i \leq n\}$ and $\{y_j : 1 \leq i \leq n\}$ of $\ker(\phi)^G$ and $\mathrm{Hom}_{\mathbb{Z}'}(\mathrm{cok}(\phi)_G, \mathbb{Z}')$ respectively for which one has*

$$\det((\rho_\phi(x_i, y_j))_{1\leq i,j\leq n}) = \epsilon(u)\epsilon(\det(D))\det(A) \in I_G^n/I_G^{n+1} \tag{3.9}$$

where u, D and A are as in (3.8).

Proof. One can show that $\epsilon(\det(D))$, and hence also $\epsilon(u)\epsilon(\det(D))$, is a unit of $\mathbb{Z}'$. The key point however is to show that there are ordered $\mathbb{Z}'$-bases of $\ker(\phi)^G$ and $\mathrm{Hom}_{\mathbb{Z}'}(\mathrm{cok}(\phi)_G, \mathbb{Z}')$ with respect to which the matrix of ρ_ϕ is equal to the image in $\mathrm{M}_n(I_G/I_G^2)$ of the matrix A in (3.7). To do this one observes that β_ϕ is equal to the connecting homomorphism which arises when applying the Snake lemma to the following commutative diagram (in which both rows and the third column are exact and the first column is a complex)

$$\begin{array}{ccccccccc}
 & & & & & & 0 & & \\
 & & & & & & \downarrow & & \\
 & & & & & & \ker(\phi)^G & & \\
 & & & & & & \downarrow & & \\
0 & \longrightarrow & I_G \otimes_{\mathfrak{A}'} F & \xrightarrow{\subseteq} & F & \xrightarrow{\mathrm{Tr}_G} & F^G & \longrightarrow & 0 \\
 & & \downarrow{\scriptstyle \mathrm{id}\otimes\phi} & & \downarrow{\scriptstyle \phi} & & \downarrow{\scriptstyle \phi^G} & & \\
0 & \longrightarrow & I_G \otimes_{\mathfrak{A}'} F & \xrightarrow{\subseteq} & F & \xrightarrow{\mathrm{Tr}_G} & F^G & \longrightarrow & 0 \\
 & & \downarrow{\scriptstyle \rho} & & & & & & \\
 & & I_G/I_G^2 \otimes_{\mathbb{Z}'} \mathrm{cok}(\phi)_G & & & & & &
\end{array}$$

where ρ is the natural projection. To compute explicitly the connecting homomorphism in this diagram one then uses the matrix representation of ϕ given in (3.7). □

Remark 3.6. In keeping with the philosophy of Mazur and Tate [**33**] and Gross [**23**], the canonical isomorphism $I_G/I_G^2 \cong \mathbb{Z}' \otimes G$ allows one to regard the pairing ρ_ϕ as taking values in the quotient $\mathbb{Z}' \otimes G$ of G.

1.4

Given the congruences (3.8) and (3.9) the connection between Conjecture 2.5 and the congruences conjectured by Gross in [**23**] is provided by the next result.

Up until this point the only property of the class τ_S' that we have used is that it is perfect. However, the proof of the following result depends critically upon the precise class field theoretical definition of τ_S'.

Theorem 3.7. *With respect to the natural identifications $\ker(\phi)^G \cong (\mathcal{O}_{k,S}^\times)'$ and $\mathrm{Hom}_{\mathbb{Z}'}(\mathrm{cok}(\phi)_G, \mathbb{Z}') \cong \mathrm{Hom}_{\mathbb{Z}'}(X_{k,S}', \mathbb{Z}') \cong X_{k,S}'$ and $I_G/I_G^2 \cong \mathbb{Z}' \otimes G$ the pairing ρ_ϕ coincides with the image under the projection $G \to \mathbb{Z}' \otimes G$ of the G-valued reciprocity pairing introduced by Gross in* [**23**].

The proof of Theorem 3.7 relies on the fact that local reciprocity maps are defined as cup products and hence can also be interpreted as connecting homomorphisms, thus providing a link to the homomorphism β_ϕ. For full details of the deduction from Conjecture 2.5 of the refined class number formula of Gross (as well as several other explicit conjectures including the Rubin-Stark Conjecture) see [**5**].

Remark 3.8. It will be clear from the above sketch that much of the argument needed to establish a link between Conjecture 2.5 and the congruences studied by Gross in [**23**] is homological algebra of a fairly universal nature. In particular, if one replaces the role of the complex $F^\bullet$ used above by a suitable 'Selmer complex' of the kind defined by Nekovář in [**34**, §11], then essentially the same approach can be used to relate the refined Birch and Swinnerton-Dyer Conjecture that is formulated by Mazur and Tate in [**33**] to a different special case of the ETNC. For more details of this important example see [**7**].

2. Higher derivatives of L-series and annihilators of class groups

In this final section we review some work in progress.[1]

We first recall Brumer's conjectural generalization of Stickelberger's Theorem. To do this we write μ_K for the group of roots of unity in the global field K.

Conjecture 3.9. *(Brumer) For each x in $\mathrm{Ann}_{\mathbb{Z}[G]}(\mu_K)$ and each non-empty subset S' of S which contains all archimedean places of k (if any) one has*

$$x\theta_S(0) \in \mathrm{Ann}_{\mathbb{Z}[G]}(\mathrm{Cl}(\mathcal{O}_{S'})).$$

Conjecture 3.9 is clearly not interesting if, as is often the case, $\theta_S(0) = 0$. However we shall now discuss how Conjecture 2.5 suggests a natural refinement of Stark's Conjecture that is in effect an analogue of Conjecture 3.9 for the value at $z = 0$ of higher derivatives of $\theta_S(z)$.

We recall that if ψ is any element of $\mathrm{Hom}_G(U_S, X_S)$, then (the abelian case of) Stark's Conjecture asserts that

$$\theta_S^*(0)R(\psi_\mathbb{Q}) \in \mathbb{Q}[G] \tag{3.10}$$

(cf. the proof of Proposition 2.12).

We consider the following possible refinement of (3.10).

Question 3.10. *Let r be any non-negative integer for which the function $\theta_S^r(z) := s^{-r}\theta_S(z)$ is holomorphic at $z = 0$. Does the element $x\theta_S^r(0)R(\psi_\mathbb{Q})$ belong to $\mathrm{Ann}_{\mathbb{Z}[G]}(\mathrm{Cl}(\mathcal{O}_S))$ for every x in $\mathrm{Ann}_{\mathbb{Z}[G]}(\mu_K)$ and every ψ in $\mathrm{Hom}_G(U_S, X_S)$?*

One can show that an affirmative answer to Question 3.10 would imply the validity of the Rubin-Stark Conjecture and hence also of Conjecture 3.9. However an affirmative answer to Question 3.10 would not itself imply the full validity of (3.10) because $\theta_S^r(0)$ does not in general coincide with the leading term $\theta_S^*(0)$.

The connection between Question 3.10 and Conjecture 2.5 is explained by the next result. For any prime p we shall say that the 'answer to Question 3.10_p is affirmative' if for every x in $\mathrm{Ann}_{\mathbb{Z}[G]}(\mu_K)$ and every ψ in $\mathrm{Hom}_G(U_S, X_S)$ the element $x\theta_S^r(0)R(\psi_\mathbb{Q})$ belongs to $\mathrm{Ann}_{\mathbb{Z}[G]}(\mathrm{Cl}(\mathcal{O}_S))_{(p)}$.

Theorem 3.11. *Assume that Conjecture 2.5 is valid. Then the answer to Question 3.10_p is affirmative in both of the following cases.*

(i) *There are r places in S which split completely in K/k and the G-module $\mu_{K,(p)}$ has finite projective dimension.*

[1]Note added in proof: the results discussed here, as well as natural generalisations of these results to non-abelian Galois extensions, are considered in detail in [**9**].

(ii) *p does not divide $|G|$.*

By combining this result with Remark 2.9 we obtain the following unconditional result.

Corollary 3.12. *If either p does not divide $|G|$ or there are r places in S which split completely in K/k and the G-module $\mu_{K,(p)}$ has finite projective dimension, then the answer to Question 3.10$_p$ is affirmative in each of the following cases.*

- *K is a finite abelian extension of $\mathbb{Q}$.*
- *k contains an imaginary quadratic field F in which p splits and is coprime to the class number and the extension K/F is abelian.*
- *K is a global function field.*

Remark 3.13. In [**32**] Macias Castillo builds on work of Rubin and Sands to show that the answer to Question 3.10 is affirmative for all quadratic extensions K/k and also for a wide range of multi-quadratic extensions. Macias Castillo also explains the connection between Question 3.10 and recent work of Popescu and Emmons and of Buckingham and Snaith.

The following results give some indication of the interest of Question 3.10 in the context of extensions of totally real number fields.

Proposition 3.14. *Let K/k be a finite abelian extension of totally real number fields and write $\mathcal{O}$ for the ring of algebraic integers in K. Let S be any finite set of places of k which contains all archimedean places, all places which ramify in K/k and at least one non-archimedean place. Fix a prime p. Then if the answer to Question 3.10$_p$ is affirmative for K/k, then for every ψ in $\mathrm{Hom}_G(U_S, X_S)$ and every g in G one has*

$$2|G|(g-1)\theta_S^{[k:\mathbb{Q}]}(0)R(\psi_{\mathbb{Q}}) \in \mathrm{Ann}_{\mathbb{Z}[G]}(\mathrm{Cl}(\mathcal{O}))_{(p)}.$$

Proof. We write S_∞ for the set of archimedean places of k. For each place v in $S \setminus S_\infty$ we write Id_v for the group of fractional ideals of K above v. Then the definition of $\mathrm{Cl}(\mathcal{O}_S)$ implies that there is an exact sequence of G-modules

$$\bigoplus_{v\in S\setminus S_\infty} \mathrm{Id}_v \to \mathrm{Cl}(\mathcal{O}) \to \mathrm{Cl}(\mathcal{O}_S) \to 0. \tag{3.11}$$

We set $r := |S_\infty| = [k:\mathbb{Q}]$ and note that, since $|S| > |S_\infty| = r$ and each place of S_∞ splits completely in K/k, the explicit formula (3.4) for the algebraic order of each function $L_S(\overline{\chi}, z)$ at $z=0$ implies that the function $\theta_S^r(z)$ is holomorphic at $z=0$ and moreover that if $\chi \neq 1_G$ and $e_\chi\theta_S^r(0)\neq 0$, then one has $e_\chi(\mathrm{Id}_{v,\mathbb{C}}) = 0$ for every $v \in S\setminus S_\infty$. In particular therefore, if we set $e := \sum_\chi e_\chi$ where χ runs over all elements of $G^\wedge$ such that $e_\chi\theta_S^r(0)\neq 0$, then one has $\theta_S^r(0) = e\theta_S^r(0)$ and, since both $|G|e \in \mathbb{Z}[G]$ and $(g-1)e_{1_G} = 0$, also

$$|G|(g-1)e \in \mathrm{Ann}_{\mathbb{Z}[G]}(\bigoplus_{v\in S\setminus S_\infty} \mathrm{Id}_v). \tag{3.12}$$

Next we note that $\mu_K = \{\pm 1\}$ so that the element $x = 2$ belongs to $\mathrm{Ann}_{\mathbb{Z}[G]}(\mu_K)$. Hence, if the answer to Question 3.10$_p$ is affirmative for K/k and with $r = [k:\mathbb{Q}]$, then for every ψ in $\mathrm{Hom}_G(U_S, X_S)$ one has $2\theta_S^r(0)R(\psi_{\mathbb{Q}}) \in \mathrm{Ann}_{\mathbb{Z}[G]}(\mathrm{Cl}(\mathcal{O}_S))_{(p)}$. By combining this containment with the equality $\theta_S^r(0) = e\theta_S^r(0)$ and the containment (3.12) one deduces from the exact sequence (3.11) that the element

$$\begin{aligned} 2|G|(g-1)\theta_S^r(0)R(\psi_{\mathbb{Q}}) &= 2|G|(g-1)e\theta_S^r(0)R(\psi_{\mathbb{Q}}) \\ &= (|G|(g-1)e)(2\theta_S^r(0)R(\psi_{\mathbb{Q}})) \end{aligned}$$

belongs to $\mathrm{Ann}_{\mathbb{Z}[G]}(\mathrm{Cl}(\mathcal{O}))_{(p)}$, as required. □

In the next result we refer to the 'p-adic Stark conjecture at $s = 1$' of Serre (see Remark 3.16 below).

Corollary 3.15. *Let K/k be an abelian extension of totally real fields. Let S be any finite set of places of k which contains all archimedean places, all places which ramify in K/k and at least one non-archimedean place. Let p be an odd prime which does not divide $[K:k]$. Then if the 'p-adic Stark conjecture at $s = 1$' is valid for K/k, then for every ψ in $\mathrm{Hom}_G(U_S, X_S)$ and every g in G one has*

$$(g-1)\theta_S^{[k:\mathbb{Q}]}(0)R(\psi_{\mathbb{Q}}) \in \mathrm{Ann}_{\mathbb{Z}[G]}(\mathrm{Cl}(\mathcal{O}_K))_{(p)}.$$

PROOF. The key point of this proof is that if the p-adic Stark conjecture at $s = 1$ is valid for K/k, then one can show that the answer to Question 3.10$_p$ is affirmative. Then, since p is both odd and coprime to $|G|$, the factor $2|G|$ is a unit of $\mathbb{Z}_{(p)}$ and so the result follows directly from Proposition 3.14. □

Remark 3.16. The 'p-adic Stark conjecture at $s = 1$' is discussed by Tate in [**40**, Chap. VI, §5] where it is attributed to Serre [**37**]. In fact there are imprecisions in the discussion of [**40**, Chap. VI, §5] but a precise statement of the conjecture can be found in [**13**, Conj. 5.2]. The conjecture predicts an explicit formula for the leading term at $s = 1$ of p-adic Artin L-functions of p-adic characters ψ of G and it is known that the validity of Leopoldt's Conjecture for K at p is equivalent to the validity of the p-adic Stark conjecture at $s = 1$ for all permutation characters ψ of G (cf. [**13**, Rem. 5.4]). Corollary 3.15 therefore constitutes a natural extension of the main result of Büyükboduk in [**14**].

Bibliography

[1] N. Aoki, J. Lee, K-S. Tan, *A refinement for a conjecture of Gross*, preprint 2005.

[2] M. F. Atiyah, C. T. C. Wall, *Cohomology of Groups*, In: *'Algebraic Number Theory'*, J. W. S. Cassels, A. Fröhlich (eds.), Academic Press, London, 1967.

[3] W. Bley, *On the equivariant Tamagawa number conjecture for abelian extensions of a quadratic imaginary field*, Doc. Math. **11** (2006) 73-118.

[4] S. Bloch, K. Kato, *L-functions and Tamagawa numbers of motives*, In: 'The Grothendieck Festschrift' vol. 1, Progress in Math. **86**, Birkhäuser, Boston, (1990) 333-400.

[5] D. Burns, *Congruences between derivatives of abelian L-functions at $s = 0$*, Inventiones Math. **169** (2007) 451-499.

[6] D. Burns, *Perfecting the nearly perfect*, Pure App. Math. Q. **4** (2008) 1041-1058 (Jean-Pierre Serre Special Issue, Part I).

[7] D. Burns, *Leading terms and values of equivariant motivic L-functions*, Pure App. Math. Q. **6** (2010) 83-172 (John Tate Special Issue, Part II).

[8] D. Burns, *Congruences between derivatives of geometric L-functions (with an appendix by D. Burns, K. F. Lai and K-S. Tan)*, to appear in Inventiones Math.

[9] D. Burns, *On derivatives of Artin L-series*, submitted for publication.

[10] D. Burns, M. Flach, *On Galois structure invariants associated to Tate motives*, Amer. J. Math. **120** (1998) 1343–1397.

[11] D. Burns, M. Flach, *Tamagawa numbers for motives with (non-commutative) coefficients*, Doc. Math. **6** (2001), 501–570.

[12] D. Burns, C. Greither, *On the Equivariant Tamagawa Number Conjecture for Tate motives*, Inventiones Math. **153** (2003) 303-359.

[13] D. Burns, O. Venjakob, *Leading terms of Zeta isomorphisms and p-adic L-functions in non-commutative Iwasawa theory*, Doc. Math., Extra Volume (2006) 165-209.

[14] K. Büyükboduk, *Kolyvagin systems of Stark units*, J. Reine Angew. Math. **631** (2009) 85–107.

[15] T. Chinburg, *On the Galois structure of algebraic integers and S-units*, Inventiones Math. **74** (1983) 321–349.

[16] T. Chinburg, *Exact sequences and Galois module structure*, Ann. Math. **121** (1985) 351–376.

[17] C. W. Curtis, I. Reiner, *Methods of Representation Theory, Vol. I and II*, John Wiley and Sons, New York, 1987.

[18] S. Dasgupta, *Shintani zeta-functions and Gross-Stark units for totally real fields*, Duke Math. J. **143** (2008) 225-279.

[19] M. Flach, *The equivariant Tamagawa number conjecture: a survey (with an appendix by C. Greither)*, Contemp. Math. **358** (2004) 79-126.

[20] M. Flach, *On the cyclotomic main conjecture for the prime* 2, to appear in J. Reine u. Angew. Math.

[21] J.-M. Fontaine, B. Perrin-Riou, *Autour des conjectures de Bloch et Kato: cohomologie galoisienne et valeurs de fonctions L*, In: Motives (Seattle) Proc. Symp. Pure Math. **55**, I, (1994) 599-706.

[22] T. Fukaya, K. Kato, *A formulation of conjectures on p-adic zeta functions in non-commutative Iwasawa theory*, Proc. St. Petersburg Math. Soc. Vol. XII, 1–85, Amer. Math. Soc. Transl. Ser. 2, **219**, Amer. Math. Soc., Providence, RI, 2006.

[23] B. H. Gross, *On the value of abelian L-functions at $s = 0$*, J. Fac. Sci. Univ. Tokyo, Sect. IA, Math., **35** (1988) 177-197.

[24] K. W. Gruenberg, J. Ritter, A. Weiss, *A local approach to Chinburg's root number conjecture*, Proc. London Math. Soc. **79** (1999) 47–80.

[25] P. J. Hilton, U. Stammbach, *A course in homological algebra* (Springer-Verlag, New York, 1997).

[26] K. Kato, *Iwasawa theory and p-adic Hodge theory*, Kodai Math. J. **16** no 1 (1993) 1-31.

[27] K. Kato, *Lectures on the approach to Iwasawa theory of Hasse-Weil L-functions via B_{dR}, I*, in: *Arithmetic algebraic geometry (Trento, 1991)*, Lecture Notes in Math. **1553** (Springer, Berlin, 1993) 50–163.

[28] S. Y. Kim, *On the Equivariant Tamagawa Number Conjecture for Quaternion Fields*, Ph. D. thesis, King's College London, 2003.

[29] G. Kings, *An Introduction to the Equivariant Tamagawa Number Conjecture: the Relation to the Birch-Swinnerton-Dyer Conjecture*, in this volume.

[30] F. Knudsen, D. Mumford, *The projectivity of the moduli space of stable curves I: Preliminaries on 'det' and 'Div'*, Math. Scand. **39** (1976) 19-55.

[31] S. Lichtenbaum, *The Weil-étale topology on schemes over finite fields*, Compositio Math. **141** (2005) 689-702.

[32] D. Macias Castillo, *On higher order Stickelberger type theorems*, preprint 2009.

[33] B. Mazur, J. Tate, *Refined Conjectures of the Birch and Swinnerton-Dyer Type*, Duke Math. J. **54** (1987) 711-750.

[34] J. Nekovář, *Selmer Complexes*, Astérisque **310**, S.M.F., Paris, 2006.

[35] C. D. Popescu, *Base Change for Stark-Type Conjectures 'over $\mathbb{Z}$'*, J. Reine u. Angew. Math. **542** (2002) 85-111.

[36] K. Rubin, *A Stark Conjecture 'over $\mathbb{Z}$' for abelian L-functions with multiple zeros*, Ann. Inst. Fourier **46** (1996) 33-62.

[37] J-P. Serre, *Sur le résidue de la fonction zêta p-adique d'un corps de nombres*, C.R. Acad. Sci. Paris **278** (1978) 183-188.

[38] J. Tate, *The cohomology groups of tori in finite Galois extensions of number fields*, Nagoya Math. J. **27** (1966) 709–719.

[39] J. Tate, *Global Class Field Theory*, In: *'Algebraic Number Theory'*, J. W. S. Cassels, A. Fröhlich (eds.), Academic Press, London, 1967.

[40] J. Tate, *Les conjectures de Stark sur les fonctions L d'Artin en $s = 0$* (Birkhäuser, Boston, 1984).

[41] J. Tate, *Letter to J. Lee*, 22 July 1997.

[42] J. Tate, *Refining Gross's conjecture on the values of abelian L-functions*, Contemp. Math. **358** (2004) 189-192.

[43] O. Venjakob, *From the Birch and Swinnerton-Dyer Conjecture to non-commutative Iwasawa theory via the Equivariant Tamagawa Number Conjecture - a survey*, in '*L*-functions and Galois Representations', Proceedings of the 2004 Durham Symposium, pp. 333-380, Cambridge Univ. Press, 2007.

Part II

Birch and Swinnerton-Dyer Conjecture

Introduction to Elliptic Curves

Alice Silverberg

IAS/Park City Mathematics Series
Volume 18, 2009

Introduction to Elliptic Curves

Alice Silverberg

Introduction

What are some reasons to study elliptic curves? Solving equations is a classical problem with a long history. Starting with the simplest equations, we know that linear and quadratic equations are easy to solve. However, there are still many interesting unanswered questions about cubic equations. In addition, there are important applications of elliptic curves to cryptography. There are also important applications of elliptic curves within mathematics, notably to the proof of Fermat's Last Theorem. By studying elliptic curves, one learns about deep connections among arithmetic, algebra, geometry, and complex analysis.

We give a very brief introduction to elliptic curves. Some places to look for more information include Silverman's books [**9, 10**], Chapter 4 of Shimura's book [**6**], two papers [**1, 8**] from the 1999 IAS/PCMI Graduate Summer Program, and the paper [**5**] on ranks, and the references therein.

1. Definitions

We begin with some basic definitions.

Definition 1.1. An *elliptic curve* E over a field F is a smooth projective curve with an affine equation of the form

$$y^2 + a_1xy + a_3y = x^3 + a_2x^2 + a_4x + a_6 \tag{1}$$

where the a_i's are in F. An equation of this form is called a *generalized Weierstrass equation.*

Recall that the points in projective space $\mathbb{P}^n(F)$ are the equivalence classes in $F^{n+1} - \{(0,\ldots,0)\}$ with respect to the equivalence relation $(x_0,\ldots,x_n) \approx (\lambda x_0,\ldots,\lambda x_n)$ with $\lambda \in F^\times$. The projective curve corresponding to the affine equation (1) is given by the homogeneous equation

$$y^2z + a_1xyz + a_3yz^2 = x^3 + a_2x^2z + a_4xz^2 + a_6z^3. \tag{2}$$

Smooth means that there is no point on the curve at which all partial derivatives vanish.

Department of Mathematics, University of California, Irvine, CA 92697-3875
E-mail address: asilverb@math.uci.edu

The author was partially supported by the National Science Foundation under grant CNS-0831004 and the National Security Agency under grant H98230-07-1-0039. The author thanks Yuri G. Zarhin for helpful comments, and Alex Abatzoglou for proofreading the paper.

Key words and phrases.
Elliptic curves, abelian varieties

If $\mathrm{char}(F) \neq 2$ or 3, then an invertible change of variables puts equation (1) in the form

$$y^2 = x^3 + ax + b$$

with $a, b \in F$. Such an equation is called a *Weierstrass equation.*

Exercise 1.2. A projective curve $y^2z = x^3 + axz^2 + bz^3$ is smooth if and only if $2(4a^3 + 27b^2) \neq 0$.

Exercise 1.3. Over which fields is $y^2 = x^3 - x$ an elliptic curve?

The points (x, y, z) on the projective curve (2) are the points $(x, y, 1)$, where (x, y) is a solution to (1), along with the point $(0, 1, 0)$. The point $(0, 1, 0)$ is called the *point at infinity*, and is denoted O_E or O. It can be viewed as the point where all vertical lines meet.

Next we give two definitions that turn out to be equivalent to Definition 1.1.

Definition 1.4. An *elliptic curve* over a field F is a smooth projective plane cubic over F with a point whose coordinates are in F.

Definition 1.5. An *elliptic curve* over a field F is a smooth projective curve of genus one with a point whose coordinates are in F.

The discriminant and j-invariant are useful quantities attached to an elliptic curve.

Definition 1.6. If E is an elliptic curve with equation $y^2 = x^3 + ax + b$, then the discriminant $\Delta(E)$ and j-invariant $j(E)$ are defined as follows:

$$\Delta(E) = -16(4a^3 + 27b^2), \qquad j(E) = \frac{(-48a)^3}{\Delta(E)}.$$

2. Group Law

The group law on an elliptic curve is based on the principle that a line meets a cubic in three points. The group (or "addition") law is defined so that three collinear points on the curve "sum" to the point at infinity O_E, which is the identity element for the group law. Figure 1 illustrates the group law.

If an elliptic curve E has equation $y^2 = x^3 + ax + b$, then the group law can be expressed algebraically as follows. If $P = (x_1, y_1)$ and $Q = (x_2, y_2)$, then $P + Q = (x_3, y_3)$ where

$$\begin{aligned} x_3 &= \lambda^2 - x_1 - x_2, \\ y_3 &= \lambda(x_1 - x_3) - y_1, \\ \lambda &= \begin{cases} (y_2 - y_1)/(x_2 - x_1) & \text{if } P \neq Q, \\ (3x_1^2 + a)/2y_1 & \text{if } P = Q. \end{cases} \end{aligned}$$

The inverse of the point P is the point $(x_1, -y_1)$, denoted $-P$, and O_E $(= (\infty, \infty))$ is the identity element.

With this group law, the points on E with coordinates in F, along with the point at infinity, form an abelian group, denoted $E(F)$. When F is a number field, $E(F)$ is called the *Mordell-Weil group* of E over F.

Figure 2 shows the Mordell-Weil group over $\mathbb{Q}$ of the elliptic curve $y^2 = x^3 - x$. (That $E(\mathbb{Q})$ has only four elements is non-trivial.)

FIGURE 1. Group Law

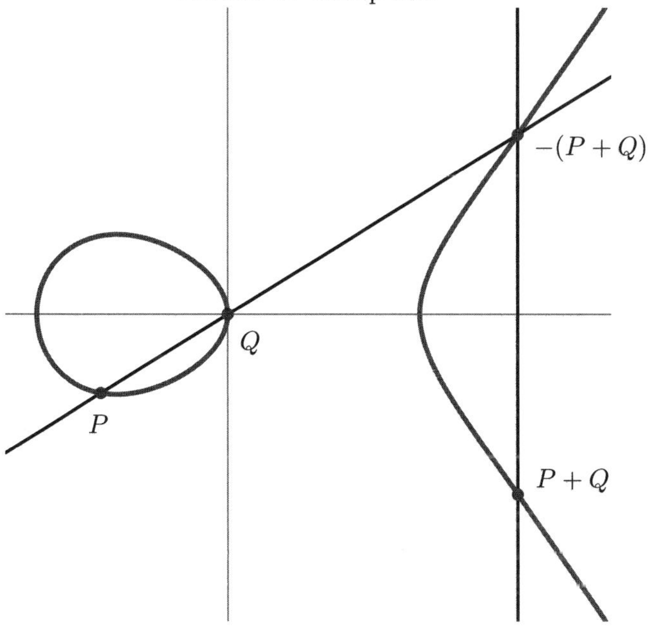

FIGURE 2. $y^2 = x^3 - x$

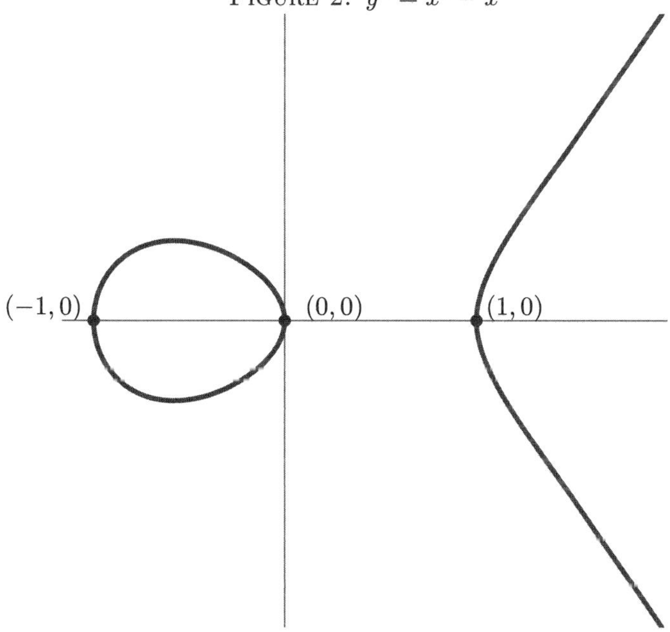

$$E(\mathbb{Q}) = \{(0,0), (1,0), (-1,0), O\} \cong \mathbb{Z}/2\mathbb{Z} \times \mathbb{Z}/2\mathbb{Z}$$

3. N-torsion

Let $\overline{F}$ denote an algebraic closure of the field F.

Definition 3.1. If E is an elliptic curve over F, then

$$E[N] := \{P \in E(\overline{F}) : NP = O_E\}.$$

Fact 3.2. If $\mathrm{char}(F) \nmid N$, then $E[N] \cong \mathbb{Z}/N\mathbb{Z} \times \mathbb{Z}/N\mathbb{Z}$.

Note that if the elliptic curve is in Weierstrass form $y^2 = x^3 + ax + b$, then the points of order 2 are the points whose y-coordinate is 0.

Exercise 3.3. If E is the elliptic curve $y^2 = x^3 - x$ over $\mathbb{Q}$, then

$$E[2] = \{(0,0), (1,0), (-1,0), O\} \cong \mathbb{Z}/2\mathbb{Z} \times \mathbb{Z}/2\mathbb{Z},$$

$$E[4] = \langle (i, i-1), (1-\sqrt{2}, 2-\sqrt{2}) \rangle \cong \mathbb{Z}/4\mathbb{Z} \times \mathbb{Z}/4\mathbb{Z}.$$

4. Elliptic Curves over $\mathbb{C}$

If E is an elliptic curve over $\mathbb{C}$, then as a complex Lie group $E(\mathbb{C})$ is isomorphic to $\mathbb{C}/L$ for a lattice $L = \mathbb{Z}\tau + \mathbb{Z}$ with τ in the complex upper half plane.

FIGURE 3. Elliptic curves over $\mathbb{C}$

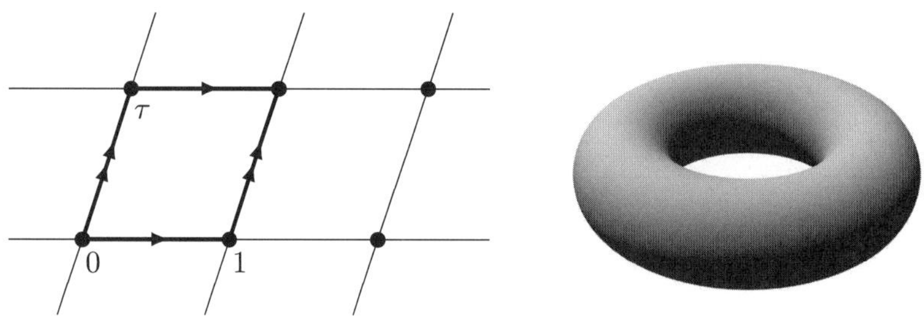

It easily follows that

$$E[N] \cong \tfrac{1}{N}L/L \cong \mathbb{Z}/N\mathbb{Z} \times \mathbb{Z}/N\mathbb{Z},$$

which agrees with Fact 3.2.

5. Elliptic Curves over Number Fields

Mordell-Weil Theorem. *If E is an elliptic curve over a number field F, then the abelian group $E(F)$ is finitely generated.*

Therefore:

$$E(F) \cong \mathbb{Z}^r \times \text{ finite group}.$$

The finite group is called the *torsion subgroup* of $E(F)$ and is denoted $E(F)_{\mathrm{tors}}$. The non-negative integer r is called the *rank* of $E(F)$.

5.1. The Torsion Subgroup

The next result gives a simple algorithm for finding all the torsion points (i.e., the elements in the torsion subgroup) for elliptic curves over $\mathbb{Q}$.

Nagell-Lutz Theorem (1930's). *If $E : y^2 = x^3 + ax + b$ is an elliptic curve with $a, b \in \mathbb{Z}$, and $(x, y) \in E(\mathbb{Q})_{\text{tors}} - \{O_E\}$, then x and y are integers, and either $y = 0$ or y^2 is a divisor of $4a^3 + 27b^2$.*

Exercise 5.1. Show that for $E : y^2 = x^3 + 4x$,

$$E(\mathbb{Q})_{\text{tors}} = \{O, (0,0), (2,4), (2,-4)\} \cong \mathbb{Z}/4\mathbb{Z}.$$

Exercise 5.2. Show that for $E : y^2 = x^3 + 1$,

$$E(\mathbb{Q})_{\text{tors}} = \{O, (-1,0), (0,1), (0,-1), (2,3), (2,-3)\} \cong \mathbb{Z}/6\mathbb{Z}.$$

The next result gives a strong bound on the size of the torsion subgroups for elliptic curves over $\mathbb{Q}$.

Mazur Theorem (1977). *If E is an elliptic curve over $\mathbb{Q}$, then $E(\mathbb{Q})_{\text{tors}}$ is isomorphic to one of the following 15 groups:*

$$\begin{array}{ll} \mathbb{Z}/N\mathbb{Z} & \text{for } N = 1, \ldots, 10 \text{ or } 12, \\ \mathbb{Z}/2\mathbb{Z} \times \mathbb{Z}/2N\mathbb{Z} & \text{for } N = 1, 2, 3, \text{ or } 4. \end{array}$$

Each of these groups occurs infinitely often.

The following result is a generalization to number fields.

Theorem 5.3 (Merel-Oesterlé-Parent, 1990's). *Suppose E is an elliptic curve defined over a number field F, and $m = [F : \mathbb{Q}]$. Then $\#E(F)_{\text{tors}}$ is bounded above by a constant depending only on m. More precisely, if $E(F)$ has a point of order p^n where p is a prime number, then*

$$p^n \leq 129(5^m - 1)(3m)^6.$$

5.2. Ranks

In 1901, Henri Poincaré pointed out the importance of the rank in the classification of rational cubics. In 1922, Mordell stated, "Mathematicians have been familiar with very few questions for so long a period with so little accomplished in the way of general results, as that of finding the rational [points on elliptic curves]." Most major open questions about elliptic curves today have something to do with the rank.

Example 5.4. If E is the elliptic curve $y^2 = x^3 - x$ over $\mathbb{Q}$, then

$$E(\mathbb{Q}) = \langle (1,0), (0,0) \rangle \cong \mathbb{Z}/2\mathbb{Z} \times \mathbb{Z}/2\mathbb{Z}$$

so the rank is 0.

Example 5.5. If E is the elliptic curve $y^2 = x^3 - 2x$ over $\mathbb{Q}$, then

$$E(\mathbb{Q}) = \langle (-1,1), (0,0) \rangle \cong \mathbb{Z} \times \mathbb{Z}/2\mathbb{Z}$$

so the rank is 1.

Example 5.6. If E is the elliptic curve $y^2 = x^3 - 17x$ over $\mathbb{Q}$, then

$$E(\mathbb{Q}) = \langle (-1,4), (9,24), (0,0) \rangle \cong \mathbb{Z} \times \mathbb{Z} \times \mathbb{Z}/2\mathbb{Z}$$

so the rank is 2.

Open Questions 5.7. Some open problems about ranks of elliptic curves over $\mathbb{Q}$ are:

(i) We still do not know an algorithm that is guaranteed to find the rational points on elliptic curves over $\mathbb{Q}$.
(ii) In particular, there is no known algorithm guaranteed to determine the rank.
(iii) It is not known which integers can occur as ranks.
(iv) It is not known if the set of ranks is unbounded.

There are similar questions about elliptic curves over other number fields.

Note that in [**4**], Manin gives an algorithm for finding the Mordell-Weil groups of elliptic curves over $\mathbb{Q}$ that is conditional on the Birch and Swinnerton-Dyer Conjecture.

The Parity Conjecture, which is a consequence of the Conjecture of Birch and Swinnerton-Dyer, implies that half of the elliptic curves over $\mathbb{Q}$ have even rank and half have odd rank. Some people believe the stronger statement that density $\frac{1}{2}$ (in a suitable sense) of the elliptic curves over $\mathbb{Q}$ have rank 0, density $\frac{1}{2}$ have rank 1, and the density of elliptic curves over $\mathbb{Q}$ with rank ≥ 2 is zero.

Many people believe that ranks of elliptic curves over $\mathbb{Q}$ are unbounded. In 2006 Elkies [**2**] found an elliptic curve over $\mathbb{Q}$ of rank at least 28; this is the largest known rank of an elliptic curve over $\mathbb{Q}$.

6. Elliptic Curves over Finite Fields

Elliptic curves over finite fields are used in elliptic curve cryptography and pairing-based cryptography. The security of elliptic curve and pairing-based cryptosystems depends on the difficulty of certain problems about elliptic curves and their Weil and Tate pairings. An important open problem in cryptography is to understand how difficult these "hard problems" are.

Fact 6.1. If E is an elliptic curve over a finite field $\mathbb{F}_q$, then

$$E(\mathbb{F}_q) \cong \mathbb{Z}/m\mathbb{Z} \times \mathbb{Z}/n\mathbb{Z}$$

where m and n are positive integers and m is a divisor of n.

Hasse Bound. *If E is an elliptic curve over $\mathbb{F}_q$, then*

$$(\sqrt{q}-1)^2 \leq \#E(\mathbb{F}_q) \leq (\sqrt{q}+1)^2.$$

7. Homomorphisms

Definition 7.1. A *homomorphism* $f : E_1 \to E_2$ of elliptic curves over the same field F is a morphism (of the algebraic varieties) that takes O_{E_1} to O_{E_2}. Write $\mathrm{Hom}_F(E_1, E_2)$ for the homomorphisms that are defined over F and $\mathrm{Hom}(E_1, E_2)$ for the homomorphisms that are defined over $\overline{F}$. An *endomorphism* is a homomorphism from E to itself. Denote $\mathrm{Hom}(E, E)$ by $\mathrm{End}(E)$.

Fact 7.2. If $f : E_1 \to E_2$ is a homomorphism as above, then the induced map $f : E_1(F) \to E_2(F)$ on the groups of points is a group homomorphism.

Example 7.3. For every $N \in \mathbb{Z}$, the map from E to itself defined by $P \mapsto N \cdot P$ is an endomorphism, and this gives an injection

$$\mathbb{Z} \hookrightarrow \mathrm{End}(E).$$

Fact 7.4. The possibilities for the ring $\mathrm{End}(E)$ are:

(i) $\mathbb{Z}$,
(ii) an order in an imaginary quadratic field, or
(iii) a maximal order in a definite quaternion algebra over $\mathbb{Q}$.

The third does not happen in characteristic 0. The first does not happen over finite fields.

Example 7.5. If E is an elliptic curve over a finite field $\mathbb{F}_q$, then

$$\phi_q : (x, y) \mapsto (x^q, y^q)$$

is an endomorphism, called the *Frobenius endomorphism.*

Example 7.6. Suppose E is $y^2 = x^3 - x$. Then $I : (x, y) \mapsto (-x, \sqrt{-1}y)$ is an automorphism. If the ground field has characteristic 0 or characteristic $p \equiv 1 \pmod 4$, then:

$$\mathrm{End}(E) = \mathbb{Z} + \mathbb{Z}I \cong \mathbb{Z}[i].$$

If the ground field has characteristic $p \equiv 3 \pmod 4$, then $\phi_p^2 = -p$, $\phi_p \circ I = -I \circ \phi_p$, and

$$\begin{aligned}\mathrm{End}(E) &= \mathbb{Z} + \mathbb{Z}I + \mathbb{Z}\left(\frac{1+\phi_p}{2}\right) + \mathbb{Z}\left(\frac{1+\phi_p}{2}\right) \circ I \\ &\cong \mathbb{Z}\begin{pmatrix} 1 & 0 \\ 0 & 1 \end{pmatrix} + \mathbb{Z}\begin{pmatrix} 0 & 1 \\ -1 & 0 \end{pmatrix} + \mathbb{Z}\begin{pmatrix} \frac{1+\sqrt{-p}}{2} & 0 \\ 0 & \frac{1-\sqrt{-p}}{2} \end{pmatrix} + \mathbb{Z}\begin{pmatrix} 0 & \frac{1-\sqrt{-p}}{2} \\ \frac{-1-\sqrt{-p}}{2} & 0 \end{pmatrix}.\end{aligned}$$

Definition 7.7. *Isogeny* can be defined using any of the following equivalent definitions:

(i) An *isogeny* of elliptic curves is a non-zero homomorphism.
(ii) An *isogeny* of elliptic curves is a surjective homomorphism.
(iii) An *isogeny* of elliptic curves is a homomorphism with finite kernel.

If there is an isogeny from E_1 to E_2 then there is an isogeny from E_2 to E_1. We say that E_1 and E_2 are *isogenous.* Being isogenous is an equivalence relation.

An *isomorphism* is an invertible isogeny. Two elliptic curves over F are isomorphic over $\overline{F}$ if and only if they have the same j-invariant. A stronger statement is the following. If E is $y^2 = x^3 + ax + b$ and E' is $y^2 = x^3 + a'x + b'$, then an isomorphism $f : E \to E'$ is of the form

$$f(x, y) = (\lambda^2 x, \lambda^3 y)$$

where $a' = \lambda^4 a$, $b' = \lambda^6 b$, and $\lambda \in \bar{F}^\times$. Note that f is defined over F if and only if $\lambda \in F^\times$.

8. Supersingular and Ordinary

A useful concept, for both number theory and cryptography, is that of supersingularity.

Definition 8.1. Suppose E is an elliptic curve over $\mathbb{F}_{p^r}$ with p prime. Then E is *supersingular* if and only if any of the following equivalent statements holds:

(i) $\#E(\mathbb{F}_{p^r}) \equiv 1 \pmod p$,
(ii) $E[p] = \{O\}$,
(iii) $\mathrm{End}(E)$ is non-commutative,

(iv) $\operatorname{End}(E)$ is an order in a quaternion algebra over $\mathbb{Q}$.

Otherwise, E is called *ordinary*.

If E is an elliptic curve over $\mathbb{F}_{p^r}$ with p prime, then E is ordinary if and only if any of the following equivalent statements holds:

(i) $\#E(\mathbb{F}_{p^r}) \not\equiv 1 \pmod{p}$,
(ii) $E[p] \cong \mathbb{Z}/p\mathbb{Z}$,
(iii) $\operatorname{End}(E)$ is commutative,
(iv) $\operatorname{End}(E)$ is an order in an imaginary quadratic field.

Example 8.2. By Example 7.6, the curve $y^2 = x^3 - x$ over $\mathbb{F}_{p^r}$ is ordinary if $p \equiv 1 \pmod{4}$ and is supersingular if $p \equiv 3 \pmod{4}$. In the latter case,

$$\#E(\mathbb{F}_{p^r}) = \begin{cases} p^r + 1 & \text{if } r \text{ is odd,} \\ ((-p)^{r/2} - 1)^2 & \text{if } r \text{ is even.} \end{cases}$$

9. Mod N representations

Let F^s be a separable closure of F and let $G_F = \operatorname{Gal}(F^s/F)$. If E is an elliptic curve defined over F, then G_F acts on $E[N]$, and this action induces the mod N representation

$$\rho_{E,N} : G_F \to \operatorname{Aut}(E[N]).$$

Note that $\operatorname{Aut}(E[N]) \cong \operatorname{GL}_2(\mathbb{Z}/N\mathbb{Z})$ if $\operatorname{char}(F) \nmid N$.

Example 9.1. If E is $y^2 = x^3 - x$ over $\mathbb{Q}$, then

$$\rho_{E,2}(\sigma) = \begin{pmatrix} 1 & 0 \\ 0 & 1 \end{pmatrix}$$

for all $\sigma \in G_F$, since $E[2] \subseteq E(\mathbb{Q})$.

Exercise 9.2. Suppose E is $y^2 = x^3 - x$ over $\mathbb{Q}$. Since $E[4] \subset \mathbb{Q}(i, \sqrt{2})$, the mod 4 representation

$$\rho_{E,4} : G_{\mathbb{Q}} \to \operatorname{Aut}(E[4])$$

factors through $\operatorname{Gal}(\mathbb{Q}(i, \sqrt{2})/\mathbb{Q}) \cong \mathbb{Z}/2\mathbb{Z} \times \mathbb{Z}/2\mathbb{Z}$. This Galois group is generated by σ and τ such that $\sigma(i) = -i$, $\sigma(\sqrt{2}) = \sqrt{2}$, $\tau(i) = i$, and $\tau(\sqrt{2}) = -\sqrt{2}$. Show that:

$$\rho_{E,4}(\sigma) = \begin{pmatrix} -1 & 0 \\ 2 & 1 \end{pmatrix}, \quad \rho_{E,4}(\tau) = \begin{pmatrix} 1 & 2 \\ 0 & 1 \end{pmatrix} \in \operatorname{GL}_2(\mathbb{Z}/4\mathbb{Z})$$

with respect to the basis $\{(i, i-1), (1-\sqrt{2}, 2-\sqrt{2})\}$ for $E[4]$.

10. Tate Modules and the Isogeny Theorem

If E is an elliptic curve and ℓ is a prime number, with respect to the maps

$$E[\ell^{n+1}] \to E[\ell^n], \quad P \mapsto \ell P$$

define

$$T_\ell(E) := \varprojlim_n E[\ell^n], \qquad V_\ell(E) := T_\ell(E) \otimes_{\mathbb{Z}_\ell} \mathbb{Q}_\ell.$$

Then $T_\ell(E)$ is a free $\mathbb{Z}_\ell$-module and $V_\ell(E)$ is a $\mathbb{Q}_\ell$-vector space.

If E is defined over a field F, then G_F acts on $V_\ell(E)$, and this action induces the *ℓ-adic representation*:

$$\rho_{E,\ell^\infty} : G_F \to \operatorname{Aut}(V_\ell(E)).$$

Note that if $\mathrm{char}(F) \neq \ell$, then $V_\ell(E) \cong \mathbb{Q}_\ell^2$ (as $\mathbb{Q}_\ell$-vector spaces) and $\mathrm{Aut}(V_\ell(E)) \simeq \mathrm{GL}_2(\mathbb{Q}_\ell)$.

Isogeny Theorem (Tate, Parshin, Faltings,...). *Suppose F is a finitely generated extension of a number field or of a finite field, and E_1 and E_2 are elliptic curves defined over F. Then the following are equivalent:*

(i) *E_1 and E_2 are isogenous over F,*
(ii) *ρ_{E_1,ℓ^∞} and ρ_{E_2,ℓ^∞} are isomorphic for every $\ell \neq \mathrm{char}(F)$,*
(iii) *ρ_{E_1,ℓ^∞} and ρ_{E_2,ℓ^∞} are isomorphic for one $\ell \neq \mathrm{char}(F)$.*

In other words, elliptic curves over F are determined (up to isogeny) by their ℓ-adic representations. See [**11**] for a survey of the Isogeny Theorem. (See also §3 of [**7**] for a brief summary of the history of the Isogeny Theorem.)

11. Reduction of Elliptic Curves

If E is an elliptic curve over a number field F, then E can be reduced modulo a prime ideal $\mathfrak{p}$ of $\mathcal{O}_F$, where $\mathcal{O}_F$ is the ring of integers of F. If $\mathfrak{p} \nmid \Delta(E)$, this gives an elliptic curve $\tilde{E}$ over the finite field $\mathcal{O}_F/\mathfrak{p}$.

Exercise 11.1. Take the elliptic curve $y^2 = x^3 - x$ over $\mathbb{Q}$ and reduce the coefficients modulo 7. This gives an elliptic curve $\tilde{E}$ over $\mathbb{F}_7$. Compute $\tilde{E}(\mathbb{F}_7)$.

12. Conjecture of Birch and Swinnerton-Dyer

The Conjecture of Birch and Swinnerton-Dyer tells us that an analytic object, namely the L-function of an elliptic curve over a number field, encodes certain arithmetic information about the Mordell-Weil group. In particular, the Birch and Swinnerton-Dyer Conjecture relates the size of the Mordell-Weil group of an elliptic curve over a number field with the numbers of points on the reductions of the curve. See [**3**] for more on the Birch and Swinnerton-Dyer Conjecture. Figure 4 shows the conjecture's originators around the time they were formulating the conjecture.

13. Abelian Varieties

Abelian varieties are higher-dimensional generalizations of elliptic curves.

Definition 13.1. An *abelian variety* is a connected projective group variety.

Remark 13.2. Note the the definition doesn't say that abelian varieties are *abelian* groups. The fact that abelian varieties are commutative groups is a theorem.

Examples 13.3.

(i) The one-dimensional abelian varieties are exactly the elliptic curves.
(ii) Products of abelian varieties are abelian varieties.
(iii) Jacobian varieties of curves of genus g are abelian varieties of dimension g.
(iv) For example, if C is a smooth plane curve of degree n, then C has genus $\frac{(n-1)(n-2)}{2}$ and its Jacobian variety has dimension $\frac{(n-1)(n-2)}{2}$.
(v) In particular, the Jacobian variety of the projective curve $x^n + y^n = z^n$ is an abelian variety of dimension $\frac{(n-1)(n-2)}{2}$.

FIGURE 4. Peter Swinnerton-Dyer and Bryan Birch in 1961

Some of the results that are true about elliptic curves are also true for abelian varieties. Examples include the Mordell-Weil Theorem and the Isogeny Theorem.

Mordell-Weil Theorem. *The group of points on an abelian variety over a number field is a finitely generated abelian group.*

Many of the facts we know about elliptic curves are open questions for higher-dimensional abelian varieties. The Torsion Conjecture is one example:

Torsion Conjecture. *If A is an abelian variety of dimension d defined over a number field F, then $\#A(F)_{\text{tors}}$ is bounded above by a constant depending only on d and F.*

For every dimension greater than one, the Torsion Conjecture is open even when $F = \mathbb{Q}$.

Strong Torsion Conjecture. *If A is an abelian variety of dimension d defined over a number field F of degree m, then $\#A(F)_{\text{tors}}$ is bounded above by a constant depending only on d and m.*

The Strong Torsion Conjecture is sometimes called the Uniform Boundedness Conjecture.

Many open questions about elliptic curves make sense and are open (though sometimes different) for abelian varieties.

Open Questions 13.4. Here are a few other well known open questions about abelian varieties:

(i) Distributions of ranks, including boundedness or unboundedness.
(ii) Conjecture of Birch and Swinnerton-Dyer.
(iii) Difficulty of the Discrete Log Problem (cryptography).

Bibliography

[1] J. P. Buhler, *Elliptic curves, modular forms, and applications*, in Arithmetic algebraic geometry (Park City, UT, 1999), 5–81, IAS/Park City Math. Ser. **9**, Amer. Math. Soc., Providence, RI, 2001.

[2] N. Elkies, *Z^28 in E(Q), etc.*, Number Theory Listserv posting, May 3, 2006, `http://listserv.nodak.edu/cgi-bin/wa.exe?A2=ind0605&L=nmbrthry&T=0&F=&S=&P=50`

[3] B. Gross, *Introduction to the Birch and Swinnerton-Dyer Conjecture*, this volume.

[4] Ju. I. Manin, *Cyclotomic fields and modular curves*, Russian Math. Surveys **26** (1971), no. 6, 7–78 (English translation of Uspehi Mat. Nauk **26** (1971), no. 6 (162), 7–71).

[5] K. Rubin, A. Silverberg, *Ranks of elliptic curves*, Bull. Amer. Math. Soc. (N.S.) **39** (2002), no. 4, 455–474.

[6] G. Shimura, Introduction to the arithmetic theory of automorphic functions, Reprint of the 1971 original, Publications of the Mathematical Society of Japan **11**, Kanô Memorial Lectures **1**, Princeton University Press, Princeton, NJ, 1994.

[7] A. Silverberg, *Galois representations attached to points on Shimura varieties*, Séminaire de Théorie des Nombres, Paris, 1990–91, 221–240, Progr. Math. **108**, Birkhäuser Boston, Boston, MA, 1993.

[8] A. Silverberg, *Open questions in arithmetic algebraic geometry*, in Arithmetic algebraic geometry (Park City, UT, 1999), 83–142, IAS/Park City Math. Ser. **9**, Amer. Math. Soc., Providence, RI, 2001.

[9] J. H. Silverman, The arithmetic of elliptic curves, Corrected reprint of the 1986 original, Graduate Texts in Mathematics **106**, Springer-Verlag, New York, 1992.

[10] J. H. Silverman, Advanced Topics in the Arithmetic of Elliptic Curves, Graduate Texts in Mathematics **151**, Springer-Verlag, New York, 1994.

[11] Yu. G. Zarkhin, A. N. Parshin, *Finiteness Problems in Diophantine Geometry*, Amer. Math. Soc. Transl. Series 2 **143** (1989), 35–102.

Lectures on the Conjecture of Birch and Swinnerton-Dyer

Benedict H. Gross

IAS/Park City Mathematics Series
Volume 18, 2009

Lectures on the Conjecture of Birch and Swinnerton-Dyer

Benedict H. Gross

Introduction

In these lectures, I will present the conjecture of Birch and Swinnerton-Dyer, for an elliptic curve E over a global field k. The first lecture studies various arithmetic invariants of the curve, focusing primarily on the Mordell-Weil group $E(k)$. In the second lecture, I introduce the L-function, describe its analytic properties (both known and expected), and give a precise statement of the conjecture for the leading term in its Taylor expansion at the point $s = 1$. In the third lecture, I discuss the advances that have been made towards a proof.

Carl Erickson and Jack Thorne helped to prepare these notes from a preliminary version of my lectures, adding many examples and references. In particular, the material in the Appendices was developed by them. I want to thank them for all their hard work, and to thank Curt McMullen, Alice Silverberg, and Mark Reeder for their comments.

Department of Mathematics, Harvard University, 1 Oxford Street, Cambridge, MA 02138
E-mail address: gross@math.harvard.edu

LECTURE 1

The Mordell-Weil Theorem

In this lecture we sketch the proof of the Mordell-Weil theorem for elliptic curves over global fields. We also introduce some of the arithmetic invariants that appear in the conjecture of Birch and Swinnerton-Dyer, such as the regulator of the height pairing and the order of the Tate-Shafarevich group.

1. Torsion on elliptic curves

Let k be a field. An elliptic curve E over k is a non-singular curve of genus 1, together with a distinguished k-rational point $\mathcal{O}_E$. The Riemann-Roch theorem implies that linear series $\|3 \cdot \mathcal{O}_E\|$ embeds E as a non-singular cubic curve in $\mathbb{P}^2$, with the distinguished point $\mathcal{O}_E$ the intersection with the line at infinity. The other points of E are solutions of a generalized Weierstrass equation

$$y^2 + a_1xy + a_3y = x^3 + a_2x^2 + a_4x + a_6,$$

whose coefficients a_i all lie in k. This equation defines a non-singular curve if and only if a certain polynomial $\Delta = \Delta(a_1, \ldots, a_6)$ (the discriminant of the generalized Weierstrass equation) in the coefficients does not vanish.

If K/k is any field extension then the chord and tangent process turns $E(K)$, the set of points $P = (x, y)$ of E with coordinates in K, into an abelian group, with the zero element being $\mathcal{O}_E$. The addition law has the following property:

$$P + Q + R = 0 \Leftrightarrow P, Q, R \text{ are collinear.}$$

For a point $P = (x, y)$, the equality $P + (-P) + \mathcal{O}_E = \mathcal{O}_E$ tells us that

$$-P = (x, -y - a_1x - a_3).$$

In particular,

$$2P = \mathcal{O}_E \Leftrightarrow P = -P \Leftrightarrow 2y + a_1x + a_3 = 0. \tag{1.1}$$

A short calculation shows that this is equivalent to asking that x satisfy the cubic equation $4x^3 + b_2x^2 + 2b_4x + b_6 = 0$, where the b_i are certain polynomial functions in the a_i. The discriminant of this cubic equation is equal to $16 \cdot \Delta$. If k is of characteristic 2, this polynomial factors as $(a_1x + a_3)^2$; otherwise, as $\Delta \neq 0$ the roots of this polynomial are distinct and

$$E[2](K) \simeq \mathbb{Z}/2\mathbb{Z} \times \mathbb{Z}/2\mathbb{Z}$$

for any field K/k that contains these roots (If A is an abelian group, then $A[n]$ denotes the n-torsion of A).

Let k^s denote a separable closure of k. If the characteristic of k is zero, and $k^s \hookrightarrow \mathbb{C}$, then we have

$$E[n](k) \subset E[n](k^s) = E[n](\mathbb{C}) \cong \mathbb{Z}/n\mathbb{Z} \times \mathbb{Z}/n\mathbb{Z}.$$

Indeed, there exists a lattice $\Lambda \subset \mathbb{C}$ and an analytic isomorphism $E(\mathbb{C}) \cong \mathbb{C}/\Lambda$ of complex Lie groups, which implies the second isomorphism above. To prove the first equality, one shows that there exists a polynomial Q_n in two variables, with coefficients in k, with the property that $nP = \mathcal{O}_E$ if and only if $Q_n(x,y) = 0$. Therefore, if $\sigma \in \mathrm{Aut}(\mathbb{C}/k)$ and $nP = \mathcal{O}_E$, then P^σ is also an n-torsion point. In particular, P has only finitely many conjugates over k; so its coordinates must in fact be algebraic over k.

In general, we have the following result.

Proposition 1.1. *Let n be a positive integer which is not divisible by the characteristic of k. Then there is an isomorphism*

$$E[n](k^s) \cong \mathbb{Z}/n\mathbb{Z} \times \mathbb{Z}/n\mathbb{Z}.$$

Moreover, all n-torsion points of E are defined over the field k^s, and multiplication by n is surjective on $E(k^s)$.

On the other hand, if k is of characteristic 2 then we have seen that $E[2](k)$ is either a cyclic group of order 2 (if $a_1 \neq 0$), or is trivial (if $a_1 = 0$). Indeed, the non-trivial 2-torsion points correspond to roots of the polynomial $a_1 x + a_3 = 0$.

A standard reference for the results described above, which does not assume a large amount of algebraic geometry, is [**52**]. For proofs using scheme-theoretic language, in the more general setting of abelian varieties, see [**42**].

2. Galois cohomology

We continue with E/k as before. Let $n > 1$ be an integer prime to the characteristic, and fix a choice of separable closure k^s. There is an exact sequence of $\mathrm{Gal}(k^s/k)$-modules

$$0 \longrightarrow E[n] \longrightarrow E \xrightarrow[\times n]{} E \longrightarrow 0.$$

When we write E or $E[n]$ here, we mean the points over k^s. Taking the long exact sequence in Galois cohomology, we get

$$E(k) \xrightarrow[\times n]{} E(k) \longrightarrow H^1(k, E[n]) \longrightarrow H^1(k, E) \xrightarrow[\times n]{} H^1(k, E).$$

This gives us a short exact sequence

$$0 \longrightarrow E(k)/nE(k) \xrightarrow[\delta]{} H^1(k, E[n]) \longrightarrow H^1(k, E)[n] \longrightarrow 0.$$

The map δ sends the equivalence class of P to the cohomology class of the cocycle $\sigma \mapsto Q^\sigma - Q$, where Q is any point in $E(k^s)$ with the property that $nQ = P$.

Now suppose that k is a global field. We can use the above exact sequence to define two important groups, as follows. For every place v of k we have a commutative diagram with exact rows:

$$\begin{array}{ccccccccc}
0 & \longrightarrow & E(k)/nE(k) & \xrightarrow[\delta]{} & H^1(k, E[n]) & \longrightarrow & H^1(k, E)[n] & \longrightarrow & 0 \\
 & & \downarrow & & \downarrow & & \downarrow & & \\
0 & \longrightarrow & E(k_v)/nE(k_v) & \xrightarrow[\delta]{} & H^1(k_v, E[n]) & \longrightarrow & H^1(k_v, E)[n] & \longrightarrow & 0
\end{array}$$

In the first row, the symbol E represents the Galois module $E(k^s)$, while in the second row E represents $E(k_v^s)$.

Definition 1.2. The *n-Selmer group* is the group of those elements of $H^1(k, E[n])$ whose restriction to $H^1(k_v, E[n])$ lies in the image of the group $E(k_v)/nE(k_v)$ for all places v. In other words, we define

$$\mathrm{Sel}_n(k, E) = \ker\left(H^1(k, E[n]) \to \prod_v H^1(k_v, E)\right).$$

The *Tate-Shafarevich group* is defined by

$$Ш(k, E) = \ker\left(H^1(k, E) \to \prod_v H^1(k_v, E)\right).$$

We then have an exact sequence

$$0 \longrightarrow E(k)/nE(k) \longrightarrow \mathrm{Sel}_n(k, E) \longrightarrow Ш(k, E)[n] \longrightarrow 0.$$

Using classical theorems of algebraic number theory (the finiteness of the S-class group, and the finite generation of the S-unit group) one can show that the n-Selmer group is always finite. We prove this for $n = 2$ in the next section. It is somewhat more difficult to calculate the n-Selmer group exactly, but this can be done in principle, and we illustrate this for some cases with $n = 2$ in Appendix B. The fact that $\mathrm{Sel}_n(k, E)$ is always finite proves that the two groups $E(k)/nE(k)$ and $Ш(k, E)[n]$ are both finite, for any integer $n \geq 1$. The first finiteness statement is called the *weak Mordell-Weil theorem.* A proof of this theorem in the general case can be found in Ch. VIII of [**52**], or Ch. 4 of [**39**].

Tate [**54**] conjectured that the group $Ш(k, E)$ is always *finite.* This is essentially a conjecture on the existence of k-rational points, which should account for most of the size of the Selmer groups. A formula for the hypothetical order $\#Ш(k, E)$ appears in the conjecture of Birch and Swinnerton-Dyer (cf. Conjecture 2.10). In all cases so far in which the finiteness of the group $Ш(k, E)$ has been proved, the full conjecture has been proved along with it.

3. A 2-descent

We continue with the notation of the previous section, and specialize to the case that $n = 2$. We will assume that the characteristic of k is not equal to 2, and that all of the points of $E[2]$ are defined over k. The curve E can then be given by an equation of the form

$$y^2 = (x - e_1)(x - e_2)(x - e_3),$$

which has $\Delta = 16\prod(e_i - e_j)^2$. If B denotes the group $k^\times/(k^\times)^2$, written multiplicatively, we have an isomorphism

$$H^1(k, E[2]) \cong \{(b_1, b_2, b_3) \in B^3 : b_1 \cdot b_2 \cdot b_3 = 1\}.$$

Lemma 1.3. *If $P = (x, y)$ in $E(k)$ has $y \neq 0$, the map δ is given by the formula $\delta(P) = (x - e_1, x - e_2, x - e_3)$.*

The proof is given in [**52**], which also explains how to extend the map δ to all of $E(k)$, using the fact that the product $b_1 \cdot b_2 \cdot b_3 = 1$.

Let us now suppose that k is a global field, that is, a number field or the function field of a non-singular projective curve defined over a finite field. Let S be the set of places v which are either archimedean, or divide 2, or such that the e_i are not all integral at v, or such that $v(\Delta) \neq 0$. Note that S is a finite set. (We note that the curve E has good reduction at all places v which are not in the finite set S - cf. Appendix A).

Lemma 1.4. *For any $v \notin S$, and for any point $P = (x, y)$ in $E(k_v)$, $v(x - e_i)$ is even for all i.*

PROOF. Suppose first that $v(x - e_i) < 0$ for one, and hence for all, i. Then $v(x-e_i)$ does not depend on i and the defining equation of E shows that $3v(x-e_i) = 2v(y)$, giving the result.

Suppose instead that $v(x - e_i) > 0$ for one i. Since $v(\Delta) = 0$, we have $v(e_i - e_j) = v(x - e_j - x + e_i) = 0$ for $j \neq i$, so $v(x - e_j) = 0$. Hence $v(x - e_i) = 2v(y)$. $\square$

The above lemma shows that $E(k)/2E(k)$ injects into the subgroup of B_1^3 consisting of those elements with $v(\alpha_i)$ even whenever v is not in S. But this subgroup is finite: let $k^\times(S, 2)/(k^\times)^2 \subset k^\times/(k^\times)^2$ be the subgroup of elements with even valuation outside of S, and let A_S be the ring of S-integers of k; that is, the subring of k consisting of elements having non-negative valuation at all places outside S. Then there is an exact sequence

$$1 \longrightarrow A_S^\times/(A_S^\times)^2 \longrightarrow k^\times(S,2)/(k^\times)^2 \longrightarrow \mathrm{Pic}(A_S)[2] \longrightarrow 1$$

where the map to the class group $\mathrm{Pic}(A_S)$ takes an element α in $k^\times(S, 2)$ generating the principal ideal $(\alpha) = I^2$ to the class of the ideal I. The outside two terms are both finite, by the finiteness of the class group $\mathrm{Pic}(A_S)$ and the finite generation of the unit group $A_S^\times$ of A_S. It follows that $E(k)/2E(k)$ is finite.

As an example, consider the curve given by the equation

$$E : y^2 = x(x - 1)(x - t)$$

over the field $k = \mathbb{F}_q(t)$, of discriminant $16t^2(t-1)^2$, where q is an odd prime power. In this case S consists of the places $t = 0, 1$ and ∞. An initial estimate for the structure of the 2-Selmer group is given by

$$\mathrm{Sel}_2(k, E) \subset (\mathbb{Z}/2\mathbb{Z})^6.$$

The group on the right is the one obtained by applying the local conditions only outside S: that is, it is the group

$$\ker\left(H^1(k, E[n]) \to \prod_{v \notin S} H^1(k_v, E)\right).$$

Taking into account the local conditions at every place, we obtain the final result:

$$\mathrm{Sel}_2(k, E) \cong (\mathbb{Z}/2\mathbb{Z})^2.$$

This is worked out in Appendix B, along with the computation of the 2-Selmer group for the twisted curve

$$E' : t(t - 1)y^2 = x(x - 1)(x - t),$$

which becomes isomorphic to E over the quadratic extension $k(\sqrt{t(t-1)})$. This curve also has bad reduction at the set $S = 0, 1, \infty$ and the initial estimate given by considering places outside S is exactly the same. By completing the 2-descent at the primes of bad reduction, we obtain the formula

$$\mathrm{Sel}_2(k, E') \cong \begin{cases} (\mathbb{Z}/2\mathbb{Z})^3 & \text{if } -1 \text{ is not a square in } \mathbb{F}_q^\times; \\ (\mathbb{Z}/2\mathbb{Z})^4 & \text{if } -1 \text{ is a square in } \mathbb{F}_q^\times. \end{cases}$$

4. Heights

We continue to assume that E is an elliptic curve over the global field k. Let n be a non-zero integer. The finiteness of $E(k)/nE(k)$ is a necessary condition for the finite generation of $E(k)$, but by no means a sufficient one. For example, $E(\mathbb{C})/nE(\mathbb{C})$ is trivial for all $n \neq 0$. To extract a proof of the full Mordell-Weil theorem from the weak version, we introduce the Néron-Tate height. This is a pairing $\langle , \rangle : E(k) \times E(k) \to \mathbb{R}$ such that:

(1) $\langle , \rangle$ is biadditive and symmetric.
(2) For all $P \in E(k), \langle P, P\rangle \geq 0$.
(3) For all $M > 0$, the set of points P such that $\langle P, P\rangle \leq M$ is finite.

Remark. The existence of such a pairing has the immediate corollary that the torsion group $E(k)_{\text{tors}}$ is finite. In fact, one can show that a point P is torsion if and only if $\langle P, P\rangle = 0$.

We will give the definition of the height pairing on the Jacobian of a curve in lecture 3. In the case that $k = \mathbb{Q}$, an approximation to the height can be given as follows. Taking a point to its x coordinate gives a map from E to $\mathbb{P}^1$, defined over $\mathbb{Q}$. Writing $x(P) = a/b$ with a, b coprime integers, we define the logarithmic height to be $h(P) = \log\max(|a|, |b|)$. Then the difference $\langle P, P\rangle - h(P)$ is bounded as P ranges over $E(\mathbb{Q})$. Tate observed that the sequence $s_n = h(nP)/n^2$ is Cauchy, and converges to the value $\langle P, P\rangle$.

With the height pairing in hand, we can deduce the full Mordell-Weil theorem.

Theorem 1.5. *Let k be a global field, E/k an elliptic curve. Then $E(k)$ is a finitely generated abelian group.*

PROOF. Choose an integer $n > 1$, and let the points $P_1, \dots, P_r$ be representatives for the finite group $E(k)/nE(k)$. Let $M = \max_i(\langle P_i, P_i\rangle)$, and let T be the set of points P such that $\langle P, P\rangle \leq M$. This is a finite set.

In fact, the points in T generate $E(k)$. Otherwise, there exists a point Q of minimal height which is not in the span of T. Since the P_i represent the cosets of $nE(k)$, there exist an i and a point R in $E(k)$ such that $Q - P_i = nR$. Since Q, P_i and the origin are not all collinear in the Euclidean space $E(k) \otimes \mathbb{R}$, we find

$$\langle Q - P_i, Q - P_i\rangle < 4 \cdot \langle Q, Q\rangle,$$

so

$$n^2 \cdot \langle R, R\rangle < 4 \cdot \langle Q, Q\rangle.$$

By the minimality of the height of Q, the point R must lie in the span of T. This implies that the same is true of $Q = nR + P_i$, which is the desired contradiction. □

With this theorem we can define an important invariant of elliptic curves over global fields.

Definition 1.6. Let n be the rank of the group $E(k)$ and let $P_1, \dots, P_n$ be a basis for a free subgroup of finite index I in $E(k)$. The *regulator of* E is defined to be the positive real number

$$R(E/k) = \det(\langle P_i, P_j\rangle)/I^2.$$

One can check that this definition does not depend on the choice of the points P_i. For example, if $E(k)$ is finite, then $R(E/k) = 1/\#E(k)^2$.

We will see in lecture 3 that when E is an elliptic curve over a global function field $k = \mathbb{F}_q(X)$, then $\langle P, Q\rangle$ is a rational multiple of $\log q$, where the denominator is controlled by the places of bad reduction. The regulator $R(E/k)$ is then a positive rational multiple of $(\log q)^n$.

The theory of heights is developed for elliptic curves in [**52**]. A rather direct approach, treating just the case when $k = \mathbb{Q}$, can be found in [**11**]. For a discussion of Néron's theory of local heights on curves, see [**25**], or the discussion in the third lecture.

LECTURE 2

Conjectures on L-functions

Before stating the conjecture of Birch and Swinnerton-Dyer, we must define the L-function of an elliptic curve over a global field k. It is given as an Euler product, the factors being indexed by the finite places of k. We begin by defining the incomplete L-function, which omits the finitely many places at which E has bad reduction. (For a discussion of the notion of reduction of elliptic curves, as well as a general discussion of elliptic curves over local fields, see Appendix A.) We then define the complete L-function, describe its analytic properties, and state the conjecture of Birch and Swinnerton-Dyer. We end with some examples over function fields and number fields, where the implications of this conjecture are explored.

1. The incomplete L-function

Let E be an elliptic curve defined over the global field k. Let S be the finite set of places of k consisting of the infinite places and the places where E has bad reduction. This means that for all places $v \notin S$, there is a model of E with the coefficients a_i lying in the local ring A_v at v and with the discriminant $\Delta = \Delta_v$ a unit in A_v. Let π_v be a uniformizing element in A_v and let $A_v/\pi_v A_v = \mathbb{F}_v$ be the finite residue field, of cardinality q_v. For $v \notin S$, we get an elliptic curve $\widetilde{E}_v$ over the residue field $\mathbb{F}_v$.

The following theorem is fundamental for elliptic curves over finite fields.

Theorem 2.1 (Hasse). *Write $\#\widetilde{E}_v(\mathbb{F}_v) = 1 + q_v - a_v$, where $v \notin S$. Then*

$$a_v^2 \leq 4q_v,$$

so the roots of the polynomial

$$h_v(t) = 1 - a_v t + q_v t^2 = (1 - \alpha_v t)(1 - \bar{\alpha}_v t)$$

are complex conjugates. Furthermore, if $F_n/\mathbb{F}_v$ is the unique field extension of degree $n \geq 1$, then

$$\#\widetilde{E}_v(F_n) = (q_v^n + 1) - (\alpha_v^n + \bar{\alpha}_v^n).$$

PROOF. See [**52**], Thm. V.1.1. □

These point counts on E over finite places are the key ingredient in defining its corresponding L-function. The polynomial $h_v(t)$ is the reciprocal of the *formal local L-factor* of an elliptic curve over a finite field:

$$L(\widetilde{E}_v/\mathbb{F}_v, t) = h_v(t)^{-1}.$$

We substitute q_v^{-s} for t to get the *local L-factor*

$$L_v(E/k, s) = L(\widetilde{E}_v/\mathbb{F}_v, q_v^{-s}) = (1 - a_v q_v^{-s} + q_v^{1-2s})^{-1}. \tag{2.1}$$

Definition 2.2. With notation as above, the *incomplete L-function* of E is

$$L_S(E/k,s) = \prod_{v\notin S} L_v(E/k,s). \tag{2.2}$$

Both this infinite product and the Dirichlet series $L_S(E/k,s) = \sum_{n\geq 1} a_n n^{-s}$ with integral coefficients which is formally assembled from it converge absolutely in the half plane $\mathrm{Re}(s) > 3/2$. This follows from the estimate in Hasse's theorem and comparison with the Dedekind zeta-function of k. Note that when k is a function field, this Dirichlet series is of the form $\sum_{m\geq 0} b_m q^{-ms}$ where q is the cardinality of the constant field of k.

The formal Euler product (2.2) determines E up to isogeny over k. See Appendix C for this fact and a discussion of the l-adic Tate module $T_l E$ of E.

2. The L-function of an elliptic curve

The general formalism of L-functions suggests that the L-function of an elliptic curve E over a global field k should have a meromorphic continuation to the whole of $\mathbb{C}$ and satisfy a functional equation. Before this can be true we must add some factors to the incomplete L-function, corresponding to the infinite places and the finite places of bad reduction.

Recall that S is the set of infinite places, together with the places where E has bad reduction. For finite $v \in S$, we define the local L-factor

$$L_v(s) = \begin{cases} 1 & \text{if } E \text{ has additive reduction at } v; \\ (1-q_v^{-s})^{-1} & \text{if } E \text{ has split multiplicative reduction at } v; \\ (1+q_v^{-s})^{-1} & \text{if } E \text{ has non-split multiplicative reduction at } v. \end{cases} \tag{2.3}$$

A common feature of local L-factors at good and bad finite places is that

$$L_v(1) = \frac{q_v}{\#\widetilde{E}_v^{\mathrm{ns}}(\mathbb{F}_v)}, \tag{2.4}$$

where $\widetilde{E}_v^{\mathrm{ns}}$ is the smooth locus of the reduction of a minimal model for E over A_v.

We can now define the global L-function.

Definition 2.3. The *L-function* of E/k is

$$L(E/k,s) = \prod_{v\nmid\infty} L_v(E/k,s). \tag{2.5}$$

When k is a number field, we also add factors at the infinite places, setting

$$\Lambda(E/k,s) = \big((2\pi)^{-s}\Gamma(s)\big)^{[k:\mathbb{Q}]} L(E/k,s).$$

Remark. The definition of the local factors in the Euler product of Artin L-functions forms the prototype for general arithmetic L-functions. These conventions show that the common features of the L-factors $L_v(E/k,s)$ at good and bad places go beyond their values at $s=1$. Namely, if $\rho : \mathrm{Gal}(k^s/k) \to GL_n(\mathbb{C}) = GL(V)$ is an Artin representation, the L-function attached to it is

$$L(\rho,s) = \prod_{\text{finite } v} (1 - q_v^{-s}\,\mathrm{Frob}_v \mid V^{I_v})^{-1},$$

where I_v is an inertia group at v, V^{I_v} denotes the invariants of I_v, and Frob_v is a geometric Frobenius element for v. Likewise, the L-function of an elliptic curve may be written

$$L(E/k, s) = \prod_{\text{finite } v} (1 - q_v^{-s}\,\mathrm{Frob}_v \mid (V_\ell E^\vee)^{I_v})^{-1}.$$

Here $(V_\ell E^\vee)$ is the ℓ-adic cohomology of E, for a prime ℓ which is not equal to the characteristic of k. To see the compatibility at good places of this formulation of $L(E/k,s)$ with our definition, see Appendix C. In particular, compare the characteristic polynomial of arithmetic Frobenius in Equation (C.1) with our definition of L-factors at good places.

The following two quantities measure bad reduction.

Definition 2.4. The *minimal discriminant* of an elliptic curve E over a number field (resp. global function field) k is the integral ideal of k (resp. positive divisor of the curve associated to the function field k) defined by

$$\mathcal{D}(E/k) = \prod_{v \nmid \infty} \mathfrak{p}_v^{v(\Delta_v)},$$

where Δ_v is the discriminant of a minimal equation for E/k_v and $\mathfrak{p}_v$ is the prime ideal (resp. prime divisor) associated to the finite place v.

Remark. Suppose that k is a number field, with ring of integers A. We say that a Weierstrass model for E is a *global minimal model* if it is a minimal model at all finite places v of k. Such a model exists if and only if a certain projective A-module of rank 1, corresponding to the Néron differentials, has trivial class in $\mathrm{Pic}(A)$. If such a model exists, then the discriminant ideal $\mathcal{D}(E/k)$ is principal, generated by the discriminant Δ of a global minimal model. For more details, see [**52**], Ch. VIII, §8.

Definition 2.5. The *conductor* of E is the integral ideal or positive divisor given by

$$N(E/k) = \prod_{v \nmid \infty} \mathfrak{p}_v^{f_v},$$

where

$$f_v = \begin{cases} 0 & \text{if } E \text{ has good reduction at } v \\ 1 & \text{if } E \text{ has multiplicative reduction at } v \\ 2+\delta_v & \text{if } E \text{ has additive reduction at } v, \end{cases} \tag{2.6}$$

and where δ_v is a non-negative integer depending on the action of wild inertia at v on $T_\ell E$. It is zero whenever the characteristic of v is not equal to 2 or 3.

Remark. In fact, the conductor of E is the Artin conductor of the Tate module of E [**49**]. It is related to $\mathcal{D}(E/k)$ by Ogg's formula [**44**]

$$f_v = \mathrm{ord}_v(\mathcal{D}(E/k)) + 1 - m_v,$$

where m_v is the number of irreducible components of the Néron model of E at v.

The following theorem is due to Grothendieck, and uses étale cohomology [**29**].

Theorem 2.6. *Let $k = \mathbb{F}_q(X)$ be the function field of the geometrically irreducible, non-singular curve X of genus g over the field $\mathbb{F}_q$, and let E be an elliptic curve over k. Then $L(E/k, s)$ is a rational function of q^{-s} of degree $\deg N(E/k)+4g-4$, and satisfies the functional equation*

$$L(E/k, s) = \pm A^{1-s} L(E/k, 2-s),$$

with $A = q^{\deg N(E/k)+4g-4}$. Moreover, if E is not a constant curve over k, then $L(E/k, s)$ is a polynomial in q^{-s} with constant coefficient 1.

We recall that X is geometrically irreducible if the algebraic closure of $\mathbb{F}_q$ in k is equal to $\mathbb{F}_q$. A constant curve E/k is one that arises from extension of scalars of an elliptic curve over $\mathbb{F}_q$ (cf. Lecture 3, §1). The rationality of $L(E/k, s)$ follows from the Lefschetz fixed point theorem, and the functional equation from a version of Poincaré duality. A sketch of the proof can be found in Appendix D.

In the case of elliptic curves over number fields, we only have the following.

Conjecture 2.7. *Let k be a number field, and E/k an elliptic curve. Then the complex analytic function $\Lambda(E/k, s)$ on the right half plane $\mathrm{Re}(s) > 3/2$ admits an analytic continuation to the entire complex plane and satisfies the functional equation*

$$\Lambda(E/k, s) = \pm A^{1-s} \Lambda(E/k, 2-s), \tag{2.7}$$

where A is the product of the absolute norm of $N(E/k)$ with the square of the discriminant of k.

Wiles and Taylor proved this conjecture in the important case when $k = \mathbb{Q}$ and the conductor $N(E/k)$ is square-free [**63**], [**59**]. Their methods were extended to cover all elliptic curves over $\mathbb{Q}$ [**6**]. Some other cases when k is a totally real number field are known.

Remark. If Conjecture 2.7 is true, then the sign $\pm$ in the functional equation, which is known in general as a *root number*, is given by

$$\pm = (-1)^{\mathrm{ord}_{s=1} L(E/k,s)}, \tag{2.8}$$

since the reflection point of the functional equation is $s = 1$. We note that the global sign is a product of local root numbers, all of which are ± 1 and almost all of which are $+1$ [**27**].

An overview of the topics of this section, with further references, may be found in [**52**], §C.16.

3. Periods

We continue with the notations of the previous section. The conjecture of Birch and Swinnerton-Dyer connects the behavior of $L(E/k, s)$ near the point $s = 1$ to the rank of $E(k)$, and to some further arithmetic invariants of E. In the previous lecture, we defined several of these invariants, namely the Tate-Shafarevich group $\text{Ш}(k, E)$ and the regulator $R(E/k)$. It remains to define the global period of E.

Recall that an elliptic curve has an invariant differential that is unique up to scalar multiple. For each place v of k let ω_v be a non-zero invariant differential for E/k_v. Together with the choice of a Haar measure dx_v on the additive group of k_v, this gives us a Haar measure $|\omega_v|$ on the compact group $E(k_v)$.

Lemma 2.8 (Tate [**57**]). *Let E be an elliptic curve over k_v where v is a finite place of k, and choose a minimal Weierstrass model for E. Let ω_v be an invariant differential, defined over A_v, which does not vanish modulo $\pi_v A_v$, and let dx_v be the Haar measure on k_v which gives A_v volume 1. Then*

$$\int_{E(k_v)} |\omega_v| = \frac{[E(k_v) : E^0(k_v)]}{L_v(1)}.$$

PROOF. See Appendix A, where the subgroup $E^0(k_v)$ is defined. □

Choose a non-zero invariant differential ω on E/k; it gives an invariant differential which we denote ω_v on E/k_v for all places v. Also, choose a decomposition

$$dx = \otimes_v dx_v \tag{2.9}$$

of the Haar measure dx on the adèles $\mathbb{A}$ of k with $\int_{\mathbb{A}/k} dx = 1$ (so dx_v is a Haar measure on k_v).

Definition 2.9. The *period* of E/k is

$$P(E/k) = P(\omega) = \prod_{v\nmid\infty} \left(L_v(1) \cdot \int_{E(k_v)} |\omega_v| \right) \cdot \prod_{v|\infty} \int_{E(k_v)} |\omega_v|.$$

To see that this quantity is well defined we make the following observations. First, the conditions of Lemma 2.8 are satisfied for almost all finite v. That is: $\int_{A_v} dx_v = 1$, ω_v is defined over A_v, and ω_v does not vanish modulo π_v. Second, we note that $E^0 \neq E$ for only finitely many finite places v, so that by Lemma 2.8 the product defining $P(E/k)$ has almost all terms equal to 1.

It remains to show that the period is independent of the choice of invariant differential. This is the case, because a rescaling $\omega \mapsto \lambda\omega$ by $\lambda \in k^\times$ modifies P by the factor

$$\prod_v |\lambda|_v = 1,$$

by the product formula. Therefore $P(E/k)$ is an invariant of the elliptic curve E/k.

Remark. The observations above amount to an introduction to adelic integration in the particular case of an elliptic curve. In this setting the $L_v(1)$ play the role of the "convergence factors" of [**62**], Ch. 2.

Example. Let k be a function field of genus g, with constant field $\mathbb{F}_q$. Then

$$P(E/k) = \prod_{\text{bad } v} [E(k_v) : E^0(k_v)] / \left(q^{g-1+\frac{1}{12}\deg \mathcal{D}(E/k)} \right). \tag{2.10}$$

The factor $q^{-(g-1)}$ reflects the volume $\int_{\mathbb{A}/k} dx = q^{g-1}$ of $\mathbb{A}/k$ under the usual product measure ([**62**], §2.1.3), and the factor $q^{-\frac{1}{12}\deg \mathcal{D}(E/k)}$ stems from the obstruction to the existence of an everywhere integral Néron differential.

Example. Let k be a number field. If there exists a Néron differential ω over the ring of integers A of k, which has non-zero reduction modulo every prime, then

$$P(E/k) = \prod_{v|\infty} \int_{E(k_v)} |\omega|_v \cdot \prod_{v \text{ bad}} [E(k_v) : E^0(k_v)] \cdot |D_k|^{-1/2}, \tag{2.11}$$

where D_k is the discriminant of k. The discriminant appears because under the usual product measure the compact quotient $\mathbb{A}/k$ has volume $|D_k|^{1/2}$ ([**62**], §2.1.3).

As the above formulae show, the global period can be computed from local invariants of the curve E and global invariants (g and D_k) of the field k.

4. The conjecture of Birch and Swinnerton-Dyer

We now have all of the definitions in place in order to state the precise form of the Birch and Swinnerton-Dyer conjecture.

Conjecture 2.10. *Let E/k be an elliptic curve over a global field, and assume that $L(E/k, s)$ has a meromorphic continuation to a neighborhood of the point $s = 1$.*

(1) *If n is the rank of the finitely generated abelian group $E(k)$, then*
$$\operatorname{ord}_{s=1} L(E/k, s) = n.$$
(2) *Let $c(E/k)$ be the leading term in the Taylor expansion at $s = 1$, that is*
$$L(E/k, s) \sim c(E/k) \cdot (s-1)^n \quad \text{as } s \to 1.$$
Then $c(E/k) = P(E/k) \cdot R(E/k) \cdot \#\text{Ш}(k, E)$.

For another discussion of this conjecture, with a reward attached, see [**64**]. Note that the formula for the leading term assumes Tate's conjecture that the Tate-Shafarevich group $\text{Ш}(k, E)$ is finite.

A naïve original motivation for this conjecture is that one might expect a larger number of rational points to yield larger numbers of points on reductions modulo finite places. Then the partial products of the infinite product
$$L_S(E/k, 1) \text{``}=\text{''} \prod_{\text{good } v} L_v(E/k, 1) = q_v / \#\widetilde{E}_v(\mathbb{F}_v)$$
would tend to be relatively smaller. The original conjecture and the numerical evidence that led to it are recorded in [**4**].

Remark. A much weaker form of Conjecture 2.10 is the parity conjecture, that $\pm$, the sign of the functional equation of E is equal to $(-1)^{\text{rank}(E(k))}$ [**18**]. Compare the remark after Conjecture 2.7.

Remark. One knows (cf. Appendix C) that the L-function is an isogeny invariant of E/k. The individual terms $P(E/k)$, $R(E/k)$, and $\#\text{Ш}(k, E)$ which occur in the conjecture of Birch and Swinnerton-Dyer are isomorphism invariants, and can change when we modify E by an isogeny. However, their product is an isogeny invariant [**10**], [**58**], [**41**].

Remark. Bloch [**5**] defines an extension of commutative algebraic groups
$$1 \longrightarrow \mathbb{G}_m^n \longrightarrow G \longrightarrow E \longrightarrow 1$$
over k, where n is the rank of $E(k)$. The existence of such an extension is a consequence of the theorem of Serre-Rosenlicht [**48**], VII, Thm. 6, which gives a natural isomorphism between the groups $\operatorname{Ext}_k(E, \mathbb{G}_m)$ and $\operatorname{Pic}^0(E)(k)$, the latter being naturally identified with the k-rational points of E.

The L-function for G is the product
$$L(G/k, s) = L(E/k, s)\zeta_k(s)^n,$$
where $\zeta_k(s)$ is the Dedekind zeta-function of k. Therefore we expect $L(G/k, s)$ to be regular and non-zero at $s = 1$. Bloch shows that $G(k)$ is discrete and co-compact in $G(\mathbb{A})$, where $\mathbb{A}$ is the ring of adeles of k, and calculates the volume of the

quotient group $G(\mathbb{A})/G(k)$ in terms of the period and regulator of E/k. He shows that the Birch and Swinnerton-Dyer conjecture for E would follow from standard conjectures on Tamagawa numbers, applied to the algebraic group G.

5. Examples of elliptic curves over function fields

We present two examples of non-constant elliptic curves where the conjecture of Birch and Swinnerton-Dyer is known to be true. In both cases, $X = \mathbb{P}^1$ and $k = \mathbb{F}_q(X) = \mathbb{F}_q(t)$.

Example. Suppose that q is odd, and consider the elliptic curve E/k given by the equation

$$y^2 = x(x-1)(x-t), \text{ of discriminant } \Delta = 16t^2(t-1)^2.$$

Making the change of variable $v = 1/t$ gives the equation

$$y^2 = x(x-v)(x-v^2), \text{ of discriminant } \Delta' = 16v^8(v-1)^2.$$

A change of variable changes the valuation of the discriminant at each place by a multiple of 12; it follows that the above equations are both minimal at every place at which they are integral. We see that E has multiplicative reduction at the places $t = 0, 1$ and additive reduction at $t = \infty$, and the conductor and minimal discriminant are given by (cf. Equation (2.6))

$$N(E/k) = (0) + (1) + 2(\infty) \text{ and } \mathcal{D}(E/k) = 2(0) + 2(1) + 8(\infty).$$

An application of Tate's algorithm shows that the reduction at the bad places $t = 0, 1, \infty$ are of Kodaira types I_2, I_2 and I_2^* respectively [**57**].

To calculate the period $P(E/k)$, we need to know the quantities $\deg \mathcal{D}(E/k) = 12$, the genus $g(X) = 0$, and the indices $[E(k_v) : E^0(k_v)]$ at the places of bad reduction $t = 0, 1, \infty$, which are $2, 2$, and 4, respectively, cf. [**57**]. Using Equation (2.10), we find that

$$P(E/k) = \prod_{\text{bad } v} [E(k_v) : E^0(k_v)] = 2_0 \cdot 2_1 \cdot 4_\infty = 16.$$

To calculate the regulator $R(E/k)$, we note that

$$E(k) = \{\mathcal{O}_E, (0,0), (1,0), (t,0)\}.$$

This can be deduced from the computation of the Selmer group, along with knowledge of the torsion subgroup of $E(k)$, which is easy to calculate. Hence

$$R(E/k) = \#E(k)^{-2} = 1/16.$$

Finally, since the degree of $N(E/k)$ is 4, Proposition D.1 implies that $L(E/k, s) = 1$. Now, the Birch and Swinnerton Dyer conjecture is known for this elliptic curve, since E is the generic fiber of a pencil of elliptic curves on a rational surface (cf. the next lecture). It follows that

$$1 = L(E/k, 1) = P(E/k) \cdot R_E \cdot \#\text{Ш}(k, E) = \#\text{Ш}(k, E),$$

so the Tate-Shafarevich group is trivial. The fact that the 2-torsion in $\text{Ш}(k, E)$ is trivial follows from the descent calculation of Appendix B.

Example. Assume that $q = p^f$ where p is an odd prime. Let E'/k be the elliptic curve defined by

$$E' : t(t-1)y^2 = x(x-1)(x-t).$$

The minimal regular model of E' over X is a K3 surface over $\mathbb{F}_q$. In this case, the conjecture of Birch and Swinnerton-Dyer is also known to be true. We now examine its consequences.

A similar computation to the one above reveals that

$$N(E'/k) = 2(0) + 2(1) + 2(\infty) \text{ and } \mathcal{D}(E'/k) = 8(0) + 8(1) + 8(\infty).$$

Proposition D.1 then implies that $L(E'/k, s)$ is a polynomial of degree 2 in q^{-s}. After a calculation (cf. section 8.8 of [**33**]) one finds that this polynomial is

$$L(E'/k, s) = (1 - \alpha q^{-s})(1 - \beta q^{-s}),$$

with

$$\text{(2.12)} \qquad \begin{cases} \alpha = (a+bi)^{2f}, & \beta = \bar{\alpha} \qquad \text{if } p \equiv 1 \pmod 4, \\ \alpha = p^f = q, & \beta = (-p)^f \quad \text{if } p \equiv 3 \pmod 4. \end{cases}$$

If $p \equiv 1 \pmod 4$ then a, b are the unique positive integers such that a is odd and $p = a^2 + b^2$. Let $\gamma = a + bi$, and write $\gamma^f = A + Bi$.

In the case that $p \equiv 1 \pmod 4$, we evaluate the L-function at $s = 1$ to get

$$L(E'/k, 1) = \frac{\left|\gamma^f - \bar{\gamma}^f\right|^2}{q} \neq 0.$$

Equation (2.10) implies that $P(E'/k) = (4 \cdot 4 \cdot 4)/q$, since $[E'(k_v) : E'^0(k_v)] = 4$ at the places v of k where $t = 0, 1,$ and ∞ and $\deg \mathcal{D}(E'/k) = 24$. Since the L-function does not vanish at $s = 1$, the group $E'(k)$ is finite. In fact, $\#E'(k) = 4$ and $R(E'/k) = 1/16$. Then the Birch and Swinnerton-Dyer conjecture implies that

$$\#Ш(k, E') = \frac{\left|\gamma^f - \bar{\gamma}^f\right|^2}{4} = B^2.$$

The fact that $Ш(k, E')$ has even order is consistent with the descent calculation in lecture 1.

Suppose that $p \equiv 3 \pmod 4$. Then Equation (2.12) implies that $L(E'/k, s)$ vanishes at $s = 1$ to order 1 if f is odd and to order 2 if f is even. We now split our discussion into cases according to the parity of f.

Let f be odd. One finds that $L(E'/k, 1) = 0$ and $L'(E'/k, 1) = 2 \log q$, so by the Birch and Swinnerton-Dyer conjecture $E'(k)$ has rank 1. As before, $P(E'/k) = 64/q$. If we take P to be a generator modulo torsion then $E'(k) = \mathbb{Z}P \oplus (\mathbb{Z}/2\mathbb{Z})^2$, so

$$R(E'/k) = \langle P, P \rangle / 4^2,$$

which as we mentioned at the end of the previous lecture is equal to $r \log q / 4^2$ where r is rational. In fact $r \in \frac{1}{2}\mathbb{Z}$; the height pairing takes on integral values except for contributions from places of bad reduction, the denominators of which can be bounded explicitly in terms of the Kodaira type, cf. lecture 3.

Applying the Birch and Swinnerton-Dyer conjecture, we see that

$$2 \log q = L'(E'/k, 1) = c(E'/k) = \frac{4r \log q}{q} \#Ш(k, E'),$$

so $\#Ш(k, E') \cdot 2r = q$. The existence of the Cassels-Tate pairing (cf. [**9**], [**58**]) implies that $Ш(k, E')$ has square order, and consequently, for all odd f, $Ш(k, E')$

is a p-group and the denominator of r is equal to 2. In fact, Elkies has shown us that $r = \frac{p}{2}$, for all odd f, so Ш(k, E') is non-trivial once $f > 1$.

We now consider the case when f is even. Observe that $L(E'/k, s)$ vanishes to order 2 at $s = 1$ with leading term equal to $(\log q)^2$, so that by the Birch and Swinnerton-Dyer conjecture, the Mordell-Weil group $E'(k)$ has rank two. Just as in the case where f is odd, $P(E'/k) = 64/q$, and the height pairing on generators modulo torsion takes values in $\frac{1}{2}\mathbb{Z}\log q$. Since the regulator is a determinant of a 2-by-2 matrix of such pairings, $R(E'/k)$ has the form $r(\log q)^2/4^2$ where $r \in \frac{1}{4}\mathbb{Z}$.

Applying the Birch and Swinnerton-Dyer conjecture, we see that

$$(\log q)^2 = \frac{L''(E'/k,1)}{2} = c(E'/k) = \frac{4r(\log q)^2}{q}\#\text{Ш}(k,E'),$$

so $\#\text{Ш}(k, E') \cdot 4r = q$. Arguing as in the odd case, we may conclude that the denominator of r is equal to 4 and that Ш(k, E') is a p-group. In fact, Elkies has shown us that $r = \frac{p^2}{4}$, for all even f, so Ш(k, E') is non-trivial once $f > 2$.

6. Examples of elliptic curves over number fields

Example. Let $E/\mathbb{Q}$ be the elliptic curve

$$E : y^2 + y = x^3 - x$$

of conductor $N = 37$. We will calculate some of the invariants of E to show what the Birch and Swinnerton-Dyer conjecture implies.

Let v be a place of good reduction. By Proposition A.3, the reduction map $E(\mathbb{Q}) \to \widetilde{E}(\mathbb{F}_v)$ is injective on n-torsion when n is prime to v. In this case we have

$$\#\widetilde{E}(\mathbb{F}_2) = 5 \text{ and } \#\widetilde{E}(\mathbb{F}_3) = 7,$$

and so $E(\mathbb{Q})$ is torsion-free. Since $P = (0, 0) \in E(\mathbb{Q})$ is not equal to $\mathcal{O}_E$, the rank of E over $\mathbb{Q}$ must be at least one. In fact, one can check that $E(\mathbb{Q}) = \mathbb{Z}P$ via a descent argument similar to the one outlined in the previous lecture.

Next, we will calculate the period of E. The factors $[E(\mathbb{Q}_p) : E^0(\mathbb{Q}_p)]$ are trivial at all places of good reduction. Applying Tate's algorithm and looking at the table in [**57**], noting that E has multiplicative reduction at 37 with $\mathrm{ord}_{37}(j(E)) = -1$, shows that $E = E^0$ at $p = 37$ as well. Equation (2.11) now gives $P(E/k) = \int_{E(\mathbb{R})} |\omega|$. The BSD conjecture predicts that $L(E/\mathbb{Q}, 1) = 0$ and

$$L'(E/\mathbb{Q},1) = \#\text{Ш}(\mathbb{Q},E)\cdot\langle P,P\rangle\cdot\int_{E(\mathbb{R})}|\omega|, \tag{2.13}$$

where ω is a Néron differential on $E/\mathbb{Q}$.

The sign in the functional equation of $L(E/\mathbb{Q}, s)$ is equal to -1, so the L-function vanishes to odd order at $s = 1$. One can compute the value $L'(E/\mathbb{Q}, 1)$, which is non-zero. Hence the L-function vanishes to order exactly 1, i.e. E has analytic rank 1. Via the work of Kolyvagin [**34**] and Gross-Zagier [**28**], the Birch and Swinnerton-Dyer conjecture is known for curves of analytic rank 0 and 1, up to some ambiguity in the leading term (cf. lecture 3). In this case, the full conjecture is known to be true. We have

$$L'(E/\mathbb{Q},1) = \langle P,P\rangle\cdot\int_{E(\mathbb{R})}|\omega|$$

and the group Ш$(\mathbb{Q}, E)$ is trivial.

Example. Let $E/\mathbb{Q}$ be the elliptic curve

$$E : y^2 + y = x^3 - 7x + 6, \tag{2.14}$$

of conductor $N = 5077$. Brumer and Kramer found that this is the elliptic curve over $\mathbb{Q}$ of rank 3 with smallest conductor, making it an interesting case study for elliptic curves of higher rank. It also has 36 points with integral coordinates on the minimal model. It has no non-trivial torsion because its reductions modulo 2 and 3 are the same as in the above example. Also, as in the previous example, $E(\mathbb{Q}_p) = E^0(\mathbb{Q}_p)$ for all p because $\mathrm{ord}_{5077}(j(E)) = -1$.

In [**8**] the authors make a detailed study of the invariants of E, which together with the conjecture of Birch and Swinnerton-Dyer would imply that $\text{Ш}(\mathbb{Q}, E) = 1$. We now review these computations. Via descent and height calculations, one can show that the points $P_1 = (0, 2), P_2 = (1, 0)$, and $P_3 = (2, 0)$ generate $E(\mathbb{Q})$, i.e.

$$E(\mathbb{Q}) = \mathbb{Z}P_1 + \mathbb{Z}P_2 + \mathbb{Z}P_3.$$

A calculation of the real period $\int_{E(\mathbb{R})} |\omega|$, the regulator $R(E/\mathbb{Q})$, and the L-function around $s = 1$ suggests that $L(E/\mathbb{Q}, s)$ vanishes to order 3 around $s = 1$ with leading term

$$c(E/k) \approx R(E/\mathbb{Q}) \cdot \int_{E(\mathbb{R})} |\omega| = 1.73185\ldots,$$

accurate to 28 decimal places. (In fact, we can show that the order of vanishing is equal to 3, using the argument in the next paragraph.) This calculation suggests strongly that $\text{Ш}(\mathbb{Q}, E)$ is trivial, but we still don't know if this group is finite.

Remark. These examples of elliptic curves with rank 3 were important in finding the first effective lower bound on class numbers of positive definite binary quadratic forms, completing the program of Goldfeld [**24**], [**23**]. Namely, the Gross-Zagier formula [**28**] implies that for an elliptic curve $E/\mathbb{Q}$ whose L-function has a odd-order zero at $s = 1$, there exists a computable point $P \in E(\mathbb{Q})$ such that

$$L'(E/\mathbb{Q}, 1) = c\langle P, P\rangle,$$

where c is a positive constant. If the point P is torsion, $\langle P, P\rangle = 0$ and the first derivative vanishes. This argument applied to the elliptic curve defined in equation (2.14) allows one to show that $L(E/\mathbb{Q}, s)$ vanishes to order 3 at $s = 1$.

LECTURE 3

Progress to date

In this lecture, we review the progress that has been made towards a proof of the conjecture of Birch and Swinnerton-Dyer. In the function field case, this work is related to the conjecture of Artin and Tate [**58**] for the Néron-Severi group of a surface over a finite field. We give some examples, both for constant and non-constant curves. In the number field case, most of the progress has come from the study of special points on Shimura curves over totally real fields, and the images of certain divisors supported on these points in the Jacobian. These ideas, due to Shimura [**50**] and Birch [**2, 3**], represent a new chapter in the theory of complex multiplication. We end with a summary of Néron's theory of local heights on curves, which is essential in the calculation of the heights of these divisor classes.

1. Results over function fields

Let $\mathbb{F}_q$ be a finite field with $q = p^f$ elements, and X a non-singular, geometrically irreducible, projective curve, defined over $\mathbb{F}_q$. Let $k = \mathbb{F}_q(X)$ be the function field of X.

Theorem 3.1 (M. Artin - Tate [**58**]). *Let E be an elliptic curve, defined over k. Then*

$$\mathrm{ord}_{s=1} L(E/k, s) \geq \mathrm{rank}\, E(k),$$

and the following are equivalent:

(1) *Equality holds above.*
(2) *$Ш(k, E)$ is finite.*
(3) *$Ш(k, E)[l^\infty]$ is finite for a single prime l.*
(4) *The conjecture of Birch and Swinnerton-Dyer is true for the curve E over k.*

The curve E can be considered as the generic fiber of an elliptic surface $S/\mathbb{F}_q$ which is fibered over X. In this case $E(k)$ is closely connected to the *Néron-Severi group $NS(S)$* of S. Motivated by this, Artin and Tate formulated a more general conjecture, similar in spirit to the conjecture of Birch and Swinnerton-Dyer, relating the rank of the Néron-Severi group and some other geometric invariants of a general surface S to its zeta function. In this setting, the cohomological Brauer group $H^2(S, \mathbb{G}_m)$ plays the role of the Tate-Shafarevich group.

In [**58**] Tate gives the translation of the above theorem into this language and a proof of the "prime-to-p" part of the conjecture, using l-adic étale cohomology for $l \neq p$. Milne completed the proof in [**38**], using more sophisticated cohomology theories capable of handling the complications at the prime p. We give a brief sketch.

We recall that for a smooth projective surface over an algebraically closed field, the Néron-Severi group is defined as the group of divisors modulo algebraic equivalence, cf. [**31**]. It is a finitely generated abelian group. We take S to be the minimal regular model of E over X. Then $NS(S)$ is defined to be the image of $\mathrm{Pic}(S)$ in $NS(S \otimes \overline{\mathbb{F}}_q)$. We then have injections

$$E(k) \otimes \mathbb{Q}_l \hookrightarrow NS(S) \otimes \mathbb{Q}_l \hookrightarrow H^2(S \otimes \overline{\mathbb{F}}_q, \mathbb{Q}_l)(1),$$

the latter arrow coming from the cycle map in étale cohomology. We remark that the first arrow is a slight modification of the obvious map, cf. [**51**]. Moreover, we have an exact sequence for every $l \neq p$:

$$0 \longrightarrow NS(S) \otimes \mathbb{Q}_l \longrightarrow (H^2(S \otimes \overline{\mathbb{F}}_q, \mathbb{Q}_l)(1))^{G_{\mathbb{F}_q}} \longrightarrow V_l Ш(k,E) \longrightarrow 0.$$

Here $V_l Ш(k,E) = \mathrm{Hom}(\mathbb{Q}_l/\mathbb{Z}_l, Ш(k,E)) \otimes_{\mathbb{Z}_l} \mathbb{Q}_l$ is the analogue of the l-adic Tate module for the torsion group $Ш(k,E)$. It vanishes if and only if $Ш(k,E)[l^\infty]$ is finite. We note that the map $E(k) \otimes \mathbb{Q}_l \to NS(S) \otimes \mathbb{Q}_l$ need not be surjective, but its cokernel is well understood. In particular, there is a formula

$$\mathrm{rank}\, NS(S) = \mathrm{rank}\, E(k) + 2 + \sum_{t \in X} (m_t - 1),$$

where m_t is the number of irreducible components in the fiber above t. We refer to [**51**] for more details.

The zeta-function of S is given by the formula

$$\prod_i P_i(S, q^{-s})^{(-1)^{i+1}}, \text{ where } 0 \le i \le 4 \text{ and } P_i(S,t) = \det(1 - \mathrm{Frob}\, t | H^i(S, \mathbb{Q}_l)).$$

Tate's geometric conjecture can be rephrased in terms of the order of vanishing of $P_2(S, q^{-s})$ at $s = 1$, because Deligne has proved that the eigenvalues of Frob on $H^i(S, \mathbb{Q}_l)$ have absolute value $q^{i/2}$. The order of vanishing is equal to the multiplicity of q as an eigenvalue of Frobenius on H^2, which gives the first inequality in the theorem. Moreover, equality holds if and only if the l-primary part of the Tate-Shafarevich group is finite. Deligne also proved that the polynomials $P_i(S,T)$ are independent of l. These facts give the equivalence of the first 3 statements; and the equivalence of the last follows from a more involved calculation, once $Ш(k,E)$ is known to be finite.

2. Constant curves

We continue with the notation of the previous section, and specialize to the case where E/k is a constant curve; that is, E arises by extension of scalars from a curve over $\mathbb{F}_q$. In this case the surface S can be taken to be $E \times_{\mathbb{F}_q} X$.

First, we note that $E(k)_{\mathrm{tors}} = E(\mathbb{F}_q)$. Indeed, the torsion points of E are defined over an algebraic extension of $\mathbb{F}_q$, and $\overline{\mathbb{F}}_q \cap k = \mathbb{F}_q$.

Next, suppose that P is a point of infinite order in $E(k)$. This point can be viewed as a section $s : X \to S$; since the section s is not constant, it induces a dominant morphism $\pi : X \to E$, of finite degree. We have the formula

$$\langle P, P \rangle = (2 \deg \pi) \log q.$$

This follows from the fact that the height pairing on $E(k)$ is induced by the intersection pairing on S.

Giving the image of P in the quotient $L = E(k)/E(\mathbb{F}_q)$ is the same as giving the map π, up to translation by elements of $E(\mathbb{F}_q)$. Since any map from X to E factors through the Jacobian J of X [**40**]), we can identify

$$L = E(k)/E(\mathbb{F}_q) = \mathrm{Hom}_{\mathbb{F}_q}(J, E).$$

We view L as an integral even lattice, that is, a free $\mathbb{Z}$-module with a positive definite integer-valued quadratic form (the height divided by $2\log q$). Lattices arising in this manner are called *Mordell-Weil lattices*, and have many interesting properties. These have been studied by Elkies, Shioda and others, cf. [**20**], [**51**].

By Equation (D.3),

$$L(E/k,s) = \frac{\det(1 - \operatorname{Frob} q^{-s} | H^1(E,\mathbb{Q}_l) \otimes H^1(X,\mathbb{Q}_l))}{(1 - aq^{-s} + q^{1-2s})(1 - aq^{1-s} + q^{3-2s})},$$

where a is the trace of Frobenius on the l-adic Tate module of E. We have identified $H^1(X,\mathbb{Q}_l) = V_l J^\vee$, and $H^1(E,\mathbb{Q}_l) = V_l E^\vee$, cf. [**40**], Cor. 9.6, and

$$V_l J^\vee \otimes V_l E^\vee \cong \operatorname{Hom}(V_l J, V_l E)(-1).$$

Hence the rank of $E(k)$, which is equal to the rank of the free quotient $L = \operatorname{Hom}_{\mathbb{F}_q}(J,E)$, is also equal to the multiplicity of q as an eigenvalue of Frobenius on $V_l J^\vee \otimes V_l E^\vee$, by Tate's theorem for abelian varieties over finite fields (Equation (C.2)). Comparison with the above equation for $L(E/k,s)$ gives

$$\operatorname{ord}_{s=1} L(E/k,s) = \operatorname{rank} E(k),$$

and by Theorem 3.1 the full Birch and Swinnerton-Dyer conjecture is true for the curve E.

Let us write $V_l E^\vee \otimes V_l J^\vee = U \oplus V$, where U is the q-eigenspace of Frobenius, and V is the sum of the other eigenspaces. Then the leading term of the L-series at $s = 1$ is given by

$$\frac{(\log q)^n \det(1 - \operatorname{Frob} q^{-1}|V)}{\#E(k)^2_{\mathrm{tors}} q^{-1}} = \frac{1}{q^{g-1}} \frac{\det(L)(\log q)^n}{\#E(k)^2_{\mathrm{tors}}} \#\text{Ш}(k,E).$$

Here we have applied the conjecture of Birch and Swinnerton-Dyer. The period is given by $P(E/k) = 1/q^{g-1}$, where g is the genus of the curve X, and the regulator is given by $R(E/k) = \det(L) \cdot (\log q)^n / \#E(k)^2_{\mathrm{tors}}$ where n is the rank. Canceling some terms gives the final identity

$$q^g \det(1 - \operatorname{Frob} q^{-1}|V) = \det(L) \#\text{Ш}(k,E).$$

The simplest situation is when $V = 0$, so the characteristic polynomial of Frobenius on $V_l E^\vee \otimes V_l J^\vee$ is $(t-q)^{4g}$. We are going to investigate a special case where this happens.

Example. Suppose that q is a square, and write $q = q_0^2$. We choose E so that both eigenvalues of the arithmetic $\mathbb{F}_q$-Frobenius are equal to $-q_0$; this can be done, for example, by taking E to be a supersingular curve with $q_0 + 1$ points over the finite field with q_0 elements (cf. Equation (C.1)). The existence of such curves is a consequence of Honda-Tate theory [**56**]; they are obtained as the reduction of elliptic curves with complex multiplication over number fields.

Let X be the non-singular curve defined by the Fermat equation

$$x^{q_0+1} + y^{q_0+1} + z^{q_0+1} = 0$$

in $\mathbb{P}^2$. This curve has genus $g = q_0(q_0-1)/2$. A simple point count shows that $\#X(\mathbb{F}_q) = q_0^3 + 1$. On the other hand, the Lefschetz fixed point theorem in étale cohomology gives the number of points as

$$\sum_{i=0}^{2} (-1)^i \operatorname{Tr}(\operatorname{Frob}|H^i(X \otimes \overline{\mathbb{F}}_q, \mathbb{Q}_l)) = q + 1 - \operatorname{Tr}(\operatorname{Frob}|H^1(J,\mathbb{Q}_l)).$$

This can be equal to q_0^3+1 if and only if all the eigenvalues of Frobenius are equal to $-q_0$, since the eigenvalues have absolute value q_0 under any embedding $\overline{\mathbb{Q}}_l \hookrightarrow \mathbb{C}$ [**39**].

We therefore have $V = 0$, and $L = \mathrm{Hom}_{\mathbb{F}_q}(J, E)$ is a lattice of rank $4g = 2q_0(q_0-1)$. The Birch and Swinnerton-Dyer conjecture then gives

$$\det L \cdot \#\text{Ш}(k,E) = q^g = q_0^{q_0(q_0-1)}.$$

Write $q_0 = p^f$. Then $\text{Ш}(k,E)$ is trivial if and only if $f \le 2$ [**26**], and Dummigan [**19**] has studied the order of $\text{Ш}(k,E)$ in the case that $f > 2$. This allows one to compute the determinant of the Mordell-Weil lattice, and in some cases identify its isomorphism type. Here is a table for some small values of q_0; note that $f \le 2$ in these cases.

q_0	2	3	4
rank L	4	12	24
det L	2^2	3^6	4^{12}
L	D_4	CT_{12}	$2\Lambda_{24}$

We obtain the D_4 root lattice, the Coxeter-Todd lattice of rank 12 and a multiple of the Leech lattice of rank 24. For example, a point count can be used to bound the degree of maps $X \to E$ from below, and to show that there are no vectors of norm 2 once $q_0 > 2$. Since the Coxeter-Todd lattice is the unique 12-dimensional even integral lattice of determinant 3^6 with no roots, this identifies L when $q_0 = 3$.

3. Non-constant curves

We briefly describe which cases of the Birch and Swinnerton-Dyer conjecture are known for general elliptic curves over function fields. Continuing with the notation of the first section, we let E/k be an elliptic curve, and $\pi : S \to X$ the minimal regular model of E. We have an isomorphism $(\pi_*\omega_{S/X})^{\otimes 12} \cong \mathcal{O}_X(\mathcal{D}(E/k))$, where $\mathcal{D}(E/k)$ is the minimal discriminant of E/k. One can show that $\deg \mathcal{D}(E/k)/12 = -(\mathcal{O}_E, \mathcal{O}_E)$, which can take on any non-negative value. E is a constant curve if and only if $(\mathcal{O}_E, \mathcal{O}_E) = 0$.

In the case that $X = \mathbb{P}^1$, the low values of $\deg \mathcal{D}(E/k)$ give (via the classification of surfaces):

$$\begin{aligned} \deg \mathcal{D}(E/k) &= 0 \quad S = E \times \mathbb{P}^1 \text{ is a ruled surface;} \\ &= 12 \quad S \text{ is a rational surface;} \\ &= 24 \quad S \text{ is a K3 surface.} \end{aligned}$$

The conjecture of Birch and Swinnerton-Dyer is known in these cases. The case of rational surfaces is due to Milne [**37**]. The case of a K3 surface is due to Artin and Swinnerton-Dyer [**1**]. However, the conjecture of Artin and Tate remains open for general elliptic surfaces as well as for surfaces of general type.

Ulmer has made progress recently by studying cases where the surface S is dominated by a product of curves. He has constructed explicit examples of non-constant curves E over rational function fields where the rank is arbitrarily large [**61**]. There is also an analogue of the theory of special points on modular curves (presented below) in the function field setting, which involves the study of Drinfeld modules of ranks 1 and 2. Using this analogy, one should be able to show that the conjecture of Birch and Swinnerton-Dyer is true whenever the order of the L-function of E/k at $s = 1$ is ≤ 1 [**7**], [**60**].

4. Results over number fields

Let k be a number field, and E/k an elliptic curve. As we saw in lecture 2, the conjecture of Birch and Swinnerton-Dyer cannot be formulated until one knows that the L-function of E/k can be analytically continued to a neighborhood of the point $s = 1$. This continuation has been established for some elliptic curves, which we now describe.

A Shimura curve X over $\mathbb{C}$ is the quotient of the upper half plane by an arithmetic Fuchsian group. These groups are constructed from quaternion algebras B over totally real number fields k, which are split at precisely one real place v of k. The associated system of Shimura curves has a canonical model over the field k, embedded in $\mathbb{C}$ via the place v [**13**], [**15**], [**66**].

Definition 3.2. We say an elliptic curve E over a totally real number field k is *modular* if it is isogenous over k to a factor of the Jacobian of a Shimura curve.

When $k = \mathbb{Q}$, it is known that all elliptic curves are modular [**6**]. Here one only needs the classical modular curves, which are the Shimura curves associated to the split quaternion algebra B of two-by-two matrices over $\mathbb{Q}$.

Theorem 3.3 (Gross-Zagier [**28**], Kolyvagin [**34**], Zhang [**66**]). *Let E be a modular elliptic curve over a totally real number field k. Then $L(E/k, s)$ is entire and satisfies Conjecture 2.7. If we assume further that* $\operatorname{ord}_{s=1} L(E/k, s) \leq 1$, *then*

(1) $\operatorname{rank}(E(k)) = \operatorname{ord}_{s=1} L(E/k, s)$;
(2) $Ш(k, E)$ *is finite; and*
(3) *the Birch and Swinnerton-Dyer conjecture is true at almost all primes.*

This theorem is due to Gross-Zagier and Kolyvagin for $k = \mathbb{Q}$ and to Zhang in the general case. By the statement "is true at almost all primes", we mean that the ratio $c(E/k)/(R(E/k) \cdot P(E/k))$ is a non-zero rational number, and that for all rational primes p outside of a specified finite set depending on the curve E, the p-part of the order of $Ш(k, E)$ is equal to the p-part of this ratio. In many cases, one can refine Kolyvagin's method to prove that the specified finite set is empty.

The crux of the proof is the exploitation of special points on Shimura curves, and the divisor classes of degree zero which are supported on these points in the Jacobian. The special points are defined over (abelian extensions of) the CM fields K with totally real subfield k, which embed in the quaternion algebra B. We give a sketch in the case that $k = \mathbb{Q}$ and the quaternion algebra B is split, where the special points are called *Heegner points*.

Assume that E has conductor $N = N(E/\mathbb{Q})$. Then the proof that E is modular exhibits a weight 2 newform f for the group $\Gamma_0(N)$ with integer Fourier coefficients such that $L(f, s) = L(E/\mathbb{Q}, s)$. The space of cusp forms of weight 2 for $\Gamma_0(N)$ is naturally identified with the space of holomorphic differentials on $X_0(N)$. This implies that E is isogenous to a quotient of the Jacobian $J_0(N)$ over $\mathbb{Q}$ [**17**].

The composition of the canonical map $X_0(N) \to J_0(N)$, obtained by sending the cusp ∞ to the zero point of $J_0(N)$, with the projection $J_0(N) \to E$ gives a dominant map $\pi : X_0(N) \to E$ defined over $\mathbb{Q}$. We let ω be the unique invariant differential on E over $\mathbb{Q}$ which satisfies $\pi^*(\omega) = \omega_f = 2\pi i f(z)dz$ [**36**].

The complement of the cusps in $X_0(N)$ is a coarse moduli space of elliptic curves (A, A') related by a cyclic N-isogeny. Let K be an imaginary quadratic number field of discriminant D in which all the prime factors of N are split. There exists an integral ideal $\mathfrak{n}$ of K such that $\gcd(\mathfrak{n}, \bar{\mathfrak{n}}) = 1$ and $\mathfrak{n} \cdot \bar{\mathfrak{n}} = (N)$, and the complex elliptic curves given by the lattice quotients $A = \mathbb{C}/\mathcal{O}_K$ and $A' = \mathbb{C}/\mathfrak{n}^{-1}$ are related by the isogeny $x : A \to A'$ with cyclic kernel of order N. By the theory of complex multiplication (cf. [**53**], Ch. II) this Heegner point x is defined over the Hilbert class field H of K, that is, $x \in X_0(N)(H)$.

Recall that $\pi(x)$ is the projection of the class $[x] - [\infty] \in J_0(N)(H)$ to $E(H)$, and let

$$P = \sum_{\sigma \in \mathrm{Gal}(H/K)} \pi(x)^\sigma \in E(K),$$

the addition occurring in $E(H)$. Birch asked when P had infinite order in the group $E(K)$. Based on extensive computations, he conjectured that this question was related to the non-vanishing of the first derivative of $L(E/K, s)$ at $s = 1$. It is proved in [**28**] that

$$L'(E/K, 1) = \frac{\langle P, P \rangle_K}{\sqrt{|D|}} \cdot \int_{E(\mathbb{C})} |\omega|.$$

The proof of this identity follows from a comparison of the calculation of the first derivative, using Rankin's integral formula for the L-function, with the calculation of the global height of P, using Néron's theory of local heights on the curve $X_0(N)$, discussed below.

In particular, P has infinite order if and only if the first derivative $L'(E/K, 1)$ is not equal to 0. If P has infinite order, it is not infinitely divisible in the finitely generated abelian group $E(K)$. Kolyvagin [**34**] shows that the Selmer group $\mathrm{Sel}_p(K, E)$ has dimension 1 over $\mathbb{Z}/p\mathbb{Z}$ for almost all primes p, and is generated by the image $\delta(P)$, where $\delta : E(K)/pE(K) \to \mathrm{Sel}_p(K, E)$ is the map defined in Lecture 1. At the remaining primes, he shows that that the p-primary part of the Tate-Shafarevich group $Ш(K, E)$ is finite. Combining this argument with some non-vanishing results for the L-series of quadratic twists gives the result in the theorem when the order of the L-function of E over $\mathbb{Q}$ is either 0 or 1.

5. Local and global heights

Let k be a global field. Néron's theory of local heights (cf. [**43**], [**28**]) expresses the canonical height pairing on the Jacobian J of a curve X/k as a sum

$$\langle \alpha, \beta \rangle = \sum_v \langle a, b \rangle_v$$

of local terms, almost all of which are zero. Here we take a, b to be divisors of degree zero on X with disjoint supports representing the classes α and β. The local symbols $\langle , \rangle_v$ are characterized by the property that they are bi-additive, symmetric, continuous, and satisfy

$$\langle a, \mathrm{div}(f) \rangle_v = \log |f(a)|_v = \log \prod_x |f(x)|_v^{m_x},$$

where $a = \sum_x m_x \cdot x$ and f is a rational function on X.

When v is archimedean, the local height pairing $\langle , \rangle_v$ is defined using potential theory [**28**]. When v is a non-archimedean place, the local pairing can be defined using intersection theory as follows. Let $\mathcal{E}$ be the minimal regular model of E over $\mathrm{Spec}\, A_v$, cf. [**12**], [**35**].

We write $\mathcal{F} = \sum a_i \mathcal{F}_i$, where $\mathcal{F}$ is the special fiber of $\mathcal{E}$, the $\mathcal{F}_i$ are its irreducible components, and the a_i are their respective multiplicities. The intersection pairing

$(,)$ in the special fiber has the property that $(\mathcal{F}, \mathcal{F}_i)$ is zero for each i. Furthermore, it is negative definite on the quotient group $\sum_i \mathbb{Z}\mathcal{F}_i/\mathbb{Z}\mathcal{F}$, so is non-degenerate over $\mathbb{Q}$.

To define the local height in this case, take a divisor $a = \sum m_x \cdot x$ on E of degree 0, and its closure A in $\mathcal{E}$. It suffices to treat the case that a is pointwise rational over k_v (i.e. if $m_x \neq 0$, then x has coordinates in k_v).

By the non-degeneracy of the intersection pairing in the special fiber, we can find a $\mathbb{Q}$-divisor μ, whose support lies in the special fiber, such that $(A - \mu, \mathcal{F}_i) = 0$ for each i. Given a degree 0 divisor b, rational over k_v, relatively prime to a, we take its closure B in $\mathcal{E}$, and define

$$\langle a, b\rangle_v = (A - \mu, B) \cdot \log \#(A_v/\pi_v A_v).$$

Using the classification of the special fibers of the minimal regular models of elliptic curves, and their intersection matrices (cf. the table in [**57**]), one can find bounds on the denominators of the local pairing, and thus of the global height pairing for function fields. This fact was used in the previous lecture. A detailed computation of the denominators, corresponding to the classification of special fibers, can be found in [**51**].

Appendices

A. Reduction modulo v

Let k_v be a field complete with respect to a discrete valuation v. For us these will arise as the completions at finite places of global fields. From now on we write A_v for the ring of integers, π_v for a choice of uniformizer of this discrete valuation ring, and $\mathbb{F}_v$ for the residue field $A_v/\pi_v A_v$.

Let E be an elliptic curve over k_v. Choose an equation for E such that the coefficients a_i are all in A_v, and such that the valuation $v(\Delta(a_1, \ldots, a_6))$ of the discriminant is minimal with respect to this property. This is called a *minimal equation* for E. Write $\bar{a}_i$ for the image of a_i in $\mathbb{F}_v$. One can show that the reduced curve

$$y^2 + \bar{a}_1 xy + \bar{a}_3 y = x^3 + \bar{a}_2 x^2 + \bar{a}_4 x + \bar{a}_6$$

does not depend on the choice of minimal equation (up to isomorphism over the residue field). It therefore has the right to be called the reduction of E modulo v, and we will write it as $\widetilde{E}_v$, or $\widetilde{E}$ if no confusion will result.

Definition A.1. If $v(\Delta) = 0$, then $\widetilde{E}$ will be non-singular, and so will be an elliptic curve over $\mathbb{F}_v$. In this case, we say that E has *good reduction at v*. If $v(\Delta) > 0$ then the reduced curve will be singular, in which case we say that E has *bad reduction*. However, the non-singular locus $\widetilde{E}^{ns}$ can still be made into a group variety using the same chord-tangent process, with the point at infinity still playing the role of zero element. There are three possibilities.

(1) $\widetilde{E}$ is a cuspidal cubic curve. In this case, $\widetilde{E}^{ns}$ is isomorphic to the additive group $\mathbb{G}_a$.
(2) $\widetilde{E}$ is a nodal cubic curve, and the tangent lines at the node are defined over the residue field. In this case, $\widetilde{E}^{ns}$ is isomorphic to the multiplicative group $\mathbb{G}_m$.
(3) $\widetilde{E}$ is a nodal cubic curve, and the tangent lines at the node are defined only over a quadratic extension of the residue field. In this case, $\widetilde{E}^{ns}$ is isomorphic to the twisted form of the multiplicative group corresponding to this quadratic extension.

We speak respectively of *additive, split multiplicative, and non-split multiplicative reduction.*

Example. Consider the curve

$$E : y^2 = x(x-1)(x-t)$$

over the function field $k = \mathbb{F}_q(t)$ of genus 0, where q is odd. This curve has discriminant $\Delta = 16 \cdot t^2 \cdot (t-1)^2$. Since a change of variables changes the valuation of Δ at each place by a multiple of 12, the above equation is minimal at all places except $t = \infty$. At the place $t = 1$, the curve E has split multiplicative reduction. At the place $t = 0$ the curve E has multiplicative reduction, and the tangents at the node are rational if and only if -1 is a square in $\mathbb{F}_q$. Finding a suitable model for the curve at $t = \infty$, one can check that the curve E has additive reduction at $t = \infty$.

One of the reasons that the notion of reduction is useful is the existence of a natural reduction map $\rho : E(k_v) \to \widetilde{E}(\mathbb{F}_v)$. Choose a minimal equation for E over A_v. Given a point $P = [X : Y : Z]$ with coordinates in k_v, one can assume that each coordinate is integral, and that at least one of X, Y, Z has valuation 0. The point $\rho(P) = [\bar{X}, \bar{Y}, \bar{Z}]$ then lies on the reduced curve. There are some important subgroups associated to the map ρ.

Definition A.2. We write

$$E^0(k_v) = \rho^{-1}(\widetilde{E}^{ns}(\mathbb{F}_v))$$

for the inverse image of the smooth locus on the reduced curve (if E has good reduction, then $E(k_v) = E^0(k_v)$). We write

$$E^1(k_v) = \rho^{-1}(\mathcal{O}_{\widetilde{E}})$$

for the inverse image of the point at infinity on the reduced curve.

Proposition A.3. *The restriction of ρ to $E^0(k_v)$ is a homomorphism, which surjects onto $\widetilde{E}^{ns}(\mathbb{F}_v)$.*

Moreover, if n is prime to the residue characteristic and E has good reduction at v, then the restriction of ρ to the n-torsion of $E(k_v)$ is also injective, and induces an isomorphism

$$E(k_v)[n] \cong \widetilde{E}^{ns}(\mathbb{F}_v)[n].$$

PROOF. The surjectivity follows from a version of Hensel's lemma. The fact that ρ is a homomorphism follows on showing that the chord-tangent process behaves well with respect to reduction.

There is a complete description of the kernel of the reduction map in terms of the formal group attached to E, giving the addition law in a neighborhood of the identity. It is known that all the torsion in the kernel has order a power of the residue characteristic; this gives the second part of the proposition. For the definition of the associated formal group, see Ch. IV of [**52**] or Thm. 4.2 in [**57**]. An introduction to one-dimensional formal groups in general can be found in [**22**]. □

The following lemma is useful in the computation of periods.

Lemma A.4 (Tate [**57**]). *Let E be an elliptic curve over k_v where v is a finite place of k, and choose a minimal Weierstrass model for E. Let ω_v be an invariant differential, defined over A_v, which does not vanish modulo $\pi_v A_v$, and let dx_v be the Haar measure on k_v which gives A_v volume 1. Then*

$$\int_{E(k_v)} |\omega_v| = \frac{[E(k_v) : E^0(k_v)]}{L_v(1)}.$$

PROOF. We begin with some notation. We may assume that E is given by a minimal equation

$$E : y^2 + a_1xy + a_3y = x^3 + a_2x^2 + a_4x + a_6,$$

and that the differential is of the form

$$\omega_v = \frac{dx}{2y + a_1x + a_3}.$$

We let $z = -x/y$. Then z is a local parameter at the point $\mathcal{O}_E$, and working in the formal completion at the origin gives that $dz = \omega_v(1 + O(z))$.

The proof of the lemma is based on the existence of the formal group of E, alluded to in the proof of the Proposition A.3. Completing along the origin $z = 0$ of E gives a formal group F with the property that $(\pi_v)_F$ (that is, the set $\pi_v A_v$ endowed with the group law defined by F) is isomorphic to $E^1(k_v)$. In particular, viewing $E(k_v)$ as a v-adic analytic manifold, $z : E^1(k_v) \to \pi_v A_v$ forms a chart. It follows that

$$\int_{E^1(k_v)} |\omega_v| = \int_{\pi_v A_v} |dz(1 + O(z)| = \frac{1}{[A_v : \pi_v A_v]} \int_{A_v} |dz| = \frac{1}{q_v},$$

using the translation invariance of Haar measure and the fact that $|1 + O(z)| = 1$ when $z \in \pi_v A_v$, by the ultrametric property of the v-adic absolute value. We recall that q_v is the cardinality of the residue field.

Write

$$N_v = [E^0(k_v) : E^1(k_v)] = \#\widetilde{E}_v^{\mathrm{ns}}(\mathbb{F}_v).$$

Applying the translation invariance of ω_v now gives

$$\int_{E(k_v)} |\omega_v| = \frac{[E(k_v) : E^0(k_v)]N_v}{q_v},$$

and re-arranging gives the result. □

B. Descent computations

One of the reasons for introducing the n-Selmer group is that it is effectively computable. In this appendix we make a complete computation of the 2-Selmer groups of some elliptic curves referred to in the first lecture.

We will again consider the case where E is an elliptic curve over a global field k, not of characteristic 2, and all the points of $E[2]$ have coordinates in k. In this case, E can be written in the form

$$E : y^2 = (x - e_1)(x - e_2)(x - e_3),$$

and we continue to write $T_i = (e_i, 0)$.

Let S denote the set of places of k which divide 2 or where E has bad reduction, together with the infinite places. Lemma 1.4 shows that we have injections

$$E(k)/2E(k) \hookrightarrow \mathrm{Sel}_2(k, E) \hookrightarrow (k^\times(S, 2)/(k^\times)^2)^2$$

the latter group being finite. In order to compute the 2-Selmer group exactly we must compute the local conditions at v imposed by the image of $E(k_v)/2E(k_v)$ in $H^1(k_v, E[2])$, for every place v.

In fact, the group on the right above is the one obtained by imposing the local conditions at places not in S. In other words,

$$(k^\times(S, 2)/(k^\times)^2)^2 = \ker\left(H^1(k, E[2]) \to \prod_{v \notin S} H^1(k, E)\right).$$

It remains therefore to compute the conditions at the finitely many places $v \in S$. This can be done using the explicit description of the map δ given earlier.

Proposition B.1. *A triple*

$$(b_1, b_2, b_3) \in (k^\times(S,2)/(k^\times)^2)^3$$

not lying in $\delta(E[2])$ lies in $\mathrm{Sel}_2(k, E)$ if and only if for every $v \in S$ there exists a solution (z_1, z_2, z_3) in k_v with $z_1 z_2 z_3 \neq 0$ to the following pair of equations:

$$b_1 z_1^2 - b_2 z_2^2 = e_2 - e_1$$
$$b_1 z_1^2 - b_3 z_3^2 = e_3 - e_1.$$

PROOF. Suppose that (b_1, b_2, b_3) does lie in the 2-Selmer group. Without loss of generality, we may suppose that $b_1 b_2 = b_3$. We need to produce, for every $v \in S$, a solution to the equations

$$\begin{aligned} b_1 z_1^2 &= x - e_1 \\ b_2 z_2^2 &= x - e_2 \\ b_3 z_3^2 &= x - e_3 \\ y^2 &= (x - e_1)(x - e_2)(x - e_3) \end{aligned}$$

with values in k_v. Then one has

$$b_1 z_1^2 = e_2 - e_1 + b_2 z_2^2 = e_3 - e_1 + b_3 z_3^2,$$

proving necessity. On the other hand, given a solution to the equations in the statement of the proposition, one can take

$$x = b_1 z_1^2 + e_1 \text{ and } y = b_1 b_2 z_1 z_2 z_3,$$

and then (x, y) is a point in $E(k_v)$ mapping to our given triple. □

The computation of the 2-Selmer group is therefore reduced to the problem of deciding whether finitely many curves have points defined over local fields. This is a simple matter, using Hensel's lemma. We illustrate with an example.

Example. Take $k = \mathbb{F}_q(t)$ with q odd, and let E be the elliptic curve

$$E : y^2 = x(x-1)(x-t).$$

Here the computation of the 2-Selmer group splits into two cases, depending on whether or not -1 is a square in $\mathbb{F}_q^\times$, although we will see that the final answer is the same:

$$\mathrm{Sel}_2(k, E) = (\mathbb{Z}/2\mathbb{Z})^2.$$

Suppose first that -1 is a square, and let α be a non-square in $\mathbb{F}_q^\times$. The above curve has bad reduction at the places $t = 0, 1$ and ∞, and good reduction elsewhere. We can therefore take $S = \{0, 1, \infty\}$, and the pair (b_1, b_2) can take the values $(\alpha^a t^b (t-1)^c, \alpha^d t^e (t-1)^f)$, where $a, \ldots, f$ are allowed to take the values 0 or 1.

Before beginning the descent computation we note a shortcut which is often useful. At each place v of k, we have the filtration

$$E(k_v) \supset E^0(k_v) \supset E^1(k_v).$$

Write $C = E(k_v)/E^1(k_v)$. Since $E^1(k_v)$ can be represented as the set of points of a formal group in $\pi_v A_v$, where multiplication by 2 is an isomorphism ([**52**], Prop. IV.2.3(b)), $E(k_v)/2E(k_v)$ is isomorphic to $C/2C$.

Now, C has a natural filtration whose quotients are

$$E(k_v)/E^0(k_v) \text{ and } E^0(k_v)/E^1(k_v) \cong \widetilde{E}_v^{ns}(\mathbb{F}_v).$$

The latter group is easily understood, and the first group can be computed using Tate's algorithm to compute the Néron model of an elliptic curve, cf. [**57**] or [**53**], Ch. IV. Once enough is known about the group $C/2C$, it can sometimes be shown that the elements of $E[2]$ give a complete set of representatives, immediately giving the local conditions at v.

In the language of the above references, the curve E has multiplicative reduction of Kodaira type I_2 at the places 0 and 1, and additive reduction of Kodaira type I_2^* at infinity. Looking at the table on page 46 of [**57**] (also Table 15.1 in [**52**]) shows that when $u = \infty$,

$$E(k_u)/E^0(k_u) \cong \mathbb{Z}/2\mathbb{Z} \times \mathbb{Z}/2\mathbb{Z}.$$

The group $\widetilde{E}_u^{ns}(\mathbb{F}_u)$ has no 2-torsion, so in this case we see that

$$E(k_u)/2E(k_u)$$

is an abelian group of the same type.

Now, computing the map δ on the 2-torsion of E locally at the place u gives

$$\begin{array}{lcl} \mathcal{O}_E & \mapsto & (1,1) \\ (0,0) & & (t,-1) \equiv (t,1) \\ (1,0) & & (1,1-t) \equiv (1,t) \\ (t,0) & & (t,t-1) \equiv (t,t), \end{array}$$

where the congruences are all taken $(\text{mod } (k_u^\times)^2)$, so we are in the case described above. Applying these local conditions implies that a set of representatives for the 2-Selmer group is contained in the set of elements of the form $(t^b(t-1)^c, t^e(t-1)^f)$. At this point it is instructive to draw a table:

b_2 \ b_1	1	t	$t-1$	$t(t-1)$
1	$\mathcal{O}_E$	$(0,0)$		
t				
$t-1$	$(1,0)$	$(t,0)$		
$t(t-1)$				

The entries above indicate that a point in $E(k)$ maps to the given pair (b_1, b_2), which must therefore lie in the 2-Selmer group. To decide whether the remaining entries lie in the 2-Selmer group, we argue using the proposition above. Note that the conditions at ∞ have already been imposed, so we need only check for points on the corresponding curves in the completions at 0 and 1.

Consider the pair $(b_1, b_2) = (t-1, 1)$. This corresponds to the pair of equations

$$(t-1)z_1^2 - z_2^2 = 1$$

and

$$(t-1)z_1^2 - (t-1)z_3^2 = t.$$

Let w be the normalized discrete valuation at $t-1$. If (z_1, z_2, z_3) is a solution in k_w with $z_1 z_2 \neq 0$, then the first equation shows that $w(z_1) \geq 0$. Then the second equation gives $1 + 2w(z_3) = 0$, a contradiction. It follows that the system of equations has no solution in k_w, and hence $(t-1, 1)$ is not in the 2-Selmer group.

Using the fact that if $x \in \mathrm{Sel}_2(k, E)$ but $y \notin \mathrm{Sel}_2(k, E)$ then $xy \notin \mathrm{Sel}_2(k, E)$, we arrive at the following table.

b_2 \ b_1	1	t	$t-1$	$t(t-1)$
1	O_E	$(0,0)$	X	X
t				
$t-1$	$(1,0)$	$(t,0)$	X	X
$t(t-1)$				

A cross indicates an element that does not lie in the 2-Selmer group. One can now check that neither of the pairs $(1, t)$ or $(t-1, t)$ lie in the 2-Selmer group, giving the following table.

b_2 \ b_1	1	t	$t-1$	$t(t-1)$
1	O_E	$(0,0)$	X	X
t	X	X	X	X
$t-1$	$(1,0)$	$(t,0)$	X	X
$t(t-1)$	X	X	X	X

It follows that

$$\mathrm{Sel}_2(k, E) = E(k)/2E(k) \cong E[2] = (\mathbb{Z}/2\mathbb{Z})^2.$$

In particular, the curve E has no rational points except for 2-torsion (it is a simple matter to check that E has no torsion except for 2-torsion). In the case that -1 is not a square in $\mathbb{F}_q^\times$ one can proceed similarly, aided by the fact that $E[2]$ now generates $E(k_w)/2E(k_w)$ at the place w used above. Hence $E(k) \cong (\mathbb{Z}/2\mathbb{Z})^2$ and $\text{Ш}(k, E)[2] = 0$.

Example. Let

$$E' : t(t-1)y^2 = x(x-1)(x-t).$$

This is a twist of E, in the sense that it becomes isomorphic to E over the separable quadratic extension obtained by adjoining $\sqrt{t(t-1)}$, via the substitution $\sqrt{t(t-1)}y \mapsto y$. This curve E' now has additive reduction of type I_2^* at the three places $t = 0, 1$ and ∞, and good reduction elsewhere. One can check that the natural map $E'[2] \to E'(k_v)/2E'(k_v)$ is an isomorphism at these 3 places, allowing one to immediately calculate the 2-Selmer group. One obtains

$$\mathrm{Sel}_2(k, E') \cong \begin{cases} (\mathbb{Z}/2\mathbb{Z})^3 & \text{if} -1 \text{ is not a square in } \mathbb{F}_q^\times; \\ (\mathbb{Z}/2\mathbb{Z})^4 & \text{if} -1 \text{ is a square in } \mathbb{F}_q^\times. \end{cases}$$

C. The Tate module and the isogeny theorem

Definition C.1. Let l be a prime and let E/k be an elliptic curve. The *l-adic Tate module* of E is

$$T_l E = \varprojlim_n E[l^n],$$

the limit being taken over the natural numbers with respect to the homomorphisms

$$E[l^{n+1}] \xrightarrow[\times l]{} E[l^n].$$

Observe that T_lE has a natural profinite topology, and a continuous action of $\mathrm{Gal}(k^s/k)$. When l is prime to the characteristic, Proposition 1.1 implies that T_lE is a free $\mathbb{Z}_l$-module of rank 2. We will often use the 2-dimensional vector space $V_lE = T_lE \otimes_{\mathbb{Z}_l} \mathbb{Q}_l$, over $\mathbb{Q}_l$. This has a continuous action of $\mathrm{Gal}(k^s/k)$, and gives (after a choice of basis) a 2-dimensional l-adic Galois representation

$$\rho_{E,l} : \mathrm{Gal}(k^s/k) \to \mathrm{GL}_2(\mathbb{Q}_l).$$

When k is a global field, one can show that this representation is unramified away from the finite collection of places $S(l)$ of k, which divide l or where the curve E has bad reduction. For each finite place $v \notin S(l)$ let ϕ_v denote a choice of arithmetic Frobenius element in $G_{S(l)}$, the Galois group of the maximal extension of k which is unramified outside of $S(l)$. Then $\rho_{E,l}(\phi_v)$ has characteristic polynomial

$$X^2 - a_vX + q_v \text{ for } v \notin S. \tag{C.1}$$

The element ϕ_v is defined only up to conjugacy in $G_{S(l)}$ but the characteristic polynomial is independent of this choice. Since $S(l)$ is finite, the Chebotarev density theorem implies that the set $\{\phi_v\}_{v\notin S}$ is dense in $G_{S(l)}$. By continuity, the terms in the Euler product defining the incomplete L-function determine the characteristic polynomials of $\rho_{E,l}(\sigma)$ for all $\sigma \in G_{S(l)}$. In particular, knowledge of the Euler product of $L_S(E/k, s)$ determines the representation $\rho_{E,l}$ up to isomorphism. We have used here the fact that $\rho_{E,l}$ is always semi-simple, cf. [**47**].

Tate conjectured that the natural map

$$\mathrm{Hom}_k(E, E') \otimes \mathbb{Z}_l \to \mathrm{Hom}_{\mathrm{Gal}(k^s/k)}(T_lE, T_lE'), \tag{C.2}$$

which was shown by Weil to be injective ([**52**],Thm. 7.4), is an isomorphism, whenever k is finitely generated over the prime field. He proved this for abelian varieties in the case when k is a finite field [**55**], and Serre proved it for elliptic curves when k is a number field, under the assumption that E has a place of multiplicative reduction [**47**]. Faltings proved the full conjecture for all abelian varieties over number fields [**21**], [**16**]. In the function field case, Paršin [**45**] proved the theorem for elliptic curves, and Zarhin [**65**] proved it for abelian varieties over arbitrary finitely generated extensions of finite fields except in characteristic two, though now all cases are complete. In particular, two elliptic curves over a global field k are isogenous if and only if their associated l-adic representations V_l are isomorphic, and the Euler product defining the incomplete L-function is a complete isogeny invariant of an elliptic curve.

More information on the topics in this appendix may be found in [**52**], Ch. 3, §7.

D. L-functions of elliptic curves over function fields

Let k be the function field of an irreducible non-singular projective curve X of genus g over the finite field $\mathbb{F}_q$ of q elements. We will assume that $\overline{\mathbb{F}}_q \cap k = \mathbb{F}_q$, or equivalently that X is geometrically irreducible (cf. [**35**], Cor. 3.2.14).

Proposition D.1. *Let E/k be an elliptic curve. Then we can write*

$$L(E/k, s) = M(q^{-s})$$

where $M(x)$ is a rational function of degree

$$\deg M = 4g - 4 + \deg N(E/k). \tag{D.1}$$

If E/k is non-constant, then $M(x)$ is a polynomial with integer coefficients, and constant coefficient equal to 1.

The proof uses étale cohomology, cf. [**14**], [**39**]. The L-functions of elliptic curves can be expressed in terms of characteristic polynomials of Frobenius actions on appropriate étale cohomology groups.

Definition D.2. Let U be a geometrically irreducible non-singular curve over $\mathbb{F}_q$. Let X be the (unique) non-singular completion of U, cf. [**31**] I.6. Then $k = \mathbb{F}_q(U) = \mathbb{F}_q(X)$. Write S for the set of places of k which correspond to points of $X - U$, and let K_S be the maximal separable extension of k which is unramified outside S, inside a suitable separable closure of k. We call the group $\mathrm{Gal}(K_S/k)$ the *étale fundamental group* $\pi_1(U)$ of U. Let l be a prime not dividing q. A *lisse l-adic sheaf* $\mathcal{F}$ on U is a continuous finite-dimensional l-adic representation of $\pi_1(U)$.

We have adopted a simplified definition here in order to avoid unnecessary complications. For the general definition, we refer to [**30**] or [**39**]. As examples of lisse l-adic sheaves, we have the trivial representation $\mathbb{Q}_l$ of $\pi_1(X)$, and the representation $V_l(E)$ of $\pi_1(U)$, where E is an elliptic curve over $k = \mathbb{F}_q(X)$ and U is the complement of the set S of places of bad reduction. By the criterion of Néron-Ogg-Shafarevich [**49**], this representation is unramified at places of good reduction, hence it is lisse on U. In Grothendieck's theory, we will often work with the dual representation $V_l E^\vee$ of $\pi_1(U)$, which is the cohomology in degree 1 of the family of curves over U. The Weil pairing gives an isomorphism with the Tate twist

$$V_l(E)^\vee \cong V_l(E)(-1).$$

We note that the arithmetic Frobenius $\phi : x \mapsto x^q$ in the Galois group acts by the scalar q on $\mathbb{Q}_l(1)$, so the geometric Frobenius $\mathrm{Frob} = \phi^{-1}$ acts by q on the twist $\mathbb{Q}_l(-1)$.

Definition D.3. Let $\mathcal{F}$ be a lisse l-adic étale sheaf on a non-singular geometrically irreducible curve $U/\mathbb{F}_q$, and let U^0 denote the set of closed points of U. Then the *formal L-function of $\mathcal{F}$* is given by

$$L(U, \mathcal{F}, t) = \prod_{y \in U^0} \det\left(1 - t^{\deg y}\,\mathrm{Frob}_y \mid \mathcal{F}\right).$$

Here Frob_y denotes the geometric Frobenius at the place v corresponding to y. The L-function $L(U, \mathcal{F}, s)$ is given by substituting q^{-s} for t.

Let us now sketch the proof of Proposition D.1. In order to avoid technical complications we will make some simplifying assumptions along the way. For example, we will assume that q is not a power of 2 or 3 to avoid dealing with wild ramification.

PROOF. **Step 1: The bad factors.** Let S be the set of places of bad reduction of E/k. We have

$$L(E/k, s) = \prod_v L_v(E/k, s) = L_S(E/k, s) \cdot \prod_{v \in S} L_v(E/k, s).$$

First we will discuss the latter factor, coming from places of bad reduction. Recall the L-factors at places of bad reduction, given in Equation (2.3), and compare them to the formula for the conductor $N(E/k)$ of E given in Definition 2.5. Our assumption on q implies that the factors δ_v in Equation (2.6) vanish, hence

$$\prod_{v\in S} L_v(E/k,t) \text{ is a rational function of degree } \deg N(E/k) - 2\cdot \#S.$$

To complete the proof, it suffices to show that the factor $L_S(E/k,t)$ is a rational function of degree $4g-4+2\cdot\#S$.

Step 2. An equality of L-functions. Write U for the complement of S in X, and consider V_lE as a representation of the étale fundamental group $\pi_1(U)$, i.e. as a lisse l-adic étale sheaf on U. Therefore, by the equality of the characteristic polynomial of $\rho_{E,l}(\phi_v)$ in Equation (C.1) with the definition of an L-factor in Equation (2.1), one obtains an identity of L-functions

$$L_S(E/k,t) = L(U, V_lE^\vee, t).$$

where $V_lE^\vee$ is the dual representation, on the first degree cohomology of E.

Step 3: Grothendieck's work on L-functions. In this step we split into two cases, depending on whether the set S of places of bad reduction is empty or not. A good general reference on this material is [**32**].

Case $S \neq \emptyset$. The curve U is affine since S is not empty, and V_lE is lisse. These conditions imply, by a version of the Lefschetz trace formula [**39**], that

$$L(U, V_lE^\vee, t) = \frac{\det(1-t\,\mathrm{Frob} \mid H^1_c(U\times_{\mathbb{F}_q}\overline{\mathbb{F}}_q, V_lE^\vee))}{\det(1-t\,\mathrm{Frob} \mid H^2_c(U\times_{\mathbb{F}_q}\overline{\mathbb{F}}_q, V_lE^\vee))},$$

where the cohomology is étale cohomology with compact support. Here Frob continues to denote the geometric Frobenius. The denominator is trivial, because $H^2_c(V_lE^\vee)=0$. Indeed, it is Poincaré dual to $H^0(V_lE)(-1)$. But the group $H^0(V_lE)$ is equal to the invariants of $\mathrm{Gal}(k^s/\overline{\mathbb{F}}_q(X))$ on V_lE. Since E/k is non-constant, there are only a finite number of l-torsion points defined over $\overline{\mathbb{F}}_q(X)$, and these invariants are trivial.

The cohomology with compact support on an affine curve always vanishes in all degrees not 1 or 2. We have shown that it is also trivial in degree 2 in this case. Hence the rank of $H^1_c(U\times_{\mathbb{F}_q}\overline{\mathbb{F}}_q, V_lE^\vee)$ is the negative of the Euler characteristic of this cohomology. We have shown that $L_S(E/k,t)$ is a polynomial. To show that its degree is equal to $4g-4+2\#S$ it remains to show that the Euler characteristic $\chi(H^*(U, V_lE^\vee))$ is equal to $4-4g-2\#S$. We can calculate this Euler characteristic using ordinary étale cohomology, which is Poincaré dual to cohomology with compact support.

Case $S=\emptyset$. In this case $X=U$. For a lisse étale sheaf over the smooth geometrically irreducible projective curve X, the determination of the L-function is even more straightforward because as $X/\mathbb{F}_q$ is proper, there is no concern with compact support. Grothendieck's theory implies that

$$L(X, V_lE^\vee, t) = \prod_{0\le i\le 2} \det(1-t\,\mathrm{Frob} \mid H^i(X\times_{\mathbb{F}_q}\overline{\mathbb{F}}_q, V_lE^\vee))^{-(-1)^i}.$$

Therefore it again suffices to show that the Euler characteristic of the étale cohomology of $V_lE^\vee$ is $4-4g$.

Step 4: Raynaud's Euler Characteristic Formula. This formula, given in [**46**], gives the Euler characteristic of the cohomology of l-adic sheaves as the product of the Euler characteristic of the base U with the rank of the sheaf, modified by terms that measure the wild ramification at points in S. There is no wild ramification on X in our case, because we have assumed that q is not a power of 2 or 3. Thus

$$\chi(H^*(U, V_l E^\vee)) = \chi(U) \cdot \dim_{\mathbb{Q}_l} V_l E^\vee.$$

As we saw above, $V_l E^\vee$ is of dimension 2 and the Euler characteristic of U is given by $\chi(X) - \#S = 2 - 2g - \#S$. This shows that $\chi(H^*(U, V_l E^\vee)) = 4 - 4g - 2\#S$, which is what was needed in both of the cases in Step 3. □

Remark. Assume that E is constant, $E = A \times_{\mathbb{F}_q} X$ for some an elliptic curve $A/\mathbb{F}_q$. If Y is a smooth projective variety over $\mathbb{F}_q$ of dimension d, we can define its zeta function as

$$Z(Y,t) = \prod_{y \in Y^0} (1 - t^{\deg y}) = L(Y, \mathbb{Q}_l, t), \tag{D.2}$$

which by the Lefschetz trace formula [**29**] is equal to

$$Z(Y,t) = \prod_{i=0}^{2d} \det(1 - t \operatorname{Frob} \mid H^i(Y, \mathbb{Q}_l))^{(-1)^{i+1}}.$$

In particular, we find that

$$Z(A,t) = \frac{1 - at + qt^2}{(1-t)(1-qt)} = \frac{L_v(E/k,t)^{-1}}{(1-t)(1-qt)}$$

where $a = q + 1 - \#A(\mathbb{F}_q)$ and

$$Z(X,t) = \frac{M(t)}{(1-t)(1-qt)}$$

where $M(t)$ is a polynomial of degree $2g(X)$. By the Künneth formula and the general relation between zeta and L-functions

$$L(E/k,t) = \frac{\det(1 - t \operatorname{Frob} \mid H^1(A) \otimes H^1(X))}{\det(1 - t \operatorname{Frob} \mid (H^1(A) \otimes H^0(X)) \oplus (H^1(A) \otimes H^2(X)))}.$$

As the Frobenius eigenvalues on the 1-dimensional $\mathbb{Q}_l$-vector spaces $H^0(X)$ and $H^2(X)$ are 1 and q respectively, we find that

$$L(E/k,s) = \frac{\det(1 - q^{-s} \operatorname{Frob} \mid H^1(A) \otimes H^1(X))}{(1 - aq^{-s} + q^{1-2s})(1 - aq^{1-s} + q^{3-2s})}. \tag{D.3}$$

Note that this is a rational function of degree $4g - 4$ in q^{-s}, as expected.

Bibliography

[1] M. Artin and H.P.F. Swinnerton-Dyer, *The Shafarevich-Tate conjecture for pencils of elliptic curves on K3 surfaces*, Invent. Math. **20** (1973), 249–266.

[2] B.J. Birch, *Elliptic curves and modular functions*, Symposia Mathematica **IV** (1970), 27–32.

[3] ———, *Heegner points of elliptic curves*, Symposia Mathematica **XV** (1975), 441–445.

[4] B.J. Birch and H.P.F. Swinnerton-Dyer, *Notes on elliptic curves. II*, J. Reine Angew. Math. **218** (1965), 79–108. MR 0179168 (31 #3419)

[5] S. Bloch, *A note on height pairings, Tamagawa numbers, and the Birch and Swinnerton-Dyer conjecture*, Invent. Math. **58** (1980), no. 1, 65–76. MR 570874 (81m:14030)

[6] C. Breuil, B. Conrad, F. Diamond, and R. Taylor, *On the modularity of elliptic curves over $\mathbb{Q}$: wild 3-adic exercises*, J. Amer. Math. Soc. **14** (2001), no. 4, 843–939 (electronic). MR 1839918 (2002d:11058)

[7] M.L. Brown, *Heegner modules and elliptic curves*, Lecture Notes in Mathematics, vol. 1849, Springer-Verlag, Berlin, 2004.

[8] J.P. Buhler, B.H. Gross, and D.B. Zagier, *On the conjecture of Birch and Swinnerton-Dyer for an elliptic curve of rank* 3, Math. Comp. **44** (1985), no. 170, 473–481. MR 777279 (86g:11037)

[9] J.W.S. Cassels, *Arithmetic on curves of genus* 1. *IV. Proof of the Hauptvermutung*, J. Reine Angew. Math. **211** (1962), 95–112. MR 0163915 (29 #1214)

[10] ———, *Arithmetic on curves of genus 1. VIII. On conjectures of Birch and Swinnerton-Dyer*, J. Reine Angew. Math. **217** (1965), 180–199.

[11] ———, *Lectures on elliptic curves*, London Mathematical Society Student Texts, vol. 24, Cambridge University Press, Cambridge, 1991.

[12] T. Chinburg, *Minimal models for curves over Dedekind rings*, Arithmetic geometry (Storrs, Conn., 1984), Springer, New York, 1986, pp. 309–326.

[13] P. Deligne, *Travaux de Shimura*, Séminaire Bourbaki, 23ème année (1970/71), Exp. No. 389, Springer, Berlin, 1971, pp. 123–165. Lecture Notes in Math., Vol. 244.

[14] ———, *Cohomologie étale*, Lecture Notes in Mathematics, Vol. 569, Springer-Verlag, Berlin, 1977, Séminaire de Géométrie Algébrique du Bois-Marie SGA $4\frac{1}{2}$, Avec la collaboration de J. F. Boutot, A. Grothendieck, L. Illusie et J. L. Verdier.

[15] ———, *Variétés de Shimura: interprétation modulaire, et techniques de construction de modèles canoniques*, Automorphic forms, representations and L-functions (Proc. Sympos. Pure Math., Oregon State Univ., Corvallis, Ore., 1977), Part 2, Proc. Sympos. Pure Math., XXXIII, Amer. Math. Soc., Providence, R.I., 1979, pp. 247–289.

[16] ———, *Preuve des conjectures de Tate et de Shafarevitch (d'après G. Faltings)*, Astérisque (1985), no. 121-122, 25–41, Seminar Bourbaki, Vol. 1983/84.

[17] F. Diamond and J. Shurman, *A first course in modular forms*, Graduate Texts in Mathematics, vol. 228, Springer-Verlag, New York, 2005. MR 2112196 (2006f:11045)

[18] T. Dokchister and V. Dokchister, *Root numbers and parity of ranks of elliptic curves*, ArXiv: 0906.1815 (2009).

[19] N. Dummigan, *The determinants of certain Mordell-Weil lattices*, Amer. J. Math. **117** (1995), no. 6, 1409–1429.

[20] N.D. Elkies, *Mordell-Weil lattices in characteristic* 2. *I. Construction and first properties*, Internat. Math. Res. Notices (1994), no. 8, 343 ff., approx. 18 pp. (electronic).

[21] G. Faltings, *Endlichkeitssätze für abelsche Varietäten über Zahlkörpern*, Invent. Math. **73** (1983), no. 3, 349–366. MR 718935 (85g:11026a)

[22] A. Fröhlich, *Formal groups*, Lecture Notes in Mathematics, No. 74, Springer-Verlag, Berlin, 1968.

[23] D. Goldfeld, *The class number of quadratic fields and the conjectures of Birch and Swinnerton-Dyer*, Ann. Scuola Norm. Sup. Pisa Cl. Sci. (4) **3** (1976), no. 4, 624–663. MR 0450233 (56 #8529)

[24] ———, *Gauss's class number problem for imaginary quadratic fields*, Bull. Amer. Math. Soc. (N.S.) **13** (1985), no. 1, 23–37. MR 788386 (86k:11065)

[25] B.H. Gross, *Local heights on curves*, Arithmetic geometry (Storrs, Conn., 1984), Springer, New York, 1986, pp. 327–339.

[26] ———, *Group representations and lattices*, J. Amer. Math. Soc. **3** (1990), no. 4, 929–960.

[27] ———, *L-functions at the central critical point*, Motives, Proc. Sympos. Pure Math., Volume 55, Part 1, Amer. Math. Soc., Providence, R.I., 1991, pp. 527–535.

[28] B.H. Gross and D.B. Zagier, *Heegner points and derivatives of L-series*, Invent. Math. **84** (1986), no. 2, 225–320. MR 833192 (87j:11057)

[29] A. Grothendieck, *Formule de Lefschetz et rationalité des fonctions L*, Séminaire Bourbaki, Vol. 9, Soc. Math. France, Paris, pp. Exp. No. 279, 41–55.

[30] ———, *Séminaire de géométrie algébrique du Bois Marie - 1960-61 - Revêtements étales et groupe fondamental - (SGA 1)*, Lecture Notes in Mathematics, No. 224, Springer-Verlag, Berlin, New York, 1971.

[31] R. Hartshorne, *Algebraic geometry*, Graduate Texts in Mathematics, vol. 52, Springer-Verlag, New York, 1977.

[32] N.M. Katz, *L-functions and monodromy: four lectures on Weil II*, Adv. Math. **160** (2001), no. 1, 81–132. MR 1831948 (2002c:11066)

[33] ———, *Twisted L-functions and monodromy*, Annals of Math Studies, vol. 150, Princeton University Press, New Jersey, 2002.

[34] V. A. Kolyvagin, *Finiteness of $E(\mathbf{Q})$ and $SH(E,\mathbf{Q})$ for a subclass of Weil curves*, Izv. Akad. Nauk SSSR Ser. Mat. **52** (1988), no. 3, 522–540, 670–671. MR 954295 (89m:11056)

[35] Q. Liu, *Algebraic geometry and arithmetic curves*, Oxford Graduate Texts in Mathematics, vol. 6, Oxford University Press, Oxford, 2002.

[36] B. Mazur and H.P.F. Swinnerton-Dyer, *Arithmetic of Weil curves*, Invent. Math. **25** (1974), 1–61.

[37] J.S. Milne, *The Brauer group of a rational surface*, Invent. Math. **11** (1970), 304–307.

[38] ———, *On a conjecture of Artin and Tate*, Ann. of Math. (2) **102** (1975), no. 3, 517–533.

[39] ———, *Étale cohomology*, Princeton Mathematical Series, vol. 33, Princeton University Press, Princeton, N.J., 1980.

[40] ———, *Jacobian varieties*, Arithmetic geometry (Storrs, Conn., 1984), Springer, New York, 1986, pp. 167–212.

[41] ———, *Arithmetic duality theorems*, second ed., BookSurge, LLC, 2006.

[42] D. Mumford, *Abelian varieties*, Tata Institute of Fundamental Research Studies in Mathematics, No. 5, Published for the Tata Institute of Fundamental Research, Bombay, 1970.

[43] A. Néron, *Quasi-fonctions et hauteurs sur les variétés abéliennes*, Ann. of Math. (2) **82** (1965), 249–331.

[44] A.P. Ogg, *Elliptic curves and wild ramification*, Amer. J. Math. **89** (1967), 1–21. MR 0207694 (34 #7509)

[45] A.N. Paršin, *Minimal models of curves of genus 2, and homomorphisms of abelian varieties defined over a field of finite characteristic*, Izv. Akad. Nauk SSSR Ser. Mat. **36** (1972), 67–109. MR 0316456 (47 #5003)

[46] M. Raynaud, *Caractéristique d'Euler-Poincaré d'un faisceau et cohomologie des variétés abéliennes*, Séminaire Bourbaki, Vol. 9, Soc. Math. France, Paris, 1995, pp. Exp. No. 286, 129–147. MR 1608794

[47] J-P. Serre, *Abelian l-adic representations and elliptic curves*, McGill University lecture notes written with the collaboration of Willem Kuyk and John Labute, W. A. Benjamin, Inc., New York-Amsterdam, 1968. MR 0263823 (41 #8422)

[48] ———, *Algebraic groups and class fields*, Graduate Texts in Mathematics, vol. 117, Springer-Verlag, New York, 1988, Translated from the French.

[49] J-P. Serre and J. Tate, *Good reduction of abelian varieties*, Ann. of Math. (2) **88** (1968), 492–517. MR 0236190 (38 #4488)

[50] G. Shimura, *Introduction to the arithmetic theory of automorphic functions*, Publications of the mathematical society of Japan, No. 11, Iwanami Shoten and Princeton University Press, Princeton, 1971.

[51] T. Shioda, *On the Mordell-Weil lattices*, Comment. Math. Univ. St. Paul. **39** (1990), no. 2, 211–240.

[52] J.H. Silverman, *The arithmetic of elliptic curves*, Graduate Texts in Mathematics, vol. 106, Springer-Verlag, New York, 1986.

[53] ———, *Advanced topics in the arithmetic of elliptic curves*, Graduate Texts in Mathematics, vol. 151, Springer-Verlag, New York, 1994.

[54] J. Tate, *Duality theorems in galois cohomology over number fields*, Proceedings of the International Congress of Mathematicians, Stockholm, 1962, pp. 288–295.

[55] ———, *Endomorphisms of abelian varieties over finite fields*, Invent. Math. **2** (1966), 134–144. MR 0206004 (34 #5829)

[56] ———, *Classes d'isogénie des variétés abeliénnes sur un corps fini*, Séminaire Bourbaki, Vol. 1968/69, Exp. No. 347 – 363, Springer-Verlag, Berlin, 1971, pp. 95–110. Lecture Notes in Math., Vol. 179.

[57] ———, *Algorithm for determining the type of a singular fiber in an elliptic pencil*, Modular functions of one variable, IV (Proc. Internat. Summer School, Univ. Antwerp, Antwerp, 1972), Springer, Berlin, 1975, pp. 33–52. Lecture Notes in Math., Vol. 476.

[58] ———, *On the conjectures of Birch and Swinnerton-Dyer and a geometric analog*, Séminaire Bourbaki 1965/66, Vol. 9, Soc. Math. France, Paris, 1995, pp. 415–440, Exp. No. 306.

[59] R. Taylor and A. Wiles, *Ring-theoretic properties of certain Hecke algebras*, Ann. of Math. (2) **141** (1995), no. 3, 553–572.

[60] D. Ulmer, *Elliptic curves and analogies between number fields and function fields*, Heegner points and Rankin L-series, vol. 49, MSRI Publications, Berkeley, 2004, pp. 285–315.

[61] ———, *L-functions with large analytic rank and abelian varieties with large algebraic rank over function fields*, Invent. Math. **167** (2007), 379–408.

[62] A. Weil, *Adeles and algebraic groups*, Progress in Mathematics, vol. 23, Birkhäuser Boston, Mass., 1982, With appendices by M. Demazure and Takashi Ono. MR 670072 (83m:10032)

[63] A. Wiles, *Modular elliptic curves and Fermat's last theorem*, Ann. of Math. (2) **141** (1995), no. 3, 443–551.

[64] ———, *The Birch and Swinnerton-Dyer conjecture*, The millennium prize problems, Clay Math. Inst., Cambridge, MA, 2006, pp. 31–41. MR 2238272

[65] Ju.G. Zarhin, *Abelian varieties in characteristic p*, Mat. Zametki **19** (1976), no. 3, 393–400. MR 0422287 (54 #10278)

[66] S. Zhang, *Heights of Heegner points on Shimura curves*, Ann. of Math. (2) **153** (2001), no. 1, 27–147. MR 1826411 (2002g:11081)

Elliptic Curves over Function Fields

Douglas Ulmer

IAS/Park City Mathematics Series
Volume 18, 2009

Elliptic Curves over Function Fields

Douglas Ulmer

Introduction

These are the notes from a course of five lectures at the 2009 Park City Math Institute. The focus is on *elliptic curves* over function fields over *finite fields*. In the first three lectures, we explain the main classical results (mainly due to Tate) on the Birch and Swinnerton-Dyer conjecture in this context and its connection to the Tate conjecture about divisors on surfaces. This is preceded by a "Lecture 0" on background material. In the remaining two lectures, we discuss more recent developments on elliptic curves of large rank and constructions of explicit points in high rank situations.

A great deal of this material generalizes naturally to the context of curves and Jacobians of *any genus* over function fields over *arbitrary ground fields*. These generalizations were discussed in a course of 12 lectures at the CRM in Barcelona in February, 2010, and will be written up as a companion to these notes, see [Ulm11]. Unfortunately, theorems on unbounded ranks over function fields are currently known only in the context of finite ground fields.

Finally, we mention here that very interesting theorems of Gross-Zagier type exist also in the function field context. These would be the subject of another series of lectures and we will not say anything more about them in these notes.

It is a pleasure to thank the organizers of the 2009 PCMI for the invitation to speak, the students for their interest, enthusiasm, and stimulating questions, and the "elder statesmen"—Bryan Birch, Dick Gross, John Tate, and Yuri Zarhin—for their remarks and encouragement. Thanks also to Keith Conrad for bringing the fascinating historical articles of Roquette [Roq06] to my attention. Last but not least, thanks are due as well to Lisa Berger, Tommy Occhipinti, Karl Rubin, Alice Silverberg, Yuri Zarhin, and an anonymous referee for their suggestions and TEXnical advice.

School of Mathematics, Georgia Institute of Technology, Atlanta, GA 30332
E-mail address: douglas.ulmer@math.gatech.edu

LECTURE 0

Background on curves and function fields

This "Lecture 0" covers definitions and notations that are probably familiar to many readers and that were reviewed very quickly during the PCMI lectures. Readers are invited to skip it and refer back as necessary.

1. Terminology

Throughout, we use the language of schemes. This is necessary to be on firm ground when dealing with some of the more subtle aspects involving non-perfect ground fields and possibly non-reduced group schemes. However, the instances where we use any hard results from this theory are isolated and students should be able to follow readily the main lines of discussion, perhaps with the assistance of a friendly algebraic geometer.

Throughout, a *variety* over a field F is a separated, reduced scheme of finite type over $\operatorname{Spec} F$. A *curve* is a variety purely of dimension 1 and a *surface* is a variety purely of dimension 2.

2. Function fields and curves

Throughout, p will be a prime number and $\mathbb{F}_q$ will denote the field with q elements with q a power of p. We write $\mathcal{C}$ for a smooth, projective, and absolutely irreducible curve of genus g over $\mathbb{F}_q$ and we write $K = \mathbb{F}_q(\mathcal{C})$ for the function field of $\mathcal{C}$ over $\mathbb{F}_q$. The most important example is when $\mathcal{C} = \mathbb{P}^1$, the projective line, in which case $K = \mathbb{F}_q(\mathcal{C}) = \mathbb{F}_q(t)$ is the field of rational functions in a variable t over $\mathbb{F}_q$.

We write v for a closed point of $\mathcal{C}$, or equivalently for an equivalence class of valuations of K. For each such v we write $\mathcal{O}_{(v)}$ for the local ring at v (the ring of rational functions on $\mathcal{C}$ regular at v), $\mathfrak{m}_v \subset \mathcal{O}_{(v)}$ for the maximal ideal (those functions vanishing at v), and $\kappa_v = \mathcal{O}_{(v)}/\mathfrak{m}_v$ for the residue field at v. The extension $\kappa_v/\mathbb{F}_q$ is finite and we set $\deg(v) = [\kappa_v : \mathbb{F}_q]$ and $q_v = q^{\deg(v)}$ so that $\kappa_v \cong \mathbb{F}_{q_v}$.

For example, in the case where $\mathcal{C} = \mathbb{P}^1$, the "finite" places of $\mathcal{C}$ correspond bijectively to monic irreducible polynomials $f \in \mathbb{F}_q[t]$. If v corresponds to f, then $\mathcal{O}_{(v)}$ is the set of ratios g/h where $g, h \in \mathbb{F}_q[t]$ and f does not divide h. The maximal ideal $\mathfrak{m}_v$ consists of ratios g/h where f does divide g, and the degree of v is the degree of f as a polynomial in t. There is one more place of K, the "infinite" place $v = \infty$. The local ring consists of ratios g/h with $g, h \in \mathbb{F}_q[t]$ and $\deg(g) \leq \deg(h)$. The maximal ideal consists of ratios g/h where $\deg(g) < \deg(h)$ and the degree of $v = \infty$ is 1. The finite and infinite places of $\mathbb{P}^1$ give all closed points of $\mathbb{P}^1$.

We write K^{sep} for a separable closure of K and let $G_K = \operatorname{Gal}(K^{sep}/K)$. We write $\overline{\mathbb{F}}_q$ for the algebraic closure of $\mathbb{F}_q$ in K^{sep}. For each place v of K we have the decomposition group D_v (defined only up to conjugacy), its normal subgroup the inertia group $I_v \subset D_v$, and Fr_v the (geometric) Frobenius at v, a canonical generator of the quotient $D_v/I_v \cong \mathrm{Gal}(\overline{\mathbb{F}}_q/\mathbb{F}_q)$ that acts as $x \mapsto x^{q_v^{-1}}$ on the residue field at a place w dividing v in a finite extension $F \subset K^{sep}$ unramified over v.

General references for this section and the next are [Gol03], [Ros02], and [Sti09].

3. Zeta functions

Let $\mathcal{X}$ be a variety over the finite field $\mathbb{F}_q$. Extending the notation of the previous section, if x is a closed point of $\mathcal{X}$, we write κ_x for the residue field at x, q_x for its cardinality, and $\deg(x)$ for $[\kappa_x : \mathbb{F}_q]$.

We define the Z and ζ functions of $\mathcal{X}$ via Euler products:

$$Z(\mathcal{X},T) = \prod_x \left(1 - T^{\deg(x)}\right)^{-1}$$

and

$$\zeta(\mathcal{X},s) = Z(\mathcal{X},q^{-s}) = \prod_x \left(1 - q_x^{-s}\right)^{-1}$$

where the products are over the closed points of $\mathcal{X}$. It is a standard exercise to show that

$$Z(\mathcal{X},T) = \exp\left(\sum_{n\geq 1} N_n \frac{T^n}{n}\right)$$

where N_n is the number of $\mathbb{F}_{q^n}$-valued points of $\mathcal{X}$. It follows from a crude estimate for the number of $\mathbb{F}_{q^n}$ points of $\mathcal{X}$ that the Euler product defining $\zeta(\mathcal{X},s)$ converges in the half plane $\mathrm{Re}(s) > \dim \mathcal{X}$.

If $\mathcal{X}$ is smooth and projective, then it is known that $Z(\mathcal{X},T)$ is a rational function of the form

$$\frac{\prod_{i=0}^{\dim \mathcal{X}-1} P_{2i+1}(T)}{\prod_{i=0}^{\dim \mathcal{X}} P_{2i}(T)}$$

where $P_0(T) = (1-T)$, $P_{2\dim\mathcal{X}}(T) = (1 - q^{\dim \mathcal{X}}T)$, and for all $0 \leq i \leq 2\dim\mathcal{X}$ $P_i(T)$ is a polynomial with integer coefficients and constant term 1. We denote the inverse roots of P_i by α_{ij} so that

$$P_i(T) = \prod_j (1 - \alpha_{ij}T)$$

The inverse roots α_{ij} of $P_i(T)$ are algebraic integers that have absolute value $q^{i/2}$ in every complex embedding. (We say that they are *Weil numbers of size* $q^{i/2}$.) It follows that $\zeta(\mathcal{X},s)$ has a meromorphic continuation to the whole s plane, with poles on the lines $\mathrm{Re}\, s \in \{0,\dots,\dim\mathcal{X}\}$ and zeroes on the lines $\mathrm{Re}\, s \in \{1/2,\dots,\dim\mathcal{X} - 1/2\}$. This is the analogue of the Riemann hypothesis for $\zeta(\mathcal{X},s)$.

It is also known that the set of inverse roots of $P_i(T)$ (with multiplicities) is stable under $\alpha_{ij} \mapsto q^i/\alpha_{ij}$ and that for $i \leq \dim\mathcal{X}$, $P_{2d-i}(T) = P_i(q^{d-i}T)$. Thus $\zeta(\mathcal{X},s)$ satisfies a functional equation when s is replaced by $\dim\mathcal{X} - s$.

In the case where $\mathcal{X}$ is a curve, $P_1(T)$ has degree $2g$ ($g =$ the genus of $\mathcal{C}$) and has the form

$$P_1(T) = 1 + \cdots + q^gT^{2g} = \prod_{j=1}^{2g}(1 - \alpha_{1j}T).$$

Thus $\zeta(\mathcal{C},s)$ has simple poles for $s \in \frac{2\pi i}{\log q}\mathbb{Z}$ and $s \in 1 + \frac{2\pi i}{\log q}\mathbb{Z}$ and its zeroes lie on the line $\mathrm{Re}\, s = 1/2$.

For a fascinating history of the early work on zeta functions and the Riemann hypothesis for curves over finite fields, see [Roq06] and parts I and II of that work.

4. Cohomology

Assume that $\mathcal{X}$ is a smooth projective variety over $k = \mathbb{F}_q$. We write $\overline{\mathcal{X}}$ for $\mathcal{X} \times_{\mathbb{F}_q} \overline{\mathbb{F}}_q$. Note that $G_k = \operatorname{Gal}(\overline{\mathbb{F}}_q/\mathbb{F}_q)$ acts on $\overline{\mathcal{X}}$ via the factor $\overline{\mathbb{F}}_q$.

Choose a prime $\ell \neq p$. We have ℓ-adic cohomology groups $H^i(\overline{\mathcal{X}}, \mathbb{Q}_\ell)$ which are finite-dimensional $\mathbb{Q}_\ell$-vector spaces and which vanish unless $0 \leq i \leq 2 \dim \mathcal{X}$.

Functoriality in $\overline{\mathcal{X}}$ gives a continuous action of $\operatorname{Gal}(\overline{\mathbb{F}}_q/\mathbb{F}_q)$. Since the geometric Frobenius ($\operatorname{Fr}_q(a) = a^{q^{-1}}$) is a topological generator of $\operatorname{Gal}(\overline{\mathbb{F}}_q/\mathbb{F}_q)$, the characteristic polynomial of Fr_q on $H^i(\overline{\mathcal{X}}, \mathbb{Q}_\ell)$ determines the eigenvalues of the action of $\operatorname{Gal}(\overline{\mathbb{F}}_q/\mathbb{F}_q)$; in fancier language, it determines the action up to semi-simplification.

An important result (inspired by [Wei49] and proven in great generality in [SGA5]) says that the factors P_i of $Z(\mathcal{X}, t)$ are characteristic polynomials of Frobenius:

$$P_i(T) = \det(1 - T \operatorname{Fr}_q | H^i(\overline{\mathcal{X}}, \mathbb{Q}_\ell)). \tag{4.1}$$

From this point of view, the functional equation and Riemann hypothesis for $Z(\mathcal{X}, T)$ are statements about duality and purity.

To discuss the connections, we need more notation. Let $\mathbb{Z}_\ell(1) = \varprojlim_n \mu_{\ell^n}(\overline{\mathbb{F}}_q)$ and $\mathbb{Q}_\ell(1) = \mathbb{Z}_\ell(1) \otimes_{\mathbb{Z}_\ell} \mathbb{Q}_\ell$, so that $\mathbb{Q}_\ell(1)$ is a one-dimensional $\mathbb{Q}_\ell$-vector space on which $\operatorname{Gal}(\overline{\mathbb{F}}_q/\mathbb{F}_q)$ acts via the ℓ-adic cyclotomic character. More generally, for $n > 0$ set $\mathbb{Q}_\ell(n) = \mathbb{Q}_\ell(1)^{\otimes n}$ (n-th tensor power) and $\mathbb{Q}_\ell(-n) = \operatorname{Hom}(\mathbb{Q}_\ell(n), \mathbb{Q}_\ell)$, so that for all n, $\mathbb{Q}_\ell(n)$ is a one-dimensional $\mathbb{Q}_\ell$-vector space on which $\operatorname{Gal}(\overline{\mathbb{F}}_q/\mathbb{F}_q)$ acts via the nth power of the ℓ-adic cyclotomic character.

We have $H^0(\overline{\mathcal{X}}, \mathbb{Q}_\ell) \cong \mathbb{Q}_\ell$ (with trivial Galois action) and $H^{2\dim \mathcal{X}}(\overline{\mathcal{X}}, \mathbb{Q}_\ell) \cong \mathbb{Q}_\ell(\dim \mathcal{X})$. The functional equation follows from the fact that we have a canonical non-degenerate, Galois equivariant pairing

$$H^i(\overline{\mathcal{X}}, \mathbb{Q}_\ell) \times H^{2\dim \mathcal{X} - i}(\overline{\mathcal{X}}, \mathbb{Q}_\ell) \to H^{2\dim \mathcal{X}}(\overline{\mathcal{X}}, \mathbb{Q}_\ell) \cong \mathbb{Q}_\ell(\dim \mathcal{X}).$$

Indeed, the non-degeneracy of this pairing implies that if α is an eigenvalue of Fr_q on $H^i(\overline{\mathcal{X}}, \mathbb{Q}_\ell)$, then $q^{\dim \mathcal{X}}/\alpha$ is an eigenvalue of Fr_q on $H^{2\dim \mathcal{X} - i}(\overline{\mathcal{X}}, \mathbb{Q}_\ell)$.

The Riemann hypothesis in this context is the statement that the eigenvalues of Fr_q on $H^i(\overline{\mathcal{X}}, \mathbb{Q}_\ell)$ are algebraic integers with absolute value $q^{i/2}$ in every complex embedding.

See [SGA4$\frac{1}{2}$] or [Mil80] for an overview of étale cohomology and its connections with the Weil conjectures.

5. Jacobians

5.1. Picard and Albanese properties

We briefly review two (dual) universal properties of the Jacobian of a curve that we will need. See [Mil86b] for more details.

We assume throughout that the curve $\mathcal{C}$ has an $\mathbb{F}_q$-rational point x, i.e., a closed point with residue field $\mathbb{F}_q$. If T is another connected variety over $\mathbb{F}_q$ with an $\mathbb{F}_q$-rational point t, a *divisorial correspondence* between (C, x) and (T, t) is an invertible sheaf $\mathcal{L}$ on $C \times_{\mathbb{F}_q} T$ such that $\mathcal{L}|_{C \times t}$ and $\mathcal{L}|_{x \times T}$ are trivial. Two divisorial correspondences are equal when they are isomorphic as invertible sheaves. Note that the set of divisorial correspondences between $(\mathcal{C}, x)$ and (T, t) forms a group under tensor product and is thus a subgroup of $\mathrm{Pic}(\mathcal{C} \times T)$. We write

$$\mathrm{DivCorr}((\mathcal{C}, x), (T, t)) \subset \mathrm{Pic}(\mathcal{C} \times T)$$

for this subgroup. One may think of a divisorial correspondence as giving a family of invertible sheaves on C: $s \mapsto \mathcal{L}|_{C \times s}$.

Let $J = J_{\mathcal{C}}$ be the Jacobian of $\mathcal{C}$ and write 0 for its identity element. Then J is a g-dimensional abelian variety over $\mathbb{F}_q$ and it carries the "universal divisorial correspondence with C." More precisely, there is a divisorial correspondence $\mathcal{M}$ between (C, x) and $(J, 0)$ such that if S is another connected variety over $\mathbb{F}_q$ with $\mathbb{F}_q$-rational point s and $\mathcal{L}$ is a divisorial correspondence between (C, x) and (S, s), then there is a unique morphism $\phi : S \to J$ sending s to 0 such that $\mathcal{L} = \phi^*\mathcal{M}$. (Of course $\mathcal{M}$ depends on the choice of base point x, but we omit this from the notation.)

It follows that there is a canonical morphism, the Abel-Jacobi morphism, $AJ : \mathcal{C} \to J$ sending x to 0. Intuitively, this corresponds to the family of invertible sheaves parameterized by $\mathcal{C}$ that sends $y \in \mathcal{C}$ to $\mathcal{O}_{\mathcal{C}}(y - x)$. More precisely, let $\Delta \subset \mathcal{C} \times C$ be the diagonal, let

$$D = \Delta - x \times \mathcal{C} - \mathcal{C} \times x,$$

and let $\mathcal{L} = \mathcal{O}_{\mathcal{C} \times \mathcal{C}}(D)$ which is a divisorial correspondence between (C, x) and itself. The universal property above then yields the morphism $AJ : \mathcal{C} \to J$. It is known that AJ is a closed immersion and that its image generates J as an algebraic group.

The second universal property enjoyed by J (or rather by AJ) is the Albanese property: it is universal for maps to abelian varieties. More precisely, if A is an abelian variety and $\phi : \mathcal{C} \to A$ is a morphism sending x to 0, then there is a unique homomorphism of abelian varieties $\psi : J \to A$ such that $\phi = \psi \circ AJ$.

Combining the two universal properties gives a useful connection between correspondences and homomorphisms: Suppose $\mathcal{C}$ and $\mathcal{D}$ are curves over $\mathbb{F}_q$ with rational points $x \in \mathcal{C}$ and $y \in \mathcal{D}$. Then we have an isomorphism

$$\mathrm{DivCorr}((\mathcal{C}, x), (\mathcal{D}, y)) \cong \mathrm{Hom}(J_{\mathcal{C}}, J_{\mathcal{D}}). \tag{5.1.1}$$

Intuitively, given a divisorial correspondence on $\mathcal{C} \times \mathcal{D}$, we get a family of invertible sheaves on $\mathcal{D}$ parameterized by $\mathcal{C}$ and thus a morphism $\mathcal{C} \to J_{\mathcal{D}}$. The Albanese property then gives a homomorphism $J_{\mathcal{C}} \to J_{\mathcal{D}}$. We leave the precise version as an exercise, or see [Mil86b, 6.3]. We will use this isomorphism later to understand the Néron-Severi group of a product of curves.

5.2. The Tate module

Let A be an abelian variety of dimension g over $\mathbb{F}_q$, for example the Jacobian of a curve of genus g. (See [Mil86a] for a brief introduction to abelian varieties and [Mum08] for a much more complete treatment.) Choose a prime $\ell \neq p$. Let $A[\ell^n]$ be the set of $\overline{\mathbb{F}}_q$ points of A of order dividing ℓ^n. It is a group isomorphic to $(\mathbb{Z}/\ell^n\mathbb{Z})^{2g}$ with a linear action of $\mathrm{Gal}(\overline{\mathbb{F}}_q/\mathbb{F}_q)$. We form the inverse limit

$$T_\ell A = \varprojlim_n A[\ell^n]$$

where the transition maps are given by multiplication by ℓ. Let $V_\ell A = T_\ell A \otimes_{\mathbb{Z}_\ell} \mathbb{Q}_\ell$, a $2g$-dimensional $\mathbb{Q}_\ell$-vector space with a linear action of $\mathrm{Gal}(\overline{\mathbb{F}}_q/\mathbb{F}_q)$. It is often called the *Tate module* of A.

According to Roquette, what we now call the Tate module seems to have first been used in print by Deuring [Deu40] as a substitute for homology in his work on correspondences on curves. It appears already in a letter of Hasse from 1935, see [Roq06, p. 36].

The following proposition is the modern interpretation of the connection between homology and torsion points.

Proposition 5.2.1. *Let A be an abelian variety over a field k and let ℓ be a prime not equal to the characteristic of k. Let $V_\ell A$ be the Tate module of A and $(V_\ell A)^*$ its dual as a $G_k = \mathrm{Gal}(k^{sep}/k)$-module.*

- *There is a canonical isomorphism of G_k-modules*
$$(V_\ell A)^* \cong H^1(A \times \overline{k}, \mathbb{Q}_\ell).$$
- *If A is the Jacobian of a curve $\mathcal{C}$ over k, then*
$$H^1(A \times \overline{k}, \mathbb{Q}_\ell) \cong H^1(\mathcal{C} \times \overline{k}, \mathbb{Q}_\ell).$$

For a proof of part 1, see [Mil86a, 15.1] and for part 2, see [Mil86b, 9.6].

Exercises 5.2.2. These exercises are meant to make the Proposition more plausible.

(1) Show that if $A(\mathbb{C})$ is a complex torus $\mathbb{C}^g/\Lambda$, then the singular homology $H_1(A(\mathbb{C}), \mathbb{Q}_\ell)$ is canonically isomorphic to $V_\ell A(\mathbb{C})$. (Hint: Use the universal coefficient theorem to show that $H_1(A(\mathbb{C}), \mathbb{Z}/\ell^n\mathbb{Z}) \cong \Lambda/\ell^n\Lambda$.)
(2) (Advanced) Let $\mathcal{C}$ be a smooth projective curve over an algebraically closed field k. Let ℓ be a prime not equal to the characteristic of k. Use geometric class field theory (as in [Ser88]) to show that unramified Galois covers $\mathcal{C}' \to \mathcal{C}$ equipped with an isomorphism $\mathrm{Gal}(\mathcal{C}'/\mathcal{C}) \cong \mathbb{Z}/\ell\mathbb{Z}$ are in bijection with elements of $\mathrm{Hom}(J_\mathcal{C}[\ell], \mathbb{Z}/\ell\mathbb{Z})$. (Make a convention to deal with the trivial homomorphism.) This suggests that $H^1(\mathcal{C}, \mathbb{Z}/\ell\mathbb{Z})$ "should be" $\mathrm{Hom}(J_\mathcal{C}[\ell], \mathbb{Z}/\ell\mathbb{Z})$ and $H_1(C, \mathbb{Z}/\ell\mathbb{Z})$ "should be" $J_\mathcal{C}[\ell]$. The reason we only have "should be" rather than a theorem is that a non-trivial Galois cover $\mathcal{C}' \to \mathcal{C}$ is never locally constant in the Zariski topology. This is a prime motivation for introducing the étale topology.

6. Tate's theorem on homomorphisms of abelian varieties

As usual, let k be a finite field and let A and B be two abelian varieties over k. Choose a prime ℓ not equal to the characteristic of k and form the Tate modules $V_\ell A$ and $V_\ell B$. Any homomorphism of abelian varieties $\phi : A \to B$ induces a homomorphism of Tate modules $\phi_* : V_\ell A \to V_\ell B$ and this homomorphism commutes with the action of $G_k = \mathrm{Gal}(\overline{k}/k)$ on the Tate modules. We get an induced homomorphism $\mathrm{Hom}_k(A, B) \otimes \mathbb{Q}_\ell \to \mathrm{Hom}_{G_k}(V_\ell A, V_\ell B)$. Tate's famous result [Tat66a] asserts that this is an isomorphism:

Theorem 6.1. *The map $\phi \mapsto \phi_*$ induces an isomorphism of $\mathbb{Q}_\ell$-vector spaces:*
$$\mathrm{Hom}_k(A, B) \otimes \mathbb{Q}_\ell \tilde{\to} \mathrm{Hom}_{G_k}(V_\ell A, V_\ell B).$$

We also mention [Zar08] which gives a different proof and a strengthening with finite coefficients.

We will use Tate's theorem in Theorem 12.1 of Lecture 2 to understand the divisors on a product of curves in terms of homomorphisms between their Jacobians.

LECTURE 1

Elliptic curves over function fields

In this lecture we discuss the basic facts about elliptic curves over function fields over finite fields. We assume the reader has some familiarity with elliptic curves over global fields such as $\mathbb{Q}$ or number fields, as explained, e.g., in [Sil09], and we will focus on aspects specific to characteristic p. The lecture ends with statements of the main results known about the conjecture of Birch and Swinnerton-Dyer in this context.

1. Elliptic curves

1.1. Definitions

We write $k = \mathbb{F}_q$ for the finite field of cardinality q and characteristic p and we let K be the function field of a smooth, projective, absolutely irreducible curve $\mathcal{C}$ over k.

An *elliptic curve* over K is a smooth, projective, absolutely irreducible curve of genus 1 over K equipped with a K-rational point O that will serve as the origin of the group law.

All the basic geometric facts, e.g., of [Sil09, Ch. III and App. A], continue to hold in the context of function fields. We review a few of them to establish notation, but will not enter into full details.

Using the Riemann-Roch theorem, an elliptic curve E over K can always be presented as a projective plane cubic curve defined by a Weierstrass equation, i.e., by an equation of the form

$$Y^2Z + a_1XYZ + a_3YZ^2 = X^3 + a_2X^2Z + a_4XZ^2 + a_6Z^3 \tag{1.1.1}$$

where $a_1, \dots, a_6 \in K$. The origin O is the point at infinity $[0:1:0]$. We often give the equation in affine form:

$$y^2 + a_1xy + a_3y = x^3 + a_2x^2 + a_4x + a_6 \tag{1.1.2}$$

where $x = X/Z$ and $y = Y/Z$.

The quantities $b_2, \dots, b_8, c_4, c_6, \Delta, j$ are defined by the usual formulas ([Sil09, III.1] or [Del75]). Since E is smooth, by the following exercise $\Delta \neq 0$.

Remark/Exercises 1.1.3. The word "smooth" in the definition of an elliptic curve means that the morphism $E \to \operatorname{Spec} K$ is smooth. Smoothness of a morphism can be tested via the Jacobian criterion (see, e.g., [Har77, III.10.4] or [Liu02, 4.3.3]). Show that the projective plane cubic (1.1.1) is smooth if and only if $\Delta \neq 0$. Because the ground field K is not perfect, smoothness is strictly stronger than the requirement that E be regular, i.e., that its local rings be regular local rings (cf. [Liu02, 4.2.2]). For example, show that the projective cubic defined by $Y^2Z = X^3 - tZ^3$ over $K = \mathbb{F}_p(t)$ with $p = 2$ or 3 is a regular scheme, but is not smooth over K.

Definitions 1.1.4. Let E be an elliptic curve over K.

(1) We say E is *constant* if there is an elliptic curve E_0 defined over k such that $E \cong E_0 \times_k K$. Equivalently, E is constant if it can be defined by a Weierstrass cubic (1.1.1) where the $a_i \in k$.
(2) We say E is *isotrivial* if there exists a finite extension K' of K such that E becomes constant over K'. Note that a constant curve is isotrivial.
(3) We say E is *non-isotrivial* if it is not isotrivial. We say E is *non-constant* if it is not constant.

Remark/Exercises 1.1.5. Show that E is isotrivial if and only if $j(E) \in k$. Suppose that E is isotrivial, so that E becomes constant over a finite extension K' and let k' be the field of constants of K' (the algebraic closure of k in K'). *A priori*, the definition of isotrivial says that there is an elliptic curve E_0 over k' such that $E \times_K K' \cong E_0 \times_{k'} K'$. Show that we may take K' to have field of constants k and E_0 to be defined over k. Show also that we may take K' to be separable and of degree dividing 24 over K.

Exercise 1.1.6. For any elliptic curve E over K, the functor on K-algebras $L \mapsto \mathrm{Aut}_L(E \times L)$ is represented by a group scheme $\underline{\mathrm{Aut}}(E)$. (Concretely, this means there is a group scheme $\underline{\mathrm{Aut}}(E)$ such that for any K-algebra L, $\mathrm{Aut}_L(E \times L)$ is $\underline{\mathrm{Aut}}(E)(L)$, the group of L-valued points of $\underline{\mathrm{Aut}}(E)$.) Show that $\underline{\mathrm{Aut}}(E)$ is an étale group scheme. Equivalently, show that any element of $\mathrm{Aut}_{\overline{K}}(E)$ is defined over a separable extension of K. (This is closely related to the previous exercise.)

1.2. Examples

Let $K = \mathbb{F}_p(t)$ with $p > 3$ and define elliptic curves

$$\begin{aligned} E_1: &\quad y^2 = x^3 + 1 \\ E_2: &\quad y^2 = x^3 + t^6 \\ E_3: &\quad y^2 = x^3 + t \\ E_4: &\quad y^2 = x^3 + x + t. \end{aligned}$$

Then $E_1 \cong E_2$ over K and both are constant, E_3 is isotrivial and non-constant, whereas E_4 is non-isotrivial.

For more examples, let $K = \mathbb{F}_p(t)$ (with p restricted as indicated) and define

$$\begin{aligned} (p \neq 3) &\qquad E_5: \quad y^2 + ty = x^3 \\ (p \neq 2) &\qquad E_6: \quad y^2 = x^3 + tx \\ (p \text{ arbitrary}) &\qquad E_7: \quad y^2 + xy + ty = x^3 \\ (p \text{ arbitrary}) &\qquad E_8: \quad y^2 + xy = x^3 + tx \\ (p \text{ arbitrary}) &\qquad E_9: \quad y^2 + xy = x^3 + t. \end{aligned}$$

Then E_5 and E_6 are isotrivial and non-constant whereas E_7, E_8, and E_9 are non-isotrivial.

2. Frobenius

If X is a scheme of characteristic p, we define the *absolute Frobenius* morphism $\mathrm{Fr}_X : X \to X$ as usual: It is the identity on the underlying topological space and raises functions to the p th power. When $X = \operatorname{Spec} K$, Fr_X is just the map of schemes induced by the ring homomorphism $K \to K$, $a \mapsto a^p$.

Suppose as usual that K is a function field and let E be an elliptic curve over K. Define a new elliptic curve $E^{(p)}$ over K by the fiber product diagram:

$$\begin{array}{ccc} E^{(p)} = \operatorname{Spec} K \times_{\operatorname{Spec} K} E & \longrightarrow & E \\ \downarrow & & \downarrow \\ \operatorname{Spec} K & \xrightarrow{\operatorname{Fr}} & \operatorname{Spec} K \end{array}$$

More concretely, if E is presented as a Weierstrass cubic as in equation (1.1.2), then $E^{(p)}$ is given by the equation with a_i replaced by a_i^p. The universal property of the fiber product gives a canonical morphism $\operatorname{Fr}_{E/K}$, the *relative Frobenius*:

$$\begin{array}{ccccc} E & \xrightarrow{\operatorname{Fr}_{E/K}} & E^{(p)} & \longrightarrow & E \\ & \searrow & \downarrow & & \downarrow \\ & & \operatorname{Spec} K & \xrightarrow{\operatorname{Fr}} & \operatorname{Spec} K \end{array}$$

By definition $\operatorname{Fr}_{E/K}$ is a morphism over K. In terms of Weierstrass equations for E and $E^{(p)}$ as above, it is just the map $(x, y) \mapsto (x^p, y^p)$.

It is evident that $\operatorname{Fr}_{E/K}$ is an isogeny, i.e., a surjective homomorphism of elliptic curves, and that its degree is p. We define $V = V_{E/K}$ to be the dual isogeny, so that $V_{E/K} \circ \operatorname{Fr}_{E/K} = [p]$, multiplication by p on E.

Note that $j(E^{(p)}) = j(E)^p$ so that if E is non-isotrivial, E and $E^{(p)}$ are not isomorphic. Thus, using Frobenius and its iterates, we see that there are infinitely many non-isomorphic elliptic curves isogenous to any non-isotrivial E. This is in marked contrast to the situation over number fields (cf. [Fal86]).

Lemma 2.1. *Let E be an elliptic curve over K. Then $j(E)$ is a p-th power in K if and only if there exists an elliptic curve E' over K such that $E \cong E'^{(p)}$.*

PROOF. We sketch a fancy argument and pose as an exercise a more down-to-earth proof. Obviously if there is an E' with $E \cong E'^{(p)}$, then $j(E) = j(E'^{(p)}) = j(E')^p \in K^p$. Conversely, suppose $j(E) \in K^p$ and choose an elliptic curve E'' such that $j(E'')^p = j(E)$. It follows that $E''^{(p)}$ is isomorphic to E over a finite separable extension of K. In other words, E is the twist of $E''^{(p)}$ by a cocycle in $H^1(G_K, \operatorname{Aut}_{K^{sep}}(E''^{(p)}))$. But there is a canonical isomorphism $\operatorname{Aut}_{K^{sep}}(E''^{(p)}) \cong \operatorname{Aut}_{K^{sep}}(E'')$ and twisting E'' by the corresponding element of

$$H^1(G_K, \operatorname{Aut}_{K^{sep}}(E'')) \cong H^1(G_K, \operatorname{Aut}_{K^{sep}}(E''^{(p)}))$$

we obtain an elliptic curve E' with $E'^{(p)} \cong E$. □

Exercise 2.2. Use explicit equations, as in [Sil09, Appendix A], to prove the lemma.

3. The Hasse invariant

Let F be a field of characteristic p and E an elliptic curve over F. Let $\mathcal{O}_E$ be the sheaf of regular functions on E and let Ω^1_E be the sheaf of Kähler differentials on E. The coherent cohomology group $H^1(E, \mathcal{O}_E)$ is a one-dimensional F-vector space and is Serre dual to the space of invariant differentials $H^0(E,\Omega^1_E)$. Choose a non-zero differential $\omega\in H^0(E,\Omega^1_E)$ and let η be the dual element of $H^1(E,\mathcal{O}_E)$. The absolute Frobenius Fr_E induces a (p-linear) homomorphism:

$$\mathrm{Fr}_E^*:H^1(E,\mathcal{O}_E)\to H^1(E,\mathcal{O}_E).$$

We define an element $A=A(E,\omega)$ of F by requiring that $\mathrm{Fr}_E^*(\eta)=A(E,\omega)\eta$. This is the *Hasse invariant* of E. It has weight $p-1$ in the sense that $A(E,\lambda^{-1}\omega)=\lambda^{p-1}A(E,\omega)$ for all $\lambda\in F^\times$.

Suppose E is given by a Weierstrass equation (1.1.2) and $\omega=dx/(2y+a_1x+a_3)$. If $p=2$, then $A(E,\omega)=a_1$. If $p>2$, choose an equation with $a_1=a_3=0$. Then $A(E,\omega)=$ the coefficient of x^{p-1} in $(x^3+a_2x^2+a_4x+a_6)^{(p-1)/2}$. These assertions follow from [KM85, 12.4] where several other calculations of A are also presented.

Recall that E/K is *ordinary* if the group of p-torsion points $E(\overline{K})[p]\neq 0$ and *supersingular* otherwise. It is known that E is supersingular if and only if $A(E,\omega)=0$ (e.g., [KM85, 12.3.6 and 12.4]) and in this case $j(E)\in\mathbb{F}_{p^2}$ (e.g., [KM85, proof of 2.9.4]). (Alternatively, one may apply [Sil09, V.3.1] to E over $\overline{K}$.) In particular, if E is supersingular, then it must also be isotrivial.

4. Endomorphisms

The classification of endomorphism rings in [Sil09, III.9] goes over verbatim to the function field case: $\mathrm{End}_{\overline{K}}(E)$ is either $\mathbb{Z}$, an order in an imaginary quadratic number field, or an order in a quaternion algebra over $\mathbb{Q}$ ramified exactly at ∞ and p. The quaternionic case occurs if and only if E is supersingular, and the imaginary quadratic case occurs if and only if $j(E)$ is in $\overline{\mathbb{F}}_p$ and E is not supersingular ([Sil09, V.3.1 and Exer. V.5.8]).

In particular, if E is non-isotrivial, then $\mathrm{End}_{\overline{K}}(E)=\mathrm{End}_K(E)=\mathbb{Z}$.

5. The Mordell-Weil-Lang-Néron theorem

We write $E(K)$ for the group of K-rational points of E and we call $E(K)$ the Mordell-Weil group of E over K. Lang and Néron (independently) generalized the classical Mordell-Weil theorem to the function field context:

Theorem 5.1. *Assume that $K=\mathbb{F}_q(\mathcal{C})$ is the function field of a curve over a finite field and let E be an elliptic curve over K. Then $E(K)$ is a finitely generated abelian group.*

(The theorems of Lang and Néron apply much more generally to any abelian variety A over a field K that is finitely generated over its "constant field" k, but one has to take care of the "constant part" of A. See [Ulm11] for details.)

We will not give a detailed proof of the MWLN theorem here, but will mention two strategies. One is to follow the method of proof of the Mordell-Weil (MW) theorem over a number field. Choose a prime number $\ell\neq p$. By an argument very similar to that in [Sil09, Ch. VIII] one can show that $E(K)/\ell E(K)$ is finite (the "weak Mordell-Weil theorem") by embedding it in a Selmer group and showing that the Selmer group is finite by using the two fundamental finiteness results of algebraic number theory (finiteness of the class group and finite generation of the unit group) applied to Dedekind domains in K. One can then introduce a theory of heights exactly as in [Sil09] and show that the MW theorem follows from the weak MW theorem and finiteness properties of heights. See the original paper of Lang and Néron [LN59] for the full details. A complete treatment in modern language has been given by Conrad [Con06].

One interesting twist in the function field setting comes if one takes $\ell = p$ above. It is still true that the Selmer group for p is finite, but one needs to use the local restrictions at all places; the maximal abelian extension of exponent p unramified outside a finite but non-empty set of places is not finite and so one needs some control on ramification at every place. See [Ulm91] for a detailed account of p-descent in characteristic p.

A second strategy of proof, about which we will say more in Lecture 3, involves relating the Mordell-Weil group of E to the Néron-Severi group of a closely related surface $\mathcal{E}$. In fact, finite generation of the Néron-Severi group (known as the "theorem of the base") is equivalent to the Lang-Néron theorem. A direct proof of the theorem of the base was given by Kleiman in [SGA6, XIII]. See also [Mil80, V.3.25].

6. The constant case

It is worth pausing in the general development to look at the case of a constant curve E. Recall that K is the function field $k(\mathcal{C})$ of the curve $\mathcal{C}$ over $k = \mathbb{F}_q$. Suppose E_0 is an elliptic curve over k and let $E = E_0 \times_k K$.

Proposition 6.1. *We have a canonical isomorphism*

$$E(K) \cong \mathrm{Mor}_k(\mathcal{C}, E_0)$$

where Mor_k *denotes morphisms of varieties over* k *(=morphisms of* k*-schemes). Under this isomorphism,* $E(K)_{tor}$ *corresponds to the subset of constant morphisms.*

PROOF. By definition, $E(K)$ is the set of K-morphisms

$$\mathrm{Spec}\, K \to E = E_0 \times_k K.$$

By the universal property of the fiber product, these are in bijection with k-morphisms $\mathrm{Spec}\, K \to E_0$. Since $\mathcal{C}$ is a smooth curve, any k-morphism $\mathrm{Spec}\, K \to E_0$ extends uniquely to a k-morphism $\mathcal{C} \to E_0$. This establishes a map $E(K) \to \mathrm{Mor}_k(\mathcal{C}, E_0)$. If $\eta : \mathrm{Spec}\, K \to \mathcal{C}$ denotes the canonical inclusion, composition with η $(\phi \mapsto \phi \circ \eta)$ induces a map $\mathrm{Mor}_k(\mathcal{C}, E_0) \to E(K)$ inverse to the map above. This establishes the desired bijection and this bijection is obviously compatible with the group structures.

Since k is finite, it is clear that a constant morphism goes over to a torsion point. Conversely, if $P \in E(K)$ is torsion, say of order n, then the image of the corresponding $\phi : \mathcal{C} \to E_0$ must lie in the set of n-torsion points of E_0, a discrete set, and this implies that ϕ is constant. □

For example, if K is rational (i.e., $\mathcal{C} = \mathbb{P}^1$ so that $K = k(t)$), then $E(K) = E_0(k)$.

Corollary 6.2. *Let* $J_\mathcal{C}$ *be the Jacobian of* $\mathcal{C}$*. We have canonical isomorphisms*

$$E(K)/E(K)_{tor} \cong \mathrm{Hom}_{k-av}(J_\mathcal{C}, E_0) \cong \mathrm{Hom}_{k-av}(E_0, J_\mathcal{C}).$$

PROOF. The Albanese property of the Jacobian of $\mathcal{C}$ (Subsection 5.1 of Lecture 0) gives a surjective homomorphism

$$\mathrm{Mor}_k(\mathcal{C}, E_0) \to \mathrm{Hom}_{k-av}(J_\mathcal{C}, E_0).$$

This homomorphism sends non-constant (and therefore surjective) morphisms to non-constant (surjective) homomorphisms, so its kernel consists exactly of the constant morphisms. The second isomorphism in the statement of the corollary follows from the fact that Jacobians are self-dual. □

By Poincaré complete reducibility [Mil86a, 12.1], $J_{\mathcal{C}}$ is isogenous to a product of simple abelian varieties. Suppose $J_{\mathcal{C}}$ is isogenous to $E_0^m \times A$ and A admits no non-zero morphisms to E_0. We say that "E_0 appears in $J_{\mathcal{C}}$ with multiplicity m." Then it is clear from the corollary that $E(K)/E(K)_{tor} \cong \mathrm{End}_k(E_0)^m$ and so the rank of $E(K)$ is m, $2m$, or $4m$.

Tate and Shafarevich used these ideas to exhibit isotrivial elliptic curves over $F = \mathbb{F}_p(t)$ of arbitrarily large rank. Indeed, using Tate's theorem on isogenies of abelian varieties over finite fields (reviewed in Section 6 of Lecture 0) and a calculation of zeta functions in terms of Gauss sums, they were able to produce a hyperelliptic curve $\mathcal{C}$ over $\mathbb{F}_p$ whose Jacobian is isogenous to $E_0^m \times A$ where E_0 is a supersingular elliptic curve and the multiplicity m is as large as desired. If $K = \mathbb{F}_p(\mathcal{C})$, E is the constant curve $E = E_0 \times F$, and E' is the twist of E by the quadratic extension K/F, then $\mathrm{Rank}\, E'(F) = \mathrm{Rank}\, E(K)$ and so $E'(F)$ has large rank by the analysis above. See the original article [TS67] for more details and a series of articles by Elkies (starting with [Elk94]) for a beautiful application to the construction of lattices with high packing densities.

7. Torsion

An immediate corollary of the MWLN theorem is that $E(K)_{tor}$ is finite. In fact, $E(K)_{tor}$ is isomorphic to a group of the form

$$\mathbb{Z}/m\mathbb{Z} \times \mathbb{Z}/n\mathbb{Z}$$

where m divides n and p does not divide m. (See for example [Sil09, Ch. 3].) One can also see using the theory of modular curves that every such group appears for a suitable K and E.

In another direction, one can give uniform bounds on torsion that depend only on crude invariants of the field K.

Indeed, in the constant case, $E(K)_{tor} \cong E_0(\mathbb{F}_q)$ which has order bounded by $(q^{1/2}+1)^2$. In the isotrivial case, there is a finite extension K' with the same field of constants $k = \mathbb{F}_q$ over which E becomes constant. Thus $E(K)_{tor} \subset E(L)_{tor}$ again has cardinality bounded by $(q^{1/2}+1)^2$.

We now turn to the non-isotrivial case.

Proposition 7.1. *Assume that E is non-isotrivial and let $g_{\mathcal{C}}$ be the genus of $\mathcal{C}$. Then there is a finite (and effectively calculable) list of groups—depending only on $g_{\mathcal{C}}$ and p—such that for any non-isotrivial elliptic curve E over K, $E(K)_{tor}$ appears on the list.*

Proof. (Sketch) First consider the prime-to-p torsion subgroup of $E(K)$. It has the form $G = \mathbb{Z}/m\mathbb{Z} \times \mathbb{Z}/n\mathbb{Z}$ where $m|n$ and $p \nmid m$. There is a modular curve $X(m,n)$, irreducible and defined over $\mathbb{F}_p(\mu_m)$, that is a coarse moduli space for elliptic curves with subgroups isomorphic to G. We get a morphism $\mathcal{C} \to X(m,n)$ which is non-constant (because E is non-isotrivial) and therefore surjective. The Riemann-Hurwitz formula then implies that $g_{\mathcal{C}} \geq g_{X(m,n)}$. But the genus of $X(m,n)$ goes to infinity with n. Indeed, $g_{X(m,n)} \geq g_{X(1,n)}$ and standard genus formulae ([Miy06, 4.2]) together with crude estimation show that the latter is bounded below by

$$1+\frac{n^2}{24\zeta(2)}-\frac{n\log_2 n}{4}.$$

This shows that for a fixed value of $g_{\mathcal{C}}$, only finitely many groups G as above can appear as $E(K)_{tor}$.

The argument for p-torsion is similar, except that ones uses the Igusa curves $Ig(p^n)$ (cf. [KM85, Ch. 12]). If $E(K)$ has a point of order p^n, we get a non-constant morphism $\mathcal{C}\to Ig(p^n)$ and the genus of $Ig(p^n)$ is asymptotic to $p^{2n}/48$ [Igu68, p. 96]. □

This proposition seems to have been rediscovered repeatedly over the years. The first reference I know of is [Lev68].

Since the genus of a function field is an analogue of the discriminant (more precisely q^{2g-2} is an analogue of the absolute value of the discriminant of a number field), the proposition is an analogue of bounding $E(K)_{tor}$ in terms of the discriminant of a number field K. One could ask for a strengthening where torsion is bounded by "gonality", i.e., by the smallest degree of a non-constant map $\mathcal{C}\to\mathbb{P}^1$. This would be an analogue of bounding $E(K)_{tor}$ in terms of the degree of a number field K, as in the theorems of Mazur, Kamienny, and Merel [Mer96]. This is indeed possible and can be proven by mimicking the proof of the proposition, replacing bounds on the genus of the modular curve with bounds on its gonality. See [Poo07] for the best results currently known on gonality of modular curves.

Exercise 7.2. Compute the optimal list mentioned in the proposition for $g=0$. (This is rather involved.) Note that the optimal list in fact depends on p. Indeed, $\mathbb{Z}/11\mathbb{Z}$ is on the list if and only if $p=11$.

One can be very explicit about p-torsion:

Proposition 7.3. *Suppose that E is a non-isotrivial elliptic curve over K. Then $E(K)$ has a point of order p if and only if $j(E)\in K^p$ and $A(E,\omega)$ is a $(p-1)$st power in $K^\times$.*

Note that whether $A(E,\omega)$ is a $(p-1)$st power is independent of the choice of the differential ω.

Proof. Let $E\xrightarrow{\mathrm{Fr}}E^{(p)}\xrightarrow{V}E$ be the standard factorization of multiplication by p into Frobenius and Verschiebung. Recall (e.g., [Ulm91, 2.1]) that $A(E,\omega)$ is a $(p-1)$st power in K if and only if $\ker\mathrm{Fr}\cong\mu_p$ if and only if $\ker V\cong\mathbb{Z}/p\mathbb{Z}$ if and only if there is a non-trivial p-torsion point in $E^{(p)}(K)$.

Now suppose that $P\in E(K)$ is a non-trivial p-torsion point. Then $\mathrm{Fr}(P)$ is a non-trivial p-torsion point in $E^{(p)}(K)$ and so $A(E,\omega)$ is a $(p-1)$st power in K. Let E' be the quotient of E by the cyclic subgroup generated by P: $E'=E/\langle P\rangle$. Since $\langle P\rangle$ is in the kernel of multiplication by p, we have a factorization of multiplication by p:

$$[p]:E\to E'\to E.$$

Since $E\to E'$ is étale of degree p and $[p]$ is inseparable of degree p^2, we have that $E'\to E$ is purely inseparable of degree p. But an elliptic curve in characteristic p has a unique inseparable isogeny of degree p (namely the quotient by the unique connected subgroup of order p, the kernel of Frobenius) so we have an identification $E = E'^{(p)}$. By 2.1, $j(E) \in K^p$.

Conversely, suppose $A(E, \omega)$ is a $(p-1)$st power and $j(E) \in K^p$. Let E' be the elliptic curve such that $E'^{(p)} \cong E$. Given a differential ω on E, there is a differential ω' on E' such that $A(E, \omega) = A(E', \omega')^p$ (as can be seen for example by using Weierstrass equations). It follows that $A(E', \omega')$ is also a $(p-1)$st power in K. Thus we have a non-trivial point of order p in $E'^{(p)}(K) = E(K)$. □

Part of the proposition generalizes trivially by iteration: if $E(K)$ has a point of order p^n, then $j(E) \in K^{p^n}$. A full characterization of p^n torsion seems harder—the condition that $A(E, \omega)$ be a $(p-1)$st power is closely related to the equations defining the Igusa curve $Ig(p)$ ([KM85, 12.8]), but we do not have such explicit equations for $Ig(p^n)$ when $n > 1$.

8. Local invariants

Let E be an elliptic curve over K and let v be a place of K. A model (1.1.2) for E with coefficients in the valuation ring $\mathcal{O}_{(v)}$ is said to be *integral at* v. The valuation of the discriminant Δ of an integral model is a non-negative integer and so there are models where this valuation takes its minimum value. Such models are *minimal integral models at* v.

Choose a model for E that is minimal integral at v:

$$y^2 + a_1xy + a_3y = x^3 + a_2x^2 + a_4x + a_6.$$

Let $\overline{a}_i \in \kappa(v)$ be the reductions of the coefficients and let E_v be the plane cubic

$$y^2 + \overline{a}_1xy + \overline{a}_3y = x^3 + \overline{a}_2x^2 + \overline{a}_4x + \overline{a}_6 \tag{8.1}$$

over the residue field κ_v. It is not hard to check using Weierstrass equations that the isomorphism type of the reduced cubic (8.1) is independent of the choice of minimal model.

If the discriminant of a minimal integral model at v has valuation zero, i.e., is a unit at v, then the reduced equation defines an elliptic curve over κ_v. If the minimal valuation is positive, then the reduced curve is singular. We distinguish three cases according to the geometry of the reduced curve.

Definition 8.2.

(1) If E_v is a smooth cubic, we say E has *good reduction at* v.
(2) If E_v is a nodal cubic, we say E has *multiplicative reduction at* v. If the tangent lines at the node are rational over $\kappa(v)$ we say the reduction is *split multiplicative* and if they are rational only over a quadratic extension, we say the reduction is *non-split multiplicative*.
(3) If E_v is a cuspidal cubic, we say E has *additive reduction*.

Define an integer a_v as follows:

$$a_v = \begin{cases} q_v + 1 - \#E_v(\kappa_v) & \text{if } E \text{ has good reduction at } v \\ 1 & \text{if } E \text{ has split multiplicative reduction at } v \\ -1 & \text{if } E \text{ has non-split multiplicative reduction at } v \\ 0 & \text{if } E \text{ has additive reduction at } v \end{cases} \tag{8.3}$$

Exercise 8.4. To make this definition less *ad hoc*, note that in the good reduction case, the numerator of the ζ-function of the reduced curve is $1 - a_v q_v^{-s} + q_v^{1-2s}$. Show that in the bad reduction cases, the ζ-function of the reduced curve is

$$\frac{1 - a_v q_v^{-s}}{(1 - q_v^{-s})(1 - q_v^{1-s})}.$$

In the good reduction case, the results about zeta functions and étale cohomology reviewed in Lecture 0, Sections 3 and 4 imply the "Hasse bound": $|a_v| \le 2\sqrt{q_v}$.

There are two more refined invariants in the bad reduction cases: the Néron model and the conductor. The local exponent of the conductor at v, denoted n_v is defined as

$$n_v = \begin{cases} 0 & \text{if } E \text{ has good reduction at } v \\ 1 & \text{if } E \text{ has multiplicative reduction at } v \\ 2 + \delta_v & \text{if } E \text{ has additive reduction at } v \end{cases} \tag{8.5}$$

Here δ_v is a non-negative integer that is 0 when $p > 3$ and is ≥ 0 when $p = 2$ or 3. We refer to [Tat75] for a definition and an algorithm to compute δ_v.

The (global) conductor of E is defined to be the divisor $\mathfrak{n} = \sum_v n_v[v]$. Its degree is $\deg \mathfrak{n} = \sum_v n_v \deg v$.

The Néron model will be discussed in Lecture 3 below.

Exercise 8.6. Mimic [Sil09, Ch. VII] to define a filtration on the points of E over a completion K_v of K. Show that the prime-to-p part of $E(K)_{tor}$ maps injectively into $E(K_v)/E(K_v)_1$. Relate $E(K_v)/E(K_v)_1$ to the special fiber of the Néron model of E at v. As in the classical case, this gives an excellent way to bound the prime-to-p part of $E(K)_{tor}$.

9. The L-function

We define the L-function of E/K as an Euler product:

$$L(E,T) = \prod_{\text{good } v} \left(1 - a_v T^{\deg v} + q_v T^{2\deg v}\right)^{-1} \prod_{\text{bad } v} \left(1 - a_v T^{\deg v}\right)^{-1} \tag{9.1}$$

and

$$L(E,s) = L(E, q^{-s}).$$

(Here T is a formal indeterminant and s is a complex number. Unfortunately, there is no standard reasonable parallel of the notations Z and ζ to distinguish the function of T and the function of s.) Because of the Hasse bound on the size of a_v, the product converges absolutely in the region $\operatorname{Re} s > 3/2$, and as we will see below, it has a meromorphic continuation to all s.

When E is constant it is elementary to calculate $L(E,s)$ in terms of the zeta-functions of E_0 and $\mathcal{C}$.

Exercise 9.2. Suppose that $E = E_0 \times_k K$. Write the ζ-functions of E_0 and $\mathcal{C}$ as rational functions:

$$\zeta(E_0, s) = \frac{\prod_{i=1}^{2}(1 - \alpha_i q^{-s})}{(1 - q^{-s})(1 - q^{1-s})}$$

and

$$\zeta(\mathcal{C}, s) = \frac{\prod_{j=1}^{2g_{\mathcal{C}}}(1 - \beta_j q^{-s})}{(1 - q^{-s})(1 - q^{1-s})}.$$

Prove that

$$L(E,s)=\frac{\prod_{i,j}(1-\alpha_i\beta_jq^{-s})}{\prod_{i=1}^{2}(1-\alpha_iq^{-s})\prod_{i=1}^{2}(1-\alpha_iq^{1-s})}.$$

Thus $L(E,s)$ is a rational function in q^{-s} of degree $4g_{\mathcal{C}}-4$, it extends to a meromorphic function of s, and it satisfies a functional equation for $s\leftrightarrow 2-s$. Its poles lie on the lines $\operatorname{Re} s=1/2$ and $\operatorname{Re} s=3/2$ and its zeroes lie on the line $\operatorname{Re} s=1$.

Although the proofs are much less elementary, these facts extend to the non-constant case as well:

Theorem 9.3. *Suppose the E is a non-constant elliptic curve over K. Let $\mathfrak{n}$ be the conductor of E. Then $L(E,s)$ is a polynomial in q^{-s} of degree $N=4g_{\mathcal{C}}-4+\deg\mathfrak{n}$, it satisfies a functional equation for $s\leftrightarrow 2-s$, and its zeroes lie on the line $\operatorname{Re} s=1$. More precisely,*

$$L(E,s)=\prod_{i=1}^{N}(1-\alpha_iq^{-s})$$

where each α_i is an algebraic integer of absolute value q in every complex embedding. The collection of α_i (with multiplicities) is invariant under $\alpha_i\mapsto q^2/\alpha_i$.

The theorem is a combination of results of Grothendieck, Deligne, and others. We will sketch a proof of it in Lecture 4.

Note that in all cases $L(E,s)$ is holomorphic at $s=1$. In the non-constant case, its order of vanishing at $s=1$ is bounded above by N and it equals N if and only if $L(E,s)=(1-q^{1-s})^N$.

10. The basic BSD conjecture

This remarkable conjecture connects the analytic behavior of the function $L(E,s)$, constructed from local data, to the Mordell-Weil group, a global invariant.

Conjecture 10.1 (Birch and Swinnerton-Dyer).

$$\operatorname{Rank} E(K)=\operatorname{ord}_{s=1}L(E,s)$$

The original conjecture was stated only for elliptic curves over $\mathbb{Q}$ [BSD65] but it is easily seen to make sense for abelian varieties over global fields. There is very strong evidence in favor of it, especially for elliptic curves over $\mathbb{Q}$ and abelian varieties over function fields. See [Gro10, Lecture 3, §4] for a summary of the best theoretical evidence in the number field case. We will discuss what is known for elliptic curves in the function field case later in this course. See Section 12 for statements of the main results and [Ulm11] for a discussion of the case of higher dimensional abelian varieties over function fields.

11. The Tate-Shafarevich group

We define the Tate-Shafarevich group of E over K as

$$Ш(E/K)=\ker\left(H^1(K,E)\to\prod_v H^1(K_v,E)\right).$$

Here the cohomology groups can be taken to be Galois cohomology groups:

$$H^1(K,E)=H^1(G_K,E(K^{sep}))$$

and similarly for $H^1(K_v, E)$; or they can be taken as étale or flat cohomology groups of Spec K with coefficients in the sheaf associated to E. The flat cohomology definition is essential for proving finer results on p-torsion in $Ш(E/K)$.

Exercise 11.1. Show that the group $H^1(K, E)$ (and therefore $Ш(E/K)$) is torsion. Hint: Show that given a class $c \in H^1(K, E)$, there is a finite Galois extension L/K such that c vanishes in $H^1(L, E)$.

The refined BSD conjecture relates the leading coefficient of $L(E, s)$ at $s = 1$ to invariants of E including heights, Tamagawa numbers, and the order of $Ш(E/K)$. In particular, the conjecture that $Ш(E/K)$ is finite is included in the refined BSD conjecture. We will not discuss that conjecture in these lectures, so we refer to [Gro10] and [Ulm11] for more details.

12. Statements of the main results

Much is known about the BSD conjecture over function fields. We start with general results.

Theorem 12.1. *Let E be an elliptic curve over a function field K. Then we have:*

(1) $\operatorname{Rank} E(K) \leq \operatorname{ord}_{s=1} L(E, s)$.
(2) *The following are equivalent:*
   - $\operatorname{Rank} E(K) = \operatorname{ord}_{s=1} L(E, s)$
   - $Ш(E/K)$ *is finite*
   - *for any one prime number ℓ ($\ell = p$ is allowed), the ℓ-primary part $Ш(E/K)_{\ell^\infty}$ is finite.*
(3) *If K'/K is a finite extension and if the BSD conjecture holds for E over K', then it holds for E over K.*

The theorem was proven by Tate [Tat66b] and Milne [Mil75] and we will sketch a proof in Lecture 3. When the equivalent conditions of Item 2 hold, it turns out that the refined BSD conjecture automatically follows. (This is also due to Tate and Milne and will be discussed in detail in [Ulm11].)

We now state several special cases where the conjecture is known to be true. As will be seen in the sequel, they all ultimately reduce either to Tate's theorem on isogenies of abelian varieties over finite fields (Theorem 6.1 of Lecture 0) or to a theorem of Artin and Swinnerton-Dyer on $K3$ surfaces [ASD73].

Theorem 12.2. *If E is an isotrivial elliptic curve over a function field K, then the BSD conjecture holds for E.*

Recall that a constant curve is also isotrivial.

To state the next result, we make an *ad hoc* definition. If E is an elliptic curve over $K = \mathbb{F}_q(t)$ we define the *height* h of E to be the smallest non-negative integer such that E can be defined by a Weierstrass equation (1.1.1) where the a_i are all polynomials and $\deg(a_i) \leq hi$. For example, the curves E_1 and E_2 in Subsection 1.2 have height $h = 0$ and the other curves $E_3, \ldots E_9$ there all have height $h = 1$. See Section 4 of Lecture 3 below for a more general definition.

Theorem 12.3. *Suppose that $K = k(t)$ and that E is an elliptic curve over K of height $h \leq 2$. Then the BSD conjecture holds for E.*

Note that this case overlaps the preceding one since an elliptic curve over $k(t)$ is constant if and only if its height is zero (cf. Proposition 4.1 in Lecture 3).

The following case is essentially due to Shioda [Shi86]. To state it, consider a polynomial f in three variables with coefficients in k which is the sum of exactly 4 non-zero monomials, say

$$f=\sum_{i=1}^{4}c_i\prod_{j=1}^{3}x_j^{e_{ij}}$$

where the $c_i \in k$ are non-zero. Set $e_{i4}=1-\sum_{j=1}^{3}e_{ij}$ and let A be the 4×4 integer matrix $A=(e_{ij})$. If $\det A\not\equiv 0 \pmod p$, we say that f *satisfies Shioda's condition.* Note that the condition is independent of the order of the variables x_j.

Theorem 12.4. *Suppose that $K=k(t)$ and that E is an elliptic curve over K. Suppose that E is birational to a plane curve $V(f)\subset \mathbb{A}^2_K$ where f is a polynomial in $k[t,x,y]\subset K[x,y]$ which is the sum of exactly 4 non-zero monomials and which satisfies Shioda's condition. Then the BSD conjecture holds for E.*

For example, the theorem applies to the curves E_4, E_7, E_8, and E_9 of Subsection 1.2 over $K=\mathbb{F}_q(t)$ for any prime power q. It applies more generally to these curves when t is replaced by t^d for any d prime to p. Note that when d is large, the height of the curve is also large, and so we get cases of BSD not covered by Theorem 12.3.

Finally we state another more recent and ultimately much more flexible special case due to Lisa Berger [Ber08].

Theorem 12.5. *Suppose that $K=k(t)$ and that E is an elliptic curve over K. Suppose that E is birational to a plane curve of the form*

$$f(x)=t^d g(y)$$

where f and g are rational functions of one variable and d is prime to p. Then the BSD conjecture holds for E.

Here one should clear denominators to interpret the equation $f=t^dg$ (or work in a Zariski open subset of the plane). For example, if $f(x)=x(x-1)$ and $g(y)=y^2/(1-y)$ then we have the plane curve over $K=k(t)$ defined by

$$x(x-1)(1-y)=t^dy^2$$

which turns out to be birational to

$$y^2+xy+t^dy=x^3+t^dx^2.$$

13. The rest of the course

The remainder of these lectures will be devoted to sketching the proofs of most of the main results and applying them to construct elliptic curves of large rank over function fields.

More precisely, in Lecture 2 we will review facts about surfaces and the Tate conjecture on divisors. This is a close relative of the BSD conjecture.

In Lecture 3 we will explain the relationship between the BSD and Tate conjectures and use it to prove the part of Theorem 12.1 related to $\ell\neq p$ as well as most of the other theorems stated in the previous section.

In Lecture 4 we will recall a general result on vanishing of L-functions in towers and combine it with the results above to obtain many elliptic curves of arbitrarily large rank.

In Lecture 5, we will give other applications of these ideas to ranks of elliptic curves and explicit points.

LECTURE 2

Surfaces and the Tate conjecture

1. Motivation

Consider an elliptic curve E/K and suppose that $K = k(t)$ and that we choose an equation for E as in Lecture 1, equation (1.1.2) where the a_i are in $k[t]$. Then (1.1.2), viewed in $K[x, y]$, defines an affine open subset of an elliptic curve E. But if we view it as an equation in $k[t, x, y]$, it defines an affine surface with a projection to the affine t line. The generic fiber of this projection is the affine curve just mentioned.

With a little more work (discussed in the next lecture), for any E over $K = k(\mathcal{C})$ we can define a smooth projective surface $\mathcal{E}$ over k with a morphism $\pi : \mathcal{E} \to \mathcal{C}$ whose generic fiber is E. Obviously there will be close connection between the arithmetic of $\mathcal{E}$ and that of E. Although $\mathcal{E}$ has higher dimension than E, it is defined over the finite field k and as a result we have better control over its arithmetic. Pursuing this line of inquiry leads to the main theorems stated at the end of the previous section.

In this lecture, we discuss the relevant facts and conjectures about surfaces over finite fields. In the next lecture we will look carefully at the connections between $\mathcal{E}$ and E and deduce the main classical theorems.

There are many excellent references for the general theory of surfaces, including [Bea96], [BHPV04], and [Băd01]. We generally refer to [Băd01] below since it works throughout over a field of arbitrary characteristic.

2. Surfaces

Let $k = \mathbb{F}_q$ be a finite field of characteristic p. As always, by a surface over k we mean a purely 2-dimensional, separated, reduced scheme of finite type over k. Such a scheme is automatically quasi-projective and is projective if and only if it is complete [Băd01, 1.28]. Since k is perfect, a surface $\mathcal{X}$ is a regular scheme if and only if $\mathcal{X} \to \operatorname{Spec} k$ is a smooth morphism (e.g., [Liu02, 4.3.3, Exer. 3.24]). We sloppily say that "$\mathcal{X}$ is smooth" if these conditions hold. Resolution of singularities is known for surfaces: For any surface $\mathcal{X}$, there is a proper birational morphism $\tilde{\mathcal{X}} \to \mathcal{X}$ with $\tilde{\mathcal{X}}$ smooth. (We may even take this morphism to be a composition of normalizations and blow ups at closed points [Lip78]. See also [Art86] for a nice exposition.) Therefore, every surface is birational to a smooth projective surface. In the cases of interest to us, this can be made very explicit in an elementary manner.

Throughout we assume that $\mathcal{X}$ is a smooth, projective, absolutely irreducible surface over k and we assume that $\mathcal{X}(k)$ is non-empty, i.e., $\mathcal{X}$ has a k-rational point.

3. Divisors and the Néron-Severi group

We give a lightning review of divisors and equivalence relations on divisors. See, for example, [Har77, V.1] for more details.

3.1. Divisor classes

A (Weil) *divisor* is a finite formal $\mathbb{Z}$-linear combination of reduced, closed, codimension 1 subvarieties of $\mathcal{X}$:

$$D=\sum a_Z Z.$$

In other words, the set of divisors is the free abelian group on the reduced, closed, codimension 1 subvarieties on $\mathcal{X}$.

If Z is a reduced, closed subvariety of $\mathcal{X}$ of codimension 1, there is an associated valuation

$$\mathrm{ord}_Z : k(\mathcal{X})^\times \to \mathcal{Z}$$

that sends a rational function to its order of zero or pole along Z.

A rational function f on $\mathcal{X}$ has a divisor:

$$\mathrm{Div}(f)=\sum_Z \mathrm{ord}_Z(f)Z.$$

A divisor D is said to be *linearly equivalent to zero* if there is a rational function f such that $\mathrm{Div}(f) = D$. Two divisors D and D' are linearly equivalent if their difference $D - D'$ is linearly equivalent to zero.

The group of divisors modulo those linear equivalent to zero is the *divisor class group $DivCl(\mathcal{X})$*. It is a fundamental invariant of $\mathcal{X}$.

3.2. The Picard group

Let $\mathrm{Pic}(\mathcal{X})$ be the Picard group of $\mathcal{X}$, i.e., the group of isomorphism classes of invertible sheaves on $\mathcal{X}$ with group law given by the tensor product. There is a cohomological calculation of $\mathrm{Pic}(\mathcal{X})$:

$$\mathrm{Pic}(\mathcal{X}) \cong H^1(\mathcal{X},\mathcal{O}_{\mathcal{X}}^\times).$$

The map sending a divisor D to the invertible sheaf $\mathcal{O}_{\mathcal{X}}(D)$ induces an isomorphism $DivCl(\mathcal{X})\tilde{\to}\mathrm{Pic}(\mathcal{X})$.

3.3. The Néron-Severi group

As usual, we write $\overline{\mathcal{X}}$ for $\mathcal{X}\times_k\overline{k}$. We first introduce the notion of algebraic equivalence for divisors on $\overline{\mathcal{X}}$. Intuitively, two divisors D and D' are algebraically equivalent if they lie in a family parameterized by a connected variety (which we may take to be a smooth curve). More precisely, if T is a smooth curve over $\overline{k}$ and $\mathcal{D}\subset\mathcal{X}\times_{\overline{k}}T$ is a divisor that is flat over T, then we get a family of divisors on $\mathcal{X}$ parameterized by T: $t\in T$ corresponds to $\mathcal{X}\times\{t\}\cap\mathcal{D}$. Two divisors D_1 and D_2 on $\mathcal{X}$ are algebraically equivalent if they lie in such a family, i.e., if there is a curve T and a divisor $\mathcal{D}$ as above and two points t_1 and $t_2\in T(\overline{k})$ such that $D_i=\mathcal{X}\times_{\overline{k}}\{t_i\}\cap\mathcal{D}$. (*A priori*, to ensure transitivity of this relation we should use chains of equivalences (see [Har77, Exer. V.1.7]) but see [Ful84, 10.3.2] for an argument that shows the definition works as is.) Note that linear equivalence is algebraic equivalence where T is restricted to be $\mathbb{P}^1$ ([Har77, Exer. V.1.7]) and so algebraic equivalence is weaker than linear equivalence.

The group of divisors on $\overline{\mathcal{X}}$ modulo those algebraically equivalent to zero is the *Néron-Severi* group $\mathrm{NS}(\overline{\mathcal{X}})$. A classical (and difficult) theorem, the "theorem of the base," says that $\mathrm{NS}(\overline{\mathcal{X}})$ is finitely generated. See [LN59] and [SGA6, XIII.5.1] for proofs and Lecture 3 below for more discussion. See also [Con06] for a modern discussion of the results in [LN59].

Since linear equivalence is weaker than algebraic equivalence, $\mathrm{NS}(\overline{\mathcal{X}})$ is a quotient of $\mathrm{Pic}(\overline{\mathcal{X}})$.

We define $\mathrm{NS}(\mathcal{X})$ to be the image of $\mathrm{Div}(\mathcal{X})$ in $\mathrm{NS}(\overline{\mathcal{X}})$ or equivalently the image of $\mathrm{Pic}(\mathcal{X})$ in $\mathrm{NS}(\overline{\mathcal{X}})$. Thus $\mathrm{NS}(\mathcal{X})$ is again a finitely generated abelian group. As we will see, it is of arithmetical nature.

Exercise 3.3.1. Let $G_k = \mathrm{Gal}(\overline{k}/k)$. Show that $\mathrm{NS}(\mathcal{X})$ is the group of invariants $\mathrm{NS}(\overline{\mathcal{X}})^{G_k}$. You will need to use that k is a finite field.

4. The Picard scheme

We define $\mathrm{Pic}^0(\mathcal{X})$ as the kernel of the surjection $\mathrm{Pic}(\mathcal{X}) \to \mathrm{NS}(\mathcal{X})$. In order to understand this group better, we will introduce more structure on the Picard group. The main fact we need to know is that the group $\mathrm{Pic}^0(\mathcal{X} \times \overline{k})$ is the set of points on an abelian variety and is therefore a divisible group. (I.e., for every class $c \in \mathrm{Pic}^0(\mathcal{X} \times \overline{k})$ and every positive integer n, there is a class c' such that $n\,c' = c$.) Readers willing to accept this assertion can skip the rest of this section.

The Picard group $\mathrm{Pic}(\mathcal{X})$ is the set of k-points of a group scheme. More precisely, under our hypotheses on $\mathcal{X}$ there is a group scheme called the *Picard scheme* and denoted $\underline{\mathrm{Pic}}_{\mathcal{X}/k}$ which is locally of finite type over k and represents the relative Picard functor. This means that if $T \to S = \mathrm{Spec}\, k$ is a morphism of schemes and $\pi_T : \mathcal{X}_T := \mathcal{X} \times_{\mathrm{Spec}\, k} T \to T$ is the base change then

$$\underline{\mathrm{Pic}}_{\mathcal{X}/k}(T) = \frac{\mathrm{Pic}(\mathcal{X}_T)}{\pi_T^* \,\mathrm{Pic}(T)}.$$

Here the left hand side is the group of T-valued points of $\underline{\mathrm{Pic}}_{\mathcal{X}/k}$. See [Kle05] for a thorough and detailed overview of the Picard scheme, and in particular [Kle05, 9.4.8] for the proof that there is a scheme representing the relative Picard functor as above.

We write $\underline{\mathrm{Pic}}^0_{\mathcal{X}/k}$ for the connected component of $\underline{\mathrm{Pic}}_{\mathcal{X}/k}$ containing the identity. Under our hypotheses, $\underline{\mathrm{Pic}}^0_{\mathcal{X}/k}$ is a geometrically irreducible projective group scheme over k [Kle05, 9.5.3, 9.5.4]. It may be non-reduced. (See examples in [Igu55] and [Ser58] and a full analysis of this phenomenon in [Mum66].) We let $\mathrm{PicVar}_{\mathcal{X}/k} = \left(\underline{\mathrm{Pic}}^0_{\mathcal{X}/k}\right)_{red}$, the *Picard variety* of $\mathcal{X}$ over k, which is an abelian variety over k.

If k' is a field extension of k, we have

$$\mathrm{Pic}^0(\mathcal{X}_{k'}) = \underline{\mathrm{Pic}}^0_{\mathcal{X}/k}(k') = \mathrm{PicVar}_{\mathcal{X}/k}(k')$$

so that $\mathrm{Pic}^0(\mathcal{X}_{k'})$ is the set of points of an abelian variety.

By [Kle05, 9.5.10], $\underline{\mathrm{Pic}}^0_{\mathcal{X}/k}(k) = \mathrm{Pic}^0(\mathcal{X})$, in other words, the class of a divisor in $\underline{\mathrm{Pic}}(\mathcal{X})$ lies in $\underline{\mathrm{Pic}}^0(\mathcal{X})$ if and only if the divisor is algebraically equivalent to 0.

5. Intersection numbers and numerical equivalence

There is an intersection pairing on the Néron-Severi group:

$$\mathrm{NS}(\mathcal{X}) \times \mathrm{NS}(\mathcal{X}) \to \mathbb{Z}$$

which is bilinear and symmetric. If D and D' are divisors, we write $D.D'$ for their intersection pairing.

There are two approaches to defining the pairing. In the first approach, one shows that given two divisors, there are divisors in the same classes in $\mathrm{NS}(\mathcal{X})$ (or even the same classes in $\mathrm{Pic}(\mathcal{X})$) that meet transversally. Then the intersection number is literally the number of points of intersection. The work in this approach is to prove a moving lemma and then show that the resulting pairing is well defined. See [Har77, V.1] for the details.

In the second approach, one uses coherent cohomology. If $\mathcal{L}$ is an invertible sheaf on $\mathcal{X}$, let

$$\chi(\mathcal{L}) = \sum_{i=0}^{2}(-1)^i \dim_k H^i(\mathcal{X},\mathcal{L})$$

be the coherent Euler characteristic of $\mathcal{L}$. Then define

$$D.D' = \chi(\mathcal{O}_\mathcal{X}) - \chi(\mathcal{O}_\mathcal{X}(-D)) - \chi(\mathcal{O}_\mathcal{X}(-D')) + \chi(\mathcal{O}_\mathcal{X}(-D-D')).$$

One checks that if C is a smooth irreducible curve on $\mathcal{X}$, then $C.D = \deg \mathcal{O}_\mathcal{X}(D)|_C$ and that if C and C' are two distinct irreducible curves on $\mathcal{X}$ meeting transversally, then $C.C'$ is the sum of local intersection multiplicities. See [Bea96, I.1-7] for details. (Nowhere is it used in this part of [Bea96] that the ground field is $\mathbb{C}$.)

Two divisors D and D' are said to be *numerically equivalent* if $D.D'' = D'.D''$ for all divisors D''. If $\mathrm{Num}(\mathcal{X})$ denotes the group of divisors in $\mathcal{X}$ up to numerical equivalence, then we have surjections

$$\mathrm{Pic}(\mathcal{X}) \twoheadrightarrow \mathrm{NS}(\mathcal{X}) \twoheadrightarrow \mathrm{Num}(\mathcal{X})$$

and so $\mathrm{Num}(\mathcal{X})$ is a finitely generated group. It is clear from the definition that $\mathrm{Num}(\mathcal{X})$ is torsion-free and so we can insert $\mathrm{NS}(\mathcal{X})/tor$ (Néron-Severi modulo torsion) into this chain:

$$\mathrm{Pic}(\mathcal{X}) \twoheadrightarrow \mathrm{NS}(\mathcal{X}) \twoheadrightarrow \mathrm{NS}(\mathcal{X})/tor \twoheadrightarrow \mathrm{Num}(\mathcal{X}).$$

6. Cycle classes and homological equivalence

There is a general theory of cycle classes in ℓ-adic cohomology, see for example [SGA4$\frac{1}{2}$, [Cycle]]. In the case of divisors, things are much simpler and we can construct a cycle class map from the Kummer sequence.

Indeed, consider the short exact sequence of sheaves on $\overline{\mathcal{X}}$ for the étale topology:

$$0 \to \mu_{\ell^n} \to \mathbb{G}_m \xrightarrow{\ell^n} \mathbb{G}_m \to 0.$$

(The sheaves μ_{ℓ^n} and $\mathbb{G}_m$ are perfectly reasonable sheaves in the Zariski topology on $\mathcal{X}$, but the arrow in the right is not surjective in that context. We need to use the étale topology or a finer one.) Taking cohomology, we get a homomorphism

$$Pic(\overline{\mathcal{X}})/\ell^n = H^1(\overline{\mathcal{X}},\mathbb{G}_m)/\ell^n \to H^2(\overline{\mathcal{X}},\mu_{\ell^n}).$$

Since $\mathrm{Pic}^0(\overline{\mathcal{X}})$ is a divisible group, we have $\mathrm{NS}(\overline{\mathcal{X}})/\ell^n = \mathrm{Pic}(\overline{\mathcal{X}})/\ell^n$ and so taking an inverse limit gives an injection

$$\mathrm{NS}(\overline{\mathcal{X}}) \otimes \mathbb{Z}_\ell \to H^2(\overline{\mathcal{X}},\mathbb{Z}_\ell(1)).$$

Composing with the natural homomorphism $\mathrm{NS}(\mathcal{X}) \to \mathrm{NS}(\overline{\mathcal{X}})$ gives our cycle class map

$$\mathrm{NS}(\mathcal{X}) \to \mathrm{NS}(\mathcal{X}) \otimes \mathbb{Z}_\ell \to H^2(\overline{\mathcal{X}},\mathbb{Z}_\ell(1)). \tag{6.1}$$

We declare two divisors to be (ℓ-)*homologically equivalent* if their classes in $H^2(\overline{\mathcal{X}}, \mathbb{Z}_\ell(1))$ are equal. (We will see below that this notion is independent of ℓ.) The group of divisors modulo homological equivalence will (temporarily) be denoted $\mathrm{Homol}(\mathcal{X})$. It will turn out to be a finitely generated free abelian group.

The intersection pairing on $NS(\mathcal{X})$ corresponds under the cycle class map to the cup product on cohomology. This means that a divisor that is homologically equivalent to zero is also numerically equivalent to zero. Thus we have a chain of surjections:

$$\mathrm{Pic}(\mathcal{X}) \twoheadrightarrow \mathrm{NS}(\mathcal{X}) \twoheadrightarrow \mathrm{NS}(\mathcal{X})/tor \twoheadrightarrow \mathrm{Homol}(\mathcal{X}) \twoheadrightarrow \mathrm{Num}(\mathcal{X}).$$

7. Comparison of equivalence relations on divisors

A theorem of Matsusaka [Mat57] asserts that the surjection

$$\mathrm{NS}(\mathcal{X})/tor \twoheadrightarrow \mathrm{Num}(\mathcal{X})$$

is in fact an isomorphism. Thus

$$\mathrm{NS}(\mathcal{X})/tor \cong \mathrm{Homol}(\mathcal{X}) \cong \mathrm{Num}(\mathcal{X})$$

and these groups are finitely generated, free abelian groups. Since $\mathrm{NS}(\mathcal{X})$ is finitely generated, $\mathrm{NS}(\mathcal{X})_{tor}$ is finite.

In all of the examples we will consider, $\mathrm{NS}(\mathcal{X})$ is torsion free. (In fact, for an elliptic surface with a section, the surjection $\mathrm{NS}(\mathcal{X}) \to \mathrm{Num}(\mathcal{X})$ is always an isomorphism, see [SS10, Theorem 6.5].) So to understand $\mathrm{Pic}(\mathcal{X})$ we have only to consider the finitely generated free abelian group $\mathrm{NS}(\mathcal{X})$ and the group $\mathrm{Pic}^0(\mathcal{X})$, which is (the set of points of) an abelian variety.

Exercise 7.1. In the case of a surface $\mathcal{X}$ over the complex numbers, use the cohomology of the exponential sequence

$$0 \to \mathbb{Z} \to \mathcal{O}_{\mathcal{X}} \overset{\exp}{\longrightarrow} \mathcal{O}_{\mathcal{X}}^{\times} \to 0$$

to analyze the structure of $\mathrm{Pic}(\mathcal{X})$.

8. Examples

8.1. $\mathbb{P}^2$

It is well known (e.g., [Har77, II.6.4]) that two curves on $\mathbb{P}^2$ are linearly equivalent if and only if they have the same degree. It follows that $\mathrm{Pic}(\mathbb{P}^2) = \mathrm{NS}(\mathbb{P}^2) \cong \mathbb{Z}$.

8.2. $\mathbb{P}^1 \times \mathbb{P}^1$

By [Har77, II.6.6.1], two curves on $\mathbb{P}^1 \times \mathbb{P}^1$ are linearly equivalent if and only if they have the same bi-degree. It follows that $\mathrm{Pic}(\mathbb{P}^1 \times \mathbb{P}^1) = \mathrm{NS}(\mathbb{P}^1 \times \mathbb{P}^1) \cong \mathbb{Z}^2$.

8.3. Abelian varieties

If $\mathcal{X}$ is an abelian variety (of any dimension g), then $\mathrm{Pic}^0(\mathcal{X})$ is the dual abelian variety and $\mathrm{NS}(\mathcal{X})$ is a finitely generated free abelian group of rank between 1 and $4g^2$. See [Mum08] for details.

8.4. Products of curves

Suppose that $\mathcal{C}$ and $\mathcal{D}$ are smooth projective curves over k with k-rational points $x \in \mathcal{C}$ and $y \in \mathcal{D}$. By definition (see Subsection 5.1 of Lecture 0), the group of divisorial correspondences between $(\mathcal{C}, x)$ and $(\mathcal{D}, y)$ is a subgroup of $\operatorname{Pic}(\mathcal{C} \times \mathcal{D})$ and it is clear that

$$\begin{aligned}\operatorname{Pic}(\mathcal{C} \times \mathcal{D}) &\cong \operatorname{Pic}(\mathcal{C}) \times \operatorname{Pic}(\mathcal{D}) \times \operatorname{DivCorr}\left((\mathcal{C}, x), (\mathcal{D}, y)\right) \\ &\cong \operatorname{Pic}^0(\mathcal{C}) \times \operatorname{Pic}^0(\mathcal{D}) \times \mathbb{Z}^2 \times \operatorname{DivCorr}\left((\mathcal{C}, x), (\mathcal{D}, y)\right).\end{aligned}$$

Moreover, as we saw in Lecture 0,

$$\operatorname{DivCorr}\left((\mathcal{C}, x), (\mathcal{D}, y)\right) \cong \operatorname{Hom}(J_{\mathcal{C}}, J_{\mathcal{D}})$$

is a discrete group. It follows that

$$\operatorname{Pic}^0(\mathcal{C} \times \mathcal{D}) \cong \operatorname{Pic}^0(\mathcal{C}) \times \operatorname{Pic}^0(\mathcal{D}) \tag{8.4.1}$$

and

$$\operatorname{NS}(\mathcal{C} \times \mathcal{D}) \cong \mathbb{Z}^2 \times \operatorname{Hom}(J_{\mathcal{C}}, J_{\mathcal{D}}). \tag{8.4.2}$$

This last isomorphism will be important for a new approach to elliptic curves of high rank over function fields discussed in Lecture 5.

8.5. Blow ups

Let $\mathcal{X}$ be a smooth projective surface over k and let $\pi : \mathcal{Y} \to \mathcal{X}$ be the blow up of $\mathcal{X}$ at a closed point $x \in \mathcal{X}$ so that $E = \pi^{-1}(x)$ is a rational curve on $\mathcal{Y}$. Then we have canonical isomorphisms

$$\operatorname{Pic}(\mathcal{Y}) \cong \operatorname{Pic}(\mathcal{X}) \oplus \mathbb{Z} \quad \text{and} \quad \operatorname{NS}(\mathcal{Y}) \cong \operatorname{NS}(\mathcal{X}) \oplus \mathbb{Z}$$

where in both groups the factor $\mathbb{Z}$ is generated by the class of E. See [Har77, V.3.2].

8.6. Fibrations

Let $\mathcal{X}$ be a smooth projective surface over k, $\mathcal{C}$ a smooth projective curve over k, and $\pi : \mathcal{X} \to \mathcal{C}$ a non-constant morphism. Assume that the induced extension of function fields $k(\mathcal{C}) \hookrightarrow k(\mathcal{X})$ is separable and $k(\mathcal{C})$ is algebraically closed in $k(\mathcal{X})$. Then for every closed point $y \in \mathcal{C}$, the fiber $\pi^{-1}(y)$ is connected, and it is irreducible for almost all y. Write F for the class in $\operatorname{NS}(\mathcal{X})$ of the fiber over a k-rational point y of $\mathcal{C}$. (This exists because we assumed that $\mathcal{X}$ has a k-rational point.) We write $\langle F \rangle$ for the subgroup of $\operatorname{NS}(\mathcal{X})$ generated by F.

It is clear from the definition of $\operatorname{NS}(\mathcal{X})$ that if y' is another closed point of $\mathcal{C}$, then the class in $\operatorname{NS}(\mathcal{X})$ of $\pi^{-1}(y')$ is equal to $(\deg y')F$.

Now suppose that $z \in \mathcal{C}$ is a closed point such that $\pi^{-1}(z)$ is reducible, say

$$\pi^{-1}(z) = \sum_{i=1}^{f_z} n_i Z_i$$

where the Z_i are the irreducible components of $\pi^{-1}(z)$ and the n_i are their multiplicities in the fiber. Then a consideration of intersection multiplicities (see for example [Sil94, III.8]) shows that for any integers m_i,

$$\sum_i m_i Z_i \in \langle F \rangle \subset \operatorname{NS}(\mathcal{X})$$

if and only if there is a rational number α such that $m_i = \alpha n_i$ for all i. More precisely, the intersection pairing restricted to the part of $\operatorname{NS}(\mathcal{X})$ generated by the classes of the Z_i is negative semi-definite, with a one-dimensional kernel spanned by integral divisors that are rational multiples of the whole fiber. It follows that the subgroup of $\mathrm{NS}(\mathcal{X})/\langle F\rangle$ generated by the classes of the Z_i has rank $f_z - 1$. It is free of this rank if the gcd of the multiplicities n_i is 1.

It also follows that if D is a divisor supported on a fiber of π and D' is another divisor supported on other fibers, then $D = D'$ in $\mathrm{NS}(\mathcal{X})/\langle F\rangle$ if and only if $D = D' = 0$ in $\mathrm{NS}(\mathcal{X})/\langle F\rangle$.

Define $L^2\,\mathrm{NS}(\mathcal{X})$ to be the subgroup of $\mathrm{NS}(\mathcal{X})$ generated by all components of all fibers of π over closed points of $\mathcal{C}$. By the above, it is the direct sum of the $\langle F\rangle$ and the subgroups of $\mathrm{NS}(\mathcal{X})/\langle F\rangle$ generated by the components of the various fibers. Thus we obtain the following computation of the rank of $L^2\,\mathrm{NS}(\mathcal{X})$.

Proposition 8.6.1. *For a closed point y of $\mathcal{C}$, let f_y denote the number of irreducible components in the fiber $\pi^{-1}(y)$. Then the rank of $L^2\,\mathrm{NS}(\mathcal{X})$ is*

$$1 + \sum_y (f_y - 1).$$

If for all y the greatest common divisor of the multiplicities of the components in the fiber of π over y is 1, then $L^2\,\mathrm{NS}(\mathcal{X})$ is torsion-free.

9. Tate's conjectures T_1 and T_2

Tate's conjecture T_1 for $\mathcal{X}$ (which we denote $T_1(\mathcal{X})$) characterizes the image of the cycle class map:

Conjecture 9.1 ($T_1(\mathcal{X})$). *For any prime $\ell \neq p$, the cycle class map induces an isomorphism*

$$\mathrm{NS}(\mathcal{X}) \otimes \mathbb{Q}_\ell \to H^2(\overline{\mathcal{X}}, \mathbb{Q}_\ell(1))^{G_k}$$

We will see below that $T_1(\mathcal{X})$ is equivalent to the apparently stronger integral statement that the cycle class induces an isomorphism

$$\mathrm{NS}(\mathcal{X}) \otimes \mathbb{Z}_\ell \to H^2(\overline{\mathcal{X}}, \mathbb{Z}_\ell(1))^{G_k}$$

We will also see that $T_1(\mathcal{X})$ is independent of ℓ which is why we have omitted ℓ from the notation.

Since G_k is generated topologically by Fr_q, we have

$$H^2(\overline{\mathcal{X}}, \mathbb{Q}_\ell(1))^{G_k} = H^2(\overline{\mathcal{X}}, \mathbb{Q}_\ell(1))^{Fr_q=1} = H^2(\overline{\mathcal{X}}, \mathbb{Q}_\ell)^{Fr_q=q}.$$

The injectivity of the cycle class map implies that

$$\mathrm{Rank}\,\mathrm{NS}(\mathcal{X}) \leq \dim_{\mathbb{Q}_\ell} H^2(\overline{\mathcal{X}}, \mathbb{Q}_\ell)^{Fr_q=q}$$

and $T_1(\mathcal{X})$ is the statement that these two dimensions are equal.

The second Tate conjecture relates the zeta-function to divisors. Recall that $\zeta(\mathcal{X}, s)$ denotes the zeta function of $\mathcal{X}$, defined in Lecture 0, Section 3.

Conjecture 9.2 ($T_2(\mathcal{X})$). *We have*

$$\mathrm{Rank}\,\mathrm{NS}(\mathcal{X}) = -\operatorname{ord}_{s=1} \zeta(\mathcal{X}, s)$$

Note that by the Riemann hypothesis, the poles of $\zeta(\mathcal{X}, s)$ at $s = 1$ come from $P_2(\mathcal{X}, q^{-s})$. More precisely, using the cohomological formula (4.1) of Lecture 0 for P_2, we have that the order of pole of $\zeta(\mathcal{X}, s)$ at $s = 1$ is equal to the multiplicity of q as an eigenvalue of Fr_q on $H^2(\overline{\mathcal{X}}, \mathbb{Q}_\ell)$.

Thus we have a string of inequalities

$$\text{Rank}\,\text{NS}(\mathcal{X}) \leq \dim_{\mathbb{Q}_\ell} H^2(\overline{\mathcal{X}}, \mathbb{Q}_\ell)^{Fr_q=q} \leq -\operatorname{ord}_{s=1}\zeta(\mathcal{X}, s). \tag{9.3}$$

Conjecture $T_1(\mathcal{X})$ is that the first inequality is an equality and conjecture $T_2(\mathcal{X})$ is that the leftmost and rightmost integers are equal. It follows trivially that $T_2(\mathcal{X})$ implies $T_1(\mathcal{X})$. Tate proved the reverse implication.

Proposition 9.4. *The conjectures $T_1(\mathcal{X})$ and $T_2(\mathcal{X})$ are equivalent. In particular, $T_1(\mathcal{X})$ is independent of ℓ.*

PROOF. First note that the intersection pairing on $\text{NS}(\mathcal{X})$ is non-degenerate, so we get an isomorphism

$$\text{NS}(\mathcal{X}) \otimes \mathbb{Q}_\ell \cong \text{Hom}(\text{NS}(\mathcal{X}), \mathbb{Q}_\ell).$$

On the other hand, the cup product on $H^2(\overline{\mathcal{X}}, \mathbb{Q}_\ell(1))$ is also non-degenerate (by Poincaré duality), so we have

$$H^2(\overline{\mathcal{X}}, \mathbb{Q}_\ell(1)) \cong \text{Hom}(H^2(\overline{\mathcal{X}}, \mathbb{Q}_\ell(1)), \mathbb{Q}_\ell).$$

If we use a superscript G_k to denote invariants and a subscript G_k to denote coinvariants, then we have a natural homomorphism

$$H^2(\overline{\mathcal{X}}, \mathbb{Q}_\ell(1))^{G_k} \to H^2(\overline{\mathcal{X}}, \mathbb{Q}_\ell(1))_{G_k}$$

which is an isomorphism if and only if the subspace of $H^2(\overline{\mathcal{X}}, \mathbb{Q}_\ell(1))$ where Fr_q acts by 1 is equal to the whole of the generalized eigenspace for the eigenvalue 1. As we have seen above, this holds if and only if we have

$$\dim_{\mathbb{Q}_\ell} H^2(\overline{\mathcal{X}}, \mathbb{Q}_\ell)^{Fr_q=q} = -\operatorname{ord}_{s=1}\zeta(\mathcal{X}, s).$$

Now consider the diagram

$$\begin{array}{ccccc}
\text{NS}(\mathcal{X}) \otimes \mathbb{Q}_\ell & & = & & \text{Hom}(\text{NS}(\mathcal{X}), \mathbb{Q}_\ell) \\
\downarrow h & & & & \uparrow h^* \\
H^2(\overline{\mathcal{X}}, \mathbb{Q}_\ell(1))^{G_k} & \xrightarrow{f} & H^2(\overline{\mathcal{X}}, \mathbb{Q}_\ell(1))_{G_k} & = & \text{Hom}(H^2(\overline{\mathcal{X}}, \mathbb{Q}_\ell(1))^{G_k}, \mathbb{Q}_\ell).
\end{array}$$

The lower right arrow is an isomorphism by elementary linear algebra. The maps h and h^* are the cycle map and its transpose and they are isomorphisms if and only if $T_1(\mathcal{X})$ holds. One checks that the diagram commutes ([Tat66b, p. 24] or [Mil75, Lemma 5.3]) and so $T_1(\mathcal{X})$ implies that f is an isomorphism. Thus $T_1(\mathcal{X})$ implies $T_2(\mathcal{X})$. □

We remark that the equality of $\dim_{\mathbb{Q}_\ell} H^2(\overline{\mathcal{X}}, \mathbb{Q}_\ell)^{Fr_q=q}$ and $-\operatorname{ord}_{s=1}\zeta(\mathcal{X}, s)$ would follow from the semi-simplicity of Fr_q acting on $H^2(\overline{\mathcal{X}}, \mathbb{Q}_\ell)$ (or even from its semisimplicity on the $Fr_q = q$ generalized eigenspace). This is a separate "standard" conjecture (see for example [Tat94]); it does not seem to imply $T_1(\mathcal{X})$.

10. T_1 and the Brauer group

We define the (cohomological) Brauer group $\text{Br}(\mathcal{X})$ by

$$\text{Br}(\mathcal{X}) = H^2(\mathcal{X}, \mathbb{G}_m) = H^2(\mathcal{X}, \mathcal{O}_{\mathcal{X}}^\times)$$

(with respect to the étale or finer topologies). Because $\mathcal{X}$ is a smooth proper surface over a finite field, the cohomological Brauer group is isomorphic to the usual Brauer group (defined in terms of Azumaya algebras) and it is known to be a torsion group.

(See [Mil80, IV.2] and also three fascinating articles by Grothendieck collected in [Gro68].) Artin and Tate conjectured in [Tat66b] that $\operatorname{Br}(\mathcal{X})$ is finite.

Similarly, define

$$\operatorname{Br}(\overline{\mathcal{X}}) = H^2(\overline{\mathcal{X}}, \mathbb{G}_m) = H^2(\overline{\mathcal{X}}, \mathcal{O}_{\mathcal{X}}^{\times}).$$

This group is torsion but need not be finite.

Taking the cohomology of the exact sequence

$$0 \to \mu_{\ell^n} \to \mathbb{G}_m \xrightarrow{\ell^n} \mathbb{G}_m \to 0$$

as in Section 6, we have an exact sequence

$$0 \to \operatorname{NS}(\overline{\mathcal{X}})/\ell^n \to H^2(\overline{\mathcal{X}}, \mu_{\ell^n}) \to \operatorname{Br}(\overline{\mathcal{X}})_{\ell^n} \to 0. \tag{10.1}$$

Taking G_k-invariants and then the inverse limit over powers of ℓ, we obtain an exact sequence

$$0 \to \operatorname{NS}(\mathcal{X}) \otimes \mathbb{Z}_\ell \to H^2(\overline{\mathcal{X}}, \mathbb{Z}_\ell(1))^{G_k} \to T_\ell \operatorname{Br}(\mathcal{X}) \to 0.$$

Since $\operatorname{Br}(\mathcal{X})_\ell$ is finite, $T_\ell \operatorname{Br}(\mathcal{X})$ is zero if and only if the ℓ-primary part of $\operatorname{Br}(\mathcal{X})$ is finite. It follows that the ℓ part of the Brauer group is finite if and only if $T_1(\mathcal{X})$ for ℓ holds if and only if the integral version of $T_1(\mathcal{X})$ for ℓ holds. In particular, since $T_1(\mathcal{X})$ is independent of ℓ, if $\operatorname{Br}(\mathcal{X})[\ell^\infty]$ is finite for one ℓ, then $\operatorname{Br}(\mathcal{X})[\ell^\infty]$ is finite for all $\ell \neq p$. It is even true, although more difficult to prove, that $T_1(\mathcal{X})$ is equivalent to the finiteness of $\operatorname{Br}(\mathcal{X})$.

Theorem 10.2. *$T_1(\mathcal{X})$ holds if and only if $\operatorname{Br}(\mathcal{X})$ is finite if and only if there is an ℓ ($\ell = p$ allowed) such that the ℓ-primary part of $\operatorname{Br}(\mathcal{X})$ is finite.*

Proof. We sketch the proof of the prime-to-p part of this assertion following [Tat66b] and refer to [Mil75] for the full proof. We already noted that the ℓ-primary part of $\operatorname{Br}(\mathcal{X})$ is finite for one $\ell \neq p$ if and only if $T_1(\mathcal{X})$ holds. To see that almost all ℓ-primary parts vanish, we consider the following diagram, which is an integral version of the diagram in the proof of Proposition 9.4:

$$\begin{array}{ccccc}
\operatorname{NS}(\mathcal{X}) \otimes \mathbb{Z}_\ell & \xrightarrow{e} & \operatorname{Hom}(\operatorname{NS}(\mathcal{X}) \otimes \mathbb{Z}_\ell, \mathbb{Z}_\ell) & = & \operatorname{Hom}(\operatorname{NS}(\mathcal{X}) \otimes \mathbb{Q}_\ell/\mathbb{Z}_\ell, \mathbb{Q}_\ell/\mathbb{Z}_\ell) \\
\downarrow{\scriptstyle h} & & & & \uparrow{\scriptstyle g^*} \\
H^2(\overline{\mathcal{X}}, \mathbb{Z}_\ell(1))^{G_k} & \xrightarrow{f} & H^2(\overline{\mathcal{X}}, \mathbb{Z}_\ell(1))_{G_k} & = & \operatorname{Hom}(H^2(\overline{\mathcal{X}}, (\mathbb{Q}_\ell/\mathbb{Z}_\ell)(1))^{G_k}, \mathbb{Q}_\ell/\mathbb{Z}_\ell)
\end{array}$$

Here e is induced by the intersection form, h is the cycle class map, f is induced by the identity map of $H^1(\overline{\mathcal{X}}, \mathbb{Z}_\ell(1))$ and g^* is the transpose of a map

$$g : \operatorname{NS}(\mathcal{X}) \otimes \mathbb{Q}_\ell/\mathbb{Z}_\ell \to H^2(\overline{\mathcal{X}}, (\mathbb{Q}_\ell/\mathbb{Z}_\ell)(1))$$

obtained by taking the direct limit over powers of ℓ of the first map in equation (10.1).

We say that a homomorphism $\phi : A \to B$ of $\mathbb{Z}_\ell$-modules is a *quasi-isomorphism* if it has a finite kernel and cokernel. In this case, we define

$$z(\phi) = \frac{\#\ker(\phi)}{\#\operatorname{coker}(\phi)}.$$

It is easy to check that if $\phi_3 = \phi_2\phi_1$ (composition) and if two of the maps ϕ_1, ϕ_2, ϕ_3 are quasi-isomorphisms, then so is the third and we have $z(\phi_3) = z(\phi_2)z(\phi_1)$.

In the diagram above, if we assume $T_1(\mathcal{X})$, then h is an isomorphism. The map e is induced from the intersection pairing and is a quasi-isomorphism and

$z(e)$ is (the ℓ part of) the order of the torsion subgroup of $\mathrm{NS}(\mathcal{X})$ divided by (the ℓ part of) discriminant of the intersection form. We saw above that under the assumption of $T_1(\mathcal{X})$, the map f is a quasi-isomorphism and it turns out that $z(f)$ is essentially (the ℓ part of) the leading term of the zeta function $\zeta(\mathcal{X}, s)$ at $s = 1$. In particular, under $T_1(\mathcal{X})$, e, f, and h are isomorphisms for almost all ℓ. The same must therefore be true of g^*. By taking G_k-invariants and a direct limit over powers of ℓ in equation (10.1), one finds that $z(g^*)$ is equal to the order of $\mathrm{Br}(\mathcal{X})[\ell^\infty]$ and so this group is trivial for almost all ℓ. This completes our sketch of the proof of the theorem. □

The sketch above has all the main ideas needed to prove that the prime-to-p part of the Artin-Tate conjecture on the leading coefficient of the zeta function at $s = 1$ follows from the Tate conjecture $T_1(\mathcal{X})$. The p-part is formally similar although more delicate. To handle it, Milne replaces the group in the lower right of the diagram with the larger group $\mathrm{Hom}(H^2(\mathcal{X}, (\mathbb{Q}_p/\mathbb{Z}_p)(1)), \mathbb{Q}_p/\mathbb{Z}_p)$. The z invariants of the maps to and from this group turn out to have more p-adic content that is related to the term $q^\alpha(\mathcal{X})$ in the Artin-Tate leading coefficient conjecture. We refer to [Mil75] for the full details and to [Ulm11] for a discussion of several related points, including the case $p = 2$ (excluded in Milne's article, but now provable due to improved p-adic cohomology) and higher dimensional abelian varieties.

11. The descent property of T_1

If $\tilde{\mathcal{X}} \to \mathcal{X}$ is the blow up of $\mathcal{X}$ at a closed point, then $T_1(\tilde{\mathcal{X}})$ is equivalent to $T_1(\mathcal{X})$. Indeed, under blowing up both the rank of $\mathrm{NS}(\cdot)$ and the dimension of $H^2(\cdot, \mathbb{Q}_\ell(1))^{G_k}$ increase by one. (See Example 8.5 above.) In fact:

Proposition 11.1. *$T_1(\mathcal{X})$ is invariant under birational isomorphism. More generally, if $\mathcal{X} \to \mathcal{Y}$ is a dominant rational map, then $T_1(\mathcal{X})$ implies $T_1(\mathcal{Y})$.*

Proof. We give simple proof of the case where $\mathcal{X}$ and $\mathcal{Y}$ are surfaces. See [Tat94] for the general case.

First, we may assume $\mathcal{X} \dashrightarrow \mathcal{Y}$ is a morphism. Indeed, let $\tilde{\mathcal{X}} \to \mathcal{X}$ be a blow up resolving the indeterminacy of $\mathcal{X} \dashrightarrow \mathcal{Y}$, i.e., so that the composition $\tilde{\mathcal{X}} \to \mathcal{X} \dashrightarrow \mathcal{Y}$ is a morphism. As we have seen above $T_1(\mathcal{X})$ implies $T_1(\tilde{\mathcal{X}})$ so we may replace $\mathcal{X}$ with $\tilde{\mathcal{X}}$ and show that $T_1(\mathcal{Y})$ holds.

So now suppose that $\pi : \mathcal{X} \to \mathcal{Y}$ is a dominant morphism. Since the dimensions of $\mathcal{X}$ and $\mathcal{Y}$ are equal, π must be generically finite, say of degree d. But then the push forward and pull-back maps on cycles present $\mathrm{NS}(\mathcal{Y}) \otimes \mathbb{Q}_\ell$ as a direct factor of $NS(\mathcal{X}) \otimes \mathbb{Q}_\ell$; they also present $H^2(\overline{\mathcal{Y}}, \mathbb{Q}_\ell(1))$ as a direct factor of $H^2(\overline{\mathcal{X}}, \mathbb{Q}_\ell(1))$. The cycle class maps and Galois actions are compatible with these decompositions and since by assumption $NS(\mathcal{X}) \otimes \mathbb{Q}_\ell \tilde{\to} H^2(\overline{\mathcal{X}}, \mathbb{Q}_\ell(1))^{G_k}$, we must also have $NS(\mathcal{Y}) \otimes \mathbb{Q}_\ell \tilde{\to} H^2(\overline{\mathcal{Y}}, \mathbb{Q}_\ell(1))^{G_k}$, i.e., $T_1(\mathcal{Y})$. □

Note that the dominant rational map $\mathcal{X} \dashrightarrow \mathcal{Y}$ could be a ground field extension, or even a purely inseparable morphism.

12. Tate's theorem on products

In this section we sketch how T_1 for products of curves follows from Tate's theorem on endomorphisms of abelian varieties over finite fields.

Theorem 12.1 (Tate). *Let $\mathcal{C}$ and $\mathcal{D}$ be curves over k and set $\mathcal{X} = \mathcal{C} \times_k \mathcal{D}$. Then $T_1(\mathcal{X})$ holds.*

PROOF. Extending k if necessary, we may assume that $\mathcal{C}$ and $\mathcal{D}$ both have rational points. Fix rational base points x and y (which we will mostly omit from the notation below). Recall from Subsection 8.4 that

$$\mathrm{NS}(\mathcal{C} \times \mathcal{D}) \cong \mathbb{Z}^2 \times \mathrm{DivCorr}(\mathcal{C}, \mathcal{D}) \cong \mathbb{Z}^2 \times \mathrm{Hom}(J_{\mathcal{C}}, J_{\mathcal{D}}).$$

By the Künneth formula,

$$\begin{aligned} H^2(\overline{\mathcal{X}}, \mathbb{Q}_\ell) &\cong \left(H^2(\overline{\mathcal{C}}, \mathbb{Q}_\ell) \otimes H^0(\overline{\mathcal{D}}, \mathbb{Q}_\ell)\right) \oplus \left(H^0(\overline{\mathcal{C}}, \mathbb{Q}_\ell) \otimes H^2(\overline{\mathcal{D}}, \mathbb{Q}_\ell)\right) \\ &\quad \oplus \left(H^1(\overline{\mathcal{C}}, \mathbb{Q}_\ell) \otimes H^1(\overline{\mathcal{D}}, \mathbb{Q}_\ell)\right) \\ &\cong \mathbb{Q}_\ell(-1) \oplus \mathbb{Q}_\ell(-1) \oplus \left(H^1(\overline{\mathcal{C}}, \mathbb{Q}_\ell) \otimes H^1(\overline{\mathcal{D}}, \mathbb{Q}_\ell)\right) \end{aligned}$$

Twisting by $\mathbb{Q}_\ell(1)$ and taking invariants, we have

$$H^2(\overline{\mathcal{X}}, \mathbb{Q}_\ell(1))^{G_k} = {\mathbb{Q}_\ell}^2 \oplus \left(H^1(\overline{\mathcal{C}}, \mathbb{Q}_\ell) \otimes H^1(\overline{\mathcal{D}}, \mathbb{Q}_\ell)(1)\right)^{G_k}.$$

Under the cycle class map, the factor $\mathbb{Z}^2$ of $\mathrm{NS}(\mathcal{X})$ (corresponding to $\mathcal{C} \times \{y\}$ and $\{x\} \times \mathcal{D}$) spans the factor ${\mathbb{Q}_\ell}^2$ of $H^2(\overline{\mathcal{X}}, \mathbb{Q}_\ell(1))^{G_k}$ (corresponding to $H^2 \otimes H^0$ and $H^0 \otimes H^2$ in the Kunneth decomposition). Thus what we have to show is that the cycle class map induces an isomorphism

$$\mathrm{Hom}(J_{\mathcal{C}}, J_{\mathcal{D}}) \otimes \mathbb{Q}_\ell \tilde{\to} \left(H^1(\overline{\mathcal{C}}, \mathbb{Q}_\ell) \otimes H^1(\overline{\mathcal{D}}, \mathbb{Q}_\ell)(1)\right)^{G_k}$$

But $H^1(\overline{\mathcal{D}}, \mathbb{Q}_\ell)(1) \cong H^1(\overline{\mathcal{D}}, \mathbb{Q}_\ell)^* \cong V_\ell(J_{\mathcal{D}})$ and $H^1(\overline{\mathcal{C}}, \mathbb{Q}_\ell) \cong V_\ell(J_{\mathcal{C}})^*$ ($* = \mathbb{Q}_\ell$-linear dual). Thus

$$\left(H^1(\overline{\mathcal{C}}, \mathbb{Q}_\ell) \otimes H^1(\overline{\mathcal{D}}, \mathbb{Q}_\ell)(1)\right)^{G_k} \cong (V_\ell(J_{\mathcal{C}})^* \otimes V_\ell(J_{\mathcal{D}}))^{G_k} \cong \mathrm{Hom}_{G_k}(V_\ell(J_{\mathcal{C}}), V_\ell(J_{\mathcal{D}})).$$

Thus the needed isomorphism is

$$\mathrm{Hom}(J_{\mathcal{C}}, J_{\mathcal{D}}) \otimes \mathbb{Q}_\ell \tilde{\to} \mathrm{Hom}_{G_k}(V_\ell(J_{\mathcal{C}}), V_\ell(J_{\mathcal{D}}))$$

and this is exactly the statement of Tate's theorem (Lecture 0, Theorem 6.1). This completes the proof of the theorem. □

Remarks 12.2.

(1) A variation of the argument above, using Picard and Albanese varieties, shows that T_1 for a product $\mathcal{X} \times \mathcal{Y}$ of varieties of any dimension follows from T_1 for the factors.

(2) It is worth noting that Tate's conjecture T_1 (and the proof of it for products of curves) only characterizes the image of in ℓ-adic cohomology of $\mathrm{NS}(\mathcal{X}) \otimes \mathbb{Z}_\ell$, not the image of $\mathrm{NS}(\mathcal{X})$ itself. This should be contrasted with the Lefschetz (1,1) theorem, which characterizes the image of $\mathrm{NS}(\mathcal{X})$ in deRham cohomology when the ground field is $\mathbb{C}$.

13. Products of curves and DPC

Assembling the various parts of this lecture gives the main result:

Proposition 13.1. *Let $\mathcal{X}$ be a smooth, projective surface over k. If there is a dominant rational map*

$$\mathcal{C} \times_k \mathcal{D} \dashrightarrow \mathcal{X}$$

from a product of curves to $\mathcal{X}$, then the Tate conjectures $T_1(\mathcal{X})$ and $T_2(\mathcal{X})$ hold.

Indeed, by Theorem 12.1, we have $T_1(\mathcal{C}\times\mathcal{D})$ and then by Proposition 11.1 we deduce $T_1(\mathcal{X})$. By Proposition 9.4, $T_2(\mathcal{X})$ follows as well.

We say that "$\mathcal{X}$ is dominated by a product of curves (DPC)." The question of which varieties are dominated by products of curves has been studied by Schoen [Sch96]. In particular, over any field there are surfaces that are not dominated by products of curves. Nevertheless, as we will see below, the collection of DPC surfaces is sufficiently rich to give some striking results on the Birch and Swinnerton-Dyer conjecture.

LECTURE 3

Elliptic curves and elliptic surfaces

We keep our standard notations throughout this lecture: p is a prime, $k = \mathbb{F}_q$ is the finite field of characteristic p with q elements, $\mathcal{C}$ is a smooth, projective, absolutely irreducible curve over k, $K = k(\mathcal{C})$ is the function field of $\mathcal{C}$, and E is an elliptic curve over K.

1. Curves and surfaces

In this section we will construct an elliptic surface $\mathcal{E} \to \mathcal{C}$ canonically associated to an elliptic curve E/K. More precisely, we give a constructive proof of the following result:

Proposition 1.1. *Given an elliptic curve E/K, there exists a surface $\mathcal{E}$ over k and a morphism $\pi : \mathcal{E} \to \mathcal{C}$ with the following properties: $\mathcal{E}$ is smooth, absolutely irreducible, and projective over k, π is surjective and relatively minimal, and the generic fiber of π is isomorphic to E. The surface $\mathcal{E}$ and the morphism π are uniquely determined up to isomorphism by these requirements.*

Here "the generic fiber of π" means $\mathcal{E}_K$, the fiber product:

$$\begin{array}{ccc} \mathcal{E}_K := \eta \times_{\mathcal{C}} \mathcal{E} & \longrightarrow & \mathcal{E} \\ \big\downarrow & & \big\downarrow \pi \\ \eta = \operatorname{Spec} K & \longrightarrow & \mathcal{C} \end{array}$$

"Relatively minimal" means that if $\mathcal{E}'$ is another smooth, absolutely irreducible, projective surface over k with a surjective morphism $\pi' : \mathcal{E}' \to \mathcal{C}$, then any birational morphism $\mathcal{E} \to \mathcal{E}'$ commuting with π and π' is an isomorphism. Relative minimality is equivalent to the condition that there are no rational curves of self-intersection -1 in the fibers of π (i.e., to the non-existence of curves in fibers that can be blown down).

Remarks 1.2. The requirements on $\mathcal{E}$ and π imply that π is flat and projective and that all geometric fibers of π are connected. These properties of π will be evident from the explicit construction below. It follows that $\pi_*\mathcal{O}_{\mathcal{E}} \cong \mathcal{O}_{\mathcal{C}}$ and more generally that π is "cohomologically flat in dimension zero," meaning that for every morphism $T \to C$ the base change

$$\pi_T : \mathcal{E}_T = \mathcal{E} \times_{\mathcal{C}} T \to T$$

satisfies $\pi_{T*}\mathcal{O}_{\mathcal{E}_T} = \mathcal{O}_T$.

Uniqueness in Proposition 1.1 follows from general results on minimal models, in particular [Lic68, Thm. 4.4]. See [Chi86] and [Liu02, 9.3] for other expositions.

We first give a detailed construction of a (possibly singular) "Weierstrass surface" $\mathcal{W} \to \mathcal{C}$ and then resolve singularities to obtain $\mathcal{E} \to \mathcal{C}$.

More precisely, the proposition follows from the following two results.

Proposition 1.3. *Given an elliptic curve E/K, there exists a surface $\mathcal{W}$ over k and a morphism $\pi_0 : \mathcal{W} \to \mathcal{C}$ with the following properties: $\mathcal{W}$ is normal, absolutely irreducible, and projective over k, π_0 is surjective, each of its fibers is isomorphic to an irreducible plane cubic, and its generic fiber is isomorphic to E.*

This proposition is elementary, but does not seem to be explained in detail in the literature, so we give a proof below.

Proposition 1.4. *There is an explicit sequence of blow ups (along closed points and curves in $\mathcal{W}$) yielding a proper birational morphism $\sigma : \mathcal{E} \to \mathcal{W}$ where the surface $\mathcal{E}$ and the composed morphism $\pi = \pi_0 \circ \sigma : \mathcal{E} \to \mathcal{W} \to \mathcal{C}$ have the properties mentioned in Proposition 1.1.*

PROOF OF PROPOSITION 1.3. Choose a Weierstrass equation for E:

$$y^2 + a_1xy + a_3y = x^3 + a_2x^2 + a_4x + a_6 \tag{1.5}$$

where the a_i are in $K = k(\mathcal{C})$. Recall that we have defined the notion of a minimal integral model at a place v of K: the a_i should be integral at v and the valuation at v of Δ should be minimal subject to the integrality of the a_i. Clearly, there is a non-empty Zariski open subset $U \subset \mathcal{C}$ such that for every closed point $v \in U$, the model (1.5) is a minimal integral model.

Let $\mathcal{W}_1$ be the closed subset of $\mathbb{P}^2_U := \mathbb{P}^2_k \times_k U$ defined by the vanishing of

$$Y^2Z + a_1XYZ + a_3YZ^2 - \left(X^3 + a_2X^2Z + a_4XZ^2 + a_6Z^3\right) \tag{1.6}$$

where X, Y, Z are the standard homogeneous coordinates on $\mathbb{P}^2_k$. Then $\mathcal{W}_1$ is geometrically irreducible and there is an obvious projection $\pi_1 : \mathcal{W}_1 \to U$ (the restriction to $\mathcal{W}_1$ of the projection $\mathbb{P}^2_U \to U$). The fiber of π_1 over a closed point v of U is the plane cubic

$$Y^2Z + a_1(v)XYZ + a_3(v)YZ^2 = X^3 + a_2(v)X^2Z + a_4(v)XZ^2 + a_6(v)Z^3$$

over the residue field κ_v at v. The generic point η of $\mathcal{C}$ lies in U and the fiber of π_1 at η is E/K.

There are finitely many points in $\mathcal{C} \setminus U$ and we must extend the model $\mathcal{W}_1 \to U$ over each of these points. Choose one of them, call it w, and choose a model of E that is integral and minimal at w. In other words, choose a model of E

$$y'^2 + a_1'x'y' + a_3'y' = x'^3 + a_2'x'^2 + a_4'x' + a_6' \tag{1.7}$$

where the $a_i' \in K$ are integral at w and the valuation at w of the discriminant Δ is minimal. The new coordinates are related to the old by a transformation

$$(x, y) = (u^2x' + r, u^3y' + su^2x' + t) \tag{1.8}$$

with $u \in K^\times$ and $r, s, t \in K$. Let U' be a Zariski open subset of $\mathcal{C}$ containing w on which all of the a_i' are integral and the model (1.7) is minimal. Let $\mathcal{W}'$ be the geometrically irreducible closed subset of $\mathbb{P}^1_{U'}$ defined by the vanishing of

$$Y'^2Z' + a_1'X'Y'Z' + a_3'Y'Z'^2 - \left(X'^3 + a_2'X'^2Z' + a_4'X'Z'^2 + a_6'Z'3\right)$$

with its obvious projection $\pi' : \mathcal{W}' \to U'$. On the open set $V = U \cap U'$, u is a unit and the change of coordinates (1.8), or rather its projective version

$$(X, Y, Z) = (u^2X' + rZ, u^3Y' + su^2X' + tZ', Z')$$

defines an isomorphism between $\pi_1^{-1}(V)$ and $\pi'^{-1}(V)$ compatible with the projections. Glueing $\mathcal{W}_1$ and $\mathcal{W}'$ along this isomorphism yields a new surface $\mathcal{W}_2$ equipped with a projection $\pi_2 : \mathcal{W}_2 \to U_2$ where $U_2 = U \cup U'$. Note that U_2 is strictly larger than U. Moreover π_2 is surjective, its geometric fibers are irreducible projective plane cubics, and its generic fiber is E.

We now iterate this construction finitely many times to extend the original model over all of $\mathcal{C}$. We arrive at a surface $\mathcal{W}$ equipped with a proper, surjective morphism $\pi : \mathcal{W} \to \mathcal{C}$ whose geometric fibers are irreducible plane cubics and whose generic fiber is E. Since $\mathcal{C}$ is projective over k, so is $\mathcal{W}$. Since $\mathcal{W}$ is obtained by glueing reduced, geometrically irreducible surfaces along open subsets, it is also reduced and geometrically irreducible. Since it has only isolated singular points, by Serre's criterion it is normal.

This completes the proof of Proposition 1.3. □

Note that the closure in $\mathcal{W}$ of the identity element of E is a divisor on $\mathcal{W}$ which maps isomorphically to the base curve $\mathcal{C}$. We write $s_0 : \mathcal{C} \to \mathcal{W}$ for the inverse morphism. This is the *zero section* of π_0. In terms of the coordinates on $\mathcal{W}_1$ used in the proof above, it is just the map $t \mapsto ([0, 1, 0], t)$.

Discussion of Proposition 1.4. The algorithm mentioned in the Proposition is the subject of Tate's famous paper [Tat75]. His article does not mention blowing up, but the steps of the algorithm nevertheless give the recipe for the blow ups needed. The actual process of blowing up is explained in detail in [Sil94, IV.9] so we will not give the details here. Rather, we explain why there is a simple algorithm, following [Con05].

First note that the surface $\mathcal{W}$ is reduced and irreducible and so has no embedded components. Also, it has isolated singularities. (They are contained in the set of singular points of fibers of π_0.) By Serre's criterion, $\mathcal{W}$ is thus normal. Moreover, and this is the key point, its singularities are *rational double points*. (See [Art86] for the definition and basic properties of rational singularities and [Bǎd01, Chapters 3 and 4] for many more details. See [Con05, Section 8] for the fact that the singularities of a minimal Weierstrass model are rational.) This implies that the blow up of $\mathcal{W}$ at one of its singular points is again normal (so has isolated singularities) and again has at worst rational double points. An algorithm to desingularize is then simply to blow up at a singular point and iterate until the resulting surface is smooth. Given the explicit nature of the equations defining $\mathcal{W}$, finding the singular points and carrying out the blow ups is straightforward.

In fact, Tate's algorithm also calls for blowing up along certain curves. (This happens at steps 6 and 7.) This has the effect of dealing with several singular points at the same time, so is more efficient, but it is not essential to the success of the algorithm.

This completes our discussion of Proposition 1.4. See below for a detailed example covering a case not treated explicitly in [Sil94]. □

Conrad's article [Con05] also gives a coordinate-free treatment of integral minimal models of elliptic curves.

It is worth remarking that Tate's algorithm and the possible structures of the bad fibers are essentially the same in characteristic p as in mixed characteristic. On the other hand, for non-perfect residue fields k of characteristic $p \leq 3$, there are more possibilities for the bad fibers, in both equal and mixed characteristics—see [Szy04].

The zero section of $\mathcal{W}$ lifts uniquely to a section which we again denote by $s_0 : \mathcal{C} \to \mathcal{E}$.

2. The bundle ω and the height of $\mathcal{E}$

We construct an invertible sheaf on $\mathcal{C}$ as follows, using the notation of the proof of Proposition 1.1. Take the trivial invertible sheaf $\mathcal{O}_U$ on U with its generating section 1_U. At each stage of the construction, extend this sheaf by glueing $\mathcal{O}_U$ and $\mathcal{O}_{U'}$ over $U \cap U'$ by identifying 1_U and $u^{-1}1_{U'}$ where u is the function appearing in the change of coordinates (1.8).

The resulting invertible sheaf ω has several other descriptions. For example, the sheaf of relative differentials $\Omega^1_{\mathcal{E}/\mathcal{C}}$ is invertible on the locus of $\mathcal{E}$ where $\pi : \mathcal{E} \to \mathcal{C}$ is smooth (in particular in a neighborhood of the zero section) and, more or less directly from the definition, ω can be identified with $s_0^*(\Omega^1_{\mathcal{E}/\mathcal{C}})$. Using relative duality theory, ω can also be identified with the inverse of $R^1\pi_*\mathcal{O}_{\mathcal{E}}$. Finally, since $\mathcal{W}$ as only rational singularities, ω is also isomorphic to $R^1\pi_{0*}\mathcal{O}_{\mathcal{W}}$

One may identify the coefficients a_i of the Weierstrass equation locally defining $\mathcal{W}$ with sections of ω^i. Using this point of view, $\mathcal{W}$ can be identified with a closed subvariety of a certain $\mathbb{P}^2$-bundle over $\mathcal{C}$. Namely, let V be the locally free $\mathcal{O}_{\mathcal{C}}$ module of rank three

$$V = \omega^2 \oplus \omega^3 \oplus \mathcal{O}_{\mathcal{C}} \tag{2.1}$$

(where the exponents denote tensor powers). If $\mathbb{P}V$ denotes the projectivization of V over $\mathcal{C}$, a $\mathbb{P}^2$ bundle over $\mathcal{C}$, then $\mathcal{W}$ is naturally the closed subset of $\mathbb{P}E$ defined locally by the vanishing of Weierstrass equations as in (1.6).

Exercises 2.2. Verify the identifications and assertions in this section. In the case where $\mathcal{C} = \mathbb{P}^1$, so $K = k(t)$, check that $\omega = \mathcal{O}_{\mathbb{P}^1}(h)$ where h is the smallest positive integer such that E has a model (1.5) where the a_i are in $k[t]$ and $\deg a_i \leq hi$.

Exercises 2.3. Check that c_4, c_6, and Δ define *canonical* sections of ω^4, ω^6, and ω^{12} respectively, independent of the choice of equation for E. If $p = 2$ or 3, check that b_2 defines a canonical section of ω^2 and that $c_4 = b_2^2$ and $c_6 = -b_2^3$. If $p = 2$, check that a_1 defines a canonical section of ω and that $b_2 = a_1^2$. Note that since positive powers of ω have non-zero sections, the degree of ω is non-negative.

Definition 2.4. The *height* of $\mathcal{E}$, denoted h, is defined by $h = \deg(\omega)$, the degree of ω as an invertible sheaf on $\mathcal{C}$.

Note that if E/K is constant (in the sense of Lecture 1) then the height of the corresponding $\mathcal{E}$ is 0.

3. Examples

The case when $\mathcal{C} = \mathbb{P}^1$ is particularly simple. First of all, one may choose a model (1.5) that is integral and minimal simultaneously at every finite v, i.e., for every $v \in \mathbb{A}^1_k$. Indeed, start with any model and change coordinates so that the

a_i are in $k[t]$. If w is a finite place where this model is not minimal, it is possible (because $k[t]$ is a PID) to choose a change of coordinates

$$(x, y) = (u^2x' + r, u^3y' + su^2x' + t)$$

where $r, s, t, u \in k[t][1/w]$ and u a unit yielding a model that is minimal at w. Such a change of coordinates does not change the minimality at any other finite place. Thus after finitely many steps, we have a model integral and minimal at all finite places. (This argument would apply for any K and any Dedekind domain $R \subset K$ which is a PID, yielding a model with the $a_i \in R$ that is minimal at all $v \in \operatorname{Spec} R$.)

Focusing attention at $t = \infty$, there is a change of coordinates (1.8) with $u = t^{-h}$ yielding a model integral and minimal at ∞. (Here h is minimal so that $\deg(a_i) \leq hi$.) So the bundle $\omega = \mathcal{O}(h) = \mathcal{O}(h\infty)$.

As a very concrete example, consider the curve

$$y^2 = x(x+1)(x+t^d)$$

over $\mathbb{F}_p(t)$ where $p > 2$ and d is not divisible by p. Since $\Delta = 16t^{2d}(t^d-1)^2$, this model is integral and minimal at all non-zero finite places. It is also minimal at zero as one may see by noting that c_4 and c_6 are units at 0. At infinity, the change of coordinates

$$(x, y) = (t^{2h}x', t^{3h}y')$$

with $h = \lceil d/2 \rceil$ yields a minimal integral model. Thus $\omega = \mathcal{O}(h)$.

Working with Tate's algorithm shows that E has I_2 reduction at the d-th roots of unity, I_{2d} reduction at $t = 0$, and either I_{2d}^* or I_{2d} reduction at infinity depending on whether d is odd or even.

Since the case of I_n reduction is not treated explicitly in [Sil94], we give more details on the blow ups needed to resolve the singularity over $t = 0$. In terms of the coordinates on $\mathcal{W}_1$ used in the proof of Proposition 1.4 we can consider the affine surface defined by

$$x^3 + (t^d + 1)x^2 + t^dx - y^2 = 0$$

which is an open neighborhood of the singularity at $x = y = t = 0$. If $d = 1$, then the tangent cone is the irreducible plane conic defined by $x^2 + tx - y^2 = 0$. The singular point thus blows up into a smooth rational curve and it is easy to check that the resulting surface is smooth in a neighborhood of the fiber $t = 0$. Now assume that $d > 1$. Then the tangent cone is the reducible conic $x^2 - y^2 = 0$ and so the singular point blows up into two rational curves meeting at one point. More precisely, the blow up is covered by three affine patches. In one of them, the surface upstairs is

$$tx_1^3 + (t^d + 1)x_1^2 + t^{d-1}x_1 - y_1^2 = 0$$

and the morphism is $x = tx_1$, $y = ty_1$. The exceptional divisor is the reducible curve $t = x_1^2 - y_1^2 = 0$ and the point of intersection of the components $t = x_1 = y_1 = 0$ is again a double point. Considering the other charts shows that there are no other singular points in a neighborhood of $t = 0$ and that the exceptional divisor meets the original fiber over $t = 0$ in two points. We now iterate this process $d-1$ times, introducing two new components at each stage. After $d-1$ blow ups, the interesting part of our surface is given by

$$t^{d-1}x_{d-1}^3 + (t^d + 1)x_{d-1}^2 + tx_{d-1} - y_{d-1}^2 = 0.$$

At this last stage, blowing up introduces one more component meeting the two components introduced in the preceding step at one point each. The (interesting part of the) surface is now

$$t^d x_d^3 + (t^d + 1)x_d^2 + x_d - y_d^2 = 0$$

which is regular in a neighborhood of $t = 0$. Thus we see that the fiber over $t = 0$ in $\mathcal{E}$ is a chain of $2d$ rational curves, i.e., a fiber of type I_{2d}.

The resolution of the singularities over points with $t^d = 1$ is similar but simpler because only one blow up is required. At $t = \infty$, if d is even then the situation is very similar to that over $t = 0$ and the reduction is again of type I_{2d}. If d is odd, the reduction is of type I_{2d}^*. We omit the details in this case since it is treated fully in [Sil94].

Exercise 3.1. In the table in Tate's algorithm paper [Tat75] (and the slightly more precise version in [Sil09, p. 448]), the last three rows have restrictions on p. Give examples showing that these restrictions are all necessary for the discriminant and conductor statements, and for the statement about j in the I_n^*, $p = 2$ case. Show that the other assertions about the j-invariant are correct for all p.

4. $\mathcal{E}$ and the classification of surfaces

It is sometimes useful to know how $\mathcal{E}$ fits into the Enriques-Kodaira classification of surfaces. In this section only, we replace k with $\overline{k}$ and write $\mathcal{E}$ for what elsewhere is denoted $\overline{\mathcal{E}}$.

Recall that the height of $\mathcal{E}$ is defined as $h = \deg \omega$.

Proposition 4.1. *$\omega \cong \mathcal{O}_\mathcal{C}$ if and only if E is constant. If $h = \deg(\omega) = 0$, then E is isotrivial.*

PROOF. It is obvious that if E is constant, then $\omega \cong \mathcal{O}_\mathcal{C}$. Conversely, suppose $\omega \cong \mathcal{O}_\mathcal{C}$. Then the construction of $\pi_0 : \mathcal{W} \to \mathcal{C}$ in Proposition 1.3 yields an irreducible closed subset of $\mathbb{P}^2_\mathcal{C}$ (because the $\mathbb{P}^2$-bundle $\mathbb{P}V$ in (2.1) is trivial):

$$\mathcal{W} \subset \mathbb{P}^2_\mathcal{C} = \mathbb{P}^2_k \times_k \mathcal{C}.$$

Let $\sigma : \mathcal{W} \to \mathbb{P}^2_k$ be the restriction of the projection $\mathbb{P}^2_\mathcal{C} \to \mathbb{P}^2_k$. Then σ is not surjective (since most points in the line at infinity $Z = 0$ are not in the image) and so its image has dimension < 2. Considering the restriction of σ to a fiber of π_0 shows that the image of σ is in fact an elliptic curve E_0 and then it is obvious from dimension considerations that

$$\mathcal{W} = \pi_0^{-1}(\pi_0(\mathcal{W})) = E_0 \times \mathcal{C}.$$

It follows that E, the generic fiber of π_0, is isomorphic to $E_0 \times \operatorname{Spec} K$, i.e., that E is constant.

Now assume that $h = 0$. Then Δ is a non-zero global section of the invertible sheaf ω^{12} on $\mathcal{C}$ of degree 0. Thus ω^{12} is trivial. It follows that there is a finite unramified cover of $\mathcal{C}$ over which ω becomes trivial and so by the first part, E becomes constant over a finite extension, i.e., E is isotrivial. □

Note that E being isotrivial does not imply that $h = 0$.

Exercise 4.2. Give an example of a non-constant E of height zero. Hint: Consider the quotient of a product of elliptic curves by a suitable free action of a group of order two.

Proposition 4.3. *The canonical bundle of $\mathcal{E}$ is $\Omega^2_{\mathcal{E}} \cong \pi^*\left(\Omega^1_{\mathcal{C}} \otimes \omega\right)$.*

Here we are using that $\mathcal{E} \to \mathcal{C}$ has a section and therefore no multiple fibers. The proof, which we omit, proceeds by considering $R^1\pi_*\mathcal{O}_{\mathcal{E}}$ and using relative duality. See for example [Băd01, 7.15].

We now consider several cases:

If $2g_{\mathcal{C}} - 2 + h > 0$, then it follows from the Proposition that the dimension of $H^0(\mathcal{E}, (\Omega^2)^{\otimes n})$ grows linearly with n, so $\mathcal{E}$ has Kodaira dimension 1.

If $2g_{\mathbb{C}} - 2 + h = 0$, then the Kodaira dimension of $\mathcal{E}$ is zero and there are two possibilities: (1) $g_{\mathcal{C}} = 1$ and $h = 0$; or (2) $g_{\mathcal{C}} = 0$ and $h = 2$. In the first case, there is an unramified cover of $\mathcal{C}$ over which $\mathcal{E}$ becomes constant and so $\mathcal{E}$ is the quotient of a product of two elliptic curves. These surfaces are sometimes called "bi-elliptic." In the second case, $\Omega^2_{\mathcal{E}} = \mathcal{O}_{\mathcal{E}}$ and $H^1(\mathcal{E}, \mathcal{O}_{\mathcal{E}}) = H^0(\mathcal{C}, \omega^{-1}) = 0$ and so $\mathcal{E}$ is a K3 surface.

If $2g_{\mathcal{C}} - 2 + h < 0$, then the Kodaira dimension of $\mathcal{E}$ is $-\infty$ and there are again two possibilities: (1) $g_{\mathcal{C}} = 0$ and $h = 1$, in which case $\mathcal{E}$ is a rational surface by Castelnuovo's criterion; or (2) $g_{\mathcal{C}} = 0$ and $h = 0$, in which case E is constant and $\mathcal{E}$ is a ruled surface $E_0 \times \mathcal{C} = E_0 \times \mathbb{P}^1$.

5. Points and divisors, Shioda-Tate

If D is an irreducible curve on $\mathcal{E}$, then its generic fiber

$$D.E := D \times_{\mathcal{C}} E$$

is either empty or is a closed point of E. The former occurs if and only if D is supported in a fiber of π. In the latter case, the residue degree of $D.E$ is equal to the generic degree of $D \to \mathcal{C}$. Extending by linearity, we get homomorphism

$$\operatorname{Div}(\mathcal{E}) \to \operatorname{Div}(E)$$

whose kernel consists of divisors supported in the fibers of π.

There is a set-theoretic splitting of this homomorphism, induced by the map sending a closed point of E to its scheme-theoretic closure in $\mathcal{E}$. However, this is not in general a group homomorphism.

Let $L^1 \operatorname{Div}(\mathcal{E})$ be the subgroup of divisors D such that the degree of $D.E$ is zero and let $L^2 \operatorname{Div}(\mathcal{E})$ be subgroup such that $D.E = 0$. We write $L^i \operatorname{Pic}(\mathcal{E})$ and $L^i \operatorname{NS}(\mathcal{E})$ $(i = 1, 2)$ for the images of $L^i(\mathcal{E})$ in $\operatorname{Pic}(\mathcal{E})$ and $\operatorname{NS}(\mathcal{E})$ respectively.

The Shioda-Tate theorem relates the Néron-Severi group of $\mathcal{E}$ to the Mordell-Weil group of E:

Theorem 5.1. *If $\mathcal{E} \to \mathcal{C}$ is non-constant, $D \mapsto D.E$ induces an isomorphism*

$$\frac{L^1 \operatorname{NS}(\mathcal{E})}{L^2 \operatorname{NS}(\mathcal{E})} \cong E(K)$$

If $\mathcal{E} \to \mathcal{C}$ is constant, we have

$$\frac{L^1 \operatorname{NS}(\mathcal{E})}{L^2 \operatorname{NS}(\mathcal{E})} \cong E(K)/E(k)$$

This theorem seems to have been known to the ancients (Lang, Néron, Weil, ...) and was stated explicitly in [Tat66b] and in papers of Shioda. A detailed proof in a more general context is given in [Shi99]. Note however that in [Shi99] the ground field is assumed to be algebraically closed. See [Ulm11] for the small modifications needed to treat finite k.

It is obvious that $NS(\mathcal{E})/L^1NS(\mathcal{E})$ is infinite cyclic. We saw in Example 8.6 of Lecture 2 that $L^2NS(\mathcal{E})$ is free abelian of rank $1+\sum_v(f_v-1)$. So as a corollary of the theorem, we have the following rank formula, known as the Shioda-Tate formula:

$$\operatorname{Rank} E(K) = \operatorname{Rank} NS(\mathcal{E}) - 2 - \sum_v (f_v - 1) \tag{5.2}$$

For more on the geometry of elliptic surfaces and elliptic curves over function fields, with an emphasis on rational and K3 surfaces, I recommend [SS10].

6. L-functions and Zeta-functions

We are going to relate the L-function of E and the zeta function of $\mathcal{E}$. We note that from the definition, $Z(\mathcal{E},T)$ depends only on the underlying set of closed points of $\mathcal{E}$ and we may partition this set using the map π.

We have

$$\begin{aligned} Z(\mathcal{E},T) &= \prod_{\text{closed } x\in\mathcal{E}} \left(1-T^{\deg(x)}\right)^{-1} \\ &= \prod_{\text{closed } y\in\mathcal{C}} \prod_{x\in\pi^{-1}(y)} \left(1-T^{\deg(x)}\right)^{-1} \\ &= \prod_{\text{closed } y\in\mathcal{C}} Z(\pi^{-1}(y), T^{\deg(y)}) \end{aligned}$$

For y such that $\pi^{-1}(y)$ is a smooth elliptic curve, we know that

$$Z(\pi^{-1}(y),T) = \frac{(1-a_yT+q_yT^2)}{(1-T)(1-q_yT)}$$

and the numerator here is the factor that enters into the definition of $L(E,T)$.

To complete the calculation, we need an analysis of the contribution of the bad fibers. We consider the fiber $\pi^{-1}(y)$ as a scheme of finite type over the residue field κ_y, the field of q_y elements. As such, it has irreducible components. Its "geometric components" are the components of the base change to $\overline{\kappa}_y$; these are defined over some finite extension of κ_y.

For certain reduction types (I_n, I_n^* ($n\geq 0$), IV and IV^*) it may happen that all the geometric components are defined over κ_y, in which case we say the reduction is "split", or it may happen that some geometric components are only defined over a quadratic extension of κ_y, in which case we say the reduction is "non-split." This agrees with the standard usage in the case of I_n reduction and may be non-standard in the other cases.

Proposition 6.1. *The zeta function of the a singular fiber of π has the form*

$$\begin{aligned} Z(\pi^{-1}(y),T) &= \frac{(1-T)^a(1+T)^b}{(1-q_yT)^f(1+q_yT)^g} \\ &= \frac{1}{(1-T)(1-q_yT)}\frac{(1-T)^{a+1}(1+T)^b}{(1-q_yT)^{f-1}(1+q_yT)^g} \end{aligned}$$

where the integers a, b, f, and g are determined by the reduction type at y and are given in the following table:

	a	b	f	g
split I_n	0	0	n	0
non-split I_n, n *odd*	-1	1	$(n+1)/2$	$(n-1)/2$
non-split I_n, n *even*	-1	1	$n/2+1$	$(n-2)/2$
split I_n^*	-1	0	$5+n$	0
non-split I_n^*	-1	0	$4+n$	1
II	-1	0	1	0
II^*	-1	0	9	0
III	-1	0	2	0
III^*	-1	0	8	0
split IV	-1	0	3	0
non-split IV	-1	0	2	1
split IV^*	-1	0	7	0
non-split IV^*	-1	0	3	4

Exercise 6.2. Use an elementary point-counting argument to verify the proposition. In particular, check that the number of components of $\pi^{-1}(y)$ that are rational over κ_y is f and that the order of pole at $T = q_y^{-1}$ of

$$Z(\pi^{-1}(y), T)(1-T)(1-q_yT)$$

is $f-1$.

Using the Proposition and the definition of the L-function (in Lecture 1, equation (9.1)) we find that

$$L(E,T) = \frac{Z(\mathcal{C},T)Z(\mathcal{C},qT)}{Z(\mathcal{E},T)} \prod_{\text{bad } v} \frac{(1-T)^{a_v+1}(1+T)^{b_v}}{(1-q_vT^{\deg(v)})^{f_v-1}(1+q_vT^{\deg(v)})^{g_v}} \tag{6.3}$$

where a_v, b_v, f_v and g_v are the invariants defined in the Proposition at the place v. Using the Weil conjectures (see Section 3 of Lecture 0), we see that the orders of $L(E,s)$ and $\zeta(\mathcal{E},s)$ at $s=1$ are related as follows:

$$\operatorname{ord}_{s=1} L(E,s) = -\operatorname{ord}_{s=1}\zeta(\mathcal{E},s) - 2 - \sum_v (f_v - 1). \tag{6.4}$$

Remark 6.5. This simple approach to evaluating the order of zero of the L-function does not yield the important fact that $L(E,T)$ is a polynomial in T when E is non-constant, nor does it yield the Riemann hypothesis for $L(E,T)$.

For a slightly more sophisticated (and less explicit) comparison of ζ-functions and L-functions in a more general context, see [Gor79].

7. The Tate-Shafarevich and Brauer groups

The last relationship between E and $\mathcal{E}$ we need concerns the Tate-Shafarevich and Brauer groups.

Theorem 7.1. *Suppose that E is an elliptic curve over $K = k(\mathcal{C})$ and $\mathcal{E} \to \mathcal{C}$ is the associated elliptic surface as in Proposition 1.1. Then there is a canonical isomorphism*

$$\operatorname{Br}(\mathcal{E}) \cong Ш(E/K).$$

The proof of this result, which is somewhat involved, is given in [Gro68, Section 4]. The main idea is simple enough: one computes $\mathrm{Br}(\mathcal{E}) = H^2(\mathcal{E}, \mathbb{G}_m)$ using the morphism $\pi : \mathcal{E} \to \mathcal{C}$ and a spectral sequence. Using that the Brauer group of a smooth, complete curve over a finite field vanishes, one finds that the main term is $H^1(\mathcal{C}, R^1\pi_*\mathbb{G}_m)$. Since $R^1\pi_*\mathbb{G}_m$ is the sheaf associated to the relative Picard group, it is closely related to the sheaf on $\mathcal{C}$ represented by the Néron model of E. This provides a connection with the Tate-Shafarevich group which leads to the theorem.

See [Ulm11] for more details about this and the closely related connection between $H^2(\overline{\mathcal{E}}, \mathbb{Z}_\ell(1))^{G_k}$ and the ℓ-Selmer group of E.

8. The main classical results

We are now in a position to prove the theorems of Section 12 of Lecture 1. For convenience, we restate Theorem 12.1 and a related result.

Theorem 8.1. *Suppose that E is an elliptic curve over $K = k(\mathcal{C})$ and $\mathcal{E} \to \mathcal{C}$ is the associated elliptic surface as in Proposition 1.1.*

(1) *BSD holds for E if and only if T_2 holds for $\mathcal{E}$.*
(2) $\operatorname{Rank} E(K) \leq \operatorname{ord}_{s=1} L(E, s)$.
(3) *The following are equivalent:*
- $\operatorname{Rank} E(K) = \operatorname{ord}_{s=1} L(E, s)$
- *$Ш(E/K)$ is finite*
- *for any one prime number ℓ ($\ell = p$ is allowed), the ℓ-primary part $Ш(E/K)_{\ell^\infty}$ is finite.*

(4) *If K'/K is a finite extension and if the BSD conjecture holds for E over K', then it holds for E over K.*

Proof. Comparing (5.2) and (6.4), we have that

$$\operatorname{Rank} E(K) - \operatorname{ord}_{s=1} L(E, s) = \operatorname{Rank} NS(\mathcal{E}) + \operatorname{ord}_{s=1} \zeta(\mathcal{E}, s).$$

Since BSD is the assertion that the left hand side is zero and T_2 is the assertion that the right hand side is zero, these conjectures are equivalent.

By Theorem 9.3 of Lecture 2, the right hand side is ≤ 0 and therefore so is the left. This gives the inequality $\operatorname{Rank} E(K) \leq \operatorname{ord}_{s=1} L(E, s)$.

The statements about $Ш(E/K)$ follow from Theorem 7.1 ($Ш(E/K) \cong \mathrm{Br}(\mathcal{E})$), the equivalence of BSD and $T_2(\mathcal{E})$, and Theorem 10.2 of Lecture 2.

The last point follows from the equivalence of BSD and $T_2(\mathcal{E})$ and Proposition 11.1 of Lecture 2. □

Proofs of Theorems 12.2 and 12.3 of Lecture 1. Theorem 12.2 of Lecture 1 concerns isotrivial elliptic curves. By the last point of Theorem 8.1 above, it suffices to show that BSD holds for constant curves. But if E is constant, then $\mathcal{E}$ is a product of curves, so the Tate conjecture for $\mathcal{E}$ follows from Theorem 12.1 of Lecture 2. The first point of Theorem 8.1 above then gives BSD for E.

Theorem 12.3 of Lecture 1 concerns elliptic curves over $k(t)$ of low height. By the discussion in Section 4, if $E/k(t)$ has height ≤ 2 then $\mathcal{E}$ is a rational or K3 surface. (Strictly speaking, this is true only over a finite extension of k, but the last point of Theorem 8.1 allows us to make this extension without loss of generality.) But $T_2(\mathcal{X})$ for a rational surface follows from Proposition 13.1 of Lecture 2. For E

such that $\mathcal{E}$ is a K3 surfaces, Artin and Swinnerton-Dyer proved the finiteness of Ш(E/K) (and therefore BSD) in [ASD73]. □

9. Domination by a product of curves

Combining part 1 of Theorem 8.1 with Proposition 13.1 of Lecture 2, we have the following.

Theorem 9.1. *Let E be an elliptic curve over K with associated surface $\mathcal{E}$. If $\mathcal{E}$ is dominated by a product of curves, then BSD holds for E.*

Theorem 12.4 ("four monomials") and Berger's theorem 11.1 are both corollaries of Theorem 9.1, as we will explain in the remainder of this lecture.

10. Four monomials

We recall Shioda's conditions. Suppose that $f \in R = k[x_1, x_2, x_3]$ is the sum of exactly four non-zero monomials:

$$f = \sum_{i=1}^{4} c_i \prod_{j=1}^{3} x_j^{e_{ij}}$$

where $c_i \in k$ and the e_{ij} are non-negative integers. Let $e_{i4} = 1 - \sum_{j=1}^{3} e_{ij}$ and form the 4×4 matrix $A = (e_{ij})$. Assuming that $\det(A) \neq 0$ (in $\mathbb{Z}$), let δ be the smallest positive integer such that there is a 4×4 integer matrix B with $AB = \delta I_{4\times 4}$. We say that f *satisfies Shioda's 4-monomial condition* if $\delta \neq 0$ in k, i.e., if $p \not| \delta$. The following exercise shows that this is equivalent to the definition in Lecture 1.

Exercises 10.1. Show that a prime ℓ divides δ if and only if it divides $\det(A)$. Show that if we change the definition of e_{i4} to $e_{i4} = d - \sum_{j=1}^{3} e_{ij}$ for some other non-zero integer d and define δ_d using the new $A = (e_{ij})$, then δ_1 divides δ_d for all d. I.e., $d = 1$ is the optimal choice to minimize δ.

Exercise 10.2. With c_i and e_{ij} as above, show that the system of equations

$$\prod_{j-1}^{4} d_j^{e_{ij}} = c_i^{-1} \qquad i = 1, \dots, 4$$

has a solution with $d_j \in \overline{\mathbb{F}}_q$, $j = 1, \dots, 4$.

Proof of Theorem 12.4 of Lecture 1. Briefly, the hypotheses imply that the associated elliptic surface $\mathcal{E} \to \mathbb{P}^1$ is dominated by a Fermat surface (of degree δ) and thus by a product of Fermat curves (of degree δ). Thus Theorem 9.1 implies that BSD holds for E.

In more detail, note that $\mathcal{E}$ is birational to the affine surface $V(f) \subset \mathbb{A}^3_k$. So it will suffice to show that $V(f)$ is dominated by a product of curves. To that end, it will be convenient to identify $k[t, x, y]$ and $R = k[x_1, x_2, x_3]$ by sending $t \mapsto x_1$, $x \mapsto x_2$ and $y \mapsto x_3$, so that f becomes

$$f = \sum_{i=1}^{4} c_i \prod_{j=1}^{3} x_j^{e_{ij}}.$$

Exercise 10.2 implies that, after extending k if necessary, we may change coordinates ($x_j \mapsto d_j x_j$) so that the coefficients c_i are all 1. Then the matrix A defines rational a map ϕ from $V(f)$ to the Fermat surface of degree 1

$$F_1^2 = \{y_1 + y_2 + y_3 + y_4 = 0\} \subset \mathbb{P}^3_k,$$

namely $\phi^*(y_i) = \prod_{j=1}^4 x_j^{e_{ij}}$. Similarly, the matrix B defines a rational map ψ from the Fermat surface of degree δ

$$F_\delta^2 = \{z_1^\delta + z_2^\delta + z_3^\delta + z_4^\delta = 0\} \subset \mathbb{P}^3_k$$

to $V(f)$, namely $\psi^*(x_i) = \prod_{j=1}^4 z_j^{B_{ij}}$. The composition of these maps is the standard projection from F_δ^2 to F_1^2, namely $y_i \mapsto z_i^\delta$ and so both maps are dominant.

Finally, Shioda and Katsura [SK79] showed that F_δ^2 is dominated by the product of Fermat curves $F_\delta^1 \times F_\delta^1$. Thus, after extending k, $\mathcal{E}$ is dominated by a product of curves and Theorem 9.1 finishes the proof. □

As we will explain below, this Theorem can be combined with results on analytic ranks to give examples of elliptic curves over $\mathbb{F}_p(t)$ with arbitrarily large Mordell-Weil rank. (In fact, similar ideas can be used to produce Jacobians of every dimension with large rank. For this, see [Ulm07] and also [Ulm11].)

Unfortunately, Theorem 12.4 is very rigid—as one sees in the proof, varying the coefficients in the 4-nomial f does not vary the isomorphism class of $\mathcal{E}$ over $\overline{\mathbb{F}}_q$ and so we get only finitely many non-isomorphic elliptic curves over $\overline{\mathbb{F}}_p(t)$. Berger's construction, explained in the next subsection, was motivated by a desire to overcome this rigidity and give *families* of examples of curves where one knows the BSD conjecture.

11. Berger's construction

Berger gave a much more flexible construction of surfaces that are dominated by a product of curves in a tower. More precisely, we note that if $\mathcal{E} \to \mathbb{P}^1$ is an elliptic surface and $\phi : \mathbb{P}^1 \to \mathbb{P}^1$ is the morphism with $\phi^*(t) = u^d$ (corresponding to the field extension $k(u)/k(t)$ with $u^d = t$), then it is not in general the case that the base changed surface

$$\begin{array}{ccc} \mathcal{E}' = \mathcal{E} \times \mathbb{P}^1_k & \longrightarrow & \mathbb{P}^1_k \\ \downarrow & & \downarrow \\ \mathbb{P}^1_k & \xrightarrow{\ \phi\ } & \mathbb{P}^1_k \end{array}$$

is dominated by a product of curves. Berger's construction gives a rich class of curves for which DPC *does* hold in every layer of a tower of coverings. We restate Theorem 12.5 from Lecture 1 in a slightly different (but visibly equivalent) form.

Theorem 11.1. *Let E be an elliptic curve over $K = k(t)$ and assume that there are rational functions $f(x)$ and $g(y)$ on $\mathbb{P}^1_k$ such that E is birational to the curve $V(f(x) - tg(y)) \subset \mathbb{P}^1_K \times \mathbb{P}^1_K$. Then the BSD conjecture holds for E over the field $k(u) = k(t^{1/d})$ for all d prime to p.*

PROOF. Clearing denominators we may interpret $f(x) - tg(y)$ as defining a hypersurface $\mathcal{X}$ in the affine space $\mathbb{A}^3$ with coordinates x, y, and t and it is clear that the elliptic surface $\mathcal{E} \to \mathbb{P}^1$ associated to E is birationally isomorphic to $\mathcal{X}$.

On the other hand, $\mathcal{X}'$ is visibly birational to $\mathbb{P}^1 \times \mathbb{P}^1$ since we may eliminate t. Thus $\mathcal{X}$ and $\mathcal{E}$ are dominated by a product of curves. This checks the case $d = 1$.

For larger d, note that the elliptic surface $\mathcal{E}_d \to \mathbb{P}^1$ associated to $E/k(u)$ is birational to the hypersurface $\mathcal{X}_d$ in $\mathbb{A}^3_k$ defined by $f(x) - u^d g(y)$. Berger showed by a fundamental group argument, generalizing [Sch96], that $\mathcal{X}_d$ is dominated by a product of curves, more precisely, by a product of covers of $\mathbb{P}^1$. (For her argument to be correct, π_1 should be replaced by the prime-to-p fundamental group $\pi_1^{p'}$ throughout.) This was later made more explicit in [Ulm09a], where it was observed that $\mathcal{X}_d$ is dominated by a product of two explicit covers of $\mathbb{P}^1$.

More precisely, let $\mathcal{C}_d$ and $\mathcal{D}_d$ be the covers of $\mathbb{P}^1_k$ defined by $z^d = f(x)$ and $w^d = g(y)$. Then there is a rational map from $\mathcal{C}_d \times \mathcal{D}_d$ to the hypersurface $\mathcal{X}_d$, namely

$$(x, z, y, w) \mapsto (x, y, u = z/w).$$

This is clearly dominant and so $\mathcal{X}_d$ and $\mathcal{E}$ are dominated by products of curves.

Applying Theorem 9.1 finishes the proof. $\square$

Note that there is a great deal of flexibility in the choice of data for Berger's construction. As an example, take $f(x) = x(x - a)/(x - 1)$ and $g(y) = y(y - 1)$ where $a \in \mathbb{F}_q$ is a parameter. Then if $a \neq 1$, the curve $f(x) = tg(y)$ in $\mathbb{P}^1 \times \mathbb{P}^1$ has genus 1 and a rational point. A simple calculation shows that it is birational to the Weierstrass cubic

$$y^2 + txy - ty = x^3 - tax^2 + t^2ax.$$

Theorem 11.1 implies that this curve satisfies the BSD conjecture over $\mathbb{F}_{q^n}(t^{1/d})$ for all n and all d prime to p. Varying q and a we get infinitely many curves for which BSD holds at every layer of a tower.

We will give more examples and discuss further applications of the idea behind Berger's construction in Lectures 4 and 5.

LECTURE 4

Unbounded ranks in towers

In order to prove results on analytic ranks in towers, we need a more sophisticated approach to L-functions. In this lecture we explain Grothendieck's approach to L-functions over function fields and then use it and a new linear algebra lemma to find elliptic curves with unbounded analytic and algebraic ranks in towers of function fields.

1. Grothendieck's analysis of L-functions

1.1. Galois representations

As usual, we let $K = k(\mathcal{C})$ be the function field of a curve over a finite field k and $G_K = \mathrm{Gal}(K^{sep}/K)$ its Galois group. As in Lecture 0, Section 2, we write D_v, I_v, and Fr_v for the decomposition group, inertia group, and (geometric) Frobenius at a place v of K.

We fix a prime $\ell \neq p$ and consider a representation

$$\rho : G_K \to \mathrm{GL}(V) \cong \mathrm{GL}_n(\overline{\mathbb{Q}}_\ell) \tag{1.1.1}$$

on a finite-dimensional $\overline{\mathbb{Q}}_\ell$ vector space. We make several standing assumptions about ρ.

First, we always assume ρ is continuous and unramified away from a finite set of places of K. By a compactness argument (see [KS99, 9.0.7]) , it is possible to define ρ over a finite extension L of $\mathbb{Q}_\ell$, i.e., there is a representation

$$\rho' : G_K \to \mathrm{GL}_n(L)$$

isomorphic to ρ over $\overline{\mathbb{Q}}_\ell$. Nothing we say will depend on the field of definition of ρ and we will generally not distinguish between ρ and isomorphic representations defined over subfields of $\overline{\mathbb{Q}}_\ell$.

We also always assume that ρ is pure of integral weight w, i.e., for all v where ρ is unramified, the eigenvalues of $\rho(\mathrm{Fr}_v)$ are Weil numbers of size $q_v^{w/2}$.

Finally, we sometimes assume that ρ is "symplectically self-dual of weight w." This means that on the space V where ρ acts, there is an G_K-equivariant, alternating pairing with values in $\overline{\mathbb{Q}}_\ell(-w)$.

1.2. Conductors

The Artin conductor of ρ is a divisor on $\mathcal{C}$ (a formal sum of places of K) and is a measure of its ramification. We write $\mathrm{Cond}(\rho) = \mathfrak{n} = \sum_v n_v[v]$. To define the local coefficients, fix a place v of K and let $G_i \subset I_v$ be the higher ramification groups at v (in the lower numbering). Then define

$$n_v = \sum_{i=0}^{\infty} \frac{1}{[G_0 : G_i]} \dim V/V^{G_i}.$$

Here V^{G_i} denotes the subspace of V invariant under G_i. It is clear that $n_v = 0$ if and only if ρ is unramified at v. If ρ is tamely ramified at v (i.e., G_1 acts trivially), then $n_v = \dim V/V^{G_0} = \dim V/V^{I_v}$. In general, the first term of the sum above is the *tame conductor* and the rest of the sum is the *Swan conductor*. We refer to [Mil80, V.2] and also [Ser77, §19] for an alternative definition and more discussion about the conductor, including the fact that the local coefficients n_v are integers.

1.3. L-functions

Let us fix an isomorphism $\overline{\mathbb{Q}}_\ell \cong \mathbb{C}$ so that we may regard eigenvalues of Frobenius on ℓ-adic representations as complex numbers. Having done this, a representation (1.1.1) gives rise to an L-function, defined as an Euler product:

$$L(\rho, T) = \prod_v \det\left(1 - T\,\mathrm{Fr}_v \,|V^{I_v}\right) \tag{1.3.1}$$

and $L(\rho, s) = L(\rho, q^{-s})$. The product is over the places of K, the exponent I_v denotes the subspace of elements invariant under the inertia group I_v, and Fr_v is a Frobenius element at v.

Because of our assumption that ρ is pure of weight w, the product defining $L(\rho, s)$ converges absolutely and defines a holomorphic function in the region $\mathrm{Re}\, s > w/2 + 1$.

It is clear from the definition that if ρ and σ are Galois representations then $L(\rho \oplus \sigma, s) = L(\rho, s)L(\sigma, s)$ and $L(\rho(n), s) = L(\rho, s - n)$.

It is also clear that $L(\rho_{triv}, s) = \zeta(\mathcal{C}, s)$. and so $L(\rho_{triv}(n), s) = \zeta(\mathcal{C}, s - n)$.

Exercise 1.3.2. Prove that if ρ factors through $G_K \to G_k$, so that Fr_v goes to $\alpha^{\deg v}$, then

$$L(\rho, T) = Z(\mathcal{C}, \alpha T)$$

is a twisted version of the zeta function of $\mathcal{C}$. Compare with Exercise 9.2 of Lecture 1. Note that a representation factors through $G_K \to G_k$ if and only if it is trivial on $G_{\overline{k}K}$, so this exercise fills in the missing cases in the following theorem.

Theorem 1.3.3. *Suppose that ρ is a representation of G_K (satisfying the standing hypotheses of Subsection 1.1) that contains no copies of the trivial representation when restricted to $G_{\overline{k}K}$. Then there is a canonically defined $\overline{\mathbb{Q}}_\ell$-vector space $H(\rho)$ with continuous G_k action such that*

$$L(\rho, s) = \det\left(1 - q^{-s}\,\mathrm{Fr}_q \,|H(\rho)\right).$$

The dimension of $H(\rho)$ is $\deg(\rho)(2g_{\mathcal{C}} - 2) + \deg \mathfrak{n}$ where $\mathfrak{n}$ is the conductor of ρ.

Proof. (Sketch) The representation $\rho : G_K \to \mathrm{GL}(V)$ gives rise to a constructible sheaf $\mathcal{F}_\rho$ on $\mathcal{C}$. In outline: ρ is essentially the same thing as a lisse sheaf $\mathcal{F}_U$ on the open subset $j : U \hookrightarrow \mathcal{C}$ over which ρ is unramified. We defined $\mathcal{F}_\rho$ as the push-forward $j_*\mathcal{F}_U$. For each closed point v of $\mathcal{C}$, the stalk of ρ at v is V^{I_v}.

Let $H^i(\overline{\mathcal{C}}, \mathcal{F})$ be the étale cohomology groups of $\mathcal{F}$. They are finite dimensional $\overline{\mathbb{Q}}_\ell$ vector spaces and give continuous representations of G_k.

The Grothendieck-Lefschetz fixed point formula says that for each finite extension $\mathbb{F}_{q^n}$ of $k \cong \mathbb{F}_q$, we have

$$\sum_{x \in \mathcal{C}(\mathbb{F}_{q^n})} \mathrm{Tr}(Fr_x|\mathcal{F}_x) = \sum_{i=0}^{2} (-1)^i\,\mathrm{Tr}\left(Fr_{q^n}|H^i(\overline{\mathcal{C}}, \mathcal{F})\right).$$

On the left hand side, the sum is over points of C with values in $\mathbb{F}_{q^n}$ and the summand is the trace of the action of the Frobenius at x on the stalk of $\mathcal{F}$ at a geometric point over x.

Multiplying both sides by T^n/n, summing over $n \geq 1$, and exponentiating, one finds that

$$L(\rho, T) = \prod_{i=0}^{2} \det\left(1 - T\operatorname{Fr}_q \left| H^i(\overline{\mathcal{C}}, \mathcal{F})\right.\right)^{(-1)^{i+1}}.$$

Now $H^0(\overline{\mathcal{C}}, \mathcal{F})$ and $H^2(\overline{\mathcal{C}}, \mathcal{F})$ are isomorphic respectively to the invariants and coinvariants of V under $G_{\overline{k}K}$ and so under our hypotheses on ρ, $H^i(\overline{\mathcal{C}}, \mathcal{F})$ vanishes for $i = 0, 2$. Thus we have

$$L(\rho, s) = \det\left(1 - q^{-s}\operatorname{Fr}_q \left| H(\rho)\right.\right)$$

where $H(\rho) = H^1(\overline{\mathcal{C}}, \mathcal{F})$.

The dimension formula comes from an Euler characteristic formula proven by Raynaud and sometimes called the Grothendieck-Ogg-Shafarevich formula. It says

$$\sum_{i=0}^{2}(-1)^i \dim H^i(\overline{\mathcal{C}}, \mathcal{F}) = \deg(\rho)(2 - 2g_{\mathcal{C}}) - \deg(\operatorname{Cond}(\rho)).$$

Since H^0 and H^2 vanish, this gives the desired dimension formula. □

Obviously we have omitted many details. I recommend [Mil80, V.1 and V.2] as a compact and readable source for several of the key points, including passing from ℓ-torsion sheaves to ℓ-adic sheaves, the conductor, and the Grothendieck-Ogg-Shafarevich formula. See [Mil80, VI.13] for the Grothendieck-Lefschetz trace formula.

Remark/Exercise 1.3.4. If we are willing to use a virtual representation of G_k in place of a usual representation, then the Theorem has a more elegant restatement which avoids singling out representations that are trivial when restricted to $G_{\overline{k}K}$. State and prove this generalization.

Exercise 1.3.5. Check that we have the Artin formalism formula: if F/K is a finite separable extension and ρ is a representation of G_F, then

$$L(\rho, s) = L(\operatorname{Ind}_{G_F}^{G_K} \rho, s).$$

Note that the left hand side is an Euler product on F with almost all factors of some degree, say N, whereas the right hand side is an Euler product on K, with almost all factors of degree $N[F:K]$. The equality can be taken to be an equality of Euler products, where that on the left is grouped according to the places of K.

1.4. Functional equation and Riemann hypothesis

Theorem 1.3.3 shows that the L-function of ρ has an analytic continuation to the entire s plane (meromorphic if we allow ρ to have trivial factors over $\overline{k}K$). In this section we deduce other good analytic properties of $L(\rho, s)$.

Theorem 1.4.1. *Suppose (in addition to the standing hypotheses) that ρ is symplectically self-dual of weight w. Then $L(\rho, s)$ satisfies a functional equation*

$$L(\rho, w + 1 - s) = \pm q^{N(s-(w+1)/2)} L(\rho, s)$$

where $N = (2g_{\mathcal{C}} - 2)\deg(\rho) + \deg(\operatorname{Cond}(\rho))$. The zeroes of ρ lie on the line $\operatorname{Re} s =$ $(w+1)/2$.

PROOF. (Sketch) We use the notation of the proof of Theorem 1.3.3. The functional equation comes from a symmetric pairing

$$H(\rho) \times H(\rho) \to H^2(\overline{\mathcal{C}}, \overline{\mathbb{Q}}_\ell(-w)) \cong \overline{\mathbb{Q}}_\ell(-w-1).$$

(Symmetric because ρ is skew-symmetric and $H = H^1$.) That there is such a pairing is not as straightforward as it looks, because we defined the sheaf $\mathcal{F}$ as a push forward $j_*\mathcal{F}_U$ where $j : U \hookrightarrow \mathcal{C}$ is a non-empty open set over which ρ is unramified and $\mathcal{F}_U$ is the lisse sheaf on U corresponding to ρ. It is well-known that j^* identifies $H^1(\overline{\mathcal{C}}, \mathcal{F})$ with the image of the "forget supports" map

$$H^1_c(\overline{U}, \mathcal{F}_U) \to H^1(\overline{U}, \mathcal{F}_U)$$

from compactly supported cohomology to usual cohomology. (This is often stated, but the only proof I know of in the literature is [Ulm05, 7.1.6].) The cup product

$$H^1_c(\overline{U}, \mathcal{F}_U) \times H^1(\overline{U}, \mathcal{F}^*_U) \to H^2_c(\overline{U}, \overline{\mathbb{Q}}_\ell) \cong \overline{\mathbb{Q}}_\ell(-1)$$

then induces a pairing on $H^1(\overline{\mathcal{C}}, \mathcal{F})$ via the above identification. Poincaré duality shows that the pairing is non-degenerate and so $H(\rho)$ is orthogonally self-dual of weight $w+1$.

The location of the zeroes is related to the eigenvalues of Frobenius on $H(\rho) = H^1(\overline{\mathcal{C}}, \mathcal{F})$ and these are Weil numbers of size q^{w+1} by Deligne's purity theorem [Del80]. I recommend the Arizona Winter School 2000 lectures of Katz (published as [Kat01]) for a streamlined proof of Deligne's theorem in the generality needed here. □

2. The case of an elliptic curve

Next, we apply the results of the previous section to elliptic curves. Throughout, E will be an elliptic curve over a function field $K = k(\mathcal{C})$ over a finite field k of characteristic p.

2.1. The Tate module

We consider the Tate module of E. More precisely, fix a prime $\ell \neq p$ and let

$$T_\ell E = \varprojlim_n E(\overline{K})[\ell^n] \quad \text{and} \quad V_\ell E = T_\ell E \otimes_{\mathbb{Z}_\ell} \mathbb{Q}_\ell.$$

Let ρ_E be the representation of G_K on the dual vector space $V_\ell^* = \operatorname{Hom}(V_\ell E, \mathbb{Q}_\ell) \cong H^1(\overline{E}, \mathbb{Q}_\ell)$. Then ρ_E is two-dimensional and continuous and (by the criterion of Ogg-Néron-Shafarevich, see [ST68, Thm. 1]) it is unramified outside the (finite) set of places where E has bad reduction.

At every place v of K where E has good reduction, we have

$$\det(1 - \rho(\mathrm{Fr}_v)T) = 1 - a_v T + q_v T^2$$

where a_v is defined as in (8.3) by $\#E_v(\kappa_v) = 1 - a_v + q_v$. This follows from the smooth base change theorem [Mil80, VI.4] and the cohomological description of the zeta function of the reduction, as in Section 4 of Lecture 0. Thus ρ is pure of weight $w = 1$.

The Weil pairing induces an alternating, G_k-equivariant pairing $V_\ell E \times V_\ell E \to \mathbb{Q}_\ell(-1)$ and so ρ is symplectically self-dual of weight 1.

If E is constant, then ρ_E factors through $G_K \to G_k$ and since G_k is abelian, ρ_E is the direct sum of two characters. More precisely, if $E \cong E_0 \times_k K$ and

$1-aT+qT^2=(1-\alpha_1T)(1-\alpha_2T)$ is the numerator of the Z-function of E_0, then ρ_E is the sum of the two characters that send Fr_v to $\alpha_i^{\deg v}$.

If E is non-isotrivial, then ρ_E restricted to $G_{\overline{k}K}$ has no trivial subrepresentations. One way to see this is to use a slight generalization of the MWLN theorem, according to which $E(\overline{k}K)$ is finitely generated (when E is non-isotrivial). Thus its ℓ-power torsion is finite and this certainly precludes a trivial subrepresentation in $\rho|_{G_{\overline{k}K}}$. In fact, by a theorem of Igusa [Igu59], $\rho|_{G_{\overline{k}K}}$ is contains an open subgroup of $\mathrm{SL}_2(\mathbb{Z}_\ell)$ so is certainly irreducible, even absolutely irreducible.

Exercise 2.1.1. Show that if E is isotrivial but not constant, then ρ_E restricted to $G_{\overline{k}K}$ has no trivial subrepresentation. Hint: E is a twist of a contant curve $E'=E_0\times_k K$. Relate the action of G_K on the Tate module of E to its action on that of E' and show that there exists an element $\sigma\in G_{\overline{k}K}$ that acts on $V_\ell E$ via a non-trivial automorphism of E. But a non-trivial automorphism has only finitely many fixed points.

We can summarize this discussion as follows.

Proposition 2.1.2. *Let ρ be the action of G_K on the Tate module $V_\ell E$ of E. Then ρ is continuous, unramified outside a finite set of places of K, and is pure and symplectically self-dual of weight 1. If E is non-constant, then $\rho|_{G_{\overline{k}K}}$ has no trivial subrepresentations.*

The conductor of ρ_E as defined in the previous section is equal to the conductor of E as mentioned in Section 8 of Lecture 1. This was proven by Ogg in [Ogg67].

2.2. The L-function

Applying the results of the previous section, we get a very satisfactory analysis of the L-function of E. Since we know everything about the constant case by an elementary analysis (cf. exercise 9.2 of Lecture 1), we restrict to the non-constant case.

Theorem 2.2.1. *Let E be a non-constant elliptic curve over $K=k(\mathcal{C})$ and let q be the cardinality of k. Let $\mathfrak{n}$ be the conductor of E. Then $L(E,s)$ is a polynomial in q^{-s} of degree $N=4g_{\mathcal{C}}-4+\deg(\mathfrak{n})$. Its inverse roots are Weil numbers of size q and it satisfies a functional equation*

$$L(E,2-s)=\pm q^{N(s-1)}L(E,s).$$

Combining the Theorem with Theorem 12.1, we obtain the following.

Corollary 2.2.2. *The rank of $E(K)$ is bounded above by $N=4g_{\mathcal{C}}-4+\deg(\mathfrak{n})$. If equality holds, then $L(E,s)=(1-q^{1-s})^N$.*

The sign in the functional equation can be computed as a product of local factors. This can be seen via the connection with automorphic forms (a connection which is outside the scope of these lectures) or, because we are in the function field situation, directly via cohomological techniques. See [Lau84] for the latter.

3. Large analytic ranks in towers

3.1. Statement of the theorem

We give a general context in which one obtains large analytic ranks by passing to layers of a suitable tower of function fields.

As usual, let p be a prime and q a power of p. Let $K=\mathbb{F}_q(t)$, for each d not divisible by p, set $F_d=\mathbb{F}_q(t^{1/d})\cong\mathbb{F}_q(u)$, and $K_d=\mathbb{F}_q(\mu_d)(t^{1/d})\cong\mathbb{F}_q(\mu_d)(u)$.

Suppose that E is an elliptic curve over K. Let $\mathfrak{n}$ be the conductor of E and let

$$\mathfrak{n}'=\mathfrak{n}-\dim(V_\ell E/V_\ell E^{I_0})[0]-\dim(V_\ell E/V_\ell E^{I_\infty})[\infty].$$

This is the conductor of E except that we have removed the tame part at $t=0$ and $t=\infty$.

Theorem 3.1.1. *Let E be an elliptic curve over K and define $\mathfrak{n}'$ as above. Suppose that $\deg\mathfrak{n}'$ is odd. Then the analytic rank of E over F_d (and K_d) is unbounded as d varies. More precisely, there exists a constant c depending only on E such that if d has the form $d=q^n+1$, then*

$$\operatorname{ord}_{s=1}L(E/F_d,s)\geq\frac{d}{2n}-c=\frac{q^n+1}{2n}-c.$$

and

$$\operatorname{ord}_{s=1}L(E/K_d,s)\geq d-c=q^n+1-c$$

This theorem is proven in detail in [Ulm07, §2-4]. We will sketch the main lines of the argument below.

3.2. A linear algebra lemma

Our analytic rank results ultimately come from the following odd-looking result of linear algebra.

Proposition 3.2.1. *Let V be a finite-dimensional vector space with subspaces W_i indexed by $i\in\mathbb{Z}/a\mathbb{Z}$ such that $V=\oplus_{i\in\mathbb{Z}/a\mathbb{Z}}W_i$. Let $\phi:V\to V$ be an invertible linear transformation such that $\phi(W_i)=W_{i+1}$ for all $i\in\mathbb{Z}/a\mathbb{Z}$. Suppose that V admits a non-degenerate, ϕ-invariant symmetric bilinear form $\langle,\rangle$. Suppose that a is even and $\langle,\rangle$ induces an isomorphism $W_{a/2}\cong W_0^*$ (the dual vector space of W_0). Suppose also that $N=\dim W_0$ is odd. Then the polynomial $1-T^a$ divides $\det(1-\phi T|V)$.*

We omit the proof of this proposition, since it is not hard and it appears in two forms in the literature already. Namely, embedded in [Ulm05, 7.1.11ff] is a matrix-language proof of the proposition, and a coordinate-free proof is given in [Ulm07, §2].

3.3. Sketch of the proof of Theorem 3.1.1

For simplicity, we assume that E is non-isotrivial. (If $p>3$ and E is isotrivial, then the theorem is vacuous because all of the local conductor exponents n_v are even.) Let ρ be the representation of G_K on $V=H^1(\overline{E},\mathbb{Q}_\ell)=(V_\ell E)^*$ and let ρ_d be the restriction of ρ to G_{F_d}. Then by Grothendieck's analysis, we have

$$L(E/F_d,s)=\det\left(1-\operatorname{Fr}_q q^{-s}|H(\rho_d)\right).$$

Here $H(\rho_d)$ is an H^1 on the rational curve whose function field is $F_d=\overline{\mathbb{F}}_q(u)=\overline{\mathbb{F}}_q(t^{1/d})$.

The projection formula in cohomology (a parallel of the Artin formalism 1.3.5) implies that

$$H(\rho_d)\cong H(\operatorname{Ind}_{G_{F_d}}^{G_K}\rho)\cong H(\rho\otimes\operatorname{Ind}_{G_{F_d}}^{G_K}\mathbf{1})$$

where **1** denotes the trivial representation. Since the cohomology H is computed on $\overline{\mathbb{P}}^1_u$ (the $\mathbb{P}^1$ with coordinate u, with scalars extended to $\overline{\mathbb{F}}_q$) and $\overline{\mathbb{P}}^1_u \to \overline{\mathbb{P}}^1_t$ is Galois with group μ_d, we have

$$H(\rho_d) \cong \bigoplus_{j=0}^{d-1} H(\rho \otimes \chi^j)$$

where χ is a character of $\mathrm{Gal}(\overline{\mathbb{F}}_q(u)/\overline{\mathbb{F}}_q(t))$ of order exactly d.

Now the decomposition displayed above is not preserved by Frobenius. Indeed Fr_q sends $H(\rho \otimes \chi^j)$ to $H(\rho \otimes \chi^{qj})$. Thus we let $o \subset \mathbb{Z}/d\mathbb{Z}$ denote an orbit for multiplication by q and we regroup:

$$H(\rho_d) \cong \bigoplus_{o \subset \mathbb{Z}/d\mathbb{Z}} \left(\bigoplus_{j \in o} H(\rho \otimes \chi^j) \right).$$

We write V_o for the summand indexed by an orbit $o \subset \mathbb{Z}/d\mathbb{Z}$ in the last display and a_o for the cardinality of o. As we will see presently, the hypotheses of the theorem imply that Proposition 3.2.1 applies to most of the V_o and for each one where it does, we get a zero of the L-function. Before we do that, there is one small technical point to take care of: The linear algebra proposition requires that V be literally self-dual (not self-dual with a weight) and it implies that 1 is an eigenvalue of ϕ on V. To get the eigenvalue q that we need, we should twist ρ by $-1/2$ (which is legitimate once we have fixed choice of square root of q) so that it has weight 0, apply the lemma, and twist back to get the desired zero. We leave the details of these points to the reader.

Assuming we have made the twist just mentioned, we need to check which V_o are self-dual. Since ρ is self-dual, Poincaré duality gives a non-degenerate pairing on $H(\rho_d)$ which puts $H(\rho \otimes \chi^j)$ in duality with $H(\rho \otimes \chi^{-j})$. Thus if $d = q^n + 1$ for some $n > 0$, then all of the orbits o will yield a self-dual V_o. Possibly two of these orbits have odd order (those through 0 and $d/2$, which have order 1) and all of the other have a_o even. Moreover, for the orbits of even order, setting $W_{o,i} = H(\rho \otimes \chi^{q^i j_o})$ for some fixed $j_o \in o$, we have

$$V_o \cong \bigoplus_{i=0}^{a_o - 1} W_{o,i}$$

with $W_{o,i}$ and $W_{o,i+a_0/2}$ in duality.

The last point that we need is that $W_{o,i}$ should be odd-dimensional. The hypothesis on $\mathfrak{n}'$ implies that for all characters χ^j of sufficiently high order (depending only on E), the conductor of $\rho \otimes \chi^j$ is odd. The Grothendieck-Ogg-Shafarevich dimension formula (mentioned at the end of the proof of Theorem 1.3.3) then implies that for all orbits o consisting of characters of high order, $H(\rho \otimes \chi^{j_o})$ has odd dimension.

The linear algebra proposition 3.2.1 now implies that for $d = q^n + 1$ and for most orbits $o \subset \mathbb{Z}/d\mathbb{Z}$, 1 is an eigenvalue of Fr_q on V_o (and q is an eigenvalue of Fr_q on the corresponding factor of $H(\rho_d)$). Since each of these orbits has size $\leq 2n$, there is a constant c such that the number of "good" orbits is $\geq d/2n$. Thus

$$\mathrm{ord}_{s=1} L(E/F_d, s) \geq \frac{d}{2n} - c$$

for a constant c depending only on E.

To get the assertions over K_d, note that in passing from F_d to K_d, each factor $(1-q^{a_o}T^{a_o})$ of $L(E/F_d,T)$ becomes $(1-qT)^{a_o}$ and so

$$\operatorname{ord}_{s=1} L(E/K_d,s) \geq d-c$$

for another c independent of E.

This completes our discussion of Theorem 3.1.1. We refer to [Ulm07, §2-4] for more details. □

3.4. Examples

It is easy to see that the hypotheses in Theorem 3.1.1 are not very restrictive and that high analytic ranks are in a sense ubiquitous. The following rephrasing of the condition in the theorem should make this clear.

Exercise 3.4.1. Prove that if $p>3$ and E is an elliptic curve over K, then Theorem 3.1.1 guarantees that E has unbounded analytic rank in the tower F_d if the number of geometric points of $\mathbb{P}^1_{\mathbb{F}_q}$ over which E has multiplicative reduction is odd.

Corollary 3.4.2. *Let p be any prime number, $K=\mathbb{F}_p(t)$, and let E be one of the curves E_7, E_8, or E_9 defined in Subsection 1.2 of Lecture 1. Then*

$$\operatorname{ord}_{s=1} L(E/\mathbb{F}_p(t^{1/d}),s)$$

is unbounded as d varies through integers prime to p

PROOF. If $p>3$, then one sees immediately by considering the discriminant and j-invariant that E has one finite, non-zero place of multiplicative reduction and is tame at 0 and ∞, thus it satisfies the hypotheses of Theorem 3.1.1. If $p=2$ or 3, one checks using Tate's algorithm that E has good reduction at all finite non-zero places and is tame at zero, but the wild part of the conductor at ∞ is odd and so the theorem again applies. □

For another example, take the Legendre curve

$$y^2 = x(x-1)(x-t)$$

over $\mathbb{F}_p(t)$, $p>2$. It is tame at 0 and ∞ and has exactly one finite, non-zero place of multiplicative reduction.

4. Large algebraic ranks

4.1. Examples via the four-monomial theorem

Noting that the curves E_7, E_8, and E_9 are defined by equations involving exactly four monomials, we get a very nice result on algebraic ranks.

Theorem 4.1.1. *Let p be any prime number, $K=\mathbb{F}_p(t)$, and let E be one of the curves E_7, E_8, or E_9 defined in Subsection 1.2 of Lecture 1. Then for all d prime to p and all powers q of p, the Birch and Swinnerton-Dyer conjecture holds for E over $K_d=\mathbb{F}_q(t^{1/d})$. Moreover, the rank of $E(\mathbb{F}_p(t^{1/d}))$ is unbounded as d varies.*

PROOF. This follows immediately from Corollary 3.4.2 and Theorem 12.4 of Lecture 1 as soon as we note that E/K_d is defined by an equation satisfying Shioda's conditions. □

Similar ideas can be used to show that for every prime p and every genus $g > 0$, there is an explicit hyperelliptic curve C over $\mathbb{F}_p(t)$ such that the Jacobian of C satisfies BSD over $\mathbb{F}_q(t^{1/d})$ for all q and d and has unbounded rank in the tower $\mathbb{F}_p(t^{1/d})$. This is the main theorem of [Ulm07].

4.2. Examples via Berger's construction

As we pointed out in Lecture 3, the Shioda 4-monomial construction is rigid—varying the coefficients does not lead to families that vary geometrically. Berger's thesis developed a new construction with parameters that leads to families of curves for which the BSD conjecture holds in a tower of fields. This together with the analytic ranks result 3.1.1 gives examples of families of elliptic curves with unbounded ranks.

To make this concrete, we quote the first example with parameters from [Ber08] that, together with the analytic rank construction 3.1.1, gives rise to unbounded analytic and algebraic ranks.

Theorem 4.2.1 (Berger). *Let $k = \mathbb{F}_q$ be a finite field of characteristic p and let $a \in \mathbb{F}_q$ with $a \neq 0, 1, 2$. Let E be the elliptic curve over $K = \mathbb{F}_q(t)$ defined by*

$$y^2 + a(t-1)xy + a(t^2-t)y = x^3 + (2a+1)tx^2 + a(a+2)t^2x + a^2t^3.$$

Then for all d prime to p the BSD conjecture holds for E over $\mathbb{F}_q(t^{1/d})$. Moreover, for every q and a as above, the rank of $E(\mathbb{F}_q(t^{1/d}))$ is unbounded as d varies.

PROOF. This is an instance of Berger's construction (Theorem 11.1 of Lecture 3). Indeed, let $f(x) = x(x-a)/(x-1)$ and $g(y) = y(y-a)/(y-1)$. Then $V(f - tg) \subset \mathbb{P}^1_K \times \mathbb{P}^1_K$ is birational to E, which is a smooth elliptic curve for all $a \neq 0, 1$. Berger's Theorem 11.1 of Lecture 3 shows that E satisfies BSD over the fields $\mathbb{F}_q(t^{1/d})$.

The discriminant of E is

$$\Delta = a^2(a-1)^4t^4(t-1)^2\left(a^2t^2 - (2a^2 - 16a + 16)t + a^2\right).$$

Assume first that $p > 3$. One checks that Δ is relatively prime to c_4 so that the zeroes of Δ are places of multiplicative reduction. Since the discriminant (in t) of the quadratic factor $a^2t^2 - (2a^2 - 16a + 16)t + a^2$ is $-64(a-1)(a-2)^2$ we see that there are three finite, non-zero geometric points of multiplicative reduction. Since $p > 3$, the reduction at 0 and ∞ is tame and so $\mathfrak{n}'$ (defined as in Subsection 3.1 of Lecture 4) has degree 3. Thus by Theorem 3.1.1 of Lecture 4, E has unbounded analytic ranks in the tower $\mathbb{F}_q(t^{1/d})$ and thus also unbounded algebraic ranks by the previous paragraph on BSD.

If $p = 2$ or 3, one needs to use Tate's algorithm to compute $\mathfrak{n}'$, which again turns out to have degree 3. We leave the details of this computation as a pleasant exercise for the reader. □

LECTURE 5

More applications of products of curves

In the last part of Lecture 4, we chose special curves E and used a domination $\mathcal{C}\times\mathcal{D}\dashrightarrow\mathcal{E}$ of the associated surface to deduce the Tate conjecture for $\mathcal{E}$ and thus the BSD conjecture for E. This yields an *a priori* equality of analytic and algebraic ranks. We then used other, cohomological, methods (namely the analytic ranks theorem) to compute the analytic rank.

It turns out to be possible to use domination by a product of curves and geometry to prove directly results about algebraic ranks and explicit points. We sketch some of these applications in this lecture.

1. More on Berger's construction

Let k be a field (not necessarily finite), $K=k(t)$, and $K_d=k(t^{1/d})=k(u)$. Recall that in Berger's construction we start with rational curves $\mathcal{C}=\mathbb{P}^1_k$ and $\mathcal{D}=\mathbb{P}^1_k$ and rational functions $f(x)$ on $\mathcal{C}$ and $g(y)$ on $\mathcal{D}$. We get a curve in $\mathbb{P}^1_K\times\mathbb{P}^1_K$ defined by $f(x)-tg(y)=0$ and we let E be the smooth proper model over K of this curve. (Some hypotheses are required for this to exist, but they are weaker than our standing hypotheses below.) The genus of E was computed by Berger in [Ber08, Theorem 3.1]. All the examples we consider will be of genus 1 and will have a K-rational point.

We establish more notation to state a precise result. Let us assume for simplicity all the zeroes and poles of f and g are k-rational. Write

$$\operatorname{div}(f)=\sum_{i=1}^{k}a_iP_i-\sum_{i'=1}^{k'}a'_{i'}P'_{i'}\quad\text{and}\quad\operatorname{div}(g)=\sum_{j=1}^{\ell}b_jQ_j-\sum_{j'=1}^{\ell'}b'_{j'}Q'_{j'}\tag{1.1}$$

with $a_i,a'_{i'},b_i,b'_{j'}$ positive integers and P_i, $P'_{i'}$, Q_j, and $Q'_{j'}$ distinct k-rational points. Let

$$m=\sum_{i=1}^{k}a_i=\sum_{i'=1}^{k'}a'_{i'}\quad\text{and}\quad n=\sum_{j=1}^{\ell}b_j=\sum_{j'=1}^{\ell'}b'_{j'}.$$

As standing hypotheses, we assume that: (i) all the multiplicities a_i, $a'_{i'}$, b_j, and $b'_{j'}$ are prime to the characteristic of k; and (ii) $\gcd(a_1\dots,a_k,a'_1,\dots,a'_{k'})=\gcd(b_1\dots,b_\ell,b'_1,\dots,b'_{\ell'})=1$.

Under these hypotheses, Berger computes that the genus of E is

$$g_E=(m-1)(n-1)-\sum_{i,j}\delta(a_i,b_j)-\sum_{i',j'}\delta(a'_{i'},b'_{j'})\tag{1.2}$$

where $\delta(a,b)=(ab-a-b+\gcd(a,b))/2$.

From now on we assume that we have chosen the data f and g so that E has genus 1. Two typical cases are where f and g are quadratic rational functions with simple zeroes and poles, or where f and g are cubic polynomials. There is always a K-rational point on E; for example, we may take a point where x and y are zeroes of f and g.

Let $\mathcal{E}_d \to \mathbb{P}^1$ be the elliptic surface over k attached to E/K_d. It is clear that $\mathcal{E}_d$ is birational to the closed subset of $\mathbb{P}^1_k \times \mathbb{P}^1_k \times \mathbb{P}^1_k$ (with coordinates x, y, u) defined by the vanishing of $f(x) - u^d g(y)$. We saw in Section 11 of Lecture 3 that $\mathcal{E}$ is dominated by a product of curves and we would now like to make this more precise.

Recall that we defined covers $\mathcal{C}_d \to \mathcal{C} = \mathbb{P}^1$ and $\mathcal{D}_d \to \mathcal{D} = \mathbb{P}^1$ by the equations $z^d = f(x)$ and $w^d = g(y)$. Note that there is an action of μ_d, the d-th roots of unity, on $\mathcal{C}_d$ and on $\mathcal{D}_d$.

Proposition 1.3. *The surface $\mathcal{E}_d$ is birationally isomorphic to the quotient surface $(\mathcal{C}_d \times \mathcal{D}_d)/\mu_d$ where μ_d acts diagonally.*

PROOF. We have already noted that $\mathcal{E}_d$ is birational to the zero set $\mathcal{X}$ of $f(x) - u^d g(y)$ in $\mathbb{P}^1_k \times \mathbb{P}^1_k \times \mathbb{P}^1_k$. Define a rational map from $\mathcal{C}_d \times \mathcal{D}_d$ to $\mathcal{X}$ by sending (x, z, y, w) to $(x, y, u = z/w)$. It is clear that this map factors through the quotient $(\mathcal{C}_d \times \mathcal{D}_d)/\mu_d$. Since the map is generically of degree d, it induces a birational isomorphism between $(\mathcal{C}_d \times \mathcal{D}_d)/\mu_d$ and $\mathcal{X}$. Thus $(\mathcal{C}_d \times \mathcal{D}_d)/\mu_d$ is birationally isomorphic to $\mathcal{E}_d$. □

In the next section we will explain how this birational isomorphism can be used to compute the Néron-Severi group of $\mathcal{E}_d$ and the Mordell-Weil group $E(K_d)$.

2. A rank formula

We keep the notation and hypotheses of the preceding subsection. Consider the base $\mathbb{P}^1_k$, the one corresponding to K, with coordinate t. For each geometric point x of this $\mathbb{P}^1_k$, let f_x be the number of components in the fiber of $\mathcal{E} \to \mathbb{P}^1$ over x. For almost all x, $f_x = 1$ and its value at any point can be computed using Tate's algorithm.

Define two constants c_1 and c_2 by the formulae

$$c_1 = \sum_{x \neq 0, \infty} (f_x - 1)$$

and

$$c_2 = (k-1)(\ell-1) + (k'-1)(\ell'-1).$$

Here the sum is over geometric points of $\mathbb{P}^1_k$ except $t = 0$ and $t = \infty$ and k, k', ℓ, and ℓ' are the numbers of distinct zeroes and poles of f and g (cf. equation (1.1)). Note that c_1 and c_2 depend only on the data defining E/K, not on d.

Theorem 2.1. *Suppose that k is algebraically closed and that d is relatively prime to all of the multiplicities a_i, $a'_{i'}$, b_j, and $b'_{j'}$ and to the characteristic of k. Then we have*

$$\operatorname{Rank} E(K_d) = \operatorname{Rank} \operatorname{Hom}(J_{\mathcal{C}_d}, J_{\mathcal{D}_d})^{\mu_d} - c_1 d + c_2.$$

Here $\operatorname{Hom}(\cdots)^{\mu_d}$ signifies the homomorphisms commuting with the actions of μ_d on the two Jacobians induced by its action on the curves.

SKETCH OF PROOF. In brief, we use the birational isomorphism

$$(\mathcal{C}_d \times \mathcal{D}_d)/\mu_d \dashrightarrow \mathcal{E}_d$$

to compute the rank of the Néron-Severi group of $\mathcal{E}_d$ and then use the Shioda-Tate formula to compute the rank of $E(K_d)$.

More precisely, we saw in Lecture 2, Subsection 8.4 that the Néron-Severi group of the product $\mathcal{C}_d \times \mathcal{D}_d$ is isomorphic to $\mathbb{Z}^2 \times \operatorname{Hom}(J_{\mathcal{C}_d}, J_{\mathcal{D}_d})$. It follows easily that the Néron-Severi group of the quotient $(\mathcal{C}_d \times \mathcal{D}_d)/\mu_d$ is isomorphic to $\mathbb{Z}^2 \times \operatorname{Hom}(J_{\mathcal{C}_d}, J_{\mathcal{D}_d})^{\mu_d}$.

One then keeps careful track of the blow-ups needed to pass from $(\mathcal{C}_d \times \mathcal{D}_d)/\mu_d$ to $\mathcal{E}_d$. The effect of blow-ups on Néron-Severi is quite simple and was noted in Subsection 8.5 of Lecture 2. This is the main source of the term c_2 in the formula.

Finally, one computes the rank of $E(K_d)$ using the Shioda-Tate formula, as in Section 5 of Lecture 3. This step is the main source of the term $c_1 d$.

The hypothesis that k is algebraically closed is not essential for any of the above, but it avoids rationality questions that would greatly complicate the formula.

For full details on the proof of this theorem (in a more general context) see [Ulm09a, Section 6]. □

3. First examples

One of the first examples is already quite interesting. We give a brief sketch and refer to [Ulm09a] for more details.

With notation as in Section 1, we take $f(x) = x(x-1)$ and $g(y) = y^2/(1-y)$. The genus formula (1.2) shows that E has genus 1. In fact, the change of coordinates $x = -y/(x+t)$, $y = -x/t$ brings it into the Weierstrass form

$$y^2 + xy + ty = x^3 + tx^2.$$

We remark in passing that if the characteristic of k is not 2, E has multiplicative reduction at $t = 1/16$ and good reduction elsewhere away from 0 and ∞. Thus by the analytic rank result of Lecture 2, when k is finite, say $k = \mathbb{F}_p$ and $p > 3$, we expect E to have unbounded analytic rank in the tower $\mathbb{F}_p(t^{1/d})$. (In fact a more careful analysis gives the same conclusion for every p.)

Now assume that k is algebraically closed. To compute the constant c_1, one checks that (for k of any characteristic) E has exactly one irreducible component over each geometric point of $\mathbb{P}^1_k$. Thus $c_1 = 0$. It is immediate from the definition that $c_2 = 0$. Thus our rank formula yields

$$\operatorname{Rank} E(K_d) = \operatorname{Rank} \operatorname{Hom}(J_{\mathcal{C}_d}, J_{\mathcal{D}_d})^{\mu_d}.$$

Next we note that there is an isomorphism $\phi : \mathcal{C}_d \to \mathcal{D}_d$ sending (x, z) to $(y = 1/x, w = 1/z)$. This isomorphism *anti-commutes* with the μ_d action: Let ζ_d be a primitive d-th root of unity and write $[\zeta_d]$ for its action on curves or Jacobians. Then $\phi \circ [\zeta_d] = [\zeta_d^{-1}] \circ \phi$. Using ϕ to identify $\mathcal{C}_d$ and $\mathcal{D}_d$, our rank formula becomes

$$\operatorname{Rank} E(K_d) = \operatorname{Rank} \operatorname{End}(J_{\mathcal{C}_d})^{anti-\mu_d}$$

where "$\operatorname{End}(\cdots)^{anti-\mu_d}$" denotes those endomorphisms anti-commuting with μ_d in the sense above.

Suppose that k has characteristic zero. Then a consideration of the (faithful) action of $\operatorname{End}(J_{\mathcal{C}_d})$ on the differentials $H^0(J_{\mathcal{C}_d}, \Omega^1)$ shows that $\operatorname{End}(J_{\mathcal{C}_d})^{anti-\mu_d} = 0$ for all d (see [Ulm09a, 7.6]). We conclude that for k of characteristic zero, the rank of $E(K_d)$ is zero for all d.

Now assume that k has characteristic p (and is algebraically closed). If we take d of the form p^f+1 then we get many elements of $\mathrm{End}(J_{\mathcal{C}_d})^{anti-\mu_d}$. Namely, we consider the Frobenius Fr_{p^f} and compute that

$$\mathrm{Fr}_{p^f}\circ[\zeta_d]=[\zeta_d^{p^f}]\circ\mathrm{Fr}_{p^f}=[\zeta_d^{-1}]\circ\mathrm{Fr}_{p^f}\,.$$

The same computation shows that $\mathrm{Fr}_{p^f}\circ[\zeta_d^i]$ anticommutes with μ_d for all i. It turns out that there are two relations among these endomorphism in $\mathrm{End}(J_{\mathcal{C}_d})$ if $p>2$ and just one relation if $p=2$ (see [Ulm09a, 7.8-7.10]). Thus we find that, for d of the special form $d=p^f+1$,

$$\mathrm{Rank}\,E(\overline{\mathbb{F}}_p(t^{1/d}))=\begin{cases}d-2 & \text{if } p>2\\ d-1 & \text{if } p=2.\end{cases}$$

The reader may enjoy checking that this is in exact agreement with what the analytic rank result (Theorem 3.1.1 of Lecture 4) predicts.

Somewhat surprisingly, there are *more* values of d for which we get high ranks. A natural question is to identify all pairs (p,d) such that $E(\overline{\mathbb{F}}_p(t^{1/d})$ has "new" rank, i.e, points of infinite order not coming from smaller values of d. The exact set of pairs (p,d) for which we get high rank is mysterious. There are "systematic" cases (such as (p,p^f+1), as above, or $(p,2(p-1)))$ and other cases that may be sporadic. This is the subject of ongoing research so we will not go into more detail, except to note that the example in Section 5 below is relevant to this question.

4. Explicit points

The main ingredients in the rank formula of Section 2 are the calculation of the Néron-Severi group of a product of curves in terms of homomorphisms of Jacobians and the Shioda-Tate formula. Tracing through the proof leads to a homomorphism

$$\mathrm{Hom}(J_{\mathcal{C}_d},J_{\mathcal{D}_d})^{\mu_d}\cong\mathrm{DivCorr}(\mathcal{C}_d,\mathcal{D}_d)\to L^1\,\mathrm{NS}(\mathcal{E}_d)\to\frac{L^1\,\mathrm{NS}(\mathcal{E}_d)}{L^2\,\mathrm{NS}(\mathcal{E}_d)}\cong E(K_d).$$

For elements of $\mathrm{Hom}(J_{\mathcal{C}_d},J_{\mathcal{D}_d})^{\mu_d}$ where we can find an explicit representation in $\mathrm{DivCorr}(\mathcal{C}_d,\mathcal{D}_d)$, the geometry of Berger's construction leads to explicit points in $E(K_d)$. This applies notably to the endomorphisms $\mathrm{Fr}_{p^f}\circ[\zeta_d^i]$ appearing in the analysis of the first example above. Indeed, these endomorphisms are represented in $\mathrm{DivCorr}(\mathcal{C}_d,\mathcal{D}_d)$ by the graphs of Frobenius composed with the automorphisms $[\zeta_d^i]$ of $\mathcal{C}_d$.

Tracing through the geometry leads to remarkable explicit expressions for points in $E(K_d)$. The details of the calculation are presented in [Ulm09a, §8] so we will just state the results here, and only in the case $p>2$.

Theorem 4.1. *Let $p>2$, $k=\overline{\mathbb{F}}_p$ and $K=k(t)$. Let E be the elliptic curve*

$$y^2+xy+ty=x^3+tx^2$$

over K. Let $q=p^f$, $d=q+1$, $K_d=k(t^{1/d})$, and

$$P(u)=\left(\frac{u^q(u^q-u)}{(1+4u)^q},\frac{u^{2q}(1+2u+2u^q)}{2(1+4u)^{(3q-1)/2}}-\frac{u^{2q}}{2(1+4u)^{q-1}}\right).$$

Then the points $P_i=P(\zeta_d^it^{1/d})$ for $i=0,\dots,d-1$ lie in $E(K_d)$ and they generate a finite index subgroup of $E(K_d)$, which has rank $d-2$. The relations among them are that $\sum_{i=0}^{d-1}P_i$ and $\sum_{i=0}^{d-1}(-1)^iP_i$ are torsion.

It is elementary to check that the points lie in $E(K_d)$. To check their independence and the relations by elementary means, one may compute the height pairing on the lattice they generate. It turns out to be a scaling of the direct sum of two copies of the $A^*_{(d-2)/2}$ lattice. Since we know from the previous section that $E(K_d)$ has rank $d-2$, the explicit points generate a subgroup of finite index. As another check that they have finite index, we could compute the conductor of E—it turns out to have degree $d+2$—and apply Corollary 2.2.2 of Lecture 4. All this is explained in detail in [Ulm09a, §8].

5. Another example

We keep the notation and hypotheses of Sections 1 and 2. For another example, assume that $k=\overline{\mathbb{F}}_p$ with $p>2$. Let $f(x)=x/(x^2-1)$ and $g(y)=y(y-1)$. The curve $f(x)-tg(y)=0$ has genus 1 and the change of coordinates $x=(x'+t)/(x'-t)$, $y=-y'/2tx'$ brings it into the Weierstrass form

$$y'^2+2tx'y'=x'^3-t^2x'.$$

This curve, call it E, has multiplicative reduction of type I_1 at the places dividing t^2+4, good reduction at other finite, non-zero places, and tame reduction at $t=0$ and $t=\infty$. We find that the constants c_1 and c_2 are both zero and that

$$\operatorname{Rank} E(\overline{\mathbb{F}}_p(t^{1/d}))=\operatorname{Rank}\operatorname{Hom}(J_{\mathcal{C}_d},J_{\mathcal{D}_d})^{\mu_d}.$$

Recall that the curves $\mathcal{C}_d$ and $\mathcal{D}_d$ are defined by the equations

$$z^d=f(x)=\frac{x}{x^2-1}\quad\text{and}\quad w^d=g(y)=y(y-1).$$

Consider the morphism $\phi:\mathcal{C}_d\to\mathcal{D}_d$ defined by $\phi^*(y)=1/(1-x^2)$ and $\phi^*(w)=z^2$. It is obviously not constant and so induces a surjective homomorphism $\phi_*:J_{\mathcal{C}_d}\to J_{\mathcal{D}_d}$.

The homomorphism ϕ_* clearly does not commute with the action of μ_d. Indeed, if ζ_d denotes a primitive d-th root of unity and $[\zeta_d]$ its action on one of the Jacobians, we have $\phi_*\circ[\zeta_d]=[\zeta_d^2]\circ\phi_*$. (This formula already holds at the level of the curves $\mathcal{C}_d$ and $\mathcal{D}_d$.)

Now let us assume that d has the form $d=2p^f-1$ and consider the map $\phi\circ\mathrm{Fr}_{p^f}:\mathcal{C}_d\to\mathcal{D}_d$. Then we find that

$$(\phi\circ\mathrm{Fr}_{p^f})_*\circ[\zeta_d]=[\zeta_d^{2p^f}]\circ(\phi\circ\mathrm{Fr}_{p^f})_*=[\zeta_d]\circ(\phi\circ\mathrm{Fr}_{p^f})_*$$

in $\operatorname{Hom}(J_{\mathcal{C}_d},J_{\mathcal{D}_d})$, in other words that $(\phi\circ\mathrm{Fr}_{p^f})_*$ commutes with the μ_d action. Similarly $([\zeta_d^i]\circ\phi\circ\mathrm{Fr}_{p^f})_*$ commutes with the μ_d action for all i.

Further analysis of the homomorphisms $([\zeta_d^i]\circ\phi\circ\mathrm{Fr}_{p^f})_*$ in $\operatorname{Hom}(J_{\mathcal{C}_d},J_{\mathcal{D}_d})^{\mu_d}$ (along the lines of [Ulm09a, 7.8]) shows that they are almost independent; more precisely, they generate a subgroup of rank $d-1$. Thus we find (for d of the form $d=2p^f-1$) that the rank of $E(k(t^{1/d}))$ is at least $d-1$.

The reader may find it a pleasant exercise to write down explicit points in this situation, along the lines of the discussion in Section 4 and [Ulm09a, §8].

6. Further developments

There have been further developments in the area of rational points on curves and Jacobians over function fields. To close, we mention three of them.

In the examples of Sections 3 and 5, the set of d that are "interesting," i.e., for which we get high rank over K_d, depends very much on p, the characteristic of k. In his thesis (University of Arizona, 2010), Tommy Occhipinti gives, for every p, remarkable examples of elliptic curves E over $\mathbb{F}_p(t)$ such that for *all* d prime to p we have

$$\operatorname{Rank} E(\overline{\mathbb{F}}_p(t^{1/d})) \geq d.$$

The curves come from Berger's construction where f and g are generic degree two rational functions. The rank inequality comes from the rank formula in Theorem 2.1 and the Honda-Tate theory of isogeny classes of abelian varieties over finite fields.

In the opposite direction, the author and Zarhin have given examples of curves of every genus over $\mathbb{C}(t)$ such that their Jacobians have bounded rank in the tower of fields $\mathbb{C}(t^{1/\ell^n})$ where ℓ is a prime. See [UZ10].

Finally, after some encouragement by Dick Gross at PCMI, the author produced explicit points on the Legendre curve over the fields $\mathbb{F}_p(\mu_d)(t^{1/d})$ where d has the form p^f+1 and proved in a completely elementary way that they give Mordell-Weil groups of unbounded rank. In fact, this construction is considerably easier than that of Tate and Shafarevich [TS67] and could have been found in the 1960s. See [Ulm09b].

It appears that this territory is rather fertile and that there is much still to be discovered about high ranks and explicit points on curves and Jacobians over function fields. Happy hunting!

Bibliography

[Art86] M. Artin, *Lipman's proof of resolution of singularities for surfaces*, Arithmetic geometry (Storrs, Conn., 1984), Springer, New York, 1986, pp. 267–287. ↑235, 249

[ASD73] M. Artin and H. P. F. Swinnerton-Dyer, *The Shafarevich-Tate conjecture for pencils of elliptic curves on K3 surfaces*, Invent. Math. **20** (1973), 249–266. ↑231, 257

[Băd01] L. Bădescu, *Algebraic surfaces*, Universitext, Springer-Verlag, New York, 2001. Translated from the 1981 Romanian original by Vladimir Maşek and revised by the author. ↑235, 249, 253

[BHPV04] W. P. Barth, K. Hulek, C. A. M. Peters, and A. Van de Ven, *Compact complex surfaces*, Second, Ergebnisse der Mathematik und ihrer Grenzgebiete. 3. Folge. A Series of Modern Surveys in Mathematics [Results in Mathematics and Related Areas. 3rd Series. A Series of Modern Surveys in Mathematics], vol. 4, Springer-Verlag, Berlin, 2004. ↑235

[Bea96] A. Beauville, *Complex algebraic surfaces*, Second, London Mathematical Society Student Texts, vol. 34, Cambridge University Press, Cambridge, 1996. Translated from the 1978 French original by R. Barlow, with assistance from N. I. Shepherd-Barron and M. Reid. ↑235, 238

[Ber08] L. Berger, *Towers of surfaces dominated by products of curves and elliptic curves of large rank over function fields*, J. Number Theory **128** (2008), 3013–3030. ↑232, 269, 271

[BSD65] B. J. Birch and H. P. F. Swinnerton-Dyer, *Notes on elliptic curves. II*, J. Reine Angew. Math. **218** (1965), 79–108. ↑230

[Chi86] T. Chinburg, *Minimal models for curves over Dedekind rings*, Arithmetic geometry (Storrs, Conn., 1984), Springer, New York, 1986, pp. 309–326. ↑247

[Con05] B. Conrad, *Minimal models for elliptic curves* (2005). Preprint. ↑249

[Con06] B. Conrad, *Chow's K/k-image and K/k-trace, and the Lang-Néron theorem*, Enseign. Math. (2) **52** (2006), 37–108. ↑225, 237

[Del75] P. Deligne, *Courbes elliptiques: formulaire d'après J. Tate*, Modular functions of one variable, IV (Proc. Internat. Summer School, Univ. Antwerp, Antwerp, 1972), Lecture Notes in Mathematics, vol. 476, Springer, Berlin, 1975, pp. 53–73. ↑221

[Del80] P. Deligne, *La conjecture de Weil. II*, Inst. Hautes Études Sci. Publ. Math. **52** (1980), 137–252. ↑264

[Deu40] M. Deuring, *Arithmetische Theorie der Korrespondenzen algebraischer Funktionenkörper. II*, J. Reine Angew. Math. **183** (1940), 25–36. ↑219

[Elk94] N. D. Elkies, *Mordell-Weil lattices in characteristic 2. I. Construction and first properties*, Internat. Math. Res. Notices (1994), 343–361. ↑226

[Fal86] G. Faltings, *Finiteness theorems for abelian varieties over number fields*, Arithmetic geometry (Storrs, Conn., 1984), Springer, New York, 1986, pp. 9–27. Translated from the German original [Invent. Math. **73** (1983), no. 3, 349–366; ibid. **75** (1984), no. 2, 381; MR 85g:11026ab] by Edward Shipz. ↑223

[Ful84] W. Fulton, *Intersection theory*, Ergebnisse der Mathematik und ihrer Grenzgebiete (3) [Results in Mathematics and Related Areas (3)], vol. 2, Springer-Verlag, Berlin, 1984. ↑236

[Gol03] D. M. Goldschmidt, *Algebraic functions and projective curves*, Graduate Texts in Mathematics, vol. 215, Springer-Verlag, New York, 2003. ↑216

[Gor79] W. J. Gordon, *Linking the conjectures of Artin-Tate and Birch-Swinnerton-Dyer*, Compositio Math. **38** (1979), 163–199. ↑255

[Gro10] B. Gross, *Lectures on the conjecture of Birch and Swinnerton-Dyer* (2010). Notes from a course at the 2009 Park City Mathematics Institute. ↑230, 231

[Gro68] A. Grothendieck, *Le groupe de Brauer. III. Exemples et compléments*, Dix Exposés sur la Cohomologie des Schémas, North-Holland, Amsterdam, 1968, pp. 88–188. ↑243, 256

[Har77] R. Hartshorne, *Algebraic geometry*, Graduate Texts in Mathematics, No. 52, Springer-Verlag, New York, 1977. ↑221, 236, 238, 239, 240

[Igu55] J-I. Igusa, *On some problems in abstract algebraic geometry*, Proc. Nat. Acad. Sci. U. S. A. **41** (1955), 964–967. ↑237

[Igu59] J-I. Igusa, *Fibre systems of Jacobian varieties. III. Fibre systems of elliptic curves*, Amer. J. Math. **81** (1959), 453–476. ↑265

[Igu68] J-I. Igusa, *On the algebraic theory of elliptic modular functions*, J. Math. Soc. Japan **20** (1968), 96–106. ↑227

[Kat01] N. M. Katz, *L-functions and monodromy: four lectures on Weil II*, Adv. Math. **160** (2001), 81–132. ↑264

[KM85] N. M. Katz and B. Mazur, *Arithmetic moduli of elliptic curves*, Annals of Mathematics Studies, vol. 108, Princeton University Press, Princeton, NJ, 1985. ↑224, 227, 228

[KS99] N. M. Katz and P. Sarnak, *Random matrices, Frobenius eigenvalues, and monodromy*, American Mathematical Society Colloquium Publications, vol. 45, American Mathematical Society, Providence, RI, 1999. ↑261

[Kle05] S. L. Kleiman, *The Picard scheme*, Fundamental algebraic geometry, 2005, pp. 235–321. ↑237

[LN59] S. Lang and A. Néron, *Rational points of abelian varieties over function fields*, Amer. J. Math. **81** (1959), 95–118. ↑225, 237

[Lau84] G. Laumon, *Les constantes des équations fonctionnelles des fonctions L sur un corps global de caractéristique positive*, C. R. Acad. Sci. Paris Sér. I Math. **298** (1984), 181–184. ↑265

[Lev68] M. Levin, *On the group of rational points on elliptic curves over function fields*, Amer. J. Math. **90** (1968), 456–462. ↑227

[Lic68] S. Lichtenbaum, *Curves over discrete valuation rings*, Amer. J. Math. **90** (1968), 380–405. ↑247

[Lip78] J. Lipman, *Desingularization of two-dimensional schemes*, Ann. Math. (2) **107** (1978), 151–207. ↑235

[Liu02] Q. Liu, *Algebraic geometry and arithmetic curves*, Oxford Graduate Texts in Mathematics, vol. 6, Oxford University Press, Oxford, 2002. Translated from the French by Reinie Erné, Oxford Science Publications. ↑221, 235, 247

[Mat57] T. Matsusaka, *The criteria for algebraic equivalence and the torsion group*, Amer. J. Math. **79** (1957), 53–66. ↑239

[Mer96] L. Merel, *Bornes pour la torsion des courbes elliptiques sur les corps de nombres*, Invent. Math. **124** (1996), 437–449. ↑227

[Mil75] J. S. Milne, *On a conjecture of Artin and Tate*, Ann. of Math. (2) **102** (1975), 517–533. ↑231, 242, 243, 244

[Mil80] J. S. Milne, *Etale cohomology*, Princeton Mathematical Series, vol. 33, Princeton University Press, Princeton, N.J., 1980. ↑217, 225, 243, 262, 263, 264

[Mil86a] J. S. Milne, *Abelian varieties*, Arithmetic geometry (Storrs, Conn., 1984), Springer, New York, 1986, pp. 103–150. ↑218, 219, 226

[Mil86b] J. S. Milne, *Jacobian varieties*, Arithmetic geometry (Storrs, Conn., 1984), Springer, New York, 1986, pp. 167–212. ↑217, 218, 219

[Miy06] T. Miyake, *Modular forms*, English, Springer Monographs in Mathematics, Springer-Verlag, Berlin, 2006. Translated from the 1976 Japanese original by Yoshitaka Maeda. ↑227

[Mum66] D. Mumford, *Lectures on curves on an algebraic surface*, With a section by G. M. Bergman. Annals of Mathematics Studies, No. 59, Princeton University Press, Princeton, N.J., 1966. ↑237

[Mum08] D. Mumford, *Abelian varieties*, Tata Institute of Fundamental Research Studies in Mathematics, vol. 5, Published for the Tata Institute of Fundamental Research, Bombay, 2008. With appendices by C. P. Ramanujam and Yuri Manin, Corrected reprint of the second (1974) edition. ↑218, 239

[Ogg67] A. P. Ogg, *Elliptic curves and wild ramification*, Amer. J. Math. **89** (1967), 1–21. ↑265

[Poo07] B. Poonen, *Gonality of modular curves in characteristic p*, Math. Res. Lett. **14** (2007), 691–701. ↑227

[Roq06] P. Roquette, *The Riemann hypothesis in characteristic p, its origin and development. III. The elliptic case*, Mitt. Math. Ges. Hamburg **25** (2006), 103–176. ↑213, 217, 219

[Ros02] M. Rosen, *Number theory in function fields*, Graduate Texts in Mathematics, vol. 210, Springer-Verlag, New York, 2002. ↑216

[Sch96] C. Schoen, *Varieties dominated by product varieties*, Internat. J. Math. **7** (1996), 541–571. ↑246, 259

[SS10] M. Schütt and T. Shioda, *Elliptic surfaces*, Algebraic Geometry in East Asia (Seoul, 2008), Advanced Studies in Pure Mathematics, vol. 60, Amer. Math. Soc., Providence, RI, 2010, pp. 51 –160. ↑239, 254

[SGA4$\frac{1}{2}$] *Cohomologie étale*, Lecture Notes in Mathematics, vol. 569, Springer-Verlag, Berlin, 1977. Séminaire de Géométrie Algébrique du Bois-Marie SGA 4$\frac{1}{2}$, Avec la collaboration de J. F. Boutot, A. Grothendieck, L. Illusie et J. L. Verdier. ↑217, 238

[SGA5] *Cohomologie l-adique et fonctions L*, Lecture Notes in Mathematics, Vol. 589, Springer-Verlag, Berlin, 1977. Séminaire de Géometrie Algébrique du Bois-Marie 1965–1966 (SGA 5), Edité par Luc Illusie. ↑217

[SGA6] *Théorie des intersections et théorème de Riemann-Roch*, Lecture Notes in Mathematics, Vol. 225, Springer-Verlag, Berlin, 1971. Séminaire de Géométrie Algébrique du Bois-Marie 1966–1967 (SGA 6), Dirigé par P. Berthelot, A. Grothendieck et L. Illusie. Avec la collaboration de D. Ferrand, J. P. Jouanolou, O. Jussila, S. Kleiman, M. Raynaud et J. P. Serre. ↑225, 237

[Ser58] J.-P. Serre, *Sur la topologie des variétés algébriques en caractéristique p*, Symposium internacional de topología algebraica International symposium on algebraic topology, Universidad Nacional Autónoma de México and UNESCO, Mexico City, 1958, pp. 24–53. ↑237

[Ser77] J.-P. Serre, *Linear representations of finite groups*, Graduate Texts in Mathematics, Vol. 42, Springer-Verlag, New York, 1977. Translated from the second French edition by Leonard L. Scott. ↑262

[Ser88] J.-P. Serre, *Algebraic groups and class fields*, Graduate Texts in Mathematics, vol. 117, Springer-Verlag, New York, 1988. Translated from the French. ↑219

[ST68] J.-P. Serre and J. Tate, *Good reduction of abelian varieties*, Ann. of Math. (2) **88** (1968), 492–517. ↑264

[Shi86] T. Shioda, *An explicit algorithm for computing the Picard number of certain algebraic surfaces*, Amer. J. Math. **108** (1986), 415–432. ↑232

[Shi99] T. Shioda, *Mordell-Weil lattices for higher genus fibration over a curve*, New trends in algebraic geometry (Warwick, 1996), London Math. Soc. Lecture Notes, vol. 264, Cambridge Univ. Press, Cambridge, 1999, pp. 359–373. ↑253

[SK79] T. Shioda and T. Katsura, *On Fermat varieties*, Tôhoku Math. J. (2) **31** (1979), 97–115. ↑258

[Sil94] J. H. Silverman, *Advanced topics in the arithmetic of elliptic curves*, Graduate Texts in Mathematics, vol. 151, Springer-Verlag, New York, 1994. ↑240, 249, 251, 252

[Sil09] J. H. Silverman, *The arithmetic of elliptic curves*, Second edition, Graduate Texts in Mathematics, vol. 106, Springer, Dordrecht, 2009. ↑221, 223, 224, 226, 229, 252

[Sti09] Henning Stichtenoth, *Algebraic function fields and codes*, Second edition, Graduate Texts in Mathematics, vol. 254, Springer-Verlag, Berlin, 2009. ↑216

[Szy04] M. Szydlo, *Elliptic fibers over non-perfect residue fields*, J. Number Theory **104** (2004), 75–99. ↑250

[Tat66a] J. T. Tate, *Endomorphisms of abelian varieties over finite fields*, Invent. Math. **2** (1966), 134–144. ↑219

[Tat66b] J. T. Tate, *On the conjectures of Birch and Swinnerton-Dyer and a geometric analog*, Séminaire Bourbaki, Vol. 9, Soc. Math. France, Paris, 1966. Exp. No. 306, pp. 415–440. ↑231, 242, 243, 253

[Tat75] J. T. Tate, *Algorithm for determining the type of a singular fiber in an elliptic pencil*, Modular functions of one variable, IV (Proc. Internat. Summer School, Univ. Antwerp, Antwerp, 1972), Lecture Notes in Mathematics, vol. 476, Springer, Berlin, 1975, pp. 33–52. ↑229, 249, 252

[Tat94] J. T. Tate, *Conjectures on algebraic cycles in l-adic cohomology*, Motives (Seattle, WA, 1991), Proc. Sympos. Pure Math., vol. 55, Amer. Math. Soc., Providence, RI, 1994, pp. 71–83. ↑242, 244

[TS67] J. T. Tate and I. R. Shafarevich, *The rank of elliptic curves*, Soviet Math. Dokl. **8** (1967), 917–920. ↑226, 276

[Ulm91] D. Ulmer, *p-descent in characteristic p*, Duke Math. J. **62** (1991), 237–265. ↑225, 227

[Ulm05] D. Ulmer, *Geometric non-vanishing*, Invent. Math. **159** (2005), 133–186. ↑264, 266

[Ulm07] D. Ulmer, *L-functions with large analytic rank and abelian varieties with large algebraic rank over function fields*, Invent. Math. **167** (2007), 379–408. ↑258, 266, 268, 269

[Ulm09a] D. Ulmer, *On Mordell-Weil groups of Jacobians over function fields* (2009). Preprint. ↑259, 273, 274, 275

[Ulm09b] D. Ulmer, *Explicit points on the Legendre curve* (2009). Preprint. ↑276

[Ulm11] D. Ulmer, *Curves and Jacobians over function fields* (2011). Notes from a course at the CRM, Barcelona. In preparation. ↑213, 224, 230, 231, 244, 253, 256, 258

[UZ10] D. Ulmer and Y. G. Zarhin, *Ranks of Jacobians in towers of function fields*, Math. Res. Lett. **17** (2010), 637–645. ↑276

[Wei49] A. Weil, *Numbers of solutions of equations in finite fields*, Bull. Amer. Math. Soc. **55** (1949), 497–508. ↑217

[Zar08] Y. G. Zarhin, *Homomorphisms of abelian varieties over finite fields*, Higher-dimensional geometry over finite fields, NATO Sci. Peace Secur. Ser. D Inf. Commun. Secur., vol. 16, IOS, Amsterdam, 2008, pp. 315–343. ↑219

Heegner's Proof

Bryan Birch

IAS/Park City Mathematics Series
Volume 18, 2009

Heegner's Proof

Bryan Birch

Preface

When Peter Swinnerton-Dyer and I told Weil and Shimura about our conjecture, they both told us to test it on those elliptic curves that were conveniently parametrised by modular functions; so clearly we needed to learn about the arithmetic properties of classical modular functions. Peter was busy at the time, designing an operating system for the Cambridge computer TITAN, so it was up to me. For the elementary theory there were various texts available, notably Bob Gunning's Princeton notes, but for the deeper theory the best available account was still Weber's Algebra; for more recent stuff one had to read the original papers of Hecke and others. Fortunately for me, Oliver Atkin was appointed to a research position at the ATLAS Laboratory near Oxford at just the right moment. Oliver was a fine mathematician, who regarded the various modular functions introduced by Klein and Weber as his personal friends; he greeted highspeed computers with delight, and it was he more than anyone else who instituted the modern computation of modular functions. He published relatively few papers, but his published papers are only a small part of his contribution: he was inconveniently ahead of his time, and preferred to put his results on the web! I am very grateful for him for teaching me about some of his friends; he died just before Christmas, and this seems an appropriate opportunity to pay tribute.

A little after my first attempt to read Weber, I encountered Heegner's famous paper, containing his solution of Gauss' tenth discriminant problem. His solution had the reputation of being wrong because results he quoted from Weber were not properly justified, but I was more immediately interested in his theorems about elliptic curves having rational points. These results were highly relevant to our conjecture and the parts of Weber that he quoted were clearly stated; it was hard to believe that he had got them wrong. Clearly, I had to understand Weber's book before using Heegner's results; and of course it turned out that Heegner was in essence entirely correct. In particular, Heegner points will play an important role towards the end of Rubin's lectures.

In this lecture, I wanted to give an account of Heegner's proof that Gauss' list of complex quadratic fields with class number 1 is complete, starting from first principles: I was reminiscing in the style of the 1960's. There was too much material for a single lecture. In this writeup, Section 1 is an introduction to the j-function in five easy steps, more geometric than the usual treatments. Section

---

Maths Institute, 24 St Giles', Oxford and Heilbronn Institute, University of Bristol
E-mail address: birch@maths.ox.ac.uk
I would like to thank the organisers for their invitation.

2 of the lecture was a breathless sightseeing tour in three more 'steps' through Weber's theory of complex multiplication, which I hoped would help to prepare my audience for later material; this tour has largely been replaced by references to Vatsal's admirable lectures in this volume. After this background, Section 3 contains Heegner's theory.

1. Preliminaries.

We will use notations $\mathbb{H}$ for the upper half plane $\{z = x + iy | y > 0\}$, and $\overline{\mathbb{H}}$ for its 'completion' $\mathbb{H} \cup \mathbb{Q} \cup \{i\infty\}$. Note that x, y will almost always mean the real and imaginary parts of z. $\Gamma(1)$ will denote the modular group $SL_2(\mathbb{Z})$, acting on $\mathbb{H}$ and on $\overline{\mathbb{H}}$ by

$$z \mapsto z\gamma = (az + b)/(cz + d).$$

G will usually be a subgroup of $\Gamma(1)$ of finite index. We will write q short for $e^{2\pi iz}$. I apologise for making my mappings act on the left and writing quotients on the right, and apologise even more if I get it wrong.

1.1. Step 1. Fundamental region.

A fundamental region for the action of $\Gamma(1)$ on $\mathbb{H}$ is

$$F : \{z = x + iy \mid -\frac{1}{2} \leq x \leq \frac{1}{2}, x^2 + y^2 \geq 1\},$$

with the edges $x = \pm\frac{1}{2}$ identified by $z \mapsto z + 1$ and the two halves of the arc $|z| = 1$ identified by $z \mapsto -1/z$; we see that $\mathbb{H}/\Gamma(1)$ is a topological sphere with puckers above i and $\rho := \frac{1}{2}(-1 + \sqrt{-3})$ and a hole at $i\infty$. By taking local parameters $(z - i)^{1/2}$ at i, $(z - \rho)^{1/3}$ at ρ, $q = e^{2\pi iz}$ at $i\infty$ and of course z everywhere else we smooth out the puckers, fill in the hole, and convert $\overline{\mathbb{H}}/\Gamma(1)$ into a Riemann sphere. Put briefly, $\overline{\mathbb{H}}/\Gamma(1)$ has genus zero; for future convenience, we will call it $X(1)$.

1.2. Step 2. $j(z)$.

We are interested in functions invariant by finite index subgroups G of $\Gamma(1)$ that are meromorphic on $\mathbb{H}$, and on $\overline{\mathbb{H}}$ in the sense that they have Fourier expansions at the 'cusps' in $\mathbb{Q} \cup \{i\infty\}$; these functions form a field, the *field of modular functions for* G. Since we have turned $\overline{\mathbb{H}}/\Gamma(1)$ into a Riemann sphere, the Riemann mapping theorem tells us that the field of modular functions for $\Gamma(1)$ is simply $\mathbb{C}(j(z))$, where we may normalise the function $j(z)$ so that j has a simple pole at $i\infty$ with residue 1, and $j(\rho) = 0$; so j has a Fourier series

$$j(z) = q^{-1} + \sum_{0}^{\infty} c_n q^n.$$

1.3. Step 3. $\Gamma_0(N)$.

For positive integer N, we define the subgroup $\Gamma_0(N)$ as the subgroup of $\Gamma(1)$ consisting of maps $z \mapsto (az + b/cz + d)$ with c divisible by N; $\Gamma_0(N)$ obviously has finite index, and we check that

$$\Gamma_0(N) = \Gamma(1) \cap D_N^{-1}\Gamma(1)D_N$$

where $D_N : z \mapsto Nz$. So $D_N^{-1}\Gamma(1)D_N$ is the subgroup of the Mobius group conjugate to $\Gamma(1)$ that leaves $j(Nz)$ invariant. We see that the field of modular functions for $\Gamma_0(N)$ is precisely $\mathbb{C}(j(z), j(Nz))$.

1.4. Step 4. $X_0(N)$ and $F_N(X,Y)$.
The subgroup $\Gamma_{(}N)$ has finite index in $\Gamma(1)$; we will denote its index by $\psi(N)$; it is not hard to check that ψ is a multiplicative function with $\psi(p^r) = (p+1)p^{r-1}$ when p^r is a prime power. The field $\mathbb{C}(j(z), j(Nz))$ is a finite extension of $\mathbb{C}(j(z))$ of degree $\psi(N)$ so there is a polynomial $F_N(X,Y)$, of degree ψ in each of X, Y, so far determined only up to a scalar factor, such that $F_N(j(z), j(Nz)) = 0$ identically. So $X_0(N) := \overline{\mathbb{H}}/\Gamma_0(N)$ is an algebraic curve with affine equation $F_N(X,Y) = 0$. We may if we like call $X_0(N)$ the modular curve and $F_N(X,Y) = 0$ the modular equation.

Let us write W_N for the involution $W_N : z \leftrightarrow -1/Nz$; then $W_N\Gamma_0(N) = \Gamma_0(N)W_N$, so W_N is an involution on $X_0(N)$ that switches $j(z)$ and $j(Nz)$. So the equation $F_N(X,Y) = 0$ is the same as $F_N(Y,X) = 0$.

I should immediately stress that $F_N(X,Y) = 0$ is a bad model for $X_0(N)$, even when projectivised. The particular model $F_N(X,Y) = 0$ has unnecessary singularities; and (health warning!) the polynomial $F_N(X,Y)$ has unnecessarily many terms and colossal coefficients; you should on no account attempt to compute it directly. There are much better models for $X_0(N)$.

1.5. Step 5. $X_0(2)$ and $X_0(3)$. Integrality.
It is time to do some computation. A fundamental region for $\Gamma_0(2)$ is bounded by the vertical lines $x = \pm\frac{1}{2}$, which are identified, and by arcs of the two circles of radius $\frac{1}{2}$ centred at $z = -\frac{1}{2}$ and at $z = \frac{1}{2}$, which too are identified. It is easy to see that $X_0(2) = \overline{\mathbb{H}}/\Gamma_0(2)$ has genus zero; it has cusps at 0 and $i\infty$. Using the Riemann mapping theorem again, we see that the field of modular functions for $\Gamma_0(2)$ is $\mathbb{C}(t_2(z))$ where we may normalise the function $t_2(z)$ by giving it a pole with residue 1 at $i\infty$ and taking $t_2(0) = 0$. It is a matter of routine to work out the relation between j and t_2 from the geometry of the covering of $X(1)$ by $X_0(2)$; we give the gory details in the next paragraph.

The index of $\Gamma_0(2)$ in $\Gamma(1)$ is 3, so $X_0(2)$ is a triple cover of $X(1)$. The branch-points of the cover are above i, ρ and ∞; $X_0(2)$ has a single triple point above ρ, a single cusp at ∞ and a double cusp at 0 above the cusp of $X(1)$, and $\{i, i, \frac{1}{2}(1+i)\}$ above i. Let us write $t_2(\rho) = A$, $t_2(\frac{1}{2}(1+i)) = B$ and $t_2(i) = C$. Then $j(z)$ is in the field $\mathbb{C}(t_2(z))$, and the covering of the cusp tells us that $j = p(t_2)/t_2^2$ where p is a cubic polynomial. Since j and t_2 both have residue 1 at ∞, the cubic must be monic; and since t_2 takes the value A three times above ρ, that is to say above $j = 0$, it follows that

$$j(z) = (t_2(z) - A)^3/t^2.$$

Since t_2 takes the values B, C, C above i, it follows that

$$j(z) - j(i) = (t_2(z) - B)(t_2(z) - C)^2/t_2(z)^2,$$

so we have an identity in $T = t_2(z)$, namely

$$(T - A)^3 - j(i)T^2 = (T - B)(T - C)^2.$$

We need one more piece of information. Remember the involution W_2; it switches $t_2(z)$ and $t_2(-1/2z)$, so it fixes their product, so $t_2(z)t_2(-1/2z) = t_2(\frac{1}{2}(1+i))^2 = B^2$ for all z. So

$$j(2z) = j(-1/2z) = (t_2(-1/2z) - A)^3/t_2(-1/2z)^2 = (B^2 - At_2(z))^3/B^4t_2(z)$$

and we see that $-A^3 = B^4$ since the leading term of the Fourier expansion must be just q^2 on both sides. We have $A^3 = BC^2$, $3A^2 = c^2 + 2BC$, $-A^3 = B^4$ and $j(i) = -3A + B + 2C$; the only possibility is $A = -256$, $B = -64$, $C = 512$ and $j(i) = 1728$.

So our modular equation is $j(z) = (T+256)^3/T^2$, $j(2z) = (T+16)^3/T$; incidentally we have shown that $j(i)$ has to be 1728.

A similar computation gives a modular equation for $X_0(3)$, namely $j(z) = (S+3)^3(S+27)/S$, $j(3z) = (S'+3)^3(S'+27)/S'$ where $SS' = 3^6$.

Theorem 1. *The Fourier coefficients c_n of $j(z)$ are rational integers.*

Suppose that t_2 has Fourier series $t_2(z) = q + \sum_0^\infty c_n q^{-n}$; then the coefficient of q^{1-n} in the relation

$$j(2z) = t_2(z)^2 + 3.2^6 t_2(z) + 3.2^{12} + 2^{18} t_2(z)^{-1}$$

tells us that $2d_n \in \mathbb{Z}[c_0, ..., c_{n-1}, d_0, ..., d_{n-1}]$, and the coefficient of q^{-n} in

$$j(z) = (t_2(z) + 256)^3 / t_2(z)^2$$

tells us that $c_n \in \mathbb{Z}[c_0, ..., c_{n-1}, d_0, ..., d_n]$. It follows by induction that $\{d_n\}$, $\{c_n\}$ are all in $\mathbb{Z}[\frac{1}{2}]$.

Similarly, using the modular equation for $X_0(3)$, we see that $\{c_n\}$ are all in $\mathbb{Z}[\frac{1}{3}]$. It follows that the c_n are all integers.

If p is prime, it follows from Theorem 1 that the coefficients of the series $j(pz)$ are congruent modulo p to the coefficients of the series $j(z)^p$; so it is reasonable to suppose that $F_p(X,Y)$ modulo p is divisible by $(X - Y^p)$ and by $(X^p - Y)$. We know that the degree of F_p is $p+1$ in each variable, so Kronecker was led to state his congruence

$$F_p(X,Y) \equiv (X - Y^p)(X^p - Y) \bmod p.$$

It is a not particularly difficult exercise to make this argument into a correct elementary proof, since $F_p(j(p\omega), j(\omega))$ is a power series in q, and every coefficient involved is integral. However, Vatsal's §7 contains a modern proof of a result that is more precise in important ways.

2. Weber.

2.1. Step 6. Quotation.

So far, I have been honest; but now I am forced to cheat. The polynomial $F_N(X,Y)$ may be defined more explicitly, and Vatsal gives the details. Putting it rather too briefly, we consider a product $\prod(X - j(\alpha_i z))$ where the α_i run over a standard set of representatives, defined in Vatsal's Prop 5.1, for the two-by-two matrices of determinant N, modulo $\Gamma(1)$ (these are in one-to-one correspondence with the cosets of $\Gamma(1)/\Gamma_0(N)$). The coefficients of the powers of X are symmetric functions in the $j(\alpha_i z)$, so are invariant by $\Gamma(1)$. Accordingly, they may be expressed in terms of $j(z)$; so our product turns out to be a polynomial $\Phi_N(X, j(z))$ with $\Phi_N(j(Nz), j(z)) = 0$. Φ_N has the same degree as F_N so we may identify $\Phi_N(X,Y)$ as the 'correct' multiple of $F_N(X,Y)$. Clearly, $\Phi_N(X,Y)$ is monic in X, and because of Theorem 1 it has integral coefficients. We have

Theorem 2. *There is a polynomial $\Phi_N(X,Y)$, monic in X and with integral coefficients, which gives the same equation for $X_0(N)$ as in Step 4.*

This is not all. By further explicit computation one may evaluate the coefficient of the leading term of the Fourier series for $\prod(j(z) - j(z\alpha_i))$; this is the same as the coefficient of the leading term of $\Phi_N(X, X)$. If N is not a square, it turns out to be an algebraic unit. As of course it is rational, we are led to the following result (stated by Vatsal as the second part of his Theorem 5.5):

Theorem 3. *If N is not a square then $\Phi_N(X, X)$ is monic.*

2.2. Step 7. Complex multiplication.

If ω is a complex quadratic surd then $j(\omega)$ is a quadratic integer; this extraordinary fact is the basis of complex multiplication. Given Theorems 2 and 3, it is not difficult. In more detail, suppose that ω is complex quadratic, so that there are coprime integers A, B, C with $B^2 - 4AC < 0$ such that $A\omega^2 + B\omega + C = 0$. We write $-\Delta(\omega) := B^2 - 4AC$, and call $-\Delta(\omega)$ the *discriminant* of ω; then

Theorem 4. *If ω is complex quadratic with discriminant $-\Delta$ then whenever $4N-\Delta$ is a square $j(\omega)$ is a root of $\Phi_N(X, X) = 0$. In particular, $j(\omega)$ is an algebraic integer.*

Indeed, suppose that $A\omega^2 + B\omega + C = 0$ with $4N + B^2 - 4AC = E^2$ then we need to show that there is an ω' $\Gamma(1)$-equivalent to ω such that $N\omega'$ is equivalent to ω'; then $j(\omega) = j(\omega')$ and $j(\omega')$ is a root of $\Phi_N(X, X) = 0$. This can be done by direct assault, but Vatsal's approach via his Lemmas 5.6 and 5.7 is neater.

2.3. Step 8. The class polynomial.

If $-\Delta$ is a complex discriminant, that is to say a negative integer with Δ congruent to either 0 or 3 modulo 4, then we define the *class polynomial* $H_\Delta(X)$ as the gcd of the polynomials $\Phi_N(X, X)$ where N runs over all N such that $4N - \Delta$ is a square. Certainly, $j(\omega)$ is a root of H_Δ if $\Delta(\omega) = \Delta$.

Theorem 5. *The polynomial $H_\Delta(X)$ is irreducible. The splitting field of H_Δ over $\mathbb{Q}(\omega)$ is $\mathbb{Q}(\omega, j(\omega))$ and the Galois group $Gal(\mathbb{Q}(\omega, j(\omega)/\mathbb{Q}(\omega))$ is isomorphic to the group of primitive ideal classes of the ring $R(-\Delta) := \mathbb{Z}[\frac{1}{2}(-\Delta + \sqrt{-\Delta})]$.*

It follows that the degree of H_Δ is the class number $h(-\Delta)$ of $R(-\Delta)$. Note that I have not restricted to the case where R is the full ring of integers of the complex quadratic field; for Heegner's application it is important to keep the general case where $R(-\Delta)$ is an order in the ring of integers of $\mathbb{Q}(\sqrt{-\Delta})$. The difficult part of the theorem (and it really is delicate) is to prove the irreducibility of H; in Vatsal's lectures, this occupies §6 and §7; the strategy is Weber's, but the details in Weber's book are much harder to follow, as class field theory was not available to him: he was inventing a particular case of it! Once we know irreducibility, the rest is easy. It is natural that the roots of H should be the $j(\omega)$ where the ω's run through inequivalent quadratic surds with discriminant $-\Delta$, which of course match up with the classes of binary quadratic forms with discriminant $-\Delta$.

We may describe the action of the Galois group as an action of the ideal class group of $R(-\Delta)$. If $A\omega^2 + B\omega + C = 0$, match $\omega = (-B + \sqrt{-\Delta})/2A$ with the ideal (class) $\mathbb{A} = (A, \frac{1}{2}(-B + \sqrt{-\Delta}))$ of the ring R, and write $j(\mathbb{A}) = j(\omega)$. Then the ideal class group acts on the roots $j(\mathbb{A})$ of $H_\Delta(X)$ by $\sigma_\mathbb{B} : j(\mathbb{A}) \mapsto j(\mathbb{B}^{-1}\mathbb{A})$.

I will write $K(-\Delta)$ for this field $\mathbb{Q}(\omega, j(\omega))$; it is the *Weber class field* for $R(-\Delta)$. An important thing to remember is that $K(-\Delta)$ is an **Abelian** extension of $\mathbb{Q}(\omega)$.

For a fixed discriminant $-\Delta$, the fields $K(-m^2\Delta)$ give a tower of Weber class fields, all abelian over $\mathbb{Q}(\sqrt{-\Delta})$.

3. Heegner.

3.1. Application 1. Heegner points of $X_0(N)$.

Given Theorem 5, these are easy to define. Suppose that $\Delta(\omega) = \Delta(N\omega)$; then $j(\omega)$ and $j(N\omega)$ are both in the field $K(-\Delta)$, so $(j(\omega), j(N\omega))$ is a point of $X_0(N)$ with coordinates in $K(-\Delta)$; this is apt to be valuable, as the field is smaller than one might expect.

In particular, $\Delta(\omega) = \Delta(N\omega)$ when ω satisfies $NA\omega^2 + B\omega + C = 0$, which may be achieved with $\Delta = 4NAC - B^2$ much smaller than N.

One may also obtain Heegner points in a tower of extensions, as noted at the end of Step 8.

If there is a covering map $\phi : X_0(N) \to C$ to some other curve then we get Heegner points on C; this has been particularly useful when C is an elliptic curve.

3.2. Application 2. Weber's Zoo.

We need a standard not very difficult Lemma.

Lemma 1. *Suppose that l is prime and that $-\Delta$ is a discriminant; then $h(-l^2\Delta) = w(l, -\Delta)h(-\Delta)$, where the index $w(l, -\Delta) = l - \left(\frac{l}{-\Delta}\right)$.*

Here, $\left(\frac{l}{-\Delta}\right)$ is just the quadratic residue symbol.

In particular, if 3 does not divide Δ then $w(3, -\Delta)$ is 2 or 4, and if Δ is odd then $w(2, -\Delta)$ is 1 or 3.

Weber was interested in the 'complex multiplication' properties of other functions beside $j(z)$, including in particular $\gamma_3(z)$, $\gamma_2(z)$ and $\sigma(z)$. These satisfy

$$\gamma_3(z)^3 = j(z),$$

and we choose the cube root so that $\gamma_3(z)$ is real when $j(z)$ is;

$$\gamma_2(z)^2 = j(z) - 1728;$$

$$j(z) = (\sigma^{24}(z) - 16)^3/\sigma^{24}(z);$$

we should adjust σ by a root of unity so that it is real on the line $x = -\frac{1}{2}$.

There are lots of wonderful identities between functions of modular type. In more detail, one defines $\eta(z) := q^{1/24}\prod_1^\infty(1 - q^n)$. You may know that $\Delta(z) := \eta^{24}(z)$ is a modular form of weight 12 for $\gamma(1)$, that is to say $\Delta(z)dz^6$ is invariant by $\Gamma(1)$. Now σ is given by $\sigma(z) = \sqrt{2}\eta(2z)/\eta(z)$, up to a root of unity. Referring to Step 5, we note that σ^{24} is equal to the function $-2^{12}/t_2$ on $X_0(2)$, and t_2 really was real and negative on $x = -\frac{1}{2}$, so adjusting σ to be real was legal!

These three functions are all invariant by congruence subgroups of $\Gamma(1)$ of various levels; I must try to say briefly what this means: a function has level M if it is invariant by the subgroup $\Gamma(M)$ consisting of maps $z \mapsto (az+b)/(cz+d)$ with b, c divisible by M and a, d congruent to 1 modulo M. Essentially, a function of level M may be expressed in terms of various functions $j((Az+B)/(Cz+D))$ with $AD - BC = M$; so if f has level M and ω has discriminant $-\Delta$ then $f(\omega)$ may be expressed in terms of j-functions evaluated at points of discriminant $M^2\Delta$.

The functions $\gamma_3, \gamma_2, \sigma$ have levels $3, 2, 48$ respectively; σ^3 has level 16.

Lemma 2. *If 3 does not divide $\Delta(\omega)$ then $\gamma_3(\omega) \in K(-\Delta(\omega))$; in other words, $j(\omega)$ is a perfect cube.*

For since γ_3 has level 3, $\gamma_3(\omega) \in K(-9\Delta)$; but by Lemma 1 this is an extension of $K(-\Delta)$ of degree 2 or 4, and $\gamma_3(\omega)$ is obviously cubic over $j(\omega)$.

In a similar way we can prove

Lemma 3. *If $\Delta(\omega)$ is odd then $\gamma_2(\omega) \in K(-\Delta)$.*

Weber went on to prove many similar results for $\sigma(\omega)$. I will restrict myself to the case where ω is $\omega_0 := \frac{1}{2}(-1+\sqrt{-p})$, with p a prime congruent to 3 modulo 8. This happens to be the case that is needed for Heegner's solution of the class number 1 problem. Note that $j(\omega_0)$, $\gamma_3(\omega_0)$, $\sigma^2(\omega_0)$, $j(2\omega_0)$ and $\sqrt{-p}\gamma_2(\omega_0)$ are all real.

Let us write K_1 for the field $K(-p) = \mathbb{Q}(\omega_0, j(\omega_0))$, and K_2 for $\mathbb{Q}(\omega_0, j(2\omega_0))$; by Lemma 1, K_2 is a cubic extension of K_1. It is important to note that $j(2\omega_0)$ generates K_2 over K_1, and $j(2z)$ is a rational function of $t_2(z) = -2^{12}/\sigma^{24}(z)$, so $\sigma^{24}(\omega_0)$ generates K_2 over K_1, it is *not* in K_1. By the relation $(\sigma^{24}-16)^3 = \sigma^{24}j$, we see that $\mathrm{Norm}_{K_2/K_1}(\sigma^{24}) = 2^{12}$.

Lemma 4. *Let p prime to 3 be congruent to 3 modulo 8, fix $\omega_0 = \frac{1}{2}(-1+\sqrt{-p})$, and define the fields K_1, K_2 as above. Then*

(i) $\sigma^{12}(\omega_0) \in K_2$, $\mathrm{Norm}_{K_2/K_1}\sigma^{12} = 2^6$.
(ii) $\sigma^{6}(\omega_0) \in K_2$, $\mathrm{Norm}_{K_2/K_1}\sigma^{6} = 8$.
(iii) $\sigma^{2}(\omega_0) \in K_2$, $\mathrm{Norm}_{K_2/K_1}\sigma^{2} = 2$.

Though we will be using the lemma only when p is prime, it is true without any such restriction.

The key to (i) is that σ^{12} is abelian over K_1 and $\sigma^{24} \in K_2$, a proper cubic extension of K_1. A quadratic extension of K_2 that is abelian over K_1 has to be of the form $K_2(\alpha)$ with $\alpha^2 \in K_1$, so if σ^{12} is not in K_2 it has to be of the form $\sigma^{12} = \beta + \alpha\gamma$ with $\beta, \gamma \in K_2$, γ not 0, and $\alpha^2 \in K_1$, α not in K_1. Squaring, we see that $\beta\gamma = 0$ so $\beta = 0$ and $\sigma^{12} = \gamma\alpha$, $\sigma^{24} = \gamma^2\alpha^2$. Taking K_2/K_1 norms gives $2^{12} = \mathrm{Norm}(\gamma)^2\alpha^6$, so α^2 is equal to the square of $2^{-6}\mathrm{Norm}(\gamma)\alpha^2$, which is patently in K_1; so α is in K_1, a contradiction, So σ^{12} has to be in K_2.

Now that we know that $\sigma^{12} \in K_2$ it is clear that its norm has to be 2^6. I have given the proof in detail because when Heegner's proof was rejected by his contemporaries, this is one of the steps that was doubted.

Once (i) has been proved the proof of (ii) is just the same; but we cannot go on to prove that $\sigma^3 \in K_2$ because the norm of σ^6 is 8, not a square.

(Exercise: what *can* we say about $\sigma^3(\omega_0)$?)

We know that $\sigma^8 = (\sigma^{24} - 16)/\gamma_3$, so part (iii) follows from (ii) and Lemma 2.

3.3. Application 3. Gauss' conjecture.

In this subsection, we retain the notations ω_0, K_1 and K_2 from the previous subsection.

We want to show that the only complex quadratic fields with class number 1 are the ones with discriminants in Gauss' list

$$\{-3, -4, -7, -8, -11, -19, -43, -67, -163\}.$$

It is elementary to show that there are no more except possibly fields with discriminant $-p$, where p is a prime congruent to 3 modulo 8. Saying that the field has class number 1 is the same as saying that there is just one class of binary quadratic forms with discriminant $-p$, which (see Step 8) is the same as saying that the class polynomial H_p has degree 1, which is the same as saying that $K_1 \cap \mathbb{R}$ is $\mathbb{Q}$; then $K_2 \cap \mathbb{R}$ is a cubic extension of $\mathbb{Q}$, and $j(\omega_0)$ is a rational integer.

So we ask, for which primes p congruent to 3 does all this happen?

Given the information in the previous subsection, Heegner found this surprisingly easy to decide.

Write $V := \sigma^2(\omega_0)$ for short. By Lemma 4, $V \in K_2$, and we also know that V is real and that it is an algebraic integer and that its norm is 2, so it satisfies an equation

$$V^3 - AV^2 + BV - 2 = 0$$

with A,B rational integers. But we also know that

$$V^{12} - \gamma_3(\omega_0)V^4 - 16 = 0,$$

where $\gamma_3(\omega_0)$ is a rational integer too. In order that the coefficient of V^8 be zero, there has to be a relation between A and B; it turns out to be

$$2B^2 + 8A - 4BA^2 + A^4 = 0.$$

This Diophantine equation has only finitely many integral solutions, and it is not difficult to find them.

Exercise 1. *Show that the only integral solutions of* $2B^2 + 8A - 4BA^2 + A^4 = 0$ *are* $(A, B) = (0,0), (-2,0), (-2,8), (2,4), (-4,4)$ *and* $(-4,28)$.

These solutions correspond, in order, to the discriminants $-p$ with

$$p = 3, 11, 67, 19, 43, 163.$$

How do we check this?

Given (A, B), we work out D so that V^3-AV^2+BV-2 divides $V^{12}+DV-16$; then $\gamma_3(\omega_0) = -D$, $j(\omega_0) - 1728 = -(D^3 + 1728)$ and by Lemma 3, $D^3 + 1728$ should turn out to be a square multiple of 3. The arithmetic is rather horrid, unless one finds moderately large numbers beautiful. (One could of course give a formula for D in terms of A, B, but it isn't efficient.)

For instance, V^3+2V^2+8V-2 divides $V^{12}+12{\cdot}440V^4-16$, giving $\gamma_3/12 = -440$ and $(j - 1728)/1728 = -(440^3 + 1) = -3 \cdot 7^2 \cdot 31^2 \cdot 67$.

3.4. Application 4. Points on an elliptic curve.

The Heegner points described in Application 1 were not those ever described by Heegner. It seems worthwhile (since we have got so near) to sketch just one of Heegner's actual examples. We keep the same notation, $p > 3$ is a prime congruent to 3 modulo 8, and ω_0, K_1, K_2 are as before. We will show that

$$E : Y^2 = X(X^2 + p^2)$$

has infinitely many rational points.

Indeed, $j - 1728 = (\sigma^{24} - 64)(\sigma^{24} + 8)^2/\sigma^{24}$, so

$$\sigma^{24} - 64 = \left(\sigma^{12}\gamma_2/(\sigma^{24} + 8)\right)^2.$$

Evaluating this gives a point on the genus 1 curve

$$C : -pu^2 = v^4 - 64$$

with $u, v \in K_2 \cap \mathbb{R}$. But we know that the class number of $\mathbb{Q}(\sqrt{-p})$ is odd, and $[K_2 : K_1] = 3$, so $[K_2 \cap \mathbb{R} : \mathbb{Q}]$ is odd, so C has an odd divisor which is defined over $\mathbb{Q}$. It has an obvious divisor at ∞ of degree 2, so it has a degree 1 divisor defined over $\mathbb{Q}$. So it has a rational point.

But C is a non-trivial 2-covering of the elliptic curve E, which accordingly has infinitely many rational points.

Complex Multiplication: a Concise Introduction

Vinayak Vatsal

IAS/Park City Mathematics Series
Volume 18, 2009

Complex Multiplication: a Concise Introduction

Vinayak Vatsal

1. Introduction

The goal of these notes is to provide a very brief and elementary introduction to the basic results of the theory that is known as complex multiplication. It is our hope that these notes will provide the beginner with the confidence and motivation to read complete or much more sophisticated treatments found elsewhere, notably in [**19**], [**2**], [**14**], or [**21**]. Our treatment here follows rather closely the discussion in [**14**]; however, we have reorganized the material where it seems appropriate, and also omitted some of the proofs where reproducing them would amount to transcription, or where it seems advisable to focus instead on the main features of the arguments.

We begin by remarking that the term complex multiplication may mean different things, depending on the context. For instance, it may refer to any of the following

(1) Elliptic curves with complex multiplications ("extra" endomorphisms),
(2) The generation of class fields of imaginary quadratic fields,
(3) The theory of modular forms associated to positive definite binary quadratic forms,
(4) Hecke L-functions associated to imaginary quadratic fields,
(5) Construction of special points on the moduli spaces associated to elliptic curves.
(6) Construction of special points or cycles on Shimura varieties.

In these notes we will talk about (1), (2), and (5). For an introduction to (3) and (4), the reader may consult [**17**], and for a discussion of (6), we refer the reader to [**13**].

We have made no attempt in this article to ascribe the results with historical accuracy. We will simply indicate here that the results here were proven over many years by many people, notably Kronecker, Weber, Hasse, Deuring, and Shimura, and that the names that have become attached over the years to some of the principal objects should not be considered as reflective of their origins. For instance, it was pointed out by the referees that the objects called Heegner points here, and in many other places in the literature, do not actually include any of the points considered by Heegner in his work on the class number problem. (For a discussion of Heegner's work, we refer to the article of Birch in this volume.) Similarly, the so-called Kronecker congruence discussed below may not have been proven by

Department of Mathematics, University of British Columbia, 1984 Mathematics Road, Vancouver, BC, V6T1Z2
E-mail address: vatsal@math.ubc.ca

Kronecker in the form in which it is most often stated. Nevertheless, we will in this article use terminology which seems to have become standard.

We would also like to thank the referees for their careful examination of the original manuscript.

2. Preliminaries

For the purposes of this article, an elliptic curve is a complex torus $E = E_\Lambda = \mathbf{C}/\Lambda$ where $\Lambda \cong \mathbf{Z}^2$ is a lattice in $\mathbf{C}$, so that $\Lambda \otimes \mathbf{R} = \mathbf{C}$.

We define the Weierstrass $\wp$-function for E (or Λ) as

$$\wp(z;\Lambda) = \frac{1}{z^2} + \sum_{\alpha\in\Lambda,\alpha\neq 0} \left(\frac{1}{(z-\alpha)^2} - \frac{1}{\alpha^2}\right).$$

Then the Weierstrass equation for E_Λ is

$$(\wp'(z))^2 = 4\wp(z)^3 - g_2(\Lambda)\wp(z) - g_3(\Lambda)$$

where

$$g_2(\Lambda) = 60 \sum_{\alpha\neq 0,\alpha\in\Lambda} \alpha^{-4}$$

and

$$g_3(\Lambda) = 140 \sum_{\alpha\neq 0,\alpha\in\Lambda} \alpha^{-6}.$$

Concretely, this means that the point $(X, Y) = (\wp(z;\Lambda), \wp'(z;\Lambda))$ lies on the locus of zeroes of the equation

$$Y^2 = 4X^3 - g_2(\Lambda)X - g_3(\Lambda).$$

Finally, we introduce the Δ-function and the j-invariant:

$$\Delta(\Lambda) = g_2(\Lambda)^3 - 27g_3(\Lambda)^2,$$

and

$$j(\Lambda) = 1728\frac{g_2(\Lambda)^3}{\Delta(\Lambda)}.$$

With these definitions, we see that $\alpha^{2k}g_k(\alpha\Lambda) = g_k(\Lambda)$ so that g_k is homogeneous of weight $2k$ as a function of lattices. Similarly, Δ has weight 12 and j has weight zero.

It will be useful in what follows to reformulate the definitions of the g_k, Δ, and j in terms of a variable τ in the upper half plane, rather than in terms of the lattice Λ. Thus if F is a function on the set of lattices, we will follow the conventional abuse of notation and write $F(\tau)$ to denote the function of τ deduced from F by evaluation at $\Lambda_\tau = \mathbf{Z} + \mathbf{Z}\tau$, so that $F(\tau) \stackrel{\text{def}}{=} F(\Lambda_\tau)$. In the rest of this article, we will view the functions g_k, Δ, and j as functions of the variable τ rather than of lattices. If $\Lambda = \Lambda_\tau$ as above, then we write E_τ instead of E_{Λ_τ} to denote the corresponding elliptic curve.

We will not develop any of the analytic theory of the functions g_k, Δ, j here. Rather, we simply collect the properties that we need. For the statements, let us set $q = q(\tau) = \exp(2\pi i\tau)$. Then we have the following facts:

- The functions $\Delta(\tau)$ and $g_k(\tau)$ are invariant under $z \mapsto z+1$ and have Fourier expansions in q with complex coefficients.
- The Fourier coefficients of g_k are rational multiples of $(2\pi i)^{-2k}$.

- The Fourier coefficients of Δ are integer multiples of $(2\pi i)^{-12}$.
- If $z = a\tau + b$ with a, b fixed in $\mathbf{Q}$, then as a function of τ, the function $(2\pi i)^2\wp(z,\tau)$ has a finite-tailed expansion in $q^{1/M}$ for some M whose coefficients are cyclotomic integers.
- The function j is holomorphic on the upper half plane and has a simple pole at infinity. Furthermore, j is invariant under the composition action of $SL_2(\mathbf{Z})$, and we have the expansion $j(\tau) = \frac{1}{q} + P(q)$ where $P(q) \in \mathbf{Z}[[q]]$ valid for the imaginary part of τ positive. The main point to note here is that the Fourier coefficients of j turn out to be *integers*; this fact is nontrivial, and will be important later. It is clear however that the coefficients in question are *rational*: from properties listed above for the Fourier coefficients of Δ and the g_k, the powers of πi all cancel.

The proofs of these statements may be found in [**14**], Chapter 4, §2. We note also that the function j is an (analytic) isomorphism invariant of the elliptic curve E_τ: we have $E_\tau \cong E_{\tau'}$ if and only if $\tau = \gamma\tau'$ for $\gamma \in SL_2(\mathbf{Z})$, which holds if and only if $j(\tau) = j(\tau')$.

3. The basic definition

Let $E = \mathbf{C}/\Lambda$ be an elliptic curve over $\mathbf{C}$. We say that E has complex multiplication, or CM for short, if $\mathrm{End}(E) \neq \mathbf{Z}$.

Example 3.1. Suppose $\Lambda = \mathbf{Z} + \tau\mathbf{Z} = \mathcal{O}_K$ is the ring of integers in an imaginary quadratic field K. Then clearly $\alpha\Lambda \subset \Lambda$ for any $\alpha \in \mathcal{O}_K$, so $z \mapsto \alpha z$ defines an endomorphism of E which is not multiplication by an integer on E unless $\alpha \in \mathbf{Z}$. For instance, we can take $\Lambda = \mathbf{Z} + \mathbf{Z}i$ to be the ring of Gaussian integers. In this case we have a model for E given by

$$Y^2 = X^3 + X.$$

This E has j-invariant 1728. In general there are very few CM elliptic curves defined over $\mathbf{Q}$, as we will see below. Note also that the map $(X, Y) \mapsto (-X, iY)$ is an endomorphism of E, which we view as being multiplication by i.

Theorem 3.2 ([**14**], Theorem 6, pg 14)**.** *Suppose that $E_i = \mathbf{C}/\Lambda_i$ are elliptic curves, for $i = 1, 2$. Suppose that there is a complex analytic homomorphism $f : E_1 \to E_2$. Then there is a complex number β with $\beta\Lambda_1 \subset \Lambda_2$ such that f is induced by the map $z \mapsto \beta z$ on $\mathbf{C}$.*

From now on we fix an imaginary quadratic field K. Then suppose that $\Lambda \subset K$ is a lattice. This means Λ is a subgroup isomorphic as a group to $\mathbf{Z}^2$ and that $\Lambda \otimes \mathbf{Q} = K$.

The set of $\alpha \in \mathbf{C}$ with $\alpha\Lambda \subset \Lambda$ is a subring $R \subset \mathcal{O}_K$, such that R has finite index in $\mathcal{O}_K$ (such an R is called an order of K) and the lattice Λ is called a fractional R-ideal. The theory of the rings R and their fractional ideals is very similar to that of the ring O_K of integers in K (the maximal order of K) and ordinary fractional ideals.

Theorem 3.3 ([**14**], Theorem 3, pg 91)**.** *Let R be an order in the imaginary quadratic field K. Then there exists a unique positive integer c such that $R = \mathbf{Z} + c\mathcal{O}_K$. The integer c is called the conductor of R and we write $R = \mathcal{O}_c$. The integer c is the index of R in $\mathcal{O}_K$ as an abelian group.*

Definition 3.4. Let Λ denote a lattice in K and let $R = \mathcal{O}_c$ denote the largest order in K such that $\alpha\Lambda \subset \Lambda$ for $\alpha \in R$. By the above theorem, $R = \mathcal{O}_c$ for some c, and we call the integer c the conductor of Λ.

Theorem 3.5 ([**14**], pp 19–20). *Suppose $E = \mathbf{C}/\Lambda$ is an elliptic curve over $\mathbf{C}$ with complex multiplication. Then there exists $\beta \in \mathbf{C}$ such that $\beta\Lambda$ is a lattice in some imaginary quadratic field K.*

Thus, by replacing $\mathbf{C}/\Lambda$ by the isomorphic curve $\beta\mathbf{C}/\beta\Lambda = \mathbf{C}/\beta\Lambda$, we may assume that the lattice Λ is a lattice in an imaginary quadratic field. From now on we will consider elliptic curves that are isomorphic to ones of the form $\mathbf{C}/\Lambda$ where Λ is a lattice of conductor c, for some positive c.

Remark 3.6. Let E be an elliptic curve which admits an abstract embedding $\iota : R \to \mathrm{End}(E)$, for an order R in K. Then for each $\alpha \in R$, we have an endomorphism $[\alpha]$ induced by the given embedding. On the other hand, if ω is an invariant differential on E, then $[\alpha]^*\omega = \mu(\alpha)\omega$ for a complex number $\mu(\alpha)$. It is clear that μ is a homomorphism from R to $\mathbf{C}$, and we may view R as a subring of $\mathbf{C}$ by this identification. We will make this identification without comment in the sequel.

Our first task is to classify the CM elliptic curves up to isomorphism (at least over $\mathbf{C}$). The most obvious way to group these elliptic curves is by the conductors. Thus let c denote a positive integer and consider the set of isomorphism classes of complex elliptic curves $\mathbf{C}/\Lambda$ where Λ is a lattice with associated order $R = \mathcal{O}_c$. The basic case is that of the maximal order, namely, curves of conductor 1.

Theorem 3.7. *The CM elliptic curves of conductor 1 are in bijective correspondence with the elements of the ideal class group of $\mathcal{O}_K$. In particular, there are h_K nonisomorphic CM elliptic curves of conductor 1, where h_K is the class number of K.*

PROOF. Let $E = \mathbf{C}/\Lambda$ denote a CM elliptic curve of conductor 1. By definition, Λ is a lattice in K which is stable under multiplication by $\mathcal{O}_K$. Thus Λ defines a fractional ideal in K. One checks without difficulty that the class of Λ modulo principal ideals depends only on the isomorphism class of E, and that every ideal class of $\mathcal{O}_K$ is obtained in this way from some E. It remains only to show that if E and E' give the same ideal class, then they are isomorphic. But this follows from Theorem 3.2. □

Now we consider the case of curves of conductor $c > 1$, which is very similar. The relevant facts are as follows. The proofs of the theorems are elementary but tedious and are mostly left as exercises. The reader may also consult Lang, Chapter 8.

We set $R = \mathcal{O}_c \subset \mathcal{O}_K$. We say that a lattice Λ in K is an R-lattice if it is stable under multiplication by R. We say that Λ is principal if there exists some $\alpha \in K$ such that $\Lambda = \alpha R$. Note that a principal R-lattice has conductor c. The product of R-lattices is clearly an R-lattice. We say that an R-lattice Λ is invertible if there exists another R-lattice Λ' such that $\Lambda\Lambda'$ is a principal R-lattice. Note that any ideal $\mathfrak{a}$ in the ring R is automatically an R-lattice. If $\mathfrak{a}$ has conductor c, we say that $\mathfrak{a}$ is a proper R-ideal.

Theorem 3.8 ([**14**], Corollary, pg 91). *Let Λ denote a lattice of K of conductor c. Then Λ is invertible as a lattice over $R = \mathcal{O}_c$. Conversely, any invertible R-lattice has conductor c. The set of lattices of conductor c form a multiplicative group.*

Definition 3.9. Let $\mathfrak{a}$ denote an ideal in R. We say that $\mathfrak{a}$ is prime to c if $\mathfrak{a}+cR = R$, or $\mathfrak{a}+c\mathcal{O}_K = R$. (These two conditions are equivalent.) Let $I_{K,c}$ denote the set of ideals of $\mathcal{O}_K$ that are prime to the ideal $c\mathcal{O}_K$ and write $I_{R,c}$ for the ideals of R that are prime to c.

Theorem 3.10 ([**14**], Theorem 4, pg 92). *There is a bijection between $I_{K,c}$ and $I_{R,c}$ given by $\mathfrak{a} \mapsto \mathfrak{a} \cap R$ whose inverse is given by $\mathfrak{a} \mapsto \mathfrak{a}\mathcal{O}_K$. Any ideal of R that is prime to c has conductor c and is a proper R-ideal.*

Theorem 3.11. *Let Λ be an R-lattice of conductor c and let m be a positive integer. Then there exists an ideal $\mathfrak{a} \subset R$ such that $\mathfrak{a} = \alpha\Lambda$ and $\mathfrak{a}$ is prime to m.*

PROOF. See [**14**], Theorem 5, pg 93. The proof is computational, but the reader is invited to find a better one! □

Definition 3.12. Let I_c denote the set of lattices of conductor c and P_c denote the set of principal R-lattices (which are automatically of conductor c). We let $G_c = I_c/P_c$ and call this the group of ideal classes of R.

In view of the theorem preceding the definition, one sees that every element of G_c has a representative that is prime to the conductor c. Thus we may replace I_c and P_c in the definition by the corresponding sets of ideals prime to c. Using this, we can recover G_c as a generalized ideal class group of the full ring of integers in K.

Definition 3.13. Let $P_{\mathbf{Z}}(c)$ denote the set of principal ideals of $\mathcal{O}_K$ which have a generator α such that α is congruent to a rational integer modulo $c\mathcal{O}_K$, and such that α is prime to c. Let $I(c)$ denote the set of ideals of $\mathcal{O}_K$ prime to c.

Theorem 3.14 ([**14**], pg 94). *We have an isomorphism $I(c)/P_{\mathbf{Z}}(c) \cong G_c$ given by $\mathfrak{a} \mapsto \mathfrak{a} \cap R$.*

Note now that $P_{\mathbf{Z}}(c)$ contains the subgroup $P_1(c)$ of ideals satisfying the further condition that there is generator congruent to 1 (mod c). Thus we have $I(c) \supset P_{\mathbf{Z}}(c) \supset P_1(c)$, and this leads quickly to a formula for the order of G_c:

Theorem 3.15 ([**14**], Theorem 7, pg 95). *The order h_c of G_c is given by*

$$h_c = h_K \frac{c}{(\mathcal{O}_K^\times : R^\times)} \prod_{p|c} (1 - \chi_K(p)/p),$$

where χ_K denotes the quadratic character associated to K.

As far as CM elliptic curves are concerned, we finally have the following predictable result, whose proof we leave as an exercise:

Theorem 3.16. *The CM elliptic curves of conductor c are in bijective correspondence with elements of the group G_c, the bijection being induced by sending an R-ideal $\mathfrak{a}$ of conductor c to the elliptic curve $\mathbf{C}/\mathfrak{a}$. There are h_c nonisomorphic CM elliptic curves of conductor c.*

We will eventually prove that the j-invariants of the elliptic curves of conductor c generate the class field of K corresponding to the generalized ideal class group G_c. For the moment, we observe that it is easy to see directly the the CM elliptic curves have j-invariants that are algebraic.

Proposition 3.17. *Let E denote an elliptic curve with complex multiplication. Then the j-invariant of E is algebraic.*

PROOF. Let σ be any automorphism of $\mathbf{C}$. Since the endomorphism ring of E^σ is the same as the endomorphism ring of E, we see that the set of conjugates of E is finite, contained in the corresponding set of curves of fixed conductor. It follows that the j-invariant of E is algebraic. □

4. Localization and class field theory

It will not surprise the reader that the foregoing results on ideals and lattices may be somewhat more simply attacked by adelic methods. Thus let Λ be a lattice of conductor c in K and for a rational prime p, let $L_p = \Lambda \otimes \mathbf{Z}_p$ denote the completion of Λ at p. Then L_p is a lattice in the completion K_p of K at p, and if $p \nmid c$, we find that L_p is a principal lattice over the discrete valuation ring $R_p = \mathcal{O}_{K,p}$ where the subscript p indicates completion again. It turns out that for quadratic fields this property holds even if $p|c$ (Lang attributes this to Ihara).

Theorem 4.1 ([14] pg 98). *Let p be arbitrary. Then the lattice L_p is principal over the ring R_p.*

Given a lattice Λ, it is clear that $\Lambda_p = \mathcal{O}_{K,p}$ for almost all p (since elements of the finitely generated $\mathbf{Z}$-module Λ have only finitely many denominators). Furthermore, if we are given for each p a lattice $M_p \subset K_p$ such that $M_p = \mathcal{O}_{K,p}$ for all but finitely many p, there exists a unique lattice $M \subset K$ such that $M_p = M \otimes \mathbf{Z}_p$ for all p. Thus, if Λ is a lattice of K and $s = (s_p)$ is an idele of K, it makes sense to define the lattice $s\Lambda$ as the unique lattice M with $M_p = s_p\Lambda_p$ for all finite primes p. In particular, the action of the ideles $s = (1, \ldots, 1, s_\infty)$ where $s_\infty \in K_\infty^\times = \mathbf{C}^\times$ is trivial. This action of the ideles on lattices will be important later.

As an exercise, the reader might try to define the idele class group corresponding to the ideal class group G_c defined above, and to unwind the action of the appropriate idele class group on the lattices of conductor c.

Finally, we recall the class fields corresponding to the ideal class group $G_c = I(c)/P_{\mathbf{Z}}(c)$ and $G_{1,c} = I(c)/P_{1,c}$. The former is the ring class field $K(c)/K$ of conductor c: it is characterized by the fact that the primes splitting completely in $K(c)$ are precisely those which are principal, with a generator congruent to a rational integer modulo c. As for the latter, it is the ray class field $K[c]/K$: it is characterized by the fact the the primes splitting completely are those which are principal and have a generator congruent to 1 modulo c.

5. The modular equation

Now we want to prove that the j-invariant of a CM elliptic curve is an algebraic integer, and prepare for the proof that these algebraic integers generate certain class fields of K. Recall that the j-invariant is an invariant function for the group $\Gamma = SL_2(\mathbf{Z})$ which is holomorphic on the upper half plane and has a Fourier expansion $q^{-1} + \ldots$ at infinity. Let n be a positive integer, and let Δ_n^* denote the set of integral 2×2 matrices with determinant n and entries with no common factor. Then one has the following elementary result:

Proposition 5.1 ([**14**], pg 51–52). *We have the decomposition* $\Delta_n^* = \Gamma \begin{pmatrix} n & 0 \\ 0 & 1 \end{pmatrix} \Gamma$ *and* $\Delta_n^* = \bigcup \Gamma\alpha_i$ *where* α_i *runs over matrices of the form*

$$\alpha_i = \begin{pmatrix} a & b \\ 0 & d \end{pmatrix}$$

with $0 < a, 0 \leq b < d$, $(a, b, d) = 1$, *and* $ad = n$.

Lemma 5.2. *Suppose* f *is a holomorphic function on the upper half plane which is invariant under the action of* Γ *by fractional linear transformations and which is meromorphic at infinity. Then* f *is a polynomial in the function* $j(z)$, *with coefficients in the* **Z**-*module generated by the Fourier coefficients of* f.

PROOF. Let us write $f = c_{-m}q^{-m} + \ldots$. Then $f - c_{-m}j^m$ has the properties in the statement of the theorem, and has a pole of order at most $m - 1$ at infinity. Repeating this process, we find a polynomial P in j with coefficients that are linear combinations of the coefficients of f such that $f - P(j)$ vanishes at infinity and is holomorphic on the upper half plane. It follows that $f - P(j)$ is identically zero, which proves our assertion. □

Now let n be a positive integer and let α_i be representatives for the right cosets of Δ_n^* as in the proposition above. Consider the formal product

$$\Phi_n(X) = \prod (X - j \circ \alpha_i).$$

Here the $j \circ \alpha_i$ denotes composition of the function j with the fractional linear transformation defined by α_i. This is a polynomial in the variable X with coefficients that are holomorphic functions on the upper half plane.

Lemma 5.3. *The coefficients of* $\Phi_n(X)$ *are invariant under the action of* Γ. *They are meromorphic at infinity and holomorphic on the upper half plane.*

PROOF. By definition, the coefficients of Φ_n are elementary symmetric functions of the $j \circ \alpha_i$. Since the right action of Γ permutes the cosets $\Gamma\alpha_i$ in Δ_n^*, the first assertion follows. It is clear that each function $j \circ \alpha_i$ is holomorphic on the upper half plane since j is so, and the meromorphy at infinity comes from the explicit formula for $\alpha_i = \begin{pmatrix} a & b \\ 0 & d \end{pmatrix}$, which shows in fact that $j \circ \alpha_i$ has a Laurent expansion in $q^{1/d}$ (and that it has coefficients in $\mathbf{Q}(\zeta_d)$). □

Corollary 5.4. *The coefficients of* $\Phi_n(X)$ *are polynomials in* j *with coefficients lying in* **Z**. *Thus we may write* $\Phi_n(X) = \Phi_n(X, j) \in \mathbf{Z}[X, j]$.

PROOF. It is clear that the coefficients of $\Phi_n(X)$ are polynomials in j with coefficients lying in the cyclotomic field $\mathbf{Q}(\zeta_n)$. Thus we may view the coefficients of Φ_n as elements of the Laurent series field $\mathbf{Q}(\zeta_n)((q))$. One checks that the automorphisms of $\mathbf{Q}(\zeta_n)$ acting on the roots of unity permute the power series expansions of the functions $j \circ \alpha_i$, and this shows in fact that these power series, being expansions of functions symmetric in the $j \circ \alpha_i$, in fact have coefficients in **Q**. That these coefficients are integral follows from the fact that the coefficients of the functions j and $j \circ \alpha_i$ are so. □

Theorem 5.5. *The polynomial $\Phi_n(X, j)$ is irreducible over $\mathbf{C}(j)$, is symmetric in X and j, and if n is not a perfect square, then $\Phi_n(j, j)$ is a polynomial in j of degree greater than 1 and has leading coefficient ± 1.*

PROOF. The first assertion follows from the fact that Γ permutes the functions $j \circ \alpha_i$ transitively.

As for the symmetry, observe that $j(z/n)$ is a root of $\Phi_n(X, j)$, since one of the matrices α_i is equal to $\begin{pmatrix} 1 & 0 \\ 0 & n \end{pmatrix}$, so that $\Phi_n(j(z/n), j(z))$ is identically zero. It follows then that $\Phi(j(z), j(nz))$ is identically zero as well. Thus $j(nz)$ is a root of the polynomial $\Phi_n(j, X)$. On the other hand, taking $\alpha_i = \begin{pmatrix} n & 0 \\ 0 & 1 \end{pmatrix}$ shows that $j(nz)$ is also a root of $\Phi_n(X, j)$. Since $\Phi_n(X, j)$ is irreducible and has a common root with $\Phi_n(j, X)$ it follows that $\Phi_n(X, j)$ divides $\Phi_n(j, X)$. By Gauss' Lemma, we must have

$$\Phi_n(j, X) = g(j, X)\Phi_n(X, j)$$

in $\mathbf{Z}[j, X]$ and hence $\Phi_n(j, X) = g(j, X)g(X, j)\Phi_n(j, X)$. It follows that $g(j, X) = \pm 1$. If $g(j, X) = -1$ then we get $\Phi_n(j, j) = -\Phi_n(j, j) = 0$, which contradicts the fact that $\Phi_n(X, j)$is irreducible over $\mathbf{Z}[j]$. Thus $g = 1$ and we get the second statement.

To get the last part, assume that n is not a square, so that in the given form for the matrices α_i we have $a \neq d$. Then we have

$$j - j \circ \alpha_i = \left(\frac{1}{q} + \dots\right) - \left(\frac{1}{\zeta_d^b q^{a/d}} + \dots\right)$$

and there is no cancellation in the leading terms of the two pieces. Thus the leading coefficient of the Laurent expansion is a root of unity, and the expansion for $\Phi_n(j, j)$ starts with $\pm\frac{1}{q^m}$, since this leading coefficient is both an integer and a root of unity. Thus $\Phi_m(j, j)$ is a polynomial in j with leading coefficient ± 1 as claimed. □

A basic fact about the modular polynomial is contained in the following lemma. In fact, it is the content of this lemma that renders the modular polynomial so useful.

Lemma 5.6 ([**14**], Theorem 5, pg 59). *If τ is an element of the upper half plane and E is the elliptic curve corresponding to the lattice $\mathbf{Z} + \mathbf{Z}\tau$, then the roots of $\Phi_n(X, j(\tau))$ are precisely the j-invariants of elliptic curves E' such that there exists a cyclic isogeny $E \to E'$ of degree n.*

We also have the following simple observation, which follows from the Chebotarev density theorem:

Lemma 5.7. *Let $\tau \in K$ and let E denote the elliptic curve corresponding to the lattice $\Lambda = \mathbf{Z} + \mathbf{Z}\tau$ as above. Then there exists a prime n such that there is a cyclic endomorphism $E \to E$ of degree n.*

Granting these facts for the moment, we can give the proof of the following basic result:

Theorem 5.8. *If τ is an element of an imaginary quadratic field, then $j(\tau)$ is an algebraic integer.*

PROOF. Let $\tau \in K$ and E be as in the statement of the lemma above. Then we find that there exists a non-square n such that $j(\tau)$ satisfies the equation $\Phi_n(j(\tau), j(\tau)) = 0$. But $\Phi_n(X, X)$ is a polynomial with integer coefficients and leading coefficient ± 1, so the theorem follows. □

We now want to prove that the j-invariants generate certain class fields of K. Thus let us recall the set up so far:

- There are $h - h_c$ non-isomorphic CM curves E_i of conductor c, where h_c is the class number of the order $R = R_c = \mathcal{O}_{K,c}$.
- The curves E_i may be described as $E_i \cong \mathbf{C}/\mathfrak{a}_i$ where $\mathfrak{a}_i$ runs over a set of representatives of the ideal classes of R.
- There is an action of the ideal class group $G = G_c$ of R on the set of isomorphism classes of curves of conductor c induced by multiplication of ideals: if $\mathfrak{b}$ in an ideal of R and $E = \mathbf{C}/\mathfrak{a}$ is a curve of conductor c, then $[\mathfrak{b}] \cdot E$ is the elliptic curve $\mathbf{C}/\mathfrak{a}\mathfrak{b}$, and one sees that this passes to equivalence classes of ideals $\mathfrak{b}$ and isomorphism classes of curves.
- The value of the j-invariant on each ideal class of R is an algebraic integer.

In view of the last two items above, and the identification of G with the Galois group $\mathrm{Gal}(K(c)/K)$, it seems reasonable to hope that the j-invariants of the E_i generate $K(c)$ and that the multiplication of ideals giving the action on curves goes over to the Galois action on j-invariants. This is what we shall prove below.

6. The key ingredient: Kronecker congruence

We will treat in detail the case where $c = 1$, leaving the case of general c for the adelic formulation of our results. Thus we take $R = \mathcal{O}_K$ and G to be the usual ideal class group of K.

Let p denote a rational prime number which is split in K. Let L/K denote the Galois closure of the field generated by the j-invariants $j(\mathfrak{a}_i)$. Let $\mathfrak{a} = \mathfrak{a}_i$ be fixed. Then if $\mathfrak{p}$ is a factor of p in K, consider the elliptic curve $\mathbf{C}/\mathfrak{p}^{-1}\mathfrak{a}$. We will show that for all but finitely split primes p, we have

$$j(\mathfrak{p}^{-1}\mathfrak{a}) \equiv j(\mathfrak{a})^p \pmod{\mathfrak{P}} \tag{1}$$

where $\mathfrak{P}$ is a prime above $\mathfrak{p}$ in L. Observe that the statement above assumes the fact that the j-invariants in question are algebraic integers.

Let us assume this congruence for the moment.

Theorem 6.1. *The field L is the Hilbert class field of K.*

PROOF. Let $\mathrm{Frob}(\mathfrak{P}) \in \mathrm{Gal}(L/K)$ denote the Frobenius automorphism of $\mathfrak{P}$. Then the congruence (1) may be restated as

$$\mathrm{Frob}(\mathfrak{P})(j(\mathfrak{a})) \equiv j(\mathfrak{p}^{-1}\mathfrak{a}) \pmod{\mathfrak{P}}.$$

Since there are only finitely many ideal classes $[a_i]$ and consequently only finitely many j-invariants $j(\mathfrak{a}_i)$ to consider, we may assume that p is relatively prime to the finitely many numbers obtained by taking differences of the conjugates of the $j(\mathfrak{a}_i)$. Then the the congruence above is in fact an equality:

$$\mathrm{Frob}(\mathfrak{P})(j(\mathfrak{a})) = j(\mathfrak{p}^{-1}\mathfrak{a}).$$

But now we are done, since this implies already that all the numbers $j(\mathfrak{a}_i)$ are conjugate over K. Indeed, we can find infinitely many split primes p in any ideal class of K, so that the set $\mathfrak{p}^{-1}\mathfrak{a}$ runs over all the ideal classes. Furthermore, $\mathrm{Frob}(\mathfrak{P})$ is trivial precisely when $\mathfrak{p}$ is principal, since $\mathfrak{p}^{-1}\mathfrak{a}$ is in the same ideal class as $\mathfrak{a}$ precisely for principal $\mathfrak{p}$. Thus the primes of degree 1 in K splitting in E are the ones which are principal. It follows that L is the Hilbert class field of K. □

Corollary 6.2. *The Hilbert class field H/K of K is generated by $j(E)$ where E is any one elliptic curve of conductor* 1.

PROOF. This follows from the fact that each of the $j(\mathfrak{a}_i)$ are conjugate, since the number of distinct $j(\mathfrak{a}_i)$ is the same as the degree of H/K. □

7. Proof of the Kronecker congruence

The proof of the key congruence is not hard, although it is somewhat technical. Thus far we have been able to get away with viewing the curves E_i as being defined over $\mathbf{C}$, but the free ride is over, and we have to deal with them as elliptic curves defined over number fields and wrestle with what reduction modulo a prime actually means.

For the purposes of this article, we will proceed formally, and assume that reduction modulo a prime works the way one would like, at least if we avoid a finite set of primes of bad reduction. We will state what we need, but we emphasize that the results we use all follow from a systematic development of Néron models.

So let us briefly state what is required. Let E_1 and E_2 denote elliptic curves over a number field L and let $\mathfrak{P}$ denote a prime of L of residue characteristic p. We assume that there exist Weierstrass models for E_i/L which have the property that the corresponding discriminants are prime to $\mathfrak{P}$. Let let R be the localization of the ring of integers of L at $\mathfrak{P}$, and let $\mathbf{F}$ denote the residue field of R at $\mathfrak{P}$. Then let $\overline{E}_i$ denote the elliptic curves over $\mathbf{F}$ defined by reducing the coefficients of the given equations of the E_i. Under our assumptions, the curves $\overline{E}_i$ are elliptic curves over $\mathbf{F}$.

The key property is the following. Let $A_i = H^0(E_i, \Omega^1)_L$ denote the 1-dimensional L-space of regular differentials on E_i. Then A_i is contravariantly functorial: if there is an isogeny $\mu : E_1 \to E_2$, then there is an induced map $A_2 \to A_1$. If $E_1 = E_2$ then there is an endomorphism $A \to A$, induced by multiplication by an element $a(\mu) \in L$, since A is a 1-dimensional space.

Similarly, let $\overline{A}_i = H^0(\overline{E}_i, \Omega^1)_{\overline{\mathbf{F}}_p}$ denote the corresponding object over $\mathbf{F}$. Then each $\overline{A}_i$ is a 1-dimensional space over $\overline{\mathbf{F}}_p$, and, if $\overline{\mu}$ is an isogeny $E_1 \to E_2$, then there is an induced map $\overline{A}_2 \to \overline{A}_1$.

The key property of good reduction that we need is the following, which can be extracted from [**20**].

Theorem 7.1. *Let the hypotheses be as above. Then there exist rank* 1 *free R-modules $\mathbf{A}_i$ such that the following properties hold:*

- $A_i = \mathbf{A}_i \otimes L$
- $\overline{A}_i = \mathbf{A}_i \otimes R/\mathfrak{P}R$
- *If $\mu : E_1 \to E_2$ is an isogeny, then there is a canonical isogeny $\overline{\mu} : \overline{E}_1 \to \overline{E}_2$ called the reduction of μ, and there is a canonical map $\mathbf{A}_2 \to \mathbf{A}_1$ which is compatible via the above tensor product with the corresponding maps on A_i and $\overline{A}_i$.*
- *The degree of $\overline{\mu}$ is the same as the degree of μ.*

Corollary 7.2. *Suppose $E_1 - E_2$ and μ is an endomorphism of E. Then the induced map $H^0(E,\Omega^1)_L \to H^0(E,\Omega^1)_L$ is induced by multiplication by an element $a(\mu) \in R$. The endomorphism of $H^0(\overline{E},\Omega^1)_{\overline{\mathbf{F}}_p}$ induced by $\overline{\mu}$ is simply multiplication by the image of $a(\mu)$ in $R/\mathfrak{P}R$.*

We also need to deal with the question of rationality of the endomorphisms that we shall have to consider.

Lemma 7.3. *Let E be a CM curve. If E is defined over a field L with $K \subset L$, then any endomorphism $[\alpha]$ of E with $\alpha \in \operatorname{End}(E) \subset K$ is defined over L. In particular,* $\ker([\alpha])$ *is an L-rational subgroup.*

PROOF. Let ω denote the invariant differential on E. Then if $\sigma \in \operatorname{Gal}(\overline{\mathbf{Q}}/L)$, we have $([\alpha]^\sigma)^*\omega = ([\alpha]^\sigma)^*\omega^\sigma = ([\alpha]^*\omega)^\sigma = (\alpha\omega)^\sigma = \alpha\omega$ since $\alpha \in K$ and ω is rational over L. Thus $[\alpha]^\sigma$ and $[\alpha]$ have the same effect on differentials, so that they must be equal. □

Lemma 7.4. *Suppose E is isomorphic to $\mathbf{C}/\mathfrak{a}$ where $\mathfrak{a}$ is a proper R-ideal for $R = \operatorname{End}(E) \subset K$. Then if $\mathfrak{b} \subset \operatorname{End}(E)$ is another proper R-ideal, and E' is an elliptic curve corresponding to $\mathbf{C}/\mathfrak{b}^{-1}\mathfrak{a}$, then the isogeny $E \to E'$ induced by the inclusion $\mathfrak{a} \subset \mathfrak{b}^{-1}\mathfrak{a}$ is rational over any extension L/K such that both E and E' are defined over L.*

PROOF. The kernel of the map $\mu : \mathbf{C}/\mathfrak{b}^{-1}\mathfrak{a} \to \mathbf{C}/\mathfrak{a}$ consists precisely of the points in E which are killed by every endomorphism β for $\beta \in \mathfrak{b}$. Thus $\ker(\mu) = \bigcap \ker([\beta])$ where β runs over $\mathfrak{b}$. But we have seen above that $\ker([\beta])$ is rational over L for all β, and so $\ker(\mu)$ is rational as well. □

With all this in hand, we proceed as follows. For each ideal class $[\mathfrak{a}_i]$ of K, we pick a Weierstrass model of E_i over the Galois closure of the field generated by the $j(\mathfrak{a}_i)$ over K. We pick a split prime $\mathfrak{p}$ of K which is relatively prime to the following quantities:

- the differences of the conjugates of the $j(\mathfrak{a}_i)$,
- the discriminants of the selected Weierstrass models of the E_i,
- the discriminant of L.

Since $\mathfrak{a} \subset \mathfrak{p}^{-1}\mathfrak{a}$, we have a natural map

$$\mathbf{C}/\mathfrak{a} \to \mathbf{C}/\mathfrak{p}^{-1}\mathfrak{a}.$$

Note that this map has degree p. Let $\mathfrak{b}$ denote an ideal of $\mathcal{O}_K$ such that $\mathfrak{p}\mathfrak{b} = (\alpha)$ is principal. We may assume that $\mathfrak{b}$ is prime to p. Then multiplication by α followed by the natural projection gives a map

$$\mathbf{C}/\mathfrak{p}^{-1}\mathfrak{a} \to \mathbf{C}/\mathfrak{b}\mathfrak{a} \to \mathbf{C}/\mathfrak{a}.$$

The composite map

$$\mathbf{C}/\mathfrak{a} \to \mathbf{C}/\mathfrak{p}^{-1}\mathfrak{a} \to \mathbf{C}/\mathfrak{b}\mathfrak{a} \to \mathbf{C}/\mathfrak{a} \tag{2}$$

is clearly induced by multiplication by α, as can be seen by tracing around the effect on $\mathbf{C}$. Furthermore, the first arrow above has degree p as was previously remarked, the second arrow is an isomorphism, and the third arrow has degree equal to the norm of $\mathfrak{b}$, which is prime to p.

We now want to reduce the endomorphism $[\alpha]$ modulo $\mathfrak{P}$, and we claim that the reduction of α is inseparable in characteristic p. Note that over $\mathbf{C}$ we have that if ω is the differential dz on $E = \mathbf{C}/\Lambda$, then the pullback $[\alpha]^*\omega = \alpha\omega$. In other words, μ induces multiplication by α on the 1-dimensional complex vector space $H^0(E, \Omega^1)$ over $\mathbf{C}$. Thus $[\alpha]$ induces multiplication by α on $H^0(E, \Omega^1)_L$, so we see that in characteristic p, the reduction of $[\alpha]$ induces multiplication by $\overline{\alpha}$ on $H^0(E, \Omega^1)$ over $\overline{\mathbf{F}}_p$, and that this is zero, since α is divisible by $\mathfrak{p}$.

On the other hand, μ factors as a composite of two maps, one of degree p, and the other of degree prime to p. Since the reduction of the second map is separable, it follows that the the reduction of the former is purely inseparable and since it has degree p, it must be the Frobenius map raising to p-th powers, up to an automorphism of the reduced curve. Thus we get

$$j(\mathfrak{p}^{-1}\mathfrak{a}) \equiv j(\mathfrak{a})^p \pmod{\mathfrak{P}}$$

as required.

8. The ray class field

We now want to look at the field generated by adding the coordinates of points of order N together with the j-invariants. However, this will require a certain precision about the exact models we choose for our elliptic curves, which we have thus far been able to avoid. So let E be an elliptic curve over $\mathbf{C}$ with $E \cong \mathbf{C}/\Lambda$ for some Λ. As we have seen above, the choice of Λ leads to a Weierstrass model for E:

$$E : Y^2 = 4X^3 - g_2(\Lambda)X - g_3(\Lambda).$$

At first glance, there is no particularly canonical choice of Λ.[1] Since the coordinates of points evidently depend on the model, we are led to seek an isomorphism invariant quantity. Let us define the Weber function h_E of E as follows. If $j(E) \neq 0, 1728$

$$h_E = \frac{g_2 g_3}{\Delta} \cdot X.$$

Here we view g_2 and g_3 as constants, and X as a rational function on E. Recall also that in terms of the complex analytic incarnation of E, the function X is just the Weierstrass $\wp$ function. Thus, h_E is just a constant multiple of the $\wp$-function, once the choice of Λ is fixed. Note also that h_E vanishes identically if $g_2 g_3 = 0$; this is excluded by our assumption that $j(E) \neq 0, 1728$, since these are the values of the j-invariant corresponding to $g_2 g_3 = 0$.

If $g_3 = 0$ we put

$$h_E = \frac{g_2^2}{\Delta} X^2$$

and if $g_2 = 0$ we put

$$h_E = \frac{g_3}{\Delta} X^3.$$

Then in every case, h_E is a rational function on E.

Proposition 8.1 ([**21**], Example 5.5.1, pg 134). *The Weber function gives a rational function $E \to \mathbf{P}^1$. If u and u' are points on E, we have $h_E(u) = h_E(u')$ if and only if there is an automorphism σ of E such that $\sigma(u) = u'$.*

---

[1] Actually, there is a canonical choice, but it comes from considerations which are not particularly useful here.

We begin with the following observation. We know from the foregoing considerations that the canonical projection $E = \mathbf{C}/\mathfrak{a} \to \mathbf{C}/\mathfrak{p}^{-1}\mathfrak{a} = E^\sigma$ for $\sigma = \text{Frob}(\mathfrak{p})$ reduces to a purely inseparable map of degree p in characteristic p. Such a map is equal to Frobenius, at least up to an automorphism of $\overline{E^\sigma}$.

Lemma 8.2 ([**14**], Lemma 1, pg 126). *We may compose the natural map $E = \mathbf{C}/\mathfrak{a} \to \mathbf{C}/\mathfrak{p}^{-1}\mathfrak{a} = E^\sigma$ with an automorphism of E^σ so that the deduced map $E \to E^\sigma$ reduces precisely to the Frobenius map.*

Theorem 8.3. *Let $E = \mathbf{C}/\mathfrak{a}$ be a CM elliptic curve of conductor 1. Then the field L generated over K by $j(\mathfrak{a})$ and the numbers $h_E(u)$ for u running over the points of order N in E is the ray class field of conductor N.*

PROOF. We will show that a degree 1 prime of K splits completely in L if and only if it is principal, generated by an element congruent to 1 modulo N. Thus suppose $\mathfrak{p} = (\alpha)$ is such a prime and $\alpha \equiv 1 \pmod{N}$. We assume again that $\mathfrak{p}$ is away from a finite set of bad primes. Then let $\mathfrak{a}$ denote an ideal of K and consider the map

$$\mu : E = \mathbf{C}/\mathfrak{a} \to \mathbf{C}/\mathfrak{p}^{-1}\mathfrak{a} = E^\sigma$$

where $\sigma = \text{Frob}(\mathfrak{P})$ for a prime $\mathfrak{P}$ above $\mathfrak{p}$. Since α is principal, the previous results on the Hilbert class field show that in fact $E^\sigma = E$.

Using the lemma above, we can assume that the reduction of μ is the Frobenius endomorphism. Now if u is a point of order N on E, then u^σ is a point of order N on E^σ and u^σ reduces to $\text{Frob}(\overline{u})$ on the reduced curve $\text{Frob}(\overline{E})$. Thus $\mu(u)$ and u^σ reduce to the same point modulo $\mathfrak{P}$, so that $\mu(u) = \sigma(u)$ for each point of order N (because $(N, p) = 1$).

Then consider the composite

$$\mathbf{C}/\mathfrak{a} \to \mathbf{C}/\alpha^{-1}\mathfrak{a} \to \mathbf{C}/\mathfrak{a}$$

where the second arrow is induced by multiplication by α. The composite is the endomorphism of $\mathbf{C}/\mathfrak{a}$ induced by multiplication by α. Under our hypothesis on α, this endomorphism acts as the identity on the points of order N. Furthermore, as endomorphisms of E, the multiplication-by-α map and the the map μ are equal up to an automorphism (multiplication by a root of unity in $\mathcal{O}_K^\times$). Since the numbers $h_E(u)$ are invariant under automorphisms, it follows that σ acts as the identity on each $h_E(u)$, which shows that $\text{Frob}(\mathfrak{P})$ is trivial on L, so $\mathfrak{p}$ splits completely.

As for the converse, suppose $\mathfrak{p}$ is a prime that has trivial Frobenius in L. Then in particular $\text{Frob}(\mathfrak{P})$ is trivial in the $K(j(\mathfrak{a}))$ which is the Hilbert class field, so $\mathfrak{p} = (\alpha)$ is principal, and, with notation as above, $E^\sigma = E$ for $E = \mathbf{C}/\mathfrak{a}$ and $E^\sigma = \mathbf{C}/\mathfrak{p}^{-1}\mathfrak{a}$. Then if u is a point of order N corresponding to to $u \in K/\mathfrak{a}$, we have $h_E(u)^\sigma = h_E(u)$ since σ is trivial on L. Furthermore, we can argue as in the previous case to see that the multiplication-by-α endomorphism acts via σ on the points of order N. But now $h_E^\sigma = h_E$ since h_E is defined over $K(j(E))$ whence we get $(h_E(u))^\sigma = h_E^\sigma(u^\sigma) = h_E(u^\sigma) = h_E(\alpha u)$. It follows that $h_E(u) = h_E(\alpha u)$ for each u, so that u and αu differ by an automorphism of E. Thus there exists a root of unity ϵ (perhaps dependent on u) such that $\alpha u = \epsilon u$. Changing the generator α by this root of unity we get $\alpha u = u$, and since $E[N]$ is principal as a module over $\mathcal{O}_K/N\mathcal{O}_K$, we find that the $\alpha u = u$ for all u in $E[N]$. Thus $\alpha \equiv 1 \pmod{N}$. □

9. Shimura's adelic formulation

Let $E = \mathbf{C}/\Lambda$ denote an elliptic curve, for a lattice Λ in K. Let σ be an automorphism of K^{ab}/K which is equal to the Artin symbol $\sigma = (s, K^{ab}/K)$ for an idele s of K. Let $s^{-1}\Lambda$ be the image of Λ under s^{-1}. Then we have the following theorem, whose proof, although straightforward in view of the preceding results, is somewhat lengthy. Observe that since the ray class fields exhaust the abelian extensions of K, our foregoing results in fact determine the maximal abelian extension of K. Then we have the following theorem, which is a variant of the results presented in [**19**], Chapter 4.

Theorem 9.1 ([**14**], Theorem 3, pg 129). *We have $\mathbf{C}/s^{-1}\Lambda = E^{\sigma}$. Furthermore, if $u \in K/\Lambda$ is a torsion point on E, then $\sigma(u) \in E^{\sigma}$ is represented by the class of $s^{-1}u$ in $K/s^{-1}\Lambda$.*

Remark 9.2. The theorem above is sometimes called the Shimura reciprocity law. Its significance lies in the fact that it identifies the Galois action on elliptic curves and their torsion points with the arithmetic action of the ideles on the corresponding lattices.

Corollary 9.3. *Let c denote a positive integer. Let $[\mathfrak{a}_i]$ denote a set of representatives for the ideal classes of the order of conductor c. Then the numbers $j(\mathfrak{a}_i)$ are conjugate over K, and generate the ring class field of conductor c.*

PROOF. This is immediate from the above theorem. One simply identifies the ideal class group of R with $K^{\times}\backslash K(\mathbf{A})^{\times}/U$ where $U = \mathbf{Q}(\mathbf{A})^{\times}\prod_v R_v^{\times} \cdot \mathbf{C}^{\times}$, for $R = \mathcal{O}_c$, and observes that the space of proper R-lattices modulo principal R-lattices is a principal homogeneous space for this group. According to the theorem, the action of the ideles on lattices goes over to the Galois action on j-invariants, so the corollary follows. □

10. CM points on modular curves

Let us recall the definition of the modular curve $X_0(N)$. Complex analytically, we know that the open modular curve $Y_0(N)$ is the quotient of the upper half plane by the group $\Gamma_0(N)$ of matrices $\begin{pmatrix} a & b \\ c & d \end{pmatrix}$ where $c \equiv 0 \pmod{N}$, and that $X_0(N)$ is the compactification of $Y_0(N)$ by adding a finite number of cusps (see [**17**]). The main facts that we will use are that

- $X_0(N)$ has the structure of an algebraic curve over $\mathbf{Q}$, and that
- The non-cuspidal points of $X_0(N)$ correspond to a certain moduli problem for elliptic curves and cyclic isogenies of degree N.

The canonical reference for the algebraic and moduli space structure on the modular curve is [**9**]. However, the reader will find a more accessible discussion in the introductory paper [**7**].

Remark 10.1. A plane model for the algebraic curve $X_0(N)$ is provided by the modular polynomial $\Phi_N(X, j)$ defined previously. However, this model is highly singular, and has extremely large coefficients.

Remark 10.2. The non-cuspidal complex points of $X_0(N)$ are simply the quotient space $\Gamma_0(N)\backslash\mathcal{H}$. A simple argument (see [**20**], Section 13 of Appendix C) shows that for each $\tau \in Y_0(N)(\mathbf{C}) = \Gamma_0(N)\backslash\mathcal{H} \subset X_0(N)$, there exists a pair of elliptic curves E, E' over $\mathbf{C}$ together with a cyclic isogeny $E \to E'$. In terms of the algebraic structure of a curve over $\mathbf{Q}$ on $X_0(N)$, it turns out that a given τ corresponds to a point rational over a subfield $L \subset \mathbf{C}$ precisely when the curves E and E' and the associated isogeny may be chosen to be defined over L.

The key fact that we will need is that CM points give rise to points on modular curves that are defined over abelian extensions of the corresponding imaginary quadratic field. The key assumption that makes this work is the so-called *Heegner hypothesis*, which may be stated as follows.

Definition 10.3. Let K be a fixed imaginary quadratic field and let $N > 1$ be a square-free integer. Then we say that N satisfies the Heegner hypothesis relative to K if there exists an ideal $\mathfrak{n} \subset \mathcal{O}_K$ of norm N. Equivalently, N satisfies the Heegner hypothesis if and only if every prime divisor of N is either split or ramified in K. To simplify matters, we will assume in the remainder of this article that N is prime to the discriminant D of K, so that all primes dividing N are split in K.

Now let $E = \mathbf{C}/\mathfrak{a}$ denote an elliptic curve with CM by the order $R = \mathcal{O}_c$ in K. Assume further that N is prime to the conductor c of E. if we fix an ideal $\mathfrak{n}'$ of $\mathcal{O}_K$ of norm N and set $\mathfrak{n} = R \cap \mathfrak{n}'$, then $R/\mathfrak{n} = \mathbf{Z}/N\mathbf{Z}$ and there is a cyclic isogeny

$$\mathbf{C}/\mathfrak{a} \to \mathbf{C}/\mathfrak{n}^{-1}\mathfrak{a}.$$

Since both source and target have conductor c, we see that each is defined over the ring class field $K(c)$. Furthermore, by Lemma 7.4, the isogeny above is also defined over $K(c)$. Thus we obtain a point P of $X_0(N)$ defined over the field $K(c)$. The point P is called a Heegner point.

Remark 10.4. The reader will note that in fact there is a choice of ideal $\mathfrak{n}$ of norm N implicit in the definition of the Heegner point above. In fact, there 2^r choices of $\mathfrak{n}$ where r is the number of prime divisors of N. The 2^r points so defined are related by way of the Atkin-Lehner operators acting on $X_0(N)$ – see [**8**].

Remark 10.5. If P is the point corresponding to the fractional R-ideal $\mathfrak{a}$ and the choice of $\mathfrak{n}$ as above, then the the complex conjugate of the point P corresponds to the choice of data $\bar{\mathfrak{a}}$ and $\bar{\mathfrak{n}}$, where the bar acting on a lattice in K denotes the action of complex conjugation acting as $\mathrm{Gal}(K/\mathbf{Q})$. In view of the theorems proven in the previous sections for the action of $\mathrm{Gal}(K(c)/K)$ in terms of ideals, we see also that if $\sigma \in \mathrm{Gal}(K(c)/K)$, then P^σ is also a Heegner point corresponding to a suitable choice of R-ideal $\mathfrak{a}$.

To proceed further, we need to introduce the Jacobian variety $J_0(N)$ of the curve $X_0(N)$. Then $J_0(N)$ is an abelian variety over $\mathbf{Q}$, and we have $J_0(N)(\mathbf{C}) = \mathrm{Pic}^0(X_0(N))$ is the group of divisors of degree 0 on $X_0(N)$ modulo divisors of functions. We have an embedding $X_0(N) \hookrightarrow J_0(N)$ which sends the point Q to the class of the divisor $Q - \infty$ where ∞ denotes the cusp at infinity. In particular, the Heegner points P defined above give rise to points on $J_0(N)$ defined over the ring class field $K(c)$. By abuse of notation (and by convention!) we will continue to write P for the Heegner point, even when we are thinking of the image of the point in $X_0(N)$ inside the Jacobian.

Now consider a modular elliptic curve E of level N over $\mathbf{Q}$. Then there is a nonconstant map $X_0(N) \to J_0(N) \to E$ defined over $\mathbf{Q}$, and the image of P is a rational point of E over the field $K(c)$. If h_c denotes the degree of $K(c)$ over K, then in fact we get h_c points in the set $E(K(c))$, corresponding to the h_c distinct CM elliptic curves with conductor c and a fixed choice of $\mathfrak{n}$. Note that it is not clear whether these points are distinct, or even non-trivial. Choice of a different ideal $\mathfrak{n}'$ of norm N does not give anything more: as noted above, the choices of $\mathfrak{n}$ are related by way of the Atkin-Lehner operators, and these act as ± 1 on E.

Let us now collect a few elementary facts about the extension $K(c)/K$. We know that $K(c)/K$ is an abelian extension, and in view of the definition, one checks that it is normal over $\mathbf{Q}$, and that complex conjugation (identified with $\mathrm{Gal}(K/\mathbf{Q})$) acts on $\mathrm{Gal}(K(c)/K)$ by inversion. Thus we have an exact sequence as follows:

$$0 \to \mathrm{Gal}(K(c)/K \to \mathrm{Gal}(K(c)/\mathbf{Q}) \to \mathrm{Gal}(K/\mathbf{Q}) \to 0$$

where the action of $\mathrm{Gal}(K/\mathbf{Q}) =< \tau >$ on an element $\sigma \in \mathrm{Gal}(K(c)/K)$ is given by $\tau\sigma\tau = \sigma^{-1}$. It follows from this that the irreducible representations of the (finite) group $\mathrm{Gal}(K(c)/\mathbf{Q})$ have degree 1 or 2. Since $\mathrm{Gal}(K(c)/K$ has order h_c, there are h_c 1-dimensional characters of $\mathrm{Gal}(K(c)/K)$. Let ρ_χ denote the 2-dimensional representation of $\mathrm{Gal}(K(c)/\mathbf{Q})$ obtained by inducing χ. Then ρ_χ is irreducible if and only if $\chi \neq \chi^{-1} \iff \chi^2 \neq 1$. In this case we have $\rho_\chi = \rho_{\chi^{-1}}$. Finally, every irreducible 2-dimensional representation of $\mathrm{Gal}(K(c)/\mathbf{Q})$ is obtained as ρ_χ for some χ, and if $\rho_\chi = \rho_{\chi'}$ then $\chi' = \chi$ or $\chi' = \chi^{-1}$.

If $\chi^2 = 1$ and ρ_χ is reducible, then then χ extends to a 1-dimensional character η of $\mathrm{Gal}(K(c)/\mathbf{Q})$ and $\rho_\chi = \eta \oplus \eta\chi_K$, where χ_K is the nontrivial quadratic character of $\mathrm{Gal}(K(c)/\mathbf{Q})$ with kernel $\mathrm{Gal}(K(c)/K)$. (Alternatively, χ_K is the character cutting out the quadratic extension $K/\mathbf{Q}$.) Given a character χ of $\mathrm{Gal}(K(c)/K)$, we define the conductor $\mathfrak{c}(\chi)$ of χ to be the conductor of the field (in the sense of class field theory) cut out by the kernel of χ.

For χ and ρ_χ as above, we may form the theta series θ_χ defined by

$$\theta_\chi = \sum_{\mathfrak{a}} \chi(\mathrm{Rec}(\mathfrak{a}))q^{\mathrm{Norm}(\mathfrak{a})}.$$

Here the sum is taken over ideals $\mathfrak{a}$ of $\mathcal{O}_K$ which are prime to $\mathfrak{c}(\chi)$, and $\mathrm{Rec}(\mathfrak{a}) \in \mathrm{Gal}(K(c)/K)$ denotes the image of K under the Artin reciprocity map. Then θ_χ is a modular form of weight 1 and level $c(\chi) = D \cdot \mathrm{Norm}(\mathfrak{c}(\chi))$. If ρ_χ is irreducible, then θ_χ is a cuspidal newform, and if $\chi^2 = 1$, then ρ_χ is an Eisenstein series. For proofs of these assertions, we refer the reader to [**17**], Section 4.8. In the first case, we note that $\theta_\chi = \theta_\chi^{-1}$ if ρ_χ is irreducible. Thus $\theta_\chi = \theta_{\rho_\chi}$ depends only on the irreducible representation ρ_χ in this case.

Finally, we write f_E to denote the modular newform of weight 2 and level N associated to the modular elliptic curve E.

With all this in hand, let us consider the L-function of the elliptic curve E, viewed as curve over the field $K(c)$. To elaborate on this, recall that associated to E there is a family of Galois representations

$$\rho_\ell : \mathrm{Gal}(\overline{\mathbf{Q}}/\mathbf{Q}) \to GL_2(\mathbf{Q}_\ell)$$

for primes ℓ. The representation ρ_ℓ is unramified outside $N\ell$, and is characterized by the fact that $\mathrm{Tr}(\rho_\ell(\mathrm{Frob}(q)) = p + 1 - \#E(\mathbf{F}_q)$ for any prime $q \nmid N\ell$. Then we have

$$L(E, \mathbf{Q}, s) = \prod_q P_q(q^{-s}) \tag{3}$$

where $P_q(X)$ is defined by taking ℓ to be any prime away from q and setting P_q to be the characteristic polynomial of $\rho_\ell(\mathrm{Frob}(q))$ on the maximal quotient of ρ_ℓ on which the inertia group of q acts trivially. It is a theorem of Deligne that the polynomial P_q is independent of ℓ. (Note however that for $q \nmid N$, this follows from the characterization of the trace of Frobenius in the definition of ρ_ℓ.)

In a similar way, we may form the L-function $L(E, K, s)$ for an extension $K/\mathbf{Q}$, simply by restriction of the family ρ_ℓ of representations to the group $\mathrm{Gal}(\overline{\mathbf{Q}}/K)$. By definition, we have

$$L(E, \mathbf{Q}, s) = \prod_{\mathfrak{q}} P_{\mathfrak{q}}((\mathrm{Norm}(\mathfrak{q}))^{-s}) \tag{4}$$

where this time the product is taken over primes of K and the polynomial $P_{\mathfrak{q}}(X)$ is the characteristic polynomial of $\mathrm{Frob}(\mathfrak{q})$ on the maximal quotient of ρ_ℓ on which the inertia group of $\mathfrak{q}$ acts trivially.

Then the usual formalism of L-functions and induction, as in [**15**], Chapter XII, where the case of Artin L-functions is treated, yields

$$L(E, K(c), s) = \prod_{\rho} L(E \otimes \rho, \mathbf{Q}, s)^{\dim(\rho)} \tag{5}$$

where the product is taken over the irreducible representations of $\mathrm{Gal}(K(c)/\mathbf{Q})$. If ρ has dimension 2, then $\rho = \rho_\chi$ for some $\chi \neq \chi^{-1}$, and $L(E \otimes \rho, \mathbf{Q}, s)$ is the Rankin-Selberg convolution of f_E with the theta series θ_χ as defined in [**8**]. If ρ has dimension 1, then let χ be the character obtained by restricting ρ to $\mathrm{Gal}(K(c)/K)$. In this case again $L(E \otimes \rho, \mathbf{Q}, s)$ is the Rankin-Selberg convolution of f_E with the theta series θ_χ, but the theta series turns out to be the weight 1 Eisenstein series associated to the 1-dimensional character ρ.

One can compute the root number ϵ_ρ of the L-series $L(E \otimes \rho, \mathbf{Q}, s)$ as in [**5**] (see also Rohrlich's course from this conference). Then $\epsilon_\rho = \pm 1$, and, under the Heegner hypothesis, we will have $\epsilon_\rho = -1$ for every irreducible ρ of dimension 2. Observe also that, as remarked above, each such ρ uniquely determines a pair χ, χ^{-1} of characters of $\mathrm{Gal}(K(c)/K)$, where $\chi \neq \chi^{-1}$.

On the other hand, if ρ is a 1-dimensional character of $\mathrm{Gal}(K(c)/K)$ (and so is either quadratic or trivial), then so is $\rho \cdot \chi_K$ for the quadratic character χ_K of K, and $\epsilon_\rho \cdot \epsilon_{\rho \cdot \chi_K} = -1$. Note that ρ and $\rho \cdot \chi_K$ have the same restriction χ to $\mathrm{Gal}(K(c)/K)$, and that induction induces a bijection between characters χ of $\mathrm{Gal}(K(c)/K)$ of order ≤ 1 and such pairs $\rho, \rho \cdot \chi_K$.

Putting these facts together, we see that $L(E, K(c), s)$ has a zero at $s = 1$ of degree at least equal to the number of distinct characters of $\mathrm{Gal}(K(c) : K)$, which is to say, a zero of order at least $[K(c) : K]$.

Given the order of the zero of $L(E, K(c), s)$ at $s = 1$, the conjecture of Birch and Swinnerton-Dyer predicts that the rank of the finitely generated group $E(K(c))$ is at least $[K(c) : \mathbf{Q}]$. Since the Heegner point P, together with its conjugates under $\mathrm{Gal}(K(c)/K)$ (which are also Heegner points corresponding to suitable $\mathfrak{a}$ and $\mathfrak{n}$) provide a set of points of cardinality $[K(c) : K]$ defined over the correct field, one is therefore led to speculate that in fact these points are linearly independent over $\mathbf{Q}$ and generate a subgroup of rank $[K(c) : K]$ in $E(K(c))$. Furthermore, one has the following theorem, due to Kolyvagin, Bertolini-Darmon, and Zhang:

Theorem 10.6 ([**1**], [**10**], [**24**]). *Suppose the points P^σ are linearly independent over* $\mathbf{Q}$*, for* $\sigma \in \mathrm{Gal}(K(c)/K)$. *Then* $\mathrm{rank}(E(K(c)) = \mathrm{ord}_{s=1}L(E, K(c), s) = [K(c) : K]$.

To be more precise, let us consider the the set $X_{\mathbf{C}} = X(E, D, c) \otimes_{\mathbf{Z}} \mathbf{C}$. This is a finite complex vector space which inherits an action of $\mathrm{Gal}(K(c)/\mathbf{Q})$. If ρ is an irreducible representation of $\mathrm{Gal}(K(c)/\mathbf{Q})$, then write $X^\rho_{\mathbf{C}}$ for the ρ-isotypic component of $X_{\mathbf{C}}$. Recall also that ρ has dimension either 1 or 2, and that $\sum_\rho \dim(\rho) = [K(c) : \mathbf{Q}]$. Then a more precise version of the theorem of Kolyvagin-Bertolini-Darmon-Zhang may be stated as follows:

Theorem 10.7 ([**1**], [**10**], [**24**]). *Let ρ be fixed, and suppose that* $\dim_{\mathbf{C}} X^\rho \neq 0$. *Then* $\mathrm{ord}_{s=1}L(E \otimes \rho, s) = 1$*, and* $\dim_{\mathbf{C}} X^\rho = \dim(\rho)$.

Thus the Heegner points generate essentially all of the points in the group $E(K(c))$. A general heuristic, elaborated by Mazur in [**16**], predicts that this is the behavior for most choices of E, K, c, namely, that any extra vanishing of the L-function, or the presence of any additional points on E over $K(c)$, should be relatively infrequent. Our next task is therefore to state this expectation somewhat more precisely, and to do this we need a bit more notation. We will be somewhat brief in this discussion, and the reader is urged to read [**16**] in detail for a beautiful introduction to this circle of ideas.

Let us write $X(E, D, c)$ for the subgroup of $E(K(c))$ generated by the Heegner points, and let $x(E, D, c)$ denote the rank of $X(E, D, c)$. Here we recall that D is the discriminant of the quadratic field K. We also let h_D denote the absolute class number of K. Note then that the degree of $[K(c) : K]$ is the product of h_D with an elementary factor depending only on c. Then Mazur's heuristic predicts that the rank of $X(E, D, c)$ is proportional to the degree of $[K(c) : K]$, with error that is small in some absolute sense. In fact, what one expects is that $\mathrm{ord}_{s=1}L(E \otimes \rho, s) = 1$, and $\dim_{\mathbf{C}} X^\rho = \dim(\rho)$, for "most" ρ, as in the hypotheses of the above theorem. For instance, one could phrase the question as follows:

Question 10.8. Let c and E be fixed, and let the quadratic field K vary. Then determine the asymptotics of the function $x(E, D, c)/h_D$. Is it bounded, independently of c? Is it true that $x(E, D, c)$ is asymptotic to h_D as D grows?

This question seems to be very hard and is not yet resolved, but some results in this direction may be found in [**22**] (at least when $c = 1$).

Alternatively, one might fix K (and D) and let c vary. For instance, one could consider the so-called p-adic anticyclotomic tower of fields $K(p^n)$, as n tends toward infinity. In this case, let us write h_n for the degree of $K(p^n)/K$. Then we have that h_n is proportional to p^n, with constant of proportionality depending only on K and p. One is therefore led to

Question 10.9. Suppose K and p and E are fixed. Then is $x(E, D, p^n)$ asymptotic to p^n for large n?

This question turns out to be easier to resolve than the one above, and the results are quite complete. In fact, one has the following strong result

Theorem 10.10 ([**23**], [**3**], [**5**]). *Suppose K and p and E are fixed. Then the following statements hold, for all but finitely many irreducible representations ρ of* $\mathrm{Gal}(K(p^n)/K)$, *as $n \to \infty$:*

- *the image of P in X^ρ is nonzero,*
- $\mathrm{ord}_{s=1} L(E \otimes \rho, s) = 1$, *and*
- $\dim_{\mathbf{C}} X^\rho = \dim(\rho)$.

The proof of this theorem is based on a theorem of Ratner [**18**] on uniform distribution of unipotent flows on p-adic Lie groups, which, when applied to our situation, allows us to conclude that the Heegner points of level p^n are uniformly distributed in the appropriate sense, as n tends toward infinity. For a sketch of the proof, we refer the reader to Chapter 1 of [**4**].

In conclusion, we remark that there is a general notion of a CM point which is vastly more general than the construction described here. Indeed, there are general root number computations that predict the size of $E(K(p^n))$ for general E and K and p, but the construction we have described only supplies points when the Heegner hypothesis on N and D is fulfilled. To produce points in the case of general N and D, and in the case of other number fields than $\mathbf{Q}$, one cannot simply look for more complicated rational points on modular curves; rather, one has to consider Shimura curves and the CM points with which they are equipped. A construction analogous to the one we described for modular curves also produces points on Shimura curves X over the relevant class fields of K, and when there is a nonconstant map $X \to E$, the images of these points produce rational points on E that enjoy properties very similar to the ones we have treated. For a discussion of these generalizations, we refer the reader to Section 1 of [**4**].

Another direction for generalization is to varieties of higher dimension, where one has CM *cycles* rather than CM points. This has been extensively studied by Kudla, Yang, and others. One also has the conjectural special points constructed by Darmon using p-adic integration. The literature on these constructions is rather scattered, but the reader could consult (for instance) [**6**], [**11**], [**12**], or [**13**].

Bibliography

[1] M. Bertolini and H. Darmon. Kolyvagin's descent and Mordell-Weil groups over ring class fields. *J. Reine Angew. Math.*, 412:63–74, 1990.

[2] C. Cornut. Multiplication complexe.

[3] C. Cornut. Mazur's conjecture on higher Heegner points. *Invent. Math.*, 148(3):495–523, 2002.

[4] C. Cornut and V. Vatsal. Nontriviality of Rankin-Selberg L-functions and CM points. Preprint, 2004.

[5] C. Cornut and V. Vatsal. CM points and quaternion algebras. *Doc. Math.*, 10:263–309 (electronic), 2005.

[6] Henri Darmon. Heegner points, Stark-Heegner points, and values of L-series. In *International Congress of Mathematicians. Vol. II*, pages 313–345. Eur. Math. Soc., Zürich, 2006.

[7] F. Diamond and J. Im. Modular curves and modular forms. In K. Murty, editor, *Seminar on Fermat's last theorem, Toronto, 1993*, C.M.S. Conference Proceedings, pages 39–133. Amer. Math. Soc., 1995.

[8] B. Gross. Heegner points on $X_0(N)$. In R. Rankin, editor, *Modular forms*. Chichester, Ellis Horwood Company, 1984.

[9] N. Katz and B. Mazur. *Arithmetic moduli of elliptic curves*, volume 108 of *Annals of Mathematics Studies*. Princeton University Press, Princeton, NJ, 1985.

[10] V. A. Kolyvagin. Euler systems. In *The Grothendieck Festschrift, Vol. II*, volume 87 of *Progr. Math.*, pages 435–483. Birkhäuser Boston, Boston, MA, 1990.

[11] Stephen S. Kudla. Derivatives of Eisenstein series and arithmetic geometry. In *Proceedings of the International Congress of Mathematicians, Vol. II (Beijing, 2002)*, pages 173–183, Beijing, 2002. Higher Ed. Press.

[12] Stephen S. Kudla. Special cycles and derivatives of Eisenstein series. In *Heegner points and Rankin L-series*, volume 49 of *Math. Sci. Res. Inst. Publ.*, pages 243–270. Cambridge Univ. Press, Cambridge, 2004.

[13] Stephen S. Kudla, Michael Rapoport, and Tonghai Yang. *Modular forms and special cycles on Shimura curves*, volume 161 of *Annals of Mathematics Studies*. Princeton University Press, Princeton, NJ, 2006.

[14] S Lang. *Elliptic functions*. Springer Verlag, 2nd edition, 1987.

[15] Serge Lang. *Algebraic number theory*, volume 110 of *Graduate Texts in Mathematics*. Springer-Verlag, New York, second edition, 1994.

[16] B. Mazur. Modular curves and arithmetic. In *Proceedings of the International Congress of Mathematicians, Vol. 1, 2 (Warsaw, 1983)*, pages 185–211. PWN, 1984.

[17] T. Miyake. *Modular forms*. Springer Verlag, 1989.

[18] M. Ratner. Raghunathan's conjectures for Cartesian products of real and p-adic Lie groups. *Duke Math. J.*, 77(2):275–382, 1995.

[19] G. Shimura. *Introduction to the arithmetic theory of automorphic functions*. Publications of the Mathematical Society of Japan, No. 11. Iwanami Shoten, Publishers, Tokyo, 1971. Kanô Memorial Lectures, No. 1.

[20] Joseph H. Silverman. *The arithmetic of elliptic curves*, volume 106 of *Graduate Texts in Mathematics*. Springer-Verlag, New York, 1992. Corrected reprint of the 1986 original.

[21] Joseph H. Silverman. *Advanced topics in the arithmetic of elliptic curves*, volume 151 of *Graduate Texts in Mathematics*. Springer-Verlag, New York, 1994.

[22] N. Templier. Minoration du rang des courbes elliptiques sur les corps de classes de Hilbert. preprint, 2009.

[23] V. Vatsal. Special values of anticylotomic L-functions. *Duke Math J.*, 116(2):219–261, 2003.

[24] S. Zhang. Gross-Zagier formula for GL_2. *Asian J. Math.*, 5(2):183–290, 2001.

The Equivariant Tamagawa Number Conjecture and the Birch–Swinnerton-Dyer Conjecture

Guido Kings

IAS/Park City Mathematics Series
Volume 18, 2009

The Equivariant Tamagawa Number Conjecture and the Birch–Swinnerton-Dyer Conjecture

Guido Kings

Introduction

The equivariant Tamagawa number conjecture (ETNC) can be seen as a vast generalization of the analytic class number formula for number fields or of the Birch-Swinnerton-Dyer conjecture (BSD) to all motives. In fact, the BSD conjecture was one of the guiding principles in the original formulation of the Tamagawa number conjecture by Bloch and Kato in [**BK**].

In this sense, it might be a useful approach for a general introduction to the ETNC to treat the BSD conjecture as a special example. It is very instructive (and a kind of small wonder) to see how all the different factors occurring in the BSD conjecture have a cohomological interpretation.

To treat just the equivalence of ETNC with the BSD conjecture, one could confine oneself with a very modest formulation of the ETNC, in fact, as no coefficients are needed, one could do with just formulating the TNC. The idea of the current lecture at the PCMI 2009 Graduate Summer School was, however, also to show how the ETNC unifies different conjectures about special values of L-functions, most notably the Stark conjecture and its equivariant refinements.

To achieve this impossible goal, of being very concrete and extremely general at the same time, the author of the current lectures had to compromise. Instead of dealing with arbitrary non-commutative coefficients with the effect of explaining a lot of technicalities, we work here with commutative coefficients and the theory of determinants in the sense of Knudsen and Mumford (which is already intimidating enough for beginners).

As a consequence our approach does not add very much to the existing literature, especially to the excellent paper [**Fo**], except that we work with (Chow) motives instead of "motivic structures". We feel that, even if most of the ETNC has to do with the realizations of a motive, it is important not to forget that the source of all regulator maps is of a purely algebraic origin.

On the positive side we treat the case of the motive $h_1(E)(1)$ associated to an elliptic curve $E/\mathbb{Q}$ with all details, only assuming elementary facts as explained in Silverman's basic book [**Si**]. In fact to be more selfcontained, in Lecture 3, Section 5 we review the facts from the reduction theory, which we need.

It was surprising for the author of these lectures that, although well-known to the experts, the equivalence of the ETNC and the BSD conjecture seems to be not well-documented in the literature. A notable exception is [**Ve**], which treats the

Universität Regensburg, NWF I-Mathematik, 93040 Regensburg, Germany
E-mail address: guido.kings@mathematik.uni-regensburg.de

general case of abelian varieties, but omits many details and does not treat $p = 2$. We therefore hope that these notes are of some value also for the expert in this field.

For a complete account of the conjecture it is indispensable to consult [**BF**], although in the case of commutative coefficients very good introductions are [**Fo**] or [**Ka**]. The later one especially explains the connection with Iwasawa theory, which is necessary for proofs of the ETNC, but unfortunately could not be treated here do to lack of space and time.

Finally, it is a great pleasure to thank all the participants of the PCMI 2009 Graduate Summer School for their interest, their questions and their comments.

LECTURE 1

Motives, cohomology and determinants

In the first lecture we introduce some background about motives, their realizations and motivic cohomology. We introduce the L-function of a motive and finally, we give a short review of the theory of determinants for commutative rings.

1. Motives

We consider here only a very elementary theory of motives. The whole ETNC can (and should) be set up for much more advanced theories of (mixed) motives. We follow here the exposition in [**Ja**] §4, where *homological* motives are considered. A very good introduction is [**Sc1**], but there *cohomological* motives are used. The reader only interested in the case of elliptic curves, should just take notice of the existence of $h_1(E)$ in Example 1.5 and proceed to its realizations in Example 1.14.

Let $\mathcal{V}_{\mathbb{Q}}$ be the category of smooth and projective schemes over $\mathbb{Q}$. If $X, Y \in \mathcal{V}_{\mathbb{Q}}$ and $Y = \coprod Y_i$ with Y_i connected, let for $r \in \mathbb{Z}$

$$A^r(X \times Y) := \bigoplus_i CH^{\dim Y_i + r}(X \times_{\mathbb{Q}} Y_i)_{\mathbb{Q}},$$

the Chow group of $\mathbb{Q}$-linear codimension $\dim Y_i + r$ cycles. Define a composition

$$\circ : A^r(X_1 \times X_2) \times A^s(X_2 \times X_3) \to A^{r+s}(X_1 \times X_3)$$

by $f \times g \mapsto g \circ f := \mathrm{pr}_{13*}(\mathrm{pr}_{12}^*(f) \cdot \mathrm{pr}_{23}^*(g))$, where $\cdot$ is the intersection product and pr_{ij} is the projection onto the i,j component. Note that $A^0(X \times X)$ is a ring with the diagonal Δ as unit element.

Definition 1.1. The category of *Chow motives* $\mathcal{M}_{\mathbb{Q}}$ over $\mathbb{Q}$ has objects

$$M = (X, q, r),$$

where $X \in \mathcal{V}_{\mathbb{Q}}$, $q \circ q = q$ an idempotent in $A^0(X \times X)$ and $r \in \mathbb{Z}$. Here the morphisms are

$$\mathrm{Hom}_{\mathcal{M}_{\mathbb{Q}}}(M, N) := q' \circ A^{r-r'}(X \times X') \circ q,$$

where $N = (X', q', r')$.

We use the following notation (here Δ denotes the diagonal):

$$\begin{aligned} h(X) &:= (X, \Delta, 0), \quad \text{the } \textit{motive} \text{ of } X \\ \mathbb{Q}(n) &:= (\mathrm{Spec}\mathbb{Q}, \Delta, n), \quad \text{the } \textit{Tate motive} \\ M(n) &:= (X, q, r+n), \quad \text{the } n\textit{-fold Tate twist.} \end{aligned}$$

The functor $\mathcal{V}_{\mathbb{Q}} \to \mathcal{M}_{\mathbb{Q}}$, which sends X to $h(X)$ and $f : X \to Y$ to its graph $\Gamma_f \subset X \times_{\mathbb{Q}} Y$ is *covariant* (this means that our motives are homological, other conventions are possible). In particular, for a morphism $f : X \to Y$ one gets maps

$$\Gamma_f : h(X) \to h(Y)$$
$$\Gamma_f^t : h(Y) \to h(X)(\dim X - \dim Y).$$

Definition 1.2. For $M = (X, q, r)$ and $N = (X', q', r') \in \mathcal{M}_{\mathbb{Q}}$ let

$$M \otimes_{\mathbb{Q}} N := (X \times_{\mathbb{Q}} X', q \times_{\mathbb{Q}} q', r + r') \quad \text{the } \textit{product}$$
$$M^\vee := (X, q^t, \dim X - r) \quad \text{the } \textit{dual},$$

where q^t is the image of q under the map, which interchanges the two factors in $X \times_{\mathbb{Q}} X$. If $r = r'$, we define the direct sum by

$$M \oplus N := (X \coprod X', q \oplus q', r).$$

For arbitrary direct sums we refer to [**Sc1**] 1.14 but remember that there cohomological motives are used.

Remark 1.3. The product $M \otimes_{\mathbb{Q}} N$ of motives is not the good one as it is not compatible with the product of the realizations of M and N. The problem is that the cup-product in cohomology is graded commutative, whereas the above product is commutative.

Definition 1.4. Let $A/\mathbb{Q}$ be a $\mathbb{Q}$-algebra, then M has *coefficients in* A, if $\mathrm{End}_{\mathcal{M}_{\mathbb{Q}}}(M)$ admits a ring homomorphism

$$\phi : A^{\mathrm{opp}} \to \mathrm{End}_{\mathcal{M}_{\mathbb{Q}}}(M).$$

This strange looking definition is necessary, if one wants that A (and not A^{opp}) acts on the realizations of M. This is a consequence of using homological motives. How this action comes up naturally, see Example 1.6 below. Note that the dual motive $M^\vee$ has coefficients in A^{opp}, where $a \in A$ acts via $\phi(a)^t$.

Example 1.5. Consider an elliptic curve $E/\mathbb{Q}$ with unit section $e : \mathrm{Spec}\mathbb{Q} \to E$ and the idempotents $q_0 := E \times e$ and $q_2 := e \times E$ and let $q_1 := \Delta - q_0 - q_2$. Then

$$h(E) = h_0(E) \oplus h_1(E) \oplus h_2(E),$$

where $h_i(E) := (E, q_i, 0)$. One has $h_2(E) \cong \mathbb{Q}(-1)$ and

$$(h_1(E)(1))^\vee(1) = h_1(E)(1).$$

It is shown in [**Sc1**] Proposition 4.5. that the natural map

$$\mathrm{End}_{\mathbb{Q}}(E) \otimes_{\mathbb{Z}} \mathbb{Q} \to \mathrm{End}_{\mathcal{M}_{\mathbb{Q}}}(h_1(E)),$$

which sends $\phi \mapsto \Gamma_\phi$ is an isomorphism.

Example 1.6. Let $K/\mathbb{Q}$ be a Galois extension with $G := \mathrm{Gal}(K/\mathbb{Q})$ and $X := \mathrm{Spec}K$. Then G acts from the right on X and one gets a ring homomorphism

$$\mathbb{Q}[G]^{\mathrm{opp}} \to \mathrm{End}(h(\mathrm{Spec}K)) = A^0(X \times X),$$

where $\mathbb{Q}[G]$ is the group ring of G with coefficients in $\mathbb{Q}$. It is not too difficult to see that this is in fact a ring-isomorphism. One has

$$h(\mathrm{Spec}K)^\vee = h(\mathrm{Spec}K).$$

Remark 1.7. If K is any field, one has also the notion of motives over K, replacing $\mathcal{V}_{\mathbb{Q}}$ by $\mathcal{V}_K$ If $K/\mathbb{Q}$ is a finite field extension, one has two functors

$$\mathrm{Res}_{K/\mathbb{Q}} : \mathcal{M}_K \to \mathcal{M}_{\mathbb{Q}} \qquad \text{resp.} \qquad \times_{\mathbb{Q}} K : \mathcal{M}_{\mathbb{Q}} \to \mathcal{M}_K,$$

which are called *restriction* and *extension* of scalars, respectively. Here $\mathrm{Res}_{K/\mathbb{Q}} M = (X, q, r)$ now considered over $\mathbb{Q}$ and $M \times_{\mathbb{Q}} K = (X \times_{\mathbb{Q}} K, q \times_{\mathbb{Q}} K, r)$. The ETNC can be formulated for motives over number fields $K/\mathbb{Q}$ but it is compatible with restriction of scalars as defined in Remark 1.7, so that it is enough to consider motives in $\mathcal{M}_{\mathbb{Q}}$.

Example 1.8. With the notion of restriction and extension of scalars, the above Example 1.6 can be generalized to arbitrary motives $M \in \mathcal{M}_{\mathbb{Q}}$ in the following way. Let $K/\mathbb{Q}$ be as above a Galois extension with $G := \mathrm{Gal}(K/\mathbb{Q})$ and consider

$$M[G] := \mathrm{Res}_{K/\mathbb{Q}}(M \times_{\mathbb{Q}} K) \cong \mathrm{Res}_{K/\mathbb{Q}}(M \otimes h(\mathrm{Spec} K)).$$

It is straightforward to check that $M[G]$ has coefficients in $\mathbb{Q}[G]$. Example 1.6 is the special case $M = \mathbb{Q}(0)$.

2. Realizations

Let $M = (X, q, r) \in \mathcal{M}_{\mathbb{Q}}$ and *fix* $i \in \mathbb{Z}_{\geq 0}$. We are going to define the i-th realizations of M but we *suppress* i from the notations.

We first need to define an action of correspondences $q \in A^{\dim X}(X \times X)$ on a (twisted) cohomology theory.

Definition 1.9. Let $H^{\cdot}$ be a (twisted) cohomology theory for $\mathcal{V}_{\mathbb{Q}}$, which admits cycle classes and has a product $\cup$ compatible with cycle classes. Let $X \in \mathcal{V}_{\mathbb{Q}}$ and $q \in A^{\dim X}(X \times X)$, then define

$$q^* : H^{\cdot}(X, *) \to H^{\cdot}(X, *)$$

by $q^*(\xi) := \mathrm{pr}_{2,*}(\mathrm{cl}(q) \cup \mathrm{pr}_1^* \xi)$, where $\mathrm{cl}(q) \in H^{2 \dim X}(X \times X, \dim X)$ is the cycle class of q.

The general rule for the i-th realization of M in the twisted cohomology theory $H^{\cdot}$ is

$$H^i(M, j) := q^* H^i(X, j + r)$$

the image under q^* of $H^i(X, j + r)$. We now describe the realizations needed to formulate the ETNC explicitly.

Definition 1.10. The (i-th)*Betti realization* of M is

$$M_B := q^* H^i_{\mathrm{sing}}(X(\mathbb{C}), \mathbb{Q}(r)),$$

with its (pure) $\mathbb{Q}$-Hodge structure of weight $w = i - 2r$ and the following involution F_∞. Complex conjugation on $\mathbb{C}$ defines a homeomorphism $\iota : X(\mathbb{C}) \to X(\mathbb{C})$ and on $\mathbb{Q}(r) := (2\pi i)^r \mathbb{Q} \subset \mathbb{C}$ it acts via $(-1)^r$. We let $F_\infty : M_B \to M_B$ be the composition of these two involutions.

We denote by

$$M_B^+ = M_B^{F_\infty = 1}$$

the subspace, which is fixed by F_∞.

Definition 1.11. The (i-th) *de Rham realization* of M is

$$M_{\mathrm{dR}} := q^* H^i_{\mathrm{dR}}(X/\mathbb{Q}) = q^* H^i(\Omega^{\cdot}_{X/\mathbb{Q}})$$

together with the shifted Hodge filtration, i.e.,

$$\mathrm{Fil}^n M_{\mathrm{dR}} := q^* \mathrm{Fil}^{n+r} H^i_{\mathrm{dR}}(X/\mathbb{Q}) = q^* \mathrm{Im}(H^i(\Omega^{\geq n+r}_{X/\mathbb{Q}}) \to H^i(\Omega^{\cdot}_{X/\mathbb{Q}})).$$

The *tangent space* of M is by definition

$$t(M) := M_{\mathrm{dR}}/\mathrm{Fil}^0 M_{\mathrm{dR}}.$$

Fix an algebraic closure $\overline{\mathbb{Q}}$ of $\mathbb{Q}$ and let $G_{\mathbb{Q}} := \mathrm{Gal}(\overline{\mathbb{Q}}/\mathbb{Q})$ be the absolute Galois group of $\mathbb{Q}$.

Definition 1.12. Let p be a prime number. The (i-th) *p-adic realization* of M is

$$M_p := q^* H^i_{\text{ét}}(X \times_{\mathbb{Q}} \overline{\mathbb{Q}}, \mathbb{Q}_p(r))$$

with its continuous $G_{\mathbb{Q}}$-action. Here $\mathbb{Q}_p(r)$ is the one dimensional $\mathbb{Q}_p$ vector space on which $G_{\mathbb{Q}}$ acts via the r-th power of the cyclotomic character.

Note that if M has coefficients in A, the realizations $M_B, M_{\mathrm{dR}}, M_p$ and $t(M)$ are A-modules.

Remark 1.13. For the dual motive $M^\vee$ we use the following convention. If we consider the i-th realization of M, then we consider the $(2\dim X - i)$-th realization of $M^\vee$. Note that if M has coefficients in A, then $M^\vee$ has coefficients in A^{opp}.

The following example is crucial for this lecture:

Example 1.14. We will use the following notations throughout these lectures. Let $E/\mathbb{Q}$ be an elliptic curve and fix $E(\mathbb{C}) \cong \mathbb{C}/\Gamma$. Denote by

$$T_pE := \varprojlim_n E[p^n](\overline{\mathbb{Q}}) \qquad \text{resp.} \qquad V_pE := T_E \otimes_{\mathbb{Z}_p} \mathbb{Q}_p$$

its Tate-module and its Tate-module tensor $\mathbb{Q}_p$. Consider $M = h_1(E)(1)$ and $i = 1$ (all other realizations are trivial). Indeed, the reader should check that q_1^* is zero on $H^i_?$ for $i \neq 1$ and the identity on $H^1_?$ for $? = B, \mathrm{dR}, p$. We identify

$$M_B = H^1(E(\mathbb{C}), \mathbb{Q}(1)) = \mathrm{Hom}(\Gamma, \mathbb{Z}(1)) \otimes_{\mathbb{Z}} \mathbb{Q} \cong \Gamma \otimes_{\mathbb{Z}} \mathbb{Q}$$

via the intersection pairing $\Gamma \times \Gamma \to \mathbb{Z}(1)$. Note that the connected component $E(\mathbb{R})^0$ of the Lie group $E(\mathbb{R})$ defines a generator

$$\mathrm{cl}_{E(\mathbb{R})^0} \in H_1(E(\mathbb{C}), \mathbb{Z})^+ \cong \Gamma^+.$$

The de Rham realization has a filtration

$$0 \to H^0(E, \Omega^1_{E/\mathbb{Q}}) \to H^1_{\mathrm{dR}}(E/\mathbb{Q}) \to H^1(E, \mathcal{O}_E) \to 0,$$

with $\mathrm{Fil}^1 H^1_{\mathrm{dR}}(E/\mathbb{Q}) = H^0(E, \Omega^1_{E/\mathbb{Q}}) = \mathrm{Fil}^0 M_{\mathrm{dR}}$. We identify

$$t(M) = H^1(E, \mathcal{O}_E) \cong \mathrm{Lie}E.$$

For later use, we fix a basis $\omega \in H^0(E, \Omega^1_{E/\mathbb{Q}})$ and the dual $\omega^\vee$ in $\mathrm{Lie}E$ as follows: Suppose that $E/\mathbb{Q}$ is written in global minimal Weierstraß equation

$$E : y^2 + a_1xy + a_3y = x^3 + a_2x^2 + a_4x + a_6,$$

then we let

$$\omega := \frac{dx}{2y + a_1x + a_3}$$

and let $\omega^\vee \in \mathrm{Lie}E$ be its dual. We identify the p-adic realization M_p of M with V_pE

$$M_p = \mathrm{Hom}(V_pE, \mathbb{Q}_p) \cong V_pE$$

via the Weil pairing $V_pE \times V_pE \to \mathbb{Q}_p(1)$. Note that $T_pE \subset V_pE$ is a $G_{\mathbb{Q}}$-stable $\mathbb{Z}_p$-lattice and that one has a natural isomorphism of $\mathbb{Z}_p$-modules $\Gamma \otimes_{\mathbb{Z}} \mathbb{Z}_p \cong T_pE$.

Example 1.15. Let $K/\mathbb{Q}$ be a finite field extension and consider $M = h(\mathrm{Spec}K)$. Then the realizations of $M(n)$ are

$$M(n)_B = \bigoplus_{\tau:K\to\mathbb{C}} \mathbb{Q}(n) \qquad M(n)_{\mathrm{dR}} = K \qquad M(n)_p = \bigoplus_{\tau:K\to\overline{\mathbb{Q}}} \mathbb{Q}_p(n).$$

Here the Hodge filtration on $M(n)_{\mathrm{dR}}$ is $\mathrm{Fil}^i M(n) = M(n)$ for $i \leq -n$ and $\mathrm{Fil}^i M(n) = 0$ for $i > -n$.

The period map for M generalizes the periods for Riemann surfaces. The embedding of $\mathbb{C}$ into the de Rham complex $\Omega^\cdot$ and a GAGA result give a comparison isomorphism $H^i_B(X(\mathbb{C}), \mathbb{Q}(j)) \otimes_{\mathbb{Q}} \mathbb{C} \cong H^i_{\mathrm{dR}}(X/\mathbb{Q}) \otimes_{\mathbb{Q}} \mathbb{C}$. In particular, there is a comparison isomorphism $M_B \otimes_{\mathbb{Q}} \mathbb{C} \cong M_{\mathrm{dR}} \otimes_{\mathbb{Q}} \mathbb{C}$ between the Betti and the de Rham realization, which is equivariant for the action of $F_\infty \otimes c$ on the left and $1 \otimes c$ on the right. Here c denotes complex conjugation. This gives

$$(M_B \otimes_{\mathbb{Q}} \mathbb{C})^{F_\infty \otimes c = 1} \cong M_{\mathrm{dR}} \otimes_{\mathbb{Q}} \mathbb{R}.$$

Definition 1.16. The *period map* is the composition of the inclusion of $M^+_{B,\mathbb{R}}$ into $(M_B \otimes_{\mathbb{Q}} \mathbb{C})^{F_\infty \otimes c=1}$ and the projection $M_{\mathrm{dR}} \otimes_{\mathbb{Q}} \mathbb{R} \to t(M) \otimes_{\mathbb{Q}} \mathbb{R}$:

$$\alpha_M : M^+_{B,\mathbb{R}} \to t(M)_{\mathbb{R}}.$$

Here we have written $M_{\mathrm{dR},\mathbb{R}} := M_{\mathrm{dR}} \otimes_{\mathbb{Q}} \mathbb{R}$ and $M^+_{B,\mathbb{R}} := M^+_B \otimes_{\mathbb{Q}} \mathbb{R}$.

Example 1.17. Let $M = h_1(E)(1)$ and $\mathrm{cl}_{E(\mathbb{R})^0} \in M^+_B$, $\omega^\vee \in \mathrm{Lie}E$ be the elements fixed in Example 1.14. Then the *period* Ω_∞ of E is the real number defined by

$$\alpha_M(\mathrm{cl}_{E(\mathbb{R})^0}) = \Omega_\infty \omega^\vee.$$

One has $\Omega_\infty = \int_{E(\mathbb{R})^0} \omega$.

Remark 1.18. The period map α_M behaves well under duality. One has a perfect pairing

$$\mathrm{coker}\alpha_M \times \ker\alpha_{M^\vee(1)} \to \mathbb{R},$$

which induces isomorphisms

$$(\mathrm{coker}\alpha_M)^\vee \cong \ker\alpha_{M^\vee(1)} \quad \text{and} \quad (\ker\alpha_M)^\vee \cong \mathrm{coker}\alpha_{M^\vee(1)}.$$

3. Motivic cohomology

Let $M = (X, q, r) \in \mathcal{M}_{\mathbb{Q}}$ and *fix* $i \in \mathbb{Z}_{\geq 0}$.

Definition 1.19. Suppose that X has a proper, flat, regular model $\mathcal{X}/\mathbb{Z}$ (i.e. $\mathcal{X} \times_{\mathbb{Z}} \mathbb{Q} = X$). Then the (unramified) *motivic cohomology* $H^i_{\mathrm{mot}}(\mathbb{Z}, M)$ is defined as

$$H^0_{\text{mot}}(\mathbb{Z}, M) := \begin{cases} 0 & \text{if } i \neq 2r \\ q^*CH^r(X)_{\mathbb{Q}}/CH^r(X)^0_{\mathbb{Q}} & \text{if } i = 2r. \end{cases}$$

$$H^1_{\text{mot}}(\mathbb{Z}, M) := \begin{cases} q^*Im(K_{2r-i-1}(\mathcal{X})^{(r)}_{\mathbb{Q}} \to K_{2r-i-1}(X)^{(r)}_{\mathbb{Q}}) & \text{if } i \neq 2r-1 \\ q^*CH^r(X)^0_{\mathbb{Q}} & \text{if } i = 2r-1. \end{cases}$$

Here $CH^r(X)^0_{\mathbb{Q}} \subset CH^r(X)_{\mathbb{Q}}$ are the cycles homologically equivalent to zero. and $K_{2r-i-1}(X)^{(r)}_{\mathbb{Q}}$ is the r-th Adams eigenspace of the algebraic K-theory of X.

This is independent of the choice of $\mathcal{X}$ (see [**Sch**] page 13). Note that for weight $w = i - 2r > -1$ one has

$$H^1_{\text{mot}}(\mathbb{Z}, M) = 0. \tag{1.1}$$

Remark 1.20. If one does not want to assume the existence of a proper, flat, regular model one can use also use a definition given by Scholl [**Sc2**], which uses alterations.

Conjecture 1 (Finite dimension). *Suppose that M has coefficients in A. Then the groups $H^i_{\text{mot}}(\mathbb{Z}, M)$ for $i = 0, 1$ are A-modules of finite rank.*

Example 1.21. Let $E/\mathbb{Q}$ be an elliptic curve, $M = h_1(E)(1)$ and $i = 1$, then Conjecture 1 is true. More precisely,

$$H^0_{\text{mot}}(\mathbb{Z}, M) = 0$$
$$H^1_{\text{mot}}(\mathbb{Z}, M) = CH^1(E)^0_{\mathbb{Q}} \cong \text{Pic}^0(E/\mathbb{Q})_{\mathbb{Q}} \cong E(\mathbb{Q})_{\mathbb{Q}}.$$

Another case where Conjecture 1 is known are the motives of number fields.

Example 1.22. Let $K/\mathbb{Q}$ be a finite field extension with ring of integers $\mathcal{O}_K$. Then Spec$\mathcal{O}_K$ is a proper, flat, regular model of Spec$K$ and for $M = h(\text{Spec}K)$ (only $i = 0$ is interesting) one gets

$$H^0_{\text{mot}}(\mathbb{Z}, M) = \mathbb{Q} \qquad H^0(\mathbb{Z}, M(n)) = 0 \text{ for } n \neq 0$$

and

$$H^1_{\text{mot}}(\mathbb{Z}, M(1)) = \mathcal{O}_K^\times \otimes_{\mathbb{Z}} \mathbb{Q}.$$

By a result of Borel [**Bo**] one knows moreover:

$$\dim_{\mathbb{Q}} H^1_{\text{mot}}(\mathbb{Z}, M(n)) = \begin{cases} r_2 & n \text{ even } n \geq 1 \\ r_1 + r_2 & n \text{ odd } n \geq 1, \end{cases}$$

where r_1 and r_2 are the number of real (resp. complex) embeddings of K.

4. L-functions

Starting from this section, we let $M = (X, q, r) \in \mathcal{M}_{\mathbb{Q}}$ be a motive over $\mathbb{Q}$ with coefficients in a finite dimensional, semi-simple and commutative algebra $A/\mathbb{Q}$ and we *fix* $i \in \mathbb{Z}_{\geq 0}$. Note that A is a product of fields.

We need Fontaine's $D_{\text{cris}}(M_p) := (B_{\text{cris}} \otimes_{\mathbb{Q}_p} M_p)^{G_{\mathbb{Q}_p}}$, which is a finite dimensional $\mathbb{Q}_p$-vector space with a Frobenius endomorphism ϕ.

Definition 1.23. For each finite place v of $\mathbb{Q}$ let

$$L_v(M_p,T) := \begin{cases} \det_{A_{\mathbb{Q}_p}}(1-\mathrm{Frob}_v^{-1}T, M_p^{I_v}) & \text{if } v \neq p \\ \det_{A_{\mathbb{Q}_p}}(1-\phi T, D_{\mathrm{cris}}(M_p)) & \text{if } v = p. \end{cases}$$

be the local Euler factor at v. This is a polynomial in $A_{\mathbb{Q}_p}[T]$. Here I_v is the inertia group and Frob_v is the (arithmetic) Frobenius endomorphism.

Conjecture 2 (Independence of p). *The polynomial $L_v(M_p,T)$ lies in $A[T]$ and is independent of p.*

Definition 1.24. The *L-function* of M is the formal Euler product for $s \in \mathbb{C}$

$$L(M,s) := \prod_v L_v(M_p, v^{-s})^{-1}.$$

If we assume Conjecture 2, then $L(M,s)$ actually converges for $Res \gg 0$ (by the Weil conjectures) and defines an element in $A \otimes_{\mathbb{Q}} \mathbb{C}$. Conjecture 2 also implies that for real s

$$L(M,s) \in A \otimes_{\mathbb{Q}} \mathbb{R}$$

(see [**BF**] Lemma 8). Note that

$$L(M(n),s) = L(M,s+n).$$

We are interested in a conjecture concerning the special value of $L(M,s)$ at $s = 0$. For this we need:

Conjecture 3 (Meromorphic continuation). *Assuming Conjecture 2, the function $s \mapsto L(M,s)$ has a meromorphic continuation to $s = 0$.*

Conjecturally, the order of vanishing of $L(M,s)$ at $s = 0$ should be determined by the rank of the motivic cohomology. More precisely:

Conjecture 4 (Order of vanishing). *Assume Conjecture 1 and Conjecture 3, then*

$$r_M := \mathrm{ord}_{s=0} L(M,s) = \mathrm{rk}_A H^1_{\mathrm{mot}}(\mathbb{Z}, M^\vee(1)) - \mathrm{rk}_A H^0_{\mathrm{mot}}(\mathbb{Z}, M^\vee(1))$$

(equality of locally constant functions on $\mathrm{Spec}A$*).*

The ETNC addresses the leading coefficient of the Laurent series of $L(M,s)$ at $s = 0$:

Definition 1.25. Assume Conjectures 2 and 3. Recall $r_M := \mathrm{ord}_{s=0}L(M,s)$ and define the *leading coefficient* at $s = 0$ by

$$L^*(M,0) := \lim_{s\to 0} s^{-r_M} L(M,s) \in (A \otimes_{\mathbb{Q}} \mathbb{R})^\times.$$

Example 1.26. Let $E/\mathbb{Q}$ be an elliptic curve and consider $M = h_1(E)(1)$ and $i = 1$. We know from Example 1.14 that $M_p \cong V_pE$ and it is shown in Lecture 3, Section 5 in Lemma 3.11 that $V_pE^{I_v} = V_p\widetilde{E}_v^{\mathrm{ns}}$ for $v \neq p$. Here $\widetilde{E}_v^{\mathrm{ns}}$ are the non-singular points in the reduction at v of a global minimal Weierstraß equation of E. Let

$$S := \{v \mid I_v \text{ acts non-trivially on } V_pE\}$$

be the set of places of $\mathbb{Q}$ at which V_pE is ramified (equals the set of places of bad reduction by the criterion of Neron-Ogg-Shafarevich). The computations in [**Si**] V§2 imply that for $v \neq p$ and $v \notin S$

$$L_v(M_p,T) = 1 - \mathrm{tr}(\mathrm{Frob}_v)v^{-1}T + v^{-1}T^2.$$

It also follows from loc. cit. that $\mathrm{tr}(\mathrm{Frob}_v) = v + 1 - \#\widetilde{E}_v^{\mathrm{ns}}(\mathbb{F}_v)$. For $v \neq p$ and $v \in S$ we have to distinguish additive and multiplicative reduction. In the case of additive reduction $E_v^{\mathrm{ns}}(\overline{\mathbb{F}}_v) = \overline{\mathbb{F}}_v$ (see [**Si**] VII Proposition 5.1. (c)) and hence $V_pE^{I_v} = 0$, which gives $L_v(M_p, T) = 1$. In the multiplicative case we have by loc. cit. $E_v^{\mathrm{ns}}(\overline{\mathbb{F}}_v) = \overline{\mathbb{F}}_v$ and $V_pE^{I_v}$ is a one-dimensional $\mathbb{Q}_p$ vector space. As Frob_v needs to have integral eigenvalues, we see that Frob_v acts by $+1$ or -1 in this case. This gives $L_v(M_p, T) = 1 \pm T$. In the case where Frob_v acts by -1 one has $E_v^{\mathrm{ns}}(\mathbb{F}_{v^2}) = \mathbb{F}_{v^2}^*$ and one sees $\#E_v^{\mathrm{ns}}(\mathbb{F}_v) = v + 1$. Thus we have in all cases

$$L_v(M_p, 1) = \frac{\#\widetilde{E}_v^{\mathrm{ns}}(\mathbb{F}_v)}{v}$$

and the L-function of M agrees up to a shift with the usual L-function of E, i.e.,

$$L(M, s) = L(E, s+1).$$

By the work of Wiles et al. it is known that E is modular. In particular, the L-function has a meromorphic continuation to $\mathbb{C}$ and Conjecture 3 holds. Conjecture 2 is known for $v \neq p$ by the remarks above (essentially the Weil conjectures) and for $v = p$ by the work of Saito [**Sa**], who again proves this result for the L-function of a modular form.

5. Determinants

We review the theory of determinants for commutative rings following [**KM**], for which we refer for more details and proofs.

For any ring A let $P_{\mathrm{fg}}(A)$ be the category of finitely generated, projective A-modules.

Definition 1.27. Let A be a commutative ring. The category of isomorphisms of graded invertible A-modules $\mathcal{L}\mathrm{is}_A$ has objects pairs (L, r), where L is an invertible A-module and $r : \mathrm{Spec} A \to \mathbb{Z}$ is a locally constant function. The morphisms are

$$\mathrm{Mor}_{\mathcal{L}\mathrm{is}_A}((L,r),(M,s)) := \begin{cases} \mathrm{Isom}(L, M) & \text{if } r = s \\ \emptyset & \text{if } r \neq s. \end{cases}$$

The (tensor) product is defined by

$$(L, r) \cdot (M, s) := (L \otimes_A M, r + s)$$

with commutativity constraint

$$\psi_{(L,r),(M,s)} : (L, r) \cdot (M, s) \cong (M, s) \cdot (L, r)$$

defined by $\psi_{(L,r),(M,s)}(l \otimes m) := (-1)^{r(\mathfrak{p})s(\mathfrak{p})} m \otimes l$, if $l \in L_{\mathfrak{p}}$, $m \in M_{\mathfrak{p}}$, where $\mathfrak{p} \subset A$ is a prime ideal.

Note that we follow the convention in [**BF**] in the commutativity constraint. The unit object for the product is $(A, 0)$. For each (L, r) let $L^{-1} := \mathrm{Hom}_A(L, A)$, then $(L, r)^{-1} := (L^{-1}, -r)$ is an inverse of (L, r) and one has an isomorphism

$$\mathrm{ev} : (L, r) \cdot (L^{-1}, -r) \cong (A, 0),$$

induced by the usual evaluation morphism $\mathrm{ev} : L \otimes_A \mathrm{Hom}_A(L, A) \to A$. There exists also a base change functor for ring homomorphisms.

Definition 1.28. Let $f : A \to B$ be a ring homomorphism, then the functor

$$\otimes_A B : \mathcal{L}\mathrm{is}_A \to \mathcal{L}\mathrm{is}_B$$

is defined by $(L, r) \otimes_A B := (L \otimes_A B, r \circ f^*)$ on objects and in the obvious way on morphisms. Here $f^* : \mathrm{Spec}B \to \mathrm{Spec}A$ is the induced map.

For each $P \in P_{\mathrm{fg}}(A)$ define

$$\mathrm{Det}_A(P) := (\bigwedge_A^{\mathrm{rk}_A(P)} P, \mathrm{rk}_A(P)),$$

where $\mathrm{rk}_A : P_{\mathrm{fg}}(A) \to \mathbb{Z}$ is the A-rank of P and $\bigwedge_A^{\mathrm{rk}_A(P)} P$ the exterior power over A. Note that $\mathrm{Det}_A(0) = (A, 0)$ is the unit object and that one has for a short exact sequence $0 \to P' \to P \to P'' \to 0$ of objects in $P_{\mathrm{fg}}(A)$ an isomorphism

$$\mathrm{Det}_A(P') \cdot \mathrm{Det}_A(P'') \cong \mathrm{Det}_A(P).$$

It is important to note that the base change functor $\otimes_A B$ commutes with determinants, i.e.,

$$\mathrm{Det}_A(C^\cdot) \otimes_A B \cong \mathrm{Det}_B(C^\cdot \otimes_A B).$$

For a bounded complex $C^\cdot$ of modules in $P_{\mathrm{fg}}(A)$, define $\mathrm{Det}_A(C^\cdot) \in \mathcal{L}\mathrm{is}_A$ by

$$\mathrm{Det}_A(C^\cdot) := \coprod_{i \in \mathbb{Z}} \mathrm{Det}_A^{-1}(C^i).$$

This construction extends to a functor Det_A from the category of bounded complexes $C^\cdot$ of modules in $P_{\mathrm{fg}}(A)$ with quasi-isomorphisms as morphisms to the category $\mathcal{L}\mathrm{is}_A$. In particular, one has for acyclic complexes $C^\cdot$ a canonical isomorphism

$$\mathrm{Det}_A(C^\cdot) \cong \mathrm{Det}_A(0).$$

In fact one can define determinants for more general complexes. Denote by Perf_A the category of complexes of A-modules, which are quasi-isomorphic to a bounded complex of modules in $P_{\mathrm{fg}}(A)$.

Definition 1.29. For each complex $C^\cdot \in \mathrm{Perf}_A$ fix a quasi-isomorphism $\widetilde{C}^\cdot \to C^\cdot$, where $\widetilde{C}^\cdot$ is a bounded complex of modules in $P_{\mathrm{fg}}(A)$. Then we define

$$\mathrm{Det}_A(C^\cdot) := \mathrm{Det}_A(\widetilde{C}^\cdot).$$

If $C^\cdot \in \mathrm{Perf}_A$ has the property that $H^j(C^\cdot) \in P_{\mathrm{fg}}(A)$ for all $j \in \mathbb{Z}$, one has

$$\mathrm{Det}_A(C^\cdot) \cong \prod_j \mathrm{Det}_A^{(-1)^j}(H^j(C^\cdot)).$$

Example 1.30. Let $\phi : P \cong Q$ be an isomorphism of modules in $\mathcal{L}\mathrm{is}_A$. Considered as an acyclic complex (in degree $0, 1$) one has a canonical isomorphism

$$\rho : \mathrm{Det}_A([P \to Q]) = \mathrm{Det}_A(P) \cdot \mathrm{Det}_A^{-1}(Q) \cong \mathrm{Det}_A(0),$$

which is induced by the isomorphism $\mathrm{Det}_A(\phi) : \mathrm{Det}_A(P) \cong \mathrm{Det}_A(Q)$ and the evaluation $\mathrm{Det}_A(Q) \cdot \mathrm{Det}_A^{-1}(Q) \cong \mathrm{Det}_A(0)$. If it happens that $P = Q$, then the evaluation gives directly $\mathrm{ev} : \mathrm{Det}_A([P \to P]) \cong \mathrm{Det}_A(0)$, but this differs from the above by $\det(\phi)$, one has

$$\rho = \det(\phi)\mathrm{ev}.$$

Example 1.31. Let $A = \mathbb{Z}_p$, then all modules in $P_{\text{fg}}(\mathbb{Z}_p)$ are free. Let H be a finite abelian p group considered as object in $\text{Perf}_{\mathbb{Z}_p}$. Then the determinant of H is defined by choosing a resolution

$$0 \to P \xrightarrow{\phi} Q \to H \to 0$$

with $P, Q \in P_{\text{fg}}(\mathbb{Z}_p)$ and defining

$$\text{Det}_{\mathbb{Z}_p}(H) := \text{Det}_{\mathbb{Z}_p}(Q)\text{Det}_{\mathbb{Z}_p}^{-1}(P).$$

We claim that under the embedding

$$\text{Det}_{\mathbb{Z}_p}(H) \subset \text{Det}_{\mathbb{Q}_p}(H \otimes_{\mathbb{Z}_p} \mathbb{Q}_p) = \text{Det}_{\mathbb{Q}_p}(0)$$

$\text{Det}_{\mathbb{Z}_p}(H)$ is the $\mathbb{Z}_p$-sublattice $(\#H)^{-1}\mathbb{Z}_p$. To show this choose bases $p_1, \ldots, p_r$ and $q_1, \ldots, q_r$ of P and Q respectively, then $\text{Det}_{\mathbb{Z}_p}(H) \subset \text{Det}_{\mathbb{Q}_p}(H \otimes_{\mathbb{Z}_p} \mathbb{Q}_p)$ is generated by

$$(q_1 \wedge \ldots \wedge q_r)(p_1 \wedge \ldots \wedge p_r)^{-1} \in \text{Det}_{\mathbb{Z}_p}(Q) \cdot \text{Det}_{\mathbb{Z}_p}^{-1}(P) \cong \text{Det}_{\mathbb{Z}_p}(H).$$

The identification

$$\text{Det}_{\mathbb{Q}_p}(0) = \text{Det}_{\mathbb{Q}_p}(H \otimes_{\mathbb{Z}_p} \mathbb{Q}_p) \cong \text{Det}_{\mathbb{Q}_p}(Q \otimes_{\mathbb{Z}_p} \mathbb{Q}_p) \cdot \text{Det}_{\mathbb{Q}_p}^{-1}(P \otimes_{\mathbb{Z}_p} \mathbb{Q}_p)$$

is as in Example 1.30 and maps $(q_1 \wedge \ldots \wedge q_r)(p_1 \wedge \ldots \wedge p_r)^{-1}$ to $\det(\phi)^{-1}$. It is well-known and not difficult to show that $\det(\phi)\mathbb{Z}_p = \#H\mathbb{Z}_p$, which proves the claim.

LECTURE 2

The equivariant Tamagawa number conjecture

In this lecture we formulate the Rationality Conjecture of Beilinson and Deligne and the equivariant Tamagawa number conjecture of Bloch-Kato, Fontaine-Perrin-Riou and Burns-Flach. As explained in the introduction, we restrict here to the case of commutative coefficients. For the general case we refer to [**BF**] (and to [**HK1**], [**Ve**] for a formulation with non-commutative determinants, more in the spirit of this lecture).

We let $M = (X, q, r) \in \mathcal{M}_{\mathbb{Q}}$ with coefficients in a finite-dimensional, semi-simple and commutative $A/\mathbb{Q}$ (i.e., A is a product of fields). Fix $i \in \mathbb{Z}_{\geq 0}$. We will use systematically notations like $V_{\mathbb{R}} := V \otimes_{\mathbb{Q}} \mathbb{R}$ for $\mathbb{Q}$-vector spaces V, if no confusion is possible. Similarly, we use $V_{\mathbb{Q}_p} := V \otimes_{\mathbb{Q}} \mathbb{Q}_p$ and $\mathcal{V}_{\mathbb{Q}_p} := \mathcal{V} \otimes_{\mathbb{Z}_p} \mathbb{Q}_p$, if $\mathcal{V}$ is a $\mathbb{Z}_p$-module.

1. Rationality conjecture

In this section we formulate the Rationality Conjecture of Beilinson and Deligne, which states essentially that $L^*(M, 0)$ divided by a certain period is "rational".

Recall from Definition 1.16 the period map

$$\alpha_M : M^+_{B,\mathbb{R}} \to t(M)_{\mathbb{R}}.$$

The following conjecture formulates a very deep relation between the kernel and the cokernel of the period map and the motivic cohomology.

Conjecture 5 (Fundamental exact sequence)**.** *There exists an exact sequence of $A_{\mathbb{R}}$-modules (of finite rank)*

$$0 \to H^0_{\text{mot}}(\mathbb{Z}, M)_{\mathbb{R}} \xrightarrow{c} \ker\alpha_M \xrightarrow{r_\infty^\vee} H^1_{\text{mot}}(\mathbb{Z}, M^\vee(1))^\vee_{\mathbb{R}} \to$$

$$\xrightarrow{<,>} H^1_{\text{mot}}(\mathbb{Z}, M)_{\mathbb{R}} \xrightarrow{r_\infty} \text{coker}\alpha_M \xrightarrow{c^\vee} H^0_{\text{mot}}(\mathbb{Z}, M^\vee(1))^\vee_{\mathbb{R}} \to 0,$$

where r_∞ is the Beilinson-Deligne regulator, c is the Chern-class map and $<,>$ the height pairing, $(\quad)^\vee$ denotes the dual vector space.

Note that the conjecture implies also that the motivic cohomology groups have finite dimension.

Remark 2.1. Assume that the weight $w = i - 2r$ of M is not -1. Then by the remark after Definition 1.19 either $H^1_{\text{mot}}(\mathbb{Z}, M)$ or $H^1_{\text{mot}}(\mathbb{Z}, M^\vee(1))$ is zero. If furthermore, $w \neq 0$, then the above conjecture says that there is either an isomorphism

$$\ker\alpha_M \xrightarrow{r_\infty^\vee} H^1_{\text{mot}}(\mathbb{Z}, M^\vee(1))^\vee_{\mathbb{R}}$$

or an isomorphism

$$H^1_{\text{mot}}(\mathbb{Z}, M)_{\mathbb{R}} \xrightarrow{r_\infty} \text{coker}\alpha_M.$$

Example 2.2. Let $M = h_1(E)(1)$ and $i = 1$, then Conjecture 5 holds for M. Indeed, α_M is an isomorphism, so that $\ker\alpha_M = 0 = \operatorname{coker}\alpha_M$. By definition $H^0_{\mathrm{mot}}(\mathbb{Z}, M) = 0 = H^0_{\mathrm{mot}}(\mathbb{Z}, M^\vee(1))$. Finally, the Neron-Tate height h defines a pairing $< P, Q >:= \frac{1}{2}(h(P+Q) - h(P) - h(Q))$, which induces an isomorphism

$$<,>: H^1_{\mathrm{mot}}(\mathbb{Z}, M^\vee(1))^\vee_{\mathbb{R}} = E(\mathbb{Q})^\vee_{\mathbb{R}} \cong H^1_{\mathrm{mot}}(\mathbb{Z}, M)_{\mathbb{R}} = E(\mathbb{Q})_{\mathbb{R}}.$$

Example 2.3. Let $K/\mathbb{Q}$ be a number field, then it follows from the results by Borel [**Bo**] (and the comparison of the Borel and Beilinson regulator by Beilinson) that Conjecture 5 holds for $h(\operatorname{Spec}K)(n)$.

Definition 2.4. The *fundamental line* is defined by

$$\begin{aligned}\Delta(M) :=&\operatorname{Det}_A(H^0_{\mathrm{mot}}(\mathbb{Z}, M)) \cdot \operatorname{Det}_A^{-1}(H^1_{\mathrm{mot}}(\mathbb{Z}, M)) \cdot \operatorname{Det}_A(t(M)) \\ &\operatorname{Det}_A^{-1}(H^0_{\mathrm{mot}}(\mathbb{Z}, M^\vee(1))^\vee) \cdot \operatorname{Det}_A(H^1_{\mathrm{mot}}(\mathbb{Z}, M^\vee(1))^\vee) \cdot \operatorname{Det}_A^{-1}(M_B^+).\end{aligned}$$

Example 2.5. Let $M = h_1(E)(1)$ and $i = 1$, then with Examples 1.14 and 1.21 we get

$$\Delta(M) = \operatorname{Det}_{\mathbb{Q}}^{-1}(E(\mathbb{Q})_{\mathbb{Q}}) \cdot \operatorname{Det}_{\mathbb{Q}}(E(\mathbb{Q})^\vee_{\mathbb{Q}}) \cdot \operatorname{Det}_{\mathbb{Q}}(\operatorname{Lie}E) \cdot \operatorname{Det}_{\mathbb{Q}}^{-1}(\Gamma^+_{\mathbb{Q}}).$$

Example 2.6. Let $M = h(\operatorname{Spec}K)$ and $K/\mathbb{Q}$ a Galois extension with $G = \operatorname{Gal}(K/\mathbb{Q})$ abelian, then the fundamental line $\Delta(M(n))$ is given by

$$\operatorname{Det}_{\mathbb{Q}[G]}(H^0_{\mathrm{mot}}(\mathbb{Z}, M(n))) \cdot \operatorname{Det}_{\mathbb{Q}[G]}(H^1_{\mathrm{mot}}(\mathbb{Z}, M^\vee(1-n))^\vee) \cdot \operatorname{Det}^{-1}_{\mathbb{Q}[G]}(M(n)^+_B)$$

if $n \leq 0$ and by

$$\begin{aligned}&\operatorname{Det}_{\mathbb{Q}[G]}(H^0_{\mathrm{mot}}(\mathbb{Z}, M^\vee(1-n))^\vee) \cdot \operatorname{Det}^{-1}_{\mathbb{Q}[G]}(H^1_{\mathrm{mot}}(\mathbb{Z}, M(n))) \\ &\quad \cdot \operatorname{Det}^{-1}_{\mathbb{Q}[G]}(M(n)^+_B) \cdot \operatorname{Det}_{\mathbb{Q}[G]}(M(n)_{\mathrm{dR}})\end{aligned}$$

if $n \geq 1$.

Taking $\operatorname{Det}_{A_{\mathbb{R}}}$ of the fundamental exact sequence in Conjecture 5 and of the tautological exact sequence

$$0 \to \ker\alpha_M \to M^+_{B,\mathbb{R}} \to t(M)_{\mathbb{R}} \to \operatorname{coker}\alpha_M \to 0$$

induces a canonical isomorphism

$$\theta_\infty : \Delta(M)_{\mathbb{R}} \cong \operatorname{Det}_{A_{\mathbb{R}}}(0).$$

The first part of the ETNC (the Rationality Conjecture, due to Deligne and Beilinson) can now be formulated.

Conjecture 6 (Rationality Conjecture)**.** *Assume Conjectures 2, 3 and 5. Let M be as above with coefficients in A. Consider $L^*(M,0) \in A_{\mathbb{R}}^\times$, then there exist a* zeta element

$$\zeta_A(M) \in \Delta(M)$$

such that

$$\theta_\infty(\zeta_A(M)) = (L^*(M,0))^{-1} \in A^\times_{\mathbb{R}}.$$

2. Local unramified cohomology

We review here the definition of local unramified cohomology, for more details one should consult [**Fo**] and [**BF**]. In fact, to work properly with determinants one has to use "true triangles", instead of just triangles in the derived category, so that for precise definitions we refer to [**BF**] 3.2.

Fix for each place v of $\mathbb{Q}$ an algebraic closure $\overline{\mathbb{Q}}_v$ of $\mathbb{Q}_v$ and an embedding $\overline{\mathbb{Q}} \subset \overline{\mathbb{Q}}_v$. Denote by $G_{\mathbb{Q}_v} := \mathrm{Gal}(\overline{\mathbb{Q}}_v/\mathbb{Q}_v)$ the absolute Galois group of $\mathbb{Q}_v$. For each continuous $G_{\mathbb{Q}_v}$-module V we let

$$R\Gamma(\mathbb{Q}_v, V) := C^{\cdot}(G_{\mathbb{Q}_v}, V)$$

the complex of continuous cochains of $G_{\mathbb{Q}_p}$ with values in V. Recall that $D_{\mathrm{cris}}(M_p) := (B_{\mathrm{cris}} \otimes_{\mathbb{Q}_p} M_p)^{G_{\mathbb{Q}_p}}$ is equipped with a Frobenius morphism ϕ and that $D_{\mathrm{dR}}(M_p) := (B_{\mathrm{dR}} \otimes_{\mathbb{Q}_p} M_p)^{G_{\mathbb{Q}_p}}$ has a filtration inherited from B_{dR}.

Definition 2.7. The p-adic *tangent space* is

$$t(M_p) := D_{\mathrm{dR}}(M_p)/\mathrm{Fil}^0 D_{\mathrm{dR}}(M_p).$$

Note that the comparison isomorphism (due to Faltings) between M_p and M_{dR} over $\mathbb{Q}_p$ induces an isomorphism

$$t(M_p) \simeq t(M)_{\mathbb{Q}_p}$$

Definition 2.8. The complex of *local unramified cohomology* is defined as

$$R\Gamma_f(\mathbb{Q}_v, M_p) := \begin{cases} R\Gamma(\mathbb{R}, M_p) & \text{if } v = \infty \\ (M_p^{I_v} \xrightarrow{1-\mathrm{Frob}_v^{-1}} M_p^{I_v}) & \text{if } v \neq p, \infty \\ (D_{\mathrm{cris}}(M_p) \xrightarrow{(1-\phi,\mathrm{pr})} D_{\mathrm{cris}}(M_p) \oplus t(M_p)) & \text{if } v = p. \end{cases}$$

Here $I_v \subset G_{\mathbb{Q}_v}$ is the inertia subgroup and the complexes for $v \neq \infty$ are placed in degree $0, 1$. We denote the cohomology of $R\Gamma_f(\mathbb{Q}_v, M_p)$ by $H^i_f(\mathbb{Q}_v, M_p)$.

The complex $R\Gamma_f(\mathbb{Q}_v, M_p)$ is quasi-isomorphic to a sub-complex of $R\Gamma(\mathbb{Q}_v, M_p)$ (see [**BF**] 3.2) and one defines $R\Gamma_{/f}(\mathbb{Q}_v, M_p)$ to be the cokernel, so that one has a natural short exact sequence of complexes

$$0 \to R\Gamma_f(\mathbb{Q}_v, M_p) \to R\Gamma(\mathbb{Q}_v, M_p) \to R\Gamma_{/f}(\mathbb{Q}_v, M_p) \to 0.$$

Definition 2.9. The *Bloch-Kato exponential map*

$$\exp_{\mathrm{BK}} : t(M_p) \to H^1_f(\mathbb{Q}_p, M_p)$$

is the composition of the inclusion $t(M_p) \to D_{\mathrm{cris}}(M_p) \oplus t(M_p)$, which maps $t \mapsto (0, t)$, and the surjection $D_{\mathrm{cris}}(M_p) \oplus t(M_p) \to H^1_f(\mathbb{Q}_p, M_p)$.

For computations it is decisive to have a version of unramified cohomology with integral coefficients.

Definition 2.10. Let $T_p \subset M_p$ be a $G_{\mathbb{Q}_v}$-stable $\mathbb{Z}_p$-lattice and let $u : H^1(\mathbb{Q}_v, T_p) \to H^1(\mathbb{Q}_v, M_p)$ be the natural map. Let

$$H^0_f(\mathbb{Q}_v, T_p) := H^0(\mathbb{Q}_v, T_p)$$
$$H^1_f(\mathbb{Q}_v, T_p) := \{\xi \in H^1(\mathbb{Q}_v, T_p) \mid u(\xi) \in H^1_f(\mathbb{Q}_v, M_p)\}$$
$$H^2_f(\mathbb{Q}_v, T_p) := 0.$$

For $H^i_{/f}$ let

$$H^i_{/f}(\mathbb{Q}_v, T_p) := H^i(\mathbb{Q}_v, T_p)/H^i_f(\mathbb{Q}_v, T_p).$$

Note that the torsion subgroups of $H^1_f(\mathbb{Q}_v, T_p)$ and $H^1(\mathbb{Q}_v, T_p)$ coincide by definition. If T_p is unramified at v, one can give a different description of $H^1_f(\mathbb{Q}_v, T_p)$.

Lemma 2.11. *Let T_p be unramified at $v \neq p$ (i.e., the inertia I_v acts trivially on T_p). Then*

$$H^1_f(\mathbb{Q}_v, T_p) = \ker\left(H^1(\mathbb{Q}_v, T_p) \to H^1(I_v, T_p)\right).$$

Note that under the conditions of the lemma $H^1(I_v, T_p) = \operatorname{Hom}_{\mathrm{ct}}(I_v, T_p)$ (continuous homomorphisms).

PROOF. By definition of $H^1_f(\mathbb{Q}_v, M_p)$ and because M_p is unramified, we have

$$H^1_f(\mathbb{Q}_v, M_p) = H^1(G_{\mathbb{Q}_v}/I_v, M_p) = \ker\left(H^1(\mathbb{Q}_v, M_p) \to H^1(I_v, M_p)\right).$$

This implies that $H^1_f(\mathbb{Q}_v, T_p)$ is the kernel of the composition

$$H^1(\mathbb{Q}_v, T_p) \to H^1(\mathbb{Q}_v, M_p) \to H^1(I_v, M_p).$$

As the canonical map $H^1(I_v, T_p) = \operatorname{Hom}_{\mathrm{ct}}(I_v, T_p) \to H^1(I_v, M_p) = \operatorname{Hom}_{\mathrm{ct}}(I_v, M_p)$ is injective, the commutative diagram

$$\begin{array}{ccc} H^1(\mathbb{Q}_v, T_p) & \longrightarrow & H^1(I_v, T_p) \\ \downarrow & & \downarrow \\ H^1(\mathbb{Q}_v, M_p) & \longrightarrow & H^1(I_v, M_p) \end{array}$$

shows that $H^1_f(\mathbb{Q}_v, T_p)$ is also the kernel of $H^1(\mathbb{Q}_v, T_p) \to H^1(I_v, T_p)$. □

3. Global unramified cohomology

Fix a prime number p. Denote by S a finite set of places of $\mathbb{Q}$, which contains p, ∞ and the places v, for which M_p is ramified (i.e., where I_v acts non-trivially). Let G_S be the Galois group of the maximal extension of $\mathbb{Q}$, which is unramified outside of S. For any continuous G_S-module V we let

$$R\Gamma(\mathbb{Z}_S, V) := \mathcal{C}^{\cdot}(G_S, V)$$

be the complex of continuous cochains of G_S with values in V.

Definition 2.12. The complex of *cohomology with compact support* is

$$R\Gamma_c(\mathbb{Z}_S, V) := \operatorname{Cone}\left(R\Gamma(\mathbb{Z}_S, V) \to \bigoplus_{v \in S} R\Gamma(\mathbb{Q}_v, V)\right)[-1],$$

where V is any continuous G_S-module. The complex of *global unramified cohomology* of M_p is by definition

$$R\Gamma_f(\mathbb{Q}, M_p) := \operatorname{Cone}\left(R\Gamma(\mathbb{Z}_S, M_p) \to \bigoplus_{v \in S} R\Gamma_{/f}(\mathbb{Q}_v, M_p)\right)[-1].$$

The cohomology groups of $R\Gamma_c(\mathbb{Z}_S, V)$ and $R\Gamma_f(\mathbb{Q}, M_p)$ are denoted by $H^i_c(\mathbb{Z}_S, V)$ and $H^i_f(\mathbb{Q}, M_p)$ respectively.

Putting the resulting exact triangles together, one obtains the important triangle (which can be made a true triangle, see [**BF**] 3.2)

$$R\Gamma_c(\mathbb{Z}_S, M_p) \to R\Gamma_f(\mathbb{Q}, M_p) \to \bigoplus_{v\in S} R\Gamma_f(\mathbb{Q}_v, M_p). \tag{2.1}$$

The global unramified cohomology $H^i_f(\mathbb{Q}, M_p)$ is self-dual in the following sense:

Proposition 2.13 ([**BF**] Lemma 19)**.** *One has $H^i_f(\mathbb{Q}, M_p) = 0$ for $i \neq 0,1,2,3$ and*

$$H^i_f(\mathbb{Q}, M_p) \cong H^{3-i}_f(\mathbb{Q}, M_p^\vee(1))^\vee,$$

where $(\dots)^\vee$ denotes the $\mathbb{Q}_p$ dual.

For later use, we introduce an integral structure on $H^1_f(\mathbb{Q}, M_p)$. Note that by definition

$$H^1_f(\mathbb{Q}, M_p) = \{\xi \in H^1(\mathbb{Z}_S, M_p) \mid \xi_v \in H^1_f(\mathbb{Q}_v, M_p) \text{ for all } v \in S\},$$

where ξ_v is the image of ξ in $H^1(\mathbb{Q}_v, M_p)$.

Definition 2.14. Let $T_p \subset M_p$ be a $G_\mathbb{Q}$-stable $\mathbb{Z}_p$-lattice, then define

$$\begin{aligned} H^1_f(\mathbb{Q}, T_p) &:= \{\xi \in H^1(\mathbb{Z}_S, T_p) \mid \xi_v \in H^1_f(\mathbb{Q}_v, T_p) \text{ for all } v \in S\} \\ &= \ker\left(H^1(\mathbb{Z}_S, T_p) \to \bigoplus_{v\in S} H^1_{/f}(\mathbb{Q}_v, T_p)\right), \end{aligned}$$

where $H^1_{/f}(\mathbb{Q}_v, T_p)$ is as in Definition 2.10.

For later use, we note:

Lemma 2.15. *Let $T_p \subset M_p$ be a $G_\mathbb{Q}$-stable $\mathbb{Z}_p$-lattice, then*

$$H^1_f(\mathbb{Q}, T_p) = \ker\left(H^1(\mathbb{Q}, T_p) \to \prod_v H^1_{/f}(\mathbb{Q}_v, T_p)\right).$$

PROOF. We claim that

$$H^1(\mathbb{Z}_S, T_p) = \ker\left(H^1(\mathbb{Q}, T_p) \to \prod_{v\notin S} H^1_{/f}(\mathbb{Q}_v, T_p)\right).$$

Accepting the claim, the statement of the lemma follows immediately: each $\xi \in H^1(\mathbb{Q}, T_p)$, which maps to zero in $\prod_{v\notin S} H^1_{/f}(\mathbb{Q}_v, T_p)$ comes from an element in $H^1(\mathbb{Z}_S, T_p)$. If it also maps to zero in $\bigoplus_{v\in S} H^1_{/f}(\mathbb{Q}_v, T_p)$ it is in $H^1_f(\mathbb{Q}, T_p)$. To prove the claim, let $\mathbb{Q}^S$ be the maximal extension of $\mathbb{Q}$, which is unramified outside of S and $G_{\mathbb{Q}^S} := \mathrm{Gal}(\overline{\mathbb{Q}}/\mathbb{Q}^S)$. The inflation-restriction sequence gives

$$0 \to H^1(\mathbb{Z}_S, T_p) \to H^1(\mathbb{Q}, T_p) \to H^1(\mathbb{Q}^S, T_p).$$

The last group is $\mathrm{Hom}_{\mathrm{ct}}(G_{\mathbb{Q}^S}, T_p)$ because $G_{\mathbb{Q}^S}$ acts trivially on T_p. As $G_{\mathbb{Q}^S}$ is the smallest closed normal subgroup of $G_\mathbb{Q}$, which contains I_v for all $v \notin S$, we have an injection

$$H^1(\mathbb{Q}^S, T_p) = \mathrm{Hom}_{\mathrm{ct}}(G_{\mathbb{Q}^S}, T_p) \hookrightarrow \prod_{v\notin S} \mathrm{Hom}_{\mathrm{ct}}(I_v, T_p),$$

so that the sequence

$$0 \to H^1(\mathbb{Z}_S, T_p) \to H^1(\mathbb{Q}, T_p) \to \prod_{v\notin S} \mathrm{Hom}_{\mathrm{ct}}(I_v, T_p) \tag{2.2}$$

is still exact. By Lemma 2.11 we see that $\mathrm{Hom}_{\mathrm{ct}}(I_v, T_p) = H^1_{/f}(\mathbb{Q}_v, T_p)$, which implies the claim. □

4. The equivariant Tamagawa number conjecture

The following conjecture relates the motivic cohomology to the global unramified cohomology. It contains deep conjectures of Tate and Jannsen as special cases and the finiteness of the p-torsion of the Tate-Shafarevich group.

Conjecture 7 (Regulators). *The p-adic regulators*

$$r^0_p : H^0_{\mathrm{mot}}(\mathbb{Z}, M)_{\mathbb{Q}_p} \to H^0_f(\mathbb{Q}, M_p)$$
$$r^1_p : H^1_{\mathrm{mot}}(\mathbb{Z}, M)_{\mathbb{Q}_p} \to H^1_f(\mathbb{Q}, M_p)$$

are isomorphisms for all p.

Example 2.16. Let $E/\mathbb{Q}$ be an elliptic curve and $M = h_1(E)(1)$, then Conjecture 7 for p is true if and only if $Ш(\mathbb{Q}, E)[p^\infty]$ is a finite group. This follows from Theorem 2.30 below as the regulator $E(\mathbb{Q})_{\mathbb{Q}_p} = H^1_{\mathrm{mot}}(\mathbb{Z}, M)_{\mathbb{Q}_p} \to H^1_f(\mathbb{Q}, M_p)$ is given by the Kummer map and $H^0_{\mathrm{mot}}(\mathbb{Z}, M) = 0 = H^0_f(\mathbb{Q}, M_p)$.

Remark 2.17. Let $K/\mathbb{Q}$ be a number field and $M = h(\mathrm{Spec} K)$. One can show that $r^i_p : H^i_{\mathrm{mot}}(\mathbb{Z}, M(n))_{\mathbb{Q}_p} \to H^i_f(\mathbb{Q}, M(n)_p)$ for $i = 0, 1$ are isomorphisms for all n.

We are going to define an isomorphism $\theta_p : \Delta(M)_{\mathbb{Q}_p} \cong \mathrm{Det}_{A_{\mathbb{Q}_p}}(R\Gamma_c(\mathbb{Z}_S, M_p))$. By definition of the complex $R\Gamma_f(\mathbb{Q}_v, M_p)$ (see Definition 2.8) and cancellation of determinants, we can identify

$$\iota_v : \mathrm{Det}^{-1}_{A_{\mathbb{Q}_p}}(R\Gamma_f(\mathbb{Q}_v, M_p)) \cong \begin{cases} \mathrm{Det}^{-1}_{A_{\mathbb{Q}_p}}(M_p^+) & \text{if } v = \infty \\ \mathrm{Det}_{A_{\mathbb{Q}_p}}(0) & \text{if } v \neq p, \infty \\ \mathrm{Det}_{A_{\mathbb{Q}_p}}(t(M)_{\mathbb{Q}_p}) & \text{if } v = p. \end{cases} \tag{2.3}$$

The triangle for $R\Gamma_c(\mathbb{Z}_S, M_p)$ in (2.1) gives

$$\mathrm{Det}_{A_{\mathbb{Q}_p}}(R\Gamma_c(\mathbb{Z}_S, M_p)) \cong \mathrm{Det}_{A_{\mathbb{Q}_p}}(R\Gamma_f(\mathbb{Q}, M_p)) \cdot \prod_{v\in S} \mathrm{Det}^{-1}_{A_{\mathbb{Q}_p}}(R\Gamma_f(\mathbb{Q}_v, M_p)).$$

Together with the isomorphism ι_v in (2.3) one gets

$$\mathrm{Det}_{A_{\mathbb{Q}_p}}(R\Gamma_c(\mathbb{Z}_S, M_p)) \cong \mathrm{Det}_{A_{\mathbb{Q}_p}}(R\Gamma_f(\mathbb{Q}, M_p)) \cdot \mathrm{Det}^{-1}_{A_{\mathbb{Q}_p}}(M_p^+) \cdot \mathrm{Det}_{A_{\mathbb{Q}_p}}(t(M)_{\mathbb{Q}_p}).$$

With Conjecture 7 and Proposition 2.13 the right hand side is canonically isomorphic to $\Delta(M)_{\mathbb{Q}_p}$.

Definition 2.18. For any p let

$$\theta_{p,S} : \Delta(M)_{\mathbb{Q}_p} \cong \mathrm{Det}_{A_{\mathbb{Q}_p}}(R\Gamma_c(\mathbb{Z}_S, M_p))$$

be the isomorphism defined above.

The ETNC states roughly that the zeta element $\zeta_A(M)$ from Definition 6 generates a canonical integral structure in $\mathrm{Det}_{A_{\mathbb{Q}_p}}(R\Gamma_c(\mathbb{Z}_S, M_p))$. To formulate this, we need:

Definition 2.19. Let $A/\mathbb{Q}$ be a finite dimensional, commutative and semi-simple $\mathbb{Q}$-algebra. An *order* $\mathcal{A}$ in A is a sub-algebra, which is a finitely generated $\mathbb{Z}$-module, such that $\mathcal{A} \otimes_{\mathbb{Z}} \mathbb{Q} = A$. A *projective* $\mathcal{A}$*-structure* in M, is a finitely generated projective $\mathcal{A}$-module $T_B \subset M_B$ such that $T_B \otimes_{\mathcal{A}} A = M_B$ and for all p the image of $T_p := T_B \otimes_{\mathbb{Z}} \mathbb{Z}_p$ under the comparison isomorphism

$$M_B \otimes_{\mathbb{Q}} \mathbb{Q}_p \cong M_p$$

is a Galois stable lattice in M_p.

Example 2.20. For an elliptic curve $E/\mathbb{Q}$, the subgroup

$$H^1(E(\mathbb{C}), \mathbb{Z}(1)) \subset h(E)(1)_B$$

is a projective $\mathbb{Z}$-structure. For the motive $h(\mathrm{Spec}K)(n)$, where $K/\mathbb{Q}$ is an (abelian) Galois extension with $G = \mathrm{Gal}(K/\mathbb{Q})$, the $\mathbb{Z}[G]$-module $\bigoplus_{\tau:K\to\mathbb{C}} \mathbb{Z} \subset h(\mathrm{Spec}K)_B$ is a projective $\mathbb{Z}[G]$-structure for the order $\mathbb{Z}[G] \subset \mathbb{Q}[G]$.

We can now formulate the equivariant Tamagawa number conjecture due to Bloch-Kato, Fontaine-Perrin-Riou and Burns-Flach.

Conjecture 8 (Equivariant Tamagawa number conjecture (ETNC)). *Let $M \in \mathcal{M}_{\mathbb{Q}}$ be a motive with coefficients in A and assume Conjectures 2, 3, 5, the Rationality Conjecture 6 and Conjecture 7. Let T_B be a projective $\mathcal{A}$-structure in M and let $\zeta_A(M) \in \Delta(M)$ be the zeta element defined in Conjecture 6. Let S be a finite set of places of $\mathbb{Q}$, such that M_p is unramified outside of S and $p, \infty \in S$. Then the $\mathcal{A}_{\mathbb{Z}_p}$-submodule*

$$\theta_{p,S}(\zeta_A(M))\mathcal{A}_{\mathbb{Z}_p} \subset \mathrm{Det}_{A_{\mathbb{Q}_p}}(R\Gamma_c(\mathbb{Z}_S, M_p))$$

coincides with $\mathrm{Det}_{\mathcal{A}_{\mathbb{Z}_p}}(R\Gamma_c(\mathbb{Z}_S, T_p))$, *where* $T_p := T_B \otimes_{\mathbb{Z}} \mathbb{Z}_p$.

Remark 2.21. The ETNC is independent of the choice of the projective $\mathcal{A}$-structure T_B. Indeed, for a different $\mathcal{A}$-structure T'_B we can consider $p^n T_p \subset T_p \cap T'_p$ and so reduce to the case of $T_p \subset T'_p$. We get an exact triangle

$$R\Gamma_c(\mathbb{Z}_S, T_p) \to R\Gamma_c(\mathbb{Z}_S, T'_p) \to R\Gamma_c(\mathbb{Z}_S, T'_p/T_p),$$

where T'_p/T_p is finite. By Example 1.31 the isomorphisms $\zeta_{\mathcal{A}_{\mathbb{Z}_p}}$ for $R\Gamma_c(\mathbb{Z}_S, T_p)$ and $R\Gamma_c(\mathbb{Z}_S, T'_p)$ differ by $\prod_{i=0,1,2,3}(\#H^i_c(\mathbb{Z}_S, T'_p/T_p))^{(-1)^i} = 1$, (see [**Mi**] I 2.8 and 5.1).

Remark 2.22. The ETNC is also independent of the choice of S in the sense that if it holds for S, then it also holds for a different set of places S'. We may assume $S \subset S'$. Then there is an exact triangle

$$R\Gamma_c(\mathbb{Z}_S, T_p) \to R\Gamma_c(\mathbb{Z}_{S'}, T_p) \to \bigoplus_{v \in S' \setminus S} R\Gamma(\mathbb{F}_v, T_p^{I_v}).$$

The complex $R\Gamma(\mathbb{F}_v, T_p^{I_v})$ can be represented by $[T_p^{I_v} \xrightarrow{1-\mathrm{Frob}_v^{-1}} T_p^{I_v}]$, which is mapped under ι_v to $\mathcal{A}_{\mathbb{Z}_p}$. This implies $\theta_{p,S}(\zeta_A(M))\mathcal{A}_{\mathbb{Z}_p} = \theta_{p,S'}(\zeta_A(M))\mathcal{A}_{\mathbb{Z}_p}$.

Remark 2.23. The independence of S and the compatibility of zeta elements in towers of number fields also shows that the zeta elements give rise to Euler systems, see [**Ka**] Remark 4.14 for more details.

5. Unramified cohomology for elliptic curves

In this section we compute the local and global unramified cohomology for elliptic curves.

Consider an elliptic curve $E/\mathbb{Q}$ and let T_pE and V_pE be as defined in Example 1.14. Fix a finite set of places S of $\mathbb{Q}$, such that T_pE is unramified outside of S and $p, \infty \in S$.

The Kummer sequence induces for any place v of $\mathbb{Q}$ an exact sequence

$$0 \to E(\mathbb{Q}_v)^{\wedge p} \to H^1(\mathbb{Q}_v, T_pE) \to T_pH^1(\mathbb{Q}_v, E) \to 0,$$

where $E(\mathbb{Q}_v)^{\wedge p} := \varprojlim_n E(\mathbb{Q}_v)/p^n E(\mathbb{Q}_v)$ is the p-adic completion and $T_p(\ldots) := \mathrm{Hom}(\mathbb{Q}_p/\mathbb{Z}_p, \ldots)$. Note that T_p of a finite group is zero.

We need the following fact on the exponential map

$$\exp_{\mathrm{BK}} : \mathrm{Lie} E_{\mathbb{Q}_p} \to H^1_f(\mathbb{Q}_p, V_pE)$$

and $H^0_f(\mathbb{Q}_p, V_pE)$.

Proposition 2.24 ([**BK**] Example 3.10.1 and 3.11). *One has $H^0_f(\mathbb{Q}_p, V_pE) = 0$ and the diagram*

$$\begin{array}{ccc} E_1(\mathbb{Q}_p)_{\mathbb{Q}_p} & \xrightarrow{\text{Kummer}} & H^1(\mathbb{Q}_p, V_pE) \\ {\scriptstyle \exp}\uparrow{\scriptstyle \cong} & & \uparrow \\ \mathrm{Lie}(E)_{\mathbb{Q}_p} & \underset{\cong}{\xrightarrow{\exp_{\mathrm{BK}}}} & H^1_f(\mathbb{Q}_p, V_pE) \end{array}$$

where exp *is the exponential of the formal group E_1 commutes. In particular, the Kummer map induces an isomorphism*

$$E(\mathbb{Q}_p)^{\wedge p}_{\mathbb{Q}_p} \cong H^1_f(\mathbb{Q}_p, V_pE).$$

Remark 2.25. The same result of Bloch and Kato also shows that for finite field extension $K/\mathbb{Q}_p$ and $M = h(\mathrm{Spec} K)$ the Kummer map induces an isomorphism

$$(\mathcal{O}^*_{K_p})^{\wedge p}_{\mathbb{Q}_p} \cong H^1_f(\mathbb{Q}_p, M(1)_p),$$

where we have written $K_p := K \otimes_{\mathbb{Q}} \mathbb{Q}_p$ and $\mathcal{O}_{K_p} \cong \prod_{w|p} \mathcal{O}_{K_w}$.

The next theorem gives the complete description of the local unramified cohomology of T_pE.

Theorem 2.26. *The Kummer sequence induces for all v*

$$E(\mathbb{Q}_v)^{\wedge p} \cong H^1_f(\mathbb{Q}_v, T_pE)$$

and $T_pH^1(\mathbb{Q}_v, E) \cong H^1_{/f}(\mathbb{Q}_v, T_pE)$. For all $v \neq \infty$ one has

$$H^0_f(\mathbb{Q}_v, T_pE) = 0.$$

Remark 2.27. Continuing Remark 2.25 let $T_p := \bigoplus_{\tau: K \to \overline{\mathbb{Q}}} \mathbb{Z}_p$ be a Galois stable $\mathbb{Z}_p$ lattice in M_p. Then, one can show that the Kummer sequence induces an isomorphism

$$(\mathcal{O}^*_{K_p})^{\wedge p} \cong H^1_f(\mathbb{Q}_p, T_p(1))$$

(cf. [**HK2**] Lemma A.1).

We first prove a lemma.

Lemma 2.28. *For $v \neq p, \infty$ the map $1 - \mathrm{Frob}_v^{-1}$ induces an isomorphism*

$$V_pE^{I_v} \xrightarrow{1-\mathrm{Frob}_v^{-1}} V_pE^{I_v},$$

so that

$$H^0_f(\mathbb{Q}_v, V_pE) = H^0(\mathbb{Q}_v, V_pE) = 0$$
$$H^1_f(\mathbb{Q}_v, V_pE) = 0.$$

For $v = p$ we have

$$H^0_f(\mathbb{Q}_v, V_pE) = 0.$$

Proof. As $\widetilde{E}^{\mathrm{ns}}_v(\mathbb{F}_v)$ is finite, this follows from Lecture 3, Section 5 Lemma 3.11 and Proposition 2.24. □

Proof of Theorem 2.26. By Lemma 2.28, we have $H^0_f(\mathbb{Q}_v, T_pE) = 0$ for all $v \neq \infty$. We claim that for all places $v \neq \infty$

$$E(\mathbb{Q}_v)^{\wedge p}_{\mathbb{Q}_p} \cong H^1_f(\mathbb{Q}_v, V_pE).$$

For $v = p$ this is contained in Proposition 2.24. For $v \neq p, \infty$ both sides are zero: indeed, the left hand side is zero by the structure of $E(\mathbb{Q}_v)$ recalled in Theorem 3.10 because the p-adic completion of $E_1(\mathbb{Q}_v)$ is zero, so that $E(\mathbb{Q}_v)^{\wedge p}$ is a finite group. The right hand side is zero by Lemma 2.28. Consider the commutative diagram:

$$\begin{array}{ccccccccc}
0 & \to & E(\mathbb{Q}_v)^{\wedge p} & \to & H^1(\mathbb{Q}_v, T_pE) & \to & T_pH^1(\mathbb{Q}_v, E) & \to & 0 \\
 & & \downarrow & & \downarrow & & \hookdownarrow & & \\
0 & \to & E(\mathbb{Q}_v)^{\wedge p} \otimes_{\mathbb{Z}_p} \mathbb{Q}_p & \to & H^1(\mathbb{Q}_v, V_pE) & \to & T_pH^1(\mathbb{Q}_v, E) \otimes_{\mathbb{Z}_p} \mathbb{Q}_p & \to & 0.
\end{array}$$

As $T_pH^1(\mathbb{Q}_v, E)$ is torsion free, the right vertical arrow is injective. A diagram chase shows that $E(\mathbb{Q}_v)^{\wedge p}$ is identified with the elements in $H^1(\mathbb{Q}_v, T_pE)$, which map to $E(\mathbb{Q}_v)^{\wedge p}_{\mathbb{Q}_p} \cong H^1_f(\mathbb{Q}_v, V_pE)$. This proves $E(\mathbb{Q}_v)^{\wedge p} \cong H^1_f(\mathbb{Q}_v, T_pE)$ for $v \neq \infty$. For $v = \infty$, the group $H^1(\mathbb{R}, E)$ is 2-torsion, which implies that $H^1(\mathbb{R}, E[2]) \to H^1(\mathbb{R}, E)$ is surjective. Hence, $H^1(\mathbb{R}, E)$ is finite, so that $T_pH^1(\mathbb{R}, E) = 0$. This gives

$$E(\mathbb{R})^{\wedge p} \cong H^1(\mathbb{R}, T_pE) = H^1_f(\mathbb{R}, T_pE).$$

The statement about $T_pH^1(\mathbb{Q}_v, E)$ follows from the definition of $H^1_{/f}((\mathbb{Q}_v, T_pE)$. □

To compute the global unramified cohomology $H^1_f(\mathbb{Q}, T_pE)$, we recall the definition of the Shafarevich-Tate group.

Definition 2.29. Let $E/\mathbb{Q}$ be an elliptic curve, then

$$Ш(E/\mathbb{Q}) := \ker\left(H^1(\mathbb{Q}, E) \to \prod_v H^1(\mathbb{Q}_v, E)\right)$$

is called the *Shafarevich-Tate group*.

Theorem 2.30. *The Kummer sequence identifies*

$$E(\mathbb{Q})^{\wedge p} \cong H^1_f(\mathbb{Q}, T_pE)$$

inside $H^1(\mathbb{Q}, T_pE)$ if and only if $Ш(\mathbb{Q}, E)[p^\infty]$ is a finite group.

PROOF. Consider the commutative diagram

$$\begin{array}{ccccccccc} 0 & \longrightarrow & E(\mathbb{Q})^{\wedge p} & \longrightarrow & H^1(\mathbb{Q},T_pE) & \longrightarrow & T_pH^1(\mathbb{Q},E) & \longrightarrow & 0 \\ & & \downarrow & & \downarrow & & \downarrow & & \\ 0 & \longrightarrow & \prod_v E(\mathbb{Q}_v)^{\wedge p} & \longrightarrow & \prod_v H^1(\mathbb{Q}_v,T_pE) & \longrightarrow & \prod_v T_pH^1(\mathbb{Q}_v,E) & \longrightarrow & 0. \end{array}$$

Consider the kernel-cokernel sequence (cf. [**Mi**] I Proposition 0.24) of the composition

$$H^1(\mathbb{Q},T_pE) \to T_pH^1(\mathbb{Q},E) \to \prod_v H^1(\mathbb{Q}_v,T_pE).$$

As the functor $T_p(\ldots) = \mathrm{Hom}(\mathbb{Q}_p/\mathbb{Z}_p,\ldots)$ is left exact,

$$H^1_{/f}(\mathbb{Q}_v,T_pE) = T_pH^1(\mathbb{Q}_v,E)$$

by Theorem 2.26 and using Lemma 2.15, this kernel-cokernel sequence gives an exact sequence

$$0 \to E(\mathbb{Q})^{\wedge p} \to H^1_f(\mathbb{Q},T_pE) \to T_p\text{Ш}(\mathbb{Q},E) \to 0.$$

If $\text{Ш}(\mathbb{Q},E)[p^\infty]$ is finite, we get that $T_p\text{Ш}(\mathbb{Q},E) = T_p\text{Ш}(\mathbb{Q},E)[p^\infty] = 0$. If $E(\mathbb{Q})^{\wedge p} \to H^1_f(\mathbb{Q},T_pE)$ is an isomorphism we have $T_p\text{Ш}(\mathbb{Q},E)[p^\infty] = 0$. But it is well-known that $\text{Ш}(\mathbb{Q},E)[p^N]$ is a finite group for every N (see [**Mi**] Remark 6.7), which implies that $\text{Ш}(\mathbb{Q},E)[p^\infty]$ is an extension of a finite group by divisible groups of the form $\mathbb{Q}_p/\mathbb{Z}_p$. If $T_p\text{Ш}(\mathbb{Q},E)[p^\infty] = 0$ then the divisible part of $\text{Ш}(\mathbb{Q},E)[p^\infty]$ must be zero. □

LECTURE 3

The relation to the Birch-Swinnerton-Dyer conjecture

In this last lecture we show that the equivariant Tamagawa number conjecture for the motive $M = h_1(E)(1)$ is equivalent to the Birch-Swinnerton-Dyer conjecture. At the end we review some results on elliptic curves, which were needed for the computation of the unramified cohomology.

1. The Tamagawa number conjecture and the Birch-Swinnerton-Dyer conjecture

Let $E/\mathbb{Q}$ be an elliptic curve and $M = h_1(E)(1)$ (as always $i = 1$). We denote by $r := \mathrm{rk}_{\mathbb{Z}} E(\mathbb{Q})$. Choose a basis $\underline{x} := (x_1, \ldots, x_r)$ of the free part $E(\mathbb{Q})^{\mathrm{free}}$ of $E(\mathbb{Q})$ and let $<,>$ be the Neron Tate height pairing as defined in Example 2.2.

Definition 3.1. The *regulator* of E is the real number

$$R(E/\mathbb{Q}) := \det(< x_i, x_j >)_{i,j=1,\ldots,r}. \tag{3.1}$$

Definition 3.2. Let

$$c_v := \begin{cases} \#(E(\mathbb{Q}_p)/E_0(\mathbb{Q}_p)) & \text{if } v \neq \infty \\ \#(E(\mathbb{R})/E(\mathbb{R})^0) & \text{if } v = \infty, \end{cases}$$

where $E_0(\mathbb{Q}_p)$ is the subgroup of $E(\mathbb{Q}_p)$ defined in Theorem 3.10 and $E(\mathbb{R})^0$ is the connected component of $E(\mathbb{R})$. Note that $c_v = 1$ for almost all v.

Remark 3.3. One can show that for finite v one has $c_v = \#(\mathcal{E}(\mathbb{F}_v)/\mathcal{E}^0(\mathbb{F}_v))$ where $\mathcal{E}$ is the Neron model of E and $\mathcal{E}^0 \subset \mathcal{E}$ is the connected component of the identity.

Conjecture 9 (Birch-Swinnerton-Dyer conjecture (BSD)). *Let $E/\mathbb{Q}$ be an elliptic curve, $r := \mathrm{rk}_{\mathbb{Z}} E(\mathbb{Q})$, then:*

a) $\mathrm{ord}_{s=1} L(E, s) = r$

b) *If $Ш(E/\mathbb{Q})$ is finite, then*

$$\frac{L^*(E,1)}{\Omega_\infty R(E/\mathbb{Q})} = \frac{\#Ш(E/\mathbb{Q})}{(E(\mathbb{Q})_{\mathrm{tors}})^2} \prod_v c_v,$$

where Ω_∞ is the period of E from Example 1.17.

We can now formulate the main theorem of these lectures:

Theorem 3.4 (Main theorem). *Let $E/\mathbb{Q}$ be an elliptic curve, $M = h_1(E)(1)$ and $A = \mathbb{Q}$.*

(1) *Conjectures 1, 2, 3 and 5 hold for M.*

(2) *Conjecture 4 on the order of vanishing of $L(E, s)$ at $s = 1$ is equivalent to part a) of the BSD conjecture.*

(3) *If (2) holds, Conjecture 6 (Rationality Conjecture) is equivalent to*

$$\frac{L^*(E,1)}{\Omega_\infty R(E/\mathbb{Q})} \in \mathbb{Q}^\times.$$

(4) *Conjecture 7 holds for p if and only if $\text{Ш}(E/\mathbb{Q})[p^\infty]$ is finite.*

(5) *The ETNC (Conjecture 8) for M and all p is equivalent to part b) of the BSD conjecture.*

We prove here (1)-(4). Part (5) is proven in Section 4.

Proof of (1)-(4). The statement in (1) just repeats Examples 1.21, 1.26 and 2.2. For (2) using Example 1.21, Conjecture 4 says for $M = h_1(E)(1)$ that

$$r = \text{rk}_{\mathbb{Z}} E(\mathbb{Q}) = \text{ord}_{s=0} L(M,s),$$

if we observe that $L(M,s) = L(E, s+1)$. This proves (2).

We show (3). The fundamental line of M is

$$\Delta(M) = \text{Det}_{\mathbb{Q}}^{-1}(E(\mathbb{Q})_{\mathbb{Q}}) \cdot \text{Det}_{\mathbb{Q}}(E(\mathbb{Q})_{\mathbb{Q}})^\vee \cdot \text{Det}_{\mathbb{Q}}^{-1}(M_B^+) \cdot \text{Det}_{\mathbb{Q}}(\text{Lie}E). \tag{3.2}$$

Let $\underline{x}^\vee := (x_1^\vee, \ldots, x_r^\vee)$ be the basis of $(E(\mathbb{Q})^{\text{free}})^\vee$ dual to $\underline{x} := (x_1, \ldots, x_r)$ and let $\text{cl}_{E(\mathbb{R})^0}$ and $\omega^\vee$ be the basis elements of M_B^+ and Lie$E$ respectively, defined in Example 1.14. We define $\beta \in \Delta(M)$ to be the element

$$\beta := (x_1 \wedge \ldots \wedge x_r)^{-1}(x_1^\vee \wedge \ldots \wedge x_r^\vee)(\text{cl}_{E(\mathbb{R})^0})^{-1}(\omega^\vee) \in \Delta(M). \tag{3.3}$$

A possible zeta element $\zeta_{\mathbb{Q}}(M)$ is necessarily a rational multiple of β, i.e., of the form $\zeta_{\mathbb{Q}}(M) = q\beta$, with $q \in \mathbb{Q}^\times$, Conjecture 6 is now equivalent to the fact that

$$q\theta_\infty(\beta) = (L^*(M,0))^{-1} \in \mathbb{R}^\times.$$

By definition of θ_∞ we get that

$$\theta_\infty(\beta) = (\Omega_\infty R(E/\mathbb{Q}))^{-1} \in \mathbb{R}^\times.$$

Thus

$$\frac{L^*(M,0)}{\Omega_\infty R(E/\mathbb{Q})} = q^{-1} \in \mathbb{Q}^\times.$$

Finally, (4) is the content of Example 2.16. □

2. The Selmer group and $R\Gamma_c(\mathbb{Z}_S, T_pE)$

In this section, $E/\mathbb{Q}$ is an elliptic curve and T_pE is the Tate-module of E. Fix a prime number p. We denote by S the union of the finite set of places where E has bad reduction and of p, ∞.

Definition 3.5. Fix a prime number p. The *Selmer group* of $E/\mathbb{Q}$ is defined as

$$\text{Sel}_{p^\infty}(E/\mathbb{Q}) := \ker\left(H^1(\mathbb{Q}, E[p^\infty]) \to \prod_v H^1(\mathbb{Q}_v, E)[p^\infty]) \right).$$

Note that one has an exact sequence

$$0 \to E(\mathbb{Q}) \otimes_{\mathbb{Z}} \mathbb{Q}_p/\mathbb{Z}_p \to \text{Sel}_{p^\infty}(E/\mathbb{Q}) \to \text{Ш}(E/\mathbb{Q})[p^\infty] \to 0. \tag{3.4}$$

Definition 3.6. For any abelian topological group H, we let

$$H^* := \text{Hom}_{\text{ct}}(H, \mathbb{Q}_p/\mathbb{Z}_p),$$

where $\mathbb{Q}_p/\mathbb{Z}_p$ has the discrete topology.

Now we can describe the determinant of the complex $R\Gamma_c(\mathbb{Z}_S, T_pE)$ more explicitly. First note that by (2.1) and Lemma 2.28, we can write

$$(3.5)\quad \mathrm{Det}_{\mathbb{Q}_p}(R\Gamma_c(\mathbb{Z}_S, V_pE)) \cong \mathrm{Det}_{\mathbb{Q}_p}(R\Gamma_f(\mathbb{Q}, V_pE)) \cdot \mathrm{Det}_{\mathbb{Q}_p}(H^1_f(\mathbb{Q}_p, V_pE)) \cdot \mathrm{Det}^{-1}_{\mathbb{Q}_p}(V_pE^+).$$

Theorem 3.7. *There is a canonical isomorphism*

$$\mathrm{Det}_{\mathbb{Z}_p}(R\Gamma_c(\mathbb{Z}_S, T_pE)) \cong \mathrm{Det}^{-1}_{\mathbb{Z}_p}(H^1_f(\mathbb{Q}, T_pE)) \cdot \mathrm{Det}_{\mathbb{Z}_p}(\mathrm{Sel}_{p^\infty}(E/\mathbb{Q})^*) \cdot \cdot \mathrm{Det}^{-1}_{\mathbb{Z}_p}(H^0(\mathbb{Z}_S, E[p^\infty])^*) \cdot \prod_{v\in S} \mathrm{Det}_{\mathbb{Z}_p}(E(\mathbb{Q}_v)^{\wedge p}) \cdot \mathrm{Det}^{-1}_{\mathbb{Z}_p}(T_pE^+),$$

which induces after $\mathbb{Q}_p \otimes_{\mathbb{Z}_p}$ *the isomorphism in* (3.5).

Proof. The Poitou-Tate sequence (see [**NSW**] (8.6.10)) gives for each integer $n \geq 1$ a long exact sequence of finite groups (we omit some of the first terms)

$$H^1(\mathbb{Z}_S, E[p^n]) \to \bigoplus_{v\in S} H^1(\mathbb{Q}_v, E[p^n]) \to H^1(\mathbb{Z}_S, E[p^n])^* \to$$
$$\to H^2(\mathbb{Z}_S, E[p^n]) \to \bigoplus_{v\in S} H^2(\mathbb{Q}_v, E[p^n]) \to H^0(\mathbb{Z}_S, E[p^n])^* \to 0.$$

Here we have identified $E[p^n] \cong \mathrm{Hom}(E[p^n], \mu_{p^n})$ via the Weil pairing. As the groups in this sequence are finite, taking $\varprojlim_n$ is exact. We get

$$H^1(\mathbb{Z}_S, T_pE) \to \bigoplus_{v\in S} H^1(\mathbb{Q}_v, T_pE) \to H^1(\mathbb{Z}_S, E[p^\infty])^* \to$$
$$\to H^2(\mathbb{Z}_S, T_pE) \to \bigoplus_{v\in S} H^2(\mathbb{Q}_v, T_pE) \to H^0(\mathbb{Z}_S, E[p^\infty])^* \to 0.$$

Taking the inverse limit of the Tate duality pairing in Theorem 3.12 one gets $E(\mathbb{Q}_v)^{\wedge p} \cong H^1(\mathbb{Q}_v, E)[p^\infty]^*$ and a commutative diagram

$$\begin{array}{ccc}
\bigoplus_{v\in S} E(\mathbb{Q}_v)^{\wedge p} & \xrightarrow{\cong} & \bigoplus_{v\in S} H^1(\mathbb{Q}_v, E)[p^\infty]^* \\
\downarrow & & \downarrow \\
\bigoplus_{v\in S} H^1(\mathbb{Q}_v, T_pE) & \longrightarrow & H^1(\mathbb{Z}_S, E[p^\infty])^* \\
\downarrow & & \downarrow \\
\bigoplus_{v\in S} T_pH^1(\mathbb{Q}_v, E) & \longrightarrow & \mathrm{Sel}_{p^\infty}(E/\mathbb{Q})^* \\
\downarrow & & \downarrow \\
0 & & 0,
\end{array}$$

where the right row is exact by Lemma 3.8 below. Thus, if we factor out the isomorphism $E(\mathbb{Q}_v)^{\wedge p} \cong H^1(\mathbb{Q}_v, E)[p^\infty]^*$ we get a long exact sequence

$$\begin{aligned} 0 \to & H^1_f(\mathbb{Q}, T_pE) \to H^1(\mathbb{Z}_S, T_pE) \to \bigoplus_{v\in S} T_pH^1(\mathbb{Q}_v, E) \to \\ \to & \mathrm{Sel}_{p^\infty}(E/\mathbb{Q})^* \to H^2(\mathbb{Z}_S, T_pE) \to \bigoplus_{v\in S} H^2(\mathbb{Q}_v, T_pE) \to \\ \to & H^0(\mathbb{Z}_S, E[p^\infty])^* \to 0. \end{aligned} \tag{3.6}$$

Here by Definition 2.14 and Theorem 2.26, the kernel of the map from $H^1(\mathbb{Z}_S, T_pE)$ to $\bigoplus_{v\in S} T_pH^1(\mathbb{Q}_v, E)$ is $H^1_f(\mathbb{Q}, T_pE)$. On the other hand we have by Definition 2.12 a long exact sequence for $R\Gamma_c(\mathbb{Z}_S, T_pE)$ (note that $H^0_c(\mathbb{Z}_S, T_pE) = H^0(\mathbb{Z}_S, T_pE) = 0$ and $H^0(\mathbb{Q}_v, T_pE) = 0$ for v finite)

$$\begin{aligned} 0 \to H^0(\mathbb{R}, T_pE) \to & H^1_c(\mathbb{Z}_S, T_pE) \to H^1(\mathbb{Z}_S, T_pE) \to \bigoplus_{v\in S} H^1(\mathbb{Q}_v, T_pE) \to \\ \to & H^2_c(\mathbb{Z}_S, T_pE) \to H^2(\mathbb{Z}_S, T_pE) \to \bigoplus_{v\in S} H^2(\mathbb{Q}_v, T_pE) \to \\ \to & H^3_c(\mathbb{Z}_S, T_pE) \to 0. \end{aligned} \tag{3.7}$$

Taking $\mathrm{Det}_{\mathbb{Z}_p}$ of the exact sequences in Equations (3.6) and (3.7) gives the desired result. □

Finally, we prove that the Selmer group can be described as a subgroup of $H^1(\mathbb{Z}_S, E[p^\infty])$ as was needed in the proof of Theorem 3.7

Lemma 3.8. *The inclusion $H^1(\mathbb{Z}_S, E[p^\infty]) \subset H^1(\mathbb{Q}, E[p^\infty])$ induces an isomorphism*

$$\mathrm{Sel}_{p^\infty}(E/\mathbb{Q}) := \ker\left(H^1(\mathbb{Z}_S, E[p^\infty]) \to \bigoplus_{v\in S} H^1(\mathbb{Q}_v, E)[p^\infty])\right).$$

PROOF. We claim that

$$H^1(\mathbb{Z}_S, E[p^\infty]) = \ker\left(H^1(\mathbb{Q}, E[p^\infty]) \to \prod_{v\notin S} H^1(\mathbb{Q}_v, E)[p^\infty])\right).$$

Accepting the claim, consider the kernel-cokernel sequence (cf. [**Mi**] Proposition 0.24) of the composition

$$H^1(\mathbb{Q}, E[p^\infty]) \to \prod_v H^1(\mathbb{Q}_v, E)[p^\infty]) \to \prod_{v\notin S} H^1(\mathbb{Q}_v, E)[p^\infty]),$$

which gives with the claim an exact sequence

$$0 \to \mathrm{Sel}_{p^\infty}(E/\mathbb{Q}) \to H^1(\mathbb{Z}_S, E[p^\infty]) \to \bigoplus_{v\in S} H^1(\mathbb{Q}_v, E)[p^\infty]).$$

This is the statement of the lemma. To prove the claim, note that by the same argument as in the proof of Lemma 2.15 we have an exact sequence

$$0 \to H^1(\mathbb{Z}_S, E[p^\infty]) \to H^1(\mathbb{Q}, E[p^\infty]) \to \prod_{v\notin S} H^1(I_v, E[p^\infty])$$

(recall that I_v acts trivially on $E[p^\infty]$ as $v \notin S$ and that E has good reduction outside of S). It follows from Theorem 3.10 that $E(\overline{\mathbb{Q}}_v)^{I_v} = E(\mathbb{Q}_v^{\mathrm{unr}})$ is p-divisible, so that $H^1(I_v, E[p^N]) \to H^1(I_v, E)$ is injective for every N (consider the long exact cohomology sequence of the Kummer sequence). We get a commutative diagram with exact rows

$$\begin{array}{ccccc} H^1(\mathbb{Z}_S, E[p^\infty]) & \longrightarrow & H^1(\mathbb{Q}, E[p^\infty]) & \longrightarrow & \prod_{v\notin S} H^1(I_v, E[p^\infty]) \\ \downarrow & & \downarrow & & \downarrow \\ \prod_{v\notin S} H^1(\mathbb{F}_v, E^{I_v})[p^\infty] & \longrightarrow & \prod_{v\notin S} H^1(\mathbb{Q}_v, E)[p^\infty] & \longrightarrow & \prod_{v\notin S} H^1(I_v, E)[p^\infty]. \end{array}$$

As the right vertical arrow is injective, the claim follows if we can show that $H^1(\mathbb{F}_v, E(\mathbb{Q}_v^{\mathrm{unr}}))[p^\infty] = 0$. As $E_1(\mathbb{Q}_v^{\mathrm{unr}})$ is uniquely p-divisible by Theorem 3.10 the long exact cohomology sequence of

$$0 \to E_1(\mathbb{Q}_v^{\mathrm{unr}}) \to E(\mathbb{Q}_v^{\mathrm{unr}}) \to \widetilde{E}_v^{\mathrm{ns}}(\overline{\mathbb{F}}_v) \to 0$$

shows that $H^1(\mathbb{F}_v, E(\mathbb{Q}_v^{\mathrm{unr}}))[p^\infty] \cong H^1(\mathbb{F}_v, \widetilde{E}_v^{\mathrm{ns}}(\overline{\mathbb{F}}_v))[p^\infty]$. Now we use the interpretation of $H^1(\mathbb{F}_v, \widetilde{E}_v^{\mathrm{ns}}(\overline{\mathbb{F}}_v))$ by principal homogeneous spaces (see [**Si**] X §3). We have to show that each principal homogeneous spaces C of E has an $\mathbb{F}_v$-rational point. This follows by Lang's trick: let $x \in C(\overline{\mathbb{F}}_v)$, then there exists $p \in E(\overline{\mathbb{F}}_v)$ such that $p + x = Fx$, where F is the Frobenius. As $\mathrm{id} - F$ is an isogeny of E, there exist $q \in E(\overline{\mathbb{F}}_v)$ with $p = q - Fq$. This implies $q + x = Fq + Fx = F(q + x)$, where the last equality holds because the action $E \times C \to C$ is Frobenius equivariant. Thus, $q + x \in C(\mathbb{F}_v)$. □

3. Local Tamagawa numbers

In this section we study how the integral structure $E(\mathbb{Q}_v)^{\wedge p}$ behaves under the identification (2.3):

$$\iota_v : \mathrm{Det}_{\mathbb{Q}_p}^{-1}(R\Gamma_f(\mathbb{Q}_v, V_pE)) \cong \begin{cases} \mathrm{Det}_{\mathbb{Q}_p}(0) & \text{if } v \neq p, \infty \\ \mathrm{Det}_{\mathbb{Q}_p}(\mathrm{Lie}E)_{\mathbb{Q}_p} & \text{if } v = p. \end{cases}$$

Recall that by Theorem 2.26

$$\mathrm{Det}_{\mathbb{Z}_p}^{-1}(R\Gamma_f(\mathbb{Q}_v, T_pE)) \cong \mathrm{Det}_{\mathbb{Z}_p}(E(\mathbb{Q}_v)^{\wedge p}).$$

Theorem 3.9. *With the above notations one has*

$$\iota_v(\mathrm{Det}_{\mathbb{Z}_p}(E(\mathbb{Q}_v)^{\wedge p})) = \begin{cases} {c_v}^1\mathbb{Z}_p & \text{if } v \neq p, \infty \\ c_p^{-1}\omega^\vee\mathbb{Z}_p & \text{if } v = p. \end{cases}$$

PROOF. We treat first the case $v \neq p$. Recall from Lemma 3.11 in Section 5 the exact sequence

$$0 \to T_pE^{I_v} \xrightarrow{1-\mathrm{Frob}_v^{-1}} T_pE^{I_v} \to \widetilde{E}_v^{\mathrm{ns}}(\mathbb{F}_v) \otimes_{\mathbb{Z}} \mathbb{Z}_p \to 0.$$

As $E_1(\mathbb{Q}_v)^{\wedge p} = 0$ for $v \neq p$, we have $E_0(\mathbb{Q}_v)^{\wedge p} \cong \widetilde{E}_v^{\mathrm{ns}}(\mathbb{F}_v) \otimes_{\mathbb{Z}} \mathbb{Z}_p$. This gives

$$\mathrm{Det}_{\mathbb{Z}_p}(E(\mathbb{Q}_v)^{\wedge p}) \cong \mathrm{Det}_{\mathbb{Z}_p}^{-1}(T_pE^{I_v}) \cdot \mathrm{Det}_{\mathbb{Z}_p}(T_pE^{I_v}) \cdot \mathrm{Det}_{\mathbb{Z}_p}(E(\mathbb{Q}_v)^{\wedge p}/E_0(\mathbb{Q}_v)^{\wedge p}).$$

As $\mathbb{Q}_p \otimes_{\mathbb{Z}_p} T_pE^{I_v} = V_pE^{I_v}$ it follows from the definition of ι_v that

$$\iota_v(\mathrm{Det}_{\mathbb{Z}_p}^{-1}(T_pE^{I_v}) \cdot \mathrm{Det}_{\mathbb{Z}_p}(T_pE^{I_v})) = \mathbb{Z}_p.$$

The determinant computation for finite $\mathbb{Z}_p$-modules in Example 1.31 shows that $\iota_v(\mathrm{Det}_{\mathbb{Z}_p}(E(\mathbb{Q}_v)^{\wedge p})) = c_v^{-1}\mathbb{Z}_p$ because

$$c_v^{-1}\mathbb{Z}_p = \#(E(\mathbb{Q}_v)^{\wedge p}/E_0(\mathbb{Q}_v)^{\wedge p})^{-1}\mathbb{Z}_p.$$

This settles the case $v \neq p$. We consider now the case $v = p$. Recall the exponential map $\exp_{\mathrm{BK}} : \mathrm{Lie}E_{\mathbb{Q}_p} \cong H^1_f(\mathbb{Q}_p, V_pE)$ from Definition 2.9. We claim that for the induced isomorphism

$$\exp_{\mathrm{BK}}^{-1} : \mathrm{Det}_{\mathbb{Q}_p}(H^1_f(\mathbb{Q}_p, V_pE)) \cong \mathrm{Det}_{\mathbb{Q}_p}(\mathrm{Lie}E_{\mathbb{Q}_p})$$

we have

$$\iota_p = \det(1-\phi)\exp_{\mathrm{BK}}^{-1},$$

where ϕ is the Frobenius on $D_{\mathrm{cris}}(V_pE)$. Indeed, consider the complex

$$R\Gamma_f(\mathbb{Q}_p, V_pE) = [D_{\mathrm{cris}}(V_pE) \xrightarrow{(1-\phi,\mathrm{pr})} D_{\mathrm{cris}}(V_pE) \oplus \mathrm{Lie}E_{\mathbb{Q}_p}].$$

Taking determinants, we get

$$\mathrm{Det}_{\mathbb{Q}_p}(\mathrm{Lie}E_{\mathbb{Q}_p}) \cdot \mathrm{Det}^{-1}_{\mathbb{Q}_p}[D_{\mathrm{cris}}(V_pE) \xrightarrow{(1-\phi)} D_{\mathrm{cris}}(V_pE)] \cong \mathrm{Det}^{-1}_{\mathbb{Q}_p}(R\Gamma_f(\mathbb{Q}_p, V_pE)).$$

The quasi-isomorphism $\exp_{\mathrm{BK}} : \mathrm{Lie}E_{\mathbb{Q}_p}[-1] \cong R\Gamma_f(\mathbb{Q}_p, V_pE)$ is induced by the embedding of the sub-complex $[0 \to \mathrm{Lie}E_{\mathbb{Q}_p}]$. The quotient is the acyclic complex $[D_{\mathrm{cris}}(V_pE) \xrightarrow{(1-\phi)} D_{\mathrm{cris}}(V_pE)]$. The isomorphism ι_p is obtained from the identification

$$\begin{aligned}(3.8)\quad \mathrm{triv} : \mathrm{Det}^{-1}_{\mathbb{Q}_p}[D_{\mathrm{cris}}(V_pE) \xrightarrow{(1-\phi)} D_{\mathrm{cris}}(V_pE)] &= \\ = \mathrm{Det}^{-1}_{\mathbb{Q}_p}(D_{\mathrm{cris}}(V_pE)) \cdot \mathrm{Det}_{\mathbb{Q}_p}(D_{\mathrm{cris}}(V_pE)) &= \mathrm{Det}_{\mathbb{Q}_p}(0)\end{aligned}$$

which forgets the morphism $(1-\phi)$. On the other hand the isomorphism $\exp_{\mathrm{BK}}^{-1}$ is obtained from the identification

$$\mathrm{acyclic} : \mathrm{Det}^{-1}_{\mathbb{Q}_p}[D_{\mathrm{cris}}(V_pE) \xrightarrow{(1-\phi)} D_{\mathrm{cris}}(V_pE)] \cong \mathrm{Det}_{\mathbb{Q}_p}(0)$$

which uses the fact that the complex in brackets is acyclic. As $\mathrm{triv} \circ \mathrm{acyclic}^{-1} = \det(1-\phi)$ the claim follows.

Now we study the map $\exp_{\mathrm{BK}}$. By Proposition 2.24 in Section 5, the map $\exp_{\mathrm{BK}}$ is given by the exponential map $\exp : \mathrm{Lie}E_{\mathbb{Q}_p} \to E_1(\mathbb{Q}_p) = \widehat{E}(p\mathbb{Z}_p)$ in the sense of p-adic Lie groups. Let $q = p$ if $p \neq 2$ and $q = 4$ if $p = 2$. Then by our choice of basis $\omega^\vee \in \mathrm{Lie}E$ one gets an isomorphism $\exp : q\omega^\vee\mathbb{Z}_p \cong \widehat{E}(q\mathbb{Z}_p)$ by Theorem 3.10. This gives an identification

$$\exp_{\mathrm{BK}}^{-1}(\mathrm{Det}_{\mathbb{Z}_p}(E(\mathbb{Q}_p)^{\wedge p})) = \#(E(\mathbb{Q}_p)^{\wedge p}/\widehat{E}(q\mathbb{Z}_p))^{-1}q\omega^\vee\mathbb{Z}_p.$$

Now $\widehat{E}(q\mathbb{Z}_p) = E_1(\mathbb{Q}_p)^{\wedge p}$ for $p \neq 2$ and $\widehat{E}(q\mathbb{Z}_p) \subset E_1(\mathbb{Q}_p)^{\wedge p}$ has index 2 if $p = 2$. This gives

$$\#(E(\mathbb{Q}_p)^{\wedge p}/\widehat{E}(q\mathbb{Z}_p))^{-1}q\omega^\vee\mathbb{Z}_p = \#(E(\mathbb{Q}_p)^{\wedge p}/E_0(\mathbb{Q}_p)^{\wedge p})^{-1}c_p^{-1}\#\widetilde{E}_p^{\mathrm{ns}}(\mathbb{F}_p)^{-1}p\omega^\vee\mathbb{Z}_p.$$

As Conjecture 2 holds for E by Example 1.26 for all v, we have $\det(1-\phi) = \frac{\#\widetilde{E}_p^{\mathrm{ns}}(\mathbb{F}_p)}{p}$. This gives finally

$$\iota_p(\mathrm{Det}_{\mathbb{Z}_p}(E(\mathbb{Q}_p)^{\wedge p})) = c_p^{-1}\omega^\vee\mathbb{Z}_p.$$

□

4. Proof of the Main Theorem

Here we finish the proof of the Main Theorem 3.4.

PROOF OF (5) OF THEOREM 3.4. Recall from the proof of Theorem 3.4 (3) that $\zeta_{\mathbb{Q}}(M) = q\beta$, with

$$\beta := (x_1 \wedge \ldots \wedge x_r)^{-1}(x_1^\vee \wedge \ldots \wedge x_r^\vee)(\mathrm{cl}_{E(\mathbb{R})^0})^{-1}(\omega^\vee) \in \Delta(M)$$

and that

$$q^{-1} = \frac{L^*(M,0)}{\Omega_\infty R(E/\mathbb{Q})}.$$

Let us identify the lattice

$$\text{(3.9)}\quad \mathrm{Det}_{\mathbb{Z}_p}(R\Gamma_c(\mathbb{Z}_S, T_pE)) \cong \mathrm{Det}^{-1}_{\mathbb{Z}_p}(H^1_f(\mathbb{Q}, T_pE)) \cdot \mathrm{Det}_{\mathbb{Z}_p}(\mathrm{Sel}_{p^\infty}(E/\mathbb{Q})^*) \cdot \\ \cdot \mathrm{Det}^{-1}_{\mathbb{Z}_p}(H^0(\mathbb{Z}_S, E[p^\infty])^*) \cdot \prod_{v\in S} \mathrm{Det}_{\mathbb{Z}_p}(E(\mathbb{Q}_v)^{\wedge p}) \cdot \mathrm{Det}^{-1}_{\mathbb{Z}_p}(T_pE^+),$$

of $\mathrm{Det}_{\mathbb{Q}_p}(R\Gamma_c(\mathbb{Z}_S, V_pE))$. By Theorem 2.30 and because $E(\mathbb{Q})$ is finitely generated, we have $H^1_f(\mathbb{Q}, T_pE) \cong E(\mathbb{Q})^{\wedge p} \cong E(\mathbb{Q}) \otimes_{\mathbb{Z}} \mathbb{Z}_p$. This gives

$$\mathrm{Det}^{-1}_{\mathbb{Z}_p}(H^1_f(\mathbb{Q}, T_pE)) \cong \mathrm{Det}^{-1}_{\mathbb{Z}_p}(E(\mathbb{Q})^{\mathrm{free}} \otimes_{\mathbb{Z}} \mathbb{Z}_p) \cdot \mathrm{Det}^{-1}_{\mathbb{Z}_p}(E(\mathbb{Q})[p^\infty])$$

and $\#E(\mathbb{Q})[p^\infty](x_1 \wedge \ldots \wedge x_r)^{-1}$ is a generator of this lattice. Taking the dual of the sequence in (3.4), one gets

$$0 \to Ш(E/\mathbb{Q})[p^\infty]^* \to \mathrm{Sel}_{p^\infty}(E/\mathbb{Q})^* \to (E(\mathbb{Q})^{\mathrm{free}})^\vee \otimes_{\mathbb{Z}} \mathbb{Z}_p \to 0.$$

This gives

$$\mathrm{Det}_{\mathbb{Z}_p}(\mathrm{Sel}_{p^\infty}(E/\mathbb{Q})^*) \cong \mathrm{Det}_{\mathbb{Z}_p}(Ш(E/\mathbb{Q})[p^\infty]^*) \cdot \mathrm{Det}_{\mathbb{Z}_p}((E(\mathbb{Q})^{\mathrm{free}})^\vee_{\mathbb{Z}_p})$$

and $(\#Ш(E/\mathbb{Q})[p^\infty])^{-1}(x_1 \vee \wedge \ldots \wedge x_r^\vee)$ generates this lattice. Putting these results together and using that $H^0(\mathbb{Z}_S, E[p^\infty]) = E(\mathbb{Q})[p^\infty]$, we get

$$\text{(3.10)}\quad \mathrm{Det}_{\mathbb{Z}_p}(R\Gamma_c(\mathbb{Z}_S, T_pE)) \cong \\ \frac{(\#E(\mathbb{Q})[p^\infty])^2}{\#Ш(E/\mathbb{Q})[p^\infty]}(x_1 \wedge \ldots \wedge x_r)^{-1}(x_1^\vee \wedge \ldots \wedge x_r^\vee)(\mathrm{cl}_{E(\mathbb{R})^0})^{-1} \prod_{v\in S} \mathrm{Det}_{\mathbb{Z}_p}(E(\mathbb{Q}_v)^{\wedge p})$$

inside $\mathrm{Det}_{\mathbb{Q}_p}(R\Gamma_c(\mathbb{Z}_S, V_pE))$. With Theorem 3.9, we see that $\theta_{p,S}$ maps $\prod_{v\in S} c_v^{-1}\omega^\vee\mathbb{Z}_p$ to $\prod_{v\in S} \mathrm{Det}_{\mathbb{Z}_p}(E(\mathbb{Q}_v)^{\wedge p})$. Taking everything together, this implies that

$$\frac{(\#E(\mathbb{Q})[p^\infty])^2}{\#Ш(E/\mathbb{Q})[p^\infty]} \prod_{v\in S} c_v^{-1}\theta_{p,S}(\beta)\mathbb{Z}_p = \mathrm{Det}_{\mathbb{Z}_p}(R\Gamma_c(\mathbb{Z}_S, T_pE)).$$

Thus, $\theta_{p,S}(\zeta_{\mathbb{Q}}(M))\mathbb{Z}_p = q\theta_{p,S}(\beta)\mathbb{Z}_p$ equals $\mathrm{Det}_{\mathbb{Z}_p}(R\Gamma_c(\mathbb{Z}_S, T_pE)$ if and only if

$$\frac{L^*(M,0)}{\Omega_\infty R(E/\mathbb{Q})} = q^{-1} = \frac{\#Ш(E/\mathbb{Q})[p^\infty]}{(\#E(\mathbb{Q})[p^\infty])^2} \prod_{v\in S} c_v.$$

This holds for all p if and only if the BSD conjecture holds. □

5. Appendix: Review of some results on elliptic curves

Here we collect some results on elliptic curves $E/\mathbb{Q}$, which we will need in these lectures. This section is independent from the rest of the text.

We assume that E is given by a global minimal Weierstraß equation. We denote by $\widetilde{E}_v$ the reduction of E at a finite place v and by $\widetilde{E}_v^{\mathrm{ns}}$ the set of non-singular points in $\widetilde{E}_v$. Note that this a group scheme over $\mathbb{F}_v$.

Theorem 3.10 ([**Si**] VII 2.1, 2.2, 6.1 and IV 6.4). *Let $K = \mathbb{Q}_v$ or $K = \mathbb{Q}_v^{\mathrm{unr}}$, then there are subgroups*

$$E_1(K) \subset E_0(K) \subset E(K)$$

such that $E(K)/E_0(K)$ is finite and one has an exact sequence

$$0 \to E_1(K) \to E_0(K) \to \widetilde{E}_v^{\mathrm{ns}}(k) \to 0,$$

where $k = \mathbb{F}_v$ or $k = \overline{\mathbb{F}}_v$ is the residue field of K. Moreover, $E_0(K) = E(K)$ if E has good reduction and $E_1(K) = \widehat{E}(\mathfrak{m})$ are the points of the formal group associated to E, where $\mathfrak{m}$ is the maximal ideal of K. The structure of $\widehat{E}(\mathfrak{m})$ is as follows: If the valuation w of K is normalized (i.e., $w(K^*) = \mathbb{Z}$), then the logarithm induces an isomorphism $\widehat{E}(\mathfrak{m}^r) \cong \mathfrak{m}^r$ for all $r > w(p)/(p-1)$ and for all $r \geq 1$ one has an isomorphism $\widehat{E}(\mathfrak{m}^r)/\widehat{E}(\mathfrak{m}^{r+1}) \cong \mathfrak{m}^r/\mathfrak{m}^{r+1}$.

Lemma 3.11. *For $v \neq p$ one has*

$$T_pE^{I_v} \cong T_p\widetilde{E}_v^{\mathrm{ns}}$$

and an exact sequence

$$0 \to T_pE^{I_v} \xrightarrow{1-\mathrm{Frob}_v^{-1}} T_pE^{I_v} \to \widetilde{E}_v^{\mathrm{ns}}(\mathbb{F}_v) \otimes_{\mathbb{Z}} \mathbb{Z}_p \to 0.$$

PROOF. The sequence in Theorem 3.10 for $K = \mathbb{Q}_v^{\mathrm{unr}}$ gives

$$0 \to E_1(\mathbb{Q}_v^{\mathrm{unr}}) \to E_0(\mathbb{Q}_v^{\mathrm{unr}}) \to \widetilde{E}_v^{\mathrm{ns}}(\overline{\mathbb{F}}_v) \to 0.$$

As $v \neq p$, the group $E_1(\mathbb{Q}_v^{\mathrm{unr}})$ is uniquely p-divisible so that one has for all n an isomorphism $E_0(\mathbb{Q}_v^{\mathrm{unr}})[p^n] \cong \widetilde{E}_v^{\mathrm{ns}}(\overline{\mathbb{F}}_v)[p^n]$. This implies

$$T_pE_0(\mathbb{Q}_v^{\mathrm{unr}}) \cong T_p\widetilde{E}_v^{\mathrm{ns}}(\overline{\mathbb{F}}_v) = T_p\widetilde{E}_v^{\mathrm{ns}}.$$

As $E(\mathbb{Q}_v^{\mathrm{unr}})/E_0(\mathbb{Q}_v^{\mathrm{unr}})$ is finite, one has $T_pE_0(\mathbb{Q}_v^{\mathrm{unr}}) = T_pE(\mathbb{Q}_v^{\mathrm{unr}})$ and one gets

$$T_pE^{I_v} = T_pE(\mathbb{Q}_v^{\mathrm{unr}}) = T_pE_0(\mathbb{Q}_v^{\mathrm{unr}}) \cong T_p\widetilde{E}_v^{\mathrm{ns}}.$$

This proves the first claim. For the exact sequence, consider the p^n-multiplication on the exact sequence

$$0 \to \widetilde{E}_v^{\mathrm{ns}}(\mathbb{F}_v) \to \widetilde{E}_v^{\mathrm{ns}}(\overline{\mathbb{F}}_v) \xrightarrow{1-\mathrm{Frob}_v^{-1}} \widetilde{E}_v^{\mathrm{ns}}(\overline{\mathbb{F}}_v) \to 0.$$

Then the snake lemma gives

$$0 \to \widetilde{E}_v^{\mathrm{ns}}(\mathbb{F}_v)[p^n] \to \widetilde{E}_v^{\mathrm{ns}}(\overline{\mathbb{F}}_v)[p^n] \xrightarrow{1-\mathrm{Frob}_v^{-1}} \widetilde{E}_v^{\mathrm{ns}}(\overline{\mathbb{F}}_v)[p^n] \to \widetilde{E}_v^{\mathrm{ns}}(\mathbb{F}_v)/p^n\widetilde{E}_v^{\mathrm{ns}}(\mathbb{F}_v) \to 0,$$

as $\widetilde{E}_v^{\mathrm{ns}}(\overline{\mathbb{F}}_v)$ is p-divisible. Taking $\varprojlim_n$ gives the desired sequence as T_p applied to the finite group $\widetilde{E}_v^{\mathrm{ns}}(\mathbb{F}_v)$ is 0. □

Finally, we need to recall Tate's local duality theorem:

Theorem 3.12 (see [**Mi**] I 3.4 and I 3.7). *For any place v of $\mathbb{Q}$ one has a perfect pairing*

$$E(\mathbb{Q}_v)/p^n E(\mathbb{Q}_v) \times H^1(\mathbb{Q}_v, E)[p^n] \to \mathbb{Q}_p/\mathbb{Z}_p.$$

PROOF. For finite v this is [**Mi**] I 3.4. For $v = \infty$, the result in [**Mi**] I 3.7 gives a perfect pairing

$$\pi_0(E(\mathbb{R})) \times H^1(\mathbb{R}, E) \to \mathbb{Q}_2/\mathbb{Z}_2,$$

where $\pi_0(E(\mathbb{R})) = E(\mathbb{R})/E(\mathbb{R})^0$. Note that both groups in this pairing are 2-torsion. This is clear for $H^1(\mathbb{R}, E)$ and for $\pi_0(E(\mathbb{R}))$ it follows from the fact that the trace map $E(\mathbb{C}) \to E(\mathbb{R})$ is a continuous map of Lie groups, whose image is a closed connected subgroup of $E(\mathbb{R})$, which contains $2E(\mathbb{R})$ because this is the image of $E(\mathbb{R})$ under the trace map. Thus, the image of the trace map is also open in $E(\mathbb{R})$ and hence equal to $E(\mathbb{R})^0$ and $2E(\mathbb{R}) \subset E(\mathbb{R})^0$. As a connected, compact and commutative Lie group, we have $E(\mathbb{R})^0 \cong \mathbb{R}/\mathbb{Z}$, which is divisible. It follows that for each prime p, we have $\pi_0(E(\mathbb{R}))/p^n\pi_0(E(\mathbb{R})) \cong E(\mathbb{R})/p^n E(\mathbb{R})$, which proves the statement in the theorem. □

Bibliography

[BK] S. Bloch and K. Kato, L-functions and Tamagawa numbers of motives. The Grothendieck Festschrift, Vol. I, 333–400, Progr. Math., 86, Birkhuser Boston, Boston, MA, 1990.

[Bo] A. Borel, Stable real cohomology of arithmetic groups, Ann. Sci. ENS 7 (1974) 235-272.

[BF] D. Burns and M. Flach, Tamagawa numbers for motives with (non-commutative) coefficients. Documenta Math. 6 (2001) 501-570.

[Fo] J-M. Fontaine, Valeurs spéciales des fonctions L de motifs, Séminaire Bourbaki, Exposé 751, Asterisque 206 (1992)

[FP] J-M. Fontaine and B. Perrin-Riou, Autour des conjectures de Bloch et Kato: cohomologie galoisienne et valeurs de fonctions L, In: Motives, PSPM vol. 55 I (1994) 599-706

[FK] T. Fukaya and K. Kato, A formulation of conjectures on p-adic zeta functions in non-commutative Iwasawa theory. Proceedings of the St. Petersburg Mathematical Society. Vol. XII, 1 85, Amer. Math. Soc. Transl. Ser. 2, 219, Amer. Math. Soc., Providence, RI, 2006.

[HK1] A. Huber and G. Kings, Equivariant Bloch-Kato conjecture and non-abelian Iwasawa main conjecture. Proceedings of the International Congress of Mathematicians, Vol. II (Beijing, 2002), 149–162.

[HK2] A. Huber and G. Kings, Bloch-Kato conjecture and main conjecture of Iwasawa theory for Dirichlet characters, Duke Math. J., 119 (2003) 393-464.

[Ja] U. Jannsen, Deligne homology, Hodge-$\mathcal{D}$-conjecture and motives, in: Beilinson's conjectures on special values of L-functions, Academic Press (1988).

[Ka] K. Kato, Iwasawa theory and p-adic Hodge-theory, Kodai Math. J. 16 (1993), 1-31.

[KM] F. Knudsen, D. Mumford, The projectivity of the moduli space of stable curves I: Preliminaries on 'det' and 'div', Math. Scand. (1976), 39:19– 55.

[NSW] J. Neukirch, A. Schmidt and K. Wingberg, Cohomology of number fields. 2nd Edition, Springer 2008.

[Mi] J. S. Milne, Arithmetic duality theorems, Academic Press (1986)

[Sa] T. Saito, Modular forms and p-adic Hodge theory, Invent. Math. 129 (1979) 607-620.

[Sc1] A. J. Scholl, Classical motives, In: Motives, PSPM vol. 55 I (1994) 163-187.

[Sc2] A. J. Scholl, Integral elements in K-theory and products of modular curves. The arithmetic and geometry of algebraic cycles (Banff, AB, 1998), 467–489

[Sch] P. Schneider, Introduction to the Beilinson conjectures, in: Beilinson's conjectures on special values of L-functions, Academic Press (1988).

[Si] J. H. Silverman, The arithmetic of elliptic curves, Springer (1986)

[Ve] O. Venjakob, From the Birch and Swinnerton-Dyer conjecture to non-commutative Iwasawa theory via the equivariant Tamagawa number conjecture—a survey. L-functions and Galois representations, 333–380, London Math. Soc. Lecture Note Ser., 320, Cambridge Univ. Press, Cambridge, 2007.

Part III

Analytical and Cohomological Methods

Root Numbers

David E. Rohrlich

IAS/Park City Mathematics Series
Volume 18, 2009

Root Numbers

David E. Rohrlich

Introduction

Starting with Riemann's derivation of an "explicit formula" for the number of primes below a given bound, the functional equation of an L-function has been an indispensable tool in analytic number theory, and in more recent years it has become a fundamental tool in automorphic forms as well via the method of converse theorems pioneered by Hamburger and Hecke. The present lectures are concerned with a third direction, naïve by comparison and more limited in scope, namely the use of the functional equation to determine the parity of the order of vanishing of an L-function at the center of its critical strip. While the insights gained from this type of information are often only conditional ("... granting the conjecture of Birch and Swinnerton-Dyer, we conclude that ..."), they are sometimes the first hint of interesting new phenomena in arithmetic geometry.

Given our focus in these lectures, the key invariant is the root number, and the first four lectures are devoted to issues that arise in computing it. The four lectures correspond roughly to four possibilities for the underlying representation: global of dimension one, local of dimension one, global of arbitrary dimension, and local of arbitrary dimension. The fifth lecture addresses a question which is hinted at from the outset: To what extent, or under what circumstances, should one expect the order of vanishing of an L-function at the center of its critical strip to be the smallest value permitted by its functional equation? Very little is known about this question, and our remarks are largely speculative.

The main prerequisites for the lectures are basic algebraic number theory and a familiarity with Dirichlet L-functions. Some prior encounters with L-functions of elliptic curves are also desirable. More general classes of L-functions (Hecke L-functions, Artin L-functions, motivic L-functions) will be introduced from first principles as the lectures progress, but since references to unspecified "L-functions" appear right from the beginning, it is essential to have some notion of what is being talked about, namely an absolutely convergent Dirichlet series represented by an Euler product in some right half-plane (thus holomorphic and nonvanishing there) which is known or conjectured to extend to a meromorphic function on $\mathbb{C}$ and to satisfy a functional equation modeled on the functional equation of the Riemann zeta function. An acquaintance with Dirichlet L-functions and perhaps even with L-functions of elliptic curves provides an adequate intuition for absorbing the concept in general.

Department of Mathematics and Statistics, Boston University, Boston, MA 02215
E-mail address: rohrlich@math.bu.edu

Some vocabulary from group representation theory is also a prerequisite. Our conventions are as follows. A **representation** ρ of a group G is always understood to be finite-dimensional, and if G is a topological group then ρ is understood to be continuous as well. Continuity is meaningful because the field of scalars of ρ will be either $\mathbb{C}$ (the default) or else, where explicitly indicated, a λ-adic field with its λ-adic topology or the algebraic closure of a finite field with the discrete topology. A **character** is either a one-dimensional representation or the trace of a representation of dimension greater than one, usually the former. Possible ambiguities, when they arise, will be resolved by referring to a **one-dimensional character.** A one-dimensional character is **unitary** if it takes values in the group of complex numbers of absolute value 1. Note that elsewhere in the literature, particularly in older treatments, the term *quasicharacter* is used for our "one-dimensional character" and the term *character* is reserved for our "unitary character." Also the "contragredient" of a representation ρ will be referred to as the **dual** of ρ and denoted $\rho^\vee$. To illustrate the definitions just made, note that if χ is a one-dimensional character then $\chi^\vee = \chi^{-1}$, but if χ is unitary then also $\chi^\vee = \overline{\chi}$. The trivial character of a group G will often be denoted by 1, or if $G = \mathrm{Gal}(\overline{K}/K)$ then by 1_K.

There is one simple fact about representations which comes up so often that it deserves to be emphasized at the outset: A complex representation of a profinite group is trivial on an open subgroup. To see why, observe first of all that $\mathrm{GL}_n(\mathbb{C})$ "has no small subgroups": there is an open neighborhood $\mathcal{U}$ of the identity in $\mathrm{GL}_n(\mathbb{C})$ such that the only subgroup of $\mathrm{GL}_n(\mathbb{C})$ which is contained in $\mathcal{U}$ is the trivial subgroup. This property is easily verified using the exponential map, and it actually characterizes real Lie groups among all locally compact groups (Hilbert's fifth problem). Suppose now that we are given a profinite group G and a representation ρ of G on a complex vector space V. Choose $\mathcal{U} \subset \mathrm{GL}(V)$ as above. Since the open subgroups of G form a neighborhood basis at the identity, there is an open subgroup H contained in $\rho^{-1}(\mathcal{U})$. Then $\rho(H)$ is a subgroup of $\mathcal{U}$, hence trivial.

Although we have made a point of proving this little fact about profinite groups, in the pages that follow results both large and small will usually be stated without proof. In particular we do not reproduce Tate's proof of the functional equation of Hecke L-functions or Deligne's proof of the existence of local epsilon factors. On the other hand, we do include some rather detailed proofs of statements for which a convenient reference to the literature is lacking, and occasionally we also include a proof to emphasize a point (as in the previous paragraph) or to illustrate a technique. In the end, both the omission of most proofs and the inclusion of some reflect our overall aim, which is to complement the literature rather than duplicate it and to emphasize motivation rather than foundations.

It is a pleasure to record my indebtedness to many people: to the organizers of the conference, Cristian Popescu, Karl Rubin, and Alice Silverberg; to the participants in the conference, especially Keith Conrad, Ralph Greenberg, Dick Gross, Chan-Ho Kim, Myoungil Kim, Thomas de La Rochefoucauld, Álvaro Lozano-Robledo, Rachel Newton, Sami Omar, Robert Pollack, and John Tate; and to many other people whose help I solicited along the way, especially Avner Ash, Barbara Beeton, Philippe Cassou-Noguès, Pierre Deligne, John Polking, and Fernando Rodriguez Villegas. To all of them I extend my sincere thanks.

LECTURE 1

Trivial central zeros

Like many things in mathematics, the subject of root numbers begins with a theorem of Gauss, who proved in 1805 that if p is an odd prime then

$$\sum_{j=1}^{p-1} \lambda(j) e^{2\pi i j/p} = \begin{cases} \sqrt{p} & \text{if } p \equiv 1 \pmod 4, \\ i\sqrt{p} & \text{if } p \equiv 3 \pmod 4, \end{cases} \tag{1.1}$$

where λ is the Legendre symbol at p:

$$\lambda(j) = \left(\frac{j}{p}\right).$$

A crude restatement of (1.1), and one that is much easier to prove, is that the left-hand side is a square root of $\lambda(-1)p$. But by summarizing the result in this way we lose the information that the square root at issue is the one with *positive* real or imaginary part. In other words, the delicate point in (1.1) is precisely the determination of the sign in front of the square root – the "root number."

More generally, suppose that χ is any primitive Dirichlet character, say with conductor N. The **Gauss sum** attached to χ is the quantity

$$\tau(\chi) = \sum_{j=0}^{N-1} \chi(j) e^{2\pi i j/N} \tag{1.2}$$

and the associated **root number** is given by

$$W(\chi) = \frac{\tau(\chi)}{i^m \sqrt{N}}, \tag{1.3}$$

where

$$m = \begin{cases} 0 & \text{if } \chi(-1) = 1, \\ 1 & \text{if } \chi(-1) = -1. \end{cases} \tag{1.4}$$

If χ is quadratic then $\tau(\chi)$ is once again equal to $\sqrt{N}$ or $i\sqrt{N}$ according as χ is even or odd. Equivalently, we can formulate the preceding statement as an assertion about root numbers:

Theorem 1.1. *If χ is a primitive quadratic Dirichlet character then $W(\chi) = 1$.*

In spite of our disclaimer in the introduction, we will actually give a proof of Theorem 1.1 at the end of the lecture. But to begin with let us examine the implications of the theorem for Dirichlet L-functions.

1. Nonexistence of trivial central zeros for Dirichlet L-functions

A **trivial zero** of an L-function is a zero which is immediately apparent from the functional equation. Any L-function worthy of the name has infinitely many trivial zeros, as one sees by playing off the holomorphy of the L-function in some right half-plane against the poles of $\Gamma(s)$ at nonpositive integers. For example, consider the functional equation of the Riemann zeta function:

$$\pi^{-s/2}\Gamma(s/2)\zeta(s) = \pi^{-(1-s)/2}\Gamma((1-s)/2)\zeta(1-s). \tag{1.5}$$

At $s = -2, -4, -6, \ldots$ the factor $\Gamma(s/2)$ on the left-hand side has a simple pole, whereas the right-hand side is holomorphic and nonvanishing. It follows that $\zeta(s)$ has a zero (in fact a simple zero) at the negative even integers, whence these points are trivial zeros of $\zeta(s)$.

On the other hand, $\zeta(s)$ does not have a **trivial central zero**. The latter term refers to a trivial zero of an L-function at $s = k/2$, where the functional equation of the L-function in question is a transformation law relative to $s \mapsto k-s$. In the case of $\zeta(s)$ we have $k = 1$; indeed if we write $Z(s)$ for the left-hand side of (1.5), then (1.5) becomes $Z(s) = Z(1-s)$. Thus a trivial central zero of $\zeta(s)$ would be a zero at $s = 1/2$ inherent in the equation $Z(s) = Z(1-s)$; but the latter equation says merely that the function $f(s) = Z(s+1/2)$ is even, and even functions, unlike odd functions, need not vanish at $s = 0$. Thus there is no trivial reason why $Z(s)$ must vanish at $s = 1/2$ and hence none why $\zeta(s)$ itself must vanish there. The expansion

$$\zeta(s) = (1-2^{1-s})^{-1}(1 - 2^{-s} + 3^{-s} - 4^{-s} + \ldots) \qquad (\Re(s) > 0)$$

shows that in fact $\zeta(1/2) \neq 0$.

More generally, no Dirichlet L-function has a trivial central zero. To verify this statement, consider a primitive Dirichlet character χ of conductor N. The functional equation of $L(s,\chi)$ is

$$\Lambda(s,\chi) = W(\chi)\Lambda(1-s,\overline{\chi}) \tag{1.6}$$

with $\Lambda(s,\chi) = N^{s/2}\,\Gamma_{\mathbb{R}}(s+m)L(s,\chi)$. Here $W(\chi)$ and m are as in (1.3) and (1.4) respectively, and

$$\Gamma_{\mathbb{R}}(s) = \pi^{-s/2}\Gamma(s/2). \tag{1.7}$$

We mention in passing that in addition to this "real gamma factor" there is also a "complex gamma factor"

$$\Gamma_{\mathbb{C}}(s) = 2\cdot(2\pi)^{-s}\Gamma(s), \tag{1.8}$$

and with this notation the duplication formula takes the attractive form

$$\Gamma_{\mathbb{R}}(s)\Gamma_{\mathbb{R}}(s+1) = \Gamma_{\mathbb{C}}(s). \tag{1.9}$$

Returning to the matter at hand, we consider three cases, namely (i) χ has order $\geqslant 3$, (ii) $\chi = 1$, and (iii) χ has order 2. In case (i), $L(s,\chi) \neq L(s,\overline{\chi})$, whence $\Lambda(s,\chi) \neq \Lambda(s,\overline{\chi})$ and (1.6) has no direct bearing on the possible vanishing of $L(s,\chi)$ at $s = 1/2$. In case (ii), $L(s,\chi) = \zeta(s)$, and we have already seen that $\zeta(s)$ does not have a trivial central zero. Finally, suppose that χ is quadratic. Then (1.6) becomes $\Lambda(s,\chi) = W(\chi)\Lambda(1-s,\chi)$, and since $W(\chi)$ is 1 rather than -1 the function $f(s) = \Lambda(s+1/2,\chi)$ is even rather than odd. Hence in case (iii) there is again no trivial central zero.

Of course it is one thing to say that there is no trivial reason for $L(s,\chi)$ to vanish at $s = 1/2$ and quite another to prove that $L(1/2,\chi)$ is not in fact zero. The latter problem is the subject of an extensive literature (see for example [**6**], [**20**], [**46**], [**47**], [**69**], [**72**], and [**90**]), and while the state of the art does not yet permit us to assert that $L(1/2, \chi) \neq 0$ for every Dirichlet character χ, that is certainly the conjecture to which the evidence points.

2. Hecke characters and Hecke L-functions

While they do not occur for Dirichlet L-functions, trivial central zeros of L-functions do exist. The first examples were found in 1966 by Birch and Stephens [**9**] and arose in connection with elliptic curves over $\mathbb{Q}$ with complex multiplication by an imaginary quadratic field. The L-function of such an elliptic curve is a Hecke L-function, and Hecke's functional equation allows one to exhibit cases in which an analogue of (1.6) holds but with $L(s, \chi) = L(s, \overline{\chi})$ and $W(\chi) = -1$, so that the L-function vanishes at the center of its critical strip. This phenomenon will be illustrated here not using elliptic curves with complex multiplication by $\mathbb{Q}(i)$ as in [**9**] but rather with the "$\mathbb{Q}$-curves" of Gross [**38**], for which the field of complex multiplication varies. But first of all we provide some background on Hecke characters and Hecke L-functions. Throughout, K denotes a number field and $\mathcal{O}$ its ring of integers. We also write I for the multiplicative group of nonzero fractional ideals of K and P for the subgroup of principal fractional ideals. As usual, a "prime ideal of K" is a nonzero prime ideal of $\mathcal{O}$, and an "integral ideal of K" is any nonzero ideal of $\mathcal{O}$.

2.1. Hecke characters

Given an integral ideal $\mathfrak{f}$ of K, we say that a fractional ideal $\mathfrak{a} \in I$ is relatively prime to $\mathfrak{f}$ if no prime ideal dividing $\mathfrak{f}$ occurs in the factorization of $\mathfrak{a}$ as a product of prime ideals to nonzero integral powers. The multiplicative group consisting of such $\mathfrak{a}$ will be denoted $I(\mathfrak{f})$, and the subgroup $P \cap I(\mathfrak{f})$ of principal fractional ideals in $I(\mathfrak{f})$ will be denoted $P(\mathfrak{f})$. Note by the way that if $\mathfrak{f} = \mathcal{O}$ then $I(\mathfrak{f}) = I$ and $P(\mathfrak{f}) = P$. We say that an element $\alpha \in K^\times$ is relatively prime to $\mathfrak{f}$ if $\alpha\mathcal{O} \in P(\mathfrak{f})$, and we write $K(\mathfrak{f})$ for the subgroup of $K^\times$ consisting of all such α. Thus $K(\mathfrak{f})$ is $(S^{-1}\mathcal{O})^\times$, the group of units of the localization of $\mathcal{O}$ at the multiplicative set

$$S = \bigcap_{\mathfrak{p}|\mathfrak{f}} (\mathcal{O} \smallsetminus \mathfrak{p})$$

(with $S = \mathcal{O} \smallsetminus \{0\}$ if $\mathfrak{f} = \mathcal{O}$). Given $\alpha \in K^\times$, we write $\alpha \equiv 1 \bmod^* \mathfrak{f}$ to mean that for every prime ideal $\mathfrak{p}$ dividing $\mathfrak{f}$ we have $v_\mathfrak{p}(\alpha - 1) \geqslant \mathrm{ord}_\mathfrak{p}\mathfrak{f}$, where $v_\mathfrak{p}$ is the valuation associated to $\mathfrak{p}$ and $\mathrm{ord}_\mathfrak{p}\mathfrak{f}$ the multiplicity of $\mathfrak{p}$ in $\mathfrak{f}$. More succinctly, $\alpha \equiv 1 \bmod^* \mathfrak{f}$ means $\alpha = 1 \bmod \mathfrak{f}(S^{-1}\mathcal{O})$. The set of all such α is a subgroup $K_\mathfrak{f}$ of $K(\mathfrak{f})$, and we write $P_\mathfrak{f}$ for the subgroup of $P(\mathfrak{f})$ consisting of all $\alpha\mathcal{O} \in P(\mathfrak{f})$ with $\alpha \in K_\mathfrak{f}$.

A **Hecke character of** K **to the modulus** $\mathfrak{f}$ is a group homomorphism $\chi : I(\mathfrak{f}) \to \mathbb{C}^\times$ with the following property: There exists a continuous homomorphism $\chi_\infty : (\mathbb{R} \otimes_\mathbb{Q} K)^\times \to \mathbb{C}^\times$ such that if $\alpha \in K_\mathfrak{f}$ then $\chi(\alpha\mathcal{O}) = \chi_\infty^{-1}(\alpha)$. Here we are identifying $\alpha \in K^\times$ with $1 \otimes \alpha \in (\mathbb{R} \otimes_\mathbb{Q} K)^\times$, as is often convenient. Without this identification the condition on χ reads

$$\chi(\alpha\mathcal{O}) = \chi_\infty^{-1}(1 \otimes \alpha) \qquad (\alpha \in K_\mathfrak{f}). \tag{1.10}$$

We call χ_∞ the **infinity type** of χ. Thus a Hecke character to the modulus $\mathfrak{f}$ is a character of $I(\mathfrak{f})$ which is completely determined on $P_\mathfrak{f}$ by its infinity type.

Two points should be noted. First of all, χ_∞ is a continuous homomorphism if and only its reciprocal is, so the content of the definition would not change if we omitted the exponent -1 on the right-hand side of (1.10). Nonetheless we retain it for the sake of the correspondence between Hecke characters and idele class characters to be discussed later. The second point is that the continuity of χ_∞ is an unambiguous concept, because all norms on the finite-dimensional real vector space

$$\mathbb{R} \otimes_{\mathbb{Q}} K \cong \mathbb{R}^{r_1} \times \mathbb{C}^{r_2} \tag{1.11}$$

are equivalent. Here r_1 and r_2 have the usual meanings; in fact we may specify the isomorphism in (1.11) – call it Ψ – by requiring that for $\alpha \in K$ we have

$$\Psi(1 \otimes \alpha) = (\sigma_1(\alpha), \sigma_2(\alpha), \ldots, \sigma_{r_1+r_2}(\alpha)), \tag{1.12}$$

where $\sigma_1, \ldots, \sigma_{r_1}$ are the distinct real embeddings of K and $\sigma_{r_1+1}, \ldots, \sigma_{r_1+r_2}$ are chosen from the distinct pairs of conjugate complex embeddings. Using (1.11) and (1.12), we may view χ_∞ as a continuous homomorphism $(\mathbb{R}^\times)^{r_1} \times (\mathbb{C}^\times)^{r_2} \to \mathbb{C}^\times$. Since $\Psi(K_\mathfrak{f})$ is dense in $(\mathbb{R}^\times)^{r_1} \times (\mathbb{C}^\times)^{r_2}$ we see that (1.10) determines χ_∞ uniquely.

As with Dirichlet characters, there is a notion of primitivity: A Hecke character χ to the modulus $\mathfrak{f}$ is **primitive** if there does not exist an integral ideal $\mathfrak{f}'$ properly dividing $\mathfrak{f}$ such that χ extends to a Hecke character to the modulus $\mathfrak{f}'$. Note that $I(\mathfrak{f}) \subset I(\mathfrak{f}')$, so that the definition is meaningful. Given a Hecke character χ to the modulus $\mathfrak{f}$, there exists a unique pair $(\chi', \mathfrak{f}')$ such that $\mathfrak{f}'$ is an integral ideal dividing $\mathfrak{f}$ and χ' is a primitive Hecke character to the modulus $\mathfrak{f}'$ extending χ. We call $\mathfrak{f}'$ the **conductor** of χ and χ' the **primitive Hecke character determined by** χ. Thus a Hecke character is primitive if and only its modulus equals its conductor. If χ is primitive, as we shall usually assume, then its conductor will be denoted $\mathfrak{f}(\chi)$.

2.2. Examples

The simplest examples are primitive Hecke characters χ with $\mathfrak{f}(\chi) = \mathcal{O}$, for then the requirement in (1.10) is simply that $\chi(\alpha\mathcal{O}) = \chi_\infty^{-1}(1 \otimes \alpha)$ for all $\alpha \in K^\times$. Consider for instance the **power-of-the-norm** map $\chi : I \to \mathbb{C}^\times$ given by $\chi(\mathfrak{a}) = (\mathbf{N}\mathfrak{a})^{s_0}$, where $s_0 \in \mathbb{C}$ is fixed and $\mathbf{N}\mathfrak{a}$ is the absolute norm of $\mathfrak{a}$. Viewing χ_∞ as a character $(\mathbb{R}^\times)^{r_1} \otimes (\mathbb{C}^\times)^{r_2} \to \mathbb{C}^\times$, we see that (1.10) holds with

$$\chi_\infty(u_1, u_2, \ldots, u_{r_1+r_2}) = |u_1 u_2 \cdots u_{r_1}|^{-s_0} \cdot |u_{r_1+1} u_{r_1+2} \cdots u_{r_1+r_2}|^{-2s_0}. \tag{1.13}$$

Note that apart from the trivial Hecke character (i. e. the case $s_0 = 0$), the power-of-the-norm map has infinite order.

Another example with $\mathfrak{f}(\chi) = \mathcal{O}$, this time of finite order, is an **ideal class character**, in other words a character χ of the ideal class group I/P of K: if we view χ as a character of I trivial on P then (1.10) holds with χ_∞ equal to the trivial character. Now if $\mathfrak{f}$ is a nonzero integral ideal of K then the natural map $I(\mathfrak{f})/P(\mathfrak{f}) \to I/P$ is an isomorphism, and therefore an ideal class character becomes a Hecke character to the modulus $\mathfrak{f}$ by restriction to $I(\mathfrak{f})$. In particular, if χ is any Hecke character to the modulus $\mathfrak{f}$ then so is $\chi\varphi$, where φ is an ideal class character of K. Note that $(\chi\varphi)_\infty = \chi_\infty$ and that $\chi\varphi$ is primitive if and only if χ is. The upshot is that whenever we have an example of a primitive Hecke character of a given infinity type then we automatically have h such examples, where h is the class number of K.

2.3. A nonexample

It may also be instructive to see a character of I which is not a Hecke character. Since I is the free abelian group on the nonzero prime ideals of $\mathcal{O}$, we can define a homomorphism $I \to \mathbb{C}^\times$ simply by specifying its values on prime ideals. Thus we get a character $\chi : I \to \{\pm 1\}$ (the "Liouville function" of K) by specifying that $\chi(\mathfrak{p}) = -1$ for every prime ideal $\mathfrak{p}$. Equivalently,

$$\chi(\mathfrak{a}) = (-1)^{\Omega(\mathfrak{a})},$$

where $\Omega(\mathfrak{a})$ is the total number of prime ideals (taking account of multiplicities) occurring in a factorization of $\mathfrak{a}$ into prime ideals. But the set of $\alpha \in K^\times$ such that $\Omega(\alpha\mathcal{O})$ is even and the set of α such that $\Omega(\alpha\mathcal{O})$ is odd are both dense in $(\mathbb{R}\otimes_{\mathbb{Q}} K)^\times$. Hence there does not exist a continuous homomorphism $\chi_\infty : (\mathbb{R} \otimes_{\mathbb{Q}} K)^\times \to \mathbb{C}^\times$ such that $\chi(\alpha\mathcal{O}) = \chi_\infty^{-1}(1 \otimes \alpha)$ for $\alpha \in K^\times$, and consequently χ is not a Hecke character.

2.4. Unitary Hecke characters

The L-function associated to a Hecke character is defined by a Dirichlet series, and in preparation for writing down this Dirichlet series explicitly we prove a result which will assure us that the series does converge in some right half-plane. If χ is a one-dimensional character of a group then the associated unitary character $\chi/|\chi|$ will be denoted χ_{unit}, so that $\chi = \chi_{\text{unit}} \cdot |\chi|$.

Proposition 1.1. *If χ is a Hecke character of K then there exists $c \in \mathbb{R}$ such that $|\chi| = \mathbf{N}^c$, whence*

$$\chi = \chi_{\text{unit}} \cdot \mathbf{N}^c.$$

In particular, every Hecke character is a unitary Hecke character times a real power of the norm.

Proof. Let $\mathbb{R}_{>0}$ denote the multiplicative group of positive real numbers. The point requiring proof is that a Hecke character with values in $\mathbb{R}_{>0}$ coincides with a real power of the norm. So after changing notation we may suppose that χ is a Hecke character $I(\mathfrak{f}) \to \mathbb{R}_{>0}$, and we must show that $\chi = \mathbf{N}^c$ for some $c \in \mathbb{R}$. It will suffice to see that χ_∞ has the form (1.13) with $s_0 \in \mathbb{R}$, for then we may take $c = s_0$. Indeed if c is so chosen then $\chi \cdot \mathbf{N}^{-c}$ is a character with values in $\mathbb{R}_{>0}$ which factors through the finite group $I(\mathfrak{f})/P_{\mathfrak{f}}$, and consequently $\chi \cdot \mathbf{N}^{-c}$ is trivial.

Put $\mathbf{T} = \{e^{i\theta} : \theta \in \mathbb{R}\}$, so that we have topological group isomorphisms $\mathbb{R}^\times \cong \{\pm 1\} \times \mathbb{R}_{>0}$ and $\mathbb{C}^\times \cong \mathbf{T} \times \mathbb{R}_{>0}$. Any continuous homomorphism $\mathbb{R}_{>0} \to \mathbb{R}_{>0}$ raises the elements of $\mathbb{R}_{>0}$ to some fixed real exponent, which we can of course choose to write as twice some other real exponent. Thus χ_∞ has the form

$$\chi_\infty(u_1, u_2, \ldots, u_{r_1+r_2}) = \prod_{j=1}^{r_1} |u_j|^{c_j} \cdot \prod_{j=r_1+1}^{r_1+r_2} |u_j|^{2c_j} \tag{1.14}$$

with $c_j \in \mathbb{R}$ for $1 \leqslant j \leqslant r_1 + r_2$. Now if $\varepsilon \in \mathcal{O}^\times \cap K_{\mathfrak{f}}$ then

$$\chi_\infty^{-1}(1 \otimes \varepsilon) = \chi(\varepsilon\mathcal{O}) = \chi(\mathcal{O}) = 1.$$

Hence using the notation of (1.12) we deduce from (1.14) that

$$\prod_{j=1}^{r_1} |\sigma_j(\varepsilon)|^{c_j} \cdot \prod_{j=r_1+1}^{r_1+r_2} |\sigma_j(\varepsilon)|^{2c_j} = 1. \tag{1.15}$$

But $\mathcal{O}^\times \cap K_{\mathfrak{f}}$ has finite index in $\mathcal{O}^\times$. Hence on taking the log of both sides of (1.15) and applying the Dirichlet unit theorem, we deduce that the linear form on $\mathbb{R}^{r_1+r_2}$

given by

$$(t_1, t_2, \ldots, t_{r_1+r_2}) \mapsto \sum_{j=1}^{r_1} c_j t_j + \sum_{j=r_1+1}^{r_1+r_2} 2c_j t_j$$

vanishes identically on the hyperplane $\sum_{j=1}^{r_1} t_j + \sum_{j=r_1+1}^{r_1+r_2} 2t_j = 0$. It follows that c_j is a constant c independent of j, whence (1.14) is indeed of the form (1.13) with $s_0 = c$, a real number. □

2.5. Hecke L-functions

If χ is a Dirichlet character to the modulus N then one puts $\chi(n) = 0$ whenever $\gcd(n, N) > 1$, and by virtue of this convention the Dirichlet series for $L(s, \chi)$ can be written either as a sum over integers prime to N or as a sum over all positive integers. In the same way, given a Hecke character χ to the modulus $\mathfrak{f}$, one sets $\chi(\mathfrak{a}) = 0$ whenever $\mathfrak{a}$ is not relatively prime to $\mathfrak{f}$, and one defines the associated L-series $L(s, \chi)$ by

$$L(s, \chi) = \sum_{\mathfrak{a}} \chi(\mathfrak{a})(\mathbf{N}\mathfrak{a})^{-s}, \tag{1.16}$$

where $\mathfrak{a}$ runs over all nonzero integral ideals of K or alternatively over the subset of ideals relatively prime to $\mathfrak{f}$. For example, if χ is the trivial Hecke character to the modulus $\mathcal{O}$ then $L(s, \chi)$ is the Dedekind zeta function

$$\zeta_K(s) = \sum_{\mathfrak{a}} (\mathbf{N}\mathfrak{a})^{-s} \tag{1.17}$$

of K, while if χ is more generally the power-of-the-norm character $\mathfrak{a} \mapsto (\mathbf{N}\mathfrak{a})^{s_0}$ then $L(s, \chi) = \zeta_K(s - s_0)$. For any χ, the definition (1.16) is meaningful in the sense that the given Dirichlet series converges in some right half-plane. Indeed by writing χ as in Proposition 1.1 we see that the Dirichlet series is absolutely convergent for $\Re(s) > c + 1$.

The basic analytic fact about $L(s, \chi)$, proved by Hecke, is that $L(s, \chi)$ extends to a meromorphic function on $\mathbb{C}$ which is either entire (if χ is not of the form $\mathbf{N}^{s_0}$) or holomorphic except for a simple pole at $s = s_0 + 1$ (if $\chi = \mathbf{N}^{s_0}$) and which satisfies a functional equation relative to $s \mapsto 2c + 1 - s$. We will say more about the functional equation in Lecture 2, but returning for now to the right half-plane of absolute convergence, we observe that $L(s, \chi)$ has an Euler product:

$$L(s, \chi) = \prod_{\mathfrak{p}} (1 - \chi(\mathfrak{p})\mathbf{N}(\mathfrak{p})^{-s})^{-1}, \tag{1.18}$$

where $\mathfrak{p}$ runs over the prime ideals of K or over the subset of prime ideals not dividing $\mathfrak{f}$. The fact that the Dirichlet series in (1.16) is equal to the Euler product in (1.18) is proved in much the same way as the corresponding equality for Dirichlet L-functions. In the latter case, the key fact needed is the unique factorization of positive integers into primes; in the case of Hecke L-functions one uses instead the fact that every nonzero ideal of $\mathcal{O}$ has a unique factorization into prime ideals.

2.6. Dirichlet characters as Hecke characters

The analogy between Dirichlet L-functions and Hecke L-functions is no coincidence, for in the case $K = \mathbb{Q}$ there is a bijection $\chi \mapsto \chi_{\text{Hec}}$ from the set of Dirichlet characters to the set of Hecke characters of $\mathbb{Q}$ of finite order. The map $\chi \mapsto \chi_{\text{Hec}}$ is defined as follows: Given a Dirichlet character χ to the modulus N, take $\mathfrak{f} = N\mathbb{Z}$ and set

$$\chi_{\mathrm{Hec}}(\mathfrak{a}) = \chi(a) \tag{1.19}$$

for $\mathfrak{a} \in I(\mathfrak{f})$, where a is the unique positive generator of $\mathfrak{a}$. Contemplating (1.19), we recognize that the subscript on χ_{Hec} is superfluous, because the left-hand side of (1.19) is a function of ideals whereas the right-hand side is a function of numbers. Hence without risk of confusion we can write (1.19) in the form $\chi(\mathfrak{a}) = \chi(a)$. Furthermore, on making the identification $(\mathbb{R}\otimes_{\mathbb{Q}} K)^\times = \mathbb{R}^\times$ one readily verifies that (1.10) holds with χ_∞ equal to the trivial character or the sign character $x \mapsto x/|x|$ according as χ is even or odd as a Dirichlet character. Thus χ_{Hec} is indeed a Hecke character. One can also check that $L(s, \chi) = L(s, \chi_{\mathrm{Hec}})$ and that χ is primitive if and only if χ_{Hec} is. Henceforth we drop the subscript on χ_{Hec}.

2.7. Hecke characters on principal ideals

While the defining property (1.10) of a Hecke character χ refers only to $\chi|P_{\mathfrak{f}}$, the following proposition shows that one can also deduce something about $\chi|P(\mathfrak{f})$. For an integer $n \geqslant 1$ let $\boldsymbol{\mu}_n \subset \mathbb{C}^\times$ be the subgroup of nth roots of unity.

Proposition 1.2. *Let $\chi : I(\mathfrak{f}) \to \mathbb{C}^\times$ and $\chi_\infty : (\mathbb{R}\otimes_{\mathbb{Q}} K)^\times \to \mathbb{C}^\times$ be respectively a homomorphism and a continuous homomorphism. Then χ is a Hecke character with infinity type χ_∞ if and only if there is an integer $n \geqslant 1$ and a homomorphism $\varepsilon : (\mathcal{O}/\mathfrak{f})^\times \to \boldsymbol{\mu}_n$ such that*

$$\chi(\alpha\mathcal{O}) = \varepsilon(\alpha)\chi_\infty^{-1}(1\otimes\alpha)$$

for $\alpha \in K(\mathfrak{f})$. Here ε is viewed as a character of $K(\mathfrak{f})$ by composition with

$$K(\mathfrak{f}) \longrightarrow K(\mathfrak{f})/K_{\mathfrak{f}} \longrightarrow (\mathcal{O}/\mathfrak{f})^\times,$$

the first arrow being the quotient map and the second the natural isomorphism.

PROOF. Sufficiency is immediate, because ε is trivial on $K_{\mathfrak{f}}$. To prove necessity let n be the order of $K(\mathfrak{f})/K_{\mathfrak{f}}$. If χ is a Hecke character with infinity type χ_∞ and $\alpha \in K(\mathfrak{f})$ then $\alpha^n \in K_{\mathfrak{f}}$, whence $\chi(\alpha^n\mathcal{O}) = \chi_\infty^{-1}(1\otimes\alpha^n)$ or in other words $\chi((\alpha\mathcal{O})^n) = \chi_\infty^{-1}((1\otimes\alpha)^n)$. As both χ and χ_∞ are homomorphisms it follows that $\chi(\alpha\mathcal{O}) = \varepsilon(\alpha)\chi_\infty^{-1}(1\otimes\alpha)$ with an nth root of unity $\varepsilon(\alpha)$. At the same time we see that $\varepsilon : K(\mathfrak{f}) \to \mathbb{C}^\times$ is a homomorphism trivial on $K_{\mathfrak{f}}$ and so may be viewed as a character of $K(\mathfrak{f})/K_{\mathfrak{f}} \cong (\mathcal{O}/\mathfrak{f})^\times$. □

Proposition 1.2 completes our discussion of Hecke characters in general. Next we specialize to the case of imaginary quadratic fields.

3. A family of Hecke L-functions with trivial central zeros

Let K be an imaginary quadratic field. After fixing an embedding of K in $\mathbb{C}$ we can identify $\mathbb{R}\otimes_{\mathbb{Q}} K$ with $\mathbb{C}$ and hence $(\mathbb{R}\otimes_{\mathbb{Q}} K)^\times$ with $\mathbb{C}^\times$. Thus if χ is a Hecke character of K then χ_∞ is a continuous homomorphism from $\mathbb{C}^\times$ to itself, and we can ask whether χ_∞ is the character $z \mapsto z^{-1}$. If it is then we say that χ is of type $(1,0)$. To demystify this terminology we add that if $\chi_\infty(z) = z^{-p}(\overline{z})^{-q}$ with $p, q \in \mathbb{Z}$ then χ is said to be of type (p,q).

As we have already indicated, our goal is to exhibit the trivial central zeros found by Gross [**38**] in his study of $\mathbb{Q}$-curves. As a first step we exhibit the relevant Hecke characters of K of type $(1,0)$. Let D be the absolute value of the discriminant of K, so that $K = \mathbb{Q}(\sqrt{-D})$. The set of Hecke characters at issue will be denoted $X(D)$. To define $X(D)$ precisely, let κ be the primitive quadratic Dirichlet character of conductor D given by

$$\kappa(n) = \left(\frac{-D}{n}\right), \tag{1.20}$$

where the Kronecker symbol on the right is understood to equal -1 when $n = -1$ (in other words, the Kronecker symbol is viewed as a Dirichlet character rather than as a norm residue symbol). Then $X(D)$ consists of all primitive Hecke character χ of K of type (1,0) satisfying three conditions:

(a) $\mathfrak{f}(\chi)|D^\infty$. (In other words, $\mathfrak{f}(\chi)$ divides some power of D.)
(b) $\chi(n\mathcal{O}) = \kappa(n)n$ for $n \in \mathbb{Z}$ prime to D.
(c) The values of χ on $P(\mathfrak{f}(\chi))$ lie in K.

Let Φ be the set of ideal class characters of K. If $\chi \in X(D)$ then $\chi\varphi \in X(D)$ for every $\varphi \in \Phi$, so the cardinality of $X(D)$ is a multiple of $h(D)$, the class number of K. Henceforth we assume that $D \neq 3, 4$.

Proposition 1.3.

$$|X(D)| = \begin{cases} h(D) & \textit{if } D \textit{ is odd,} \\ 0 & \textit{if } 4||D, \\ 2h(D) & \textit{if } 8|D. \end{cases}$$

PROOF. Consider characters of the form $\varepsilon : (\mathcal{O}/\mathfrak{f})^\times \to \{\pm 1\}$ with $\mathfrak{f}|D^\infty$. We impose two conditions: First, $\varepsilon(n) = \kappa(n)$ for $n \in \mathbb{Z}$ relatively prime to D, and second, ε is primitive in the usual sense that ε does not factor through $(\mathcal{O}/\mathfrak{f}')^\times$ for any ideal $\mathfrak{f}'$ properly dividing $\mathfrak{f}$. The set of such ε will be denoted E. We claim that the proposition is equivalent to the assertion

$$|E| = \begin{cases} 1 & \text{if } D \text{ is odd,} \\ 0 & \text{if } 4||D, \\ 2 & \text{if } 8|D. \end{cases} \tag{1.21}$$

In other words, we claim that $|X(D)| = |E|h(D)$.

To verify the claim, we use Proposition 1.2: The restriction to $P(\mathfrak{f}(\chi))$ of any $\chi \in X(D)$ has the form $\chi(\alpha\mathcal{O}) = \varepsilon(\alpha)\alpha$ for some character ε of $(\mathcal{O}/\mathfrak{f}(\chi))^\times$ with values in the nth roots of unity. As the values of χ on $P(\mathfrak{f}(\chi))$ lie in K and $K \neq \mathbb{Q}(\sqrt{-3}), \mathbb{Q}(\sqrt{-4})$ it follows that n can be taken to be 2. Thus we may view ε as a character $(\mathcal{O}/\mathfrak{f}(\chi))^\times \to \{\pm 1\}$, necessarily primitive since χ is primitive. Since $\chi(n\mathcal{O}) = \kappa(n)n$ for $n \in \mathbb{Z}$ prime to D we deduce that $\varepsilon \in E$, and thus we obtain a map $X(D) \to E$. The fibers of the map have cardinality $h(D)$, because there are $h(D)$ ways to extend a character of $P(\mathfrak{f}(\chi))$ to a character of $I(\mathfrak{f}(\chi))$. To see that the map $\chi \mapsto \varepsilon$ is surjective, let $\varepsilon : (\mathcal{O}/\mathfrak{f})^\times \to \{\pm 1\}$ be a a given element of E. We would like to define a character χ of $P(\mathfrak{f})$ by setting

$$\chi(\alpha\mathcal{O}) = \varepsilon(\alpha)\alpha \tag{1.22}$$

for $\alpha \in K(\mathfrak{f})$, but we must check that the right-hand side of (1.22) depends only on the principal ideal $\alpha\mathcal{O}$ and not on the choice of generator α. Since $K \neq \mathbb{Q}(\sqrt{-3}), \mathbb{Q}(\sqrt{-4})$, the only other generator is $-\alpha$; but $\varepsilon(-\alpha) = \kappa(-1)\varepsilon(\alpha) = -\varepsilon(\alpha)$

by (1.20). Hence if α is replaced by $-\alpha$ then the right-hand side of (1.22) is unchanged, so we obtain a well-defined character χ of $P(\mathfrak{f})$. Extending it arbitrarily to $I(\mathfrak{f})$ we obtain an element of $X(D)$.

We must now prove (1.21). Given $\varepsilon \in E$, write $\mathfrak{f}(\varepsilon)$ for the ideal $\mathfrak{f}$ such that ε is a primitive character of $(\mathcal{O}/\mathfrak{f})^\times$. Also, if $D||4$ then let $\mathfrak{T}$ denote the prime ideal of $\mathcal{O}$ lying over 2. We claim that if $\varepsilon \in E$ then

$$(1.23) \qquad \mathfrak{f}(\varepsilon) \text{ is divisible by } \begin{cases} \sqrt{-D}\mathcal{O} & \text{if } D \text{ is odd,} \\ \sqrt{-D}\mathfrak{T} & \text{if } 4||D, \\ 2\sqrt{-D}\mathcal{O} & \text{if } 8|D. \end{cases}$$

To verify (1.23) use the fact $\varepsilon(n) = \kappa(n)$ for $n \in \mathbb{Z}$ prime to D. It follows that D divides $\mathbb{Z} \cap \mathfrak{f}(\varepsilon)$, but one readily checks that an ideal $\mathfrak{a}$ of $\mathcal{O}$ with the property that D divides $\mathbb{Z} \cap \mathfrak{a}$ is divisible by the right-hand side of (1.23).

At this point we consider the three cases in (1.21) one by one. Suppose first that D is odd. If $\varepsilon \in E$ then $\sqrt{-D}\mathcal{O}|\mathfrak{f}(\varepsilon)$ by (1.23), but also $\mathfrak{f}(\varepsilon)|D^\infty$ by assumption. Thus if $\mathfrak{f}(\varepsilon)$ is *properly* divisible by $\sqrt{-D}\mathcal{O}$ then $(\mathcal{O}/\mathfrak{f}(\varepsilon))^\times$ is a nontrivial extension of $(\mathcal{O}/\sqrt{-D}\mathcal{O})^\times$ by a group of odd order, contradicting the fact that ε is both quadratic and primitive. It follows that $\mathfrak{f}(\varepsilon) = \sqrt{-D}\mathcal{O}$. But the natural map $(\mathbb{Z}/D\mathbb{Z})^\times \to (\mathcal{O}/\sqrt{-D}\mathcal{O})^\times$ is an isomorphism, and $\varepsilon(n) = \kappa(n)$ for $n \in (\mathbb{Z}/D\mathbb{Z})^\times$. Hence there is a unique choice for ε, and $|E| = 1$. At the same time we see that

$$(1.24) \qquad \mathfrak{f}(\chi) = \sqrt{-D}\mathcal{O}$$

for $\chi \in X(D)$, because it follows from (1.22) that $\mathfrak{f}(\chi) = \mathfrak{f}(\varepsilon)$.

Next suppose that $4||D$. Then $D = 4C$ with $C \equiv 1 \bmod 4$. If there exists an $\varepsilon \in E$, then $\sqrt{-C}\mathfrak{T}^3|\mathfrak{f}(\varepsilon)$ by (1.23); we claim that in fact

$$(1.25) \qquad \mathfrak{f}(\varepsilon) = \sqrt{-C}\mathfrak{T}^k \qquad \text{with } k = 3 \text{ or } 4.$$

To see this, we first argue as in the case D odd: Since ε is quadratic and primitive, the kernel of the reduction map $(\mathcal{O}/\mathfrak{f}(\varepsilon))^\times \to (\mathcal{O}/\sqrt{-C}\mathfrak{T}^3)^\times$ has 2-power order. As $\mathfrak{f}(\varepsilon)|D^\infty$ this already implies that $\mathfrak{f}(\varepsilon) = \sqrt{-C}\mathfrak{T}^k$ with $k \geqslant 3$. But one checks by induction that if $k \geqslant 5$ then every element of the kernel of $(\mathcal{O}/\mathfrak{T}^k)^\times \to (\mathcal{O}/\mathfrak{T}^5)^\times$ is a square in $(\mathcal{O}/\mathfrak{T}^k)^\times$, so again, the fact that ε is quadratic and primitive ensures that $k = 3, 4$, or 5. Now choose a rational integer n such that $n \equiv 5 \bmod 8$ and $n \equiv 1 \bmod C$. Then n represents the nontrivial element of the kernel of $(\mathcal{O}/\mathfrak{T}^5)^\times \to (\mathcal{O}/\mathfrak{T}^4)^\times$; but $\varepsilon(n) = \kappa(n) = 1$. Since ε is primitive, (1.25) follows.

To obtain a contradiction from (1.25), write

$$(\mathcal{O}/\sqrt{-C}\mathfrak{T}^k)^\times \cong (\mathcal{O}/\mathfrak{T}^k)^\times \times (\mathcal{O}/\sqrt{-C}\mathcal{O})^\times$$

and $\varepsilon = \varepsilon'\varepsilon''$ with quadratic characters ε' and ε'' of $(\mathcal{O}/\mathfrak{T}^k)^\times$ and $(\mathcal{O}/\sqrt{-C}\mathcal{O})^\times$ respectively. Then $\varepsilon(-1) = \kappa(-1) = -1$, but $\varepsilon''(-1) = 1$ because $C \equiv 1 \bmod 4$, so $\varepsilon'(-1) = -1$. This is a contradiction, because -1 is a square in $(\mathcal{O}/\mathfrak{T}^k)^\times$ for $k = 4$ and *a fortiori* for $k = 3$: indeed $(2 + \sqrt{-C})^2 \equiv -1 \bmod 4\mathcal{O}$.

Finally, suppose that $8|D$. Write $\mathfrak{T}$ for the prime ideal of $\mathcal{O}$ lying over 2. As in the case $4||D$, if $k \geqslant 5$, then every element of the kernel of $(\mathcal{O}/\mathfrak{T}^k)^\times \to (\mathcal{O}/\mathfrak{T}^5)^\times$ is a square in $(\mathcal{O}/\mathfrak{T}^k)^\times$. Appealing to (1.23) and arguing as before, we deduce that $\mathfrak{f}(\varepsilon) = 2\sqrt{-D}\mathcal{O}$. Now write $D = 8C$, and let $\mathfrak{C}$ be the product of the prime ideals of $\mathcal{O}$ dividing C. We have $(\mathcal{O}/2\sqrt{-D}\mathcal{O})^\times \cong (\mathcal{O}/\mathfrak{T}^5)^\times \times (\mathcal{O}/\mathfrak{C})^\times$ and a corresponding decomposition of characters $\varepsilon = \varepsilon'\varepsilon''$. Also $(\mathbb{Z}/D\mathbb{Z})^\times \cong (Z/8\mathbb{Z})^\times \times (Z/C\mathbb{Z})^\times$ and

$\kappa = \kappa'\kappa''$. Using the natural embedding of $(Z/C\mathbb{Z})^\times$ in $(\mathcal{O}/\mathfrak{C})^\times$ to identify these two groups, we have $\varepsilon'' = \kappa''$, so ε'' is uniquely determined and $|E|$ is equal to the number of possibilities for ε'. Now the natural embedding of $(\mathbb{Z}/8\mathbb{Z})^\times$ in $(\mathcal{O}/\mathfrak{T}^5)^\times$ realizes $(\mathbb{Z}/8\mathbb{Z})^\times$ as one summand in a direct sum decomposition of $(\mathcal{O}/\mathfrak{T}^5)^\times$, the complementary summand being the cyclic group of order 4 generated by the coset of $1+\sqrt{-2C}$. On the factor $(\mathbb{Z}/8\mathbb{Z})^\times$ the character ε' coincides with κ', and since ε is quadratic there are exactly two possibilities for the value of ε' on the coset of $1+\sqrt{-2C}$, namely ± 1. Thus $|E| = 2$. Furthermore, we see that

$$\mathfrak{f}(\chi) = 2\sqrt{-D}\mathcal{O} \tag{1.26}$$

for $\chi \in X(D)$, because if χ and ε are related as in (1.22) then $\mathfrak{f}(\chi) = \mathfrak{f}(\varepsilon)$. □

3.1. The functional equation

While we have not yet discussed the functional equation of Hecke L-functions over arbitrary number fields, if K is imaginary quadratic and χ a primitive Hecke character of K of type $(1,0)$ then the functional equation is easily stated:

$$\Lambda(s,\chi) = W(\chi)\Lambda(2-s,\overline{\chi}) \tag{1.27}$$

with $|W(\chi)| = 1$ and

$$\Lambda(s,\chi) = (D\mathbf{N}\mathfrak{f}(\chi))^{s/2}\Gamma_{\mathbb{C}}(s)L(s,\chi). \tag{1.28}$$

Since $\overline{\chi}$ is of type $(0,1)$, the definition of $\Lambda(s,\overline{\chi})$ is technically not covered by (1.28), but it offers no surprises:

$$\Lambda(s,\overline{\chi}) = (D\mathbf{N}\mathfrak{f}(\chi))^{s/2}\Gamma_{\mathbb{C}}(s)L(s,\overline{\chi}). \tag{1.29}$$

The appearance of $\mathfrak{f}(\chi)$ in place of $\mathfrak{f}(\overline{\chi})$ on the right-hand side of (1.29) is not a misprint; one readily checks that $\mathfrak{f}(\overline{\chi}) = \mathfrak{f}(\chi)$. Now take $\chi \in X(D)$ with D either odd or divisible by 8. We shall give explicit formulas for the factors that go into the functional equation of $L(s,\chi)$. One such factor has already been made explicit (cf. (1.24) and (1.26)):

Proposition 1.4. *If $\chi \in X(D)$ then*

$$\mathfrak{f}(\chi) = \begin{cases} \sqrt{-D}\mathcal{O} & \text{if } D \text{ is odd,} \\ 2\sqrt{-D}\mathcal{O} & \text{if } 8|D. \end{cases}$$

The root number $W(\chi)$ can also be computed. First consider the case D odd. The proof of the following proposition is as in Gross ([**38**], pp. 60 – 63) and will be reproduced in Lecture 2.

Proposition 1.5. *If D is odd and $\chi \in X(D)$ then*

$$W(\chi) = \left(\frac{2}{D}\right).$$

Next recall that if $8|D$ then $|X(D)| = 2h(D)$. A proof of the following statement (albeit a proof in a more general context) can be found in [**76**], pp. 538 – 541, and a proof in the present setting will be outlined in Exercise 2.4.

Proposition 1.6. *Suppose that $8|D$, and put*

$$X^{\pm}(D) = \{\chi \in X(D) : W(\chi) = \pm 1\}.$$

Then $|X^{\pm}(D)| = h(D)$. In fact if $\chi \in X^{\pm}(D)$ then $X^{\pm}(D) = \{\chi\varphi : \varphi \in \Phi\}$, where Φ is the set of ideal class characters of K.

In spite of Propositions 1.5 and 1.6, we cannot conclude that our "canonical" family of Hecke L-functions exhibits trivial central zeros until we have verified that $\Lambda(s,\overline{\chi}) = \Lambda(s,\chi)$. But if we think of $L(s,\chi)$ as the Dirichlet series $\sum \chi(\mathfrak{a})(\mathbf{N}\mathfrak{a})^{-s}$ then the desired identity $L(s,\overline{\chi}) = L(s,\chi)$ is an immediate consequence of the equivariance of χ with respect to complex conjugation:

Proposition 1.7. *If* $\chi \in X(D)$ *then*

$$\chi(\overline{\mathfrak{a}}) = \overline{\chi(\mathfrak{a})}$$

for $\mathfrak{a} \in I(\mathfrak{f}(\chi))$.

PROOF. Put $n = \mathbf{N}\mathfrak{a}$, so that $\mathfrak{a}\overline{\mathfrak{a}} = n\mathcal{O}$. Then

$$\chi(\mathfrak{a})\chi(\overline{\mathfrak{a}}) = \chi(n\mathcal{O}) = \kappa(n)n = n,$$

because the Kronecker symbol κ is trivial on norms from K. Thus $\chi(\mathfrak{a})\chi(\overline{\mathfrak{a}}) = \mathbf{N}\mathfrak{a}$, and it suffices to see that $\chi(\mathfrak{a})\overline{\chi(\mathfrak{a})} = \mathbf{N}\mathfrak{a}$ or in other words that

$$|\chi(\mathfrak{a})| = \sqrt{\mathbf{N}\mathfrak{a}}. \tag{1.30}$$

Now in contrast to the identity $\chi(\mathfrak{a})\chi(\overline{\mathfrak{a}}) = \mathbf{N}\mathfrak{a}$, which depended on the relation $\varepsilon(n) = \kappa(n)$, (1.30) is a general property of Hecke characters of type (1,0). In fact since both sides of (1.30) are positive, it suffices to verify that equality holds after both sides are raised to the power $h(D)|K(\mathfrak{f})/K_{\mathfrak{f}}|$, where $\mathfrak{f} = \mathfrak{f}(\chi)$. Thus we may assume that $\mathfrak{a} = \alpha\mathcal{O}$ with $\alpha \in K_{\mathfrak{f}}$. But then $\chi(\mathfrak{a}) = \alpha$ and (1.30) is immediate. $\square$

Thus if $\chi \in X(D)$ then the functional equation (1.27) becomes

$$\Lambda(s,\chi) = W(\chi)\Lambda(2-s,\chi), \tag{1.31}$$

and we can talk about trivial central zeros. (Note by the way that quite apart from Propositions 1.5 and 1.6, the fact that $W(\chi) = \pm 1$ is clear *a priori* from (1.31).) Now $-D$ is a discriminant, so if D is odd then D is 3 mod 4 and in particular either 3 or 7 mod 8. Hence Propositions 1.5 and 1.6 imply that $L(s,\chi)$ has a trivial central zero if and only if either $D \equiv 3 \bmod 8$ or else $8|D$ and $\chi \in X^-(D)$. In the remaining cases, when $D \equiv 7 \bmod 8$ or $8|D$ and $\chi \in X^+(D)$, there is no trivial reason for $L(s,\chi)$ to vanish at $s = 1$, and we can ask the same question as with Dirichlet L-functions: Is $L(1,\chi)$ in fact nonzero? Actually, even if $W(\chi) = -1$ we can ask the analogous question about $L'(1,\chi)$, for while $L(s,\chi)$ vanishes to odd order at $s = 1$, there is no trivial reason for the order of vanishing to be > 1.

Theorem 1.2.

$$\mathrm{ord}_{s=1}L(s,\chi) = \begin{cases} 0 & \textit{if } W(\chi) = 1, \\ 1 & \textit{if } W(\chi) = -1. \end{cases}$$

For the proof see Montgomery-Rohrlich [**66**] or Miller-Yang [**65**] according as $W(\chi)$ is 1 or -1. We mention just one aspect of these proofs and of others like them, namely the key role played by the fact that

$$\{\chi^\sigma : \sigma \in \mathrm{Aut}(\mathbb{C}/K)\} = \{\chi\varphi : \varphi \in \Phi\} \tag{1.32}$$

for $\chi \in X(D)$. Here χ^σ is the character defined by $\chi^\sigma(\mathfrak{a}) = \chi(\mathfrak{a})^\sigma$ for $\mathfrak{a} \in I(\mathfrak{f})$, where $\mathfrak{f} = \mathfrak{f}(\chi)$. As the values of χ on principal ideals lie in K, it follows that χ and χ^σ coincide on principal ideals, and consequently the left-hand side of (1.32) is contained in the right-hand side. Thus to prove that equality holds it suffices to see that the cardinality of the left-hand side of (1.32) is $\geqslant h(D)$. Given $\mathfrak{a} \in$

$I(\mathfrak{f})$, let n be its order in the ideal class group $I(\mathfrak{f})/P(\mathfrak{f})$; then $\mathfrak{a}^n = \alpha\mathcal{O}$ for some $\alpha \in K(\mathfrak{f})$, and consequently $\chi(\mathfrak{a})^n = \pm\alpha$. One readily deduces that $\chi(\mathfrak{a})$ generates an extension of K of degree n. Now choose ideals $\mathfrak{a}_1, \mathfrak{a}_2, \ldots, \mathfrak{a}_t \in I(\mathfrak{f})$ such that I/P is the direct sum of the cyclic subgroups generated by the classes of the ideals $\mathfrak{a}_i$. Then $h(D) = n_1 n_2 \cdots n_t$, where n_i is the order of the class of $\mathfrak{a}_i$. Given these observations, it is not hard to believe or to prove that the extension of K generated by $\chi(\mathfrak{a}_1), \chi(\mathfrak{a}_2), \ldots, \chi(\mathfrak{a}_t)$ has degree $h(D)$ over K. It follows that the left-hand side of (1.32) has cardinality $\geqslant h(D)$, whence equality holds and (1.32) follows.

The significance of (1.32) is that it meshes well with algebraicity results for special values of L-functions. In the case case $W(\chi) = 1$, results of Shimura [**87**], [**88**] imply that if $L(1, \chi) = 0$ then $L(1, \chi^\sigma) = 0$ for all $\sigma \in \mathrm{Aut}(\mathbb{C})$, whence $h(D)^{-1} \sum_{\varphi\in\Phi} L(1, \chi\varphi) = 0$ by (1.32). Similarly, in the case $W(\chi) = -1$ the Gross-Zagier formula [**39**] implies that if $L'(1, \chi) = 0$ then $h(D)^{-1} \sum_{\varphi\in\Phi} L'(1, \chi\varphi) = 0$. Thus to prove the theorem it suffices to show that $h(D)^{-1} \sum_{\varphi\in\Phi} L(1, \chi\varphi) \neq 0$ or that $h(D)^{-1} \sum_{\varphi\in\Phi} L'(1, \chi\varphi) \neq 0$ according as $W(\chi) = 1$ or $W(\chi) = -1$. The point of this reduction is that as a Dirichlet series, $L(s, \chi)$ is the sum $\sum \chi(\mathfrak{a})(\mathbf{N}\mathfrak{a})^{-s}$ over all nonzero ideals of $\mathcal{O}$, and in particular over ideals belonging to all ideal classes. By contrast, $h(D)^{-1} \sum_{\varphi\in\Phi} L(s, \chi\varphi)$ is the sum $\sum \chi(\mathfrak{a})(\mathbf{N}\mathfrak{a})^{-s}$ taken over the *principal* ideals only. Analytically the latter sum is a much more tractable object.

3.2. Gross's $\mathbb{Q}$-curves

The significance of Hecke characters of type $(1, 0)$ is that they correspond to elliptic curves with complex multiplication, and the significance of the Hecke characters $\chi \in X(D)$ is that the corresponding elliptic curves are the canonical examples of Gross's "$\mathbb{Q}$-curves." To make this precise, recall that the modular invariant j can be evaluated not only on elliptic curves but also on lattices in $\mathbb{C}$: in fact if A is an elliptic curve over $\mathbb{C}$ and $\mathcal{L}$ its period lattice relative to a nonzero regular differential then $j(A) = j(\mathcal{L})$. In particular, since we are viewing K as a subfield of $\mathbb{C}$ we may take $\mathcal{L}$ to be $\mathcal{O}$, and then an elliptic curve with invariant $j(\mathcal{O})$ has complex multiplication by $\mathcal{O}$. Putting $F = \mathbb{Q}(j(\mathcal{O}))$ and $H = K(j(\mathcal{O}))$, we see that F is the minimal field of definition for an elliptic curve with invariant $j(\mathcal{O})$ and H the minimal field of definition for its complex multiplication.

Now if D is odd then the set $X(D)$ picks out a canonical isogeny class of elliptic curves over F with invariant $j(\mathcal{O})$, any member of which will be denoted $A(D)$. Similarly, if $8|D$ then the sets $X^+(D)$ and $X^-(D)$ each pick out such isogeny classes, say with members $A^+(D)$ and $A^-(D)$ respectively. We then have

$$L(s, A(D)) = \prod_{\chi\in X(D)} L(s, \chi) \qquad (D \text{ odd}) \tag{1.33}$$

and

$$L(s, A^\pm(D)) = \prod_{\chi\in X^\pm(D)} L(s, \chi) \qquad (8|D). \tag{1.34}$$

We emphasize that the isogeny classes at issue all contain more than one isomorphism class over F, so that $A(D)$, $A^+(D)$, and $A^-(D)$ have not been specified up to isomorphism. It is possible to do so, at least in the case of $A(p)$ with a prime $p \equiv 3 \bmod 4$ $(p > 3)$, by a consideration of minimal discriminants (Gross [**38**], p. 35), but for the validity of (1.33) and (1.34) this refinement is unnecessary: the L-function of an elliptic curve depends only on its isogeny class. Incidentally, if A is any one of $A(D)$, $A^+(D)$, and $A^-(D)$ then the isogeny class of A over H is defined over $\mathbb{Q}$ in the sense that A is isogenous over H to all of its Galois conjugates. This is the reason for the term "$\mathbb{Q}$-curve."

Combining Proposition 1.7 and Theorem 1.2 with (1.33) and (1.34), and applying either Rubin's generalization [**78**] of the Coates-Wiles theorem (if $W(\chi) = 1$) or the Gross-Zagier formula [**39**] and the theorem of Kolyvagin-Logachev [**54**] supplemented by either Bump-Friedberg-Hoffstein [**14**] or Murty-Murty [**70**] (if $W(\chi) = -1$), we obtain:

Theorem 1.3. *If D is odd then the rank of $A(D)(F)$ is 0 or $h(D)$ according as D is 7 or 3 modulo 8. If D is divisible by 8 then the rank of $A^+(D)$ over F is 0 and the rank of $A^-(D)$ over F is $h(D)$.*

In the case of a prime $p \equiv 7 \bmod 8$, the fact that $A(p)(F)$ has rank 0 was proved by Gross [**38**] several years before Theorem 1.2 using descent.

3.3. Yang's simplest abelian varieties

While we have seen that $X(D) = \emptyset$ if $4||D$, the exclusion of this case was nonetheless a peculiar anomaly for several years. However Yang [**101**] has shown that the case $4||D$ can be incorporated into the theory if on the geometric side elliptic curves are replaced by abelian varieties and on the arithmetic side the requirement that the values of χ on principal ideals lie in K – condition (c) in the original definition of $X(D)$ – is replaced by conditions (c) and (d) below. Let K be an imaginary quadratic field and D the absolute value of its discriminant. We consider the set $Y(D)$ of primitive Hecke characters v of K of type (1,0) satisfying the following conditions:

(a) $\mathfrak{f}(v)|D^\infty$.
(b) $v(n\mathcal{O}) = \kappa(n)n$ for $n \in \mathbb{Z}$ prime to D.
(c) Let T be the extension of K generated by the values of v. Then $[T : K]$ is minimal subject to (a) and (b).
(d) Also $\mathbf{N}\mathfrak{f}(v)$ is minimal subject to (a) and (b).

Suppose once again that $D \neq 3, 4$. Yang associates an isogeny class of abelian varieties over K with complex multiplication by T to the Galois orbit of an element $v \in Y(D)$, and he shows that these abelian varieties are in a natural sense the "simplest" among all abelian varieties over K with complex multiplication by T. If D is odd or divisible by 8 then $Y(D) = X(D)$, and if we fix a Galois orbit of elements of this set then Yang's abelian variety B is related to Gross's $\mathbb{Q}$-curve A via Weil's restriction-of-scalars functor: $B = \mathrm{res}_{H/K}A$. (In the case where D is a prime congruent to 3 mod 4 this restriction of scalars figured prominently already in [**38**].) But if $4||D$ then B need not be the restriction of scalars of any elliptic curve over H. Nonetheless, Yang proves analogues for all of the results already mentioned for $X(D)$. The proof of Yang's analogue of Theorem 1.2 is particularly daunting, because one no longer has (1.32): the Galois conjugates of χ are not all of the form $\chi\varphi$ with $\varphi \in \Phi$.

4. An open problem

In a nutshell, the problem is to prove an analogue of Theorem 1.2 with χ replaced by a power of χ. Let w be a positive integer and take χ in $X(D)$. If w is odd then χ^w is still primitive of conductor $\mathfrak{f}(\chi)$, but if w is even then χ^w extends to a

Hecke character to the modulus $\mathcal{O}$ and so is no longer primitive. We are primarily concerned with the case where w is odd, but to carry along both cases for one moment longer, write χ_w to mean χ^w if w is odd and the primitive Hecke character determined by χ^w if w is even. Then Hecke's functional equation for $L(s,\chi_w)$ is

$$\Lambda(s,\chi_w) = W(\chi_w)\Lambda(w+1-s,\chi_w) \tag{1.35}$$

with

$$\Lambda(s,\chi_w) = \begin{cases} (D\mathbf{N}\mathfrak{f}(\chi))^{s/2}\Gamma_{\mathbb{C}}(s)L(s,\chi_w) & \text{if } w \text{ is odd} \\ D^{s/2}\Gamma_{\mathbb{C}}(s)L(s,\chi_w) & \text{if } w \text{ is even} \end{cases} \tag{1.36}$$

and

$$W(\chi_w) = \begin{cases} (-1)^{(w-1)/2}W(\chi) & \text{if } w \text{ is odd} \\ 1 & \text{if } w \text{ is even} \end{cases} \tag{1.37}$$

(see Exercise 2.5). We note in particular that the center of symmetry of the functional equation is $(w+1)/2$. Thus if w is odd then the center is an integer, and in fact a critical integer in the sense of Deligne [**24**].

Problem 1. *Suppose that w is odd and relatively prime to $h(D)$. Show that*

$$\operatorname{ord}_{s=(w+1)/2}L(s,\chi^w) = \begin{cases} 0 & \textit{if } W(\chi)=1, \\ 1 & \textit{if } W(\chi)=-1. \end{cases} \tag{1.38}$$

An example of Rodriguez Villegas shows that the assumption $\gcd(w,h(D))=1$ cannot be omitted if (1.38) is to hold without exception ([**74**], p. 437, Remark 2), but we would also like a variant of the problem in which the coprimality hypothesis is eliminated at the expense of a weaker conclusion. However to begin with let us consider the problem as stated. A basic result in this domain is the theorem of Liu and Xu [**58**], who have shown that if w is given then there exists a constant $c(w)$ such that (1.38) is satisfied for $D \geqslant c(w)$. As is common in papers of this sort (cf. [**59**], [**60**], [**75**], and [**102**]), the authors actually prove the stronger statement that (1.38) remains valid when $L(s,\chi^w)$ is replaced by $L(s,\chi^w\mu)$ with a primitive quadratic Dirichlet character μ of sufficiently small conductor d relative to D (the precise condition in [**58**] is $d \ll D^{1/12-\varepsilon}$ for any $\varepsilon > 0$; of course D must still be sufficiently large relative to w). On the other hand, if $W(\chi)=1$ then the constant $c(w)$ implicit in [**58**] is ineffective, so that the validity of (1.38) for all D is not reduced to a finite number of verifications even when w is fixed. However in the case where D is a prime p, necessarily with $p \equiv 3 \bmod 4$, Boxer and Diao [**10**] have recently proved a result which is not only effective but also remarkably tidy: If $W(\chi)=1$ and $p \geqslant 13(w-1)^2/8$ then $L((w+1)/2,\chi^w) \neq 0$. Using Proposition 1.5 and (1.37), one readily verifies that when $D=p$ the condition $W(\chi)=1$ is equivalent to $w \equiv (-1)^{(p+1)/4} \bmod 4$, so the result of Boxer and Diao is as explicit as can be. Even so, the validity of (1.38) for all D and all odd w prime to $h(D)$ remains an open problem.

If the condition $\gcd(w,h(D))=1$ is dropped then the problem is not only open but open-ended in the sense that we lack a conjecture to guide us. The most optimistic conjecture would be that (1.38) holds for all but finitely many triples (w,D,χ) with $\chi \in X(D)$, but the evidence is as yet too weak to support this hope. The underlying problem here is that most attempts to prove (1.38) begin by showing that (1.38) holds for *some* $\chi \in X(D)$, and if the characters $\chi^w\varphi$ with $\varphi \in \Phi$

are all Galois-conjugate (in other words, if χ can be replaced by χ^w in (1.32)) then one deduces that (1.38) holds for *all* $\chi \in X(D)$. The deduction here is based on Shimura [**87**] as before if $W(\chi^w) = 1$ and on Zhang [**105**] rather than Gross-Zagier [**39**] if $W(\chi^w) = -1$ with $w \geqslant 3$. But if $\gcd(w, h(D)) > 1$ then the argument falls apart, because the characters $\chi^w\varphi$ are simply not all Galois-conjugate. Let $h_w(D)$ be the order of the quotient of the ideal class group by its subgroup of elements of order dividing w. Then the size of a Galois orbit of $\{\chi^w\varphi : \varphi \in \Phi\}$ is $h_w(D)$, so the fact that (1.38) holds for one element of $X(D)$ implies only that (1.38) holds for at least $h_w(D)$ elements. By estimating $h_w(D)$ using the recent bounds of Ellenberg and Venkatesh [**28**] on torsion in ideal class groups, Masri [**60**] deduces that the number of characters $\chi \in X(D)$ satisfying (1.38) for fixed w and $D = p$ is $\gg p^\delta$ for some $\delta > 0$ (for instance if $w = 3$ then δ can be any number $< 1/6$). If $w \geqslant 5$ then the results of [**28**] and hence of [**60**] depend on the generalized Riemann hypothesis, but Masri's papers [**59**] and [**60**] also give a second method which is based on the subconvexity results of Duke, Friedlander, and Iwaniec [**27**] and is independent of the generalized Riemann hypothesis. See also Masri-Yang [**61**]. The implicit constant $c(w)$ in [**60**] such that (1.38) holds for at least p^δ characters if $p \geqslant c(w)$ is ineffective, but once again Boxer and Diao [**10**] give an effective result: If $W(\chi^w) = 1$ and $D = p \geqslant 13(w-1)^2/8$ then the number of good characters is at least $h_w(D)$.

All of this meshes well with the example of Rodriguez Villegas [**74**] mentioned above, in which $D = 59$, $w = 3$, and $L(2, \chi^3) = 0$ for some $\chi \in X(D)$, despite the fact that $W(\chi^3) = 1$. The point is that $h(D) = 3$ in this case, in line with the preceding discussion. (Incidentally, Rodriguez Villegas deduces the vanishing of $L(2, \chi^3)$ from a calculation and an *a priori* bound on the denominator of the special value.) One might imagine that in this example there are infinitely many w divisible by 3 such that $W(\chi^w) = 1$ but $L((w+1)/2, \chi^w) = 0$ for some $\chi \in X(D)$. However this possibility is excluded by a general theorem of Greenberg [**36**], valid for any imaginary quadratic field K and any primitive Hecke character of K of type $(w, 0)$. In our setting, the theorem asserts that if we fix $\chi \in X(D)$ then there are only finitely many odd positive integers w such that $W(\chi^w) = 1$ but $L((w+1)/2, \chi^w) = 0$. Pondering this statement, we realize that Problem 1 is subsumed in a larger problem, still to be formulated, in which the hypothesis $\gcd(w, h(D)) = 1$ is omitted but the conclusion (1.38) is asserted to hold for as large a set of characters as possible. An optimal description of this set is the point that remains open-ended.

5. Evaluation of the quadratic Gauss sum

For the sake of contrasting two different techniques, we give a self-contained proof of Theorem 1.1 here and a proof of a more general statement in the next lecture. To get started we record a few facts about the "Fourier transform" on $\mathbb{Z}/N\mathbb{Z}$.

Let V be the complex vector space consisting of $\mathbb{C}$-valued functions on $\mathbb{Z}/N\mathbb{Z}$. The Fourier transform on $\mathbb{Z}/N\mathbb{Z}$ is the linear automorphism $f \mapsto \hat{f}$ of V, where

$$\hat{f}(x) = \sum_{y \in \mathbb{Z}/N\mathbb{Z}} f(y) e^{2\pi i x y/N}.$$

Sometimes we write $\hat{f}$ as $f\hat{\ }$. That $f \mapsto \hat{f}$ is an automorphism follows from the identity $(f\hat{\ })\hat{\ }(x) = Nf(-x)$, which in turn is a consequence of the calculation

$$(f\hat{\ })\hat{\ }(x) = \sum_{y \in \mathbb{Z}/N\mathbb{Z}} \hat{f}(y) e^{2\pi i x y/N} = \sum_{z \in \mathbb{Z}/N\mathbb{Z}} f(z) \sum_{y \in \mathbb{Z}/N\mathbb{Z}} e^{2\pi i (x+z) y/N}$$

(observe that the inner sum is N or 0 according as z is or is not $-x$). The main fact about the Fourier transform which is needed for the proof of Theorem 1.1 is a formula for $\widehat{\chi}$, where χ is a primitive Dirichlet character of conductor N. Here we are viewing χ as a character of $(\mathbb{Z}/N\mathbb{Z})^\times$ and hence as a function on $\mathbb{Z}/N\mathbb{Z}$ via the usual convention that $\chi(x) = 0$ if the residue class $x \in \mathbb{Z}/N\mathbb{Z}$ is not invertible.

The formula that we need is

$$\widehat{\chi} = \tau(\chi)\overline{\chi}. \tag{1.39}$$

To verify (1.39), we take $x \in \mathbb{Z}/N\mathbb{Z}$ and compute $\widehat{\chi}(x)$ from the definition:

$$\widehat{\chi}(x) = \sum_{y \in (\mathbb{Z}/N\mathbb{Z})^\times} \chi(y) e^{2\pi i x y/N}. \tag{1.40}$$

There are two cases to consider, according as x is a nonunit or a unit.

Suppose first that $x \in M\mathbb{Z}/N\mathbb{Z}$, where M is a divisor of N and $M > 1$. To prove (1.39) in this case we must see that the right-hand side of (1.40) is 0. Put $G = (\mathbb{Z}/N\mathbb{Z})^\times$ and let H be the kernel of the map from $(\mathbb{Z}/N\mathbb{Z})^\times$ to $(\mathbb{Z}/(N/M)\mathbb{Z})^\times$ given by reduction modulo N/M. Write $y = gh$, where $h \in H$ and g runs over a set of coset representatives for H in G. Then $\chi(y) = \chi(g)\chi(h)$, and since χ is primitive $\chi|H$ is nontrivial. If we write the right-hand side of (1.40) as a double sum consisting of an inner sum over h and an outer sum over g, then the inner sum is 0 for each g, because $\chi|H$ is nontrivial and $e^{2\pi i x g h/N}$ is independent of h.

Next suppose that $x \in (\mathbb{Z}/N\mathbb{Z})^\times$. Then in the sum over y we can replace y by yx^{-1}, and referring to the definition (1.2) we see that $\widehat{\chi}(x) = \tau(\chi)\overline{\chi}(x)$. This completes the proof of (1.39).

An immediate corollary of (1.39), given the relation $(f\hat{\ })\hat{\ }(x) = Nf(-x)$, is that

$$\tau(\chi)\tau(\overline{\chi}) = \chi(-1)N. \tag{1.41}$$

It follows in particular that if χ is quadratic then $\tau(\chi)^2 = \chi(-1)N$. Thus the validity of (1.1) up to sign is easy, as already mentioned. Returning to the case of an arbitrary primitive χ, and observing that (1.39) and (1.40) together give $\overline{\tau(\chi)} = \chi(-1)\tau(\overline{\chi})$, we deduce from (1.41) that

$$|\tau(\chi)| = \sqrt{N}. \tag{1.42}$$

Of course both (1.41) and (1.42) can be reformulated in terms of root numbers: Applying the definition (1.3), we obtain

$$W(\chi)W(\overline{\chi}) = 1 \tag{1.43}$$

and

$$|W(\chi)| = 1 \tag{1.44}$$

respectively.

So much for general orientation. We turn now to the proof of Theorem 1.1 itself. The key step is the special case (1.1), for which we use an argument of Schur.

5.1. Schur's proof

Now take $N = p$ and write $\mathcal{F}$ for the Fourier transform $f \mapsto \hat{f}$ on V. To prove (1.1) we put

$$Q = \frac{\det \mathcal{F}}{|\det \mathcal{F}|}. \tag{1.45}$$

and compute Q in two different ways.

The first way is to use the ordered basis $\delta_0, \delta_1, \ldots, \delta_{p-1}$ for V, where $\delta_j(y) = 1$ if y is j mod p and $\delta_j(y) = 0$ otherwise. The matrix of $\mathcal{F}$ relative to this basis has $e^{2\pi ijk/p}$ as its (j,k) entry, and consequently $\det \mathcal{F}$ is a Vandermonde determinant:

$$\det \mathcal{F} = \prod_{0 \leqslant j < k \leqslant p-1} (e^{2\pi ik/p} - e^{2\pi ij/p}). \tag{1.46}$$

To compute Q we can replace $\det \mathcal{F}$ in (1.45) by any positive scalar multiple of $\det \mathcal{F}$. In particular, the factors in (1.46) with $j = 0$ can be removed from (1.46), because the factors corresponding to $(0,k)$ and $(0,p-k)$ are complex conjugates, hence their product is positive. Now the remaining factors correspond to pairs (j,k) with $1 \leqslant j < k \leqslant p-1$, and the map $(j,k) \mapsto (p-k, p-j)$ is an involution on the set of such pairs. Furthermore the fixed points are precisely the pairs $(j, p-j)$ with $1 \leqslant j \leqslant (p-1)/2$, and if (j,k) is not a fixed point then the factor corresponding to $(p-k, p-j)$ is the negative of the complex conjugate of the factor corresponding to (j,k). The upshot of these remarks is that $Q = R/|R|$ with

$$R = (-1)^{(p-1)(p-3)/4} \prod_{1 \leqslant j \leqslant (p-1)/2} (e^{-2\pi ij/p} - e^{2\pi ij/p}).$$

But $(p-1)(p-3)/4$ is even, so the factor $(-1)^{(p-1)(p-3)/4}$ can be removed. Then

$$Q = (-i)^{(p-1)/2}, \tag{1.47}$$

because $e^{-2\pi ij/p} - e^{2\pi ij/p} = -2i\sin(2\pi j/p)$ and $\sin(2\pi j/p) > 0$ for $1 \leqslant j < p/2$.

On the other hand, we obtain most of a second basis for V from the characters $\chi : \mathbb{F}_p^\times \to \mathbb{C}^\times$ of $\mathbb{F}_p^\times$. In conformity with a convention established earlier, if χ is nontrivial then we extend it to a function on $\mathbb{F}_p$ by setting $\chi(0) = 0$. We also extend the trivial character χ_0 by setting $\chi_0(0) = 1$. Now let

$$\chi_1, \overline{\chi}_1, \chi_2, \overline{\chi}_2, \ldots, \chi_{(p-3)/2}, \overline{\chi}_{(p-3)/2} \tag{1.48}$$

be an enumeration of the conjugate pairs of nontrivial nonquadratic characters of $\mathbb{F}_p^\times$. Then

$$\lambda, \delta_0, \chi_0, \chi_1, \overline{\chi}_1, \chi_2, \overline{\chi}_2, \ldots, \chi_{(p-3)/2}, \overline{\chi}_{(p-3)/2} \tag{1.49}$$

is an ordered basis for V. Let us compute the matrix of $\mathcal{F}$ relative to this basis.

Since λ is quadratic we have $\mathcal{F}\lambda = \tau(\lambda)\lambda$ by (1.39). Furthermore $\mathcal{F}\delta_0 = \chi_0$, whence $\mathcal{F}\chi_0 = p\delta_0$ by the relation $(\mathcal{F}^2 f)(x) = pf(-x)$. Thus the matrix of $\mathcal{F}$ relative to the basis (1.49) is block-diagonal: The entry $\tau(\lambda)$ in the upper left-hand corner is followed by the 2×2 block

$$B_0 = \begin{pmatrix} 0 & p \\ 1 & 0 \end{pmatrix},$$

which is followed in turn by the 2×2 blocks – here we are using (1.39) again –

$$B_j = \begin{pmatrix} 0 & \tau(\overline{\chi_j}) \\ \tau(\chi_j) & 0 \end{pmatrix}$$

for $1 \leqslant j \leqslant (p-3)/2$. Now $\det B_0 = -p$ by inspection while $\det B_j = -\chi_j(-1)p$ for $1 \leqslant j \leqslant (p-3)/2$ by (1.41). Computing modulo positive real numbers, we deduce that

$$Q = \frac{\tau(\lambda)}{\sqrt{p}}(-1)^{(p-1)/2}\prod_{j=1}^{(p-3)/2}\chi_j(-1),$$

and comparing this result with (1.47) we obtain

$$\frac{\tau(\lambda)}{\sqrt{p}} = i^{(p-1)/2}\prod_{j=1}^{(p-3)/2}\chi_j(-1).$$

Equivalently,

$$\frac{\tau(\lambda)}{\sqrt{p}} = i^{(p-1)/2}(-1)^{\nu}, \tag{1.50}$$

where ν is the number of odd characters among the χ_j $(1 \leqslant j \leqslant (p-3)/2)$. Since χ is odd if and only if $\overline{\chi}$ is odd, we can also say that ν is half the number of odd characters among the characters listed in (1.48).

If $p \equiv 1 \bmod 4$ then all $(p-1)/2$ odd characters of $\mathbb{F}_p^{\times}$ occur in (1.48), because neither χ_0 nor λ is odd. Hence $\nu = (p-1)/4$ and the right-hand side of (1.50) is 1. If $p \equiv 3 \bmod 4$ then λ is odd, and consequently only $(p-3)/2$ of the odd characters of $\mathbb{F}_p^{\times}$ occur in (1.48). Hence $\nu = (p-3)/4$ and the right-hand side of (1.50) is i. This completes Schur's proof of (1.1).

5.2. The general case

Now suppose that χ is an arbitrary primitive quadratic Dirichlet character, and let N be the conductor of χ. If N is an odd prime then Theorem 1.1 has just been proved, and if $N = 4$ or $N = 8$ then the theorem is easily verified by explicit calculation. Putting these cases aside, and keeping in mind that N is the conductor of a *primitive quadratic* Dirichlet character, we can write $N = N_1N_2$ and $\chi = \chi_1\chi_2$ with coprime integers N_1 and N_2 and primitive quadratic Dirichlet characters χ_1 and χ_2 of conductors N_1 and N_2 respectively. The numbers

$$j = j_1N_2 + j_2N_1 \qquad (0 \leqslant j_1 \leqslant N_1 - 1, \quad 0 \leqslant j_2 \leqslant N_2 - 1) \tag{1.51}$$

represent the distinct residue classes modulo N, and when j is written in this way we have $j/N = (j_1/N_1) + (j_2/N_2)$, $\chi_1(j) = \chi_1(j_1)\chi_1(N_2)$, and $\chi_2(j) = \chi_2(j_2)\chi_2(N_1)$. Hence inserting (1.51) in (1.2), we obtain

$$\tau(\chi) = \chi_1(N_2)\chi_2(N_1)\cdot\tau(\chi_1)\tau(\chi_2). \tag{1.52}$$

At this point it is convenient to write the integer m of (1.4) as $m(\chi)$ to indicate its dependence on χ. Dividing both sides of (1.52) by $\sqrt{N}\,i^{m(\chi)}$ and applying the law of quadratic reciprocity in the form

$$\chi_1(N_2)\chi_2(N_1) = i^{m(\chi_1\chi_2)-m(\chi_1)-m(\chi_2)},$$

we obtain $W(\chi) = W(\chi_1)W(\chi_2)$. Hence Theorem 1.1 follows by induction on the number of distinct prime factors of N.

6. Exercises

Exercise 1.1. We have observed that the negative even integers are trivial zeros of $\zeta(s)$. Generalize this remark in two directions:

- Determine the trivial zeros of the Dedekind zeta function $\zeta_K(s)$. Your answer will depend on the number of real and complex embeddings of the number field K. (See Theorem 2.1 for the functional equation of $\zeta_K(s)$.)
- Determine the trivial zeros of $L(s,\chi)$ for an arbitrary primitive Dirichlet character χ. Your answer will depend on the parity of χ.

The assumption that χ is primitive is natural when one talks about trivial zeros of $L(s,\chi)$, because imprimitivity perturbs the functional equation. Note however that trivial *central* zeros are unaffected: If χ is an imprimitive Dirichlet character and χ' is the primitive Dirichlet character determined by χ then $L(s,\chi)$ differs from $L(s,\chi')$ by a factor which does not vanish at $s = 1/2$. On the other hand, what happens at $s = 0$?

Exercise 1.2. Let χ be a primitive Dirichlet character of order $\geqslant 3$. To see that $L(s,\chi)$ does not have a trivial central zero, we argued that the functional equation could have no bearing on $\mathrm{ord}_{s=1/2}L(s,\chi)$ because $\chi \neq \overline{\chi}$ and hence $L(s,\chi) \neq L(s,\overline{\chi})$. Implicit in this argument is a basic analytic fact:

> *If two Dirichlet series $\sum_{n\geqslant 1} a(n)n^{-s}$ and $\sum_{n\geqslant 1} b(n)n^{-s}$ coincide as holomorphic functions in some right half-plane in which they both converge absolutely then $a(n) = b(n)$ for all $n \geqslant 1$.*

Verify this assertion by proving an equivalent statement:

> *If a Dirichlet series $\sum_{n\geqslant 1} a(n)n^{-s}$ is identically* 0 *in some right half-plane in which the series is absolutely convergent then $a(n) = 0$ for all $n \geqslant 1$.*

Then explain why Proposition 1.7 does imply that $L(s,\chi) = L(s,\overline{\chi})$ for $\chi \in X(D)$, even though in this case $\chi \neq \overline{\chi}$.

Exercise 1.3. Apart from the identities $W(\chi)W(\overline{\chi}) = 1$ and $|W(\chi)| = 1$, which appeared as equations (1.43) and (1.44) in the prolegomena to the proof of Theorem 1.1, we have said nothing at all about $W(\chi)$ when χ is a primitive Dirichlet character of order $\geqslant 3$. In this problem we assume that χ is not only primitive of order $\geqslant 3$ but also of prime conductor p. The easier case (easier in the sense that more complete results can be given) is actually the case of prime-power conductor p^ν with $\nu \geqslant 2$, for which see Exercises 1.4, 1.5, and 1.6 below.

(a) Stickelberger's theorem (see for example [**55**], p. 97, Theorem 10) gives a factorization of $\tau(\chi)^{p-1}$ as a product of prime ideals in the cyclotomic field $\mathbb{Q}(e^{2\pi i/(p-1)})$. Using this factorization, show that $W(\chi)$ is not an algebraic integer, and in particular not a root of unity, in spite of the fact that $|W(\chi^\sigma)| = 1$ for every automorphism σ of $\mathbb{C}$.

(b) (*Literature search.*) In the case where χ has order 3 or 4, Matthews [**62**], [**63**] expresses $\tau(\chi)$ in terms of values of the Weierstrass $\wp$-function at quadratic imaginary arguments. Furthermore, Heath-Brown and Patterson [**41**] prove the equidistribution (relative to Lebesgue measure on the unit circle) of the numbers $W(\chi)$ as χ runs over primitive Dirichlet characters of order 3 and prime conductor. What is known about possible generalizations of the results of Matthews and of Heath-Brown and Patterson to Dirichlet characters of orders greater than 3 or 4?

Exercise 1.4. The purpose of this exercise and the two that follow is to show that if χ is a primitive Dirichlet character of conductor p^ν with $\nu \geqslant 2$ then $W(\chi)$ is a root of unity (in contrast to part (a) of Exercise 1.3).

(a) Suppose that χ is of p-power order (and hence of order dividing $p^{\nu-1}$). Using the defining formula (1.2) and (1.42), show that $\tau(\chi)$ is an element of $\mathbb{Z}[e^{2\pi i/p^\nu}]$ of absolute value $p^{\nu/2}$ in every complex embedding of $\mathbb{Q}(e^{2\pi i/p^\nu})$. Deduce that $\tau(\chi)^2$ is a root of unity times p^ν, and conclude that $W(\chi)$ is a root of unity.

(b) Deduce that if χ is primitive of conductor 2^ν then $W(\chi)$ is a root of unity.

Exercise 1.5. With notation as in Exercise 1.4, suppose that ν is even, and put $n = \nu/2$. Observe that $\chi(1+p^n(x+y)) = \chi(1+p^n x)\chi(1+p^n y)$ for $x, y \in \mathbb{Z}/p^n\mathbb{Z}$, and deduce that $W(\chi)$ is a root of unity. (Hint: Put $G = (\mathbb{Z}/p^\nu\mathbb{Z})^\times$ and let H be the kernel of $(\mathbb{Z}/p^\nu\mathbb{Z})^\times \to (\mathbb{Z}/p^n\mathbb{Z})^\times$. Express (1.2) as a sum over $j = gh$, where $h \in H$ and g runs over a set of coset representatives for H in G. Then write the sum over j as a double sum over g and h, and show that the inner sum over h is 0 for all but one value of g.)

Exercise 1.6. With notation as in Exercise 1.4, suppose that ν is odd and hence $\geqslant 3$, and put $m = (\nu-1)/2$. Since the case $p = 2$ has already been dealt with in part (b) of Exercise 1.4, we may assume that p is odd.

(a) Using the binomial theorem, show that $(1+p^m x)^p \equiv 1 + p^{m+1}x \bmod p^\nu$ for arbitrary $x \in \mathbb{Z}$.

(b) Using the hint for Exercise 1.5 with $n = m+1$, show that there is a unique element $c \in \mathbb{Z}/\mathbb{Z}p^m$ such that $\chi(1+p^n x) = e^{2\pi i c x/p^m}$ for $x \in \mathbb{Z}/p^m\mathbb{Z}$. Deduce that $\tau(\chi)$ is a root of unity times $p^m S$, where S is the sum

$$S = \sum_{x \in \mathbb{F}_p} \chi(1+p^m x) e^{2\pi i g_0 x/p^n}$$

and $g_0 \in \mathbb{Z}/\mathbb{Z}p^n$ is any element which reduces mod p^m to $-c$.

(c) Using (a), show that $\chi(1+p^m x)e^{2\pi i g_0 x/p^n}$ is a pth root of unity, whence $S \in \mathbb{Z}[e^{2\pi i/p}]$. Show that S has absolute value $p^{1/2}$ in every complex embedding of $\mathbb{Q}(e^{2\pi i/p})$, and conclude as in Exercise 1.4 that $W(\chi)$ is a root of unity.

Exercise 1.7. (*Literature search.*) While it is widely expected that Dirichlet L-functions do not vanish at $s = 1/2$, the history of this conjecture deserves to be elucidated. Is it correct to say that the first mention of the conjecture (at least in the quadratic case) is in Chowla [**18**]? Soundararajan [**90**] notes that the nonvanishing of $L(1/2, \chi)$ would follow from the conjectured $\mathbb{Q}$-linear independence of the set

$$\{\gamma : L(1/2 + i\gamma, \chi) = 0, \ \ \gamma \geqslant 0\},$$

but what is the history of the latter conjecture?

LECTURE 2

Local formulas

In principle, we could derive the explicit formula for $W(\chi)$ in Proposition 1.5 by calculating directly from formula (45) of Hecke's original paper [**42**]. However Hecke's formula is expressed in terms of "ideal numbers," an extrinsic construction long superseded by the intrinsically defined "ideles" of Chevalley and Weil. Rather than rescue ideal numbers from desuetude, we prefer to emphasize the correspondence between Hecke characters and idele class characters and the use of Tate's local formulas.

1. The idele class group

Let K be a number field. The **ring of adeles** of K is the restricted direct product

$$\mathbb{A} = \prod_v{}' K_v, \tag{2.1}$$

where v runs over the standard set of places of K and K_v is the completion of K at v. If we wish to indicate the dependence of $\mathbb{A}$ on K then we write $\mathbb{A}_K$. The restriction (indicated by the prime) is that an element $x = (x_v)$ of the usual direct product belongs to $\mathbb{A}$ if and only if $x_v \in \mathcal{O}_v$ for all but finitely many finite v, where $\mathcal{O}_v$ is the ring of integers of K_v. Since K is naturally embedded in each of its completions, we may view it as a subring of $\mathbb{A}$ via the diagonal embedding. In other words, we identify $x \in K$ with the adele (x_v) such that $x_v = x$ for all v.

If v in (2.1) runs over the finite places only then the resulting ring $\mathbb{A}_{\text{fin}}$ is called the **ring of finite adeles** of K. Putting $\mathbb{A}_\infty = \prod_{v|\infty} K_v$, we may write the full adele ring $\mathbb{A}$ as the ordinary direct product of its finite and infinite components:

$$\mathbb{A} = \mathbb{A}_{\text{fin}} \times \mathbb{A}_\infty. \tag{2.2}$$

Of course $\mathbb{A}_\infty \cong \mathbb{R}^{r_1} \times \mathbb{C}^{r_2} \cong \mathbb{R} \otimes_{\mathbb{Q}} K$, where r_1 and r_2 have their usual meaning.

Next consider the **ring of adelic integers of** K, defined as the direct product

$$\widehat{\mathcal{O}} = \prod_{v\nmid\infty} \mathcal{O}_v \tag{2.3}$$

and viewed as a subring of $\mathbb{A}_{\text{fin}}$. We topologize $\mathbb{A}_{\text{fin}}$ by imposing two requirements:

- $\widehat{\mathcal{O}}$ is open in $\mathbb{A}_{\text{fin}}$, and the relative topology on $\widehat{\mathcal{O}}$ induced by $\mathbb{A}_{\text{fin}}$ is the usual product topology coming from (2.3).
- For each $a \in \mathbb{A}_{\text{fin}}$, the map $x \mapsto a + x$ is a homeomorphism from $\mathbb{A}_{\text{fin}}$ to itself.

One can check that there is a unique topology on $\mathbb{A}_{\text{fin}}$ satisfying these conditions and that with this topology $\mathbb{A}_{\text{fin}}$ becomes a topological ring. The topology on $\mathbb{A}$ is then the direct product topology afforded by (2.2), where $\mathbb{A}_\infty$ has its standard topology as the finite-dimensional real vector space $\mathbb{R} \otimes_{\mathbb{Q}} K$.

The multiplicative group $\mathbb{A}^\times$ of $\mathbb{A}$ is known as the **group of ideles** of K. It too is a restricted direct product:

$$\mathbb{A}^\times = {\prod_v}' K_v^\times, \tag{2.4}$$

but this time the restriction is that an element $x = (x_v)$ of the unrestricted direct product belongs to $\mathbb{A}^\times$ if and only if $x_v \in \mathcal{O}_v^\times$ for all but finitely many finite v. The topology on $\mathbb{A}^\times$ is *not* the relative topology from $\mathbb{A}$, but it can nonetheless be defined in a similar way. Indeed consider the multiplicative group of $\widehat{\mathcal{O}}$:

$$\widehat{\mathcal{O}}^\times = \prod_{v\nmid\infty} \mathcal{O}_v^\times. \tag{2.5}$$

The topology on $\mathbb{A}_{\text{fin}}$ is characterized by two properties:

- $\widehat{\mathcal{O}}^\times$ is open in $\mathbb{A}_{\text{fin}}^\times$, and the relative topology on $\widehat{\mathcal{O}}^\times$ induced by $\mathbb{A}_{\text{fin}}^\times$ is the usual product topology coming from (2.5).
- For each $a \in \mathbb{A}_{\text{fin}}^\times$, the map $x \mapsto ax$ is a homeomorphism from $\mathbb{A}_{\text{fin}}^\times$ to itself.

Once again, there is a unique topology on $\mathbb{A}_{\text{fin}}^\times$ satisfying these conditions, and with this topology $\mathbb{A}_{\text{fin}}^\times$ becomes a topological group. To topologize $\mathbb{A}^\times$ we use (2.2) to write

$$\mathbb{A}^\times = \mathbb{A}_{\text{fin}}^\times \times \mathbb{A}_\infty^\times, \tag{2.6}$$

and then we give $\mathbb{A}^\times$ the direct product topology corresponding to (2.6).

While $\mathbb{A}_\infty^\times$ can be identified either with $(\mathbb{R}\otimes_{\mathbb{Q}} K)^\times$ or with $\prod_{v|\infty} K_v^\times$, it will frequently be viewed as the subgroup of $\mathbb{A}^\times$ consisting of ideles $x = (x_v)$ such that $x_v = 1$ for $v \nmid \infty$. If $\mathbb{A}_{\text{fin}}^\times$ is similarly identified with the subgroup of $\mathbb{A}^\times$ consisting of ideles $x = (x_v)$ such that $x_v = 1$ for $v|\infty$ then (2.6) expresses $\mathbb{A}^\times$ as a direct product of two subgroups. The associated projection functions will be written $x \mapsto x_{\text{fin}}$ and $x \mapsto x_\infty$ respectively, so that $x = x_{\text{fin}}x_\infty$.

Since $K^\times$ is naturally embedded in each of its completions, we may view it as a subgroup of $\mathbb{A}^\times$ via the diagonal embedding, just as K was viewed as a subring of $\mathbb{A}$. Thus an element $x \in K^\times$ is identified with the idele (x_v) such that $x_v = x$ for all v. The quotient group $\mathbb{A}^\times/K^\times$ is called the **idele class group** of K.

2. Idele class characters

Let v be a place of K, finite or infinite, and let $p \leqslant \infty$ be the place of $\mathbb{Q}$ below v. We write $|*|_v$ for the absolute value on K_v which extends the standard absolute value $|*|_p$ on $\mathbb{Q}_p$, and we define the **local norm** $||*||_v$ on $K_v^\times$ by setting

$$||*||_v = |*|_v^{[K_v:\mathbb{Q}_p]}. \tag{2.7}$$

For example, if $K_v \cong \mathbb{C}$ then $p = \infty$ and $||*||_v = |*|_v^2$.

If v is finite then a character $\chi_v : K_v^\times \to \mathbb{C}^\times$ is **ramified** or **unramified** according as the restriction $\chi_v|\mathcal{O}_v^\times$ is nontrivial or trivial. Now $\mathcal{O}_v^\times$ is precisely the set $\{x \in K_v^\times : |x|_v = 1\}$, and if we temporarily denote this set by $\mathcal{O}_v^\times$ even when v is an infinite place then we obtain a seamless extension of the notions *ramified* and *unramified* to the infinite places: In all cases, $\mathcal{O}_v^\times$ is a subgroup of $K_v^\times$ (coinciding with $\{\pm1\}$ if v is real and with the circle group if v is complex), and in all cases we call χ_v ramified or unramified according as $\chi_v|\mathcal{O}_v^\times$ is nontrivial or trivial.

The "seamless extension" just described is often useful, particularly when one wants to distinguish between "narrow ray class characters" and "wide ray class characters" in class field theory. But this distinction is tangential to the matter at hand, and henceforth we will speak of ramified and unramified characters only at the finite places, reserving the notation $\mathcal{O}_v^\times$ for the finite places also.

By an **idele class character** of K we mean a continuous homomorphism $\chi : \mathbb{A}^\times \to \mathbb{C}^\times$ which is trivial on the diagonally embedded subgroup $K^\times$. Such a character necessarily factors as a product of local characters,

$$\chi = \prod_v{}' \chi_v, \tag{2.8}$$

where the prime indicates that χ_v is unramified for all but finitely many finite v. It is only by virtue of this last property that (2.8) has a meaning, for we interpret (2.8) to mean that if $x = (x_v) \in \mathbb{A}^\times$ then $\chi(x) = \prod_v \chi_v(x_v)$, and the product is finite precisely because for all but finitely many finite v we have $x_v \in \mathcal{O}_v^\times$ and $\chi_v|\mathcal{O}_v^\times = 1$. When χ_v is unramified we also say that χ is unramified at v.

By definition, an idele class character of K factors through the idele class group $\mathbb{A}^\times/K^\times$, whence the term *idele class character.* In fact one often identifies idele class characters with characters of $\mathbb{A}^\times/K^\times$.

As an example of an idele class character, consider the **idelic norm**, defined as the product of the local norms:

$$||x|| = \prod_v ||x_v||_v \qquad (x = (x_v) \in \mathbb{A}_K^\times). \tag{2.9}$$

This product is meaningful, because for all but finitely many finite v we have $x_v \in \mathcal{O}_v^\times$ and hence $||x_v||_v = 1$. It is immediately verified that the idelic norm is a continuous character of $\mathbb{A}^\times$, and by the so-called "Product Formula" it is trivial on $K^\times$, hence an idele class character.

2.1. Hecke characters as idele class characters

The L-function of an idele class character χ of K is defined by the formula

$$L(s, \chi) = \prod_{\substack{v\nmid\infty \\ \chi_v \text{ unram}}} (1 - \chi_v(\pi_v)q_v^{-s})^{-1}, \tag{2.10}$$

where the Euler product on the right-hand side runs over the finite places of K at which χ is unramified, q_v being the order of the residue class field of $\mathcal{O}_v$ and $\pi_v \in \mathcal{O}_v$ a uniformizer. The fact that χ_v is unramified means precisely that $\chi_v(\pi_v)$ is independent of the choice of π_v, so the right-hand side of (2.10) is well defined at least as a formal product. But in fact the product converges in some right half-plane and hence defines a holomorphic function there.

This last assertion may sound familiar, because the very same claim was made in connection with the L-function of a Hecke character. This is no coincidence: an idele class character is essentially the same thing as a primitive Hecke character. More precisely, there is a map $\chi \mapsto \chi_{\mathbb{A}}$ from Hecke characters of K to idele class characters of K which is a bijection when restricted to primitive Hecke characters. The map $\chi \mapsto \chi_{\mathbb{A}}$ comes about as follows.

Given a nonzero integral ideal $\mathfrak{f}$ of $\mathcal{O}$, let $\mathbb{A}_{\mathfrak{f}} \subset \mathbb{A}_{\text{fin}}^\times$ be the subgroup consisting of all elements $x = (x_v) \in \mathbb{A}_{\text{fin}}^\times$ such that $x_v \in 1 + \mathfrak{f}\mathcal{O}_v$ whenever $v = v_{\mathfrak{p}}$ with $\mathfrak{p}|\mathfrak{f}$.

By the Artin-Whaples approximation theorem (or simply the Chinese remainder theorem), we can write

$$\mathbb{A}^\times = K^\times \cdot \mathbb{A}_\mathfrak{f} \cdot (\mathbb{R} \otimes_\mathbb{Q} K)^\times \tag{2.11}$$

with

$$K^\times \cap (\mathbb{A}_\mathfrak{f}(\mathbb{R} \otimes_\mathbb{Q} K)^\times) = K_\mathfrak{f}. \tag{2.12}$$

Suppose now that χ is a Hecke character of K to the modulus $\mathfrak{f}$ and with infinity type χ_∞. Given $x \in \mathbb{A}^\times$, we use (2.11) to write

$$x = \alpha \cdot y \cdot r \tag{2.13}$$

with $\alpha \in K^\times$, $y \in \mathbb{A}_\mathfrak{f}$, and $r \in (\mathbb{R} \otimes_\mathbb{Q} K)^\times$. Then we set

$$\chi_\mathbb{A}(x) = \chi(\mathfrak{a}_y)\chi_\infty(r), \tag{2.14}$$

where

$$\mathfrak{a}_y = \prod_{v \nmid \infty} \mathfrak{p}_v^{\operatorname{ord}_v y_v} \tag{2.15}$$

and $\mathfrak{p}_v$ is the prime ideal of $\mathcal{O}$ underlying v. The definition of $\mathbb{A}^\times$ as a restricted direct product ensures that $\operatorname{ord}_v y_v = 0$ for all but finitely many v, whence (2.15) is meaningful. As $y \in \mathbb{A}_\mathfrak{f}$ we have $\mathfrak{a}_y \in I(\mathfrak{f})$, and therefore $\chi(\mathfrak{a}_y)$ is defined.

By (2.12), the definition (2.14) is unambiguous provided the right-hand side is trivial whenever $x = 1$, $y = \alpha_{\text{fin}}^{-1}$, and $r = \alpha_\infty^{-1}$ with $\alpha \in K_\mathfrak{f}$. In other words if $\alpha \in K_\mathfrak{f}$ then we must have $\chi(\alpha\mathcal{O}) = \chi_\infty(1 \otimes \alpha)^{-1}$. This is precisely the defining property (1.10) of a Hecke character.

By construction, $\chi_\mathbb{A}$ is trivial on $K^\times$. To see that it is continuous, put

$$\widehat{\mathcal{O}}_\mathfrak{f} = \mathbb{A}_\mathfrak{f} \cap \widehat{\mathcal{O}}^\times. \tag{2.16}$$

If y in (2.13) belongs to $\widehat{\mathcal{O}}_\mathfrak{f}$ then $\mathfrak{a}_y = \mathcal{O}$, whence (2.14) becomes $\chi_\mathbb{A}(x) = \chi_\infty(r)$. Furthermore, if x belongs to the open subgroup $\widehat{\mathcal{O}}_\mathfrak{f} \times (\mathbb{R} \otimes_\mathbb{Q} K)^\times$ of $\mathbb{A}^\times$ then we can take $y = x_{\text{fin}}$ and $r = x_\infty$ in (2.13), whence the restriction of $\chi_\mathbb{A}$ to this open subgroup is the function $x \mapsto \chi_\infty(x_\infty)$. As χ_∞ is continuous by assumption, the continuity of $\chi_\mathbb{A}$ on all of $\mathbb{A}^\times$ follows from the fact that a group homomorphism is continuous if and only if its restriction to some open subgroup is.

Thus $\chi_\mathbb{A}$ is an idele class character. A review of the construction shows that if $\mathfrak{f}$ had been replaced by an ideal divisible by $\mathfrak{f}$ then $\chi_\mathbb{A}$ would have been unchanged. It follows that $\chi_\mathbb{A}$ depends only on the primitive Hecke character determined by χ. Furthermore, once one has the notion of a "conductor" (still to come), one can verify that every idele class character has the form $\chi_\mathbb{A}$ for a unique primitive χ.

In practice, since χ and $\chi_\mathbb{A}$ can be distinguished by their arguments – ideals and ideles respectively – the subscript on $\chi_\mathbb{A}$ will usually be omitted. For example, (2.14) can be written $\chi(x) = \chi(\mathfrak{a}_y)\chi_\infty(r)$.

2.2. Local components of idele class characters

For some calculations it is useful to be able to go directly from a Hecke character χ written as in Proposition 1.2 to the local components χ_v in (2.8). The following proposition helps us to do so. Let χ be a primitive Hecke character of K of conductor $\mathfrak{f}$ and infinity type χ_∞, and let ε be the character of $(\mathcal{O}/\mathfrak{f})^\times$ such that

$$\chi(\alpha\mathcal{O}) = \varepsilon(\alpha)\chi_\infty^{-1}(1 \otimes \alpha) \tag{2.17}$$

for $\alpha \in K(\mathfrak{f})$. (As in Proposition 1.2, we are viewing ε as a character of $K(\mathfrak{f})$ via the identification $K(\mathfrak{f})/K_{\mathfrak{f}} \cong (\mathcal{O}/\mathfrak{f})^{\times}$.) By the Chinese remainder theorem we can write

$$(\mathcal{O}/\mathfrak{f})^{\times} = \prod_{\mathfrak{p}|\mathfrak{f}} (\mathcal{O}/\mathfrak{p}^{n(\mathfrak{p})})^{\times}, \tag{2.18}$$

where the product runs over the distinct prime ideals dividing $\mathfrak{f}$ and $n(\mathfrak{p})$ is the multiplicity of $\mathfrak{p}$ in $\mathfrak{f}$. There is a corresponding decomposition

$$\varepsilon = \prod_{\mathfrak{p}|\mathfrak{f}} \varepsilon_{\mathfrak{p}}, \tag{2.19}$$

where $\varepsilon_{\mathfrak{p}}$ is a character of $(\mathcal{O}/\mathfrak{p}^{n(\mathfrak{p})})^{\times}$. If $v = v_{\mathfrak{p}}$ is the place of K corresponding to $\mathfrak{p}$ then we write ε_v for the character of $\mathcal{O}_v^{\times}$ obtained by composing $\varepsilon_{\mathfrak{p}}$ with the natural map of $\mathcal{O}_v^{\times}$ onto $(\mathcal{O}/\mathfrak{p}^{n(\mathfrak{p})})^{\times}$.

Proposition 2.1. *Let $\mathfrak{p}$ be a prime ideal of K and $v = v_{\mathfrak{p}}$ the corresponding finite place. Let $\pi_v \in \mathcal{O}_v$ be a uniformizer.*

(a) *If $\mathfrak{p} \nmid \mathfrak{f}$ then χ_v is unramified and $\chi_v(\pi_v) = \chi(\mathfrak{p})$.*

(b) *If $\mathfrak{p}|\mathfrak{f}$ then $\chi_v|\mathcal{O}_v^{\times} = \varepsilon_v^{-1}$, whence χ_v is in particular ramified. Furthermore, suppose that for some $\beta \in \mathcal{O}$ the principal ideal $\beta\mathcal{O}$ is a power of $\mathfrak{p}$. Then*

$$\chi_v(\beta) = \chi_{\infty}^{-1}(\beta) \cdot \prod_{\substack{\mathfrak{q}|\mathfrak{f} \\ \mathfrak{q} \neq \mathfrak{p}}} \varepsilon_{\mathfrak{q}}(\beta),$$

where $\mathfrak{q}$ runs over prime ideals dividing $\mathfrak{f}$ but different from $\mathfrak{p}$.

PROOF. Throughout the proof, w denotes an arbitrary place of K.

(a) Given $z \in \mathcal{O}_v^{\times}$, take $x = (x_w)$ to be the idele with $x_v = z$ and $x_w = 1$ for $w \neq v$. Then we may take $\alpha = r = 1$ and $y = x$ in (2.13), whence $\mathfrak{a}_y = \mathcal{O}$. So (2.14) and (2.8) give $\chi_v(z) = \chi(x) = 1$, and we conclude that χ_v is unramified. On the other hand, choosing $x = (x_w)$ to be the idele with $x_v = \pi_v$ and $x_w = 1$ for $w \neq v$, we may again take $\alpha = r = 1$ and $y = x$ in (2.13), but this time we get $\mathfrak{a}_y = \mathfrak{p}$ and consequently $\chi_v(\pi_v) = \chi(\mathfrak{p})$.

(b) Given $z \in \mathcal{O}_v^{\times}$, take $x = (x_w)$ to be the idele with $x_v = z$ and $x_w = 1$ for $w \neq v$. Applying the Chinese remainder theorem and the notation of (2.18), we choose $\alpha \in \mathcal{O}$ so that $\alpha \equiv z \bmod \mathfrak{p}^{n(\mathfrak{p})}\mathcal{O}_v$ and also $\alpha \equiv 1 \bmod \mathfrak{q}^{n(\mathfrak{q})}$ for all prime ideals $\mathfrak{q}$ dividing $\mathfrak{f}$ but different from $\mathfrak{p}$. Then $\alpha_{\text{fin}}^{-1}x \in \mathbb{A}_{\mathfrak{f}}$, so we may take $y = \alpha_{\text{fin}}^{-1}x$ and $r = \alpha_{\infty}^{-1}$ in (2.13). Then $\mathfrak{a}_y = \alpha^{-1}\mathcal{O}$, and consequently (2.14) gives $\chi(x) = \chi(\alpha^{-1}\mathcal{O})\chi_{\infty}(1 \otimes \alpha^{-1})$. Replacing α by α^{-1} in (2.17), we deduce that $\chi(x) = \varepsilon^{-1}(\alpha)$. In view of the choice of x and α, we obtain $\chi_v(z) = \varepsilon_v^{-1}(z)$ by (2.8) and (2.19). Thus $\chi_v|\mathcal{O}_v^{\times} = \varepsilon_v^{-1}$.

Now suppose that $\beta\mathcal{O}$ is a power of $\mathfrak{p}$. Evaluating both sides of (2.8) at β gives

$$1 = \prod_{w|\mathfrak{f}\infty} \chi_w(\beta), \tag{2.20}$$

because $\chi|K^{\times}$ is trivial and χ_w is unramified for $w \nmid \mathfrak{f}\infty$. Now if $\mathfrak{q}$ is a prime ideal dividing $\mathfrak{f}$ and $\mathfrak{q} \neq \mathfrak{p}$ then $\beta \in \mathcal{O}_w^{\times}$, where w is the place corresponding to $\mathfrak{q}$. Hence we can apply the result of the previous paragraph with $\mathfrak{p}$ replaced by $\mathfrak{q}$, obtaining $\chi_w(\beta) = \varepsilon_{\mathfrak{q}}^{-1}(\beta)$. Inserting this information in (2.20), we obtain the claimed formula for $\chi_v(\beta)$. □

2.3. An example

To illustrate both the map $\chi \mapsto \chi_{\mathbb{A}}$ and the use of Proposition 2.1, consider the case where χ is the absolute norm, $\chi(\mathfrak{a}) = \mathbf{N}\mathfrak{a}$. We claim that $\chi_{\mathbb{A}}$ is $|| * ||^{-1}$, the reciprocal of the idelic norm defined by (2.9). Indeed in the notation of Proposition 1.2 we have $\mathfrak{f} = \mathcal{O}$ and $\varepsilon = 1$, and by taking $s_0 = 1$ in (1.13) we see that χ_∞ is the product of the reciprocals of the local norms at infinity. In other words, if v is an infinite place then $\chi_v = || * ||_v^{-1}$. To see that the same is true at the finite places, let $\mathfrak{p}$ be a prime ideal of K and v the corresponding finite place. By part (a) of Proposition 2.1, χ_v is the unramified character of $K_v^\times$ taking the value $\mathbf{N}\mathfrak{p}$ on any uniformizer π_v. But the local norm is also unramified, and $\mathbf{N}\mathfrak{p} = q_v = ||\pi_v||_v^{-1}$, where q_v is the order of the residue class field of K. Hence again $\chi_v = || * ||_v^{-1}$, and we conclude that $\chi_{\mathbb{A}} = || * ||^{-1}$, as claimed.

2.4. The conductor

As pointed out in the introduction, a complex representation of a profinite group is trivial on an open subgroup. The one-dimensional case of this remark underlies some verifications that have already been passed over without comment, for example the fact that every idele class character is a restricted direct product of local characters as in (2.8), or the fact that the map $\chi \mapsto \chi_{\mathbb{A}}$ from Hecke characters to idele class characters is surjective. The relevant profinite groups are $\widehat{\mathcal{O}}$ and $\widehat{\mathcal{O}}^\times$; instead of (2.3) and (2.5) we write $\widehat{\mathcal{O}} = \varprojlim_{\mathfrak{f}} \mathcal{O}/\mathfrak{f}$ and $\widehat{\mathcal{O}}^\times = \varprojlim_{\mathfrak{f}} (\mathcal{O}/\mathfrak{f})^\times$, where $\mathfrak{f}$ runs over the nonzero integral ideals of K ordered by divisibility. In particular, the expression for $\widehat{\mathcal{O}}^\times$ as an inverse limit shows that the restriction of an idele class character to $\widehat{\mathcal{O}}^\times$ is trivial on $\widehat{\mathcal{O}}_{\mathfrak{f}}$ for some $\mathfrak{f}$, where $\widehat{\mathcal{O}}_{\mathfrak{f}}$ is as in (2.16).

The same remark holds locally at every finite place v: If $v = v_{\mathfrak{p}}$ then we have $\mathcal{O}_v = \varprojlim_n \mathcal{O}/\mathfrak{p}^n$ and $\mathcal{O}_v^\times = \varprojlim_n (\mathcal{O}/\mathfrak{p}^n)^\times$, and we deduce that any character of $K_v^\times$ is trivial on $1 + \pi_v^n\mathcal{O}_v$ for some $n \geqslant 1$.

These remarks permit us to define the **conductor** of a character both locally and globally. Consider first the local case. If v is a finite place of K and χ_v a character of $K_v^\times$ then the **exponent of the conductor** of χ_v is the integer $a(\chi_v)$ defined as follows: If χ_v is unramified then $a(\chi_v) = 0$, and if χ_v is ramified then $a(\chi_v)$ is the smallest integer $n \geqslant 1$ such that χ_v is trivial on $1 + \pi_v^n\mathcal{O}_v$. The conductor of χ_v is the ideal $\pi_v^{a(\chi_v)}\mathcal{O}_v$ of $\mathcal{O}_v$. Turning now to the global setting, we have two ways of defining the conductor $\mathfrak{f}(\chi)$ of an idele class character χ of K: Either we consider integral ideals $\mathfrak{f}$ of K such that χ is trivial on $\widehat{\mathcal{O}}_{\mathfrak{f}}$, defining $\mathfrak{f}(\chi)$ to be the smallest such $\mathfrak{f}$ in terms of divisibility, or else we set

$$\mathfrak{f}(\chi) = \prod_{v \nmid \infty} \mathfrak{p}_v^{a(\chi_v)}, \tag{2.21}$$

where $\mathfrak{p}_v$ is the prime ideal determined by v. One verifies that these two definitions are equivalent to each other and also to our original definition of the conductor of a primitive Hecke character when χ is viewed as such.

3. The functional equation

Our goal now is to see how the adelic viewpoint facilitates the statement of the functional equation for Hecke L-functions. The key point is that once we think of an idele class character χ as a product of local characters using (2.8) we can define the objects appearing in the functional equation of $L(s, \chi)$ as products of local objects also. In the case of the conductor $\mathfrak{f}(\chi)$ we have already taken this step in (2.21), although the benefit gained may not yet be apparent. The real prize we anticipate is a factorization of the root number $W(\chi)$. But first we consider the factorization of $L(s,\chi)$ itself as a product of local L-factors.

3.1. L-factors

It may appear at first that there is nothing new here. Given a finite place v of K and a character χ_v of $K_v^\times$, we set

$$L(s,\chi_v) = \begin{cases} (1-\chi_v(\pi_v)q_v^{-s})^{-1} & \text{if } \chi_v \text{ is unramified} \\ 1 & \text{if } \chi_v \text{ is ramified,} \end{cases} \tag{2.22}$$

where as before, π_v is a uniformizer and q_v the order of the residue class field of $\mathcal{O}_v$. As we have already noted, the fact that χ_v is unramified means precisely that χ_v is independent of the choice of π_v. Now if χ is an idele class character of K then a comparison of (2.10) and (2.22) shows that the global L-function $L(s,\chi)$ is the product of the local L-factors:

$$L(s,\chi) = \prod_{v\nmid\infty} L(s,\chi_v). \tag{2.23}$$

Furthermore, if one thinks of χ as a primitive Hecke character then one can verify that the original definition (1.18) of $L(s,\chi)$ is equivalent to (2.10) and (2.23). (The key point is that if v is a finite place where χ is unramified and x in (2.13) is the idele with π_v at the place v and 1 at all other places then we can take $\alpha = r = 1$ and $y = x$, whence $\mathfrak{a}_y = \mathfrak{p}_v$.) So our definitions are compatible, but the introduction of local L-factors appears to add nothing new.

However from the adelic point of view it is natural to associate L-factors not only to the finite places of K but also to the infinite places, where the "L-factors" turn out to be the gamma factors in the functional equation. In fact what we have been calling $L(s,\chi)$ would in some contexts be regarded as merely "the finite part of the L-function," $L_{\text{fin}}(s,\chi)$, and the notation $L(s,\chi)$ would be reserved for the "completed L-function" $L_\infty(s,\chi)L_{\text{fin}}(s,\chi)$, where $L_\infty(s,\chi)$ is the product of the L-factors at the infinite places:

$$L_\infty(s,\chi) = \prod_{v\mid\infty} L(s,\chi_v). \tag{2.24}$$

Hence the completed L-function $L_\infty(s,\chi)L_{\text{fin}}(s,\chi)$ is the product of the L-factors $L(s,\chi_v)$ over *all* the places of K and includes both the traditional L-function $L_{\text{fin}}(s,\chi)$ and its gamma factors.

In practice we will continue to write $L(s,\chi)$ for the traditional L-function $L_{\text{fin}}(s,\chi)$, but the factorization (2.24) will be used in the statement of the functional equation of $L(s,\chi)$. Hence we need to make the local factors in (2.24) explicit.

Suppose first that v is a real place. Then $K_v = \mathbb{R}$, and the identification is unique because $\mathbb{R}$ has no nontrivial automorphisms (even as an abstract field). Thus a character χ_v of $K_v^\times$ can be identified with a character of $\mathbb{R}^\times$. But a character of $\mathbb{R}^\times$ is necessarily of the form $t \mapsto |t|^{s_0}(t/|t|)^m$ with unique numbers $s_0 \in \mathbb{C}$ and $m \in \{0,1\}$. We set

$$L(s,\chi_v) = \Gamma_{\mathbb{R}}(s+s_0+m), \tag{2.25}$$

where we recall that the real gamma factor $\Gamma_{\mathbb{R}}(s)$ is defined by (1.7).

Next suppose that v is complex. Then there are two possible identifications $K_v \cong \mathbb{C}$. Choosing one of them, we may view a character χ_v of $K_v^\times$ as a character of $\mathbb{C}^\times$. Then χ_v is necessarily of the form $z \mapsto |z|^{2s_0}(z/|z|)^m$ with unique numbers $s_0 \in \mathbb{C}$ and $m \in \mathbb{Z}$. We set

$$L(s,\chi_v) = \Gamma_{\mathbb{C}}(s + s_0 + |m|/2). \tag{2.26}$$

If we replace our chosen identification of K_v with $\mathbb{C}$ with the complex-conjugate identification then $\chi_v(z)$ is replaced by $\chi_v(\overline{z})$ and hence m by $-m$; but (2.26) stays the same. Thus $L(s,\chi_v)$ is well defined.

3.2. Hecke's theorem

Given an idele class character χ of K, put

$$\Lambda(s,\chi) = (D\mathbf{N}\mathfrak{f}(\chi))^{s/2} L_\infty(s,\chi)L(s,\chi), \tag{2.27}$$

where D is the absolute value of the discriminant of K and $L_\infty(s,\chi)$ and $L(s,\chi)$ are as in (2.24) and (2.23) respectively. Let c be as in Proposition 1.1, and put

$$\begin{cases} w = 2c \\ k = w + 1, \end{cases} \tag{2.28}$$

so that $k = 2c + 1$.

Theorem 2.1. *There is a constant $W(\chi) \in \mathbb{C}$ with $|W(\chi)| = 1$ such that*

$$\Lambda(s,\chi) = W(\chi)\Lambda(k - s, \overline{\chi}).$$

Furthermore, if χ is the trivial character then $W(\chi) = 1$.

Of course if χ is the trivial character then $L(s,\chi)$ is just the Dedekind zeta function $\zeta_K(s)$ of K, and $\Lambda(s,\chi)$ is often written as $Z_K(s)$ in this case. If $K = \mathbb{Q}$ then we will continue to write $\zeta_{\mathbb{Q}}(s)$ and $Z_{\mathbb{Q}}(s)$ simply as $\zeta(s)$ and $Z(s)$, as we did in Lecture 1.

4. Quadratic root numbers

Before saying even one word about local root numbers, we can deduce from Theorem 2.1 that root numbers of quadratic idele class characters are trivial:

Theorem 2.2. *Suppose that χ is an idele class character of K such that χ^2 is trivial. Then $W(\chi) = 1$.*

Proof. If χ is the trivial character then Theorem 2.2 is already contained in Theorem 2.1, so we may assume that χ is quadratic. We will use the identity

$$\zeta_L(s) = \zeta_K(s)L(s, \mathrm{sign}_{L/K}), \tag{2.29}$$

where L/K is a quadratic extension of number fields and $\mathrm{sign}_{L/K}$ is **the quadratic Hecke character associated to** L/K. The meaning of this last phrase is as follows. Let $\mathfrak{d}_{L/K}$ be the relative discriminant ideal of L/K. Then $\mathrm{sign}_{L/K}$: $I(\mathfrak{d}_{L/K}) \to \{\pm 1\}$ is the unique homomorphism satisfying

$$\mathrm{sign}_{L/K}(\mathfrak{p}) = \begin{cases} 1 & \text{if } \mathfrak{p} \text{ splits in } L \\ -1 & \text{if } \mathfrak{p} \text{ remains prime in } L \end{cases} \tag{2.30}$$

for prime ideals $\mathfrak{p}$ of K unramified in L. The fact that the homomorphism defined by (2.30) is actually a Hecke character of K (indeed a primitive Hecke character of conductor $\mathfrak{d}_{L/K}$) is essentially quadratic reciprocity over number fields, although a little bit of work is required to go back and forth between this statement and the classical version found for example in [**43**], p. 246. In any case, to prove the theorem we combine (2.29) with the fact that any quadratic idele class character χ of K has the form $\chi = \mathrm{sign}_{L/K}$ for some quadratic extension L of K. In other words, given χ we can write

$$\zeta_L(s) = \zeta_K(s)L(s,\chi), \tag{2.31}$$

where the quadratic extension L of K is determined by χ.

We claim that (2.31) remains valid when $\zeta_L(s)$, $\zeta_K(s)$, and $L(s,\chi)$ are replaced by their normalized versions:

$$Z_L(s) = Z_K(s)\Lambda(s,\chi). \tag{2.32}$$

To verify (2.32), write $r_{1/1}(L/K)$ for the number of real places of K which split into two real places of L and $r_{2/1}(L/K)$ for the number of real places of K which ramify into a complex place of L. Since every real place of K either splits or ramifies, the number of such places satisfies

$$r_1(K) = r_{1/1}(L/K) + r_{2/1}(L/K). \tag{2.33}$$

Also

$$\begin{cases} r_1(L) = 2r_{1/1}(L/K) \\ r_2(L) = 2r_2(K) + r_{2/1}(L/K), \end{cases} \tag{2.34}$$

because every place of L is an extension of a unique place of K. Now $\zeta_L(s)$ and $\zeta_K(s)$ are the L-functions of the trivial idele class character of L and K respectively. Hence in applying (2.24) and (2.27) we take $s_0 = m = 0$ in (2.25) and (2.26), obtaining

$$Z_L(s) = D_L^{s/2}\Gamma_{\mathbb{R}}(s)^{r_1(L)}\Gamma_{\mathbb{C}}(s)^{r_2(L)}\zeta_L(s) \tag{2.35}$$

and

$$Z_K(s) = D_K^{s/2}\Gamma_{\mathbb{R}}(s)^{r_1(K)}\Gamma_{\mathbb{C}}(s)^{r_2(K)}\zeta_K(s). \tag{2.36}$$

As for $\Lambda(s,\chi)$, the relation $\chi = \mathrm{sign}_{L/K}$ has the following consequences: If v is a complex place of K then $(s_0, m) = (0,0)$ in (2.26), while if v is a real place then $(s_0, m) = (0,0)$ or $(s_0, m) = (0,1)$ in (2.25) according as v splits or ramifies in L. As $\mathfrak{f}(\chi) = \mathfrak{d}_{L/K}$, we see that (2.27) gives

$$\Lambda(s,\chi) = (D_{L/K}D_K)^{s/2}\Gamma_{\mathbb{R}}(s)^{r_{1/1}(L/K)}\Gamma_{\mathbb{R}}(s+1)^{r_{2/1}(L/K)}\Gamma_{\mathbb{C}}(s)^{r_2(K)}L(s,\chi), \tag{2.37}$$

where $D_{L/K}$ is the absolute norm of the relative different ideal of L/K and hence also the absolute norm of $\mathfrak{d}_{L/K}$. Now compare the product of (2.36) and (2.37) with (2.35). Taking account of (2.33) and (2.34) as well as the duplication formula (1.9) and the standard relation $D_L = D_{L/K}D_K^2$, we obtain (2.32).

To deduce the theorem we apply Theorem 2.1 on both sides of (2.32), obtaining

$$Z_L(1-s) = Z_K(1-s)W(\chi)\Lambda(1-s,\chi). \tag{2.38}$$

Replacing s by $1-s$ in (2.32) and comparing the result with (2.38), we conclude that $W(\chi) = 1$. $\square$

5. Local root numbers

Let K be a number field and χ an idele class character of K. As we have already hinted, the root number $W(\chi)$ defined by Theorem 2.1 has a factorization

$$W(\chi) = \prod_{v} W(\chi_v), \tag{2.39}$$

where $W(\chi_v)$ is the **local root number** attached to χ_v and is equal to 1 for all but finitely many places v. We shall give formulas for $W(\chi_v)$ and then illustrate their use by verifying Proposition 1.5. Initially we treat these formulas as a black box, ignoring their provenance. Afterwards we fill in a number of points: the implicit dependence of the local root number on a choice of additive character, the connection with epsilon factors, and so on.

5.1. Formulas for local root numbers

We change notation, writing K_v and χ_v simply as K and χ respectively. Thus K is a finite extension of $\mathbb{Q}_p$ for some fixed $p \leqslant \infty$ and χ is a character of $K^\times$; a notation like $||*||$, for example, now refers to the *local* norm on $K^\times$. An important point about root numbers is that they see only the unitary part of a character. In other words, if we put $\chi_{\text{unit}} = \chi/|\chi|$ as before then

$$W(\chi) = W(\chi_{\text{unit}}). \tag{2.40}$$

Hence if it is convenient to do so one may assume that χ is unitary.

If K is archimedean then there is no need to do so. Indeed if K is archimedean then χ has the form $x \mapsto ||x||^{s_0}(x/|x|)^m$ with unique elements $s_0 \in \mathbb{C}$ and either $m \in \{0,1\}$ or $m \in \mathbb{Z}$ according as $K = \mathbb{R}$ or $K \cong \mathbb{C}$. The character χ is unitary if and only if $s_0 \in i\mathbb{R}$, but $W(\chi)$ depends only on m, not on s_0:

$$W(\chi) = i^{-|m|}. \tag{2.41}$$

In other words, if $K = \mathbb{R}$ and $\chi(t) = |t|^{s_0}(t/|t|)^m$ then $W(\chi) = i^{-m}$, and if $K \cong \mathbb{C}$ and $\chi(z) = |z|^{2s_0}(z/|z|)^m$ then (2.41) holds and is independent of the identification $K \cong \mathbb{C}$ chosen: the alternative identification merely replaces m by $-m$.

Now take $p < \infty$. In the nonarchimedean setting it would be a slight convenience to assume that χ is unitary, but instead we shall replace χ by χ_{unit} in the formulas themselves, so that the formulas are universally valid. As before, we write $a(\chi)$ for the exponent of the conductor of χ. Furthermore, we write d for the **exponent of the different ideal** of K. Thus if π is a uniformizer of K then $\pi^d\mathcal{O}$ is the different ideal of K over $\mathbb{Q}_p$. Now put $\mathfrak{f}(\chi) = \pi^{a(\chi)}\mathcal{O}$ and choose an element $\gamma \in \pi^{a(\chi)+d}\mathcal{O}^\times$. It follows from the definitions that the functions $x \mapsto \chi(x)$ and $x \mapsto e^{2\pi i \operatorname{tr}_{K/\mathbb{Q}_p}(x/\gamma)}$ on $\mathcal{O}^\times$ depend only on the residue class of x in $(\mathcal{O}/\mathfrak{f}(\chi))^\times$. Writing q for the order of the residue class field of K, we have

$$W(\chi) = \chi_{\text{unit}}(\gamma) \cdot q^{-a(\chi)/2} \sum_{x \in (\mathcal{O}/\mathfrak{f}(\chi))^\times} \chi^{-1}(x) e^{2\pi i \operatorname{tr}_{K/\mathbb{Q}_p}(x/\gamma)}. \tag{2.42}$$

Note that we have not bothered to write $\chi^{-1}(x)$ as $\chi_{\text{unit}}^{-1}(x)$, because the restriction of any character of $K^\times$ to $\mathcal{O}^\times$ has finite order and is therefore automatically unitary.

It is somewhat unconventional to express the nonarchimedean local root number by a single formula, as in (2.42). Normally something like (2.42) would be stated for ramified characters only, and for unramified characters one would give the separate formula

$$W(\chi) = \chi_{\text{unit}}(\gamma) \qquad (a(\chi) = 0). \tag{2.43}$$

However (2.42) actually reduces to (2.43) in the unramified case provided we agree that if $a(\chi) = 0$ and hence $\mathfrak{f}(\chi) = \mathcal{O}$ then $(\mathcal{O}/\mathfrak{f}(\chi))^\times$ has just one element, namely the coset of 1.

Having stated the local formulas, we return to the global setting and write K for a number field, χ for an idele class character of K, and $\mathfrak{f}(\chi)$ for the conductor of χ. Let $\mathfrak{D}$ denote the different ideal of K. It follows from (2.43) that if v is the finite place of K corresponding to a prime ideal $\mathfrak{p}$ of K not dividing $\mathfrak{D}\mathfrak{f}(\chi)$ then $W(\chi_v) = 1$. In particular, since $\mathfrak{p} \nmid \mathfrak{D}\mathfrak{f}(\chi)$ for all but finitely many $\mathfrak{p}$ we conclude that the product in (2.39) has only finitely many factors different from 1, as claimed.

5.2. An example

We illustrate these formulas by proving Proposition 1.5, or in other words by computing $W(\chi)$ for $\chi \in X(D)$ with D odd. Thus K is now $\mathbb{Q}(\sqrt{-D})$. By Proposition 1.4, the different ideal $\mathfrak{D} = \sqrt{-D}\mathcal{O}$ coincides with the conductor $\mathfrak{f}(\chi)$, whence (2.43) gives $W(\chi_v) = 1$ for all finite places $v \nmid \mathfrak{D}$. And since $\chi_\infty(z) = z^{-1}$ we have $W(\chi_\infty) = i^{-1}$ (take $s_0 = -1/2$ and $m = -1$ in (2.41)). Thus (2.39) gives

$$W(\chi) = i^{-1} \prod_{v|\mathfrak{D}} W(\chi_v), \tag{2.44}$$

and it remains to evaluate $W(\chi_v)$ at the places v above the prime divisors p of D.

Suppose then that $p|D$ and that $v = v_\mathfrak{p}$, where $\mathfrak{p}$ is the prime ideal above p. Then χ_v is ramified, so the appropriate local formula is (2.42). Since $a(\chi_v) = d_v = 1$ (d_v being the exponent of the local different at v) we may take the quantity γ in (2.42) to be p. Furthermore, since the natural map $(\mathbb{Z}/p\mathbb{Z})^\times \to (\mathcal{O}/\mathfrak{p})^\times$ is an isomorphism, the coset representatives x in (2.42) may be taken to be rational integers. So (2.42) becomes

$$W(\chi_v) = (\chi_v)_{\text{unit}}(p) \cdot p^{-1/2} \sum_{n \in (\mathbb{Z}/p\mathbb{Z})^\times} \chi_v^{-1}(n) e^{2\pi i(2n/p)}. \tag{2.45}$$

Now recall condition (b) in the original definition of $X(D)$: $\chi(n\mathcal{O}) = \kappa(n)n$ for $n \in \mathbb{Z}$ prime to D. Here κ is the Kronecker symbol with numerator $-D$, as before. It follows that if $\chi|P(\mathfrak{f}(\chi))$ is written as in Proposition 1.2 then $\varepsilon(n) = \kappa(n)$. Thus in the notation of Proposition 2.1, we have $\varepsilon_v(n) = \lambda(n)$ for n prime to p, where λ is the Legendre symbol at p. Consequently the proposition just cited gives $\chi_v^{-1}(n) = \lambda(n)$ for such n. Making this substitution in (2.45), and replacing the summation over n by a summation over $\overline{2}n$, where $\overline{2}$ is a representative for the multiplicative inverse of 2 in $(\mathbb{Z}/p\mathbb{Z})^\times$, we find

$$W(\chi_v) = \lambda(\overline{2})(\chi_v)_{\text{unit}}(p) \cdot p^{-1/2} \sum_{n \in (\mathbb{Z}/p\mathbb{Z})^\times} \lambda(n) e^{2\pi i(n/p)}. \tag{2.46}$$

Of course $\lambda(\overline{2}) = \lambda(2)$ since λ is quadratic. Furthermore, by taking $\beta = p$ in Proposition 2.1 we obtain

$$\chi_v(p) = p \prod_{\substack{q|D \\ q \neq p}} \left(\frac{p}{q}\right), \tag{2.47}$$

and the factor of p on the right-hand side disappears when χ_v is replaced by $(\chi_v)_{\text{unit}}$ on the left-hand side. Finally, the sum in (2.46) is evaluated by Gauss's formula (1.1). Thus (2.46) becomes

$$W(\chi_v) = i^{\delta}\left(\frac{2}{p}\right)\prod_{\substack{q|D\\ q\neq p}}\left(\frac{p}{q}\right) \tag{2.48}$$

with δ equal to 0 or 1 according as p is 1 mod 4 or 3 mod 4.

The rest is bookkeeping. Let t be the number of prime divisors of D which are congruent to 3 mod 4. Substituting (2.48) in (2.44), we get

$$W(\chi) = i^{t-1}\left(\frac{2}{D}\right)\prod_{p\neq q}\left(\frac{p}{q}\right), \tag{2.49}$$

where the product on the right-hand side of (2.49) runs over pairs (p,q) of distinct prime divisors of D. By quadratic reciprocity, this product is $(-1)^{t(t-1)/2}$. On the other hand, the odd integer $-D$ is a discriminant, hence congruent to 1 mod 4. Therefore t is odd, and we can write $i^{t-1} = (-1)^{(t-1)/2}$. Multiplying this factor by the factor $(-1)^{t(t-1)/2}$ coming from quadratic reciprocity, we obtain $(-1)^{(t^2-1)/2}$, which is 1. Thus (2.49) does give Gross's result, Proposition 1.5.

6. An open problem

The preceding example illustrates a simple point: Armed with the formulas (2.39) through (2.43), we can in principle detect trivial central zeros of Hecke L-functions whenever they exist. But do we always care? Does a trivial central zero of a Hecke L-function, or indeed of any L-function, always have arithmetic significance? Consider for example the L-functions associated to Maass forms for $\mathrm{SL}(2,\mathbb{Z})$. A theorem of Venkov [**98**] implies that half of these L-functions have a trivial central zero. What is the arithmetic significance of this fact, if any?

The L-functions associated to Maass forms for $\mathrm{SL}(2,\mathbb{Z})$ lie outside the scope of these lectures, but a satisfactory substitute is available, namely Hecke L-functions which are of "Maass type" in the sense that they coincide with the L-functions associated to certain Maass forms for congruence subgroups of $\mathrm{SL}(2,\mathbb{Z})$. Let K be a real quadratic field, viewed as a subfield of $\mathbb{R}$, and write $\alpha \mapsto \alpha'$ for the nonidentity embedding of K in $\mathbb{R}$. We will call a primitive Hecke character χ of K **equivariant** if $\chi(\mathfrak{a}') = \overline{\chi(\mathfrak{a})}$ for $\mathfrak{a} \in I(\mathfrak{f}(\chi))$. The Hecke characters of Maass type to be considered here have the form $\chi = \eta\chi_0$, where η is a primitive equivariant Hecke character of K of finite order and $\chi_0 : I \to \mathbb{C}^\times$ is the Hecke character of K defined in two steps as follows. First we define χ_0 on P by the formula

$$\chi_0(\alpha\mathcal{O}) = |\alpha/\alpha'|^{\pi i/\log\varepsilon_0} \qquad (\alpha \in K^\times), \tag{2.50}$$

where ε_0 is the fundamental unit of K. Then we extend χ_0 to a character of I arbitrarily. In Exercise 2.6 the reader is invited to verify that (2.50) gives a well-defined function on principal ideals and that any extension of (2.50) to I is an equivariant Hecke character. Now put $\chi = \eta\chi_0$, and let D be the discriminant of K. Applying Theorem 2.1, we find that the functional equation of $L(s,\chi)$ can be written $\Lambda(s,\chi) = W(\chi)\Lambda(1-s,\chi)$ with

$$\Lambda(s,\chi) = (D\mathbf{N}\mathfrak{f}(\eta))^{s/2}\Gamma_{\mathbb{R}}(s+\pi i/\log\varepsilon_0)\Gamma_{\mathbb{R}}(s-\pi i/\log\varepsilon_0)L(s,\chi). \tag{2.51}$$

However if we make the additional assumption that the conductor of η has the form $\mathfrak{f}(\eta) = N\sqrt{D}\mathcal{O}$ for some rational integer $N \geqslant 1$ then $W(\chi) = W(\eta)$, and consequently the functional equation becomes

$$\Lambda(s, \chi) = W(\eta)\Lambda(1 - s, \chi). \tag{2.52}$$

Incidentally, the equivariance of η already ensures that $\mathfrak{f}(\eta)' = \mathfrak{f}(\eta)$ and hence that $\mathfrak{f}(\eta)$ has the form $N\mathfrak{C}$ for some ideal $\mathfrak{C}$ of $\mathcal{O}$ dividing $\sqrt{D}\mathcal{O}$, so the condition $\mathfrak{f}(\eta) = N\sqrt{D}\mathcal{O}$ is relatively mild. In any case, assume that the condition is satisfied. Then (2.52) suggests the following question:

Problem 2. *Suppose that $W(\eta) = -1$. Does the resulting trivial central zero of $L(s, \chi)$ have any arithmetic significance?*

An example where the hypothesis $W(\eta) = -1$ is satisfied will be given below. Historically, the first examples of a Hecke character η of finite order for which $L(s, \eta) = L(s, \overline{\eta})$ and $W(\eta) = -1$ were given by Armitage [**3**] and Serre around 1972. At the time there was no arithmetic interpretation for such trivial central zeros, but in the case of certain quartic Hecke characters of real quadratic fields, Fröhlich [**31**] found a connection with Galois module structure: The quartic characters η considered by Fröhlich correspond to certain Galois extensions N of $\mathbb{Q}$ with Galois group the quaternion group of order 8, and Fröhlich proved that $W(\eta)$ is 1 or -1 according as $\mathcal{O}_N$ is or is not a free $\mathbb{Z}[\mathrm{Gal}(N/\mathbb{Q})]$-module. Since then a vast literature has developed relating root numbers to Galois module structure; see for example [**15**], [**16**], [**17**], [**21**], [**32**], and [**95**]. This snippet of history should caution us against dismissing Problem 2 too cavalierly.

Returning to the matter at hand, we need an example of a real quadratic field K and an equivariant Hecke character η of K with $W(\eta) = -1$ and $\mathfrak{f}(\eta)$ of the required form. Take $K = \mathbb{Q}(\sqrt{r(r+4)})$ with a prime $r > 5$ congruent to 1 mod 4. Then the discriminant of K has the form $D = rs$, where $r + 4 = sm^2$ with s square-free and $m \in \mathbb{Z}$. Let κ be the primitive quadratic Dirichlet character of conductor D given by

$$\kappa(n) = \left(\frac{n}{D}\right), \tag{2.53}$$

and define $\varepsilon : (\mathcal{O}/\sqrt{D}\mathcal{O})^\times \to \{\pm 1\}$ by composing κ (viewed as a character of $(\mathbb{Z}/D\mathbb{Z})^\times$) with the canonical identification $(\mathcal{O}/\sqrt{D}\mathcal{O})^\times \cong (\mathbb{Z}/D\mathbb{Z})^\times$. We claim that ε is trivial on the image of $\mathcal{O}^\times$ in $(\mathcal{O}/\sqrt{D}\mathcal{O})^\times$. In view of (2.53) we have at least $\varepsilon(-1) = \kappa(-1) = 1$. On the other hand, put $u = ((r+2) + \sqrt{r(r+4)})/2$. According to Katayama [**48**], u is the fundamental unit of K, so we must verify that $\varepsilon(u) = 1$ also.

To see this, write $(\mathbb{Z}/D\mathbb{Z})^\times \cong (\mathbb{Z}/r\mathbb{Z})^\times \times (\mathbb{Z}/s\mathbb{Z})^\times$, and let $\kappa = \kappa'\kappa''$ be the corresponding decomposition of κ into primitive quadratic characters of conductors r and s respectively. Also put $\mathfrak{f} = \sqrt{D}\mathcal{O}$ and let $\mathfrak{r}$ and $\mathfrak{s}$ be respectively the prime ideal of K over r and the product of the prime ideals dividing ε. Then $(\mathcal{O}/\mathfrak{f})^\times \cong (\mathcal{O}/\mathfrak{r})^\times \times (\mathcal{O}/\mathfrak{s})^\times$ and we have a corresponding decomposition of characters $\varepsilon = \varepsilon'\varepsilon''$. Recalling that $u = ((r+2) + \sqrt{r(r+4)})/2$ and $r + 4 = sm^2$, we see that $u \equiv 1 \bmod \mathfrak{r}$ and $u \equiv -1 \bmod \mathfrak{s}$, whence $\varepsilon(u) = \varepsilon'(1)\varepsilon''(-1) = \kappa''(-1)$. But $\kappa''(-1) = 1$ because $s \equiv 1 \bmod 4$, so ε is trivial on $\mathcal{O}^\times$, as claimed.

It follows that we obtain a well-defined character η of $P(\mathfrak{f})$ by setting

$$\eta(\alpha\mathcal{O}) = \varepsilon(\alpha) \qquad (\alpha \in K(\mathfrak{f})), \tag{2.54}$$

where ε is viewed as a character of $K(\mathfrak{f})$ as in Proposition 1.2. Extending η to $I(\mathfrak{f})$ arbitrarily, we obtain a primitive Hecke character of conductor $\mathfrak{f}$ which we also denote η and which is readily verified to be equivariant. A calculation shows that

$$W(\eta) = \left(\frac{2}{D}\right) \tag{2.55}$$

(Exercise 2.7). But $D = r(r+4)/m^2$ and in particular $D \equiv 5 \bmod 8$, so $W(\eta) = -1$.

7. Epsilon factors

Even a brief perusal of the literature on root numbers will reveal that our discussion has so far neglected two basic issues: the dependence of the local root number on an "additive character" and the relation between root numbers and "epsilon factors." In rectifying these omissions we shall also add a few words about Tate's global and local functional equations [**92**], which are the source of the formulas for $W(\chi)$ stated earlier. For a more thorough treatment see [**92**] or Chapter XIV of [**55**].

7.1. Additive characters

Let K be a finite extension of $\mathbb{Q}_p$ with $p \leqslant \infty$. By an **additive character** of K we mean a nontrivial unitary character $\psi : K \to \mathbb{C}^\times$. There is a canonical choice of ψ which we denote ψ^{can}. If $p = \infty$ then

$$\psi^{\text{can}}(x) = e^{-2\pi i \operatorname{tr}_{K/\mathbb{R}}(x)}, \tag{2.56}$$

and if $p < \infty$ then

$$\psi^{\text{can}}(x) = e^{2\pi i \{\operatorname{tr}_{K/\mathbb{Q}_p}(x)\}_p}. \tag{2.57}$$

Here $\{z\}_p$ is the p-adic principal part of a number $z \in \mathbb{Q}_p$: thus if $z = \sum_{n\in\mathbb{Z}} a_n p^n$ with $a_n \in \{0, 1, \dots, p-1\}$ for all n and $a_n = 0$ for $n \ll 0$ then $\{z\}_p = \sum_{n<0} a_n p^n$. Note that (2.56) can also be written

$$\psi^{\text{can}}(x) = e^{-2\pi i \{\operatorname{tr}_{K/\mathbb{R}}(x)\}_\infty}, \tag{2.58}$$

where $\{t\}_\infty$ is the fractional part of a real number t, defined by the requirements $0 \leqslant \{t\}_\infty < 1$ and $t \equiv \{t\} \bmod \mathbb{Z}$. Normally $\{t\}_\infty$ is written simply as $\{t\}$, but we have included the subscript to emphasize the analogy with (2.57).

Let χ be a character of $K^\times$. Associated to χ and to an arbitrary additive character ψ of K is a **local root number** $W(\chi, \psi)$. The definition of $W(\chi, \psi)$ will be given later, and it will turn out that our $W(\chi)$ coincides with $W(\chi, \psi^{\text{can}})$. Furthermore, any ψ has the form $\psi(x) = \psi^{\text{can}}(ax)$ for some $a \in K^\times$, and we shall see that $W(\chi, \psi) = \chi_{\text{unit}}(a) W(\chi)$. It follows that

$$W(\chi, \psi_b) = \chi_{\text{unit}}(b) W(\chi, \psi) \tag{2.59}$$

for any $b \in K^\times$, where $\psi_b(x) = \psi(bx)$.

We now switch to the global setting and change notation accordingly. Let K be a number field and $\mathbb{A}$ its ring of adeles. A **global additive character** of K is a nontrivial unitary character ψ of $\mathbb{A}$ which is trivial on K. Again there is a canonical choice ψ^{can}. It is simply the product of the local canonical choices:

$$\psi^{\text{can}}(x) = \prod_v \psi_v^{\text{can}}(x_v) \qquad (x = (x_v) \in \mathbb{A}). \tag{2.60}$$

Of course to make sense of (2.60) we must verify that $\psi_v^{\text{can}}(x_v) = 1$ for all but finitely many v. Certainly $x_v \in \mathcal{O}_v$ for all but finitely many finite v, and if $x_v \in \mathcal{O}_v$ then

$\operatorname{tr}_{K_v/\mathbb{Q}_p}(x_v) \in \mathbb{Z}_p$, where p is the residue characteristic of v. Then $\psi_v^{\mathrm{can}}(x_v) = 1$. Hence (2.60) is indeed meaningful.

To see that ψ^{can} is a global additive character of K we must still check that it is trivial on K. So suppose that $\alpha \in K$. Referring to (2.57) and (2.58), we find that the identity to be verified is

$$\sum_{p<\infty} \sum_{v|p} \{\operatorname{tr}_{K_v/\mathbb{Q}_p}(\alpha)\}_p \equiv \sum_{v|\infty} \{\operatorname{tr}_{K_v/\mathbb{R}}(\alpha)\}_\infty \pmod{\mathbb{Z}}.$$

The sum of the local traces at the places of K lying over a given place of $\mathbb{Q}$ is equal to the global trace, so putting $\beta = \operatorname{tr}_{K/\mathbb{Q}}(\alpha)$, we must show that

$$\sum_{p<\infty} \{\beta\}_p \equiv \{\beta\}_\infty \pmod{\mathbb{Z}}.$$

But this is a familiar fact: The principal part of a rational number differs from its fractional part by an integer.

Now let χ be an idele class character of K and ψ a global additive character of K. We define the global root number $W(\chi, \psi)$ to be the product of the local root numbers:

$$W(\chi, \psi) = \prod_v W(\chi_v, \psi_v). \tag{2.61}$$

However the dependence of $W(\chi, \psi)$ on ψ is illusory: We can write $\psi(x) = \psi^{\mathrm{can}}(\alpha x)$ for some $\alpha \in K^\times$, and then the right-hand side of (2.61) becomes

$$\prod_v W(\chi_v, (\psi_v^{\mathrm{can}})_\alpha) = \prod_v (\chi_v)_{\mathrm{unit}}(\alpha) W(\chi_v, \psi_v^{\mathrm{can}}) \tag{2.62}$$

by virtue of (2.59). Note that $(\chi_v)_{\mathrm{unit}} = (\chi_{\mathrm{unit}})_v$. Since χ_{unit} is an idele class character and $\alpha \in K^\times$, the right-hand side of (2.62) is simply $\prod_v W(\chi_v, \psi_v^{\mathrm{can}})$ or in other words our previous $\prod_v W(\chi_v)$. Hence we recover the definition (2.39) of the global root number which we had before the introduction of additive characters.

If we are back where we started, then what was the point of introducing additive characters in the first place? One reason is that one wants a theory which is applicable to function fields in one variable over finite fields, not just to number fields. Usually the difference between number fields and function fields is thought to be the presence or absence of archimedean places, but another difference is that the prime field of a number field is another number field, whereas the prime field of a function field is not a global field at all. It is the latter difference which forces one to consider arbitrary additive characters. Indeed for the prime field $\mathbb{Q}$ there is a canonical choice of additive character, whence for any number field there is a canonical choice via composition with trace. But a function field is of infinite degree over its prime field, and consequently a preferred additive character is lacking. In the absence of a preferred choice one is forced to consider all choices.

7.2. The global epsilon factor and the global functional equation

The fact that there is no trace from a function field to its prime field has other consequences as well. The framework within which we have been working – ideles, idele class characters, L-functions – carries over without change to function fields, and Hecke's functional equation

$$\Lambda(s, \chi) = W(\chi)\Lambda(k - s, \overline{\chi}) \tag{2.63}$$

(Theorem 2.1, with k as in (2.28)) is subsumed in Tate's functional equation

$$L(s,\chi) = \varepsilon(s,\chi)L(1-s,\chi^{-1}) \tag{2.64}$$

for the L-function associated to an idele class character χ of an arbitrary global field. Here $L(s,\chi)$ is the completed L-function associated to χ and $\varepsilon(s,\chi)$ is the **global epsilon factor**, an elementary factor of the form $a\cdot b^s$ with $a \in \mathbb{C}^\times$ and $b>0$. We define $\varepsilon(s,\chi)$ more precisely below, but right now the key point is that in Tate's more general setting, (2.63) has to be replaced by (2.64), because $\Lambda(s,\chi)$ no longer has a meaning: The definition of $\Lambda(s,\chi)$ involves the absolute discriminant, and function fields do not have absolute discriminants. Such an invariant exists for a number field K only because one can take the trace down to the prime field and obtain the canonical pairing $\langle x,y\rangle = \operatorname{tr}_{K/\mathbb{Q}}(xy)$.

Before explaining how (2.64) reduces to (2.63) in the case of number fields, we repeat that the L-function $L(s,\chi)$ in (2.64) is the *completed* L-function associated to χ. In other words, $L(s,\chi) = L_\infty(s)L_{\text{fin}}(s,\chi)$, where $L_{\text{fin}}(s,\chi)$ is the Euler product (2.10). Normally it is this Euler product itself which we denote $L(s,\chi)$, but *for the remainder of the present lecture only,* we shall use $L(s,\chi)$ to mean the completed L-function, reverting in subsequent lectures to the more traditional usage in force until now. Note that it is only in the number field case that this distinction is even an issue: If there are no infinite places then $L_\infty(s,\chi)=1$, whence $L(s,\chi)$ coincides with (2.10) by either convention.

To make a connection between (2.63) and (2.64) we must say a word about $\varepsilon(s,\chi)$, although we postpone the formal definition a little longer. Roughly speaking, epsilon factors are Gauss sums, or a generalization and renormalization of Gauss sums. For example if χ is a primitive Dirichlet character of conductor N, simultaneously viewed as the corresponding idele class character $\chi_{\mathbb{A}}$, then

$$\varepsilon(s,\chi) = \tau(\chi)/(i^m N^s) = W(\chi)N^{1/2-s}, \tag{2.65}$$

where $\tau(\chi)$ is the Gauss sum (1.2) and m is as in (1.4). The second equality in (2.65) is a consequence of (1.3). As this example illustrates, the epsilon factor of an idele class character of a number field binds the root number to the the product of the conductor and the absolute value of the discriminant, although the latter factor is of course 1 in the case of $\mathbb{Q}$. In general, if K is an arbitrary number field and χ an idele class character of K then

$$\varepsilon(s,\chi) = W(\chi)(D\mathbf{N}\mathfrak{f}(\chi))^{k/2-s} \tag{2.66}$$

(see Exercise 2.9). Granting (2.66), let us verify that (2.64) does reduce to (2.63) when K is a number field. In addition to (2.66), we will need the formula

$$L(s,\chi\cdot||*||^{s_0}) = L(s+s_0,\chi), \tag{2.67}$$

where $||*||$ is as usual the idelic norm. The validity of (2.67) follows by inspection from formulas (2.22) through (2.26).

To deduce (2.63) from (2.64), we recall first of all that when the absolute norm is viewed as an idele class character it coincides with the *reciprocal* of the idelic norm. Thus the idelic version of Proposition 1.1, given (2.28), is

$$\chi = \chi_{\text{unit}}\cdot ||*||^{-(k-1)/2}. \tag{2.68}$$

Consequently $\chi^{-1} = \overline{\chi}_{\text{unit}}\cdot||*||^{(k-1)/2}$, whence

$$L(1-s,\chi^{-1}) = L(1-s+(k-1)/2,\overline{\chi}_{\text{unit}})$$

by (2.67). Another appeal to (2.67) now gives

$$L(1-s,\chi^{-1}) = L(k-s,\overline{\chi}_{\text{unit}}\cdot ||*||^{-(k-1)/2}) = L(k-s,\overline{\chi}), \tag{2.69}$$

where the second equality follows from (2.68). On substituting (2.66) and (2.69) in (2.64) and then multiplying through by $(D\mathbf{N}\mathfrak{f}(\chi))^{s/2}$, we do indeed recover (2.63).

We come finally to the definition of $\varepsilon(s,\chi)$, where χ is now an idele class character of a global field K. Choose a global additive character ψ of K – recall this means that ψ is a nontrivial unitary character of the adele ring $\mathbb{A}$ of K, trivial on K – and let dx be the Haar measure on $\mathbb{A}$ giving the quotient $\mathbb{A}/K$ measure 1. As with any Haar measure on $\mathbb{A}$, we can write dx as a restricted direct product measure, $dx = \otimes_v dx_v$, where dx_v is a Haar measure on K_v and dx_v gives $\mathcal{O}_v$ measure 1 for all but finitely many finite v. If $\{c_v\}$ is a family of positive real numbers such that $c_v = 1$ for all but finitely many v and $\prod_v c_v = 1$, then we also have $dx = \otimes_v c_v dx_v$, so the decomposition $dx = \otimes_v dx_v$ is not unique. Neither is ψ, of course. Nonetheless, we obtain a global factor $\varepsilon(s,\chi)$ independent of any choices by setting

$$\varepsilon(s,\chi) = \prod_v \varepsilon(s,\chi_v,\psi_v,dx_v), \tag{2.70}$$

where the **local epsilon factor** $\varepsilon(s,\chi_v,\psi_v,dx_v)$ must now be defined.

7.3. The local epsilon factor and the local functional equation

Since the issue is now local, we drop the subscript v and switch to a local setting. Thus K is a local field, χ a character of $K^\times$, ψ an additive character of K, and dx a Haar measure on K. We must define $\varepsilon(s,\chi,\psi,dx)$. Once we have done so we will put

$$\varepsilon(\chi,\psi,dx) = \varepsilon(s,\chi,\psi,dx)|_{s=0} \tag{2.71}$$

and define the local root number $W(\chi,\psi)$ by the formula

$$W(\chi,\psi) = \frac{\varepsilon(\chi,\psi,dx)}{|\varepsilon(\chi,\psi,dx)|}. \tag{2.72}$$

As the notation suggests, the right-hand side of (2.72) turns out to be independent of the choice of dx.

One point to understand at the outset is that the definition of $\varepsilon(s,\chi,\psi,dx)$ involves integrals which may converge only for $\Re(s) \gg 0$. However $\varepsilon(s,\chi,\psi,dx)$ extends to an entire function of s, so that (2.71) defines $\varepsilon(\chi,\psi,dx)$ by analytic continuation. Going in the other direction, we will see that

$$\varepsilon(s,\chi,\psi,dx) = \varepsilon(\chi\cdot ||*||^s,\psi,dx); \tag{2.73}$$

of course $||*||$ now denotes the local norm (2.7), given that we are dropping the subscript v.

The integrals just mentioned appear along with $\varepsilon(s,\chi,\psi,dx)$ in Tate's local functional equation. Quite apart from the fact that Tate's method is applicable to function fields as well as number fields, the local functional equation is from the perspective of Hecke's method something completely new. The setting for the innovation is the **Schwartz space** $\mathcal{S}(K)$ of K. If K is nonarchimedean then $\mathcal{S}(K)$ consists of locally constant functions on K of compact support, while if K is $\mathbb{R}$ or $\mathbb{C}$ then $\mathcal{S}(K)$ consists of C^∞ functions f on K such that the derivatives of f of all orders (mixed partial derivatives of all orders if $K \cong \mathbb{C}$) are of rapid decay. In all cases we define the **Fourier transform** $\hat{f}$ of a function $f \in \mathcal{S}(K)$ by

$$\hat{f}(x) = \int_K f(y)\psi(xy)\,dy. \tag{2.74}$$

The definition depends on the choice of ψ, but for any choice, $\hat{f}$ is again in $\mathcal{S}(K)$.

To state the local functional equation, take $f \in \mathcal{S}(K)$ and set

$$I(s,\chi,\psi,dx,f) = \int_{K^\times} f(x)\chi(x)||x||^s \frac{dx}{||x||}. \tag{2.75}$$

Even though $I(s,\chi,\psi,dx,f)$ does not actually depend on ψ, we retain it in the notation to remind ourselves that $\hat{f}$ depends on ψ. As for the right-hand side of (2.75), the reason for writing the integrand as $f(x)\chi(x)||x||^s\,dx/||x||$ rather than simply as $f(x)\chi(x)||x||^{s-1}\,dx$ is that $dx/||x||$ is a Haar measure on $K^\times$. By a consideration of cases one can show that the integral converges for $\Re(s) \gg 0$ and extends to a meromorphic function on $\mathbb{C}$. Tate's local functional equation is the statement that

$$\frac{I(1-s,\chi^{-1},\psi,dx,\hat{f})}{L(1-s,\chi^{-1})} = \varepsilon(s,\chi,\psi,dx)\frac{I(s,\chi,\psi,dx,f)}{L(s,\chi)}, \tag{2.76}$$

where the local L-factors are as in (2.22), (2.25), and (2.26) and $\varepsilon(s,\chi,\psi,dx)$ is an entire nowhere vanishing function independent of f.

Formula (2.76) is the *definition* of $\varepsilon(\chi,\psi,dx,s)$. More precisely, one first proves that for arbitrary $f, g \in \mathcal{S}(K)$ the identity

$$I(s,\chi,\psi,dx,f)I(1-s,\chi^{-1},\psi,dx,\hat{g}) = I(s,\chi,\psi,dx,g)I(1-s,\chi^{-1},\psi,dx,\hat{f})$$

holds, and then for each χ one exhibits a choice of g such that $I(s,\chi,\psi,dx,g)$ and $I(1-s,\chi^{-1},\psi,dx,\hat{g})$ are nonzero as meromorphic functions. Finally, given such a g one defines $\varepsilon(s,\chi,\psi,dx)$ by putting

$$\varepsilon(s,\chi,\psi,dx) = \frac{L(s,\chi)I(1-s,\chi^{-1},\psi,dx,\hat{g})}{L(1-s,\chi^{-1})I(s,\chi,\psi,dx,g)}. \tag{2.77}$$

Then $\varepsilon(s,\chi,\psi,dx)$ is independent of the choice of g and (2.76) follows.

We will not delve into the details of these calculations, nor into the passage from (2.76) to (2.64), which is based on an adelic version of the Poisson summation formula. For all of this see [**92**]. However we will illustrate the use of (2.76) by determining the dependence of $\varepsilon(s,\chi,\psi,dx)$ on the parameters ψ, dx, and s, and by proving a duality relation. We will also record some formulas for $\varepsilon(s,\chi,\psi,dx)$ which arise as a by-product of the proof of (2.76) and which in the number field case are the source of the local formulas for $W(\chi)$ stated earlier.

7.4. Dependence on the parameters

To determine how $\varepsilon(s,\chi,\psi,dx)$ depends on ψ, we return to the spot where ψ entered the picture, namely the definition (2.74) of the Fourier transform. If we temporarily write $(\psi,\hat{f})$ and $(\psi_b,\hat{f})$ for the Fourier transform of f taken relative to ψ and ψ_b respectively, then (2.74) gives $(\psi_b,\hat{f})(x) = (\psi,\hat{f})(bx)$. Hence by using the invariance of $dx/||x||$ under $x \mapsto b^{-1}x$ in the integral for $I(1-s,\chi^{-1},\psi_b,dx,\hat{f})$, we see that

$$I(1-s,\chi^{-1},\psi_b,dx,\hat{f}) = \chi(b)||b||^{s-1}I(1-s,\chi^{-1},\psi,dx,\hat{f}).$$

Since the L-factors and $I(s, \chi, \psi, dx, f)$ are unaffected by the switch from ψ to ψ_b, the formula

$$\varepsilon(s, \chi, \psi_b, dx) = \chi(b)||b||^{s-1}\varepsilon(s, \chi, \psi, dx) \tag{2.78}$$

now follows from (2.76). In particular, taking $s = 0$, we obtain

$$\varepsilon(\chi, \psi_b, dx) = \chi(b)||b||^{-1}\varepsilon(\chi, \psi, dx), \tag{2.79}$$

and inserting this information in (2.72), we recover the claimed formula (2.59) for the dependence of $W(\chi, \psi)$ on ψ. Furthermore, if we redo the calculation (2.62) with (2.61) replaced by (2.70) and (2.59) by (2.78) then we find that the global epsilon factor in (2.70) is indeed independent of the global additive character.

Next we determine the dependence of $\varepsilon(s, \chi, \psi, dx)$ on dx. Any other Haar measure on K has the form $c\,dx$ with $c > 0$, and from (2.74) and (2.75) we deduce that $I(1-s, \chi^{-1}, c\,dx, \hat{f}) = c^2\, I(1-s, \chi^{-1}, dx, \hat{f})$ and $I(s, \chi, c\,dx, f) = c\,I(s, \chi, dx, f)$. The formula

$$\varepsilon(s, \chi, \psi, c\,dx) = c\,\varepsilon(s, \chi, \psi, dx) \tag{2.80}$$

now follows from (2.76). Taking $s = 0$, we find that

$$\varepsilon(\chi, \psi, c\,dx) = c\,\varepsilon(\chi, \psi, dx). \tag{2.81}$$

Thus the definition (2.72) of $W(\chi, \psi)$ is independent of dx, as claimed. Another claim also follows, namely that the global epsilon factor does not depend on the decomposition of the global Haar measure as a restricted direct product of local Haar measures.

Finally, let us verify the relation (2.73), the dependence of $\varepsilon(s, \chi, \psi, dx)$ on s. This is easy. First of all, it is immediate from the definition of $I(s, \chi, \psi, dx, f)$ that if s' is a second complex variable then

$$I(s+s', \chi, \psi, dx, f) = I(s', \chi \cdot ||*||^s, \psi, dx, f).$$

It is likewise immediate from the definition of $L(s, \chi)$ given in formulas (2.22), (2.25), and (2.26) that $L(s+s', \chi) = L(s', \chi \cdot ||*||^s)$. Hence (2.76) gives

$$\varepsilon(s+s', \chi, \psi, dx) = \varepsilon(s', \chi \cdot ||*||^s, \psi, dx).$$

Setting $s' = 0$, we obtain (2.73).

7.5. A duality relation

Given $f \in \mathcal{S}(K)$, define $f^- \in \mathcal{S}(K)$ by $f^-(x) = f(-x)$. If we fix ψ and use it to define the Fourier transform $f \mapsto \hat{f}$ on $\mathcal{S}(K)$, then the **self-dual Haar measure** dx_ψ relative to ψ is the unique Haar measure on K such that $(f^\wedge)^\wedge = f^-$. Taking $dx = dx_\psi$ and applying (2.76) a second time, we find that

$$I(s, \chi, \psi, dx_\psi, f^-) = \varepsilon(s, \chi, \psi, dx_\psi)\varepsilon(1-s, \chi^{-1}, \psi, dx_\psi)I(s, \chi, \psi, dx_\psi, f).$$

The substitution $x \mapsto -x$ in the integral $I(s, \chi, \psi, dx_\psi, f^-)$ now gives

$$\varepsilon(s, \chi, \psi, dx_\psi)\varepsilon(1-s, \chi^{-1}, \psi, dx_\psi) = \chi(-1). \tag{2.82}$$

This is the duality relation at issue.

A less transparent version of (2.82) will actually be more useful to us. We claim first of all that a local analogue of Proposition 1.1 holds:

$$\chi = \chi_{\text{unit}} \cdot ||*||^{-c} \tag{2.83}$$

with $c \in \mathbb{R}$. The assertion in (2.83) is that $|\chi| = ||*||^{-c}$ and indeed that any character of $K^\times$ with values in $\mathbb{R}_{>0}$ is a real power of $||*||$. If K is $\mathbb{R}$ or $\mathbb{C}$ this fact has already been noted, and if K is nonarchimedean then it suffices to observe that a character of $K^\times$ with values in $\mathbb{R}_{>0}$ is trivial on $\mathcal{O}^\times$ and thus determined by its value on a uniformizer, hence equal to $||*||^{-c}$ for some c. Now (2.83) implies that $\chi^{-1} = \overline{\chi} \cdot ||*||^{2c}$, whence substitution in (2.82) gives

$$\varepsilon(\chi, \psi, dx_\psi)\varepsilon(1 + 2c, \overline{\chi}, \psi, dx_\psi) = \chi(-1) \tag{2.84}$$

when we take $s = 0$ and use (2.73).

On the other hand, applying complex conjugation to both sides of (2.76) gives $\overline{\varepsilon(s, \chi, \psi, dx_\psi)} = \varepsilon(\overline{s}, \overline{\chi}, \psi_{-1}, dx_\psi)$, because $\overline{\psi} = \psi^{-1} = \psi_{-1}$. Thus

$$\overline{\varepsilon(s, \chi, \psi, dx_\psi)} = \chi(-1)\varepsilon(\overline{s}, \overline{\chi}, \psi, dx_\psi)$$

by (2.78). Taking $s = 1 + 2c$ and substituting the result in (2.84), we obtain

$$\varepsilon(\chi, \psi, dx_\psi)\overline{\varepsilon(\chi \cdot ||*||^{1+2c}, \psi, dx_\psi)} = 1 \tag{2.85}$$

after canceling $\chi(-1)$ on both sides and once again using (2.73). In spite of its apparent awkwardness, (2.85) will lead to a formula for the absolute value of $\varepsilon(\chi, \psi, dx_\psi)$.

7.6. Explicit formulas

As we have already indicated, an essential part of the proof of (2.76) is to exhibit a Schwartz function g for which $I(s, \chi, \psi, dx, g) \neq 0$. In the process one obtains the following useful information about $\varepsilon(s, \chi, \psi, dx)$.

First, if K is archimedean and dx^{can} is the self-dual measure relative to ψ^{can} then

$$\varepsilon(s, \chi, \psi^{\text{can}}, dx^{\text{can}}) = \varepsilon(\chi, \psi^{\text{can}}, dx^{\text{can}}) = W(\chi) \tag{2.86}$$

with $W(\chi)$ as in (2.41). Standard formulas for the Fourier transform on euclidean space show that dx^{can} is Lebesgue measure or twice Lebesgue measure according as $K = \mathbb{R}$ or $K \cong \mathbb{C}$. In the latter case we can also say that dx^{can} is $|dz \wedge d\overline{z}|$.

Next suppose that K is nonarchimedean. Let $\mathcal{O}$ be its ring of integers, π a uniformizer, $a(\chi)$ the exponent of the conductor of χ, and $n(\psi)$ the largest integer n such that $\psi|\pi^{-n}\mathcal{O}$ is trivial. Choose $\gamma \in \pi^{a(\chi)+n(\psi)}\mathcal{O}^\times$. If χ is unramified then

$$||\gamma||\,\varepsilon(\chi, \psi, dx) = \chi(\gamma)\int_{\mathcal{O}} dx, \tag{2.87}$$

and if χ is ramified then

$$||\gamma||\,\varepsilon(\chi, \psi, dx) = \int_{\mathcal{O}^\times} \chi^{-1}(x/\gamma)\psi(x/\gamma)dx. \tag{2.88}$$

Of course the integral in (2.87) is just meas($\mathcal{O}$), the measure of $\mathcal{O}$ relative to dx.

We note two consequences of (2.87) and (2.88). First, an inspection of these formulas shows that the effect of replacing χ by $\chi \cdot ||*||^s$ is to multiply the right-hand side of the formulas by $||\gamma||^s$. In other words

$$\varepsilon(\chi \cdot ||*||^s, \psi, dx) = q^{-(a(\chi)+n(\psi))s}\varepsilon(\chi, \psi, dx) \tag{2.89}$$

with $q = |\mathcal{O}/\pi\mathcal{O}|$. In particular, take $s = 2c + 1$, where c is as in (2.83). Inserting (2.89) in (2.85), we obtain

$$|\varepsilon(\chi, \psi, dx_\psi)|^2 = q^{(a(\chi)+n(\psi))k} \tag{2.90}$$

with $k = 2c + 1$.

The second consequence is particularly simple and depends only on (2.87): If $a(\chi) = n(\psi) = 0$ and $\text{meas}(\mathcal{O}) = 1$ then $\varepsilon(\chi, \psi, dx) = 1$. On replacing χ by $\chi \cdot ||*||^s$ and applying (2.73) we find more generally that if $a(\chi) = n(\psi) = 0$ and $\text{meas}(\mathcal{O}) = 1$ then $\varepsilon(s, \chi, \psi, dx) = 1$. If we switch to the global setting, so that K is now a global field, χ an idele class character, ψ a global additive character, and dx the Haar measure on $\mathbb{A}$ giving $\mathbb{A}/K$ measure 1, then for all but finitely many finite places v we have $a(\chi_v) = n(\psi_v) = 0$ and $\text{meas}(\mathcal{O}_v) = 1$, and thus $\varepsilon(s, \chi_v, \psi_v, dx_v) = 1$. Hence the product (2.70) defining the global epsilon factor is indeed meaningful.

7.7. The number field case

It remains to check that we do recover the formulas for $W(\chi)$ stated earlier – namely (2.41), (2.42), and (2.43) – when we take K to be a finite extension of $\mathbb{Q}_p$ ($p \leqslant \infty$) and ψ to be ψ^{can}. That (2.86) and (2.72) imply (2.41) is a tautology, so we may assume that $p < \infty$.

If χ is unramified then γ in (2.87) is just an element of valuation $n(\psi)$. Furthermore $n(\psi)$ is d, the exponent of the different ideal, because $\psi = \psi^{\text{can}}$. Hence dividing the two sides of (2.87) by their absolute values, we obtain (2.43).

To derive (2.42), let dx^{can} denote the self-dual measure on K relative to ψ^{can}. We claim first of all that dx^{can} coincides with the measure dx on K for which $\text{meas}(\mathcal{O}) = q^{-d/2}$. To justify the claim it suffices to exhibit a single nonzero element $f \in \mathcal{S}(K)$ such that $(f\hat{\ })\hat{\ } = f^-$ when the Fourier transform is computed using ψ^{can} and dx. Let f and g be the characteristic functions of $\mathcal{O}$ and $\pi^{-d}\mathcal{O}$ respectively. A straightforward calculation shows that $\hat{f} = q^{-d/2} g$ and that $\hat{g} = q^{d/2} f$, whence $(f\hat{\ })\hat{\ } = f = f^-$ and $dx^{\text{can}} = dx$, as claimed. Thus $\mathcal{O}$ has measure $q^{-d/2}$ relative to dx^{can}. It follows that each coset of $\pi^{a(\chi)}\mathcal{O}$ in $\mathcal{O}$ has measure $q^{-d/2-a(\chi)}$. But the value at $x \in \mathcal{O}^\times$ of $\chi^{-1}(x/\gamma)\psi(x/\gamma)$ depends only on the coset of x modulo $\pi^{a(\chi)}\mathcal{O}$, so we can replace the integral in (2.88) by a sum:

$$q^{-a(\chi)-d}\, \varepsilon(\chi, \psi^{\text{can}}, dx^{\text{can}}) = q^{-a(\chi)-d/2} \sum_{x \in (\mathcal{O}/\pi^{a(\chi)}\mathcal{O})^\times} \chi^{-1}(x/\gamma)\psi^{\text{can}}(x/\gamma). \tag{2.91}$$

Furthermore, we know from (2.90) that the absolute value of $\varepsilon(\chi^{\text{can}}, \psi, dx^{\text{can}})$ is $q^{(a(\chi)+d)k/2}$, where $k = 2c + 1$ and $|\chi(\pi)| = q^c$. It follows that the absolute value of the two sides of (2.91) is $q^{(a(\chi)+d)(c-1/2)}$. Dividing both sides of (2.91) by this quantity while substituting

$$\chi(\gamma) = \chi_{\text{unit}}(\gamma) q^{(a(\chi)+d)c},$$

we obtain (2.42).

8. Exercises

Exercise 2.1. Let $\mathbb{A}$ be the adele ring of a number field K. We have characterized the topology on $\mathbb{A}$ by two properties. Verify that there does in fact exist a unique topology on $\mathbb{A}$ with these properties and that a basis for the topology is given by the sets $\prod_v \mathcal{U}_v$ with $\mathcal{U}_v$ open in K_v for all v and $\mathcal{U}_v = \mathcal{O}_v$ for all but finitely many finite v. Similarly, verify that there is a unique topology on $\mathbb{A}^\times$ with the two properties claimed for it, and show that a basis for this topology is given by the sets $\prod_v \mathcal{U}_v$ with $\mathcal{U}_v$ open in $K_v^\times$ for all v and $\mathcal{U}_v = \mathcal{O}_v^\times$ for all but finitely many finite v.

Exercise 2.2. Let K be a number field, C its ideal class group, and $c(\mathfrak{D}) \in C$ the class of the different ideal $\mathfrak{D}$. Prove that $c(\mathfrak{D}) \in C^2$. (Hint: If this is not the case then there is a quadratic ideal class character χ of K such that $\chi(\mathfrak{D}) = -1$. Compute $W(\chi)$ and obtain a contradiction to Theorem 2.2.) Although the theorem that $c(\mathfrak{D}) \in C^2$ is due to Hecke, Hecke's proof on p. 261 of [**43**] is quite different. The proof outlined in the hint follows an argument of Armitage [**2**] suggested by Serre.

Exercise 2.3. Given an idele class character χ of a number field K, write $W(\chi_{\text{fin}})$ and $W(\chi_\infty)$ for the product of the local root numbers $W(\chi_v)$ taken over $v \nmid \infty$ and over $v|\infty$ respectively. Using (2.42), show that if χ' is an idele class character of K with $\mathfrak{f}(\chi') = \mathcal{O}$ then

$$W((\chi\chi')_{\text{fin}}) = \chi'_{\text{unit}}(\mathfrak{f}(\chi)\mathfrak{D})W(\chi_{\text{fin}}),$$

where $\mathfrak{D}$ is the different ideal of K. (Of course in writing $\chi'_{\text{unit}}(\mathfrak{f}(\chi)\mathfrak{D})$ we are thinking of χ'_{unit} as a Hecke character.) It follows that

$$W(\chi\chi') = \omega \cdot \chi'_{\text{unit}}(\mathfrak{f}(\chi)\mathfrak{D})W(\chi) \tag{2.92}$$

with $\omega = W((\chi\chi')_\infty)/W(\chi_\infty)$.

Exercise 2.4. This exercise outlines a proof of Proposition 1.6. Notation is as in that proposition and as in the last paragraph of the proof of Proposition 1.3.

(a) Show that that $W(\chi\varphi) = W(\chi)$ for $\chi \in X(D)$ and $\varphi \in \Phi$. (Hint: Use (2.92) with $\chi' = \varphi$.)

(b) Let $\mathfrak{T}$ be the prime ideal of $\mathcal{O}$ above 2, let E be as in the proof of Proposition 1.3, and write $E = \{\varepsilon, \delta\}$. Also let ε' and δ' be the corresponding quadratic characters of

$$(\mathcal{O}/\mathfrak{T}^5)^\times \cong (\mathbb{Z}/8\mathbb{Z})^\times \times \langle(1+\sqrt{-2C}) + \mathfrak{T}^5\rangle, \tag{2.93}$$

where the second factor on the right-hand side is the cyclic group of order 4 generated by the coset of $1+\sqrt{-2C}$. Replacing this second factor by its subgroup of order 2, we obtain a subgroup H of index 2 in $(\mathcal{O}/\mathfrak{T}^5)^\times$ such that $\varepsilon' = -\delta'$ on the complement of H and $\varepsilon' = \delta'$ on H. By direct calculation, show that

$$\sum_{h\in H} \varepsilon'(h)e^{2\pi i \operatorname{tr}_{K/\mathbb{Q}}(h/16)} = 0.$$

(To interpret $e^{2\pi i \operatorname{tr}_{K/\mathbb{Q}}(h/16)}$, replace h by any of its coset representatives in $\mathcal{O}$: The value of the exponential is independent of the coset representative because the different ideal of K is $\mathfrak{T}^3\mathfrak{C}$ and $16\mathcal{O} = \mathfrak{T}^8$.)

(c) Deduce that $\sum_{\chi\in X(D)} W(\chi) = 0$, and conclude that the sets $X^\pm(D)$ are both nonempty, whence both are of cardinality $h(D)$ by (a) and Proposition 1.3.

Exercise 2.5. Given $\chi \in X(D)$ and a positive integer w, use the local formulas and Theorem 2.1 to verify that (1.35) holds with $\Lambda(s,\chi_w)$ as in (1.36) and $W(\chi_w)$ as in (1.37). (Hint: If w is odd then (2.92) can be applied with $\chi' = \chi^{w-1}$.)

Exercise 2.6. This exercise pertains to our example of a Hecke L-function of "Maass type." Let K be a real quadratic field of discriminant D.

(a) Show that (2.50) gives a well-defined equivariant character χ_0 of P and that any extension of this character to I is an equivariant Hecke character. (Hint: To check the equivariance, observe that if $\mathfrak{a}$ is any ideal of $\mathcal{O}$ then $\mathfrak{a}\mathfrak{a}'$ is the principal ideal generated by a rational integer, namely $\mathbf{N}\mathfrak{a}$.)

(b) Let η be a primitive equivariant Hecke character of K of finite order, and assume that $\mathfrak{f}(\eta) = N\sqrt{D}\mathcal{O}$ for some integer $N \geqslant 1$. Put $\chi = \eta\chi_0$, where χ_0 is now any extension of (2.50) to I. Show that $W(\chi) = W(\eta)$. (Hint: Use (2.92) with χ and χ' replaced by η and χ_0.)

Exercise 2.7. This exercise leads to a proof of (2.55). Fix a prime $r > 5$ congruent to 1 mod 4, put $K = \mathbb{Q}(\sqrt{r(r+4)})$, and let $\mathfrak{f} = \sqrt{D}\mathcal{O}$, where D as before is the discriminant of K. Also let η be any extension of (2.54) to $I(\mathfrak{f})$.

(a) Verify that η is an equivariant Hecke character of conductor $\mathfrak{f}$ and trivial infinity type.

(b) Deduce that $W(\eta_v) = 1$ if v is a place of K not dividing $\mathfrak{f}$.

(c) Now consider the place v of K corresponding to a given prime ideal $\mathfrak{p}$ dividing $\mathfrak{f}$, and let p be the prime number below $\mathfrak{p}$. Show that

$$W(\eta_v) = i^\delta \left(\frac{2}{p}\right) \prod_{\substack{q|D \\ q \neq p}} \left(\frac{p}{q}\right)$$

with δ equal to 0 or 1 according as p is 1 mod 4 or 3 mod 4. (Note the similarity to (2.48).)

(d) Finally, use quadratic reciprocity to complete the proof of (2.55).

(e) Deduce from Theorem 2.2 that the class number of K is even.

Exercise 2.8. Let K be a number field, and let χ and χ' be idele class characters of K of relatively prime conductors.

(a) Show that the following formula of Langlands is a generalization of (2.92):

$$W(\chi\chi') = \omega \cdot \chi_{\text{unit}}(\mathfrak{f}(\chi'))\chi'_{\text{unit}}(\mathfrak{f}(\chi))W(\chi)W(\chi'), \tag{2.94}$$

where $\omega = W((\chi\chi')_\infty)/(W(\chi_\infty)W(\chi'_\infty))$.

(b) Prove (2.94).

(c) By taking $K = \mathbb{Q}$ and choosing χ and χ' appropriately, derive quadratic reciprocity from (2.94).

Exercise 2.9. Let χ be an idele class character of a number field K, and define k as in (2.28). Using (2.86), (2.89), and (2.90), derive (2.66).

LECTURE 3

Motivic L-functions

The discussion now moves to L-functions associated to Galois representations. Within this large framework Artin L-functions form a natural point of departure for one simple reason: All known methods of obtaining an L-function from a Galois representation are variants of Artin's original construction.

1. Artin representations and Artin L-functions

Let K be a number field. A representation of $\mathrm{Gal}(\overline{K}/K)$ over $\mathbb{C}$ is called an **Artin representation** of K. The requirement that a representation of a topological group be continuous is very restrictive in the case of Artin representations, because $\mathrm{Gal}(\overline{K}/K)$ with its Krull topology is the profinite group $\varprojlim_L \mathrm{Gal}(L/K)$, where L runs over finite Galois extensions of K inside $\overline{K}$, and a complex representation of a profinite group is trivial on an open subgroup. It follows that an Artin representation of K can be regarded as a representation of $\mathrm{Gal}(L/K)$ for some finite Galois extension L of K, and this is the point of view that we shall usually adopt.

Consider then a finite Galois extension L of K, a finite-dimensional complex vector space V, and an Artin representation $\rho : \mathrm{Gal}(L/K) \to \mathrm{GL}(V)$. The **Artin L-function** $L(s, \rho)$ is defined by an Euler product:

$$L(s,\rho) = \prod_{\mathfrak{p}} B_{\mathfrak{p}}((\mathbf{N}\mathfrak{p})^{-s})^{-1}, \tag{3.1}$$

where $\mathfrak{p}$ runs over nonzero prime ideals of K and $B_{\mathfrak{p}}(x)$ is the polynomial with constant term 1 defined as follows.

Given $\mathfrak{p}$, fix a prime ideal $\mathfrak{P}$ of L over $\mathfrak{p}$ and let D and I be the corresponding decomposition and inertia subgroups of $\mathrm{Gal}(L/K)$. Also write l and k for the residue class fields of $\mathfrak{P}$ and $\mathfrak{p}$ respectively. The natural action of D on l induces an isomorphism of D/I onto $\mathrm{Gal}(l/k)$, and the latter group has a canonical generator, the Frobenius automorphism. If $\sigma \in D$ is a preimage of the Frobenius automorphism under the composition of maps $D \to D/I \to \mathrm{Gal}(l/k)$ then σ is called a **Frobenius element** at $\mathfrak{P}$. While it is only σI and not σ which is uniquely determined by $\mathfrak{P}$, if we restrict attention to the **subspace of inertial invariants**

$$V^I = \{v \in V : \rho(i)(v) = v \text{ for all } i \in I\} \tag{3.2}$$

then the resulting linear automorphism $\rho(\sigma)|V^I$ of V^I is well defined. We set $B_{\mathfrak{p}}(x) = x^d P(x^{-1})$, where $P(x)$ is the characteristic polynomial of $\rho(\sigma)|V^I$ and d is the dimension of V^I. Thus

$$B_{\mathfrak{p}}(x) = \det(1 - x\rho(\sigma)|V^I). \tag{3.3}$$

Of course I is trivial unless $\mathfrak{p}$ is ramified in L, and when I is trivial, $V^I = V$. It follows that the degree of $B_{\mathfrak{p}}(x)$ is $\leqslant \dim V$ for all $\mathfrak{p}$ and is equal to $\dim V$ for all but finitely many $\mathfrak{p}$. Furthermore, since ρ is a representation of the finite group $\mathrm{Gal}(L/K)$, the eigenvalues of $\rho(\sigma)|V^I$ are roots of unity, and consequently the Euler product in (3.1) converges for $\Re(s) > 1$.

The definition (3.3) of $B_\mathfrak{p}(x)$ may appear to depend on our choice of a prime ideal $\mathfrak{P}$ over $\mathfrak{p}$, but if $\mathfrak{P}'$ is another choice then there is an element $g \in \mathrm{Gal}(L/K)$ such that $\mathfrak{P}' = g(\mathfrak{P})$, and then D, I, σ, and V^I are replaced by gDg^{-1}, gIg^{-1}, $g\sigma g^{-1}$, and $\rho(g)(V^I)$ respectively. Since characteristic polynomials are similarity invariants, (3.3) is unchanged. We also see that we can define ρ to be **unramified at $\mathfrak{p}$** (or at the corresponding place of K) if ρ is trivial on I, for then it is trivial on gIg^{-1} for all $g \in \mathrm{Gal}(L/K)$. If $\rho|I$ is nontrivial then ρ is **ramified at $\mathfrak{p}$**.

One can also define a notion of ramification at the infinite places. Let v be an infinite place of K and w a place of L over v. If v is real and w is complex then we let $I \subset \mathrm{Gal}(L/K)$ be the subgroup of order two generated by the complex conjugation corresponding to w. If v is complex or w is real then we take I to be the trivial subgroup of $\mathrm{Gal}(L/K)$. We say that ρ is **ramified** or **unramified** at v according as ρ is nontrivial or trivial on I. Note once again that ramification at v is a meaningful concept even though I may depend on w.

If S is a set of places of K then we say that ρ is **unramified outside** S if ρ is unramified at every place $v \notin S$. For example, if S contains all of the places of K which ramify in L then ρ is unramified outside S. In particular, an Artin representation is always unramified outside a finite set of places.

1.1. Idele class characters of finite order as Artin representations

The phrase *finite order* is crucial here. Idele class characters of infinite order do not correspond to Artin representations. However we do have a canonical bijection $\xi \mapsto \chi_\xi$ from one-dimensional Artin representations to idele class characters of finite order. The mechanism underlying this bijection is the Artin symbol. Since it will be convenient to think of χ_ξ as a primitive Hecke character we shall describe the Artin symbol at the level of ideals rather than at the level of ideles.

Let K be a number field and L a finite abelian extension of K, and let $\mathfrak{f}$ be a nonzero integral ideal of K which is divisible by every prime ideal of K ramified in L. Given a prime ideal $\mathfrak{p} \in I(\mathfrak{f})$ and a prime ideal $\mathfrak{P}$ of L above $\mathfrak{p}$, we can speak of *the* Frobenius element $\sigma_\mathfrak{P} \in \mathrm{Gal}(L/K)$ determined by $\mathfrak{P}$, because the inertia subgroup $I \subset \mathrm{Gal}(L/K)$ corresponding to $\mathfrak{P}$ is trivial. In fact we can write $\sigma_\mathfrak{P}$ as $\sigma_\mathfrak{p}$, because $\sigma_\mathfrak{P}$ is independent of the choice of prime ideal $\mathfrak{P}$ over $\mathfrak{p}$: Since $\mathrm{Gal}(L/K)$ is abelian we have $g\sigma_\mathfrak{P}g^{-1} = \sigma_\mathfrak{P}$ for all $g \in \mathrm{Gal}(L/K)$. We define the Artin symbol $(*, L/K)$ on prime ideals by setting

$$(\mathfrak{p}, L/K) = \sigma_\mathfrak{p}. \tag{3.4}$$

Since $I(\mathfrak{f})$ is the free abelian group on the prime ideals not dividing $\mathfrak{f}$, the map $\mathfrak{p} \mapsto \sigma_\mathfrak{p}$ extends uniquely to a homomorphism $I(\mathfrak{f}) \to \mathrm{Gal}(L/K)$, the **Artin map** or **reciprocity law map**. We write this map as $\mathfrak{a} \mapsto (\mathfrak{a}, L/K)$, where $\mathfrak{a}$ denotes an arbitrary element of $I(\mathfrak{f})$.

The Artin map is surjective, and hence it gives rise to the **Artin isomorphism** from an appropriate quotient of $I(\mathfrak{f})$ onto $\mathrm{Gal}(L/K)$. But an even deeper fact of class field theory is embedded in the next statement: If ξ is a one-dimensional character of $\mathrm{Gal}(L/K)$ and we pull it back to $I(\mathfrak{f})$ via the Artin map then the resulting character of $I(\mathfrak{f})$ is a Hecke character, necessarily of finite order. Since we have not been careful about the choice of $\mathfrak{f}$ we cannot claim that the Hecke character in question is primitive, but it certainly determines a primitive Hecke character, and it is this primitive character that we denote χ_ξ. The relation between ξ and χ_ξ is summarized in the formula

$$\chi_\xi(\mathfrak{a}) = \xi((\mathfrak{a}, L/K)) \qquad (\mathfrak{a} \in I(\mathfrak{f})). \tag{3.5}$$

If we start with a character ξ of $\mathrm{Gal}(\overline{K}/K)$ then there are many choices of a finite abelian extension L of K such that ξ factors through $\mathrm{Gal}(L/K)$, but the primitive Hecke character χ_ξ obtained is independent of the choice. This follows from the "consistency property" of the Artin symbol: the fact that if M is an intermediate field of the extension L/K then the restriction of $(\mathfrak{a}, L/K)$ to M is $(\mathfrak{a}, M/K)$. The upshot is that we may think of $\xi \mapsto \chi_\xi$ as a map from one-dimensional Artin representations of K to primitive Hecke characters of K of finite order.

This map is a bijection. The injectivity is clear from the definition, and the surjectivity is essentially one version of the "existence theorem" of class field theory. To elaborate on this point very briefly, we introduce the **narrow ray class group** $I(\mathfrak{f})/P_{\mathfrak{f},>0}$ to the modulus $\mathfrak{f}$. Here $\mathfrak{f}$ is an integral ideal of K, and $P_{\mathfrak{f},>0}$ is the subgroup of $P_\mathfrak{f}$ consisting of principal fractional ideals generated by a *totally positive* element of $K_\mathfrak{f}$ (an element which is sent to a positive number by every real embedding of K). Of course if K is totally complex then $P_{\mathfrak{f},>0} = P_\mathfrak{f}$. Now when we combine the defining property (1.10) of Hecke characters with the fact that a finite-order character of $\mathbb{C}^\times$ or $\mathbb{R}_{>0}$ is trivial, we see that if χ is a primitive Hecke character of K of finite order with $\mathfrak{f} = \mathfrak{f}(\chi)$ then χ factors through $I(\mathfrak{f})/P_{\mathfrak{f},>0}$. On the other hand, the existence theorem of class field theory assures us that there is an abelian extension $K^\mathfrak{f}$ of K, unramified outside the infinite places of K and the places dividing $\mathfrak{f}$, such that the kernel of the Artin symbol $(*, K^\mathfrak{f}/K)$ on $I(\mathfrak{f})$ is precisely $P_{\mathfrak{f},>0}$. The field $K^\mathfrak{f}$ is uniquely determined by these conditions and is called the **narrow ray class field** of K to the modulus $\mathfrak{f}$. If K is totally complex then we can omit the word *narrow*. In any case, the Artin map gives an isomorphism

$$I(\mathfrak{f})/P_{\mathfrak{f},>0} \cong \mathrm{Gal}(K^\mathfrak{f}/K). \tag{3.6}$$

Thus every primitive Hecke character χ of K of finite order with $\mathfrak{f} = \mathfrak{f}(\chi)$ has the form χ_ξ for some character ξ of $\mathrm{Gal}(K^\mathfrak{f}/K)$.

1.2. The arithmetic versus the geometric convention

The definitions (3.3) and (3.5) follow what we will call the **arithmetic convention** for Frobenius elements. The arithmetic convention is the traditional convention, and it seems like the appropriate convention to follow in the context of Artin L-functions. However there is a more recent convention, the **geometric convention**, and if we were to follow that convention then (3.5) would be replaced by

$$\chi_\xi(\mathfrak{a}) = \xi((\mathfrak{a}, L/K)^{-1}) \qquad (\mathfrak{a} \in I(\mathfrak{f})) \tag{3.7}$$

and (3.3) would be replaced by

$$B_\mathfrak{p}(x) = \det(1 - x\rho(\Phi)|V^I), \tag{3.8}$$

where Φ is an **inverse Frobenius element** at $\mathfrak{p}$, the inverse of a Frobenius element. While we are still on the topic of Artin L-functions we will continue to follow the arithmetic convention, but once we start to look at more general motivic L-functions we will make a permanent switch to the geometric convention, for reasons to be discussed later.

1.3. The Artin formalism

Almost anything one does with Artin L-functions depends on three basic properties. These properties are collectively known as the **Artin formalism.**

The first is **additivity**: If ρ and ρ' are representations of $\mathrm{Gal}(L/K)$ then

$$L(s,\rho\oplus\rho')=L(s,\rho)L(s,\rho'). \tag{3.9}$$

This is immediate from the additivity of the characteristic polynomial and the additivity of the map $V\mapsto V^I$.

The second property, which is trickier to prove, is **inductivity**. If M is an intermediate field of the Galois extension L/K and ρ is a representation of $\mathrm{Gal}(L/M)$, let $\mathrm{ind}_{M/K}\rho$ denote the representation of $\mathrm{Gal}(L/K)$ induced by ρ. Then

$$L(s,\mathrm{ind}_{M/K}\rho)=L(s,\rho). \tag{3.10}$$

Note that the left-hand side is an Artin L-function of K while the right-hand side is an Artin L-function of M.

The third property is **compatibility in dimension one**. Let ξ be a one-dimensional Artin representation and and χ_ξ the primitive Hecke character of finite order which corresponds to ξ under the identification (3.5). Then

$$L(s,\xi)=L(s,\chi_\xi). \tag{3.11}$$

This is a straightforward consequence of (1.18) (3.3), and (3.4).

There is actually a fourth property which can be considered part of the Artin formalism, namely **invariance under inflation**, although the invariance of $L(s,\rho)$ under inflation is more a property of Frobenius elements than of L-functions. Suppose that M is an intermediate field of the finite Galois extension L/K, and let ρ be a representation of $\mathrm{Gal}(M/K)$. Then ρ can be inflated to a representation of $\mathrm{Gal}(L/K)$ by composition with the canonical map $\mathrm{Gal}(L/K)\to\mathrm{Gal}(M/K)$, and we write $\mathrm{infl}_{L/M}\rho$ for the representation of $\mathrm{Gal}(L/K)$ so obtained. Then

$$L(s,\mathrm{infl}_{L/M}\rho)=L(s,\rho). \tag{3.12}$$

This follows from the fact that if $\sigma\in\mathrm{Gal}(L/K)$ is a Frobenius element at a prime ideal $\mathfrak{P}$ of L then $\sigma|M$ is a Frobenius element at the prime ideal of M lying below $\mathfrak{P}$. The consistency property of the Artin symbol mentioned earlier is just an abelian consequence of this fact. In any case, the significance of (3.12) is that if ρ is presented to us as a representation of $\mathrm{Gal}(\overline{K}/K)$ then $L(s,\rho)$ is independent of the choice of a finite Galois extension L of K such that ρ factors through $\mathrm{Gal}(L/K)$.

This concludes our recitation of the basic properties constituting the Artin formalism. We can now see that we encountered the first nontrivial instance of the Artin formalism in the previous lecture, when we looked at quadratic root numbers (Theorem 2.2). Given a quadratic extension of number fields L/K, let 1_L denote the one-dimensional character of the trivial subgroup $\mathrm{Gal}(L/L)$ of $\mathrm{Gal}(L/K)$, and let 1_K and ξ denote respectively the trivial and the nontrivial one-dimensional characters of $\mathrm{Gal}(L/K)$. Then $\mathrm{ind}_{L/K}1_L=1_K\oplus\xi$, so (3.9) and (3.10) give

$$L(s,1_L)=L(s,1_K)L(s,\xi). \tag{3.13}$$

Now it is immediate from the definitions that $L(s,1_L)=\zeta_L(s)$ and $L(s,1_K)=\zeta_K(s)$. Furthermore, on combining (3.4) with (3.5) we find that χ_ξ is the quadratic Hecke character $\mathrm{sign}_{L/K}$ defined in (2.30). It follows that (3.13) is simply the relation (2.29).

2. The functional equation

The example (3.13) was a warm-up exercise for the task now at hand, which is to deduce a functional equation for Artin L-functions from the properties listed above and the known functional equation of Hecke L-functions. The key ingredient here is Brauer's induction theorem.

2.1. Derivation of the functional equation from Brauer's theorem

Let G be a finite group. We recall that $\mathrm{Groth}(G)$, the **Grothendieck group of virtual representations of G over $\mathbb{C}$**, can be viewed as the free abelian group on the isomorphism classes of the irreducible complex representations of G, whence a **virtual representation** of G is just an integral linear combination of such isomorphism classes. Also, a representation of G is **monomial** if it is induced by a one-dimensional character of G. Given a representation ρ of G over $\mathbb{C}$, write $[\rho]$ for its class in $\mathrm{Groth}(G)$. For our purposes, the essential content of Brauer's theorem is that the classes $[\rho]$ with ρ monomial span $\mathrm{Groth}(G)$ over $\mathbb{Z}$.

Now take $G = \mathrm{Gal}(L/K)$, where L/K is a Galois extension of number fields. If ρ is any representation of G then by Brauer's theorem we can write

$$[\rho] = \sum_{(M,\xi)} n_{M,\xi}\,[\mathrm{ind}_{M/K}\xi] \tag{3.14}$$

with $n_{M,\xi} \in \mathbb{Z}$, where (M,ξ) runs over pairs consisting of an intermediate field M and a one-dimensional character ξ of $\mathrm{Gal}(L/M)$. On the other hand, by virtue of the additivity property (3.9) we can view $L(s,*)$ as a homomorphism from $\mathrm{Groth}(G)$ to the multiplicative group of nonzero meromorphic functions on the right half-plane $\Re(s) > 1$. Applying this homomorphism to both sides of (3.14), we obtain

$$L(s,\rho) = \prod_{(M,\xi)} L(s,\mathrm{ind}_{M/K}\xi)^{n_{M,\xi}}, \tag{3.15}$$

and then the inductivity and compatibility properties (3.10) and (3.11) give

$$L(s,\rho) = \prod_{(M,\xi)} L(s,\chi_\xi)^{n_{M,\xi}}. \tag{3.16}$$

Each $L(s,\chi_\xi)$ is a Hecke L-function and so extends to a meromorphic function on $\mathbb{C}$. Thus (3.16) gives the continuation of $L(s,\rho)$ to a meromorphic function on $\mathbb{C}$.

But we want more: a functional equation. For each pair (M,ξ) in (3.14), put

$$A_{M,\xi} = D_M \mathbf{N}\mathfrak{f}(\chi_\xi), \tag{3.17}$$

where D_M is the absolute value of the discriminant of M. We set

$$A(\rho) = \prod_{(M,\xi)} A_{M,\xi}^{n_{M,\xi}}, \tag{3.18}$$

$$L_\infty(s,\rho) = \prod_{(M,\xi)} L_\infty(s,\chi_\xi)^{n_{M,\xi}}, \tag{3.19}$$

$$W(\rho) = \prod_{(M,\xi)} W(\chi_\xi)^{n_{M,\xi}}, \tag{3.20}$$

and

$$\Lambda(s,\rho) = A(\rho)^{s/2} L_\infty(s,\rho) L(s,\rho). \tag{3.21}$$

Then

$$\Lambda(s,\rho) = \prod_{(M,\xi)} \Lambda(s,\chi_\xi)^{n_{M,\xi}} \tag{3.22}$$

with $\Lambda(s,\chi_\xi)$ as in (2.27). Now dualization is a well-defined operation on Groth(G) and commutes with induction, so (3.14) also gives

$$[\rho^\vee] = \sum_{(M,\xi)} n_{M,\xi}[\mathrm{ind}_{M/K}\bar{\xi}]. \tag{3.23}$$

Furthermore $\chi_{\bar{\xi}} = \overline{\chi_\xi}$. Hence the counterpart to (3.16) is

$$L(s,\rho^\vee) = \prod_{(M,\xi)} L(s,\overline{\chi_\xi})^{n_{M,\xi}}. \tag{3.24}$$

Similarly, to define $A(\rho^\vee)$ and $L_\infty(s,\rho^\vee)$ we replace $A_{M,\xi}$ by $A_{M,\bar{\xi}}$ in (3.18) and $L_\infty(s,\chi_\xi)$ by $L_\infty(s,\overline{\chi_\xi})$ in (3.19) (neither replacement actually changes anything). Finally, to define $\Lambda(s,\rho^\vee)$ we replace ρ by $\rho^\vee$ on the right-hand side of (3.21), or simply in $L(s,\rho)$. Then

$$\Lambda(s,\rho^\vee) = \prod_{(M,\xi)} \Lambda(s,\overline{\chi_\xi})^{n_{M,\xi}}, \tag{3.25}$$

whence (3.20), (3.22), (3.25), and Hecke's functional equation (Theorem 2.1) give

$$\Lambda(s,\rho) = W(\rho)\Lambda(1-s,\rho^\vee). \tag{3.26}$$

This is the functional equation of $L(s,\rho)$.

2.2. Dependence on Brauer's theorem

With the functional equation (3.26) now established, let us review the definition of the four types of quantities which appear in it: the L-function $L(s,\rho)$, the gamma factor $L_\infty(s,\rho)$, the exponential factor $A(\rho)$, and the root number $W(\rho)$. Our goal in reviewing the definitions is to distinguish between those that are "Brauer-dependent" – in other words, dependent on an expression for $[\rho]$ like (3.14) – and those that are not. The definition of $L(s,\rho)$ is of the latter type: It is both Brauer-independent and local in the sense that (3.1) makes no reference to Brauer's theorem and expresses $L(s,\rho)$ as a product of local factors (3.3) defined in an intrinsic way. The same is true of $A(\rho)$ and $L_\infty(s,\rho)$, for the Brauer-dependent global definitions (3.18) and (3.19) will eventually be replaced by the Brauer-independent local definitions (3.50) and (3.51) below. However in the case of $W(\rho)$ a Brauer-independent definition is simply not known. By itself this is not problematic: while the decomposition (3.14) of $[\rho]$ is not unique, the resulting quantity (3.20) has to be independent of the decomposition because the functional equation (3.26) can't hold with two different values of $W(\rho)$. What is problematic, however, is that the definition (3.20) is not local. True, each factor $W(\chi_\xi)$ in (3.20) can be written as a product of local root numbers, but no analogue of Tate's local functional equation is known in dimension > 1, so even if one decomposes each $W(\chi_\xi)$ into local factors and reassembles the local factors corresponding to a given place v it is not obvious that the resulting local root number $W(\rho_v)$ is independent of the decomposition of $[\rho]$. That there is in fact a well-defined purely local root number is a theorem of Langlands and Deligne [**23**], about which we will have a little more to say in Lecture 4. For now we continue with the global theory but develop it in a more general context.

3. Compatible families

The L-function of an elliptic curve without complex multiplication is neither a Hecke L-function nor an Artin L-function, and for this reason alone we need to broaden the discussion. The L-functions associated to the "compatible families" to be discussed next include all L-functions of elliptic curves, all Artin L-functions, and much else. In particular, since they include all L-functions of elliptic curves and all Artin L-functions they also include the L-functions of Hecke characters of type (1,0) of imaginary quadratic fields and the L-functions of Hecke characters of finite order. However they do not include all Hecke L-functions: the L-functions of "nonalgebraic" Hecke characters like (2.50) will now fall by the wayside.

3.1. ℓ-adic representations

Up to this point, all representations have been defined over $\mathbb{C}$. Now we consider representations with field of scalars $\mathbb{Q}_\ell$, where ℓ is a prime number. Such a representation is called an **ℓ-adic representation.**

A key difference between complex and ℓ-adic Galois representations is that the latter need not factor through the Galois group of a finite Galois extension. Nonetheless, given a number field K and a prime ideal $\mathfrak{p}$ of K we can choose a prime ideal $\mathfrak{P}$ of $\overline{K}$ over $\mathfrak{p}$ and consider the associated inertia subgroup $I \subset \mathrm{Gal}(\overline{K}/K)$ and Frobenius coset $\sigma I = I\sigma$, where $\sigma \in \mathrm{Gal}(\overline{K}/K)$ is any Frobenius element at $\mathfrak{P}$. If $\mathfrak{P}'$ is another prime ideal of $\overline{K}$ over $\mathfrak{p}$ and I' and σ' are the analogues of I and σ then there is an element of $\mathrm{Gal}(\overline{K}/K)$ which conjugates $\mathfrak{P}$ to $\mathfrak{P}'$, I to I', and σI to $\sigma' I'$. This was the property justifying certain definitions which we made for Artin representations and which therefore now go through for ℓ-adic representations as well. In particular, ρ_ℓ is **ramified** or **unramified** at $\mathfrak{p}$ according as $\rho_\ell|I$ is nontrivial or trivial, and if $\mathfrak{p} \nmid \ell$ then $B_\mathfrak{p}(x)$ is defined by analogy with (3.8):

$$B_\mathfrak{p}(x) = \det(1 - x\rho_\ell(\Phi)|V_\ell^I), \tag{3.27}$$

where $\Phi = \sigma^{-1}$ and V_ℓ is the space of ρ_ℓ. The subspace V_ℓ^I is defined as in (3.2) but with V and ρ replaced by V_ℓ and ρ_ℓ.

Note that in (3.27) we are following the geometric convention, as we shall do consistently from now on. The arithmetic convention would dictate that

$$B_\mathfrak{p}(x) = \det(1 - x\rho_\ell(\sigma)|(V_\ell)_I), \tag{3.28}$$

where $(V_\ell)_I$ is the space of **inertial coinvariants**, the quotient of V_ℓ by the subspace spanned by all expressions of the form $v - \rho_\ell(i)v$ with $v \in V_\ell$ and $i \in I$. The relation between the two conventions is that if $V_\ell^\vee$ is the dual space of V_ℓ then

$$\det(1 - x\rho_\ell^\vee(\Phi)|(V_\ell^\vee)^I) = \det(1 - x\rho_\ell(\sigma)|(V_\ell)_I). \tag{3.29}$$

To see that (3.28) generalizes Artin's original arithmetic definition (3.3), observe that V^I can be replaced by V_I in (3.3): Artin representations are complex representations of a finite group, hence semisimple.

3.2. Full compatibility

Since the coefficients of $B_\mathfrak{p}(x)$ lie in $\mathbb{Q}_\ell$ rather than in $\mathbb{C}$, it is not *a priori* meaningful to substitute $x = (\mathbf{N}\mathfrak{p})^{-s}$. But suppose that for each prime number ℓ we have an ℓ-adic representation ρ_ℓ of $\mathrm{Gal}(\overline{K}/K)$. The resulting collection $\{\rho_\ell\}$ is called a **family of ℓ-adic representations of** $\mathrm{Gal}(\overline{K}/K)$, and we say that the family is **fully compatible** if the following conditions are satisfied:

(i) There is a finite set S of prime ideals of K, independent of ℓ, such that if $\mathfrak{p} \notin S$ and $\mathfrak{p} \nmid \ell$ then ρ_ℓ is unramified at $\mathfrak{p}$.
(ii) The polynomial $B_\mathfrak{p}(x)$ in (3.27), which *a priori* has coefficients in $\mathbb{Q}_\ell$, actually has coefficients in $\mathbb{Q}$ and is independent of ℓ in the sense that $B_\mathfrak{p}(x)$ is unchanged if ℓ in (3.27) is replaced by some other rational prime ℓ' with $\mathfrak{p} \nmid \ell'$.

Although we have not made it part of the definition, it follows from (i) and (ii) that the dimension of ρ_ℓ is independent of ℓ. Indeed, given a second rational prime ℓ', we can choose a prime ideal $\mathfrak{p}$ of K such that $\mathfrak{p} \notin S$ and $\mathfrak{p} \nmid \ell\ell'$. Then $V_\ell^I = V_\ell$ and $V_{\ell'}^I = V_{\ell'}$, whence the degree of $B_\mathfrak{p}(x)$ coincides both with $\dim(V_\ell)$ and $\dim(V_{\ell'})$.

A warning is in order here: The term *fully compatible* is not a standard term, and no standard term for the concept just defined seems to exist in the literature. The usual term is *strictly compatible*, but this is a slightly weaker notion: For strict compatibility $B_\mathfrak{p}(x)$ is required to be independent of ℓ *only for* $\mathfrak{p}$ *not in* S. There is also mere *compatibility*, an even weaker concept; see Serre [**82**], pp. I-10 – I-11. While the concept that we have dubbed *full compatibility* may lack a widely accepted name, the concept itself is all over the literature; see for example [**26**], [**80**], and [**81**]. Fortunately, standard terms do exist for two other concepts: The fully compatible family $\{\rho_\ell\}$ is **integral** if the coefficients of $B_\mathfrak{p}(x)$ are rational integers, and the minimal set S satisfying (i) is the **exceptional set** of the family.

3.3. Examples

The prototypical example of a fully compatible family of integral ℓ-adic representations of $\mathrm{Gal}(\overline{K}/K)$ is the one-dimensional family $\{\omega_\ell^{-1}\}$ consisting of the duals of the ℓ-adic cyclotomic characters

$$\omega_\ell : \mathrm{Gal}(\overline{K}/K) \to \mathbb{Z}_\ell^\times.$$

Let σ denote an arbitrary element of $\mathrm{Gal}(\overline{K}/K)$ and ζ an arbitrary root of unity of ℓ-power order. Then ω_ℓ is defined by the condition

$$\sigma(\zeta) = \zeta^{\omega_\ell(\sigma)}.$$

In particular, if σ is a Frobenius element at a prime ideal $\mathfrak{p}$ of K not dividing ℓ then $\omega_\ell(\sigma) = \mathbf{N}\mathfrak{p}$. Equivalently, if Φ is an inverse Frobenius at $\mathfrak{p}$ then $\omega_\ell^{-1}(\Phi) = \mathbf{N}\mathfrak{p}$, and consequently $B_\mathfrak{p}(x) = x - \mathbf{N}\mathfrak{p}$ by (3.27). Thus because we are following the geometric convention the family $\{\omega_\ell^{-1}\}$ is integral, but not the family $\{\omega_\ell\}$. The exceptional set of both families is the empty set.

Another example is provided by any elliptic curve E over K. Let $T_\ell(E)$ be the ℓ-adic Tate module of E, and let $\rho_{E,\ell}$ be the associated representation of $\mathrm{Gal}(\overline{K}/K)$ on the space $V_\ell(E) = \mathbb{Q}_\ell \otimes_{\mathbb{Z}_\ell} T_\ell(E)$. Then the family $\{\rho_{E,\ell}\}$ is fully compatible, and the dual family $\{\rho_{E,\ell}^\vee\}$ is fully compatible and integral. We remark that the ℓ-adic cohomology groups $H_\ell^1(E)$ are dual to the spaces $V_\ell(E)$, so that $\rho_{E,\ell}^\vee$ is the representation of $\mathrm{Gal}(\overline{K}/K)$ on $H_\ell^1(E)$. The exceptional set S of $\{\rho_{E,\ell}^\vee\}$ consists of the places where E has bad reduction.

The preceding example serves as a useful mnemonic device. The spaces $V_\ell(E)$ yield integral families relative to the arithmetic convention (3.28), while the spaces $H_\ell^1(E)$ yield integral families relative to the geometric convention (3.27). Since ℓ-adic cohomology is expected to be the primary source of fully compatible families, the geometric convention has been adopted as the standard.

3.4. λ-adic representations

If χ is a Dirichlet character of order $\geqslant 3$ then there are infinitely many ℓ such that the values of χ do not lie in $\mathbb{Q}_\ell$, but we would nonetheless like to associate a compatible family to χ and indeed to any Artin representation. A similar comment applies to Hecke characters of imaginary quadratic fields. Thus we need to expand our notion of a compatible family.

Let $\mathbb{E}$ be a number field and λ a finite place of $\mathbb{E}$ of residue characteristic ℓ. A representation with field of scalars $\mathbb{E}_\lambda$ is called a **λ-adic representation**. If $\mathbb{E} = \mathbb{Q}$ then we recover the notion of an ℓ-adic representation. Now let K be a number field and ρ_λ a λ-adic representation of $\mathrm{Gal}(\overline{K}/K)$. If $\mathfrak{p}$ is a prime ideal of K and I the inertia subgroup of $\mathrm{Gal}(\overline{K}/K)$ at some prime ideal of $\overline{K}$ over $\mathfrak{p}$ then we say that ρ_λ is **ramified** or **unramified** at $\mathfrak{p}$ according as $\rho_\lambda|I$ is nontrivial or trivial. And if $\mathfrak{p} \nmid \ell$ then we set

$$B_\mathfrak{p}(x) = \det(1 - x\rho_\lambda(\Phi)|V_\lambda^I), \tag{3.30}$$

where V_λ is the space of ρ_λ.

By a **family of λ-adic representations of** $\mathrm{Gal}(\overline{K}/K)$ we mean a collection $\{\rho_\lambda\}$, where λ runs over the finite places of $\mathbb{E}$ and ρ_λ is a λ-adic representation of $\mathrm{Gal}(\overline{K}/K)$. The family $\{\rho_\lambda\}$ is **fully compatible** if two conditions hold:

(i) There is a finite set S of prime ideals of K, independent of λ, such that if $\mathfrak{p} \notin S$ and $\mathfrak{p} \nmid \ell$ (ℓ being the residue characteristic of λ) then ρ_λ is unramified at $\mathfrak{p}$.

(ii) The polynomial $B_\mathfrak{p}(x)$ in (3.30) has coefficients in $\mathbb{E}$ and is unchanged if λ in (3.30) is replaced by λ', where λ' is another finite place of $\mathbb{E}$ of residue characteristic not divisible by $\mathfrak{p}$.

The fully compatible family $\{\rho_\lambda\}$ is **integral** if the coefficients of $B_\mathfrak{p}(x)$ lie in $\mathcal{O}_\mathbb{E}$, and the **exceptional set** of the family is the minimal set S satisfying (i). We refer to the field $\mathbb{E}$ as the **coefficient field** of the family.

3.5. Compatible families of Artin and Hecke type

Let ρ be an Artin representation of K. Since any complex representation of a finite group is realizable over a number field, there is a representation of $\mathrm{Gal}(\overline{K}/K)$ on a vector space V over a number field $\mathbb{E} \subset \mathbb{C}$ such that ρ is the representation on $\mathbb{C} \otimes_\mathbb{E} V$ afforded by extension of scalars. But extension of scalars also gives a representation ρ_λ on $\mathbb{E}_\lambda \otimes_\mathbb{E} V$ for each finite place λ of $\mathbb{E}$. The resulting family $\{\rho_\lambda\}$ is fully compatible and integral, and its exceptional set is the set of places where ρ is ramified.

Next let K be an imaginary quadratic field and χ a primitive Hecke character of K of type (1,0) and conductor $\mathfrak{f}$. Let $\mathbb{E}$ be the finite extension of K generated by the values of χ. Given a prime number ℓ and a prime ideal $\mathfrak{l}$ of $\mathbb{E}$ dividing ℓ, we consider the map

$$I(\mathfrak{f}\ell^n)/P_{\mathfrak{f}\ell^n} \to \mathbb{E}(\mathfrak{l}^n)/\mathbb{E}_{\mathfrak{l}^n}$$

given by

$$\mathfrak{a}P_{\mathfrak{f}\ell^n} \mapsto \chi(\mathfrak{a})^{-1}\mathbb{E}_{\mathfrak{l}^n} \tag{3.31}$$

for $\mathfrak{a} \in I(\mathfrak{f}\ell^n)$. This map is well defined by (1.10) (note in particular that $\chi(\mathfrak{a})$ is relatively prime to ℓ because $\chi(\mathfrak{a})^r$ is relatively prime to ℓ, where $r = |I(\mathfrak{f})/P_\mathfrak{f}|$).

Making our usual identification of $\mathbb{E}(\mathfrak{l}^n)/\mathbb{E}_{\mathfrak{l}^n}$ with $(\mathcal{O}_{\mathbb{E}}/\mathfrak{l}^n)^\times$ and then taking inverse limits, we obtain a one-dimensional λ-adic representation

$$\varprojlim_n I(\mathfrak{f}\ell^n)/P_{\mathfrak{f}\ell^n} \to \mathcal{O}_\lambda^\times, \tag{3.32}$$

where λ is the place of $\mathbb{E}$ corresponding to $\mathfrak{l}$. To get a λ-adic *Galois* representation χ_λ from (3.32), we identify the left-hand side of (3.32) with a quotient of $\mathrm{Gal}(K^{\mathrm{ab}}/K)$ by using the Artin isomorphism (3.6). The resulting family $\{\chi_\lambda\}$ is fully compatible and integral, and the exceptional set consists of the places of K dividing $\mathfrak{f}$.

3.6. The L-function of a fully compatible family

Henceforth it will be convenient to view all coefficient fields as subfields of $\mathbb{C}$. Notationally it is more convenient to associate an L-function to the *isomorphism class* of a fully compatible family rather than to the family itself, so a definition is in order. Let $\{\rho_\lambda\}$ and $\{\rho'_{\lambda'}\}$ be fully compatible families of representations of $\mathrm{Gal}(\overline{K}/K)$ with coefficient fields $\mathbb{E}$ and $\mathbb{E}'$ respectively. We say that these two families are **isomorphic** if there exists a number field $\mathbb{E}''$ containing $\mathbb{E}$ and $\mathbb{E}'$ such that for every finite place λ'' of $\mathbb{E}''$, the representations $\rho_{\lambda''}$ and $\rho'_{\lambda''}$ are isomorphic over $\mathbb{E}''_{\lambda''}$. Here $\rho_{\lambda''}$ and $\rho'_{\lambda''}$ are the representations over $\mathbb{E}''_{\lambda''}$ obtained by extension of scalars from ρ_λ and $\rho'_{\lambda'}$ respectively, where λ and λ' lie below λ''.

Now let M be the isomorphism class of a fully compatible family of λ-adic representations of $\mathrm{Gal}(\overline{K}/K)$ with coefficient field $\mathbb{E}$. Since we are viewing $\mathbb{E}$ as a subfield of $\mathbb{C}$, we can define the L-function of M by analogy with (3.1):

$$L(s,M) = \prod_{\mathfrak{p}} B_{\mathfrak{p}}((\mathbf{N}\mathfrak{p})^{-s})^{-1}, \tag{3.33}$$

where $\mathfrak{p}$ runs over the prime ideals of K and $B_{\mathfrak{p}}(x)$ is as in (3.30).

The analogy with Artin L-functions extends beyond the definition to include also the properties of additivity and inductivity. Let M and M' be the isomorphism classes of two fully compatible families $\{\rho_\lambda\}$ and $\{\rho'_\lambda\}$, which by extension of scalars may be assumed to have the same coefficient field. Then $M \oplus M'$ denotes the isomorphism class of the family $\{\rho_\lambda \oplus \rho'_\lambda\}$, and

$$L(s, M\oplus M') = L(s,M)L(s,M'). \tag{3.34}$$

On the other hand, let L be a finite extension of K and let M be the isomorphism class of a fully compatible family $\{\rho_\lambda\}$ of λ-adic representations of $\mathrm{Gal}(\overline{K}/L)$. Then the family $\{\mathrm{ind}_{L/K}\rho_\lambda\}$ is also fully compatible, and its isomorphism class, which we will denote $\mathrm{ind}_{L/K}M$, satisfies

$$L(s, \mathrm{ind}_{L/K}M) = L(s,M). \tag{3.35}$$

As in the case of Artin L-functions, (3.34) is immediate from the definitions, (3.35) less straightforward.

By way of illustration, if ρ is an Artin representation and M is the isomorphism class of $\{\rho_\lambda\}$ then $L(s,M) = L(s,\rho^\vee)$, because we followed the arithmetic convention when defining Artin L-functions. On the other hand, if χ is a Hecke character of type $(1,0)$ of an imaginary quadratic field and M is the isomorphism class of $\{\chi_\lambda\}$ then the net effect of the replacement of $\chi(\mathfrak{a})$ by $\chi(\mathfrak{a})^{-1}$ in (3.31) and the replacement of σ by Φ in (3.27) is to leave the L-function unchanged: $L(s,M) = L(s,\chi)$. Finally, if E is an elliptic curve and M is the isomorphism class of $\{\rho^\vee_{E,\ell}\}$ then $L(s,M) = L(s,E)$ by virtue of (3.29) with $V_\ell = V_\ell(E)$.

3.7. Semisimplicity

Let M be the isomorphism class of a fully compatible family $\{\rho_\lambda\}$. Since characteristic polynomials are insensitive to semisimplification, $L(s, M)$ is unchanged if the representations ρ_λ are replaced by their semisimiplifications. So from this point of view there is no loss in assuming that the ρ_λ are semisimple to being with, and in fact there is something to be gained:

Proposition 3.1. *Let $\{\rho_\lambda\}$ be a fully compatible family of semisimple λ-adic representations of $\mathrm{Gal}(\overline{K}/K)$, and let M be its isomorphism class. Then M is uniquely determined by the isomorphism class of any one of the representations ρ_λ.*

PROOF. This is a simple consequence of the fact that a semisimple representation over a field of characteristic 0 is determined up to isomorphism by its character. Indeed fix places λ and λ' of the coefficient field $\mathbb{E}$ of the family, and given a prime ideal $\mathfrak{p}$ of K let $\Phi_\mathfrak{p} \in \mathrm{Gal}(\overline{K}/K)$ be an inverse Frobenius element at $\mathfrak{p}$. For all but finitely many $\mathfrak{p}$ we have $\operatorname{tr}\rho_\lambda(\Phi_\mathfrak{p}) = \operatorname{tr}\rho_{\lambda'}(\Phi_\mathfrak{p})$, because both traces coincide with the coefficient of $-x$ in $B_\mathfrak{p}(x)$. Thus $\operatorname{tr}\rho_\lambda$ and $\operatorname{tr}\rho_{\lambda'}$ coincide on a dense subset of $\mathrm{Gal}(\overline{K}/K)$. Since both are continuous each determines the other. □

Without the semisimplicity assumption the assertion is false. For example, fix a finite place λ_0 of $\mathbb{E}$, and for $\lambda \neq \lambda_0$ set $\rho_\lambda = 1_K \oplus 1_K$. We can complete $\{\rho_\lambda\}_{\lambda\neq\lambda_0}$ to a fully compatible family $\{\rho_\lambda\}$ by setting $\rho_{\lambda_0} = 1_K \oplus 1_K$ but also by setting

$$\rho_{\lambda_0}(g) = \begin{pmatrix} 1 & \log\omega_{\ell_0}(g) \\ 0 & 1 \end{pmatrix} \qquad (g \in \mathrm{Gal}(\overline{K}/K)),$$

where ℓ_0 is the residue characteristic of λ_0 and log is the ℓ_0-adic logarithm on $\mathbb{Z}_{\ell_0}$. Of course the associated L-function is $\zeta_K(s)^2$ in both cases.

3.8. Analytic desiderata

Let M be the isomorphism class of a fully compatible family of λ-adic representations of $\mathrm{Gal}(\overline{K}/K)$ with coefficient field $\mathbb{E}$ and exceptional set S. It is not at all clear that the Euler product defining $L(s, M)$ converges in some right half-plane, and without this property $L(s, M)$ is of no use to us. Hence we need to impose a condition on the **reciprocal roots** of $B_\mathfrak{p}(x)$, in other words the numbers $\alpha \in \mathbb{C}^\times$ such that $B_\mathfrak{p}(\alpha^{-1}) = 0$ (recall that $B_\mathfrak{p}(x)$ has constant term 1, so that 0 is not a root). Actually we consider two conditions: one for $\mathfrak{p} \notin S$ and one for $\mathfrak{p} \in S$. Fix $w \in \mathbb{Z}$ and let τ run over arbitrary field automorphisms of $\mathbb{C}$. The conditions are

$$(3.36) \qquad |\tau(\alpha)| = (\mathbf{N}\mathfrak{p})^{w/2} \qquad (\mathfrak{p} \notin S)$$

and

$$(3.37) \qquad |\tau(\alpha)| \leqslant (\mathbf{N}\mathfrak{p})^{w/2} \qquad (\mathfrak{p} \in S)$$

respectively. If M satisfies (3.36) and (3.37) then we say that M has **weight** w.

The point of this definition is that if M has weight w then the Euler product defining $L(s, M)$ converges for $\Re(s) > w/2 + 1$. Of course this would be true even if we required (3.36) and (3.37) only for τ equal to the identity automorphism, but by allowing $\tau \in \mathrm{Aut}(\mathbb{C})$ to be arbitrary, we compensate for the fact that we have fixed an embedding of $\mathbb{E}$ in $\mathbb{C}$. Fixing such an embedding is convenient, but we do not want our definition of *weight* to depend on the choice of embedding.

Next we want an analytic continuation and functional equation. To begin to formulate the latter, we introduce the notion of a **gamma factor of weight** w over K. This term will refer to any product of the form

$$\gamma(s) = \prod_{v\mid\infty} \gamma_v(s), \tag{3.38}$$

where $\gamma_v(s)$ is a a gamma factor of weight w over K_v in the following sense.

If w is odd, or if w is even and $K_v \cong \mathbb{C}$, then $\gamma_v(s)$ is a product of the form

$$\gamma_v(s) = \prod_{\substack{p+q=w \\ q\geqslant p\geqslant 0}} \Gamma_{\mathbb{C}}(s-p)^{h^{pq}}, \tag{3.39}$$

where p and q are nonnegative integers satisfying the stated conditions and the exponents h^{pq} are nonnegative integers. Of course if w is odd then the condition $q \geqslant p \geqslant 0$ can be replaced by $q > p \geqslant 0$.

If w is even and $K_v = \mathbb{R}$ then $\gamma_v(s)$ is a product of the form

$$\gamma_v(s) = \Gamma_{\mathbb{R}}(s-w/2)^{h^{w/2+}} \cdot \Gamma_{\mathbb{R}}(s-w/2+1)^{h^{w/2-}} \cdot \prod_{\substack{p+q=w \\ q>p\geqslant 0}} \Gamma_{\mathbb{C}}(s-p)^{h^{pq}}, \tag{3.40}$$

where as before, p and q are nonnegative integers satisfying the stated conditions and the exponents $h^{w/2+}$, $h^{w/2-}$, and h^{pq} are all nonnegative integers.

3.9. Duality

To state a functional equation we need not only gamma factors but also a notion of duality. This is straightforward: If M is the isomorphism class of $\{\rho_\lambda\}$ then the **dual** $M^\vee$ of M is the isomorphism class of $\{\rho_\lambda^\vee\}$. Given $r \in \mathbb{Z}$, we define the **r-fold Tate twist** $M(r)$ of M to be the isomorphism class of $\{\rho_\lambda \otimes \omega_\ell^r\}$, where the tensor product of ρ_λ with the ℓ-adic cyclotomic character ω_ℓ is formed by viewing the latter as a representation over $\mathbb{E}_\lambda$. Now if M has weight w then we put

$$\overline{M} = M^\vee(-w) \tag{3.41}$$

and call M **essentially self-dual** if $\overline{M} \cong M$. The notation $\overline{M}$ seems reasonable, because if $\alpha_1, \alpha_2, \ldots, \alpha_n$ are the reciprocal roots of $B_\mathfrak{p}(x)$ in $\mathbb{C}$, listed with their multiplicities, then the reciprocal roots of the counterpart to $B_\mathfrak{p}(x)$ for $M^\vee(-w)$ are the numbers $\alpha_j^{-1} \cdot (\mathbf{N}\mathfrak{p})^w$ with $1 \leqslant j \leqslant n$. But $\alpha_j^{-1} = \overline{\alpha_j}/|\alpha_j|^2 = \overline{\alpha_j}/(\mathbf{N}\mathfrak{p})^w$ by (3.36), at least if $\mathfrak{p} \notin S$, so

$$\alpha_j^{-1} \cdot (\mathbf{N}\mathfrak{p})^w = \overline{\alpha_j} \qquad (\mathfrak{p} \notin S). \tag{3.42}$$

The notation for the left-hand side of (3.41) is meant to remind us of (3.42).

To illustrate the definitions, let K be an imaginary quadratic field and consider a primitive Hecke character χ of K of type (1,0). Let $M'(\chi)$ be the isomorphism class of the associated one-dimensional family $\{\chi_\lambda\}$. Then $M'(\chi)$ is of weight one by (1.30), but $\chi \neq \overline{\chi}$ and consequently $M'(\chi)$ is not essentially self-dual. But put

$$M(\chi) = \mathrm{ind}_{K/\mathbb{Q}} M'(\chi). \tag{3.43}$$

If χ is equivariant in the sense that it satisfies the identity $\chi(\overline{\mathfrak{a}}) = \overline{\chi}(\mathfrak{a})$ then one readily verifies that $M(\chi)$ is essentially self-dual of weight 1. In particular this is the case if $\chi \in X(D)$, by Proposition 1.7. And since $L(s, M(\chi))$ coincides by (3.35) with $L(s, M'(\chi))$ and hence with $L(s,\chi)$, Theorem 1.2 remains valid with $L(s,\chi)$ replaced by $L(s, M(\chi))$.

3.10. An algebraic desideratum

There is one further requirement that we would like to impose on M. Let $\{\rho_\lambda\}$ be a representative of M and $\mathbb{E}$ its coefficient field. We will say that M satisfies **Condition $\mathbf{C_8}$** if the following property holds. Let $\mathfrak{p}$ be a prime ideal of K and D and I the decomposition and inertia subgroups of $\mathrm{Gal}(\overline{K}/K)$ associated to a prime ideal $\mathfrak{P}$ of $\overline{K}$ above $\mathfrak{p}$. Suppose that $g \in D$ is an element such that the coset of g in D/I coincides with the coset of σ^n for some Frobenius element σ at $\mathfrak{P}$ and some $n \in \mathbb{Z}$. Then we require the characteristic polynomial of $\rho_\lambda(g)$ to have coefficients in $\mathbb{E}$ and to be independent of λ for all finite places λ of $\mathbb{E}$ such that $\mathfrak{p}$ and λ are of distinct residue characteristics. The reason for referring to this property as "Condition C_8" is that it is so labeled in Serre [**81**]. We could also have referred to Problem 2 on p. 514 of Serre-Tate [**86**]. Note that Condition C_8 neither supersedes nor is superseded by full compatibility, because if $\mathfrak{p}$ belongs to the exceptional set S then Condition C_8 pertains to $\rho_\lambda(g)$ itself whereas full compatibility pertains to the restriction of $\rho_\lambda(g)$ to the space of inertial invariants.

4. Premotives

We come now to the main point. Let M be the isomorphism class of a fully compatible family of semisimple integral λ-adic representations of $\mathrm{Gal}(\overline{K}/K)$. We call M a **premotive of weight** w over K if three conditions are satisfied:

(i) M has weight w and satisfies Condition C_8.

(ii) $L(s, M)$ extends to a meromorphic function on $\mathbb{C}$ which is entire if w is odd and holomorphic everywhere except possibly at $s = w/2 + 1$ if w is even.

(iii) There is a positive integer $A(M)$, a constant $W(M) \in \mathbb{C}$ of absolute value 1, and a gamma factor $\gamma(s) = L_\infty(s, M)$ of weight w over K such that

$$\Lambda(s, M) = W(M)\Lambda(k - s, \overline{M})$$

with $\Lambda(s, M) = A(M)^{s/2}\gamma(s)L(s, M)$, $\Lambda(s, \overline{M}) = A(M)^{s/2}\gamma(s)L(s, \overline{M})$, and $k = w + 1$.

The **rank** of M is the dimension of ρ_λ for $\{\rho_\lambda\} \in M$, and if we can choose $\{\rho_\lambda\}$ to have a given number field $\mathbb{E}$ as coefficient field then we say that M **admits $\mathbb{E}$ as coefficient field**.

If M is a premotive of weight w then the integrality of the underlying representations ρ_λ implies that $w \geqslant 0$. We could broaden the definition and allow premotives of negative weight by declaring that $M((r - w)/2)$ has weight $w < 0$ if M is a premotive of weight $r \geqslant 0$ with $r \equiv w \bmod 2$. However premotives of negative weight will play no role in what follows.

What is more problematic about the definition is that by requiring (ii) and (iii) we have limited our stock of examples. Not that we are entirely without examples: The isomorphism class of $\{\omega_\ell^{-w/2}\}$ is a premotive of weight w for any even integer $w \geqslant 0$, and if K is an imaginary quadratic field, χ a Hecke character of K of type $(1, 0)$, and $M(\chi)$ and $M(\chi')$ as in (3.43), then $M(\chi)$ and $M(\chi')$ are premotives of weight 1 over K and $\mathbb{Q}$ respectively by virtue of the analytic properties of Hecke L-functions, in particular Theorem 2.1. But we cannot make the analogous claim for an arbitrary Artin representation ρ of a number field K because (ii) is unknown in general: The statement that $L(s, \rho)$ is holomorphic everywhere except possibly at $s = 1$ is the Artin conjecture. (More precisely, the Artin conjecture asserts that if ρ is irreducible and nontrivial then $L(s,\rho)$ is entire. But if ρ is irreducible and trivial then $L(s,\rho)=\zeta_K(s)$. Hence by additivity the Artin conjecture implies that for any ρ, irreducible or not, $L(s,\rho)$ is holomorphic except possibly at $s=1$.) While the Artin conjecture is known in certain cases – including for example the case of monomial representations, where it follows by inductivity from the analytic properties of Hecke L-functions, and many two-dimensional cases (Langlands [**57**], Tunnell [**97**], and Khare and Wintenberger [**50**], [**51**] with Kisin [**52**]) – the fact remains that Artin representations do not automatically provide examples of premotives. Similarly, if E is an elliptic curve over $\mathbb{Q}$ then the isomorphism class of $\{\rho_{E,\ell}^{\vee}\}$ is a premotive of weight 1 by virtue of the modularity of E (Wiles [**100**], Taylor and Wiles [**96**], and Breuil, Conrad, Diamond, and Taylor [**11**]), but if $\mathbb{Q}$ is replaced by an arbitrary number field then the assertion remains conjectural.

In spite of this objection, we have included (ii) and (iii) in the definition so as to be able to refer to the order of vanishing of $L(s,M)$ at $s=(w+1)/2$ without using the word *conjectural* at every turn. But when we are not talking about trivial central zeros (ii) and (iii) will play no role.

5. Uniqueness of the functional equation

Given that we do include the functional equation in the definition, it may seem strange that the factors $A(M)$, $W(M)$, and $L_\infty(s,M)$ are not defined more precisely. However the following proposition shows that they are in fact uniquely determined by the functional equation. Both the proposition and its proof were suggested by Théorème 4.6 on p. 514 of Deligne-Serre [**25**].

Proposition 3.2. *If $\widetilde{A}(M)$ is a positive integer, $\widetilde{W}(M)\in\mathbb{C}$ a constant of absolute value 1, and $\tilde{\gamma}(s)$ a gamma factor of weight w over K such that*

$$\widetilde{\Lambda}(s,M)=\widetilde{W}(M)\widetilde{\Lambda}(k-s,\overline{M})$$

with $\widetilde{\Lambda}(s,M)=\widetilde{A}(M)^{s/2}\tilde{\gamma}(s)L(s,M)$ *and* $\widetilde{\Lambda}(s,\overline{M})=\widetilde{A}(M)^{s/2}\tilde{\gamma}(s)L(s,\overline{M})$, *then* $\widetilde{A}(M)=A(M)$, $\widetilde{W}(M)=W(M)$, *and* $\tilde{\gamma}(s)=\gamma(s)$.

PROOF. Taking the ratio of the two functional equations, we obtain

$$(A(M)/\widetilde{A}(M))^{s/2}\frac{\gamma(s)}{\tilde{\gamma}(s)}=(W(M)/\widetilde{W}(M))(A(M)/\widetilde{A}(M))^{(k-s)/2}\frac{\gamma(k-s)}{\tilde{\gamma}(k-s)}. \tag{3.44}$$

Now it follows from the formulas (3.39) and (3.40) that the left-hand side of (3.44) is holomorphic and nonvanishing for $\Re(s)>[w/2]$ and the right-hand side for $\Re(s)<k-[w/2]$. Since $k=w+1$ we have $[w/2]<k-[w/2]$ and consequently both sides of (3.44) are entire and nonvanishing. Thus $\gamma(s)/\tilde{\gamma}(s)$ is entire and nonvanishing. If w is odd then $\gamma(s)/\tilde{\gamma}(s)$ has the form $\prod_{p=0}^{(w-1)/2}\Gamma_{\mathbb{C}}(s-p)^{n_p}$ with $n_p\in\mathbb{Z}$, and the fact that $\gamma(s)/\tilde{\gamma}(s)$ is holomorphic and nonzero at $s=(w-1)/2$ shows that $n_p=0$ for $p=(w-1)/2$. Applying this argument inductively we find that $n_p=0$ for $0\leqslant p\leqslant(w-1)/2$, whence $\tilde{\gamma}(s)=\gamma(s)$. If w is even we use the duplication formula to write $\gamma(s)/\tilde{\gamma}(s)$ in the form $\prod_{p=-1}^{w/2}\Gamma_{\mathbb{R}}(s-p)^{n_p}$, and a similar argument again gives $\tilde{\gamma}(s)=\gamma(s)$. Thus in both cases we conclude that

$$(A(M)/\widetilde{A}(M))^{s}=(W(M)/\widetilde{W}(M))(A(M)/\widetilde{A}(M))^{k/2}$$

for all s, whence $\widetilde{A}(M)=A(M)$ and $\widetilde{W}(M)=W(M)$. □

6. An open problem

The term *premotive* is just a device enabling us to talk about motivic L-functions without first talking about motives, but the terminology suggests a question:

Problem 3. *Does every premotive come from a motive?*

This is not a new question. In fact a stronger version appears as Question 2 on p. I-12 of Serre [**82**], and while the Fontaine-Mazur conjecture [**30**] is concerned with other issues, it too implies a statement about the provenance of fully compatible families which is in most respects much stronger than what we are asking for here (see [**30**], pp. 196 – 197). Furthermore, the converse of Problem 3 is also a well-known open problem. In other words, if we start with a pure motive M of weight w then it is not known in general that M is a premotive of weight w, not only because the analytic continuation and functional equation of $L(s, M)$ are not known, but also because the full compatibility of the family of λ-adic representations attached to M is not known either, nor the semisimplicity of the representations ρ_λ, nor Condition C_8.

Since Problem 3 and its converse are already well-known open problems, the only reason for drawing attention to them here is to justify the coinage *premotive* and the use of the term *motivic L-function* for the L-function associated to a premotive. By postulating a connection with motives we also justify the notations h^{pq} and $h^{p\pm}$ in (3.39) and (3.40), because if M does come from a motive then the gamma factor $L_\infty(s, M)$ that one associates to M is given by (3.38) with h^{pq} and $h^{p\pm}$ equal to the usual Hodge numbers; cf. [**24**], p. 329. More precisely, h^{pq} is the usual Hodge number and $h^{p\pm}$ is the multiplicity of the eigenvalue $(-1)^p(\pm 1)$ of the "Frobenius at infinity" – in other words, of complex conjugation – on H^{pp}.

That said, Problem 3 has little bearing on our present train of thought. We would like to define $A(M)$ and $W(M)$ as products of local factors rather than via the abstract uniqueness of a conjectural functional equation. While the quest for such a self-contained local definition will encounter a difficulty, the difficulty would not disappear if we knew that our premotive came from a motive.

7. Local factors for Artin L-functions

Earlier we mentioned that if ρ is an Artin representation of K then the factors $A(\rho)$ and $L_\infty(s, \rho)$ defined by (3.18) and (3.19) respectively could also be defined in a Brauer-independent and local way. We now record these local definitions, thereby demonstrating that in the case of Artin representations, the theory to be described in the next lecture is needed only for the sake of $W(\rho)$.

7.1. The exponential factor

Given a finite place v of K, we identify $\mathrm{Gal}(\overline{K_v}/K_v)$ with the decomposition subgroup of $\mathrm{Gal}(\overline{K}/K)$ at some place of $\overline{K}$ above v, and we put $\rho_v = \rho|\mathrm{Gal}(\overline{K_v}/K_v)$. The **conductor** of ρ is the integral ideal

$$\mathfrak{f}(\rho) = \prod_{v \nmid \infty} \mathfrak{p}_v^{a(\rho_v)} \tag{3.45}$$

of K, where $\mathfrak{p}_v$ is the prime ideal corresponding to v and $a(\rho_v)$ is the **exponent of the local conductor**, a nonnegative integer still to be defined. The definition will show that $a(\rho_v) > 0$ if and only if ρ is ramified at $\mathfrak{p}_v$. Since ρ is unramified outside a finite set it follows that the product on the right-hand side of (3.45) is finite.

The definition of $a(\rho_v)$ is a purely local matter, and thus we change our notation by dropping the subscript v: K is now a finite extension of $\mathbb{Q}_p$ with $p < \infty$, and ρ is a representation of $\mathrm{Gal}(\overline{K}/K)$. We choose a finite Galois extension L of K such that ρ factors through $\mathrm{Gal}(L/K)$, and we view ρ as a representation of $\mathrm{Gal}(L/K)$.

The definition of $a(\rho)$ involves the higher ramification subgroups I_n $(n \geqslant 0)$ of $\mathrm{Gal}(L/K)$. If $n = 0$ then $I_n = I$, the inertia subgroup of $\mathrm{Gal}(L/K)$, and in general

$$I_n = \{\sigma \in I : \sigma(x) \equiv x \pmod{\mathfrak{p}_L^{n+1}} \text{ for all } x \in \mathcal{O}_L\},$$

where $\mathcal{O}_L$ is the ring of integers of L and $\mathfrak{p}_L$ its maximal ideal. Let V be the space of ρ and V^{I_n} the subspace of vectors fixed by $\rho(I_n)$. Then

$$a(\rho) = \sum_{n \geqslant 0} (e_n/e)\dim(V/V^{I_n}), \tag{3.46}$$

where $e_n = |I_n|$ and e $(= e_0)$ is the ramification index of L over K. If n is sufficiently large then I_n is the trivial subgroup of $\mathrm{Gal}(L/K)$, so the sum is finite. Also $V^I = V$ if and only if $\rho(I)$ is the trivial subgroup of $\mathrm{GL}(V)$, so $a(\rho) = 0$ if and only if ρ is unramified, as already mentioned. That $a(\rho)$ is integral and independent of the choice of L follows from an alternative expression for $a(\rho)$ as an inner product of the character of ρ with the "Artin character" (see Chapter VI of Serre [**83**], especially Theorem 1' on p. 99 and the Corollary to Proposition 3 on p. 101).

The conductor-exponents $a(\rho)$ satisfy a modified Artin formalism, and consequently so does the global conductor (3.45). We will go directly to the global version, and thus we take K to be a number field again. The additivity of the conductor is immediate from the definitions: If ρ and ρ' are Artin representations of K then

$$\mathfrak{f}(\rho \oplus \rho') = \mathfrak{f}(\rho)\mathfrak{f}(\rho'). \tag{3.47}$$

Next let M be a finite extension of K and ρ an Artin representation of M. The counterpart to inductivity is not quite invariance under induction but rather

$$\mathfrak{f}(\mathrm{ind}_{M/K}\rho) = \mathfrak{d}_{M/K}^{\dim\rho} N_{M/K}(\mathfrak{f}(\rho)), \tag{3.48}$$

where $\mathfrak{d}_{M/K}$ is the relative discriminant ideal of the extension M/K. Finally, if ξ is a one-dimensional Artin representation and χ_ξ the corresponding idele class character of finite order, then

$$\mathfrak{f}(\xi) = \mathfrak{f}(\chi_\xi), \tag{3.49}$$

where the right-hand side is the usual conductor of an idele class character, defined as in (2.21).

Now let D be the absolute value of the discriminant of K. Given an Artin representation ρ of K, we define its **exponential factor** to be the positive integer

$$A(\rho) = D^{\dim\rho}\mathbf{N}\mathfrak{f}(\rho). \tag{3.50}$$

From (3.47), (3.48), and (3.49) we deduce the Artin formalism for the exponential factor: $A(\rho \oplus \rho') = A(\rho)A(\rho')$, $A(\inf_{M/K}\rho) = A(\rho)$, and $A(\xi) = A(\chi_\xi)$. Using these identities, one can verify that the exponential factor defined by (3.50) coincides with the Brauer-dependent quantity defined by (3.18).

7.2. The gamma factor

Given an infinite place v of K we identify the decomposition subgroup of $\mathrm{Gal}(\overline{K}/K)$ at some place above v with $\mathrm{Gal}(\overline{K_v}/K_v)$, just as in the archimedean case. Of course in the archimedean case there is no distinction between the decomposition and inertia groups: Either both are trivial or both are the group of order two generated by the relevant complex conjugation. If ρ is an Artin representation of K then we put $\rho_v = \rho|\mathrm{Gal}(\overline{K_v}/K_v)$ and define the gamma factor of ρ to be the product

$$L_\infty(s,\rho) = \prod_{v|\infty} L(s,\rho_v), \tag{3.51}$$

where the individual factors $L(s,\rho_v)$ must still be defined. Since the issue is now local, we fix v and drop the subscript on ρ_v and K_v. Thus ρ is a representation of $\mathrm{Gal}(\overline{K}/K)$ with $K = \mathbb{R}$ or $K \cong \mathbb{C}$.

The first step is to declare that $L(s,\rho\oplus\rho') = L(s,\rho)L(s,\rho')$ and thus to reduce the definition of $L(s,\rho)$ to the case where ρ is one-dimensional. Next, to a one-dimensional ρ we associate a character χ of $K^\times$ as follows: If ρ is trivial then χ is trivial, and if $K = \mathbb{R}$ and ρ is the nontrivial character of $\mathrm{Gal}(\mathbb{C}/\mathbb{R})$ then χ is the sign character of $\mathbb{R}^\times$. We put

$$L(s,\rho) = L(s,\chi). \tag{3.52}$$

Referring to (2.25) and (2.26), we see that the definition (3.52) amounts to saying that if $K \cong \mathbb{C}$ then $L(s,\rho) = \Gamma_\mathbb{C}(s)$ and if $K = \mathbb{R}$ then $L(s,\rho)$ is either $\Gamma_\mathbb{R}(s)$ or $\Gamma_\mathbb{R}(s+1)$ according as ρ is trivial or nontrivial.

8. Exercises

Exercise 3.1. Give examples of a one-dimensional Artin representation and a two-dimensional irreducible Artin representation such that the associated Artin L-function has a trivial central zero. (Hint: Exercise 2.7 with (3.5) and induction.)

Exercise 3.2. Let K be a number field and L a Galois extension of K with $\mathrm{Gal}(L/K)$ isomorphic to A_4, the alternating group on 4 letters. Let ρ be the irreducible complex representation of $\mathrm{Gal}(L/K)$ of dimension 3, unique up to isomorphism. Show that $W(\rho) = 1$. (Hint: Write ρ as a monomial representation and apply Theorem 2.2, or else use Theorem 2.2 in conjunction with (3.53) below.) Incidentally, both this exercise and Theorem 2.2 are instances of a result of Fröhlich and Queyrut [**33**] asserting that $W(\rho) = 1$ if ρ is orthogonal.

Exercise 3.3. Let K be a number field and L a finite Galois extension of K. Using the fact that $\mathrm{ind}_{L/K}1_L$ is the regular representation of $\mathrm{Gal}(L/K)$, show that

$$\zeta_L(s) = \prod_\rho L(s,\rho)^{\dim\rho}, \tag{3.53}$$

where ρ runs over the distinct isomorphism classes of irreducible complex representations of $\mathrm{Gal}(L/K)$. In particular, if one of the L-functions in (3.53) has a trivial central zero (as Exercise 3.1 shows can happen) and the others are holomorphic at $s = 1/2$ (as the Artin conjecture says they must be) then $\zeta_L(1/2) = 0$. But the zero of $\zeta_L(s)$ at $s = 1/2$ is not a trivial central zero as we have defined the term, because the functional equation of $\zeta_L(s)$ is $Z_L(s) = Z_L(1-s)$. This is arguably a reason to revise our notion of a trivial zero to take account not only of the functional equation of the given L-function but also of the functional equation of its factors.

Exercise 3.4. Let $\mathbb{E}_\lambda$ be a finite extension of $\mathbb{Q}_\ell$ and ρ_λ a λ-adic representation of a compact group G. Put $d = \dim\rho_\lambda$. Prove that ρ_λ is equivalent to a representation into $\mathrm{GL}_d(\mathcal{O}_\lambda)$, where $\mathcal{O}_\lambda$ is the ring of integers of $\mathbb{E}_\lambda$. (Hint: View ρ_λ as a continuous homomorphism $G \to \mathrm{GL}_d(\mathbb{E}_\lambda)$, and let H be the inverse image of $\mathrm{GL}_d(\mathcal{O}_\lambda)$. Then H is an open subgroup of G, hence of finite index. Let $\mathcal{L}$ be the sum of the $\mathcal{O}_\lambda$-submodules $\rho_\lambda(g)(\mathcal{O}_\lambda^d)$ of $\mathbb{E}_\lambda^d$, where g runs over a set of coset representatives for H in G. Show that $\mathcal{L}$ is a G-stable $\mathcal{O}_\lambda$-lattice in $\mathbb{E}_\lambda^d$.)

Exercise 3.5. Let $\{\rho_\ell\}$ be a fully compatible family of one-dimensional ℓ-adic representations of $\mathrm{Gal}(\overline{\mathbb{Q}}/\mathbb{Q})$ with exceptional set $S = \emptyset$. Prove that $\rho_\ell = \omega_\ell^n$ for some $n \in \mathbb{Z}$.

Exercise 3.6. Let K be a number field, and consider the map $\rho \mapsto \{\rho_\lambda\}$ which sends an Artin representation ρ of $\mathrm{Gal}(\overline{K}/K)$ to a fully compatible family of integral λ-adic representations of $\mathrm{Gal}(\overline{K}/K)$. Show that every premotive M of weight 0 over K arises from some Artin representation ρ in this way. (Hint: If M is the isomorphism class of a family $\{\rho_\lambda\}$ then the key point is to show that the image of ρ_λ is finite. By Exercise 3.4, ρ_λ may be viewed as a continuous homomorphism $\mathrm{Gal}(\overline{K}/K) \to \mathrm{GL}_d(\mathcal{O}_\lambda)$. Let $\mathbb{F}_\lambda$ be the residue class field of $\mathcal{O}_\lambda$ and ℓ the residue characteristic. Show that the reduction map $\mathrm{GL}_d(\mathcal{O}_\lambda) \to \mathrm{GL}_d(\mathbb{F}_\lambda)$ is injective on elements of order prime to ℓ.)

Exercise 3.7. (*Reading.*) Let K and $\mathbb{E}$ be number fields. Given a fully compatible family $\{\rho_\lambda\}$ over K with coefficient field $\mathbb{E}$, consider the extension F of $\mathbb{E}$ generated by all the roots of the equations $B_\mathfrak{p}(x) = 0$ as $\mathfrak{p}$ varies over prime ideals of K. A necessary and condition for F to be of finite degree over $\mathbb{E}$ is given by Khare [**49**]. By examining some special cases (e. g. elliptic curves, Artin representations) and perhaps glancing at the title of Khare's paper, try to guess what Khare's necessary and sufficient condition is. Then compare your answer to [**49**].

LECTURE 4

Local formulas in arbitrary dimension

To associate a local factor to an Artin representation one starts by restricting the Artin representation to a decomposition subgroup, but to deal with more general premotives one is forced to replace the decomposition subgroup by its close relative, the local Weil or Weil-Deligne group.

1. The local Weil and Weil-Deligne groups

Until further notice, K denotes a finite extension of $\mathbb{Q}_p$ with $p < \infty$. As usual, $\mathcal{O}$ is the ring of integers of K and π a uniformizer of $\mathcal{O}$, and we put $q = ||\pi||^{-1} = |\mathcal{O}/\pi\mathcal{O}|$. The maximal unramified extension of K inside $\overline{K}$ will be denoted K_{unr}, and any element $\sigma \in \text{Gal}(\overline{K}/K)$ which reduces to the map $x \mapsto x^q$ on $\mathcal{O}/\pi\mathcal{O}$ will be called a Frobenius element of $\text{Gal}(\overline{K}/K)$. The symbol Φ denotes the inverse of a Frobenius element, and I is the inertia group $\text{Gal}(\overline{K}/K_{\text{unr}})$.

1.1. The Weil group

As an abstract group, the **Weil group** $\text{W}(\overline{K}/K)$ of K is the union of those cosets of I in $\text{Gal}(\overline{K}/K)$ which are represented by integral powers of a Frobenius element:

$$\text{W}(\overline{K}/K) = \bigcup_{n \in \mathbb{Z}} \sigma^n I. \tag{4.1}$$

Since I is normal in $\text{Gal}(\overline{K}/K)$, the union is a subgroup of $\text{Gal}(\overline{K}/K)$, and since any two Frobenius elements differ by an element of I the definition (4.1) is independent of the choice of σ. We topologize $\text{W}(\overline{K}/K)$ by imposing two requirements:

- I is open in $\text{W}(\overline{K}/K)$, and the relative topology on I from $\text{W}(\overline{K}/K)$ coincides with its relative topology from $\text{Gal}(\overline{K}/K)$.
- For every $g \in \text{W}(\overline{K}/K)$, the map $x \mapsto gx$ is a homeomorphism from $\text{W}(\overline{K}/K)$ to itself.

These conditions determine a unique topology on $\text{W}(\overline{K}/K)$ and make $\text{W}(\overline{K}/K)$ into a topological group. The most important property of this topology, immediate from its definition, is that *an abstract group homomorphism from* $\text{W}(\overline{K}/K)$ *into another topological group is continuous if and only its restriction to I is continuous.* We also note that if L is a finite extension of K inside $\overline{K}$ then $\text{W}(\overline{K}/L)$ is an open subgroup of $\text{W}(\overline{K}/K)$ just as $\text{Gal}(\overline{K}/L)$ is an open subgroup of $\text{Gal}(\overline{K}/K)$, and if L is Galois over K then there are identifications

$$\text{W}(\overline{K}/K)/\text{W}(\overline{K}/L) \cong \text{Gal}(\overline{K}/K)/\text{Gal}(\overline{K}/L) \cong \text{Gal}(L/K). \tag{4.2}$$

However the open subgroups of $\text{Gal}(\overline{K}/K)$ are *precisely* the subgroups $\text{Gal}(\overline{K}/L)$ with L finite over K, whereas the subgroups $\text{W}(\overline{K}/L)$ of $\text{W}(\overline{K}/K)$ are merely the open subgroups *of finite index.*

A character of $W(\overline{K}/K)$ is **unramified** if its restriction to I is trivial, and it is then determined by its value on σ. In particular, there is a unique unramified character ω of $W(\overline{K}/K)$ such that $\omega(\sigma) = q$. The similarity to the notation ω_ℓ for the ℓ-adic cyclotomic character is not coincidental. We introduced ω_ℓ as a character of a global Galois group, but if we restrict to a decomposition group then we obtain a character of our local Galois group $\mathrm{Gal}(\overline{K}/K)$. Restricting further to $W(\overline{K}/K)$, and making the assumption $\ell \neq p$, we get our present ω, because both ω and $\omega_\ell | W(\overline{K}/K)$ are unramified characters taking the value q on Frobenius elements. We may think of ω as the **prime-to-p cyclotomic character.**

We write $W(\overline{K}/K)^{\mathrm{ab}}$ for the quotient of $W(\overline{K}/K)$ by the closure of its commutator subgroup, or equivalently for the quotient of $W(\overline{K}/K)$ by the intersection $\mathrm{Gal}(\overline{K}/K^{\mathrm{ab}}) \cap W(\overline{K}/K)$. The latter description realizes $W(\overline{K}/K)^{\mathrm{ab}}$ as a subgroup of $\mathrm{Gal}(K^{\mathrm{ab}}/K)$, and we shall denote this subgroup $W(K^{\mathrm{ab}}/K)$. One pleasant feature of $W(\overline{K}/K)$ that distinguishes it from $\mathrm{Gal}(\overline{K}/K)$ is that the **local Artin homomorphism** $x \mapsto (x, K^{\mathrm{ab}}/K)$ from $K^\times$ to $W(K^{\mathrm{ab}}/K)$ is an isomorphism rather than merely an injective homomorphism with dense image. Thus a one-dimensional representation ρ of $W(\overline{K}/K)$ is the same thing as a character χ of $K^\times$. In making the identification we follow the geometric convention:

$$\rho((x, K^{\mathrm{ab}}/K)) = \chi(x^{-1}) \qquad (x \in K^\times). \tag{4.3}$$

For example, when the prime-to-p cyclotomic character ω of $W(\overline{K}/K)$ is viewed as a character of $K^\times$ it coincides with $|| * ||$, the local norm on $K^\times$.

1.2. The Weil-Deligne group

With regard to the **Weil-Deligne group** $WD(\overline{K}/K)$ our point of view will be tannakian: Instead of defining $WD(\overline{K}/K)$ itself we define its representations. A **representation of** $WD(\overline{K}/K)$ is a pair $\boldsymbol{\rho} = (\rho, N)$, where ρ is a representation of $W(\overline{K}/K)$ and N a nilpotent endomorphism of the space of ρ satisfying

$$\rho(g) N \rho(g)^{-1} = \omega(g) N \tag{4.4}$$

for $g \in W(\overline{K}/K)$. Henceforth representations of $W(\overline{K}/K)$ will be viewed as the special case $N = 0$ of representations of $WD(\overline{K}/K)$. In other words, we identify a representation ρ of $W(\overline{K}/K)$ with the representation $\boldsymbol{\rho} = (\rho, 0)$ of $WD(\overline{K}/K)$.

The following proposition will be needed later and for the moment can serve to illustrate the definitions just made. Let us say that two endomorphisms of a finite-dimensional vector space V are **simultaneously triangularizable** if there is a basis for V relative to which both endomorphisms are represented by upper triangular matrices.

Proposition 4.1. *Let $\boldsymbol{\rho} = (\rho, N)$ be a representation of $WD(\overline{K}/K)$ over $\mathbb{C}$ and g any element of $W(\overline{K}/K)$. Then $\rho(g)$ and N are simultaneously triangularizable.*

PROOF. The proof is like the standard argument that commuting matrices over an algebraically closed field are simultaneously triangularizable. The only reason for reviewing the details is to illustrate the use of (4.4). In fact consider endomorphisms A and N of a finite-dimensional complex vector space V such that N is nilpotent and $AN = cNA$ with a nonzero scalar c. Write V_N for the kernel of N. We will prove that A and N are simultaneously triangularizable by induction on the dimension of V.

If $\dim V = 1$ there is nothing to prove. If $\dim V = n \geqslant 2$ then we use the identity $AN = cNA$, which shows that V_N is stable under A. Furthermore $V_N \neq \{0\}$ because N is nilpotent. Let $v_1 \in V_N$ be a nonzero eigenvector of A and W its span. Applying the inductive hypothesis to the endomorphisms of V/W determined by A and N, we obtain a basis $v_2 + W, v_3 + W, \ldots, v_n + W$ for V/W relative to which these endomorphisms are upper-triangular. Then A and N are upper-triangular relative to the basis for V consisting of $v_1, v_2, \ldots, v_n$. □

1.3. Operations on representations of the Weil-Deligne group

A representation of $\mathrm{WD}(\overline{K}/K)$ as defined above is not quite a group representation in the usual sense, so the standard operations of representation theory may require some explication. In the following definitions the field of scalars is taken to be $\mathbb{C}$, but $\mathbb{C}$ could be replaced by any field of characteristic zero.

If $\boldsymbol{\rho} = (\rho, N)$ and $\boldsymbol{\rho}' = (\rho', N')$ are representations of $\mathrm{WD}(\overline{K}/K)$ then we define their **direct sum** by

$$\boldsymbol{\rho} \oplus \boldsymbol{\rho}' = (\rho \oplus \rho', N \oplus N') \tag{4.5}$$

and their **tensor product** by

$$\boldsymbol{\rho} \otimes \boldsymbol{\rho}' = (\rho \otimes \rho', N \otimes 1' + 1 \otimes N'), \tag{4.6}$$

where 1 and $1'$ denote the identity automorphism of the space of $\boldsymbol{\rho}$ and $\boldsymbol{\rho}'$ respectively. In particular, for $s_0 \in \mathbb{C}$ we have

$$\boldsymbol{\rho} \otimes \omega^{s_0} = (\rho \otimes \omega^{s_0}, N). \tag{4.7}$$

This follows from (4.6) in view of our identification of ω with $\boldsymbol{\omega} = (\omega, 0)$.

Let $\boldsymbol{\rho}$ and $\boldsymbol{\rho}'$ be arbitrary representations of $\mathrm{WD}(\overline{K}/K)$ again. We define an **intertwining map** or **homomorphism of representations** from $\boldsymbol{\rho}$ to $\boldsymbol{\rho}'$ to be a linear map T from the space of $\boldsymbol{\rho}$ to the space of $\boldsymbol{\rho}'$ which intertwines ρ with ρ' and N with N': Thus $T\rho(g) = \rho'(g)T$ for $g \in \mathrm{WD}(\overline{K}/K)$ and $TN = N'T$. An intertwining map which is a linear isomorphism is an **isomorphism of representations**.

Given a representation $\boldsymbol{\rho}$ of $\mathrm{WD}(\overline{K}/K)$, we define its **dual** $\boldsymbol{\rho}^\vee$ by

$$\boldsymbol{\rho}^\vee = (\rho^\vee, -N^\vee). \tag{4.8}$$

Here $N^\vee$ is the transpose of N: thus if V is the space of $\boldsymbol{\rho}$ and $V^\vee$ the dual space of linear forms on V then $N^\vee(f) = f \circ N$ for $f \in V^\vee$.

Given a finite extension L of K inside $\overline{K}$, put $\mathrm{res}_{L/K}\rho = \rho|\mathrm{WD}(\overline{K}/L)$. Then

$$\mathrm{res}_{L/K}\boldsymbol{\rho} = (\mathrm{res}_{L/K}\rho, N) \tag{4.9}$$

is the **restriction** of $\boldsymbol{\rho}$ to $\mathrm{WD}(\overline{K}/L)$.

Finally, with L still a finite extension of K, take $\boldsymbol{\rho}$ to be a representation of $\mathrm{WD}(\overline{K}/L)$. To define the **induced representation** $\mathrm{ind}_{L/K}\boldsymbol{\rho}$ of $\mathrm{WD}(\overline{K}/K)$, let V be the space of $\boldsymbol{\rho}$ and put $G = \mathrm{W}(\overline{K}/K)$ and $H = \mathrm{W}(\overline{K}/L)$. Then ρ makes V into an H-module, and we can take the space of $\mathrm{ind}_{L/K}\rho$ to be $\mathbb{C}[G] \otimes_{\mathbb{C}[H]} V$. We set

$$\mathrm{ind}_{L/K}\boldsymbol{\rho} = (\mathrm{ind}_{L/K}\rho, \omega^{-1} \cdot (1 \otimes N)), \tag{4.10}$$

where the endomorphism $\omega^{-1} \cdot (1 \otimes N)$ of $\mathbb{C}[G] \otimes_{\mathbb{C}[H]} V$ is defined by the formula

$$\omega^{-1} \cdot (1 \otimes N)(g \otimes v) = \omega^{-1}(g)(g \otimes Nv) \tag{4.11}$$

for $g \in G$ and $v \in V$.

Let us verify that (4.11) gives a well-defined endomorphism of $\mathbb{C}[G] \otimes_{\mathbb{C}[H]} V$ and that the resulting pair (4.10) satisfies (4.4). To facilitate the verifications, we put a subscript on ω: Our current ω is ω_K, and the prime-to-p cyclotomic character of $\mathrm{W}(\overline{K}/L)$ will be denoted ω_L. Then $\omega_K|\mathrm{W}(\overline{K}/L) = \omega_L$. Now to see that $\omega^{-1} \cdot (1 \otimes N)$ is well defined, we must examine (4.11) when $g \otimes v$ is rewritten as $gh \otimes \rho(h)^{-1}v$ with $h \in H$. According to (4.11), we get

$$\omega^{-1} \cdot (1 \otimes N)(gh \otimes \rho(h)^{-1}v) = \omega_K^{-1}(gh)(gh \otimes N\rho(h)^{-1}v). \tag{4.12}$$

As $N\rho(h)^{-1} = \omega_L(h)\rho(h)^{-1}N$, we see that (4.11) and (4.12) are consistent.

Now put $\boldsymbol{\varphi} = \mathrm{ind}_{L/K}\boldsymbol{\rho}$, $\varphi = \mathrm{ind}_{L/K}\rho$, and $M = \omega^{-1} \cdot (1 \otimes N)$. To verify that $\boldsymbol{\varphi}$ satisfies the required identity $\varphi(g)M\varphi(g)^{-1} = \omega_K(g)M$ for $g \in G$, we compare the effect of both sides on pure tensors. Since $\varphi(g)^{-1}(g' \otimes v) = g^{-1}g' \otimes v$, (4.11) gives $M\varphi(g)^{-1}(g' \otimes v) = \omega_K(g')^{-1}\omega_K(g)(g^{-1}g' \otimes Nv)$, whence $\varphi(g)M\varphi(g)^{-1}(g' \otimes v)$ is indeed $\omega_K(g)M(g' \otimes v)$.

This completes our discussion of the standard operations. It is instructive to drop the tannakian perspective for a moment and let the cat out of the bag: As an actual group, $\mathrm{WD}(\overline{K}/K)$ is just the semidirect product $\mathbb{C} \rtimes \mathrm{W}(\overline{K}/K)$ with $gzg^{-1} = \omega(g)z$ for $g \in \mathrm{W}(\overline{K}/K)$ and $z \in \mathbb{C}$. One is supposed to view the factor $\mathbb{C}$ as the set of complex points of the algebraic group $\mathbb{G}_{\mathrm{a}}$, so a representation of $\mathbb{C} \rtimes \mathrm{W}(\overline{K}/K)$ should be algebraic and in particular holomorphic when restricted to the factor $\mathbb{C}$. Using this fact, one can show that a representation $\boldsymbol{\rho}$ of $\mathrm{WD}(\overline{K}/K)$ has the form $zg \mapsto \exp(zN)\rho(g)$, where ρ is a representation of $\mathrm{W}(\overline{K}/K)$ and N a nilpotent endomorphism of the space of ρ. This not only explains the identification of $\boldsymbol{\rho}$ with the pair (ρ, N) but also shows (after some calculation) that the preceding definitions for $\oplus$, $\otimes$, and so on are just the standard operations of representation theory applied to the group $\mathbb{C} \rtimes \mathrm{W}(\overline{K}/K)$.

1.4. The archimedean Weil group

Now suppose that $K = \mathbb{R}$ or $K \cong \mathbb{C}$. In the archimedean case there is no distinction between $\mathrm{W}(\overline{K}/K)$ and $\mathrm{WD}(\overline{K}/K)$; the notations $\mathrm{W}(\overline{K}/K)$ and $\mathrm{WD}(\overline{K}/K)$ are interchangeable. The definition is as follows. If $K \cong \mathbb{C}$ then

$$\mathrm{W}(\overline{K}/K) = \mathrm{W}(\overline{K}/\overline{K}) = K^\times \cong \mathbb{C}^\times,$$

and if $K = \mathbb{R}$ then

$$\mathrm{W}(\overline{K}/K) = \overline{K}^\times \cup J\overline{K}^\times \cong \mathbb{C}^\times \cup J\mathbb{C}^\times,$$

where $J^2 = -1$ and $JzJ^{-1} = \overline{z}$ for $z \in \overline{K}^\times \cong \mathbb{C}^\times$. In the case $K = \mathbb{R}$ we identify the subgroup $\overline{K}^\times$ of $\mathrm{W}(\overline{K}/K)$ with $\mathrm{W}(\overline{K}/\overline{K})$, which thus becomes a subgroup of index 2 in $\mathrm{W}(\overline{K}/K)$. The nontrivial coset is represented by J.

The reason that we have been so pedantic about distinguishing $\mathrm{W}(\overline{K}/K)$ from $\mathbb{C}^\times$ (when $K \cong \mathbb{C}$) or from $\mathbb{C}^\times \cup J\mathbb{C}^\times$ (when $K = \mathbb{R}$) is that we would like to identify $\mathrm{W}(\overline{K}/K)^{\mathrm{ab}}$ with $K^\times$, just as in the nonarchimedean case. Let $\pi : \mathrm{W}(\overline{K}/K) \to K^\times$ be the identity map if $K \cong \mathbb{C}$ and the map sending J to -1 and z to $|z|^2$ if $K = \mathbb{R}$. Then π factors through a map $\mathrm{W}(\overline{K}/K)^{\mathrm{ab}} \to K^\times$, and we claim that the latter map is an isomorphism. This is obvious if $K \cong \mathbb{C}$, so suppose that $K = \mathbb{R}$. One readily verifies that the commutator subgroup of $\mathrm{W}(\overline{K}/K)$ consists of all elements of the form $JzJ^{-1}z^{-1}$ with $z \in \overline{K}$. Furthermore, let $\mathbf{T}$ be the subgroup of $\overline{K}^\times$ consisting of numbers of absolute value 1. Since $JzJ^{-1}z^{-1} = \overline{z}/z$ it follows that the commutator subgroup of $\mathrm{W}(\overline{K}/K)$ is $\mathbf{T}$, and it is also easy to see that $\mathbf{T}$ is the kernel of π and that π is surjective. Thus $\mathrm{W}(\overline{K}/K)^{\mathrm{ab}} \cong K^\times$ as claimed.

Henceforth we shall identify the one-dimensional characters of $\mathrm{W}(\overline{K}/K)$ with those of $K^\times$ by putting

$$\chi(g) = \chi(\pi(g)) \qquad (g \in \mathrm{W}(\overline{K}/K)). \tag{4.13}$$

This is the archimedean analogue of (4.3). Given that we know the characters of $\mathbb{C}^\times$ and $\mathbb{R}^\times$ explicitly, we can regard (4.13) as an explicit description of the characters of $\mathrm{W}(\overline{K}/K)$. In fact we can describe not just the one-dimensional representations but all the irreducible complex representations of $\mathrm{W}(\overline{K}/K)$. If $K \cong \mathbb{C}$ then $\mathrm{W}(\overline{K}/K)$ is abelian and there are no irreducible representations of dimension > 1, and if $K = \mathbb{R}$ then $\mathrm{W}(\overline{K}/K)$ has an abelian subgroup of index 2, namely $\mathrm{W}(\overline{K}/\overline{K})$, whence any irreducible representation of $\mathrm{W}(\overline{K}/K)$ of dimension > 1 is two-dimensional, induced by a character of $\mathrm{W}(\overline{K}/\overline{K})$. Let χ be a character of $\mathrm{W}(\overline{K}/\overline{K})$ and let $\rho = \mathrm{ind}_{\overline{K}/K}\chi$ be the representation of $\mathrm{W}(\overline{K}/K)$ it induces. Identifying $\mathrm{W}(\overline{K}/\overline{K})$ with $\mathbb{C}^\times$ and writing $\chi(z) = |z|^{2s_0}(z/|z|)^m$ with $s_0 \in \mathbb{C}$ and $m \in \mathbb{Z}$, one checks that ρ is irreducible if and only if $m \neq 0$.

2. From Galois representations to Weil-Deligne representations

Let K be a number field. We seek an analogue for premotives of the map $\rho \mapsto \rho_v$ sending an Artin representation of K to its restriction to a decomposition subgroup of $\mathrm{Gal}(\overline{K}/K)$ at a given place v of K. The analogue should be a map $M \mapsto \boldsymbol{\rho}_{M,v}$, where M is a premotive over K and $\boldsymbol{\rho}_{M,v}$ a complex representation of $\mathrm{WD}(\overline{K}_v/K_v)$. What the theory of Grothendieck and Deligne provides in the first instance, however, is a purely local correspondence from λ-adic representations of $\mathrm{Gal}(\overline{K}_v/K_v)$ to λ-adic representations of $\mathrm{WD}(\overline{K}_v/K_v)$. To describe this local correspondence we make two preliminary remarks about profinite groups.

Given a prime ℓ, we say that a profinite group has order prime to ℓ if it is an inverse limit of finite groups of order prime to ℓ. The first remark is that every homomorphism (of topological groups) from a profinite group of order prime to ℓ to a pro-ℓ-group is trivial. This follows from the corresponding fact about finite groups.

Let Γ be a pro-ℓ-group and $\varphi : \mathbb{Z}_\ell \to \Gamma$ a homomorphism. The second remark is that if $\gamma = \varphi(1)$ then for arbitrary $z \in \mathbb{Z}_\ell$ we have $\varphi(z) = \gamma^z$, the point being that γ^z is meaningful as a limit even if $z \notin \mathbb{Z}$. More generally, suppose that I is a profinite group with the following property: There is a closed normal subgroup Q of I, of profinite order prime to ℓ, such that $I/Q \cong \mathbb{Z}_\ell$. If $t : I \to \mathbb{Z}_\ell$ is an epimorphism and $\varphi : I \to \Gamma$ an arbitrary homomorphism then

$$\varphi(i) = \gamma^{t(i)} \tag{4.14}$$

for $i \in I$, where $\gamma \in \Gamma$ is the image under φ of any preimage of $1 \in \mathbb{Z}_\ell$ under t.

2.1. The local correspondence

Now let K be a finite extension of $\mathbb{Q}_p$ with $p < \infty$, and write K_{tame} for the maximal tamely ramified extension of K inside $\overline{K}$. Put $P = \mathrm{Gal}(\overline{K}/K_{\mathrm{tame}})$. If we fix a uniformizer ϖ of K_{unr} then K_{tame} can be described as the compositum of all extensions of K_{unr} of the form $K_{\mathrm{unr}}(\varpi^{1/n})$ with positive integers n prime to p.

Kummer theory then gives an identification of $\mathrm{Gal}(K_{\mathrm{unr}}(\varpi^{1/n})/K_{\mathrm{unr}})$ with $\mathbb{Z}/n\mathbb{Z}$ and hence an identification

$$I/P \cong \prod_{\ell \neq p} \mathbb{Z}_\ell \tag{4.15}$$

after taking inverse limits. Since P is a pro-p-group, it follows that if $\ell \neq p$ then there is a closed normal subgroup Q of I, profinite of order prime to ℓ, such that $I/Q \cong \mathbb{Z}_\ell$. In particular, the space of homomorphisms $I \to \mathbb{Q}_\ell$ is one-dimensional over $\mathbb{Q}_\ell$. Furthermore (4.15) and our deductions from it hold with I replaced by any open subgroup I' of I and P by $P \cap I'$, because any open subgroup of the right-hand side of (4.15) is again isomorphic to the right-hand side of (4.15).

The theorem to be stated next combines results of Grothendieck and Deligne, and the proof follows the exposition of Serre-Tate ([**86**], pp. 515 – 516) and Deligne ([**23**], pp. 566 – 571). Fix a nonzero homomorphism $t_\ell : I \to \mathbb{Q}_\ell$ and a Frobenius element $\sigma \in \mathrm{W}(\overline{K}/K)$. Since t_ℓ is unique only up to a scalar multiple and σ only up to multiplication by an element of I, it is important to remark that the isomorphism class of the representation $\boldsymbol{\rho}$ constructed in the theorem below is independent of the choice of t_ℓ and σ (cf. [**23**], p. 569). However we do not bother to prove this remark, because the local issue presented by the choice of t_ℓ and σ will be overshadowed by a far more problematic global pair of choices later on. In any case, once a choice of t_ℓ has been fixed, the following identity holds for arbitrary elements g and i of $\mathrm{W}(\overline{K}/K)$ and I respectively:

$$t_\ell(gig^{-1}) = \omega(g)t_\ell(i). \tag{4.16}$$

This is proved by applying the usual Galois equivariance of the Kummer pairing to the extension $K_{\mathrm{unr}}(\varpi^{1/n})/K_{\mathrm{unr}}$ and then taking inverse limits. Note the formal resemblance of (4.16) to (4.4)!

Theorem 4.1. *Let $\mathbb{E}_\lambda$ be a finite extension of $\mathbb{Q}_\ell$ with $\ell \neq p$, and let ρ_λ be a representation of* $\mathrm{Gal}(\overline{K}/K)$ *over $\mathbb{E}_\lambda$.*

(a) *There is a unique nilpotent endomorphism N of the space of ρ_λ such that*

$$\rho_\lambda(i) = \exp(t_\ell(i)N)$$

for all i in some open subgroup of I. Furthermore, consider the function ρ on $\mathrm{W}(\overline{K}/K)$ *defined by setting*

$$\rho(g) = \exp(-t_\ell(i)N)\rho_\lambda(g)$$

for $g = i\sigma^n$ with $i \in I$ and $n \in \mathbb{Z}$. This function is a representation of $\mathrm{W}(\overline{K}/K)$ *on the space of ρ_λ, and the pair $\boldsymbol{\rho} = (\rho, N)$ is a representation of* $\mathrm{WD}(\overline{K}/K)$.

(b) *Let ρ and N be as in* (a), *and for each $g \in \mathrm{W}(\overline{K}/K)$ let $\rho^{\mathrm{ss}}(g)$ be the semisimple component of $\rho(g)$ in a multiplicative Jordan decomposition of $\rho(g)$. Then the map $g \mapsto \rho^{\mathrm{ss}}(g)$ is a semisimple representation of* $\mathrm{W}(\overline{K}/K)$ *trivial on an open subgroup of I, and the pair $\boldsymbol{\rho}^{\mathrm{ss}} = (\rho^{\mathrm{ss}}, N)$ is a representation of* $\mathrm{WD}(\overline{K}/K)$.

Proof. (a) Let $\mathcal{O}_\lambda$ be the ring of integers of $\mathbb{E}_\lambda$. By Exercise 3.4, we may think of ρ_λ as a map $\mathrm{Gal}(\overline{K}/K) \to \mathrm{GL}_d(\mathcal{O}_\lambda)$. Let Γ be the open pro-ℓ-subgroup of $\mathrm{GL}_d(\mathcal{O}_\lambda)$ consisting of matrices congruent to 1 mod ℓ^2, and put $I' = I \cap \rho_\lambda^{-1}(\Gamma)$. Then I' is an open subgroup of I and $\rho_\lambda|I'$ is a homomorphism of I' into a pro-ℓ-group. Choose $c \in \mathbb{Q}_\ell$ such that $ct_\ell|I'$ is a surjection of I' onto $\mathbb{Z}_\ell$. Applying (4.14) with $\varphi = \rho_\lambda|I'$ and $t = ct_\ell$, we see that $\rho_\lambda(i) = \gamma^{ct_\ell(i)}$ for some $\gamma \in \Gamma$ and all $i \in I'$. Let $\mathrm{M}_{d\times d}(\mathcal{O}_\lambda)$ be the set of $d \times d$ matrices with coefficients in $\mathcal{O}_\lambda$, and write

$\gamma = \exp(N)$ with $N \in \ell^2 \mathrm{M}_{d\times d}(\mathcal{O}_\lambda)$. Then $\rho_\lambda(i) = \exp(ct_\ell(i)N)$ for $i \in I'$, and it follows from (4.16) that for $g \in \mathrm{W}(\overline{K}/K)$ we have $\rho_\lambda(gig^{-1}) = \exp(c\omega(g)t_\ell(i)N)$. But if we simply conjugate the equation $\rho_\lambda(i) = \exp(ct_\ell(i)N)$ by $\rho_\lambda(g)$ then we get a second expression for $\rho_\lambda(gig^{-1})$, namely $\exp(ct_\ell(i)\rho_\lambda(g)N\rho_\lambda(g)^{-1})$. Applying the ℓ-adic logarithm to both expressions, we find that $\rho_\lambda(g)N\rho_\lambda(g)^{-1} = \omega(g)N$. This identity immediately carries over to the identity $\rho(g)N\rho(g)^{-1} = \omega(g)N$ if we define ρ as in the statement of the theorem. And by taking $g = \sigma^\nu$ with $\nu \in \mathbb{Z}$ we deduce that N is nilpotent, for if N had a nonzero eigenvalue r then N would have infinitely many eigenvalues, namely the numbers rq^ν. Replacing N by cN we preserve the nilpotence of N and the relation $\rho(g)N\rho(g)^{-1} = \omega(g)N$ and we gain the simplified formula $\rho_\lambda(i) = \exp(t_\ell(i)N)$ for $i \in I'$. This equation determines N uniquely, because the exponential is a bijection from nilpotent matrices to unipotent matrices.

To complete the proof of (a) we must check that ρ is a homomorphism. So suppose that $g = i\sigma^n$ and $g' = h\sigma^m$ with $m, n \in \mathbb{Z}$ and $h, i \in I$. Then (4.16) gives

$$\rho(gg') = \exp((-t_\ell(i) - q^n t_\ell(h))N)\rho_\lambda(gg') \tag{4.17}$$

while

$$\rho(g)\rho(g') = \exp(-t_\ell(i)N)\rho_\lambda(g)\exp(-t_\ell(h)N)\rho_\lambda(g'). \tag{4.18}$$

The identity $\rho_\lambda(g)N\rho_\lambda(g)^{-1} = \omega(g)N$ shows that the right-hand sides of (4.17) and (4.18) are equal, whence $\rho(gg') = \rho(g)\rho(g')$.

(b) Let J be the kernel of $\rho|I$. Since I is normal in $\mathrm{W}(\overline{K}/K)$ so is J. In addition, J is open in I, because it is the subgroup of I on which ρ_λ coincides with the map $i \mapsto \exp(t_\ell(i)N)$, and this subgroup is open by (a). It follows that I/J is a finite normal subgroup of $\mathrm{W}(\overline{K}/K)/J$, and consequently the action of $\mathrm{W}(\overline{K}/K)/J$ on I/J by conjugation gives a map from $\mathrm{W}(\overline{K}/K)$ to the finite group $\mathrm{Aut}(I/J)$. Hence the kernel of this map has finite index in $\mathrm{W}(\overline{K}/K)$, and there is an integer $l \geqslant 1$ such that σ^l acts trivially on I/J. Since ρ factors through $\mathrm{W}(\overline{K}/K)/J$, we deduce that $\rho(\sigma^l)$ centralizes $\rho(I)$. But $\rho(\sigma^l)$ certainly commutes with $\rho(\sigma)$, so $\rho(\sigma^l)$ centralizes the image of ρ.

Now let u be the unipotent Jordan component of $\rho(\sigma)$. Then u^l is the unipotent Jordan component of $\rho(\sigma^l)$. But the semisimple and unipotent components of an invertible matrix are polynomials in the matrix. Since $\rho(\sigma^l)$ centralizes the image of ρ it follows that u^l does too. Using the binomial series for $(1+x)^{1/l}$, we see that u is a polynomial in u^l, so we conclude that u centralizes the image of ρ.

Next consider an arbitrary element $g \in \mathrm{W}(\overline{K}/K)$, and write $g = i\sigma^n$ with $i \in I$ and $n \in \mathbb{Z}$. Let $\rho^{\mathrm{u}}(g)$ denote the unipotent Jordan component of $\rho(g)$; we claim that $\rho^{\mathrm{u}}(g) = u^n$. Since $\rho^{\mathrm{u}}(g^l) = \rho^{\mathrm{u}}(g)^l$ and unipotent automorphisms have unique unipotent lth roots, it suffices to see that $\rho^{\mathrm{u}}(g^l) = u^{nl}$. But $\rho(g^l) = \rho(i')\rho(\sigma^{nl})$ for some $i' \in I$ and $\rho(\sigma^{nl}) = \rho^{\mathrm{ss}}(\sigma^{nl})u^{nl}$, so

$$\rho(g^l) = (\rho(i')\rho^{\mathrm{ss}}(\sigma^{nl})) \cdot u^{nl}. \tag{4.19}$$

We contend that (4.19) is the multiplicative Jordan decomposition of $\rho(g^l)$, whence $\rho^{\mathrm{u}}(g^l) = u^{nl}$, as desired. As u^{nl} is unipotent and commutes with $\rho(i')\rho^{\mathrm{ss}}(\sigma^{nl})$ it suffices to see that $\rho(i')\rho^{\mathrm{ss}}(\sigma^{nl})$ is semisimple. But $\rho(i')$ is semisimple because $\rho|I$ factors through the finite group I/J, and $\rho^{\mathrm{ss}}(\sigma^{nl})$ is semisimple and commutes with $\rho(i')$. Hence $\rho(i')\rho^{\mathrm{ss}}(\sigma^{nl})$ is indeed semisimple, and we conclude that $\rho^{\mathrm{u}}(g) = u^n$.

We can now show that ρ^{ss} is a representation (necessarily trivial on J because ρ is). Given $g, g' \in \mathrm{W}(\overline{K}/K)$, write $g = i\sigma^n$ and $g' = h\sigma^m$ with $m, n \in \mathbb{Z}$ and $h, i \in I$. Then $\rho^{\mathrm{ss}}(g)\rho^{\mathrm{ss}}(g') = \rho(g)u^{-n}\rho(g')u^{-m} = \rho(gg')u^{-(n+m)} = \rho^{\mathrm{ss}}(gg')$, where the last equality follows from the fact that $gg' = i'\sigma^{n+m}$ for some $i' \in I$, whence $u^{n+m} = \rho^{\mathrm{u}}(gg')$.

To see that ρ^{ss} is semisimple we quote a general fact: A representation of a group over a field of characteristic 0 is semisimple if and only if its restriction to a subgroup of finite index is semisimple. In the case at hand, the infinite cyclic group $\langle\sigma\rangle$ generated by σ is of finite index in $\mathrm{W}(\overline{K}/K)/J$, and $\rho^{\mathrm{ss}}|\langle\sigma\rangle$ is semisimple by the very definition of ρ^{ss}.

Finally, we must check the identity $\rho^{\mathrm{ss}}(g)N\rho^{\mathrm{ss}}(g)^{-1} = \omega(g)N$. Since we already know that $\rho(g)N\rho(g)^{-1} = \omega(g)N$, it will suffice to see that u commutes with N. Denote the adjoint representation of $\mathrm{GL}_d(\mathbb{E}_\lambda)$ on $\mathrm{M}_{d\times d}(\mathbb{E}_\lambda)$ by Ad, so that

$$\mathrm{Ad}(x)(y) = xyx^{-1}$$

for $x \in \mathrm{GL}_d(\mathbb{E}_\lambda)$ and $y \in \mathrm{M}_{d\times d}(\mathbb{E}_\lambda)$. It is readily verified that $\mathrm{Ad}(x)^{\mathrm{ss}} = \mathrm{Ad}(x^{\mathrm{ss}})$ and $\mathrm{Ad}(x)^{\mathrm{u}} = \mathrm{Ad}(x^{\mathrm{u}})$. In particular, since $\mathrm{Ad}(x)^{\mathrm{u}}$ is a polynomial in $\mathrm{Ad}(x)$, we see that any eigenvector of $\mathrm{Ad}(x)$ is also an eigenvector of $\mathrm{Ad}(x^{\mathrm{u}})$. Apply the preceding remark with $x = \rho(\sigma)$ and $x^{\mathrm{u}} = u$. The relation $\rho(\sigma)N\rho(\sigma)^{-1} = qN$ shows that N is an eigenvector of $\mathrm{Ad}(\rho(\sigma))$ and hence of $\mathrm{Ad}(u)$. But $\mathrm{Ad}(u)$ is $\mathrm{Ad}(\rho(\sigma))^{\mathrm{u}}$ and therefore unipotent; its eigenvalues equal 1. Thus N is an eigenvector of $\mathrm{Ad}(u)$ with eigenvalue 1; in other words, u commutes with N. □

2.2. Characteristic polynomials

Next we examine the effect of the maps $\rho_\lambda \mapsto \boldsymbol{\rho}$ and $\boldsymbol{\rho} \mapsto \rho^{\mathrm{ss}}$ of Theorem 4.1 on the space of inertial invariants. Let V_λ denote the space of ρ_λ and V the space of $\boldsymbol{\rho}$. While V_λ and V are equal as abstract vector spaces, by using different notations we can distinguish between the subspaces V_λ^I and V^I, which need not be equal: one consists of vectors fixed by $\rho_\lambda(I)$, the other of vectors fixed by $\rho(I)$. We claim that the relation between them is

$$V_\lambda^I = V_N^I, \tag{4.20}$$

where V_N is the kernel of N and $V_N^I = V_N \cap V^I$.

To verify (4.20), we return to the relation $\rho_\lambda(i) = \exp(t_\ell(i)N)\rho(i)$ for $i \in I$. The inclusion $V_N^I \subset V_\lambda^I$ is an immediate consequence. For the reverse inclusion, recall that $\rho_\lambda(j) = \exp(t_\ell(j)N)$ for all j in some open subgroup J of I. Any element of V_λ^I is in particular fixed by J and hence by $\exp(t_\ell(j)N)$ for $j \in J$. By writing $t_\ell(j)N$ as $-\sum_{\nu\geqslant 1}(1-x)^\nu/\nu$ with $x = \exp(t_\ell(j)N)$, we see that a vector fixed by $\exp(t_\ell(j)N)$ is in the kernel of N, whence $V_\lambda^I \subset V_N$. Now the relation $\rho(g)N = \omega(g)N\rho(g)$ for $g \in \mathrm{W}(\overline{K}/K)$ shows that V_N is stable under ρ, so the inclusion $V_\lambda^I \subset V_N$ and the relation $\rho_\lambda(i) = \exp(t_\ell(i)N)\rho(i)$ for $i \in I$ together imply that $\rho_\lambda(i) = \rho(i)$ on V_λ^I. Hence $V_\lambda^I \subset V_N^I$, and (4.20) follows.

Put $\Phi = \sigma^{-1}$. Then the definition of ρ in part (a) of Theorem 4.1 gives $\rho(\Phi) = \rho_\lambda(\Phi)$, whence in particular $\rho_\lambda(\Phi)|V_\lambda^I = \rho(\Phi)|V_N^I$ by (4.20). Since characteristic polynomials are insensitive to semisimplification, we obtain:

Proposition 4.2. $\det(1 - x\rho_\lambda(\Phi)|V_\lambda^I) = \det(1 - x\rho^{\mathrm{ss}}(\Phi)|V_N^I)$.

3. An open problem

Let K be a number field again and M a premotive over K. Fix a finite place v of K and write p for its residue characteristic. We would like to claim that M determines a complex representation $\boldsymbol{\rho}_{M,v}$ of $\mathrm{WD}(\overline{K}_v/K_v)$ up to isomorphism.

There is an obvious candidate for $\boldsymbol{\rho}_{M,v}$. Let $\{\rho_\lambda\} \in M$ be a fully compatible family and $\mathbb{E}$ its coefficient field, and choose a place λ of $\mathbb{E}$ of residue characteristic $\ell \neq p$. As usual, we identify $\mathrm{Gal}(\overline{K}_v/K_v)$ with the decomposition subgroup of $\mathrm{Gal}(\overline{K}/K)$ at a place of $\overline{K}$ above v, so it is meaningful to consider the restriction $\rho_{\lambda,v} = \rho_\lambda|\mathrm{Gal}(\overline{K}_v/K_v)$. This is a λ-adic representation of $\mathrm{Gal}(\overline{K}_v/K_v)$ to which we may apply Theorem 4.1. The result is a representation $\boldsymbol{\rho}_v^{\mathrm{ss}} = (\rho_v^{\mathrm{ss}}, N_v)$ of $\mathrm{WD}(\overline{K}_v/K_v)$ over $\mathbb{E}_\lambda$. To obtain a representation over $\mathbb{C}$, fix an abstract field embedding ι of $\mathbb{E}_\lambda$ in $\mathbb{C}$. Since we regard $\mathbb{E}$ as a subfield both of $\mathbb{E}_\lambda$ and of $\mathbb{C}$, we can require ι to be the identity on $\mathbb{E}$. Extending scalars from $\mathbb{E}_\lambda$ to $\mathbb{C}$ via ι, we obtain a representation

$$\boldsymbol{\rho}_{M,v} = ((\rho_v^{\mathrm{ss}})^\iota, (N_v)^\iota) \tag{4.21}$$

of $\mathrm{WD}(\overline{K}_v/K_v)$ over $\mathbb{C}$.

Problem 4. *Up to isomorphism, is $\boldsymbol{\rho}_{M,v}$ independent of the choice of λ and ι?*

If M comes from an Artin representation or a Hecke character of type $(1,0)$ then an affirmative answer follows tautologically from the definitions, and an affirmative answer is also known if M comes from an elliptic curve or more generally from an abelian variety, cf. [**23**], p. 571. Admittedly, an Artin representation for which we do not know the Artin conjecture cannot be offered as an example of a premotive according to our definition of the term, nor can an elliptic curve over an arbitrary number field. But the analytic conditions that we have imposed on a premotive could be omitted from the definition and Problem 4 would still make prefect sense. The real issue is that an affirmative answer to Problem 4 is unknown in general even if one assumes that M comes from a motive. It should be added, however, that if we write $\boldsymbol{\rho}_{M,v} = (\rho_{M,v}, N_{M,v})$ then it is only $N_{M,v}$ which is problematic. Indeed the theory of Grothendieck and Deligne does yield the following.

Theorem 4.2. *Up to isomorphism, $\rho_{M,v}$ is independent of the choice of λ and ι.*

PROOF. Let $\boldsymbol{\rho}_v = (\rho_v, N_v)$ be the representation of $\mathrm{WD}(\overline{K}_v/K_v)$ resulting from $\rho_{\lambda,v}$ as in part (a) of Theorem 4.1, and let g denote an arbitrary element of $\mathrm{W}(\overline{K}_v/K_v)$, written as in the theorem. By Proposition 4.1 , there is a basis for the space of ρ_v relative to which the matrices of $\rho_v(g)$ and N_v are both upper triangular. Then $\exp(t_\ell(i)N_v)$ is upper triangular with all diagonal entries equal to 1, and as $\rho_{\lambda,v}(g) = \exp(t_\ell(i)N_v)\rho_v(g)$ we deduce that the characteristic polynomials of $\rho_{\lambda,v}(g)$ and $\rho_v(g)$ are equal. On the other hand, the characteristic polynomials of $\rho_v(g)$ and $\rho_v^{\mathrm{ss}}(g)$ are equal because $\rho_v^{\mathrm{ss}}(g)$ is the semisimple Jordan component of $\rho_v(g)$. By Condition C_8, the characteristic polynomial of $\rho_{\lambda,v}(g)$ has coefficients in $\mathbb{E}$ and is independent of λ, so the same is true for the characteristic polynomial of $\rho_v^{\mathrm{ss}}(g)$. Furthermore this characteristic polynomial is unchanged when ι is used to extend the field of scalars of $\rho_v^{\mathrm{ss}}(g)$ from $\mathbb{E}_\lambda$ to $\mathbb{C}$, because ι is the identity on $\mathbb{E}$. We conclude that the characteristic polynomial of $(\rho_v^{\mathrm{ss}})^\iota(g)$ is independent of the choice of λ and ι. In particular the trace of $(\rho_v^{\mathrm{ss}})^\iota$ is independent of the choice of λ and ι, and since $(\rho_v^{\mathrm{ss}})^\iota$ is semisimple we conclude that the isomorphism class of $(\rho_v^{\mathrm{ss}})^\iota$ is independent of the choices as well. $\square$

As stated, Problem 4 pertains only to the finite places of K, because it is only for finite v that we have presented a candidate for $\boldsymbol{\rho}_{M,v}$. However the Fontaine-Mazur conjectures provide a candidate for $\boldsymbol{\rho}_{M,v}$ when v is an infinite place as well, at least if $K = \mathbb{Q}$ ([**30**], p. 197, Conjecture 3a). The definition of $\boldsymbol{\rho}_{M,v}$ depends on the standing assumption in [**30**], namely potential semistability, and in this sense it does not mesh too well with the present framework. However if we simply impose potential semistability as an additional requirement on M then in principle we can take v in Problem 4 to be any place of K, finite or infinite. The choice of λ and ι remains very much an issue at the infinite places, because for us there is no distinguished prime p and of course no distinguished embedding of $\overline{\mathbb{Q}_p}$ into $\mathbb{C}$.

In the remainder of this lecture we define the local factors – the L-factor, the conductor, the root number – associated to a complex representation of the Weil-Deligne group of a finite extension of $\mathbb{Q}_p$ for $p \leqslant \infty$. If Problem 4 has an affirmative answer, then once we have made the local definitions we can put

$$L_\infty(s, M) = \prod_{v|\infty} L(s, \boldsymbol{\rho}_{M,v}), \tag{4.22}$$

$$W(M) = \prod_{v} W(\boldsymbol{\rho}_{M,v}), \tag{4.23}$$

$$\mathfrak{f}(M) = \prod_{v\nmid\infty} \mathfrak{p}_v^{a(\boldsymbol{\rho}_{M,v})}, \tag{4.24}$$

and

$$A(M) = D^{\mathrm{rk}(M)} \mathbf{N}\mathfrak{f}(M), \tag{4.25}$$

where D is the absolute value of the discriminant of K and $\mathrm{rk}(M)$ is the rank of M (recall that this is $\dim \rho_\lambda$ for $\{\rho_\lambda\} \in M$). Thus if we grant an affirmative answer to Problem 4 then $L_\infty(s, M)$, $A(M)$, and $W(M)$ will finally have an intrinsic definition, one that does not depend on the uniqueness of a conjectural functional equation (Proposition 3.2). For the sake of the overall coherence of the discussion we will also check that

$$L(s, M) = \prod_{v\nmid\infty} L(s, \boldsymbol{\rho}_{M,v}), \tag{4.26}$$

even though we already have the intrinsic definition (3.33) of $L(s, M)$. In fact by virtue of this intrinsic definition, verifying (4.26) will amount to checking that

$$B_{\mathfrak{p}}((\mathbf{N}\mathfrak{p})^{-s}) = L(s, \boldsymbol{\rho}_{M,v}), \tag{4.27}$$

where $\mathfrak{p}$ is an arbitrary prime ideal of K and $v = v_{\mathfrak{p}}$.

4. Local factors

Henceforth K is a finite extension of $\mathbb{Q}_p$ ($p \leqslant \infty$) and $\boldsymbol{\rho}$ a representation of $\mathrm{WD}(\overline{K}/K)$ over $\mathbb{C}$. We will define the local factors associated to $\boldsymbol{\rho}$.

4.1. L-factors

Suppose first that $p < \infty$, and write $\boldsymbol{\rho} = (\rho, N)$. Let V be the space of ρ and V_N the kernel of N, and put $V_N^I = V^I \cap V_N$ as before. Recall also that by virtue of (4.4), V_N^I is stable under ρ. We put

$$L(s, \boldsymbol{\rho}) = \det(1 - q^{-s}\rho(\Phi)|V_N^I)^{-1}, \tag{4.28}$$

where $\Phi \in \mathrm{W}(\overline{K}/K)$ is an inverse Frobenius element and q is the order of the residue class field of K. Since $V_N^I \subset V^I$ the definition of $L(s, \boldsymbol{\rho})$ is as usual independent of the choice of Φ. Furthermore, (4.27) is an immediate consequence of Proposition 4.2 given the definitions (3.30), (4.21), and (4.28).

Suppose next that $p = \infty$. Then there is no distinction between $\boldsymbol{\rho} = (\rho, 0)$ and ρ, and the definition of the L-factor $L(s, \boldsymbol{\rho}) = L(s, \rho)$ is a straightforward generalization of our earlier discussion of gamma factors for Artin L-functions. In fact the only new issue is that the L-factor of ρ is by definition the L-factor of the semisimplification of ρ. (Of course in the context of Artin representations, ρ is automatically semisimple.) If ρ is a semisimple representation of $\mathrm{W}(\overline{K}/K)$ then we define $L(s, \rho)$ by imposing the Artin formalism. In particular, additivity holds by fiat: in other words we declare that $L(s, \rho \oplus \rho') = L(s, \rho)L(s, \rho')$, whence it suffices to define $L(s, \rho)$ for ρ irreducible.

So suppose that ρ is irreducible. If ρ is one-dimensional then we use (4.13) to identify ρ with a character χ of $K^\times$, and we set $L(s, \rho) = L(s, \chi)$, thus defining $L(s, \rho)$ by (2.25) and (2.26). If $K = \mathbb{R}$ and ρ is two-dimensional then $\rho = \mathrm{ind}_{\overline{K}/K}\chi$ for some character χ of $\mathrm{W}(\overline{K}/\overline{K})$, and again we set $L(s, \rho) = L(s, \chi)$, as required by inductivity. Incidentally, the fact that inductivity holds even for reducible inductions follows from the duplication formula (1.9).

4.2. The exponent of the conductor

We return to the case $p < \infty$ and write $\boldsymbol{\rho} = (\rho, N)$ as before. The exponent $a(\boldsymbol{\rho})$ of the conductor $\pi^{a(\boldsymbol{\rho})}\mathcal{O}$ of $\boldsymbol{\rho}$ is a sum of two terms,

$$a(\boldsymbol{\rho}) = a(\rho) + b(\boldsymbol{\rho}), \tag{4.29}$$

and only the second term depends on N. In fact

$$b(\boldsymbol{\rho}) = \dim(V^I/V_N^I). \tag{4.30}$$

It follows in particular that if $N = 0$ then $b(\boldsymbol{\rho}) = 0$, so that (4.29) is consistent with our identification of ρ with $(\rho, 0)$. Turning to the first term, we declare first of all that $a(\rho)$ depends only on the semisimplification of ρ. We also impose additivity – in other words, we take the identity $a(\rho \oplus \rho') = a(\rho) + a(\rho')$ to be part of the definition of $a(*)$ – and so reduce the problem of defining $a(\rho)$ to the case where ρ is irreducible.

To handle this case we need the notion of a representation **of Galois type**. This term refers to any representation of $\mathrm{W}(\overline{K}/K)$ over $\mathbb{C}$ which is trivial on an open subgroup of finite index in $\mathrm{W}(\overline{K}/K)$, hence on an open normal subgroup of finite index. Since the open normal subgroups of finite index are precisely the subgroups of the form $\mathrm{W}(\overline{K}/L)$ with L a finite Galois extension of K, we see from (4.2) that a representation of Galois type is simply a representation of $\mathrm{Gal}(L/K)$ for some L. The following proposition is drawn from Deligne [**23**], p. 542. The proof (also drawn from [**23**]) recalls the first step in the proof of part (b) of Theorem 4.1.

Proposition 4.3. *Let ρ be a representation of* $\mathrm{W}(\overline{K}/K)$ *over* $\mathbb{C}$. *If ρ is irreducible then there exists $s_0 \in \mathbb{C}$ such that $\rho \otimes \omega^{s_0}$ is of Galois type.*

PROOF. Let J be the kernel of $\rho|I$. Then J is normal in $\mathrm{W}(\overline{K}/K)$, and the action of $\mathrm{W}(\overline{K}/K)$ on I by conjugation defines a map $\mathrm{W}(\overline{K}/K) \to \mathrm{Aut}(I/J)$. But I/J is a finite group, so we deduce that if $\sigma \in \mathrm{W}(\overline{K}/K)$ is a given Frobenius element then there is a positive integer n such that σ^n acts trivially on I/J. It follows that the coset $\sigma^n J$ is in the center of $\mathrm{W}(\overline{K}/K)/J$, and since ρ can be viewed as an irreducible representation of $\mathrm{W}(\overline{K}/K)/J$ we conclude that $\rho(\sigma^n)$ is scalar. Choose $s \in \mathbb{C}$ such that $\rho(\sigma^n)$ is multiplication by q^s, where $q = \omega(\sigma)$ as usual. Then $(\rho \otimes \omega^{-s/n})(\sigma^n)$ is trivial. Hence if $s_0 = -s/n$ then $\rho \otimes \omega^{s_0}$ is trivial on the open subgroup of $\mathrm{W}(\overline{K}/K)$ generated by J and σ^n, which is of finite index. □

The significance of the proposition for us is that if ρ is a representation of Galois type and hence a representation of $\mathrm{Gal}(L/K)$ for some finite Galois extension L of K then we have already seen a definition of $a(\rho)$ in the context of Artin representations: $a(\rho)$ is defined by (3.46). Hence given an arbitrary *irreducible* representation ρ of $\mathrm{W}(\overline{K}/K)$ we can put

$$a(\rho) = a(\rho \otimes \omega^{s_0}), \tag{4.31}$$

where the right-hand side is defined by (3.46) with s_0 chosen so that $\rho \otimes \omega^{s_0}$ is of Galois type. While such an s_0 is not unique (for it can be replaced by $s_0 + 2\pi i r/\log q$ with $r \in \mathbb{Q}$), an inspection of (3.46) shows that $a(\rho \otimes \omega^{s_0})$ depends only on $(\rho \otimes \omega^{s_0})|I$ and hence only on $\rho|I$. Thus Proposition 4.3 and (4.31) together assign a meaning to $a(\rho)$ for irreducible ρ and hence by semisimplification and additivity for all ρ.

4.3. Root numbers

To start with take $p < \infty$ and $\boldsymbol{\rho} = (\rho, N)$. Just as $a(\boldsymbol{\rho})$ is the sum of the two terms $a(\rho)$ and $b(\boldsymbol{\rho})$, with $a(\rho)$ independent of N, the root number $W(\boldsymbol{\rho})$ is similarly the product of two factors, the first of which is independent of N:

$$W(\boldsymbol{\rho}) = W(\rho)\Delta(\boldsymbol{\rho}). \tag{4.32}$$

Furthermore the definition of $\Delta(\boldsymbol{\rho})$, like the definition of $b(\boldsymbol{\rho})$, is straightforward: Writing V for the space of ρ and the same letter ρ for the quotient representation on V^I/V_N^I determined by ρ, we put

$$\delta(\boldsymbol{\rho}) = \det(-\rho(\Phi)|V^I/V_N^I) \tag{4.33}$$

and

$$\Delta(\boldsymbol{\rho}) = \frac{\delta(\boldsymbol{\rho})}{|\delta(\boldsymbol{\rho})|}. \tag{4.34}$$

Note that the identification of ρ with $(\rho, 0)$ is once again respected here, because if $N = 0$ then $\delta(\boldsymbol{\rho})$ is the determinant of a linear automorphism of the trivial vector space, whence $\Delta(\boldsymbol{\rho}) = \delta(\boldsymbol{\rho}) = 1$ and $W(\boldsymbol{\rho}) = W(\rho)$.

If $p = \infty$ then (4.32) is still valid provided we understand that $N = 0$. Thus $\Delta = 1$ and $W(\boldsymbol{\rho}) = W(\rho)$. Furthermore the definition of $W(\rho)$ is straightforward in the archimedean case, but the archimedean case will not be treated separately because it is included in the general case to be discussed now.

While the definition of $W(\rho)$ is indirect, the underlying strategy is simple: Define $W(\rho)$ by imposing a modified Artin formalism. This means first of all that $W(\rho)$ depends only on the semisimplification of ρ, and secondly that additivity holds – in other words that $W(\rho \oplus \rho') = W(\rho)W(\rho')$. We also impose compatibility in dimension one: If ρ is one-dimensional and we identify it with a character χ of $K^\times$ via (4.3) and (4.13), then $W(\rho) = W(\chi)$. So far this is just the standard Artin formalism. But as soon as we impose additivity and compatibility in dimension one there are simple counterexamples to inductivity. For instance take the case of the unramified quadratic extension L of a finite extension K of $\mathbb{Q}_p$, where $p < \infty$. Write $\mathrm{sign}_{L/K}$ for the nontrivial character of $\mathrm{Gal}(L/K)$. Then $\mathrm{ind}_{L/K} 1_L = 1_K \oplus \mathrm{sign}_{L/K}$, while (2.43) gives $W(1_L) = W(1_K) = 1$ and $W(\mathrm{sign}_{L/K}) = (-1)^d$ with d equal to the exponent of the different ideal of K. If the root number were to satisfy inductivity we would have $1 = (-1)^d$, a contradiction if d is odd (as it is for example if $K = \mathbb{Q}_p(\sqrt{p})$).

Thus inductivity must be modified. The modification is a weaker condition called **inductivity in dimension zero** (or "in degree zero"). Fix a finite extension K of $\mathbb{Q}_p$ with $p \leqslant \infty$, and consider the Grothendieck group $\mathrm{Groth}(\mathrm{W}(\overline{K}/K))$ of virtual representations of $\mathrm{W}(\overline{K}/K)$. If we want the root number to depend only on the semisimplification of its argument and to satisfy additivity then it becomes a function on $\mathrm{Groth}(\mathrm{W}(\overline{K}/K))$, so that $W(\rho)$ acquires a meaning even for virtual representations of $\mathrm{W}(\overline{K}/K)$. Now a virtual representation has a virtual dimension, and to impose inductivity in dimension zero is to demand that

$$W(\mathrm{ind}_{L/K}\rho) = W(\rho) \tag{4.35}$$

whenever L is a finite extension of K and ρ a virtual representation of $\mathrm{W}(\overline{K}/L)$ of dimension zero.

With this modification, the Artin formalism can be realized: By the theorem of Langlands and Deligne ([**23**], p. 535; see also Tate [**93**], [**94**]), there is a unique map $\rho \mapsto W(\rho)$ from virtual representations of $\mathrm{W}(\overline{K}/K)$ to complex numbers of absolute value 1 which satisfies additivity, compatibility in dimension one, and inductivity in dimension zero. The uniqueness of $\rho \mapsto W(\rho)$ is the easy part of the assertion, but it is worth verifying here, for the argument shows how $W(\rho)$ can be computed in practice.

What is needed for the verification is a "dimension-zero" variant of Brauer's theorem (cf. [**23**], p. 510, Proposition 1.5). Let us say that a virtual representation of a finite group G is **monomial of dimension zero** if it is induced by the difference of two one-dimensional representations of a subgroup of G. The dimension-zero version of Brauer's theorem states that in $\mathrm{Groth}(G)$ any virtual representation of G of dimension zero is an integral linear combination of monomial representations of dimension zero. Suppose now that ρ is an irreducible representation of $\mathrm{W}(\overline{K}/K)$ over $\mathbb{C}$. By Proposition 4.3 there exists $s_0 \in \mathbb{C}$ such that $\rho \otimes \omega^{s_0}$ factors through $\mathrm{Gal}(L/K)$ for some finite Galois extension L of K. Hence taking $G = \mathrm{Gal}(L/K)$ we can write

$$[\rho \otimes \omega^{s_0}] - (\dim\rho)[1_K] = \sum_{(M,\xi,\xi')} n_{M,\xi,\xi'} \mathrm{ind}_{M/K}([\xi] - [\xi']) \tag{4.36}$$

with integers $n_{M,\xi,\xi'}$, subfields M of L containing K, and one-dimensional characters ξ and ξ' of $\mathrm{Gal}(L/M)$. Tensoring both sides of (4.36) with $[\omega^{-s_0}]$, we obtain

$$[\rho] = (\dim\rho)[\omega^{-s_0}] + \sum_{(M,\xi,\xi')} n_{M,\xi,\xi'} \mathrm{ind}_{M/K}([\xi_0] - [\xi_0']) \tag{4.37}$$

in Groth(W($\overline{K}/K$)), where $\xi_0 = \xi\omega_M^{-s_0}$ and $\xi_0' = \xi'\omega_M^{-s_0}$. Here we are using the fact that our ω is really ω_K and satisfies $\omega|\mathrm{W}(\overline{K}/M) = \omega_M$. The modified Artin formalism now implies that

$$W(\rho) = W(||*||^{-s_0})^{\dim\rho} \prod_{(M,\xi,\xi')} (W(\chi_{\xi_0})/W(\chi_{\xi_0'}))^{n_{M,\xi,\xi'}}, \tag{4.38}$$

where χ_{ξ_0} and $\chi_{\xi_0'}$ are the characters of $M^\times$ corresponding to ξ_0 and ξ_0' respectively under (4.3) and (4.13). It follows from (4.38) that the modified Artin formalism does indeed determine $W(\rho)$ uniquely for irreducible ρ, and consequently, by semisimplification and additivity, for all ρ.

4.4. Epsilon factors

Just as in the one-dimensional case, the local root number is subsumed in a broader concept, the epsilon factor. Given a local field K, a complex representation $\boldsymbol{\rho}$ of WD($\overline{K}/K$), an additive character ψ of K, and a Haar measure dx on K, one puts

$$\varepsilon(\boldsymbol{\rho}, \psi, dx) = \varepsilon(\rho, \psi, dx)\delta(\boldsymbol{\rho}), \tag{4.39}$$

where $\delta(\boldsymbol{\rho})$ is as in (4.33) – note that the definition is indeed independent of ψ and dx – and $\varepsilon(\rho, \psi, dx)$ is defined by the theorem of Langlands and Deligne. In the context of epsilon factors their theorem states that the modified Artin formalism can be imposed on $\varepsilon(\rho, \psi, dx)$ and then defines $\varepsilon(\rho, \psi, dx)$ uniquely. Inductivity in dimension zero now takes account of the additive character: the requirement is that

$$\varepsilon(\mathrm{ind}_{L/K}\rho, \psi \circ \mathrm{tr}_{L/K}, d_L x) = \varepsilon(\rho, \psi, d_K x) \tag{4.40}$$

for every finite extension L of K, virtual representation ρ of W($\overline{K}/L$) of dimension zero, additive character ψ of K, and Haar measures $d_L x$ and $d_K x$ on L and K respectively. Additivity means as usual that

$$\varepsilon(\rho \oplus \rho', \psi, dx) = \varepsilon(\rho, \psi, dx)\varepsilon(\rho', \psi, dx),$$

and compatibility in dimension one means that if ρ is one-dimensional then

$$\varepsilon(\rho, \psi, dx) = \varepsilon(\chi, \psi, dx),$$

where χ is the character of $K^\times$ corresponding to ρ under (4.3) and (4.13).

Although a local functional equation is lacking in dimension > 1, one can still introduce a complex parameter s by putting

$$\varepsilon(s, \boldsymbol{\rho}, \psi, dx) = \varepsilon(\boldsymbol{\rho} \otimes \omega^s, \psi, dx) \tag{4.41}$$

and thus generalizing (2.73). One can also define $W(\boldsymbol{\rho}, \psi)$ by a straightforward generalization of (2.72):

$$W(\boldsymbol{\rho}, \psi) = \frac{\varepsilon(\boldsymbol{\rho}, \psi, dx)}{|\varepsilon(\boldsymbol{\rho}, \psi, dx)|}. \tag{4.42}$$

If K has characteristic zero then we recover $W(\boldsymbol{\rho})$ by taking $\psi = \psi^{\mathrm{can}}$. Indeed if L is a finite extension of K and ψ_L^{can} and ψ_K^{can} denote the respective canonical additive characters then $\psi_L^{\mathrm{can}} = \psi_K^{\mathrm{can}} \circ \mathrm{tr}_{L/K}$, so (4.40) and (4.42) do give (4.35). Thus the theorem of Langlands and Deligne does imply that $W(\rho)$ can be defined by imposing the modified Artin formalism.

5. Normalizations of the root number in the literature

We conclude this lecture with a brief but important caveat: In (2.72), (4.32), and (4.42) we have normalized the local and hence also the global root number to have absolute value 1, but this is far from being a standard convention. In fact what is more commonly found in the literature (see for example [**93**], p. 105) is the normalization

$$W(\rho,\psi)=\epsilon(1/2,\rho,\psi,dx_\psi), \tag{4.43}$$

which does not always give $W(\rho,\psi)$ absolute value 1. Suppose for instance that K is nonarchimedean, with residue class field of order q. Take ρ to be one-dimensional and view it as a character χ of $K^\times$ via (4.3). Applying (2.89) and (2.90), we see that if the root number is defined by (4.43) then $|W(\chi,\psi)|=q^{(a(\chi)+n(\psi))c}$, where c is as in (2.83). Hence $|W(\chi,\psi)|\neq 1$ unless χ is unitary or $a(\chi)+n(\psi)=0$.

On the other hand, while our convention that root numbers have absolute value 1 may not be widely supported in the literature, it seems to be standard in the mathematical vernacular. For example, have you ever heard anybody say that the root number of the elliptic curve $X_0(11)$ is $+\sqrt{11}$?

6. Exercises

Exercise 4.1. Let K be a finite extension of $\mathbb{Q}_p$ with $p<\infty$. Show that the open subgroups of infinite index in $\mathrm{W}(\overline{K}/K)$ are precisely the subgroups of the form $\mathrm{Gal}(\overline{K}/R)$ with R a finite extension of K_{unr} inside $\overline{K}$.

Exercise 4.2. Let E be an elliptic curve over a finite extension K of $\mathbb{Q}_p$, and let $\rho_{E,\ell}$ be the representation of $\mathrm{Gal}(\overline{K}/K)$ on $V_\ell(E)$ for some prime $\ell\neq p$. Let $\boldsymbol{\rho}=(\rho,N)$ be the representation of $\mathrm{WD}(\overline{K}/K)$ obtained from $\rho_{E,\ell}^\vee$ by applying part (a) of Theorem 4.1. Show that if E has (i) good reduction then ρ is unramified and $N=0$, (ii) bad but potentially good reduction then ρ is ramified but $N=0$, and (iii) potentially multiplicative reduction then $N\neq 0$ and $\rho\cong\chi\oplus\chi\omega$, where χ is the unique quadratic or trivial character of $\mathrm{W}(\overline{K}/K)$ such that the twist of E by χ is a Tate curve over K. Furthermore, show that $W(\boldsymbol{\rho})=1$ in case (i) and that

$$W(\boldsymbol{\rho})=\begin{cases}-1 & \text{if } \chi=1\\ 1 & \text{if } \chi \text{ is the unique unramified quadratic character of } \mathrm{W}(\overline{K}/K)\\ \chi(-1) & \text{if } \chi \text{ is ramified}\end{cases}$$

in case (iii).

Exercise 4.3. Let K be a number field with r_1 real embeddings and r_2 pairs of complex conjugate embeddings, and suppose that E is a semistable elliptic curve over K which has split multiplicative reduction at exactly s finite places of K. Put $W(E)=W(M)$, where M is the isomorphism class of $\{\rho_{E,\ell}^\vee\}$. Using Exercise 4.2, derive the classic formula $W(E)=(-1)^{r_1+r_2+s}$. (Essential information: The representation ρ_v of $\mathrm{W}(\overline{K}_v/K_v)$ associated to E at an infinite place v is as follows. Let $\chi:\mathbb{C}^\times\to\mathbb{C}^\times$ be the character $z\mapsto z^{-1}$. If v is complex and we identify $\mathrm{W}(\overline{K}_v/K_v)$ with $\mathbb{C}^\times$ then $\rho_v=\chi\oplus\overline{\chi}$. If v is real and we identify the subgroup $\mathrm{W}(\overline{K}_v/\overline{K}_v)$ of $\mathrm{W}(\overline{K}_v/K_v)$ with $\mathbb{C}^\times$ then $\rho_v=\mathrm{ind}_{\overline{K}_v/K_v}\chi$.)

Exercise 4.4. As we have already mentioned, the class of motivic L-functions, while very broad, does not include even all Hecke L-functions of number fields – for example if χ is as in (2.50) then $L(s,\chi)$ is not motivic – let alone the L-functions of arbitrary automorphic forms. Nonetheless, at the *local* level, representations of the Weil-Deligne group still serve as parameters for local components of automorphic representations. To illustrate this point, let f be a Maass form for $\mathrm{SL}(2,\mathbb{Z})$. Then the representation of $\mathrm{W}(\mathbb{C}/\mathbb{R})$ associated to f at the infinite place of $\mathbb{Q}$ is $\chi\oplus\overline{\chi}$, where χ has the form $\chi(t)=|t|^{ir}(t/|t|)^m$ with $r\in\mathbb{R}$ and $m\in\{0,1\}$. Show that the root number $W(f)$ in the functional equation of $L(s,f)$ is 1 or -1 according as $m=0$ or 1. (Hint: Given that f is a Maass form for $\mathrm{SL}(2,\mathbb{Z})$ rather than for one of its congruence subgroups, what can you infer about the conductor of the representation of $\mathrm{WD}(\overline{\mathbb{Q}}_p/\mathbb{Q}_p)$ associated to f at primes $p<\infty$?)

Exercise 4.5. (*Reading.*) Let G be a finite group and ρ a representation of G. There are theorems of Snaith [**89**] and others which express $[\rho]$ in $\mathrm{Groth}(G)$ as an integral linear combination of classes of monomial representations in a *canonical* way. Theorems of this type are call **canonical Brauer induction** theorems. Can they be used to prove the existence of local root numbers?

Exercise 4.6. Let $\boldsymbol{\rho}^{\mathrm{ss}}=(\rho^{\mathrm{ss}},N)$ be as in part (b) of Theorem 4.1, and let V be the space of $\boldsymbol{\rho}^{\mathrm{ss}}$.

(a) Show that there is a unique invariant subspace W of V such that $V=V^I\oplus W$. Here "invariant" means "stable under ρ^{ss}."

(b) Let N^* be the nilpotent endomorphism of V which coincides with N on V^I and with 0 on W. Show that the pair $\boldsymbol{\rho}^{\mathrm{ss}*}=(\rho^{\mathrm{ss}},N^*)$ is a representation of $\mathrm{WD}(\overline{K_v}/K_v)$ and that $L(s,\boldsymbol{\rho}^{\mathrm{ss}*})=L(s,\boldsymbol{\rho}^{\mathrm{ss}})$, $a(\boldsymbol{\rho}^{\mathrm{ss}*})=a(\boldsymbol{\rho}^{ss})$, and $W(\boldsymbol{\rho}^{\mathrm{ss}*})=W(\boldsymbol{\rho}^{ss})$.

Exercise 4.7. With notation as in (4.21), put

$$\boldsymbol{\rho}^*_{M,v}=((\rho_v^{\mathrm{ss}})^\iota,(N_v^*)^\iota),\tag{4.44}$$

where N_v^* is defined as in Exercise 4.6. (Note that there is an implicit choice of λ in (4.44) as well as of ι.) Assume that a condition stronger than (3.37) holds, namely that for every prime ideal $\mathfrak{p}$ in the exceptional set S of M, for every pair of reciprocal roots α and α' of $B_{\mathfrak{p}}(x)$, and for every pair of field automorphisms τ and τ' of $\mathbb{C}$, we have

$$|\tau(\alpha)|=|\tau'(\alpha')|\leqslant(\mathbf{N}\mathfrak{p})^{w/2}.\tag{4.45}$$

Prove that the isomorphism class of $\boldsymbol{\rho}^*_{M,v}$ is independent of the choice of λ and ι.

Exercise 4.8. The preceding two exercises may appear to be a partial solution to Problem 4, or at least a successful circumvention of it: If we simply replace N by N^* then the local L-factor, conductor, and root number are unaffected (Exercise 4.6) but the isomorphism class of $\boldsymbol{\rho}^*_{M,v}$ is now independent of all choices provided (4.45) holds (Exercise 4.7). The purpose of the present exercise is to show that this "solution" to Problem 4 is completely unsatisfactory.

(a) Give an example showing that (4.45) need not be satisfied even in the simple case of the motive associated to H^1 of the product of two elliptic curves.

(b) Give an example of representations $\boldsymbol{\rho}=(\rho,N)$ and $\boldsymbol{\rho}'=(\rho',N')$ for which $(N\otimes 1'+1\otimes N')^*\neq N^*\otimes 1'+1\otimes N'^*$, in contrast to (4.6).

LECTURE 5

The minimalist dichotomy

In this final lecture we would like to reflect on the following question: To what extent, or under what circumstances, should we expect the order of vanishing of a motivic L-function at the center of its critical strip to be the minimum compatible with its functional equation? To begin with we restrict our attention to essentially self-dual premotives M, so that the functional equation of $L(s, M)$ has the form $\Lambda(s, M) = W(M)\Lambda(k-s, M)$ with $k = w+1$, where $w \geqslant 0$ is the weight of M. We are then asking how likely it is that

$$\text{(5.1)} \qquad \operatorname{ord}_{s=k/2} L(s, M) = \begin{cases} 0 & \text{if } W(M) = 1, \\ 1 & \text{if } W(M) = -1. \end{cases}$$

In the case of elliptic curves E over $\mathbb{Q}$ with $W(E) = 1$ this question is discussed at length in the paper of Bektemirov, Mazur, Stein, and Watkins [**8**], who refer to the conjecture that $L(1, E) \neq 0$ with probability one as the "minimalist conjecture" for such E. Adopting their language, we shall say that the **minimalist dichotomy** holds for M, or for the associated L-function $L(s, M)$, if (5.1) is satisfied.

We have already encountered an infinite family of self-dual premotives in which the minimalist dichotomy holds for every member: the premotives $M(\chi)$ associated by (3.43) to the "canonical" characters χ. Indeed we have $L(s, M(\chi)) = L(s, \chi)$ and $W(M(\chi)) = W(\chi)$, so that (5.1) holds with $M = M(\chi)$ by Theorem 1.3. On the other hand, as a universal statement about L-functions of essentially self-dual premotives, (5.1) is simply false. In the case of elliptic curves this point is perhaps so familiar as to require no comment, but for the record, if one takes the base field to be $\mathbb{Q}$ and orders elliptic curves by their conductor then the first counterexample to (5.1) is the curve 389A1 in Cremona's tables ([**22**], p. 306). Of course the fact that one can be so precise depends on the confluence of several breakthroughs of the past quarter-century: first of all the modularity of elliptic curves over $\mathbb{Q}$ proved by Wiles [**100**], Taylor and Wiles [**96**], and Breuil, Conrad, Diamond, and Taylor [**11**], and secondly the theorems of Kolyvagin [**53**] (supplemented by Bump-Friedberg-Hoffstein [**14**] or Murty-Murty [**70**]) and Gross-Zagier [**39**], which together imply that if E is an elliptic curve over $\mathbb{Q}$ with $\operatorname{ord}_{s=1} L(s, E) \leqslant 1$ then the rank of E is $\leqslant 1$. For curves of conductor $\leqslant 999$ an inspection of Cremona's tables ([**22**], pp. 293 – 340) reveals that the converse also holds, and consequently the first curve in the tables of rank > 1 – namely the curve 389A1, which has rank 2 – it is also the first elliptic curve over $\mathbb{Q}$ for which the minimalist dichotomy fails. Using results like those in [**79**] one can produce as many other counterexamples as one likes.

Nonetheless, we can ask as in [**8**] whether the minimalist dichotomy holds for a dense set of elliptic curves over $\mathbb{Q}$. After briefly surveying what is known or conjectured about this question, we shall broaden the discussion to include more general motivic L-functions.

1. Elliptic curves

Let $\mathcal{E}$ be the set of isomorphism classes of elliptic curves E over $\mathbb{Q}$ and $\mathcal{D}$ the subset of isomorphism classes for which $L(s, E)$ satisfies the minimalist dichotomy. Write $\vartheta_{\mathcal{E}}(x)$ and $\vartheta_{\mathcal{D}}(x)$ for the number of isomorphism classes in $\mathcal{E}$ and $\mathcal{D}$ respectively which have conductor $\leqslant x$. By the **minimalist conjecture for elliptic curves over Q** we mean the hypothesis that $\lim_{x\to\infty} \vartheta_{\mathcal{D}}(x)/\vartheta_{\mathcal{E}}(x)$ exists and equals 1.

Before committing ourselves too firmly to this formulation of the problem, we should note that the answer could depend on the fact that we are counting elliptic curves using the conductor rather than some other natural invariant. Quite generally, consider a set $\mathcal{S}$ and a function $\nu : \mathcal{S} \to \mathbb{Z}_{\geqslant 0}$ such that for every $x > 0$ there are only finitely many $s \in \mathcal{S}$ with $\nu(s) \leqslant x$. We will call ν a **counting function** on $\mathcal{S}$. Writing $\vartheta_{\mathcal{S},\nu}(x)$ for the number of such s, one can consider the limit $\lim_{x\to\infty} \vartheta_{\mathcal{T},\nu}(x)/\vartheta_{\mathcal{S},\nu}(x)$ for a given subset $\mathcal{T}$ of $\mathcal{S}$, but even if this limit exists, its value may depend on ν. For example, fix an integer $n \geqslant 3$, let $\mathcal{P}$ be the set of primes, and let $\mathcal{Q}$ the subset of primes $p \equiv -1$ modulo n. If we take $\nu(p) = p$ then $\lim_{x\to\infty} \vartheta_{\mathcal{Q},\nu}(x)/\vartheta_{\mathcal{P},\nu}(x) = 1/\varphi(n)$, but if instead $\nu(p) = p^{\langle p\rangle}$, where $\langle p\rangle$ is the least positive residue of p modulo n, then $\lim_{x\to\infty} \vartheta_{\mathcal{Q},\nu}(x)/\vartheta_{\mathcal{P},\nu}(x) = 0$.

Returning to $\mathcal{E}$, and writing $\nu(E)$ for the value of ν on the isomorphism class of E, we can defend the choice $\nu(E) = N(E)$ as the only *analytic* possibility for ν – analytic in the sense that $N(E)$ appears in the functional equation of $L(s, E)$ – but on the arithmetic side there are many other possibilities: for example the absolute value of the minimal discriminant $\Delta(E)$, the Arakelov height of E, or simply the coarse height

$$\nu(E) = \min_{\substack{4a^3+27b^2\neq 0\\ E\cong E_{a,b}}} \max(|a|^3, |b|^2), \tag{5.2}$$

where the minimum is taken over all pairs of integers (a, b) such that $4a^3+27b^2 \neq 0$ and E is isomorphic to the curve $E_{a,b} : y^2 = x^3+ax+b$. Now one can argue that the arithmetic choices are less natural than $N(E)$, for they depend on the *isomorphism* class of E, whereas the validity of (5.1) depends only on the *isogeny* class. However the arithmetic choices of ν are often easier to work with, so it behooves us to know whether the choice $\nu(E) = |\Delta(E)|$, say, is equivalent to the choice $\nu(E) = N(E)$ for the purpose of evaluating the limit $\lim_{x\to\infty} \vartheta_{\mathcal{D},\nu}(x)/\vartheta_{\mathcal{E},\nu}(x)$. This does not seem like an easy question, particularly since it is not known whether $|\Delta(E)|$ is bounded by a power of $N(E)$ – in a stronger form this is Szpiro's conjecture.

But in fact one hopes for more: Not only should $\lim_{x\to\infty} \vartheta_{\mathcal{D},\nu}(x)/\vartheta_{\mathcal{E},\nu}(x)$ be the same for $\nu(E) = |\Delta(E)|$ as for $\nu(E) = N(E)$, but even the shape of the error term should be the same. To spell this out, consider the conjecture

$$\vartheta_{\mathcal{E},\nu}(x) \sim c \cdot x^{5/6}. \tag{5.3}$$

The expectation is that (5.3) holds both in the case $\nu(E) = |\Delta(E)|$ (Brumer and McGuinness [**13**]) and in the case $\nu(E) = N(E)$ (Watkins [**99**]), although the constant c may depend on ν. Next consider Conjecture 3.4 on p. 244 of [**8**] (based on the heuristics of Watkins [**99**]), which in principle gives

$$\vartheta_{\mathcal{E},\nu}(x) - \vartheta_{\mathcal{D},\nu}(x) \sim c' \cdot x^{19/24}(\log x)^{3/8} \tag{5.4}$$

both for $\nu(E) = N(E)$ and $\nu(E) = |\Delta(E)|$, although the constant c' may again depend on the choice of ν. We say "in principle" because the focus in [**8**] is on the first line of (5.1), so that the roles of $\mathcal{E}$ and $\mathcal{D}$ are actually played by the set $\mathcal{E}^+$ of isomorphism classes with $W(E) = 1$ and the subset $\mathcal{D}^+$ of isomorphism classes with $L(1, E) \neq 0$. In any case, since $19/24 < 5/6$ we obtain from (5.3) and (5.4) that $\lim_{x\to\infty} \vartheta_{\mathcal{D}}(x)/\vartheta_{\mathcal{E}}(x) = 1$, regardless of whether the implicit counting function is $|\Delta|$ or N. Henceforth the omission of the subscript ν on $\vartheta_{\mathcal{D}}(x)$ and $\vartheta_{\mathcal{E}}(x)$ indicates as before that $\nu = N$, but now with the implication that the choice of ν shouldn't matter anyway.

As the authors of [**8**] acknowledge, the numerical evidence for all of this is weak. In fact one can almost say that the minimalist conjecture for elliptic curves over $\mathbb{Q}$ is made in defiance of the available data. These data include the calculations of Brumer and McGuinness [**13**] with elliptic curves of prime conductor $< 10^8$, the calculations of Stein and Watkins [**91**] with elliptic curves of composite conductor $\leqslant 10^8$ or prime conductor $< 10^{10}$, and the calculations of the authors themselves involving selected elliptic curves of prime conductor around 10^{14}. None of these works gives much support for the minimalist dichotomy, although there is some hint that the desired numerical evidence may simply lie outside the range of computation. At least the results for 10^{14} can be regarded as a bit more supportive than those for 10^8 or 10^{10}.

Turning from the numerical to the theoretical, we find that the known results pertain less to the minimalist conjecture than to a slightly different hypothesis, the **average rank conjecture**. Originally enunciated by Goldfeld [**34**] for quadratic twists of a fixed elliptic curve, the average rank conjecture is here understood to apply to all isomorphism classes of elliptic curves over $\mathbb{Q}$ simultaneously. It asserts that if ν is any of the counting functions $\mathcal{E} \to \mathbb{Z}_{\geqslant 0}$ mentioned above then the limit

$$r_\nu(\mathcal{E}) = \lim_{x\to\infty} \frac{\sum_{\nu(E)\leqslant x} \operatorname{ord}_{s=1} L(s, E)}{\vartheta_{\mathcal{E},\nu}(x)} \tag{5.5}$$

exists and equals $1/2$. In principle the choice of ν favored in the literature is the coarse height (5.2), but even when ν is so chosen, the "average rank" r that one is likely to encounter in research papers differs from our r_ν in that both the summation in the numerator on the right-hand side of (5.5) and the implicit summation in the denominator run over *all* elliptic curves $E_{a,b}$ such that $|a|^3, |b|^2 \leqslant x$: In other words, redundancies arising from isomorphisms among the curves $E_{a,b}$ are not eliminated. Granting this point, and assuming the generalized Riemann hypothesis for L-functions of elliptic curves over $\mathbb{Q}$, one can cite the successive upper bounds $r \leqslant 23/10$ (Brumer [**12**]), $r \leqslant 2$ (Heath-Brown [**40**]), $r \leqslant 25/14$ (Young [**103**], cf. also [**104**]), and $r \leqslant 27/14$ (Baier and Zhao [**5**]). The two more recent works draw inspiration from random matrix theory (cf. Iwaniec, Luo, and Sarnak [**45**]), and while Young's bound is sharper than that of Baier and Zhao, it depends on the generalized Riemann hypothesis for Dirichlet and symmetric square L-functions, a dependence eliminated in [**5**]. See also Exercise 5.4. It should be added that the limit defining r is not actually known to exist: the results cited above are to be understood as upper bounds for the corresponding limit superior.

Quite apart from the large gap between the upper bounds for r cited above and the conjectured value $r = 1/2$, the average rank conjecture does not seem to imply anything about the minimalist conjecture unless one knows something about the equidistribution of root numbers. As before, let $\mathcal{E}^\pm \subset \mathcal{E}$ be the subset of isomorphism classes with root number ± 1. It does not appear to be known that

$\lim_{x\to\infty} \vartheta_{\mathcal{E}^{\pm}}(x)/\vartheta_{\mathcal{E}}(x) = 1/2$, let alone that

$$\vartheta_{\mathcal{E}^{\pm}}(x) = \frac{1}{2}\vartheta_{\mathcal{E}}(x) + O(x^{\gamma}) \tag{5.6}$$

with a constant $\gamma < 5/6$. However if one grants (5.6) along with (5.3) then one can show (see Exercise 5.1) that an estimate of the form

$$\sum_{N(E)\leqslant x} \mathrm{ord}_{s=1}L(s,E) = \frac{1}{2}\vartheta_{\mathcal{E}}(x) + O(x^{\gamma'}) \tag{5.7}$$

with $\gamma' < 5/6$ implies an estimate of the form

$$\vartheta_{\mathcal{D}}(x) = \vartheta_{\mathcal{E}}(x) + O(x^{\gamma''}) \tag{5.8}$$

with $\gamma'' < 5/6$, and conversely. Note that (5.7) is stronger than the assertion that $r_N(\mathcal{E}) = 1/2$ in (5.5), while (5.8) is weaker than (5.4). But if we grant (5.3) then (5.8) does imply the minimalist conjecture for elliptic curves over $\mathbb{Q}$.

2. The minimalist trichotomy

Let us attempt to formalize the idea that " with probability 1, the order of vanishing of a motivic L-function at its center of symmetry is the minimum compatible with its functional equation." This notion may simply be wrong, but without a precise formulation there is nothing to refute.

Fix a number field K and an integer $w \geqslant 0$, and put $k = w+1$, so that if M is a premotive of weight w over K then the functional equation of $L(s,M)$ is $\Lambda(s,M) = W(M)\Lambda(k-s,\overline{M})$. Since we are not restricting ourselves to essentially self-dual premotives, it is not necessarily the case that $W(M) = \pm 1$. Hence the dichotomy (5.1) should now be replaced by

$$\mathrm{ord}_{s=k/2}L(s,M) = \begin{cases} 0 & \text{if } M \not\cong \overline{M}, \\ 0 & \text{if } M \cong \overline{M} \text{ and } W(M) = 1, \\ 1 & \text{if } M \cong \overline{M} \text{ and } W(M) = -1. \end{cases} \tag{5.9}$$

Of course an equivalent but more succinct formulation would be

$$\mathrm{ord}_{s=k/2}L(s,M) = \begin{cases} 1 & \text{if } M \cong \overline{M} \text{ and } W(M) = -1, \\ 0 & \text{otherwise,} \end{cases} \tag{5.10}$$

but perhaps (5.9) is more illuminating than (5.10). We shall refer to (5.9) as the **minimalist trichotomy**.

Now fix an integer $n \geqslant 1$ and let $\mathcal{S}_{K,w,n}$ be the set of premotives over K of weight w and rank n. Let $\mathcal{T}_{K,w,n} \subset \mathcal{S}_{K,w,n}$ be the subset of elements satisfying the minimalist trichotomy, and put $\mathcal{S} = \mathcal{S}_{K,w,n}$ and $\mathcal{T} = \mathcal{T}_{K,w,n}$ for simplicity. We also take $\nu(M) = \mathbf{N}\mathfrak{f}(M)$ and omit the subscript ν on $\vartheta_{\mathcal{S},\nu}$ and $\vartheta_{\mathcal{T},\nu}$. A naïve attempt at a **minimalist conjecture for premotives** would assert that

$$\lim_{x\to\infty} \vartheta_{\mathcal{T}}(x)/\vartheta_{\mathcal{S}}(x) = 1. \tag{5.11}$$

This blunt formulation implies in particular that factorizations like (1.33) and (1.34) are relatively rare: the resulting central zeros of high multiplicity occur with density zero. Conceivably this is a reason to reject (5.11). But a more fundamental problem is that we do not know that ν is a counting function on $\mathcal{S}_{K,w,n}$. In other words, to formulate the minimalist conjecture as in (5.11) we need:

Hypothesis 1. *Given $x > 0$, there are only finitely many premotives M over K of weight w and rank n such that $\mathbf{N}\mathfrak{f}(M) \leqslant x$.*

It might be possible to reduce this statement to the following assertion about Galois representations over finite fields (cf. [**67**]), which however is also unknown:

Hypothesis 2. *Fix a prime number ℓ. If $x > 0$ is given, then there are only finitely many isomorphism classes of semisimple representations $\rho : \mathrm{Gal}(\overline{K}/K) \to \mathrm{GL}_n(\overline{\mathbb{F}}_\ell)$ with $\mathbf{N}\mathfrak{f}(\rho) \leqslant x$. Here $\mathfrak{f}(\rho)$ is defined exactly as in* (3.45), *except that the product runs only over the prime ideals of K which do not divide ℓ.*

The evidence in favor of these hypotheses includes the following points:

- Hypothesis 2 is known to hold if the image of ρ is solvable (Moon and Taguchi [**68**], p. 2530, Theorem 2), and for Hypothesis 2 to hold in general it suffices that it hold when the image of ρ is a finite simple group of Lie type in characteristic ℓ ([**68**], p. 2531, Proposition 3).
- The analogue of Hypothesis 2 over $\mathbb{C}$ – in other words, the statement that there are only finitely many isomorphism classes of n-dimensional Artin representations of K with conductor below a given bound – was proved by Ralph Greenberg (unpublished) and by Anderson, Blasius, Coleman, and Zettler [**1**]. Thus Hypothesis 1 holds for $w = 0$.
- Hypothesis 2 would follow from a suitable generalization of Serre's conjecture [**85**], as in Ash, Doud, and Pollack [**4**].

Nonetheless, if we want to formulate the minimalist conjecture without relying on unproven hypotheses then a slight modification is needed.

With K, w, and n as before, fix a number field $\mathbb{E}$ and let $\mathcal{S}_{K,\mathbb{E},w,n} \subset \mathcal{S}_{K,w,n}$ be the subset of elements which admit $\mathbb{E}$ as coefficient field. Let $\mathcal{T}_{K,\mathbb{E},w,n} \subset \mathcal{S}_{K,\mathbb{E},w,n}$ be the subset of elements satisfying the minimalist trichotomy. If we take the sets $\mathcal{S}$ and $\mathcal{T}$ in (5.11) to be $\mathcal{S}_{K,\mathbb{E},w,n}$ and $\mathcal{T}_{K,\mathbb{E},w,n}$ rather than $\mathcal{S}_{K,w,n}$ and $\mathcal{T}_{K,w,n}$ then we obtain a variant of the minimalist conjecture which is meaningful unconditionally:

Proposition 5.1. *The map $M \mapsto \mathbf{N}\mathfrak{f}(M)$ is a counting function on $\mathcal{S}_{K,\mathbb{E},w,n}$.*

To prove Proposition 5.1 we use the following result, which restates a well-known lemma of Faltings ([**29**], pp. 362 – 363, or see [**44**], p. 285). The lemma was embedded in the proof of Satz 5 of [**29**] and used there only in the case $\mathbb{E}_\lambda = \mathbb{Q}_\ell$, but the argument works for any finite extension of $\mathbb{Q}_\ell$.

Proposition 5.2. *Fix a number field K, a finite set of prime ideals S of K, a prime number ℓ, a finite extension $\mathbb{E}_\lambda$ of $\mathbb{Q}_\ell$, and an integer $n \geqslant 1$. Then there exists a finite set of prime ideals T of K, disjoint from S, with the following property: If ρ_λ is an n-dimensional semisimple representation of $\mathrm{Gal}(\overline{K}/K)$ over $\mathbb{E}_\lambda$ which is unramified outside S then ρ_λ is determined up to isomorphism by the $|T|$ values $\mathrm{tr}\, \rho_\lambda(\Phi_\mathfrak{p})$ for $\mathfrak{p} \in T$, where $\Phi_\mathfrak{p}$ denotes an inverse Frobenius element at $\mathfrak{p}$.*

Given an integer $n \geqslant 2$, let $p_{\max}(n)$ be the largest prime number dividing n. We also put $p_{\max}(1) = 1$, so that $p_{\max}(n)$ is defined for every positive integer n. Since $p_{\max}(n) \leqslant n$ for every positive integer n, Proposition 5.1 is an immediate corollary of the following statement:

Proposition 5.3. *The map $M \mapsto p_{\max}(\mathbf{N}\mathfrak{f}(M))$ is a counting function on $\mathcal{S}_{K,\mathbb{E},w,n}$.*

PROOF. Fix a prime number ℓ_0 and a place λ_0 of $\mathbb{E}$ above ℓ_0, and let $x > \ell_0$ be given. Let S be the set of prime ideals of K which lie over a rational prime $\leqslant x$. We apply Proposition 5.2 with ℓ and λ replaced by ℓ_0 and λ_0. Given $M \in \mathcal{S}_{K,\mathbb{E},w,n}$, with $p_{\max}(\mathbf{N}\mathfrak{f}(M)) \leqslant x$, choose $\{\rho_\lambda\} \in M$ with coefficient field $\mathbb{E}$; then M is determined by the isomorphism class of ρ_{λ_0} (Proposition 3.1), and ρ_{λ_0} is unramified outside S. Thus it suffices to see that there are only finitely many possibilities for the numbers $\operatorname{tr}\rho_{\lambda_0}(\Phi_\mathfrak{p})$ with $\mathfrak{p} \in T$. Now $\operatorname{tr}\rho_{\lambda_0}(\Phi_\mathfrak{p})$ is the sum of the reciprocal roots of $B_\mathfrak{p}(x)$ and is therefore an element of $\mathcal{O}_\mathbb{E}$ having absolute value $\leqslant n(\mathbf{N}\mathfrak{p})^{w/2}$ in every archimedean embedding of $\mathbb{E}$. But $\mathcal{O}_E$ has only finitely many such elements. □

3. Elliptic curves revisited

A reasonable expectation of any "minimalist conjecture for premotives" is that it should imply the minimalist conjecture for elliptic curves over $\mathbb{Q}$. However the latter conjecture will need to be reformulated if we are to make a connection. The problem is that $\mathcal{E}$ was defined to be the set of isomorphism classes of elliptic curves over $\mathbb{Q}$, not the set of isogeny classes. Isomorphism classes are the right objects to consider if ν is taken to be $|\Delta|$, but now that we have settled on $\nu = N$ it is natural to take the domain of ν to be the set of isogeny classes. In fact by the isogeny theorem [**29**] we may identify the isogeny class of an elliptic curve E over $\mathbb{Q}$ with the premotive determined by the fully compatible family $\{\rho_{E,\ell}^\vee\}$. Writing $\mathcal{I}$ for the set of all such premotives, we will henceforth take the minimalist conjecture for elliptic curves over $\mathbb{Q}$ to be the assertion that

$$\lim_{x\to\infty} \vartheta_\mathcal{J}(x)/\vartheta_\mathcal{I}(x) = 1, \tag{5.12}$$

where $\mathcal{J} \subset \mathcal{I}$ is the subset of elements satisfying the minimalist dichotomy.

It follows from the definitions that $\mathcal{I} \subset \mathcal{S}_{K,\mathbb{E},w,n}$ with $K = \mathbb{E} = \mathbb{Q}$, $w = 1$, and $n = 2$, and the question at hand is whether the preceding inclusion is actually an equality: if so, then (5.12) is simply (5.11) with $\mathcal{S} = \mathcal{S}_{\mathbb{Q},\mathbb{Q},1,2}$ and $\mathcal{T} = \mathcal{T}_{\mathbb{Q},\mathbb{Q},1,2}$. However the desired equality $\mathcal{I} = \mathcal{S}_{\mathbb{Q},\mathbb{Q},1,2}$ is at present unknown. In fact even a slightly weaker question posed by Lang and Trotter ([**56**], pp. 5 and 19) more than thirty years ago remains open. On the positive side, the Fontaine-Mazur conjecture [**30**] combined with a certain "ordinariness conjecture" (about which more in a moment) does imply that $\mathcal{I} = \mathcal{S}_{\mathbb{Q},\mathbb{Q},1,2}$. An implication in this spirit but oriented more toward the Lang-Trotter question is proved in [**77**], and for the sake of completeness we shall now give a proof of the implication as stated here. The reader is cautioned that the arguments involved – mostly variants of the arguments in [**77**], which are elementary, but also some arguments based on modular forms – temporarily lead us outside the prerequisites for the lectures. This deviation is inevitable, because the Fontaine-Mazur conjecture itself lies outside our self-imposed perimeter. Hence we simply refer the reader to pp. 190 – 191 of [**30**] for the precise statement to be used here. Similarly, we refer to pp. 97–98 of Greenberg [**37**] for the notion of an **ordinary** prime relative to a strictly compatible family $\{\rho_\ell\}$ of ℓ-adic representations of $\operatorname{Gal}(\overline{\mathbb{Q}}/\mathbb{Q})$. One feature of Greenberg's definition is that the ordinariness of p is a condition on ρ_p only (and in fact only on the restriction of ρ_p to a decomposition subgroup at p), not on ρ_ℓ for $\ell \neq p$. On the other hand, there is a second notion of ordinariness in the literature which is a condition on ρ_ℓ precisely for $\ell \neq p$. This second notion, like the first, can be elaborated in great generality, but we prefer to present it only for the type of strictly compatible family that is relevant here. Thus we take the dimension of ρ_ℓ to be two, and if $p \notin S$ (the exceptional set of the family) then we assume that $B_p(x)$ has the form $B_p(x) = 1 - a(p)x \pm px^2$ with $a(p) \in \mathbb{Z}$. In this setting we will say that p is **classically ordinary** if $p \nmid a(p)$. The **ordinariness conjecture** mentioned above is the first of the two assumptions in the following proposition:

Proposition 5.4. *Assume that classically ordinary primes are ordinary and that the Fontaine-Mazur conjecture holds. Then* $\mathcal{I} = \mathcal{S}_{\mathbb{Q},\mathbb{Q},1,2}$.

PROOF. Given a premotive $M \in S_{\mathbb{Q},\mathbb{Q},1,2}$ and a fully compatible family $\{\rho_\ell\}$ in M with coefficient field $\mathbb{Q}$, we must show that there is an elliptic curve E over $\mathbb{Q}$ such that $\{\rho_\ell\} \cong \{\rho_{E,\ell}^\vee\}$.

Let S be the exceptional set of M and let p denote an arbitrary prime not in S. The coefficients of $B_p(x)$ lie in $\mathbb{Z}$ by assumption, and the reciprocal roots of $B_p(x) = 0$ have complex absolute value $\sqrt{p}$ by (3.36). Consequently we have:

(i) $B_p(x) = 1 - a(p)x \pm px^2$ with $a(p) \in \mathbb{Z}$.
(ii) $|a(p)| < 2\sqrt{p}$.

If for all $p \notin S$ the sign $\pm$ in (i) is the plus sign then the desired conclusion follows from Theorem 1 of [**77**]. Hence it suffices to see that the occurrence of a minus sign in (i) leads to a contradiction.

Let ℓ and p be prime numbers with $p \notin S \cup \{\ell\}$, and let $\Phi_p \in \mathrm{Gal}(\overline{\mathbb{Q}}/\mathbb{Q})$ be an inverse Frobenius element at p. Since $\det \rho_\ell(\Phi_p) = \pm p$ in (i) and $\omega_\ell^{-1}(\Phi_p) = p$, the formula $\eta = (\det \rho_\ell)/\omega_\ell^{-1}$ defines a character $\eta : \mathrm{Gal}(\overline{\mathbb{Q}}/\mathbb{Q}) \to \{\pm 1\}$ independent of ℓ. Under the assumption that η is nontrivial the equation $\det \rho_\ell = \eta\omega_\ell^{-1}$ will lead to a contradiction.

View η as a primitive quadratic Dirichlet character, and write P for the set of prime numbers $p \notin S$ such that $\eta(p) = -1$. We claim that if $p \in P$ then $a(p) = 0$. To see this, write $1 - a(p)x - px^2 = (1 - \alpha x)(1 - \alpha' x)$ with $\alpha, \alpha' \in \mathbb{C}$. By (3.36) we can write $\alpha = e^{it}\sqrt{p}$ with $t \in \mathbb{R}$, and then $\alpha' = -e^{-it}\sqrt{p}$. Hence $a(p) = 2i(\sin t)\sqrt{p} \in i\mathbb{R}$. But $a(p) \in \mathbb{Z}$ by assumption, so $a(p) = 0$, as claimed.

Let $\mathcal{L}$ be the set of prime numbers ℓ such that there exist primes $p, q \in P$ with $p, q \neq \ell$ and

$$\left(\frac{p}{\ell}\right) = -\left(\frac{q}{\ell}\right). \tag{5.13}$$

We claim that $\mathcal{L}$ is a set of density 1. It suffices to see that the complement $\mathcal{M}$ of $\mathcal{L}$ has density 0. Now $\mathcal{M} = \mathcal{M}^+ \cup \mathcal{M}^-$, where $\mathcal{M}^\pm$ consists of the prime numbers ℓ such that

$$\left(\frac{p}{\ell}\right) = \pm 1 \tag{5.14}$$

for all $p \in P$ with $p \neq \ell$. But P is an infinite set (in fact a set of density 1/2), so it follows from the Chebotarev density theorem – or simply from Dirichlet's theorem – that $\mathcal{M}^+$ and $\mathcal{M}^-$ are sets of density 0. A small technical point here: In case the infinite set of conditions $\ell \neq p$ $(p \in P)$ appears to be a barrier to the application of the Chebotarev or Dirichlet theorems, choose an infinite subset P_0 of P of density 0, and let $\mathcal{M}_0^\pm$ be the set of primes $\ell \notin P_0$ such that (5.14) holds for all $p \in P_0$. If n is any positive integer and Q is any subset of P_0 of cardinality n then $\mathcal{M}_0^\pm$ is a subset of the set of primes $\ell \notin Q$ such that (5.14) holds for all $p \in Q$. But the latter set has density 2^{-n} and n is arbitrary, so we deduce that $\mathcal{M}_0^\pm$ has density 0.

As $\mathcal{M}^{\pm} \subset \mathcal{M}_0^{\pm} \cup \mathcal{P}_0$ and $\mathcal{M}_0^{\pm}$ and $\mathcal{P}_0$ are both sets of density 0, we do conclude that $\mathcal{M}^{\pm}$ has density 0 as well.

Now choose any odd prime ℓ_0 and view ρ_{ℓ_0} as a representation into $\mathrm{GL}_2(\mathbb{Z}_{\ell_0})$ (cf. Exercise 3.4). Let $\overline{\rho}_{\ell_0}$ denote the representation into $\mathrm{GL}_2(\mathbb{F}_{\ell_0})$ obtained from ρ_{ℓ_0} by reduction modulo ℓ_0. We consider the set Ξ of prime numbers $p \notin S \cup \{\ell_0\}$ such that $\overline{\rho}_{\ell_0}(\Phi_p) = 1$, the identity matrix in $\mathrm{GL}_2(\mathbb{F}_{\ell_0})$. By Chebotarev, Ξ has positive density, whence the same is true of $\Xi \cap \mathcal{L}$ because $\mathcal{L}$ has density one. Thus we can choose a prime $r \in \Xi \cap \mathcal{L}$ with $r \geqslant 5$. We claim that r is ordinary and that ρ_r is absolutely irreducible.

To see that ρ_r is ordinary, observe first of all that by construction, $\overline{\rho}_{\ell_0}(\Phi_r) = 1$, whence $\operatorname{tr} \rho_{\ell_0}(\Phi_r) \equiv 2 \bmod \ell_0$. But $\operatorname{tr} \rho_{\ell_0}(\Phi_r) = a(r)$ and ℓ_0 is odd, so we deduce that $a(r) \not\equiv 0 \bmod \ell_0$ and hence in particular that $a(r) \neq 0$. As $|a(r)| < 2\sqrt{r}$ it follows that $r \nmid a(r)$: in other words, r is classically ordinary and hence, under our hypotheses, ordinary.

To see that ρ_r is absolutely irreducible, we use the fact that $r \in \mathcal{L}$. Choose primes $p, q \in P$ with $p, q \neq r$ such that (5.13) holds with ℓ replaced by r. Then one of p and q is a square in $\mathbb{F}_r$ and the other is not. Also $a(p) = a(q) = 0$ because $p, q \in P$. Let us once again view ρ_r as a representation into $\mathrm{GL}_2(\mathbb{Z}_r)$, writing $\overline{\rho}_r$ for its reduction modulo r. Since the characteristic polynomials of $\rho_r(\Phi_p)$ and $\rho_r(\Phi_q)$ are $x^2 B_p(x^{-1})$ and $x^2 B_q(x^{-1})$ respectively, those of $\overline{\rho}_r(\Phi_p)$ and $\overline{\rho}_r(\Phi_q)$ are $x^2 - p$ and $x^2 - q$, where p and q are regarded as elements of $\mathbb{F}_r$. It follows that the eigenspaces of $\overline{\rho}_r(\Phi_p)$ and $\overline{\rho}_r(\Phi_q)$ over $\overline{\mathbb{F}}_r$ are pairs of distinct lines, but in one case the lines are rational over $\mathbb{F}_r$ and in the other case irrational. Hence a line in $\overline{\mathbb{F}}_r^2$ which is stable under $\overline{\rho}_r$ is both rational and irrational over $\mathbb{F}_r$ and therefore does not exist. Thus $\overline{\rho}_r$ is absolutely irreducible, and *a fortiori* so is ρ_r.

We are now in a position to apply the Fontaine-Mazur conjecture to ρ_r. Since r is ordinary it follows that ρ_r is semistable, and we have just seen that ρ_r is absolutely irreducible. Furthermore $\det \rho_r = \eta \omega_r^{-1}$, and consequently ρ_r is not a Tate twist of a two-dimensional Artin representation (for then $\det \rho_r$ would be a finite-order character times an *even* power of ω_r). Hence the Fontaine-Mazur conjecture implies that there is a primitive cusp form f of weight 2 and Nebentypus character η such that ρ_r is isomorphic to the semisimple representation $\rho_{f,r} : \mathrm{Gal}(\overline{\mathbb{Q}}/\mathbb{Q}) \to \mathrm{GL}_2(\mathbb{Q}_r)$ associated to f. But $a(p) = 0$ for $p \in P$, a set of density 1/2. It follows (Serre [**84**], p. 174, Corollaire 2) that f is a form of CM type. Thus the results of Ribet ([**73**], pp. 38 – 39, (4.4) and p. 40, (4.5)) imply that there is an imaginary quadratic field K and a primitive Hecke character χ of K of type $(1,0)$ such that $L(s, \chi) = L(s, f)$. It follows in particular that if p is a prime which splits in K and is relatively prime to $\mathfrak{f}(\chi)$ then $p \notin S$ and

$$a(p) = \chi(\mathfrak{p}) + \chi(\overline{\mathfrak{p}}), \tag{5.15}$$

where $\mathfrak{p}$ and $\overline{\mathfrak{p}}$ are the prime ideals of K above p.

Now let $\kappa = \mathrm{sign}_{K/\mathbb{Q}}$, viewed as a primitive quadratic Dirichlet character. Since f is a form of weight two we have $\eta(-1) = 1$, whereas $\kappa(-1) = -1$ since K is imaginary. But κ and by assumption also η are nontrivial characters, so if N is a common multiple of the conductors of η and κ then the sums $\sum_j \eta(j)$ and $\sum_j \kappa(j)$ are both 0, where j runs over invertible residue classes modulo N. Since $\eta(-1) = 1$ and $\kappa(-1) = -1$ it follows that there is also an invertible residue class c such that $\eta(c) = -1$ and $\kappa(c) = 1$. Hence by Dirichlet's theorem there are infinitely many prime numbers p relatively prime to $\mathfrak{f}(\chi)$ such that $\eta(p) = -1$ and $\kappa(p) = 1$, or in other words such that $p \in P$ and p splits in K: but these conditions imply respectively that $a(p) = 0$ and that (5.15) holds. Thus

$$\chi(\mathfrak{p}) = -\chi(\overline{\mathfrak{p}}). \tag{5.16}$$

Put $\mathfrak{F} = \mathbf{N}\mathfrak{f}(\chi)$ and choose n so that $\mathfrak{p}^n \in P_{\mathfrak{F}}$; then choose $\alpha \in K_{\mathfrak{F}}$ so that $\mathfrak{p}^n = \alpha\mathcal{O}$. By complex conjugation we also have $\overline{\mathfrak{p}}^n = \overline{\alpha}\mathcal{O}$. Both α and $\overline{\alpha}$ belong to $K_{\mathfrak{f}(\chi)}$, whence on raising both sides of (5.16) to the nth power we obtain $\alpha = (-1)^n\overline{\alpha}$. Passing to ideals we get $\mathfrak{p}^n = \overline{\mathfrak{p}}^n$ and consequently $\mathfrak{p} = \overline{\mathfrak{p}}$. Since p splits in K this is a contradiction. □

4. An open problem

Someday somebody may be able to formulate a conjecture which fully captures the notion that "motivic L-functions usually satisfy the minimalist trichotomy," but the hypotheses proposed in this lecture fall short of the mark, for two reasons.

First of all, no matter how $\mathcal{S}$ is chosen, a conjecture like (5.11) has the limitations inherent in any probabilistic statement: It cannot account for phenomena which hold for *all* or for *all but finitely many* members of a family. For example, we expect that every Dirichlet L-function satisfies the minimalist trichotomy, and Serre has conjectured more generally that the same is true for the L-function of any irreducible Artin representation of $\mathbb{Q}$ ([**35**], p. 324, Conjecture 8.24.1), but no such consequence can be deduced from (5.11). Neither can results like Theorem 1.2 or like Greenberg's theorem [**36**] on powers of Hecke characters of type $(1, 0)$.

Granting this objection, we come to a second issue, namely the choice of $\mathcal{S}$. We have portrayed $\mathcal{S}_{K,w,n}$ as the desired choice and $\mathcal{S}_{K,\mathbb{E},w,n}$ as the default choice, but neither is likely to be the right choice: A satisfactory conjecture would accommodate a wider variety of sets $\mathcal{S}$, subject only to some condition (still to be formulated) which plausibly ensures that (5.11) holds. The point is illustrated already by the two choices $\mathcal{S}_{K,w,n}$ and $\mathcal{S}_{K,\mathbb{E},w,n}$: The former is more inclusive, but without the latter we have no hope of recovering the minimalist conjecture for elliptic curves over $\mathbb{Q}$, and neither $\mathcal{S}_{K,w,n}$ nor $\mathcal{S}_{K,\mathbb{E},w,n}$ is likely to be suitable if one wants to conjecture that L-functions of essentially self-dual premotives satisfy the minimalist dichotomy with probability one. Underlying this last remark is the expectation that essentially self-dual premotives have density zero among all premotives, so that the validity of the minimalist trichotomy with probability one would say nothing about the minimalist dichotomy.

Unable to propose a compelling conjecture in general, we return to the case of elliptic curves over $\mathbb{Q}$, where we would still like to know that the minimalist conjecture is equivalent to the average rank hypothesis. If we grant (5.3) then the missing link, as already noted, is (5.6), which should hold for any of the counting functions ν mentioned earlier:

Problem 5. *Show that*

$$\vartheta_{\mathcal{E}^{\pm},\nu}(x) = \frac{1}{2}\vartheta_{\mathcal{E},\nu}(x) + O(x^{\gamma})$$

with a constant $\gamma < 5/6$.

5. Exercises

Exercise 5.1. It follows from work of Mestre [**64**] that if E is an elliptic curve over $\mathbb{Q}$ then

$$\operatorname{ord}_{s=1} L(s, E) \ll \log N(E), \tag{5.17}$$

where the implied constant is absolute. Using this estimate, prove that if one grants (5.3) and (5.6) then (5.7) and (5.8) are equivalent, as claimed.

Exercise 5.2. The case $w = 0$ of Proposition 5.3 implies that if K and $\mathbb{E}$ are fixed number fields, S a fixed finite set of prime ideals of K, and n a fixed positive integer, then there are only finitely many isomorphism classes of n-dimensional Artin representations of K which are unramified outside S and the infinite places of K and which are realizable over $\mathbb{E}$. Give a direct proof of this assertion.

Exercise 5.3. This problem refers to a conjecture of Serre already mentioned: Artin L-functions associated to irreducible Artin representations of $\mathbb{Q}$ satisfy the minimalist trichotomy.

(a) Why is the conjecture restricted to Artin representations of $\mathbb{Q}$? Why not make the same conjecture for Artin representations of arbitrary number fields? (Hint: Exercise 3.3.)

(b) (*Literature search.*) Some numerical evidence in support of Serre's conjecture can be found in the paper of Omar [**71**], where the Artin representations at issue are two-dimensional with image isomorphic to the quaternion group. Have other irreducible representations of dimension > 1 been investigated numerically?

Exercise 5.4. (*Reading.*) Using a counting function similar to (5.2), Bhargava and Shankar [**7**] have recently shown that the average rank of an elliptic curve over $\mathbb{Q}$ is $\leqslant 7/6$. While this result pertains to the Mordell-Weil rank rather than the "analytic rank" $\operatorname{ord}_{s=1} L(s, E)$, under the conjecture of Birch and Swinnerton-Dyer their work does give an upper bound for the quantity $r_\nu(\mathcal{E})$ in (5.5). What are they able to deduce about analytic ranks unconditionally?

Exercise 5.5. As we have already noted, the elliptic curve 389 A1 of [**22**] violates the minimalist dichotomy. The purpose of this problem is to give an example of a premotive M which is *not* essentially self-dual but which also violates the minimalist trichotomy.

(a) Put $K = \mathbb{Q}(\sqrt{-7})$ and $d = -118 - 18\sqrt{-7}$, and let η be the quadratic Hecke character of K associated to the extension $K(\sqrt{d})$ of K. View $A(7)$ as an elliptic curve over K and write E for the twist of $A(7)$ by η. Using [**38**], p. 82, verify that $y^2 = x^3 - 35x - 98$ is an equation for $A(7)$ over K, whence $dy^2 = x^3 - 35x - 98$ is an equation for E.

(b) Show that the point with coordinates $(x, y) = ((1 + \sqrt{-7})/2, 1)$ relative to the equation $dy^2 = x^3 - 35x - 98$ is a point of infinite order on E.

(c) Let χ be the unique element of $X(7)$, and put $\xi = \chi\eta$ and $M = M(\xi)$, so that $L(s, M) = L(s, \xi)$ and $L(s, E/K) = L(s, \xi)L(s, \bar{\xi})$. Using the Coates-Wiles theorem [**19**], deduce from (b) that $L(1, M) = 0$. But show that $L(s, \xi) \neq L(s, \bar{\xi})$, whence M is not self-dual.

Bibliography

[1] G. Anderson, D. Blasius, R. Coleman, and G. Zettler, *On representations of the Weil group with bounded conductor*, Forum Math. 6 (1994), 537 – 545.

[2] J. Armitage, *On a theorem of Hecke in number fields and function fields*, Invent. Math. 2 (1967), 238 – 246.

[3] J. Armitage, *Zeta functions with a zero at $s = 1/2$*, Invent. Math. 15 (1972), 199 – 205.

[4] A. Ash, D. Doud, and D. Pollack, *Galois representations with conjectural connections to arithmetic cohomology*, Duke Math. J. 112 (2002), 521 – 579.

[5] S. Baier and L. Zhao, *On the low-lying zeros of Hasse-Weil L-functions for elliptic curves*, Advances in Math. 219 (2008), 952 – 985.

[6] R. Balasubramanian and V. K. Murty, *Zeros of Dirichlet L-functions*, Ann. Sci. Ecole Norm. Sup. 25 (1992), 567 – 615.

[7] M. Bhargava and A. Shankar, *Ternary cubic forms having bounded invariants, and the existence of a positive proportion of elliptic curves having rank* 0. Online preprint: `http://arxiv.org/abs/1007.0052`.

[8] B. Bektemirov, B. Mazur, W. Stein, and M. Watkins, *Average ranks of elliptic curves: tension between data and conjecture*, Bull. Amer. Math. Soc. 44 (2007), 233 – 254

[9] B. J. Birch and N. M. Stephens, *The parity of the rank of the Mordell-Weil group*, Topology 5 (1966), 295 – 299.

[10] G. A. Boxer and P. Diao, *Effective nonvanishing of canonical Hecke L-functions*, Proc. Amer. Math. Soc., to appear.

[11] C. Breuil, B. Conrad, F. Diamond, and R. Taylor, *On the modularity of elliptic curves over $\mathbb{Q}$: wild 3-adic exercises*, J. Amer. Math. Soc. 14 (2001), 843-939.

[12] A. Brumer, *The average rank of elliptic curves I*, Invent. Math. 109 (1992), 445 – 472.

[13] A. Brumer and O. McGuinness, *The behavior of the Mordell-Weil group of elliptic curves*, Bull. Amer. Math. Soc. 23 (1990), 375 – 382.

[14] D. Bump, S. Friedberg, and J. Hoffstein, *Nonvanishing theorems for L-functions of modular forms and their derivatives*, Invent. Math. 102 (1990), 543 – 618.

[15] Ph. Cassou-Noguès, *Quelques théorèmes de base normale d'entiers*, Ann. Inst. Fourier 28 (1978), 1 – 33.

[16] T. Chinburg, B. Erez, G. Pappas, and M. J. Taylor, *ε-constants and the Galois structure of de Rham cohomology*, Ann. of Math. 146 (1997), 411 – 473.

[17] T. Chinburg, G. Pappas, and M. J. Taylor, *ε-constants and the Galois structure of de Rham cohomology II*, Jour. reine angew. Math. 519 (2000), 334 – 365.

[18] S. D. Chowla, *The Riemann Hypothesis and Hilbert's Tenth Problem (Mathematics and its Applications Vol. 4)*, Gordon and Breach Science Publishers (1965).

[19] J. Coates and A. Wiles, *On the conjecture of Birch and Swinnerton-Dyer*, Invent. Math. 39 (1977), 223 – 251.

[20] B. Conrey and K. Soundararajan, *Real zeros of quadratic Dirichlet L-functions*, Invent. Math. 150 (2002), 1 – 44.

[21] J. Cougnard, *Les travaux de A. Fröhlich, Ph. Cassou-Noguès et M.J. Taylor sur les bases normales*, Séminaire Bourbaki 25, 1982-1983, Exp. No. 598, 25 – 38.

[22] J. E. Cremona, *Algorithms for Modular Elliptic Curves*, Cambridge University Press (1980).

[23] P. Deligne, *Les constantes des équations fonctionelles des fonctions L*. In: *Modular Functions of One Variable, II*, Lect. Notes in Math. 349, Springer-Verlag (1973), 501–595.

[24] P. Deligne, *Valeurs de fonctions L et périodes d'intégrales*. In: *Automorphic Forms, Representations, and L-Functions*, Proc. Symp. Pure Math. Vol. 33 – Part 2, Amer. Math. Soc. (1979), 313 – 346

[25] P. Deligne and J.-P. Serre, *Formes modulaires de poids 1*, Ann. Sci. Ec. Norm. Sup. 7 (1974), 507 – 530.

[26] C. Deninger, *L-functions of mixed motives.* In: *Motives*, Proc. Symp. Pure Math. Vol. 55 – Part 1, Amer. Math. Soc. (1994), 517 – 525.

[27] W. Duke, J. Friedlander, and H. Iwaniec, *Bounds for automorphic L-functions. II*, Invent. Math. 115 (1994), 219 – 239.

[28] J. Ellenberg and A. Venkatesh *Reflection principles and bounds for class group torsion*, Int. Math. Res. Not. (2007).

[29] G. Faltings, *Endlichkeitssätze für abelsche Varietäten über Zahlkörper*, Invent. Math. 73 (1983), 349 – 366.

[30] J.-M. Fontaine and B. Mazur, *Geometric Galois representations.* In: *Elliptic Curves, Modular Forms, & Fermat's Last Theorem*, J. Coates and S.-T. Yau, eds. International Press, Cambridge MA (1997), pp. 190 – 227.

[31] A. Fröhlich, *Artin root numbers and normal integral bases for quaternion fields*, Invent. Math. 17 (1972), 143 – 166.

[32] A. Fröhlich, *Galois module structure* In: *Algebraic Number Fields*, Proceedings of the Durham Symposium, A. Fröhlich, ed., Academic Press (1977), 133 – 191.

[33] A. Fröhlich and J. Queyrut, *On the functional equation for the Artin L-function for characters of real representations*, Invent. Math. 20 (1973), 125–138.

[34] D. Goldfeld, *Conjectures on elliptic curves over quadratic fields.* In: *Number Theory, Carbondale, 1979*, Springer, Lecture Notes in Math. 751 (1979), 108 – 118.

[35] D. Goss, *Basic Structures of Function Field Arithmetic*, Ergebnisse der Math. 35, Springer (1966).

[36] R. Greenberg, *On the critical values of Hecke L-functions for imaginary quadratic fields*, Invent. Math. 79 (1985), 79 – 94.

[37] R. Greenberg, *Iwasawa theory for p-adic representations.* In: *Algebraic Number Theory – in honor of K. Iwasawa*, Advanced Studies in Pure Mathematics 17 (1989), pp. 97 – 137.

[38] B. H. Gross, *Arithmetic on Elliptic Curves with Complex Multiplication*, SLN 776, Springer-Verlag (1980).

[39] B. H. Gross and D. B. Zagier, *Heegner points and derivatives of L-series*, Invent. Math. 84 (1986), 225 – 320.

[40] D. R. Heath-Brown, *The average rank of elliptic curves*, Duke Math. J. 122 (2004), 225 – 320.

[41] D. R. Heath-Brown and S. J. Patterson, *The distribution of Kummer sums at prime arguments*, J. reine angew. Math. 310 (1979), 111 – 130.

[42] E. Hecke, *Eine neue Art von Zetafunktionen und ihre Beziehungen zur Verteilung der Primzahlen. Zweite Mitteilung*, Math. Z. 6 (1920), 11 – 51.

[43] E. Hecke, *Vorlesungen über die Theorie der algebraischen Zahlen*, Chelsea Publishing Company, 2nd ed. (1970).

[44] D. Husemöller, *Elliptic Curves*, Springer GTM 111 (1987).

[45] H. Iwaniec, W. Luo, and P. Sarnak, *Low lying zeros of families of L-functions*, Publ. Math. Inst. Hautes Études Sci. 91 (2000), 55 – 131.

[46] H. Iwaniec and P. Sarnak, *Dirichlet L-functions at the central point.* In: *Number Theory in Progress* vol. 2, de Gruyter, Berlin (1999), 941 – 952.

[47] M. Jutila, *On the mean value of $L(1/2, \chi)$ for real characters*, Analysis 1 (1981), 149 – 161.

[48] S. Katayama, *On fundamental units of real quadratic fields with norm +1*, Proc. Japan Acad. 68 Ser. A (1992), 18 – 20.

[49] C. Khare, *F-split Galois representations are potentially abelian*, Proc. Amer. Math. Soc. 131 (2003), 3021 – 3023.

[50] C. Khare and J.-P. Wintenberger, *Serre's modularity conjecture (I)*, Invent. Math. 178 (2009), 485 – 504.

[51] C. Khare and J.-P. Wintenberger, *Serre's modularity conjecture (II)*, Invent. Math. 178 (2009), 505 – 586.

[52] M. Kisin, *Modularity of 2-adic Barsotti-Tate representations*, Invent. Math. 178 (2009), 587 – 634.

[53] V. A. Kolyvagin, *Finiteness of $E(\mathbf{Q})$ and $Ш(E, \mathbf{Q})$ for a subclass of Weil curves*, Izv. Akad. Nauk SSSR Ser. Mat. 52 (1988), 522 – 540; Math. USSR Izv. 32 (1989), 523 – 541.

[54] V. A. Kolyvagin and D. Yu. Logachev, *Finiteness of the Shafarevich-Tate group and the group of rational points for some modular abelian varieties*; Algebra i Analiz 1 (1989), 171 – 196; Leningrad Math. J. 1 (1990) 1229 – 1253.

[55] S. Lang, *Algebraic Number Theory*, 2nd. ed., Springer GTM 110 (1994).

[56] S. Lang and H. Trotter, *Frobenius Distributions in GL_2-Extensions*, Springer-Verlag, LNM vol. 504 (1976).

[57] R. P. Langlands, *Base Change for* GL(2), Princeton Univ. Press, Ann. of Math. Studies 96 (1980).

[58] Chunlei Liu and Lanju Xu *The vanishing order of certain Hecke L-functions of imaginary quadratic fields*, J. of Number Thy. 108 (2004), 76-89.

[59] R. Masri, *Quantitative nonvanishing of L-series associated to canonical Hecke characters*, IMRN (2007).

[60] R. Masri, *Asymptotics for sums of central values of canonical Hecke L-series*, IMRN (2007).

[61] R. Masri and T. H. Yang, *Average values of Hecke L-functions via Galois suborbits of Heegner points*.

[62] C. R. Matthews, *Gauss sums and elliptic functions: I. The Kummer sum*, Invent. Math. 52 (1979), 163 – 185.

[63] C. R. Matthews, *Gauss sums and elliptic functions: II. The quartic sum*, Invent. Math. 54 (1979), 23 – 52.

[64] J.-F. Mestre, *Formules explicites et minorations de conducteurs de variétés algébriques*, Compositio Math. 58 (1986), 209 – 232.

[65] S. D. Miller and T. H. Yang, *Non-vanishing of the central derivative of canonical Hecke L-functions*, Math. Res. Lett. 7 (2000), 263 – 277.

[66] H. Montgomery and D. E. Rohrlich, *On the L-functions of canonical Hecke characters of imaginary quadratic fields. II* Duke Math. J. 49 (1982), 937-942.

[67] H. Moon, *Finiteness results on certain mod p Galois representations*, J. Number Theory 84 (2000), 156 – 165.

[68] H. Moon and Y. Taguchi, *Mod p Galois representations of solvable image*, Proc. Amer. Math. Soc. 129 (2001), 2529 – 2534.

[69] M. R. Murty and V. K. Murty, *Non-vanishing of L-functions and applications.* In: Progress in Mathematics 157, Birkhäuser (1997).

[70] M. R. Murty and V. K. Murty, *Mean values of derivatives of modular L-series*, Ann. of Math. 133 (1991), 447 – 475.

[71] S. Omar, *On Artin L-functions for octic quaternion fields*, Experim. Math. 10 (2001), 237 – 245.

[72] S. Omar, *Non-vanishing of Dirichlet L-functions at the central point.* In: *Algorithmic Number Theory, 8th International Symposium, ANTS VIII*, A. J. van der Poorten and A. Stein, eds., Springer LNCS 5011 (2008), 443 – 453.

[73] K. A. Ribet, *Galois representations attached to eigenforms with Nebentypus.* In: *Modular Functions of One Variable, V*, Lect. Notes in Math. 601, Springer-Verlag (1977), 17 – 52.

[74] F. Rodriguez Villegas, *Square root formulas for central values of Hecke L-series. II* Duke Math. J. 72 (1993), 431 – 440.

[75] D. E. Rohrlich, *On the L-functions of canonical Hecke characters of imaginary quadratic fields*, Duke Math. J. 47 (1980), 547 – 557.

[76] D. E. Rohrlich, *Root numbers of Hecke L-functions of CM fields*, Am. J. Math. 104 (1982), 517 – 543.

[77] D. E. Rohrlich, *Compatible families of elliptic type*, Acta Arithmetica, 142 (2010), 1 – 9.

[78] K. Rubin, *Elliptic curves with complex multiplication and the conjecture of Birch and Swinnerton-Dyer*, Invent. Math. 64 (1981), 455 – 470.

[79] K. Rubin and A. Silverberg, *Ranks of elliptic curves*, Bull. Amer. Math. Soc. 39 (2002), 455 – 474.

[80] P. Schneider, *Introduction to the Beilinson conjectures.* In: *Beilinson's Conjectures on Special Values of L-Functions*, Academic Press (1988), 1 – 33.

[81] J.-P. Serre, *Facteurs locaux de fonctions zêta des variétés algébriques (définitions et conjectures).* In: Séminaire Delange-Poitou-Pisot, 1969/70, no. 19.

[82] J.-P. Serre, *Abelian ℓ-Adic Representations and Elliptic Curves*, W. A. Benjamin, 1968.

[83] J.-P. Serre, *Local Fields*, Springer GTM 67 (1979).

[84] J.-P. Serre, *Quelques applications du théorème de densité de Chebotarev*, Publ. Math. IHES 54 (1981), 123 – 201.

[85] J.-P. Serre, *Sur les représentations modulaires de degré 2 de* $\mathrm{Gal}(\overline{\mathbf{Q}}/\mathbf{Q})$, Duke Math. J. 54 (1987), 179 – 230.

[86] J.-P. Serre and J. Tate, *Good reduction of abelian varieties*, Ann. of Math. 88 (1968), 492 – 517.

[87] G. Shimura, *The special values of the zeta functions associated with cusp forms*, Comm. Pure Appl. Math. 29 (1976), 783 – 804.

[88] G. Shimura, *On the periods of modular forms*, Math. Ann. 229 (1977), 211 – 221.

[89] V. P. Snaith, *Explicit Brauer Induction*, Cambridge Studies in Advanced Math. 40, Cambridge Univ. Press (1994).

[90] K. Soundararajan, *Nonvanishing of quadratic Dirichlet L-functions at* $s = 1/2$, Ann. of Math. 152 (2000), 447 – 488.

[91] W. Stein and M. Watkins, *A database of elliptic curves – first report.* In: *Algorithmic Number Theory (Sydney, 2002)*, Springer Lecture Notes in Comput. Sci. 2369, Springer (2002), 267 – 275.

[92] J. Tate, *Fourier analysis in number fields and Hecke's zeta functions.* In: *Algebraic Number Theory*, J. W. S. Cassels and A. Fröhlich, eds., Academic Press (1967), 305 – 347.

[93] J. Tate, *Local constants* (prepared in collaboration with C. J. Bushnell and M. J. Taylor). In: *Algebraic Number Fields*, Proceedings of the Durham Symposium, A. Fröhlich, ed., Academic Press (1977), 89 – 131.

[94] J. Tate, *Number theoretic background.* In: *Automorphic Forms, Representations, and L-Functions*, Proc. Symp. Pure Math. Vol. 33 – Part 2, Amer. Math. Society (1979), 3 – 26.

[95] M. J. Taylor, *On Fröhlich's conjecture for rings of integers of tame extensions*, Invent. Math. 63 (1981), 41 – 79.

[96] R. Taylor and A. Wiles, *Ring-theoretic properties of certain Hecke algebras*, Ann. of Math. 141 (1995), 553 – 572.

[97] J. Tunnell, *Artin's conjecture for representations of octahedral type*, Bull. AMS 5 (1981), 173 – 175.

[98] A. B. Venkov, *Selberg's trace formula for the Hecke operator generated by an involution, and the eigenvalues of the Laplace-Beltrami operator on the fundamental domain of the modular group* $\mathrm{PSL}(2,\mathbb{Z})$, Math. USSR Izvestija 12 (1978), 448 – 462.

[99] M. Watkins, *Some heuristics about elliptic curves*, Experimental Math. 17 (2008), 105 – 125.

[100] A. Wiles, *Modular elliptic curves and Fermat's last theorem*, Ann. of Math. 141 (1995), 443 – 551.

[101] T. H. Yang, *On CM abelian varietiers over imaginary quadratic fields*, Math. Ann. 329 (2004), 87 – 117.

[102] T. H. Yang, *Nonvanishing of central Hecke L-values and rank of certain elliptic curves*, Compositio Math. 117 (1999), 337-359.

[103] M. P. Young, *Low-lying zeros of families of elliptic curves*, J. Amer. Math. Soc. 19 (2005), 205 – 250.

[104] M. P. Young, *Analytic number theory and ranks of elliptic curves.* In: *Ranks of Elliptic Curves and Random Matrix Theory*, London Math. Soc. Lecture Note Ser. 341, Cambridge University Press (2007), 71 – 92.

[105] S. W. Zhang, *Heegner cycles and derivatives of L-series*, Invent. Math. 130 (19917), 99 – 152.

Euler Systems and Kolyvagin Systems

Karl Rubin

IAS/Park City Mathematics Series
Volume 18, 2009

Euler Systems and Kolyvagin Systems

Karl Rubin

Introduction

In these lectures we will give an introduction to the theory of Kolyvagin systems and Euler systems. A Kolyvagin system is a collection of cohomology classes with special properties. Our aim here is to show how Kolyvagin systems can be used to bound the size of Selmer groups, and to give an idea of when one expects Kolyvagin systems to exist. An Euler system is a (different kind of) collection of cohomology classes, that can be used to construct a Kolyvagin system.

Ideally one would like to relate a Kolyvagin system to special values of an L-function, so that the corresponding Selmer bound would involve an appropriate L value. From this point of view, we will view a Kolyvagin system as an algebraic incarnation of a special value of an L-function, while an Euler system plays the role of the (p-adic) L-function itself.

The work described in these lectures (after the background in Lecture 1) is joint with Barry Mazur, and we follow closely the presentation in [**MR1**]. Many proofs are included here, some are left as exercises, and for the rest references (mainly to [**MR1**] and [**R2**]) are given. Everything in these lectures builds on Kolyvagin's groundbreaking work [**Ko**].

The first lecture gives a quick introduction to the facts that we will need from Galois cohomology, in order to define Selmer structures and Selmer groups and to state the local and global dualities that play a crucial role in everything that follows. Lecture 2 introduces Kolyvagin systems for finite Galois modules, and gives the main applications to describing the sizes of Selmer groups. In Lecture 3 we adapt these results to Selmer groups attached to p-adic representations, and go into detail in the special cases where the p-adic representation is the Tate module of roots of unity, or of an elliptic curve.

Lecture 4 introduces Euler systems, and shows how an Euler system gives rise (via Kolyvagin's derivative construction) to a Kolyvagin system. The final lecture completes the application to ideal class groups and to Selmer groups of elliptic curves, and ends with some general speculation.

Department of Mathematics, University of California, Irvine, CA 92697-3875, USA
E-mail address: krubin@math.uci.edu
I would like to thank Sunil Chetty and Zev Klagsbrun for their comments and corrections on an early version of these notes. This material is based upon work supported by the National Science Foundation under grant DMS-0757807.

LECTURE 1

Galois cohomology

We begin with a *very* quick and selective introduction to the facts from group cohomology and Galois cohomology that will be needed for the following lectures. For basic details see [**AW, Gr, Se2**], and for some of the more advanced results see [**Se1, Mi**]. For another quick overview see the lectures of Tate [**Ta3**] from PCMI 1999. We omit most proofs, although accessible ones are often given as exercises.

1.1. G-modules.

Suppose G is a group. A *G-module* is an abelian group A with an action of G on A that respects the group operation on A. That is, there is a map

$$G \times A \longrightarrow A$$

such that, if we let ga (or sometimes a^g) denote the image of (g, a) in A, then

$$(gh)a = g(ha), \qquad g(a+b) = ga + gb$$

for $g, h \in G$ and $a, b \in A$. Define the fixed subgroup

$$A^G := \{a \in A : ga = a \text{ for every } g \in G\}.$$

Example 1.1.1. If X is an abelian group we can view X as a G-module with trivial G action, and then $X^G = X$.

Example 1.1.2. If F/K is a Galois extension of fields and $G = \mathrm{Gal}(F/K)$, then F and $F^\times$ are G-modules. More generally, if $\mathcal{H}$ is an algebraic group defined over K, then the group of F-points $\mathcal{H}(F)$ is a G-module and $\mathcal{H}(F)^G = \mathcal{H}(K)$.

Example 1.1.3. If A and B are G-modules and $\varphi : A \to B$ is a group homomorphism, we define a new group homomorphism $g\varphi$ for $g \in G$ by $(g\varphi)(a) = g(\varphi(g^{-1}a))$. This makes $\mathrm{Hom}(A, B)$ into a G-module, and $\mathrm{Hom}(A, B)^G$ is the group of G-module homomorphisms from A to B.

We say that a G module A is *co-induced* if $A \cong \mathrm{Hom}(\mathbf{Z}[G], X)$ for some abelian group X.

Exercise 1.1.4. Suppose A is a G-module, and A_0 is the G-module whose underlying abelian group is A, but whose G action is trivial. Show that the map $a \mapsto \varphi_a$, where $\varphi_a(g) = g^{-1}a$, is an injection from A to the co-induced module $\mathrm{Hom}(\mathbf{Z}[G], A_0)$.

1.2. Characterization of the cohomology groups.

For this section suppose that the group G is finite. For every G-module A, there are abelian (cohomology) groups $H^i(G, A)$ for $i \geq 0$. For an explicit definition using cocycles and coboundaries, see [**AW, Se2**]. We will omit the definition, and just make use of the following properties.

Theorem 1.2.1. *There is a unique collection of functors $H^i(G, \cdot)$ from G-modules to abelian groups, for $i \geq 0$, satisfying the following properties:*

(1) *$H^0(G, A) = A^G$ for every G-module A.*
(2) *If $0 \to A \to B \to C \to 0$ is a short exact sequence of G-modules, then there is a (functorial) long exact sequence*

$$0 \to H^0(G, A) \to H^0(G, B) \to H^0(G, C) \to H^1(G, A) \to H^1(G, B) \to \cdots$$
$$\cdots \to H^i(G, A) \to H^i(G, B) \to H^i(G, C) \to H^{i+1}(G, A) \to \cdots$$

(3) *If A is co-induced, then $H^i(G, A) = 0$ for all $i \geq 1$.*

Exercise 1.2.2. Show that the three properties above determine the cohomology groups $H^i(G, A)$ uniquely (assuming they exist). Hint: use induction and the fact (Exercise 1.1.4) that for every G-module A there is a short exact sequence $0 \to A \to B \to C \to 0$ with B co-induced.

In these lectures we will only make use of $H^i(G, A)$ for $i \leq 2$ (and mostly $i \leq 1$). When $i = 0$ the groups are described explicitly by condition (1) of Theorem 1.2.1; when $i = 1$ we have the following explicit description.

Define $C^1(G, A)$ (the 1-cochains) to be the group of (set) maps from G to A. Define subgroups of cocycles and coboundaries $B^1(G, A) \subset Z^1(G, A) \subset C^1(G, A)$ by

$$Z^1(G, A) := \{f \in C^1(G, A) : f(gh) = f(g) + g(f(h))\}$$
$$B^1(G, A) := \{f \in C^1(G, A) : \text{for some } a \in A,\ f(g) = ga - a \text{ for every } g \in G\}.$$

Exercise 1.2.3. Show that $B^1(G, A) \subset Z^1(G, A)$.

Proposition 1.2.4. $H^1(G, A) = Z^1(G, A)/B^1(G, A)$.

There is a similar definition of $H^i(G, A)$ for every i, where the cochains $C^i(G, A)$ (are) set maps from G^i to A, and $B^i(G, A) \subset Z^i(G, A) \subset C^i(G, A)$ are defined appropriately.

Example 1.2.5. Suppose that G acts trivially on A. Then $Z^1(G, A) = \text{Hom}(G, A)$ and $B^1(G, A) = 0$, so in this case we conclude from Proposition 1.2.4 that

$$H^1(G, A) = \text{Hom}(G, A).$$

Exercise 1.2.6. Suppose that G is a finite cyclic group. For every G-module A, let $A_N := \{a \in A : \sum_{g \in G} ga = 0\}$ (here "N" stands for "norm"; A_N is the kernel of the norm map $a \mapsto \sum_{g \in G} ga$).

Show that if g is a generator of G, then the map $f \mapsto f(g)$ is an injective homomorphism from $Z^1(G, A)$ to A with image A_N. Deduce that in this case

$$H^1(G, A) \cong A_N/(g - 1)A.$$

(Warning: note that this isomorphism depends on the choice of generator g.)

Exercise 1.2.7. Suppose $0 \to A \to B \to C \to 0$ is an exact sequence of G-modules, and suppose $c \in C^G$. Fix an element $b \in B$ that maps to c.

Show that the map $g \mapsto gb - b$ defines a 1-cocycle $f_c \in Z^1(G, A)$ (that depends on the choice of b). Show that $c \mapsto f_c$ induces a well-defined homomorphism $H^0(G, C) \to H^1(G, A)$ and check that with this homomorphism, the beginning of the long exact sequence of Theorem 1.2.1(2) is in fact exact.

1.3. Continuous cohomology.

Now suppose that G is a *profinite group* (see for example [**Gr**]), i.e., there is an isomorphism

$$G = \varprojlim G/U \tag{1.1}$$

where U runs over open subgroups of G of finite index. This isomorphism gives G natural topology, where we view each finite quotient G/U as a discrete topological space, so the product $\prod_U (G/U)$ is a compact group with the product topology, and then (1.1) identifies G with a closed (and hence compact) subset of $\prod_U (G/U)$.

Note that a finite group G is profinite, with the discrete topology.

If A is a G-module, we will view A as a topological group with the discrete topology, and we call A a continuous G-module if the action of G on A (i.e., the map $G \times A \to A$) is continuous.

Exercise 1.3.1. Suppose A is a G-module. Show that the following are equivalent:
(1) A is a continuous G-module,
(2) for every $a \in A$, the stabilizer of a in G is open,
(3) $A = \cup A^U$, union over open subgroups $U \subset G$.

Example 1.3.2. If F/K is an infinite Galois extension of fields, and we put $G := \mathrm{Gal}(F/K)$, then there is a natural isomorphism

$$G = \varprojlim \mathrm{Gal}(L/K)$$

inverse limit over finite Galois extensions L of K in F. Thus G is a profinite group. If $\mathcal{H}$ is an algebraic group over K as in Example 1.1.2, then

$$\mathcal{H}(F) = \cup \mathcal{H}(L) = \cup \mathcal{H}(F)^{\mathrm{Gal}(L/K)},$$

union over finite extensions L of K in F. Therefore G acts continuously on $\mathcal{H}(F)$ by Exercise 1.3.1.

It follows (for example) that when $F = K^{\mathrm{sep}}$ is a separable closure of K, the following G-modules are continuous: K^{sep}, $(K^{\mathrm{sep}})^\times$, $\boldsymbol{\mu}_{p^\infty}$ (the p-power roots of unity in K^{sep}), $E(K^{\mathrm{sep}})$ for an elliptic curve E, and $E[p^\infty]$ (the p-power torsion in $E(K^{\mathrm{sep}})$).

If A is a continuous G-module, we can define *continuous* cohomology groups $H^i(G, A)$, defined similarly to the case of finite groups G but with *continuous* cochains (that is, $C^i(G, A)$ consists of continuous maps from G^i to A). Theorem 1.2.1(1) and (2) also hold for continuous cohomology groups.

If G is finite, then all relevant maps are continuous and the continuous cohomology groups agree with the cohomology groups described in §1.2.

In the next section we will see (Proposition 1.4.7) how to describe the continuous cohomology groups in terms of the cohomology of finite groups.

Until further notice we will always assume that the group G is profinite, and G-module will mean continuous G-module, a discrete abelian group with a continuous action of G.

1.4. Change of group.

Suppose H is a closed subgroup of a profinite group G, and A is a G-module. Then A is an H-module, and $A^G \subset A^H$. For every i there is a restriction map on cochains

$\mathrm{Res} : C^i(G,A) \to C^i(H,A)$, and these maps induce *restriction maps*

$$\mathrm{Res} : H^i(G,A) \to H^i(H,A).$$

If $[G : H]$ is finite, then there is also a norm map $A^H \to A^G$, defined by $a \mapsto \sum_g ga$, summing over a set of left coset representatives of G/H. This map extends in a less obvious way to a *corestriction map*

$$\mathrm{Cor} : H^i(H,A) \to H^i(G,A)$$

for every i.

Proposition 1.4.1. *If $[G : H]$ is finite, then* $\mathrm{Cor} \circ \mathrm{Res} : H^i(G,A) \to H^i(G,A)$ *is multiplication by $[G : H]$.*

Corollary 1.4.2. *If G is finite, then for every G-module A and every $i \geq 1$, we have*

$$|G| \cdot H^i(G,A) = 0.$$

Proof. Take $H = \{1\}$ in Proposition 1.4.1. □

Exercise 1.4.3. Suppose $m \in \mathbf{Z}$. Show that if $mA = 0$, then $mH^i(G,A) = 0$ for every i. Show that if $m : A \to A$ is an isomorphism, then $mH^i(G,A) = H^i(G,A)$ for every i. Deduce that if G is finite and $|G| : A \to A$ is an isomorphism, then $H^i(G,A) = 0$ for $i \geq 1$. In particular if G and A are both finite and $\gcd(|A|,|G|) = 1$, then $H^i(G,A) = 0$ for $i \geq 1$.

Exercise 1.4.4. Suppose G is finite. Using Exercise 1.1.4, show that for $i \geq 1$, $H^i(G,A)$ can be expressed in terms of $H^{i-1}(G,C)$ for some G-module C. Use this fact, the norm map on H^0, and induction to define the corestriction map on H^i. Using this definition, prove Proposition 1.4.1.

Now suppose that H is a closed normal subgroup of G. Then A^H is a G/H-module, and for every i there is an inflation map on cochains $\mathrm{Inf} : C^i(G/H, A^H) \to C^i(G,A)$. These maps induce *inflation maps*

$$\mathrm{Inf} : H^i(G/H, A^H) \to H^i(G,A).$$

Theorem 1.4.5. *If H is a normal subgroup of G and A is a G-module, then there is an natural exact sequence*

$$\begin{aligned} 0 \to H^1(G/H, A^H) \xrightarrow{\mathrm{Inf}} H^1(G,A) \xrightarrow{\mathrm{Res}} H^1(H,A)^{G/H} \\ \to H^2(G/H, A^H) \to H^2(G,A). \end{aligned}$$

Corollary 1.4.6. *Suppose H acts trivially on A, and*

$$H^1(G/H,A) = H^2(G/H,A) = 0.$$

Then $H^1(G,A) \cong \mathrm{Hom}(H,A)^{G/H}$.

Proposition 1.4.7. *If A is a G-module, then*

$$H^i(G,A) = \varinjlim H^i(G/U, A^U)$$

direct limit over open subgroups $U \subset G$, with respect to the inflation maps.

Exercise 1.4.8. Prove Proposition 1.4.7 when $i = 1$, using the description (Proposition 1.2.4) of $H^1(G,A)$ in terms of cocycles.

Exercise 1.4.9. Use Proposition 1.4.7 to show that if A is a G-module, then $H^i(G, A)$ is a torsion group for $i \geq 1$.

Proposition 1.4.10. *Suppose* $G \cong \hat{\mathbf{Z}} := \varprojlim \mathbf{Z}/n\mathbf{Z}$, *and* A *is a torsion* G*-module.*

(1) *If* γ *is a topological generator of* G, *then evaluation of cocycles at* γ *induces an isomorphism* $H^1(G, A) \cong A/(\gamma - 1)A$.
(2) $H^i(G, A) = 0$ *for* $i \geq 2$.

PROOF. See [**Se2**, §XIII.1]. Assertion (1) is the following exercise. □

Exercise 1.4.11. Prove Proposition 1.4.10(1) using Exercise 1.2.6 and Proposition 1.4.7.

From now on, if F/K is a Galois extension and A is a $\mathrm{Gal}(F/K)$-module, we will write $H^i(F/K, A)$ in place of $H^i(\mathrm{Gal}(F/K), A)$, and when F is a separable closure K^{sep} of K we write simply $G_K := \mathrm{Gal}(K^{\mathrm{sep}}/K)$ and $H^i(K, A)$ in place of $H^i(K^{\mathrm{sep}}/K, A) = H^i(G_K, A)$.

Definition 1.4.12. Suppose K is a nonarchimedean local field K of characteristic zero, i.e., a finite extension of some $\mathbf{Q}_p$. Let $I_K \subset G_K$ denote the inertia group, $K^{\mathrm{ur}} := \bar{K}^{I_K}$ the maximal unramified extension of K, and $\varphi \in \mathrm{Gal}(K^{\mathrm{ur}}/K)$ the Frobenius automorphism. Then $\mathrm{Gal}(K^{\mathrm{ur}}/K) = G_K/I_K \cong \hat{\mathbf{Z}}$, generated by φ.

If A is a G_K-module, define the *unramified cohomology group* $H^1_{\mathrm{u}}(K, A) \subset H^1(K, A)$ by

$$H^1_{\mathrm{u}}(K, A) := \ker\big[H^1(K, A) \xrightarrow{\mathrm{Res}} H^1(K^{\mathrm{ur}}, A)\big].$$

The inflation-restriction exact sequence (Theorem 1.4.5) shows that

$$H^1_{\mathrm{u}}(K, A) = H^1(K^{\mathrm{ur}}/K, A^{I_K}). \tag{1.2}$$

If A is a G_K-module, we will say that A is *unramified* if I_K acts trivially on A.

Proposition 1.4.13. *If* A *is a finite unramified* G_K*-module, then:*

(1) $H^1_{\mathrm{u}}(K, A) = H^1(K^{\mathrm{ur}}/K, A) = A/(\varphi - 1)A$,
(2) $H^1(K, A)/H^1_{\mathrm{u}}(K, A) \cong \mathrm{Hom}(I_K, A)^{G_K}$.

PROOF. This follows from Theorem 1.4.5 and Proposition 1.4.10. □

1.5. Selmer groups.

Suppose for this section that K is a number field, and A is a G_K-module. For every place v of K, we can view the decomposition group G_{K_v} as a subgroup of G_K, so we have restriction maps

$$\mathrm{Res}_v : H^i(K, A) \longrightarrow H^i(K_v, A).$$

Definition 1.5.1. A *Selmer structure* $\mathcal{F}$ for A is a collection of distinguished subgroups

$$H^1_{\mathcal{F}}(K_v, A) \subset H^1(K_v, A)$$

for every place v of K, such that $H^1_{\mathcal{F}}(K_v, A) = H^1_{\mathrm{u}}(K_v, A)$ for all but finitely many v.

If $\mathcal{F}$ is a Selmer structure for A, we define the *Selmer group* $H^1_{\mathcal{F}}(K, A)$ by

$$H^1_{\mathcal{F}}(K, A) := \ker\big[H^1(K, A) \xrightarrow{\oplus \mathrm{Res}_v} \bigoplus_v (H^1(K_v, A)/H^1_{\mathcal{F}}(K_v, A))\big].$$

In other words, $H^1_{\mathcal{F}}(K,A)$ is the subgroup of all classes in $H^1(K,A)$ whose localization lies in $H^1_{\mathcal{F}}(K_v,A)$ for every v.

Exercise 1.5.2. Suppose $\mathcal{F}$ is a Selmer structure, and suppose Σ is a finite set of places of K containing all archimedean places, all places where A is ramified, and all v such that $H^1_{\mathcal{F}}(K_v,A) \neq H^1_{\mathrm{u}}(K_v,A)$. Let K_Σ be the maximal extension of K unramified outside of Σ. Show that

$$H^1_{\mathcal{F}}(K,A) = \ker\big[H^1(K_\Sigma/K,A) \xrightarrow{\oplus \mathrm{Res}_v} \bigoplus_{v\in\Sigma} (H^1(K_v,A)/H^1_{\mathcal{F}}(K_v,A))\big]$$

Proposition 1.5.3. *If A is finite and $\mathcal{F}$ is a Selmer structure for A, then $H^1_{\mathcal{F}}(K,A)$ is finite.*

PROOF. With notation as in Exercise 1.5.2, a standard result (for example [**Mi**, Corollary I.4.15]) shows that $H^1(K_\Sigma/K,A)$ is finite. Thus $H^1_{\mathcal{F}}(K,A)$ is finite by Exercise 1.5.2. □

In what follows, Selmer groups will be arithmetically interesting objects (ideal class groups, Selmer groups of elliptic curves, ...) and their orders should be related to values of L-functions. We will see examples of Selmer groups in the next section.

Exercise 1.5.4. If C is a G_K-quotient of B, show that a Selmer structure $\mathcal{F}$ for B induces a Selmer structure for C (which we also denote by $\mathcal{F}$), where we define $H^1_{\mathcal{F}}(K_v,C)$ to be the image of $H^1_{\mathcal{F}}(K_v,B)$ under the canonical map $H^1(K_v,B) \to H^1(K_v,C)$.

Similarly, show that if A is a G_K-submodule of B, then a Selmer structure $\mathcal{F}$ for B induces one for A, where $H^1_{\mathcal{F}}(K_v,A)$ is defined to be the inverse image of $H^1_{\mathcal{F}}(K_v,B)$ under the canonical map $H^1(K_v,A) \to H^1(K_v,B)$.

1.6. Kummer theory.

Proposition 1.6.1. *Suppose F/K is a Galois extension. Then*

$$H^i(F/K,F) = \begin{cases} K & \text{if } i=0 \\ 0 & \text{if } i>0, \end{cases} \qquad H^i(F/K,F^\times) = \begin{cases} K^\times & \text{if } i=0 \\ 0 & \text{if } i=1. \end{cases}$$

Exercise 1.6.2. Use Proposition 1.4.7 to reduce the proof of Proposition 1.6.1 to the case where F/K is finite.

Theorem 1.6.3. *For every $m>0$ prime to the characteristic of K, there is a natural isomorphism*

$$K^\times/(K^\times)^m \cong H^1(K,\boldsymbol{\mu}_m).$$

PROOF. The long exact cohomology sequence coming from the short exact sequence

$$0 \longrightarrow \boldsymbol{\mu}_m \longrightarrow (K^{\mathrm{sep}})^\times \xrightarrow{m} (K^{\mathrm{sep}})^\times \longrightarrow 0$$

begins

$$0 \longrightarrow \boldsymbol{\mu}_m(K) \longrightarrow K^\times \xrightarrow{m} K^\times \longrightarrow H^1(K,\boldsymbol{\mu}_m) \longrightarrow H^1(K,(K^{\mathrm{sep}})^\times).$$

Now the theorem follows from Proposition 1.6.1. □

Exercise 1.6.4. Show using the description from Exercise 1.2.7 that the isomorphism of Theorem 1.6.3 is given by sending $x \in K^\times$ to the cocycle $g \mapsto g(\sqrt[m]{x})/\sqrt[m]{x}$ for $g \in G_K$ and for any fixed m-th root $\sqrt[m]{x}$ of x.

Corollary 1.6.5. *If $\boldsymbol{\mu}_m \subset K$ then $K^\times/(K^\times)^m \cong \mathrm{Hom}(G_K, \boldsymbol{\mu}_m)$.*

Exercise 1.6.6. Suppose $\boldsymbol{\mu}_m \subset K$ and X is a finite subgroup of $K^\times/(K^\times)^m$. Show that the isomorphism of Corollary 1.6.5 identifies

$$X \cong \mathrm{Hom}(\mathrm{Gal}(K(\sqrt[m]{X})/K), \boldsymbol{\mu}_m),$$

where $K(\sqrt[m]{X})$ is the extension of K generated by m-th roots of all elements of X. Show that the isomorphism of Corollary 1.6.5 induces a bijection between finite subgroups of $K^\times/(K^\times)^m$ and finite abelian extensions of K of exponent dividing m.

Definition 1.6.7. If K is a nonarchimedean local field of characteristic zero, define

$$H^1_{\mathrm{f}}(K, \boldsymbol{\mu}_m) \subset H^1(K, \boldsymbol{\mu}_m)$$

to be the image under the Kummer map of Theorem 1.6.3

$$\mathcal{O}_K^\times/(\mathcal{O}_K^\times)^m \hookrightarrow K^\times/(K^\times)^m \cong H^1(K, \boldsymbol{\mu}_m)$$

where $\mathcal{O}_K$ is the ring of integers of K. If K is $\mathbf{R}$ or $\mathbf{C}$, let $H^1_{\mathrm{f}}(K, \boldsymbol{\mu}_m) = H^1(K, \boldsymbol{\mu}_m)$.

Exercise 1.6.8. Show that if K is a number field, then the collection of subgroups $\{H^1_{\mathrm{f}}(K_v, \boldsymbol{\mu}_m)\}$ is a Selmer structure (i.e., show that $H^1_{\mathrm{f}}(K_v, \boldsymbol{\mu}_m) = H^1_{\mathrm{u}}(K_v, \boldsymbol{\mu}_m)$ for all $v \nmid m\infty$).

Proposition 1.6.9. *The Selmer group $H^1_{\mathrm{f}}(K, \boldsymbol{\mu}_m)$ defined with the Selmer structure above satisfies*

$$0 \longrightarrow \mathcal{O}_K^\times/(\mathcal{O}_K^\times)^m \longrightarrow H^1_{\mathrm{f}}(K, \boldsymbol{\mu}_m) \longrightarrow \mathcal{C}_K[m] \longrightarrow 0$$

where $\mathcal{C}_K[m]$ is the m-torsion subgroup of the ideal class group of K.

Exercise 1.6.10. Prove Proposition 1.6.9 as follows. Viewing $H^1_{\mathrm{f}}(K, \boldsymbol{\mu}_m) \subset K^\times/(K^\times)^m$, show that if $x \in K^\times$ projects to an element of $H^1_{\mathrm{f}}(K, \boldsymbol{\mu}_m)$, then the fractional principal ideal $x\mathcal{O}_K$ is $\mathfrak{a}^m$ for some fractional ideal $\mathfrak{a}$. Show that sending x to the class of $\mathfrak{a}$ induces a surjective map $H^1_{\mathrm{f}}(K, \boldsymbol{\mu}_m) \twoheadrightarrow \mathcal{C}_K[m]$ with kernel $\mathcal{O}_K^\times/(\mathcal{O}_K^\times)^m$.

Theorem 1.6.3 has the following analogue for elliptic curves.

Theorem 1.6.11. *Suppose E is an elliptic curve defined over K and $m > 0$ is prime to the characteristic of K. There is an exact sequence*

$$0 \longrightarrow E(K)/mE(K) \longrightarrow H^1(K, E[m]) \longrightarrow H^1(K, E(K^{\mathrm{sep}}))[m] \longrightarrow 0.$$

Exercise 1.6.12. Prove Theorem 1.6.11 using the short exact sequence

$$0 \longrightarrow E[m] \longrightarrow E(K^{\mathrm{sep}}) \xrightarrow{\ m\ } E(K^{\mathrm{sep}}) \longrightarrow 0.$$

Show that the injection $E(K)/mE(K) \hookrightarrow H^1(K, E[m])$ is given by sending $x \in E(K)$ to the cocycle $g \mapsto gy - y$, where $y \in E(K^{\mathrm{sep}})$ satisfies $my = x$ (and the same y is used for all g).

Definition 1.6.13. If K is a nonarchimedean local field of characteristic zero, E is an elliptic curve over K, and $m > 0$, define

$$H^1_{\mathrm{f}}(K, E[m]) \subset H^1(K, E[m])$$

to be the image of the Kummer map of Theorem 1.6.11

$$E(K)/mE(K) \hookrightarrow H^1(K, E[m]).$$

If K is $\mathbf{R}$ or $\mathbf{C}$, let $H^1_{\mathrm{f}}(K, E[m]) = H^1(K, E[m])$.

Suppose now that K is a number field and E is an elliptic curve over K. If $v \nmid m\infty$ and E has good reduction at v, then $H^1_{\mathrm{f}}(K_v, E[m]) = H^1_{\mathrm{u}}(K_v, E[m])$ by [**Ca**], so the collection of subgroups $\{H^1_{\mathrm{f}}(K_v, E[m])\}$ is a Selmer structure. The corresponding Selmer group is the classical m-Selmer group of E/K, $H^1_{\mathrm{f}}(K, E[m]) = \mathrm{Sel}_m(E/K)$ sitting in an exact sequence

$$0 \longrightarrow E(K)/mE(K) \longrightarrow \mathrm{Sel}_m(E/K) \longrightarrow Ш(E/K)[m] \longrightarrow 0 \tag{1.3}$$

where $Ш(E/K)$ is the Shafarevich-Tate group of E/K.

1.7. The Brauer group.

If K is a field, $\boldsymbol{\mu}_\infty$ will denote the roots of unity in K^{sep}.

Exercise 1.7.1. Suppose K is a field. Show that for every $m > 0$ prime to the characteristic of K, there is a natural isomorphism $H^2(K, \boldsymbol{\mu}_m) \cong H^2(K, (K^{\mathrm{sep}})^\times)[m]$. Deduce that if K has characteristic zero, then $H^2(K, \boldsymbol{\mu}_\infty) \cong H^2(K, (K^{\mathrm{sep}})^\times)$.

Theorem 1.7.2. *If K is a nonarchimedean local field of characteristic zero (i.e., a finite extension of some $\mathbf{Q}_p$), then there is a canonical isomorphism*

$$\mathrm{inv}_K : H^2(K, \boldsymbol{\mu}_\infty) \xrightarrow{\sim} \mathbf{Q}/\mathbf{Z}.$$

If $K = \mathbf{R}$ or $\mathbf{C}$, there are canonical isomorphisms

$$\mathrm{inv}_{\mathbf{R}} : H^2(\mathbf{R}, \boldsymbol{\mu}_\infty) \xrightarrow{\sim} \tfrac{1}{2}\mathbf{Z}/\mathbf{Z} \subset \mathbf{Q}/\mathbf{Z},$$
$$\mathrm{inv}_{\mathbf{C}} : H^2(\mathbf{C}, \boldsymbol{\mu}_\infty) \xrightarrow{\sim} \{0\} \subset \mathbf{Q}/\mathbf{Z}.$$

PROOF. See [**Se2**, §XIII.3]. □

Theorem 1.7.3. *Suppose K is a number field. For every $m \leq \infty$ there is an exact sequence*

$$0 \longrightarrow H^2(K, \boldsymbol{\mu}_m) \xrightarrow{\oplus \mathrm{Res}_v} \bigoplus_v H^2(K_v, \boldsymbol{\mu}_m) \xrightarrow{\sum \mathrm{inv}_{K_v}} \mathbf{Q}/\mathbf{Z} \longrightarrow 0$$

where K_v is the completion of K at v, and Res_v *is the restriction map with respect to the subgroup $G_{K_v} \subset G_K$.*

PROOF. See [**AT**, Chapter 7, Theorem 8]. □

1.8. Local fields and duality.

Suppose A and B are G-modules. There is a cup product homomorphism

$$H^i(G, A) \otimes H^j(G, B) \longrightarrow H^{i+j}(G, A \otimes B).$$

We view this as a bilinear pairing on $H^i(G, A) \times H^j(G, B)$, written $(a, b) \mapsto a \cup b$. The cup product pairing has numerous properties [**AW, Se2**]; here we mention only that it commutes with restriction maps: if $H \subset G$, then $\mathrm{Res}(a) \cup \mathrm{Res}(b) = \mathrm{Res}(a \cup b)$.

Suppose now that K is a field, A is a finite G_K module, and $\boldsymbol{\mu}_\infty$ is the G_K-module of roots of unity in K^{sep}. Define the Cartier dual $A^* := \mathrm{Hom}(A, \boldsymbol{\mu}_\infty)$. Then A^* is also a (continuous) G_K-module, and there is a natural pairing

$$A \otimes A^* \longrightarrow \boldsymbol{\mu}_\infty.$$

We can compose the cup product with this pairing to get a pairing

$$H^i(K, A) \otimes H^j(K, A^*) \longrightarrow H^{i+j}(K, \boldsymbol{\mu}_\infty). \tag{1.4}$$

If K is a local field of characteristic zero, and $i+j=2$, we can compose the pairing (1.4) with the isomorphism of Proposition 1.7.2 to get a new pairing

$$H^i(K,A)\times H^j(K,A^*)\xrightarrow{\ \cup\ }H^2(K,\boldsymbol{\mu}_\infty)\xrightarrow{\ \mathrm{inv}_K\ }\mathbf{Q}/\mathbf{Z}. \tag{1.5}$$

Theorem 1.8.1 (Tate local duality). *Suppose K is a local field of characteristic zero, A is a finite G_K-module, and $0\le i\le 2$. Then (1.5) is a perfect pairing of finite groups.*

PROOF. See [**Mi**, Corollary I.2.3]. □

In the following examples and exercise, K is a local field of characteristic zero and $\mathcal{O}_K$ is its ring of integers.

Example 1.8.2. Fix $m\ge 0$, and let $A=\boldsymbol{\mu}_m$. Then $A^*=\mathbf{Z}/m\mathbf{Z}$ (with trivial G_K-action), and we have

$$H^1(K,\boldsymbol{\mu}_m)\cong K^\times/(K^\times)^m,\quad H^1(K,\mathbf{Z}/m\mathbf{Z})=\mathrm{Hom}(G_K,\mathbf{Z}/m\mathbf{Z}).$$

By Theorem 1.8.1, the pairing (1.5) gives an isomorphism (for every m)

$$\begin{aligned}K^\times/(K^\times)^m\xrightarrow{\sim}H^1(K,\boldsymbol{\mu}_m)&\xrightarrow{\sim}\mathrm{Hom}(H^1(K,\mathbf{Z}/m\mathbf{Z}),\mathbf{Q}/\mathbf{Z})\\&=\mathrm{Hom}(\mathrm{Hom}(G_K,\mathbf{Z}/m\mathbf{Z}),\mathbf{Q}/\mathbf{Z})\xrightarrow{\sim}G_K^{\mathrm{ab}}/(G_K^{\mathrm{ab}})^m.\end{aligned}$$

The inverse limit of these maps over m gives the Artin map $K^\times\to G_K^{\mathrm{ab}}$ of local class field theory.

Exercise 1.8.3. Recall that $H^1_{\mathrm{f}}(K,\boldsymbol{\mu}_m)\subset H^1(K,\boldsymbol{\mu}_m)$ is the image of the Kummer map $\mathcal{O}_K^\times/(\mathcal{O}_K^\times)^m\hookrightarrow H^1(K,\boldsymbol{\mu}_m)$. Define

$$H^1_{\mathrm{f}}(K,\mathbf{Z}/m\mathbf{Z}):=H^1_{\mathrm{u}}(K,\mathbf{Z}/m\mathbf{Z})=\mathrm{Hom}(G_K/I_K,\mathbf{Z}/m\mathbf{Z})\subset H^1(K,\mathbf{Z}/m\mathbf{Z}),$$

where $I_K\subset G_K$ is the inertia group. Show, using the local class field theory description of the isomorphism in Example 1.8.2, that $H^1_{\mathrm{f}}(K,\boldsymbol{\mu}_m)$ and $H^1_{\mathrm{f}}(K,\mathbf{Z}/m\mathbf{Z})$ are orthogonal complements of each other under the Tate pairing

$$H^1(K,\boldsymbol{\mu}_m)\times H^1(K,\mathbf{Z}/m\mathbf{Z})\to\mathbf{Q}/\mathbf{Z}.$$

Example 1.8.4. Suppose E is an elliptic curve over K, $m>0$, and $A=E[m]$. The Weil pairing $E[m]\times E[m]\to\boldsymbol{\mu}_m$ gives a canonical isomorphism $A^*\cong E[m]$. By Theorem 1.8.1, the pairing (1.5) gives a perfect pairing

$$H^1(K,E[m])\times H^1(K,E[m])\longrightarrow\mathbf{Z}/m\mathbf{Z}.$$

The subgroup $H^1_{\mathrm{f}}(K,E[m])\subset H^1(K,E[m])$ given by Definition 1.6.13 is its own orthogonal complement under this pairing [**Ta1**].

1.9. Unramified and transverse cohomology groups.

Fix for this section a nonarchimedean local field K of characteristic zero, and let $\mathcal{O}_K$ be its ring of integers. As in Definition 1.4.12, we let $I_K\subset G_K$ denote the inertia group, $K^{\mathrm{ur}}=\bar{K}^{I_K}$ the maximal unramified extension of K, and $\varphi\in\mathrm{Gal}(K^{\mathrm{ur}}/K)$ the Frobenius automorphism. Then $\mathrm{Gal}(K^{\mathrm{ur}}/K)=G_K/I_K\cong\hat{\mathbf{Z}}$, generated by φ. Recall that a G_K-module A is *unramified* if I_K acts trivially on A.

Proposition 1.9.1. *Suppose A is a finite unramified G_K-module of order prime to the residue characteristic of K. Then $H^1_{\mathrm{u}}(K,A)$ and $H^1_{\mathrm{u}}(K,A^*)$ are orthogonal complements of each other under the pairing* (1.5).

PROOF. See for example [**Mi**, Theorem I.2.6], or combine the following two exercises. □

Exercise 1.9.2. Show using Proposition 1.4.13 that if A is a finite unramified G_K-module of order prime to the residue characteristic of K, then

$$|H^1_{\mathrm{u}}(K,A)| = [H^1(K,A^*) : H^1_{\mathrm{u}}(K,A^*)].$$

You will need to use that the tame quotient of I_K is isomorphic (as a G_K-module) to $\prod_{\ell \neq p} \mathbf{Z}_\ell(1)$, where p is the residue characteristic of K and $\mathbf{Z}_\ell(1) = \varprojlim \boldsymbol{\mu}_{\ell^n}$.

Exercise 1.9.3. Assuming that the cup product commutes with inflation, show that $H^1_{\mathrm{u}}(K,A)$ and $H^1_{\mathrm{u}}(K,A^*)$ are orthogonal because the following diagram commutes:

$$\begin{array}{ccc} H^1(K,A) \times H^1(K,A^*) & \xrightarrow{\ \cup\ } & H^2(K,\boldsymbol{\mu}_\infty) \\ \uparrow{\scriptstyle \mathrm{Inf}} & & \uparrow{\scriptstyle \mathrm{Inf}} \\ H^1_{\mathrm{u}}(K,A) \times H^1_{\mathrm{u}}(K,A^*) & \xrightarrow{\ \cup\ } & H^2(K^{\mathrm{ur}}/K,\boldsymbol{\mu}_\infty) \end{array}$$

and the lower right corner is zero by Proposition 1.4.10(2).

Let $\Bbbk$ denote the residue field of K, and $q := |\Bbbk|$.

Definition 1.9.4. Suppose L/K is a totally tamely ramified extension of degree $q-1$. Then there is a canonical isomorphism $\mathrm{Gal}(L/K) \cong \Bbbk^\times$. (When $K = \mathbf{Q}_\ell$ we can take $L = \mathbf{Q}_\ell(\boldsymbol{\mu}_\ell)$.) We define the *$L$-transverse subgroup* $H^1_{\mathrm{t}}(K,A) \subset H^1(K,A)$ by

$$H^1_{\mathrm{t}}(K,A) := \ker\big[H^1(K,A) \xrightarrow{\mathrm{Res}} H^1(L,A)\big] = H^1(L/K, A^{G_L}).$$

Proposition 1.9.5. *Suppose A is a finite unramified G_K-module and $(q-1)A = 0$. Then:*

(1) $H^1_{\mathrm{t}}(K,A) \cong \mathrm{Hom}(\mathrm{Gal}(L/K), A^{\varphi=1})$.
(2) $H^1_{\mathrm{t}}(K,A) \otimes \mathrm{Gal}(L/K) \cong A^{\varphi=1}$.
(3) *There is a direct sum decomposition* $H^1(K,A) = H^1_{\mathrm{u}}(K,A) \oplus H^1_{\mathrm{t}}(K,A)$.
(4) $H^1_{\mathrm{t}}(K,A)$ *and* $H^1_{\mathrm{t}}(K,A^*)$ *are orthogonal complements of each other under the pairing* (1.5).

PROOF. See [**MR1**, Lemmas 1.2.1, 1.2.4 and Proposition 1.3.2]. Since A is unramified and I_K and G_L generate G_K, we have $A^{G_L} = A^{G_K} = A^{\varphi=1}$. This proves (1) and (2). By (1) and Proposition 1.4.13(2), $|H^1(K,A)| = |H^1_{\mathrm{u}}(K,A)| \cdot |H^1_{\mathrm{t}}(K,A)|$, so to prove (3) it is enough to show that $H^1_{\mathrm{u}}(K,A) \cap H^1_{\mathrm{t}}(K,A) = 0$, and then to prove (4) it is enough to show that $H^1_{\mathrm{t}}(K,A)$ and $H^1_{\mathrm{t}}(K,A^*)$ are orthogonal. These are left as an exercises, or see [**MR1**]. □

Definition 1.9.6. Suppose $m \mid q-1$, and let $R = \mathbf{Z}/m\mathbf{Z}$. Suppose that A is an unramified G-module that is free of finite rank over R, and that $\det(1-\varphi|A) = 0$. Consider the characteristic polynomial

$$P(x) := \det(1-\varphi x|A) \in R[x].$$

Since $P(1) = \det(1-\varphi|A) = 0$, we have $P(x) = (x-1)Q(x)$ for some $Q(x) \in R[x]$. By the Cayley-Hamilton theorem, $P(\varphi^{-1})$ annihilates A, so $Q(\varphi^{-1}) \subset A^{\varphi=1}$. Define the *unramified-transverse comparison map* ϕ^{ut} to be the composition

$$H^1_{\mathrm{u}}(K,A) \xrightarrow{\sim} A/(\varphi-1)A \xrightarrow{Q(\varphi^{-1})} A^{\varphi=1} \xrightarrow{\sim} H^1_{\mathrm{t}}(K,A) \otimes \mathrm{Gal}(L/K)$$

using the isomorphisms of Proposition 1.4.13(1) and 1.9.5(2).

Exercise 1.9.7 (Lemma 1.2.3 of [**MR1**]). Suppose $m \mid q-1$ and A is an unramified G-module that is free of finite rank over $\mathbf{Z}/m\mathbf{Z}$. Show that $A^{\varphi=1}$ is free of rank one over $\mathbf{Z}/m\mathbf{Z}$ if and only if $A/(\varphi-1)A$ is free of rank one over $\mathbf{Z}/m\mathbf{Z}$. If these conditions are satisfied, show that the map ϕ^{ut} of Definition 1.9.6 is an isomorphism and $H^1_{\mathrm{u}}(K,A)$, $H^1_{\mathrm{t}}(K,A)$, $H^1_{\mathrm{u}}(K,A^*)$, $H^1_{\mathrm{t}}(K,A^*)$, are all free of rank 1 over $\mathbf{Z}/m\mathbf{Z}$.

1.10. Comparing Selmer groups.

Suppose K is a number field and A is a finite G_K-module.

Definition 1.10.1. If $\mathcal{F}$ is a Selmer structure for A, define a Selmer structure $\mathcal{F}^*$ for the Cartier dual $A^* := \mathrm{Hom}(A, \boldsymbol{\mu}_\infty)$ by

$$H^1_{\mathcal{F}^*}(K_v, A^*) := H^1_{\mathcal{F}}(K_v, A)^\perp \subset H^1(K_v, A^*)$$

for every v, where $H^1_{\mathcal{F}}(K_v,A)^\perp$ is the orthogonal complement of $H^1_{\mathcal{F}}(K_v,A)$ in $H^1(K_v,A^*)$ under the local Tate pairing (1.5). By Proposition 1.9.1, $\mathcal{F}^*$ is a Selmer structure.

If $\mathcal{F},\mathcal{G}$ are Selmer structures for A, we will say $\mathcal{F}\subset\mathcal{G}$ if $H^1_{\mathcal{F}}(K_v,A)\subset H^1_{\mathcal{G}}(K_v,A)$ for every v. Note that if $\mathcal{F}\subset\mathcal{G}$, then

- $H^1_{\mathcal{F}}(K,A)\subset H^1_{\mathcal{G}}(K,A)$,
- $\mathcal{G}^*\subset\mathcal{F}^*$.

Theorem 1.10.2 (Global duality). *Suppose $\mathcal{G}_1,\mathcal{G}_2$ are Selmer structures for A, and $\mathcal{G}_1\subset\mathcal{G}_2$. There are exact sequences*

$$0 \longrightarrow H^1_{\mathcal{G}_1}(K,A) \longrightarrow H^1_{\mathcal{G}_2}(K,A) \xrightarrow{\oplus\mathrm{Res}_v} \bigoplus_v H^1_{\mathcal{G}_2}(K_v,A)/H^1_{\mathcal{G}_1}(K_v,A),$$

$$0 \longrightarrow H^1_{\mathcal{G}_2^*}(K,A^*) \longrightarrow H^1_{\mathcal{G}_1^*}(K,A^*) \xrightarrow{\oplus\mathrm{Res}_v} \bigoplus_v H^1_{\mathcal{G}_1^*}(K_v,A^*)/H^1_{\mathcal{G}_2^*}(K_v,A^*),$$

summing over v such that $H^1_{\mathcal{G}_1}(K_v,A)\neq H^1_{\mathcal{G}_2}(K_v,A)$. The images of the two right-hand maps are orthogonal complements of each other under the sum of the local Tate pairings (1.5).

PROOF. See [**R2**, Theorem 1.7.3] to derive this statement from the usual statement of Poitou-Tate duality ([**Ta2**, Theorem 3.1] or [**Mi**, Theorem I.4.10]). □

Exercise 1.10.3. Show that the two sequences of Theorem 1.10.2 are exact. Show that the images of the two right-hand maps are orthogonal, using Theorem 1.7.3.

The following corollary of Theorem 1.10.2 will be used to bound the size of $H^1_{\mathcal{F}}(K,A)$. Note that given $\mathcal{F}$, there will exist $\mathcal{G}\subset\mathcal{F}$ "small enough" that $H^1_{\mathcal{G}}(K,A)=0$.

Corollary 1.10.4. *Suppose $\mathcal{F}\subset\mathcal{G}$ are Selmer structures for A, and $H^1_{\mathcal{G}^*}(K,A^*)=0$. Show that*

$$|H^1_{\mathcal{F}^*}(K,A^*)| = \left|\mathrm{coker}\left[H^1_{\mathcal{G}}(K,A) \xrightarrow{\oplus\mathrm{Res}_v} \bigoplus_v H^1_{\mathcal{G}}(K_v,A)/H^1_{\mathcal{F}}(K_v,A)\right]\right|.$$

Exercise 1.10.5. Prove Corollary 1.10.4.

LECTURE 2

Kolyvagin systems

From now on, to simplify things we will only consider $K = \mathbf{Q}$. Much of what we do can be extended easily to arbitrary number fields. We follow the presentation in [**MR1**, Chapters 3 and 4], and more details can be found there.

Fix a prime p and a positive integer m, and let $R = \mathbf{Z}/p^m\mathbf{Z}$. Fix a $G_{\mathbf{Q}}$-module A that is free of finite rank over R, and a Selmer structure $\mathcal{F}$ for A.

The letter ℓ will always denote a rational prime.

2.1. Varying the Selmer structure.

As in Exercise 1.5.2, let Σ be a finite set of places of $\mathbf{Q}$ containing ∞, p, all primes where A is ramified, and all ℓ such that $H^1_{\mathcal{F}}(\mathbf{Q}_\ell, A) \neq H^1_{\mathrm{u}}(\mathbf{Q}_\ell, A)$.

Definition 2.1.1. Suppose a, b, c are pairwise relatively prime positive integers, and c is not divisible by any primes in Σ. Define a new Selmer structure $\mathcal{F}_a^b(c)$ for A by

$$H^1_{\mathcal{F}_a^b(c)}(\mathbf{Q}_v, A) := \begin{cases} H^1_{\mathcal{F}}(\mathbf{Q}_v, A) & \text{if } v \nmid abc \\ 0 & \text{if } v \mid a, \\ H^1(\mathbf{Q}_v, A) & \text{if } v \mid b, \\ H^1_{\mathrm{t}}(\mathbf{Q}_v, A) & \text{if } v \mid c. \end{cases}$$

In other words, $\mathcal{F}_a^b(c)$ is constructed by modifying $\mathcal{F}$ at the primes dividing abc, with the "strict" condition at ℓ if $\ell \mid a$, the "relaxed" condition at ℓ if $\ell \mid b$, and the transverse condition at ℓ if $\ell \mid c$. (Here the transverse subgroup $H^1_{\mathrm{t}}(\mathbf{Q}_\ell, A)$ is defined with respect to the extension $\mathbf{Q}_\ell(\boldsymbol{\mu}_\ell)/\mathbf{Q}_\ell$ in Definition 1.9.4.)

If any of a, b, or c are equal to 1, we will suppress them from the notation.

Exercise 2.1.2. Show that the dual Selmer structure $(\mathcal{F}_a^b(c))^*$ (in the sense of Definition 1.10.1) is $(\mathcal{F}^*)_b^a(c)$.

Definition 2.1.3. Let

$$\mathcal{P} := \{\ell \notin \Sigma : \ell \equiv 1 \pmod{p^m} \text{ and } A/(\mathrm{Fr}_\ell - 1)A \text{ is free of rank 1 over } R\}.$$

Let $\mathcal{N} = \mathcal{N}(\mathcal{P}) \subset \mathbf{Z}^+$ denote the set of squarefree products of primes in $\mathcal{P}$ (with the convention that $1 \in \mathcal{N}$). If $n \in \mathcal{N}$ let

$$G_n := \bigotimes_{\ell \mid n} \mathbf{F}_\ell^\times = \bigotimes_{\ell \mid n} \mathrm{Gal}(\mathbf{Q}(\boldsymbol{\mu}_\ell)/\mathbf{Q}).$$

Since each group in the tensor product is cyclic of order divisible by p^m, we see that $G_n \otimes R$ is free of rank one over R. By convention, we set $G_1 = R$.

If $\ell \mid n \in \mathcal{N}$, then $\ell - 1 = 0$ and $\det(1 - \mathrm{Fr}_\ell|A) = 0$ in R, so we can apply Proposition 1.9.5, Definition 1.9.6, and Exercise 1.9.7 to $H^1(\mathbf{Q}_\ell, A)$. In particular we will write

$$\phi_\ell^{\mathrm{ut}} : H^1_{\mathrm{u}}(\mathbf{Q}_\ell, A) \xrightarrow{\sim} H^1_{\mathrm{t}}(\mathbf{Q}_\ell, A) \otimes G_\ell$$

for the unramified-transverse isomorphism of Definition 1.9.6.

If $n\ell \in \mathcal{N}$, we can compare $H^1_{\mathcal{F}(n)}(\mathbf{Q}, A) \otimes G_n$ and $H^1_{\mathcal{F}(n\ell)}(\mathbf{Q}, A) \otimes G_{n\ell}$ by localizing at ℓ and using ϕ_ℓ^{ut}:

$$(2.1)\qquad \begin{array}{ccc} & & H^1_{\mathcal{F}(n)}(\mathbf{Q}, A) \otimes G_n \\ & & \Big\downarrow{\scriptstyle \mathrm{Res}_\ell \otimes 1} \\ & & H^1_{\mathrm{u}}(\mathbf{Q}_\ell, A) \otimes G_n \\ & & \Big\downarrow{\scriptstyle \phi_\ell^{\mathrm{ut}} \otimes 1} \\ H^1_{\mathcal{F}(n\ell)}(\mathbf{Q}, A) \otimes G_{n\ell} & \xrightarrow{\mathrm{Res}_\ell \otimes 1} & H^1_{\mathrm{t}}(\mathbf{Q}_\ell, A) \otimes G_{n\ell}. \end{array}$$

Exercise 2.1.4 (Corollary 2.3.6 of [**MR1**]). If $n \in \mathcal{N}$, show that

$$\begin{aligned} \mathrm{length}(H^1_{\mathcal{F}}(\mathbf{Q}, A)) &- \mathrm{length}(H^1_{\mathcal{F}^*}(\mathbf{Q}, A^*)) \\ &= \mathrm{length}(H^1_{\mathcal{F}(n)}(\mathbf{Q}, A)) - \mathrm{length}(H^1_{\mathcal{F}(n)^*}(\mathbf{Q}, A^*)). \end{aligned}$$

where length means length as an R-module, so $|M| = p^{\mathrm{length}(M)}$ for every finite R-module M. Hint: apply Theorem 1.10.2 with $(\mathcal{G}_1, \mathcal{G}_2) = (\mathcal{F}, \mathcal{F}^n)$ and with $(\mathcal{G}_1, \mathcal{G}_2) = (\mathcal{F}(n), \mathcal{F}^n)$, and use Exercise 1.9.7.

2.2. Kolyvagin systems.

Keep the notation of the previous section.

Definition 2.2.1. A *Kolyvagin system* for $(A, \mathcal{F})$ is a collection

$$\{\kappa_n \in H^1_{\mathcal{F}(n)}(\mathbf{Q}, A) \otimes G_n : n \in \mathcal{N}\}$$

such that if $n\ell \in \mathcal{N}$, then the images of κ_n and $\kappa_{n\ell}$ in $H^1_{\mathrm{t}}(\mathbf{Q}_\ell, A) \otimes G_{n\ell}$ in (2.1) are equal. We let $\mathbf{KS}(A, \mathcal{F})$ (or simply $\mathbf{KS}(A)$, if $\mathcal{F}$ is understood) denote the R-module of Kolyvagin systems.

We next reformulate this definition.

Definition 2.2.2. Suppose X is a graph, with vertex set V and edge set E. By a *sheaf* $\mathcal{S}$ of R-modules on X, we mean a rule assigning

- to each vertex v an R-module $\mathcal{S}(v)$,
- to each edge e an R-module $\mathcal{S}(e)$,
- to each pair (v, e) such that the vertex v is an endpoint of the edge e, an R-module homomorphism $\psi_v^e : \mathcal{S}(v) \to \mathcal{S}(e)$.

We call $\mathcal{S}(v)$ the *stalk* of $\mathcal{S}$ at v.

A *global section* of $\mathcal{S}$ is a collection $\{\kappa_v \in \mathcal{S}(v) : v \in V\}$ such that for every edge e, if v, v' are the endpoints of e, then $\psi_v^e(\kappa_v) = \psi_{v'}^e(\kappa_{v'})$. We let $\Gamma(\mathcal{S})$ denote the R-module of global sections of $\mathcal{S}$.

Definition 2.2.3. Define a graph $\mathcal{X}$ associated to $(A, \mathcal{F})$ by taking the set of vertices to be $\mathcal{N}$, and whenever $n, n\ell \in \mathcal{N}$ we join n and $n\ell$ by an edge. Define the *Selmer sheaf* $\mathcal{H} = \mathcal{H}_{(A,\mathcal{F})}$ on $\mathcal{X}$ by

- $\mathcal{H}(n) := H^1_{\mathcal{F}(n)}(\mathbf{Q}, A) \otimes G_n$ for $n \in \mathcal{N}$,

and if e is the edge joining n and $n\ell$,

- $\mathcal{H}(e) := H^1_{\mathrm{t}}(\mathbf{Q}_\ell, A) \otimes G_{n\ell}$,
- ψ^e_n is the composition of the vertical maps of (2.1),
- $\psi^e_{n\ell}$ is the horizontal map of (2.1).

Definition 2.2.1$'$. A *Kolyvagin system* for $(A, \mathcal{F})$ is a global section of the Selmer sheaf $\mathcal{H}_{(A,\mathcal{F})}$, and so $\mathbf{KS}(A, \mathcal{F}) = \Gamma(\mathcal{H}_{(A,\mathcal{F})})$.

Exercise 2.2.4. Suppose $H^1_{\mathcal{F}}(\mathbf{Q}, A) = 0$. Show that every Kolyvagin system is identically zero.

Exercise 2.2.5. Suppose $\boldsymbol{\kappa} \in \mathbf{KS}(A, \mathcal{F})$ and $\ell \in \mathcal{P}$. Fix a surjection $\xi : G_\ell \to R$. Show that the collection

$$\{(1 \otimes \xi)(\kappa_{n\ell}) : n \in \mathcal{N}, \ell \nmid n\}$$

is a Kolyvagin system for $(A, \mathcal{F}(\ell))$.

We postpone until later any discussion of how one can find Kolyvagin systems, and instead we discuss consequences of the existence of Kolyvagin systems.

2.3. Sheaves and monodromy.

Suppose for this section that $\mathcal{S}$ is a sheaf on a graph X, as in Definition 2.2.2.

Definition 2.3.1. If v and w are vertices of X, a *path* in X from v to w is a sequence of vertices and edges $(v = v_0, e_1, v_1, e_2, \ldots, e_n, v_n = w)$ in X such that for each i, e_i is an edge joining v_{i-1} and v_i. A *loop* in X (at v) is a path from v to v.

We say that $\mathcal{S}$ is *locally cyclic* if all the R-modules $\mathcal{S}(v)$, $\mathcal{S}(e)$ are cyclic and all the maps ψ^e_v are surjective.

If $\mathcal{S}$ is locally cyclic then a *surjective path* (relative to $\mathcal{S}$) from v to w is a path $(v = v_0, e_1, v_1, e_2, \ldots, e_n, v_n = w)$ in X such that for each i, the map $\psi^{e_i}_{v_i}$ is an isomorphism. We say that the vertex v is a *hub* of $\mathcal{S}$ if for every vertex w there is an $\mathcal{S}$-surjective path from v to w.

Suppose now that the sheaf $\mathcal{S}$ is locally cyclic. If P is a surjective path in X:

$$\mathcal{S}(v_0) \twoheadrightarrow \mathcal{S}(e_1) \xleftarrow{\sim} \mathcal{S}(v_1) \twoheadrightarrow \mathcal{S}(e_2) \xleftarrow{\sim} \mathcal{S}(v_2) \twoheadrightarrow \cdots \xleftarrow{\sim} \mathcal{S}(v_{n-1}) \twoheadrightarrow \mathcal{S}(e_n) \xleftarrow{\sim} \mathcal{S}(v_n)$$

$$v_0 \overset{}{\underset{e_1}{\text{———}}} v_1 \underset{e_2}{\text{———}} v_2 \text{——} \cdots \text{——} v_{n-1} \underset{e_n}{\text{———}} v_n$$

we can define a surjective map $\psi_P : \mathcal{S}(v_0) \to \mathcal{S}(v_n)$ by

$$\psi_P := (\psi^{e_n}_{v_n})^{-1} \circ \psi^{e_n}_{v_{n-1}} \circ (\psi^{e_{n-1}}_{v_{n-1}})^{-1} \circ \cdots \circ (\psi^{e_1}_{v_1})^{-1} \circ \psi^{e_1}_{v_0} \tag{2.2}$$

since all the inverted maps are isomorphisms. We will say that $\mathcal{S}$ has *trivial monodromy* if whenever v, w, w' are vertices, P, P' are surjective paths $(v, \ldots, w)$ and $(v, \ldots, w')$, and w, w' are joined by an edge e, then $\psi^e_w \circ \psi_P = \psi^e_{w'} \circ \psi_{P'} \in \mathrm{Hom}(\mathcal{S}(v), \mathcal{S}(e))$. In particular for every pair v, w of vertices and and every pair P, P' of surjective paths from v to w, we require that $\psi_P = \psi_{P'} \in \mathrm{Hom}(\mathcal{S}(v), \mathcal{S}(w))$.

Recall that $\Gamma(\mathcal{S})$ is the set of global sections of $\mathcal{S}$.

Proposition 2.3.2. *Suppose $\mathcal{S}$ is locally cyclic and v is a hub of $\mathcal{S}$.*

(1) *The map $f_v : \Gamma(\mathcal{S}) \to \mathcal{S}(v)$ defined by $\boldsymbol{\kappa} \mapsto \kappa_v$ is injective, and is surjective if and only if $\mathcal{S}$ has trivial monodromy.*
(2) *If $\boldsymbol{\kappa} \in \Gamma(\mathcal{S})$, and if u is a vertex such that $\kappa_u \neq 0$ and κ_u generates $p^k\mathcal{S}(u)$ for some $k \in \mathbf{Z}^+$, then κ_w generates $p^k\mathcal{S}(w)$ for every vertex w.*

Exercise 2.3.3. Prove Proposition 2.3.2

Definition 2.3.4. A global section $\kappa \in \Gamma(\mathcal{S})$ will be called *primitive* if for every vertex v, $\kappa(v) \in \mathcal{S}(v)$ is a generator of the R-module $\mathcal{S}(v)$.

It follows from Proposition 2.3.2 that a locally cyclic sheaf $\mathcal{S}$ with a hub has a primitive global section if and only if $\mathcal{S}$ has trivial monodromy.

The Selmer sheaf will not in general be locally cyclic. However, we will see later that under suitable hypotheses, there is a natural subsheaf $\mathcal{H}'$ of the Selmer sheaf $\mathcal{H}$ such that $\mathcal{H}'$ is locally cyclic with trivial monodromy, $\mathcal{H}'$ has hub vertices, and every global section of $\mathcal{H}$ is a global section of $\mathcal{H}'$, i.e.,

$$\mathbf{KS}(A, \mathcal{F}) = \Gamma(\mathcal{H}) = \Gamma(\mathcal{H}').$$

2.4. Hypotheses on A and $\mathcal{F}$.

From now on we will assume that all of the following hypotheses hold. Let $\bar{A} = A \otimes_R \mathbf{F}_p$. We will assume that in addition to being free of finite rank over R, A satisfies:

(H.1) $\bar{A}$ is an absolutely simple $\mathbf{F}_p[G_{\mathbf{Q}}]$-module, not isomorphic (as a $G_{\mathbf{Q}}$-module) to $\mathbf{F}_p$ or to $\boldsymbol{\mu}_p$.
(H.2) There is a $\tau \in G_{\mathbf{Q}}$ such that $\tau = 1$ on $\boldsymbol{\mu}_{p^\infty}$ and $A/(\tau - 1)A$ is free of rank one over R.
(H.3) $H^1(\mathbf{Q}(A)/\mathbf{Q}, \bar{A}) = H^1(\mathbf{Q}(A^*)/\mathbf{Q}, \bar{A}^*) = 0$, where $\mathbf{Q}(A)$ is the fixed field of the kernel of the map $G_{\mathbf{Q}} \to \mathrm{Aut}(A)$, and similarly for $\mathbf{Q}(A^*)$.
(H.4) Either $\bar{A} \not\cong \bar{A}^*$, or $p \geq 5$.

We will also assume that for every k, $0 \leq k \leq m$, the quotient Selmer structure induced by $\mathcal{F}$ on A/p^kA (via the map $A \twoheadrightarrow A/p^kA$) is the same as the subgroup Selmer structure induced by $\mathcal{F}$ on A/p^kA (via the map $p^{m-k} : A/p^kA \hookrightarrow A$), where the induced Selmer structures are defined in Exercise 1.5.4. In other words, for every v,

(H.5) (image of $H^1_{\mathcal{F}}(\mathbf{Q}_v, A)$ under $H^1(\mathbf{Q}_v, A) \to H^1(\mathbf{Q}_v, A/p^kA)$)
$= $ (inverse image of $H^1_{\mathcal{F}}(\mathbf{Q}_v, A)$ under $H^1(\mathbf{Q}_v, A/p^kA) \to H^1(\mathbf{Q}_v, A)$).

Remark 2.4.1. Hypotheses (H.1), (H.2), and (H.3) are generally requiring that the image of $G_{\mathbf{Q}}$ in $\mathrm{Aut}(A)$ is sufficiently large. For a discussion of the condition (H.5) on $\mathcal{F}$, and some sufficient conditions for it to be satisfied, see [**MR1**, §3.7]. (Note that the condition on $\mathcal{F}$ is vacuous if $m = 1$, $R = \mathbf{F}_p$.)

If $(A, \mathcal{F})$ satisfy the conditions above, then so do $(A/p^kA, \mathcal{F})$ for $0 \leq k \leq m$.

Exercise 2.4.2. Show that under the assumptions above, $\mathcal{P}$ has positive density. Hint: check that if $\ell \notin \Sigma$ and the Frobenius of ℓ in $\mathrm{Gal}(\mathbf{Q}(A, \boldsymbol{\mu}_{p^m})/\mathbf{Q})$ is in the conjugacy class of the element τ of (H.2), then $\ell \in \mathcal{P}$.

Exercise 2.4.3. Show that (H.5) is automatically satisfied for $v \notin \Sigma$.

Exercise 2.4.4 (Lemma 3.7.4 of [**MR1**]). Show that if A and $\mathcal{F}$ satisfy the assumptions above, and $n \in \mathcal{N}$, then so do $(A, \mathcal{F}(n))$.

Proposition 2.4.5. *Under the assumptions above, if $n \in \mathcal{N}$ and $0 \le k \le m$, there are isomorphisms*

(1) $H^1_{\mathcal{F}(n)}(\mathbf{Q}, A/p^k A) \xrightarrow{\sim} H^1_{\mathcal{F}(n)}(\mathbf{Q}, A[p^k]) \xrightarrow{\sim} H^1_{\mathcal{F}(n)}(\mathbf{Q}, A)[p^k]$ *induced by* $A/p^k A \xrightarrow{\sim} A[p^k] \hookrightarrow A$,

(2) $H^1_{\mathcal{F}(n)}(\mathbf{Q}, A)[p^k] = \ker\left[H^1_{\mathcal{F}(n)}(\mathbf{Q}, A) \to H^1_{\mathcal{F}(n)}(\mathbf{Q}, A/p^{m-k} A)\right]$,

(3) $H^1_{\mathcal{F}(n)^*}(\mathbf{Q}, A^*[p^k]) \xrightarrow{\sim} H^1_{\mathcal{F}(n)^*}(\mathbf{Q}, A^*)[p^k]$ *induced by* $A^*[p^k] \hookrightarrow A^*$.

Exercise 2.4.6 (Lemmas 3.5.3, 3.5.4 of [**MR1**]). Prove Lemma 2.4.5. (By Exercise 2.4.4, it is enough to do this for $n = 1$.)

Exercise 2.4.7. Show that the composition $A/p^k A \xrightarrow{\sim} A[p^k] \hookrightarrow A$ induces an inclusion $\mathbf{KS}(A/p^k A) \hookrightarrow \mathbf{KS}(A)[p^k]$. Why is not obvious that this inclusion is an isomorphism?

2.5. The core Selmer group.

Recall that $\bar{A} = A \otimes \mathbf{F}_p$.

Definition 2.5.1. If $n \in \mathcal{N}$, define

$$\lambda(n, A) := \operatorname{length}(H^1_{\mathcal{F}(n)}(\mathbf{Q}, A)) = \operatorname{length}(\mathcal{H}(n)),$$
$$\lambda(n, A^*) := \operatorname{length}(H^1_{\mathcal{F}(n)^*}(\mathbf{Q}, A^*)),$$

and similarly with A replaced by $A/p^k A$. Note that by Proposition 2.4.5,

$$\lambda(n, A) = 0 \iff \lambda(n, \bar{A}) = 0, \quad \lambda(n, A^*) = 0 \iff \lambda(n, \bar{A}^*) = 0.$$

Theorem 2.5.2. *There are integers $r, s \ge 0$, with one of them equal to zero, such that for every $n \in \mathcal{N}$ there is a noncanonical isomorphism*

$$H^1_{\mathcal{F}(n)}(\mathbf{Q}, A) \oplus R^r \cong H^1_{\mathcal{F}(n)^*}(\mathbf{Q}, A^*) \oplus R^s.$$

PROOF. To prove the existence of such an isomorphism, it is enough to show that for $1 \le k \le m$, the kernels of p^k on the left-hand side and on the right-hand side have the same order. In other words, we need to show that there is a $t \in \mathbf{Z}$ such that for every n,

$$\operatorname{length}(H^1_{\mathcal{F}(n)}(\mathbf{Q}, A)[p^k]) - \operatorname{length}(H^1_{\mathcal{F}(n)^*}(\mathbf{Q}, A^*)[p^k]) = kt.$$

By Proposition 2.4.5, this is equivalent to

$$\lambda(n, A/p^k A) - \lambda(n, A^*[p^k]) = kt.$$

By Exercise 2.1.4, the left-hand side is independent of n, so we only need to consider $n = 1$. In this case what we need follows from [**Wi**, Proposition 1.6] (see also [**MR1**, Proposition 2.3.5]) and [**MR1**, Proposition 1.1.5]. □

Definition 2.5.3. Define the *core rank* $\chi(A)$ of A (suppressing $\mathcal{F}$ from the notation) to be the nonnegative integer s of Theorem 2.5.2. We call $n \in \mathcal{N}$ a *core vertex* if either $\lambda(n, A) = 0$ or $\lambda(n, A^*) = 0$.

Exercise 2.5.4. Show the following:

(1) If $\chi(A) > 0$, then $H^1_{\mathcal{F}(n)}(\mathbf{Q}, A) \cong H^1_{\mathcal{F}(n)^*}(\mathbf{Q}, A^*) \oplus R^{\chi(A)}$ for every $n \in \mathcal{N}$.

(2) If n is a core vertex, then $H^1_{\mathcal{F}(n)}(\mathbf{Q}, A)$ is free over R of rank $\chi(A)$,
(3) If n is a core vertex and $\chi(A) > 0$ then $H^1_{\mathcal{F}(n)^*}(\mathbf{Q}, A^*) = 0$.
(4) If n is a core vertex and $\chi(A) = 0$ then $H^1_{\mathcal{F}(n)}(\mathbf{Q}, A) = 0$.

2.6. Stepping along the Selmer graph.

The next proposition is an application of the global duality (Theorem 1.10.2). Proposition 2.6.1 and Corollary 2.6.2 are crucial in many of the calculations that follow, beginning in §2.8.

Proposition 2.6.1. *Suppose $n\ell \in \mathcal{N}$. We have the following diagrams of inclusions, in which the labels on the arrows are the lengths of the corresponding cyclic cokernels.*

$$\begin{array}{ccccc} & & H^1_{\mathcal{F}^\ell(n)}(\mathbf{Q}, A) & & \\ & \overset{a}{\nearrow} & & \overset{b}{\nwarrow} & \\ H^1_{\mathcal{F}(n)}(\mathbf{Q}, A) & & & & H^1_{\mathcal{F}(n\ell)}(\mathbf{Q}, A) \\ & \underset{c}{\nwarrow} & & \underset{d}{\nearrow} & \\ & & H^1_{\mathcal{F}_\ell(n)}(\mathbf{Q}, A) & & \end{array} \qquad \begin{array}{ccccc} & & H^1_{\mathcal{F}_\ell(n)^*}(\mathbf{Q}, A^*) & & \\ & \overset{c^*}{\nearrow} & & \overset{d^*}{\nwarrow} & \\ H^1_{\mathcal{F}(n)^*}(\mathbf{Q}, A^*) & & & & H^1_{\mathcal{F}(n\ell)^*}(\mathbf{Q}, A^*) \\ & \underset{a^*}{\nwarrow} & & \underset{b^*}{\nearrow} & \\ & & H^1_{\mathcal{F}^\ell(n)^*}(\mathbf{Q}, A^*) & & \end{array}$$

These lengths satisfy
(1) $0 \le a, b, c, d, a^*, b^*, c^*, d^* \le m$,
(2) $a + c = b + d,\ a^* + c^* = b^* + d^*$,
(3) $a + a^* = b + b^* = c + c^* = d + d^* = m$,
(4) $a \ge d,\ b \ge c,\ c^* \ge b^*,\ d^* \ge a^*$.

PROOF. By definition

$$H^1_{\mathcal{F}(n)}(\mathbf{Q}, A) = \ker[H^1_{\mathcal{F}^\ell(n)}(\mathbf{Q}, A) \to H^1(\mathbf{Q}_\ell, A)/H^1_{\mathrm{u}}(\mathbf{Q}_\ell, A)],$$
$$H^1_{\mathcal{F}_\ell(n)}(\mathbf{Q}, A) = \ker[H^1_{\mathcal{F}(n)}(\mathbf{Q}, A) \to H^1_{\mathrm{u}}(\mathbf{Q}_\ell, A)],$$

and similarly for A^*. The two diagrams come from the definitions in this way. Since $H^1_{\mathrm{u}}(\mathbf{Q}_\ell, A)$, $H^1_{\mathrm{t}}(\mathbf{Q}_\ell, A)$, $H^1_{\mathrm{u}}(\mathbf{Q}_\ell, A^*)$, and $H^1_{\mathrm{t}}(\mathbf{Q}_\ell, A^*)$ are all free of rank one over R by Exercise 1.9.7, the inequalities (1) hold. The equalities (2) are immediate from the diagrams.

The equality $a + a^* = m$ follows from global duality (apply Theorem 1.10.2 with $\mathcal{G}_1 = \mathcal{F}(n)$ and $\mathcal{G}_2 = \mathcal{F}^\ell(n)$), and similarly for the other three equalities of (3). Finally, we have

$$H^1_{\mathcal{F}(n)}(\mathbf{Q}, A) \cap H^1_{\mathcal{F}(n\ell)}(\mathbf{Q}, A) = H^1_{\mathcal{F}_\ell(n)}(\mathbf{Q}, A).$$

The first two inequalities of (4) follow from this, and the other two similarly with $(A, \mathcal{F})$ replaced by $(A^*, \mathcal{F}^*)$. □

Corollary 2.6.2. *Suppose $n\ell \in \mathcal{N}$.*
(1) $|\lambda(n\ell, A) - \lambda(n, A)| \le m$ *and* $|\lambda(n\ell, A^*) - \lambda(n, A^*)| \le m$.
(2) *If the localization map $H^1_{\mathcal{F}(n)}(\mathbf{Q}, A) \to H^1_{\mathrm{u}}(\mathbf{Q}_\ell, A)$ is surjective, then*

$$H^1_{\mathcal{F}(n\ell)^*}(\mathbf{Q}, A^*) = H^1_{\mathcal{F}^\ell(n)^*}(\mathbf{Q}, A^*) \subset H^1_{\mathcal{F}(n)^*}(\mathbf{Q}, A^*).$$

(3) *The image of the composition*

$$p^{\lambda(n,A^*)}H^1_{\mathcal{F}(n)}(\mathbf{Q},A) \xrightarrow{\mathrm{loc}_\ell} H^1_{\mathrm{u}}(\mathbf{Q}_\ell,A) \xrightarrow{\phi_\ell^{\mathrm{ut}}} H^1_{\mathrm{t}}(\mathbf{Q}_\ell,A)\otimes G_\ell$$

is equal to the image of $p^{\lambda(n\ell,A^*)}H^1_{\mathcal{F}(n\ell)}(\mathbf{Q},A)\otimes G_\ell \xrightarrow{\mathrm{loc}_\ell} H^1_{\mathrm{t}}(\mathbf{Q}_\ell,A)\otimes G_\ell$.

(4) *If both localization maps*

$$H^1_{\mathcal{F}(n)}(\mathbf{Q},A)[p] \to H^1_{\mathrm{u}}(\mathbf{Q}_\ell,A), \qquad H^1_{\mathcal{F}(n)^*}(\mathbf{Q},A^*)[p] \to H^1_{\mathrm{u}}(\mathbf{Q}_\ell,A^*)$$

are nonzero, then $\lambda(n\ell,\bar{A}) = \lambda(n,\bar{A})-1$ *and* $\lambda(n\ell,\bar{A}^*) = \lambda(n,\bar{A}^*)-1$.

Proof. Consider the diagrams of Proposition 2.6.1. Assertion (1) is immediate. Recall (Exercise 1.9.7) that $H^1_{\mathrm{u}}(\mathbf{Q}_\ell,A)$, $H^1_{\mathrm{t}}(\mathbf{Q}_\ell,A)$, $H^1_{\mathrm{u}}(\mathbf{Q}_\ell,A^*)$, and $H^1_{\mathrm{t}}(\mathbf{Q}_\ell,A^*)$ are free of rank one over R, and ϕ_ℓ^{ut} is an isomorphism.

If the localization map in (2) is surjective, then $c=m$ in the left-hand diagram of Proposition 2.6.1. Therefore by (3) and (4) of that lemma, $b^*=0$, which proves (2).

To prove (3) it is enough to show that $\mathrm{length}(C_n) = \mathrm{length}(C_{n\ell})$ where C_n and $C_{n\ell}$ are the images of $p^{\lambda(n,A^*)}H^1_{\mathcal{F}(n)}(\mathbf{Q},A)$ and $p^{\lambda(n\ell,A^*)}H^1_{\mathcal{F}(n\ell)}(\mathbf{Q},A)$, respectively, under localization at ℓ. The left-hand diagram of Proposition 2.6.1 shows that

$$\mathrm{length}(C_n) = \max\{0, c-\lambda(n,A^*)\}, \qquad \mathrm{length}(C_{n\ell}) = \max\{0, d-\lambda(n\ell,A^*)\}.$$

The right-hand diagram shows that

$$\lambda(n,A^*)-\lambda(n\ell,A^*) = d^*-c^* = c-d,$$

so $\mathrm{length}(C_n) = \mathrm{length}(C_{n\ell})$.

For (4), if the localization maps in question are nonzero, then using Proposition 2.4.5 (with $k=1$) we see that the localization maps

$$H^1_{\mathcal{F}(n)}(\mathbf{Q},\bar{A}) \to H^1_{\mathrm{u}}(\mathbf{Q}_\ell,\bar{A}), \quad H^1_{\mathcal{F}(n)^*}(\mathbf{Q},\bar{A}^*) \to H^1_{\mathrm{u}}(\mathbf{Q}_\ell,\bar{A}^*)$$

are surjective. Applying Proposition 2.6.1 with A replaced by $\bar{A}$ (and $m=1$) we have $c=a^*=1$, so $b^*\le c^*=0$ and $d\le a=0$, so $\lambda(n\ell,\bar{A}^*) = \lambda(n,\bar{A}^*)-1$ and $\lambda(n\ell,\bar{A}) = \lambda(n,\bar{A})-1$. □

2.7. Choosing useful primes.

Proposition 2.7.1. *Suppose* $c\in H^1_{\mathcal{F}(n)}(\mathbf{Q},A)$ *and* $d\in H^1_{\mathcal{F}(n)^*}(\mathbf{Q},A^*)$ *are both nonzero. Then* $\{\ell\in\mathcal{P} : \mathrm{Res}_\ell(c)\ne 0 \text{ and } \mathrm{Res}_\ell(d)\ne 0\}$ *has positive density.*

Sketch of proof. (For details, see [**MR1**, Proposition 3.6.1].) First, find $\gamma\in G_{\mathbf{Q}}$ such that

(1) $\gamma=\tau$ on $\mathbf{Q}(A,\boldsymbol{\mu}_{p^k})$, where τ is as in (H.2) of §2.4,
(2) $c(\gamma)\notin(\tau-1)A$,
(3) $d(\gamma)\notin(\tau-1)A^*$.

where $c(\gamma)$ means the image of γ under any cocycle representing c, which is well-defined modulo $(\gamma-1)A=(\tau-1)A$, and similarly for $d(\gamma)$. (One needs to check that these three conditions can be satisfied simultaneously; this uses our hypotheses on A).

If $\ell\notin\Sigma$, $\ell\nmid n$, is a prime whose Frobenius (in a suitably large extension of $\mathbf{Q}$) is γ, then it follows from the first condition that $\ell\in\mathcal{P}$ (Exercise 2.4.2), from the second condition that $\mathrm{Res}_\ell(c)\in H^1_{\mathrm{u}}(\mathbf{Q}_\ell,A)\cong A/(\tau-1)A$ is nonzero, and from the third condition that $\mathrm{Res}_\ell(d) \in H^1_{\mathrm{u}}(\mathbf{Q}_\ell, A^*) \cong A^*/(\tau - 1)A^*$ is nonzero. Thus ℓ belongs to the set in question. □

Corollary 2.7.2. *Suppose $n \in \mathcal{N}$ is not a core vertex. Then there is an $\ell \in \mathcal{P}$ such that $\lambda(n\ell, \bar{A}) = \lambda(n, \bar{A}) - 1$ and $\lambda(n\ell, \bar{A}^*) = \lambda(n, \bar{A}^*) - 1$.*

If in addition $\boldsymbol{\kappa} \in \mathbf{KS}(A)$ and $\kappa_n \neq 0$, then ℓ can be chosen so $\kappa_{n\ell} \neq 0$.

PROOF. Since n is not a core vertex, $H^1_{\mathcal{F}(n)}(\mathbf{Q}, A)$ and $H^1_{\mathcal{F}(n)^*}(\mathbf{Q}, A^*)$ are both nonzero. Apply Proposition 2.7.1 with a nonzero $c \in H^1_{\mathcal{F}(n)}(\mathbf{Q}, A)[p]$ and a nonzero $d \in H^1_{\mathcal{F}(n)^*}(\mathbf{Q}, A^*)[p]$, and let $\ell \in \mathcal{P}$ be such that $\mathrm{Res}_\ell(c) \neq 0$ and $\mathrm{Res}_\ell(d) \neq 0$. For this ℓ, Corollary 2.6.2(4) shows that $\lambda(n\ell, \bar{A}) = \lambda(n, \bar{A}) - 1$ and $\lambda(n\ell, \bar{A}^*) = \lambda(n, \bar{A}^*) - 1$.

If we take c to be a multiple of κ_n, then $\mathrm{Res}_\ell(\kappa_n) \neq 0$, so $\mathrm{Res}_\ell(\kappa_{n\ell}) \neq 0$ (by definition of a Kolyvagin system), so $\kappa_{n\ell} \neq 0$. □

Corollary 2.7.3. *Suppose $n \in \mathcal{N}$. Then there is a core vertex n' and a path of length* $\min\{\lambda(n, \bar{A}), \lambda(n, \bar{A}^*)\}$ *from n to n'.*

If in addition $\boldsymbol{\kappa} \in \mathbf{KS}(A)$ and $\kappa_n \neq 0$, then n' can be chosen so $\kappa_{n'} \neq 0$.

If $n \in \mathcal{N}$, let $\omega(n)$ denote the number of prime factors of n.

Corollary 2.7.4. *Let $r = \min\{\dim_{\mathbf{F}_p} H^1_{\mathcal{F}}(\mathbf{Q}, \bar{A}), \dim_{\mathbf{F}_p} H^1_{\mathcal{F}^*}(\mathbf{Q}, \bar{A}^*)\}$. If n is a core vertex, then $\omega(n) \geq r$. There are core vertices n with $\omega(n) = r$.*

Exercise 2.7.5. Prove Corollaries 2.7.3 and 2.7.4.

Theorem 2.7.6. *If $\chi(A) = 0$, then $\mathbf{KS}(A) = 0$.*

PROOF. Suppose $\boldsymbol{\kappa} \in \mathbf{KS}(A)$ and $n \in \mathcal{N}$. If $\kappa_n \neq 0$, then by Corollary 2.7.3 there is a core vertex n' such that $\kappa_{n'} \in H^1_{\mathcal{F}(n')}(\mathbf{Q}, A)$ is nonzero. But since $\chi(A) = 0$, we have $H^1_{\mathcal{F}(n')}(\mathbf{Q}, A) = 0$ by Exercise 2.5.4(4), so this is impossible. □

Remark 2.7.7. Theorem 2.7.6 gives another proof of Exercise 2.2.4, because if $H^1_{\mathcal{F}}(\mathbf{Q}, A) = 0$ then Theorem 2.5.2 shows that $\chi(A) = 0$.

2.8. The stub subsheaf.

We define a subsheaf of the Selmer sheaf $\mathcal{H} = \mathcal{H}_{(A,\mathcal{F})}$ of Definition 2.2.3 as follows.

Definition 2.8.1. The *sheaf of stub Selmer modules* $\mathcal{H}' = \mathcal{H}'_{(A,\mathcal{F})}$ is the subsheaf of $\mathcal{H}$ defined by

- $\mathcal{H}'(n) := p^{\lambda(n,A^*)}\mathcal{H}(n) = p^{\lambda(n,A^*)} H^1_{\mathcal{F}(n)}(\mathbf{Q}, A) \otimes G_n \subset \mathcal{H}(n)$ if $n \in \mathcal{N}$,
- $\mathcal{H}'(e) := \psi^e_n(\mathcal{H}'(n)) \subset \mathcal{H}(e)$ if e is an edge joining n and $n\ell$,

and the vertex-to-edge maps $\psi^e_n, \psi^e_{n\ell}$ are the restrictions of those of $\mathcal{H}$.

By Corollary 2.6.2, $\psi^e_{n\ell}(\mathcal{H}'(n\ell)) = \psi^e_n(\mathcal{H}'(n))$, so $\mathcal{H}'$ is a well-defined sheaf and all vertex-to-edge maps are surjective.

Exercise 2.8.2. Show that for every n, $\mathcal{H}'(n)$ is free of rank $\chi(A)$ over $\mathbf{Z}/p^k\mathbf{Z}$, where $k = \max\{m - \lambda(n, A^*), 0\}$. Deduce that if $\chi(A) = 1$, then $\mathcal{H}'$ is locally cyclic.

Exercise 2.8.3. Show that if $\chi(A) = 1$, then Corollary 2.7.3 can be strengthened to say that for every vertex n, there is a *surjective path* (see Definition 2.3.1) from some core vertex to n. (Hint: in the proof of Corollary 2.7.2, take c to be a nonzero element of $p^{m-1}H^1_{\mathcal{F}(n)}(\mathbf{Q}, A)$. Show that

$$\mathrm{Res}_\ell : H^1_{\mathcal{F}(n)}(\mathbf{Q}, A) \to H^1_{\mathrm{u}}(\mathbf{Q}_\ell, A)$$

is surjective, and so

$$\mathrm{Res}_\ell : p^{\lambda(n,A^*)}H^1_{\mathcal{F}(n)}(\mathbf{Q}, A) \to p^{\lambda(n,A^*)}H^1_{\mathrm{u}}(\mathbf{Q}_\ell, A)$$

is an isomorphism.)

Theorem 2.8.4. *Suppose $\chi(A) = 1$. Then the natural inclusion $\Gamma(\mathcal{H}') \hookrightarrow \Gamma(\mathcal{H})$ is an isomorphism, i.e., if $\boldsymbol{\kappa} \in \mathbf{KS}(A)$ then $\kappa_n \in \mathcal{H}'(n)$ for every $n \in \mathcal{N}$.*

Sketch of proof. We give the proof in the special case where $m = 1$, so $R = \mathbf{F}_p$. See [**MR1**, Theorem 4.3.4] for the complete proof. (In this special case, it is not necessary to assume that $\chi(A) = 1$.)

Suppose $R = \mathbf{F}_p$ and $\boldsymbol{\kappa} \in \mathbf{KS}(A)$. If n is a core vertex, then $\mathcal{H}'(n) = \mathcal{H}(n)$ and there is nothing to prove. If n is not a core vertex, then $\mathcal{H}'(n) = 0$ and we need to show $\kappa_n = 0$.

Suppose n is not a core vertex, and $\kappa_n \neq 0$. Using Proposition 2.7.1, we can find an $\ell \in \mathcal{P}$ such that $\mathrm{Res}_\ell(\kappa_n) \neq 0$ and $\mathrm{Res}_\ell(H^1_{\mathcal{F}^*}(\mathbf{Q}, A^*)) \neq 0$. In the notation of Proposition 2.6.1, we have $a^* = 1$, so $a = 0$, so $d = 0$. But then $\mathrm{Res}_\ell(\kappa_{n\ell}) = 0$, so $\mathrm{Res}_\ell(\kappa_n) = 0$ by definition of a Kolyvagin system. This is a contradiction. □

Theorem 2.8.5. *Suppose $\chi(A) = 1$. Then the stub sheaf $\mathcal{H}'$ is locally cyclic, and every core vertex is a hub.*

A key part of the proof of Theorem 2.8.5 is the following.

Definition 2.8.6. Suppose $\chi(A) = 1$. Let $\mathcal{X}^0$ be the subgraph of $\mathcal{X}$ whose vertices are the core vertices in $\mathcal{N}$, with an edge joining n and $n\ell$ whenever both maps ψ^e_n and $\psi^e_{n\ell}$ are isomorphisms.

Let $\mathcal{H}^0$ be the restriction of the Selmer sheaf $\mathcal{H}$ to $\mathcal{X}^0$, i.e., $\mathcal{H}^0(n) = \mathcal{H}(n)$ for core vertices n, and $\mathcal{H}^0(e) = \mathcal{H}(e)$. Note that $\mathcal{H}^0$ is also the restriction of $\mathcal{H}'$ to $\mathcal{X}^0$, since $\mathcal{H}'$ and $\mathcal{H}$ agree precisely at the vertices and edges of $\mathcal{X}^0$.

Theorem 2.8.7. *The graph $\mathcal{X}^0$ is connected.*

Proof. The proof is "more of the same". See [**MR1**, Theorem 4.3.12] □

Proof of Theorem 2.8.5. We have already seen (Exercise 2.8.2) that $\mathcal{H}'$ is locally cyclic. Suppose $n \in \mathcal{N}$ and v is a core vertex; we need to show that there is a surjective path from v to n. By Exercise 2.8.3, there is a core vertex w and a surjective path from w to n. By Theorem 2.8.7, there is a path in $\mathcal{X}^0$ from v to w. But every path in $\mathcal{X}^0$ is a surjective path, so the concatenated path from v to w to n is a surjective path from v to n. □

Theorem 2.8.8 (Howard, Appendix B of [**MR1**]). *For every $n \in \mathcal{N}$, the map $\Gamma(\mathcal{H}') \to \mathcal{H}'(n)$ is surjective.*

Corollary 2.8.9. (1) *If $\chi(A) = 0$, then $\mathbf{KS}(A) = 0$.*
(2) *If $\chi(A) = 1$, then $\mathbf{KS}(A)$ is free of rank one over R.*
(3) *If $\chi(A) > 1$, then $\mathbf{KS}(A)$ is infinite.*

PROOF. The first assertion is Theorem 2.7.6.

Suppose $\chi(A) = 1$ and n is a core vertex (core vertices exist by Corollary 2.7.3). Then $\mathcal{H}(n) = \mathcal{H}'(n)$, and by Theorem 2.8.4 we have $\mathbf{KS}(A) := \Gamma(\mathcal{H}) = \Gamma(\mathcal{H}')$. Therefore Theorem 2.8.5 and Proposition 2.3.2(1) show that the map $\mathbf{KS}(A) \to \mathcal{H}(n)$ is injective, Theorem 2.8.8 shows that it is surjective, and Exercise 2.5.4(2) shows that $\mathcal{H}(n)$ is free of rank one over R.

For (3), see the following exercise. □

Exercise 2.8.10 (Proof of Corollary 2.8.9(3)). If $\chi(A, \mathcal{F}) > 0$, and $\ell \in \mathcal{P}$, use Proposition 2.6.1 to show that $\chi(\bar{A}, \mathcal{F}_\ell) = \chi(\bar{A}, \mathcal{F}) - 1$. (Note: the Selmer structure $\mathcal{F}_\ell$ for A may not satisfy condition (H.5) of §2.4, but for $\bar{A}$, (H.5) is vacuous.) Deduce that there are many different n such that $\chi(\bar{A}, \mathcal{F}_n) = 1$, and for each of these we have $\mathbf{KS}(\bar{A}, \mathcal{F}_n) \subset \mathbf{KS}(\bar{A}, \mathcal{F}) = \mathbf{KS}(A, \mathcal{F})[p]$. Now apply Corollary 2.8.9(2) to each of the $\mathbf{KS}(\bar{A}, \mathcal{F}_n)$.

We end this section with one example of how to use the results above to bound the size of a Selmer group.

Corollary 2.8.11. *Suppose $\chi(A) = 1$ and $\boldsymbol{\kappa} \in \mathbf{KS}(A)$. If $\kappa_1 \in H^1_{\mathcal{F}}(\mathbf{Q}, A)$ has order p^k, then*

$$\mathrm{length}(H^1_{\mathcal{F}^*}(\mathbf{Q}, A^*)) \le m - k.$$

PROOF. By Theorem 2.8.4(1), $\kappa_1 \in p^{\mathrm{length}(H^1_{\mathcal{F}^*}(\mathbf{Q},A^*))} H^1_{\mathcal{F}}(\mathbf{Q}, A)$, so κ_1 is killed by $p^{m-\mathrm{length}(H^1_{\mathcal{F}^*}(\mathbf{Q},A^*))}$. □

2.9. Recovering the group structure of the dual Selmer group.

For this section we suppose that A is a $G_{\mathbf{Q}}$-module, free of finite rank over $R = \mathbf{Z}/p^m\mathbf{Z}$, and $\mathcal{F}$ is a Selmer structure for A, satisfying all the hypotheses of §2.4. We assume in addition that $\chi(A) = 1$.

Proposition 2.9.1. *Suppose $\chi(A) = 1$ and $0 \le k \le m$. The natural quotient map $A \twoheadrightarrow A/p^kA$ induces a map $\mathbf{KS}(A) \twoheadrightarrow \mathbf{KS}(A/p^kA)$.*

Exercise 2.9.2. Prove Proposition 2.9.1 Hint: this is not as easy as it might seem, because the index set $\mathcal{N}$ in the definition of a Kolyvagin system is larger for A/p^kA than for A. Use Exercise 2.4.7, and show that the map $\mathbf{KS}(A/p^kA) \hookrightarrow \mathbf{KS}(A)[p^k]$ of Exercise 2.4.7 is an isomorphism if $\chi(A) = 1$.

Proposition 2.9.3. *Suppose $\chi(A) = 1$ and $\boldsymbol{\kappa} \in \mathbf{KS}(A)$. Then the following are equivalent.*

(1) *κ_n generates $\mathcal{H}'(n)$ for every $n \in \mathcal{N}$,*
(2) *κ_n generates $\mathcal{H}'(n)$ for some $n \in \mathcal{N}$ with $\mathcal{H}'(n) \neq 0$,*
(3) *$\boldsymbol{\kappa} \notin p\mathbf{KS}(A)$,*
(4) *$p^{m-1}\boldsymbol{\kappa} \neq 0$,*
(5) *$\boldsymbol{\kappa}$ generates $\mathbf{KS}(A)$,*
(6) *the image of $\boldsymbol{\kappa}$ in $\mathbf{KS}(\bar{A})$ (under the map of Proposition 2.9.1) is nonzero,*

Exercise 2.9.4. Prove Proposition 2.9.3, using Theorems 2.8.4, 2.8.5, and 2.8.8, and Proposition 2.3.2.

Definition 2.9.5. We say that $\boldsymbol{\kappa} \in \mathbf{KS}(A)$ is *primitive* if it satisfies the equivalent conditions of Proposition 2.9.3

If $\boldsymbol{\kappa}$ is primitive, we have the following sharpening of Corollary 2.8.11.

Corollary 2.9.6. *Suppose $\chi(A) = 1$ and $\boldsymbol{\kappa} \in \mathbf{KS}(A)$ is primitive. If $\kappa_1 \neq 0$ then*

$$\operatorname{length}(H^1_{\mathcal{F}^*}(\mathbf{Q}, A^*)) = m - \operatorname{length}(R\kappa_1) = \max\{i : \kappa_1 \in p^i H^1_{\mathcal{F}}(\mathbf{Q}, A)\}.$$

If $\kappa_1 = 0$ then $\operatorname{length}(H^1_{\mathcal{F}^*}(\mathbf{Q}, A^*)) \geq k$.

Proof. Let $\lambda = \operatorname{length}(H^1_{\mathcal{F}^*}(\mathbf{Q}, T^*))$. Since $\boldsymbol{\kappa}$ is primitive, we have

$$R\kappa_1 = \mathcal{H}'(1) = p^\lambda H^1_{\mathcal{F}}(\mathbf{Q}, A),$$

and $\operatorname{length}(p^\lambda H^1_{\mathcal{F}}(\mathbf{Q}, A)) = \max\{m - \lambda, 0\}$ by Exercise 2.8.2. □

We can formulate a more precise version of Corollary 2.9.6, which in some cases determines the group structure of $H^1_{\mathcal{F}^*}(\mathbf{Q}, T^*)$. Recall that $\omega(n)$ denotes the number of prime divisors of $n \in \mathcal{N}$.

Definition 2.9.7. If $\boldsymbol{\kappa} \in \mathbf{KS}(T)$ and $r \geq 0$, define

$$\partial^r(\boldsymbol{\kappa}) := \max\{k : k < m \text{ and } \kappa_n \in p^k H^1_{\mathcal{F}(n)}(\mathbf{Q}, A) \otimes G_n \\ \text{for every } n \in \mathcal{N} \text{ with } \omega(n) = r\}.$$

The *elementary divisors* of $\boldsymbol{\kappa}$ are defined by

$$e_i(\boldsymbol{\kappa}) := \partial^{i-1}(\boldsymbol{\kappa}) - \partial^i(\boldsymbol{\kappa}), \quad i \geq 1.$$

Proposition 2.9.8. *Suppose $\chi(A) = 1$, $\boldsymbol{\kappa} \in \mathbf{KS}(A)$, and let j be such that $\boldsymbol{\kappa}$ generates $p^j\mathbf{KS}(A)$. Write $H^1_{\mathcal{F}^*}(\mathbf{Q}, A^*) \cong \oplus_i \mathbf{Z}/p^{d_i}\mathbf{Z}$ with nonnegative integers $d_1 \geq d_2 \geq \cdots$. Then for every $r \geq 0$,*

$$\partial^r(\boldsymbol{\kappa}) = \min\{m, j + \sum_{i>r} d_i\}.$$

Proof. Note that since $\boldsymbol{\kappa}$ generates $p^j\mathbf{KS}(A)$, Proposition 2.3.2 shows that κ_n generates $p^j\mathcal{H}'(n) = p^{j+\lambda(n,A^*)}\mathcal{H}(n)$ for every n. Therefore

$$\partial^r(\boldsymbol{\kappa}) = \min\{m, j + \lambda(n, A^*) : n \in \mathcal{N}, \omega(n) = r\},$$

and to prove the proposition we need to show that

$$\min\{\lambda(n, A^*) : n \in \mathcal{N}, \omega(n) = r\} = \sum_{i>r} d_i. \tag{2.3}$$

This is clear when $r = 0$, since $\lambda(1, A^*) = \sum_i d_i$.

Suppose $n \in \mathcal{N}$ and $\omega(n) = r$. Consider the map

$$H^1_{\mathcal{F}^*}(\mathbf{Q}, A^*) \longrightarrow \bigoplus_{\ell \mid n} H^1_{\mathrm{f}}(\mathbf{Q}_\ell, A^*).$$

The right-hand side is free of rank r over R, so the image is a quotient of $H^1_{\mathcal{F}^*}(\mathbf{Q}, A^*)$ generated by (at most) r elements. Hence the image has length at most $\sum_{i \leq r} d_i$. Therefore the kernel has length at least $\sum_{i>r} d_i$. But by definition this kernel is $H^1_{(\mathcal{F}^*)_n}(\mathbf{Q}, A^*) \subset H^1_{\mathcal{F}(n)^*}(\mathbf{Q}, A^*)$, so we conclude that $\lambda(n, A^*) \geq \sum_{i>r} d_i$, which is one inequality of (2.3).

We will prove the opposite inequality by induction on r (for varying $\mathcal{F}$). The case $r = 0$ was proved above.

Since $\chi(A) > 0$, Theorem 2.5.2 shows that $p^{m-1}H^1_{\mathcal{F}}(\mathbf{Q}, A) \neq 0$. Fix a nonzero element $c \in p^{m-1}H^1_{\mathcal{F}}(\mathbf{Q}, A) \subset H^1_{\mathcal{F}}(\mathbf{Q}, A)[p]$. If $d_1 > 0$, then choose a nonzero element $c^* \in p^{d_1-1}H^1_{\mathcal{F}^*}(\mathbf{Q}, A^*) \subset H^1_{\mathcal{F}^*}(\mathbf{Q}, A^*)[p]$. Using Proposition 2.7.1, choose a prime $\ell \in \mathcal{P}$ such that the localization $\mathrm{Res}_\ell(c)$ is nonzero and, if $d_1 > 0$, such that $\mathrm{Res}_\ell(c^*)$ is nonzero as well.

It follows that

- the localization map $H^1_{\mathcal{F}}(\mathbf{Q}, A) \to H^1_{\mathrm{u}}(\mathbf{Q}_\ell, A)$ is surjective, and
- $H^1_{(\mathcal{F}^\ell)^*}(\mathbf{Q}, A^*) \cong \oplus_{i\geq 1}\mathbf{Z}/p^{d_{i+1}}\mathbf{Z}$.

By Theorem 2.6.2(2) we have $H^1_{\mathcal{F}(\ell)^*}(\mathbf{Q}, A^*) = H^1_{(\mathcal{F}^\ell)^*}(\mathbf{Q}, A^*)$. Thus

$$\begin{aligned}\min\{\lambda(n, A^*) : n \in \mathcal{N}, \omega(n) = r\} &\leq \min\{\lambda(n\ell, A^*) : n \in \mathcal{N}, \ell \nmid n, \omega(n) = r-1\} \\ &= \min\{\mathrm{length}_R(H^1_{\mathcal{F}(\ell)(n)^*}(\mathbf{Q}, A^*)) : n \in \mathcal{N}, \ell \nmid n, \omega(n) = r-1\} \\ &= \sum_{i>r-1} d_{i+1}\end{aligned}$$

where the last equality comes from our induction hypothesis applied to $(A, \mathcal{F}(\ell))$. This completes the proof of (2.3). □

Theorem 2.9.9. *Suppose* $\chi(A) = 1$, $\boldsymbol{\kappa} \in \mathbf{KS}(A)$, *and* $\kappa_1 \neq 0$. *Then*

$$\partial^0(\boldsymbol{\kappa}) \geq \partial^1(\boldsymbol{\kappa}) \geq \partial^2(\boldsymbol{\kappa}) \geq \cdots \geq 0,$$
$$e_1(\boldsymbol{\kappa}) \geq e_2(\boldsymbol{\kappa}) \geq e_3(\boldsymbol{\kappa}) \geq \cdots \geq 0,$$

$\partial^r(\boldsymbol{\kappa})$ *stabilizes at a value* $\partial^\infty(\boldsymbol{\kappa})$ *for large* r, $e_r(\boldsymbol{\kappa}) = 0$ *for large* r, *and*

$$H^1_{\mathcal{F}^*}(\mathbf{Q}, A^*) \cong \bigoplus_{i\geq 1}\mathbf{Z}/p^{e_i(\boldsymbol{\kappa})}\mathbf{Z}.$$

In particular $\mathrm{length}(H^1_{\mathcal{F}^*}(\mathbf{Q}, A^*)) = \partial^0(\boldsymbol{\kappa}) - \partial^\infty(\boldsymbol{\kappa})$.

Proof. If $\kappa_1 \neq 0$ then $\partial^0(\boldsymbol{\kappa}) < m$, so in Proposition 2.9.8 we have $\partial^r(\boldsymbol{\kappa}) = j + \sum_{i>r} d_i$ for every r. The theorem follows immediately. □

Remark 2.9.10. Theorem 2.9.9 shows that the elementary divisors of $\boldsymbol{\kappa}$ are independent of $\boldsymbol{\kappa}$ as long as $\kappa_1 \neq 0$. Note that $\boldsymbol{\kappa}$ is primitive if and only if $\partial^\infty(\boldsymbol{\kappa}) = 0$.

LECTURE 3

Kolyvagin systems for p-adic representations.

Suppose for this lecture that $\mathcal{A}$ is a free $\mathbf{Z}_p$-module of finite rank, with a continuous action of $G_{\mathbf{Q}}$. We assume further that $\mathcal{A}$ is ramified at only finitely many primes (i.e., for all but finitely many primes ℓ, the inertia group of every prime above ℓ acts trivially on $\mathcal{A}$). Let $\mathcal{A}^* = \mathrm{Hom}(\mathcal{A}, \boldsymbol{\mu}_{p^\infty})$. For every positive integer m we have the finite $G_{\mathbf{Q}}$-modules $A_m := A/p^m A$ and $A_m^* = \mathcal{A}^*[p^m]$. We will study Selmer groups attached to $\mathcal{A}^*$ by applying the results of the previous section to A_m and A_m^*.

Typical examples for $\mathcal{A}$ will be

$$\mathbf{Z}_p(1) := \varprojlim \boldsymbol{\mu}_{p^m}, \qquad \mathbf{Z}_p(1)^* = \mathbf{Q}_p/\mathbf{Z}_p,$$
$$T_p(E) := \varprojlim E[p^m], \; T_p(E)^* = E[p^\infty],$$

for an elliptic curve E defined over $\mathbf{Q}$.

3.1. Cohomology groups for p-adic representations.

Exercise 3.1.1. Show that $\mathcal{A}^*$ is a discrete, torsion $G_{\mathbf{Q}}$-module, and $\mathcal{A}^* = \varinjlim A_m^*$.

Since $\mathcal{A}$ is not a discrete $G_{\mathbf{Q}}$-module, our definitions in Lecture 1 do not apply. We will simply define, for any closed subgroup G of $G_{\mathbf{Q}}$,

$$H^0(G, \mathcal{A}) := \mathcal{A}^G, \qquad H^1(G, \mathcal{A}) := \varprojlim H^1(G, A_m).$$

This is the same group one would get by defining $H^1(G, \mathcal{A})$ using continuous cocycles, but that would not be true in general for $H^i(G, \mathcal{A})$ with $i > 1$. Since $\mathcal{A}^*$ is a discrete $G_{\mathbf{Q}}$-module, no modifications are necessary to define $H^i(G, \mathcal{A}^*)$.

The definitions of $H^1_{\mathrm{u}}(\mathbf{Q}_\ell, \mathcal{A})$, Selmer structures for $\mathcal{A}$, and the corresponding Selmer groups, are exactly the same as Definitions 1.4.12 and 1.5.1. If $\mathcal{F}$ is a Selmer structure for $\mathcal{A}$, then $\mathcal{F}$ induces a Selmer structure (that we also denote by $\mathcal{F}$) for every A_m, by taking $H^1_{\mathcal{F}}(\mathbf{Q}_v, A_m)$ to be the image of $H^1_{\mathcal{F}}(\mathbf{Q}_v, \mathcal{A})$ under the canonical map $H^1(\mathbf{Q}_v, \mathcal{A}) \to H^1(\mathbf{Q}_v, A_m)$. This in turn defines a Selmer structure $\mathcal{F}^*$ on every A_m^*, and we also denote by $\mathcal{F}^*$ the Selmer structure on $\mathcal{A}^*$ defined by $H^1_{\mathcal{F}^*}(\mathbf{Q}_\ell, \mathcal{A}^*) = \varinjlim H^1_{\mathcal{F}^*}(\mathbf{Q}_\ell, A_m^*)$.

Exercise 3.1.2. Show that the construction described above defines a Selmer structure on A_m for every m (i.e., show that if $\mathcal{A}$ is unramified at ℓ and $H^1_{\mathcal{F}}(\mathbf{Q}_\ell, \mathcal{A}) = H^1_{\mathrm{u}}(\mathbf{Q}_\ell, \mathcal{A})$, then $H^1_{\mathcal{F}}(\mathbf{Q}_\ell, A_m) = H^1_{\mathrm{u}}(\mathbf{Q}_\ell, A_m)$). Show that

$$H^1_{\mathcal{F}}(\mathbf{Q}, \mathcal{A}) = \varprojlim H^1_{\mathcal{F}}(\mathbf{Q}, A_m), \qquad H^1_{\mathcal{F}^*}(\mathbf{Q}, \mathcal{A}^*) = \varinjlim H^1_{\mathcal{F}^*}(\mathbf{Q}, A_m^*).$$

We will make the following assumptions on $(\mathcal{A}, \mathcal{F})$, in order to ensure that the $(A_m, \mathcal{F})$ satisfy the hypotheses of §2.4. Let $\bar{A} := \mathcal{A}/p\mathcal{A} = A_1$, so $(\bar{A})^* = \mathcal{A}^*[p]$.

(H.1) $\bar{\mathcal{A}}$ is an absolutely simple $\mathbf{F}_p[G_{\mathbf{Q}}]$-module, not isomorphic (as a $G_{\mathbf{Q}}$-module) to $\mathbf{F}_p$ or to $\boldsymbol{\mu}_p$.
(H.2) There is a $\tau \in G_{\mathbf{Q}}$ such that $\tau = 1$ on $\boldsymbol{\mu}_{p^\infty}$ and $\mathcal{A}/(\tau-1)\mathcal{A}$ is free of rank one over $\mathbf{Z}_p$.
(H.3) $H^1(\mathbf{Q}(\mathcal{A})/\mathbf{Q}, \bar{\mathcal{A}}) = H^1(\mathbf{Q}(\mathcal{A}^*)/\mathbf{Q}, \bar{\mathcal{A}}^*) = 0$, where $\mathbf{Q}(\mathcal{A})$ is the fixed field of the kernel of the map $G_{\mathbf{Q}} \to \mathrm{Aut}(\mathcal{A})$, and similarly for $\mathbf{Q}(\mathcal{A}^*)$.
(H.4) Either $\bar{\mathcal{A}} \not\cong \bar{\mathcal{A}}^*$, or $p \geq 5$.

Let Σ be a finite set of places containing ∞, p, all ℓ such that $\mathcal{A}$ is ramified at ℓ, and all ℓ such that $H^1_{\mathcal{F}}(\mathbf{Q}_\ell, \mathcal{A}) \neq H^1_{\mathrm{u}}(\mathbf{Q}_\ell, \mathcal{A})$. We will also assume that:

(H.5) For every $v \in \Sigma$, $H^1(\mathbf{Q}_v, \mathcal{A})/H^1_{\mathcal{F}}(\mathbf{Q}_v, \mathcal{A})$ is a torsion-free $\mathbf{Z}_p$-module.

Lemma 3.1.3. *If $\mathcal{A}$ and $\mathcal{F}$ satisfy the properties above, then:*

(1) *A_m and $\mathcal{F}$ satisfy the hypotheses of §2.4,*
(2) *the core rank $\chi(A_m)$ is independent of m.*

PROOF. It is immediate that property (H.i) for $\mathcal{A}$ implies the corresponding property (H.i) for every A_m, $1 \leq i \leq 4$. That (H.5) for $(\mathcal{A}, \mathcal{F})$ implies (H.5) for $(A_m, \mathcal{F})$ is [**MR1**, Lemma 3.7.1(i)], and we omit the proof.

The second assertion follows from Proposition 2.4.5 and Theorem 2.5.2; the details are left as an exercise. □

Definition 3.1.4. We define the core rank $\chi(\mathcal{A})$ to be the common value $\chi(A_m)$.

Example 3.1.5. We define a canonical Selmer structure $\mathcal{F}_{\mathrm{can}}$ on $\mathcal{A}$ by:

- if $v \in \{\infty, p\}$ then $H^1_{\mathcal{F}_{\mathrm{can}}}(\mathbf{Q}_v, \mathcal{A}) := H^1(\mathbf{Q}_v, \mathcal{A})$,
- if $v \notin \{\infty, p\}$ then $H^1_{\mathcal{F}_{\mathrm{can}}}(\mathbf{Q}_v, \mathcal{A}) := \ker\big[H^1(\mathbf{Q}_\ell, \mathcal{A}) \to H^1(\mathbf{Q}_\ell^{\mathrm{ur}}, \mathcal{A}) \otimes \mathbf{Q}_p\big]$.

Note that if $\mathcal{A}$ is unramified at ℓ, then $H^1(\mathbf{Q}_\ell^{\mathrm{ur}}, \mathcal{A}) = \mathrm{Hom}(I_{\mathbf{Q}_\ell}, \mathcal{A})$ where $I_{\mathbf{Q}_\ell}$ is the inertia group, and $\mathrm{Hom}(I_{\mathbf{Q}_\ell}, \mathcal{A})$ is torsion-free, so $H^1_{\mathcal{F}_{\mathrm{can}}}(\mathbf{Q}_\ell, \mathcal{A}) = H^1_{\mathrm{u}}(\mathbf{Q}_\ell, \mathcal{A})$ in this case.

Exercise 3.1.6. (1) Check that $\mathcal{F}_{\mathrm{can}}$ as defined above is a Selmer structure, and that it satisfies (H.5).
(2) Show that if $\ell \neq p$, $\mathcal{A}$ is unramified at ℓ, and $m > 0$, then $H^1_{\mathcal{F}_{\mathrm{can}}}(\mathbf{Q}_\ell, A_m) = H^1_{\mathrm{u}}(\mathbf{Q}_\ell, A_m)$.

Proposition 3.1.7. *For every v, the limit of the Tate pairings of Theorem 1.8.1 induces a nondegenerate pairing*

$$H^1(\mathbf{Q}_v, \mathcal{A}) \times H^1(\mathbf{Q}_v, \mathcal{A}^*) \longrightarrow \mathbf{Q}_p/\mathbf{Z}_p$$

under which $H^1_{\mathcal{F}_{\mathrm{can}}}(\mathbf{Q}_v, \mathcal{A})$ and $H^1_{\mathcal{F}^*_{\mathrm{can}}}(\mathbf{Q}_v, \mathcal{A}^*)$ are orthogonal complements. In particular:

(1) *if $v \in \{\infty, p\}$ then $H^1_{\mathcal{F}^*_{\mathrm{can}}}(\mathbf{Q}_v, \mathcal{A}^*) = 0$,*
(2) *if $v \notin \{\infty, p\}$ then $H^1_{\mathcal{F}^*_{\mathrm{can}}}(\mathbf{Q}_v, \mathcal{A}^*) = H^1_{\mathrm{u}}(\mathbf{Q}_v, \mathcal{A}^*)_{\mathrm{div}}$, the maximal divisible subgroup of $H^1_{\mathrm{u}}(\mathbf{Q}_v, \mathcal{A}^*)$.*
(3) *If $m > 0$ then $H^1_{\mathcal{F}^*_{\mathrm{can}}}(\mathbf{Q}_v, A^*_m)$ is the inverse image of $H^1_{\mathcal{F}^*_{\mathrm{can}}}(\mathbf{Q}_v, \mathcal{A}^*)$ under the map $H^1(\mathbf{Q}_v, A^*_m) \to H^1(\mathbf{Q}_v, \mathcal{A}^*)$.*

PROOF. See [**R2**, §1.3], or work out the details as an exercise. □

Exercise 3.1.8. Suppose that either $v = p$ or $H^1_{\mathrm{u}}(\mathbf{Q}_v, \mathcal{A}^*)$ is finite. Use Proposition 3.1.7 to show that there are natural isomorphisms

$$H^0(\mathbf{Q}_v, \mathcal{A}^*)/p^m H^0(\mathbf{Q}_v, \mathcal{A}^*) \xrightarrow{\sim} H^1_{\mathcal{F}^*_{\mathrm{can}}}(\mathbf{Q}_v, A^*_m).$$

Show that if in addition $H^0(\mathbf{Q}_v, \mathcal{A}^*)$ is divisible then

$$H^1_{\mathcal{F}^*_{\mathrm{can}}}(\mathbf{Q}_v, A^*_m) = 0, \qquad H^1_{\mathcal{F}_{\mathrm{can}}}(\mathbf{Q}_v, A_m) = H^1(\mathbf{Q}_v, A_m).$$

If M is a $\mathbf{Z}_p$-module, we let

$$\mathrm{rank}_{\mathbf{Z}_p}(M) := \dim_{\mathbf{Q}_p} M \otimes \mathbf{Q}_p, \quad \mathrm{corank}_{\mathbf{Z}_p}(M) := \mathrm{rank}_{\mathbf{Z}_p}\mathrm{Hom}(M, \mathbf{Q}_p/\mathbf{Z}_p).$$

Proposition 3.1.9.

$$\chi(\mathcal{A}, \mathcal{F}_{\mathrm{can}}) = \mathrm{rank}_{\mathbf{Z}_p}\mathcal{A}^- + \mathrm{corank}_{\mathbf{Z}_p} H^0(\mathbf{Q}_p, \mathcal{A}^*).$$

where $\mathcal{A}^-$ denotes the minus part of $\mathcal{A}$ for some complex conjugation.

Idea of proof. For the details of the proof, see [**MR1**, Theorem 5.2.15]. The outline is as follows. By Proposition 2.4.5 and Theorem 2.5.2, there is an integer k such that

$$\text{(3.1)} \qquad \mathrm{length}(H^1_{\mathcal{F}_{\mathrm{can}}}(\mathbf{Q}, A_m)) - \mathrm{length}(H^1_{\mathcal{F}^*_{\mathrm{can}}}(\mathbf{Q}, A^*_m)) = km$$

for every m, and $\chi(\mathcal{A}) = \max\{k, 0\}$. To prove the proposition we need to compute the left-hand side of (3.1) (in fact, computing it up to an error bounded independently of m will suffice). This can be done using [**Wi**, Proposition 1.6]. $\square$

3.2. Kolyvagin systems for $\mathcal{A}$.

Fix a $\mathbf{Z}_p[G_{\mathbf{Q}}]$-module $\mathcal{A}$ as above and a Selmer structure $\mathcal{F}$ for $\mathcal{A}$, satisfying hypotheses (H.1) through (H.5).

If $\ell \notin \Sigma$, let

$$\nu(\ell) = \max\{m : \ell \equiv 1 \pmod{p^m} \text{ and } A_m/(\mathrm{Fr}_\ell - 1)A_m \text{ is free of rank 1 over } \mathbf{Z}/p^m\mathbf{Z}\}.$$

Let $\mathcal{N} = \mathcal{N}_{\mathcal{A}}$ be the set of squarefree product of primes $\ell \notin \Sigma$, and if $n \in \mathcal{N}$ define $\nu(n) := \min\{\nu(\ell) : \ell \mid n\}$ (by convention we let $\nu(1) = \infty$).

Definition 3.2.1. A Kolyvagin system for $\mathcal{A}$ is a collection

$$\{\kappa_n \in H^1_{\mathcal{F}(n)}(\mathbf{Q}, A_{\nu(n)}) \otimes G_n : n \in \mathcal{N}\}$$

such that if $n\ell \in \mathcal{N}$, then $\mathrm{Res}_\ell(\kappa_{n\ell}) = \phi^{\mathrm{ut}}_\ell \circ \mathrm{Res}_\ell(\kappa_n)$ in $H^1_{\mathrm{t}}(\mathbf{Q}_\ell, A_{\nu(n\ell)}) \otimes G_{n\ell}$. (We understand $A_\infty = \mathcal{A}$, so $\kappa_1 \in H^1_{\mathcal{F}}(\mathbf{Q}, \mathcal{A})$.) Let $\mathbf{KS}(\mathcal{A})$ be the $\mathbf{Z}_p$-module of Kolyvagin systems for $\mathcal{A}$.

Exercise 3.2.2. Show that restricting from $\mathcal{N}$ to $\{n \in \mathcal{N} : \nu(n) \geq m\}$ induces a map $\mathbf{KS}(\mathcal{A}) \to \mathbf{KS}(A_m)$. Show that if $\chi(\mathcal{A}) = 1$, then

$$\mathbf{KS}(\mathcal{A}) = \varprojlim \mathbf{KS}(A_m),$$

inverse limit with respect to the maps of Proposition 2.9.1.

We say that $\boldsymbol{\kappa} \in \mathbf{KS}(\mathcal{A})$ is primitive if its image in $\mathbf{KS}(A_1)$ is nonzero.

Theorem 3.2.3. *Suppose $\chi(\mathcal{A}) = 1$. Then $\mathbf{KS}(\mathcal{A})$ is free of rank one over $\mathbf{Z}_p$, and every generator is primitive.*

Exercise 3.2.4. Prove Theorem 3.2.3, using the results of Lecture 2 for $\mathbf{KS}(A_m)$.

Definition 3.2.5. Suppose $\boldsymbol{\kappa} \in \mathbf{KS}(\mathcal{A})$, $\boldsymbol{\kappa} \neq 0$. We define the *order of vanishing* of $\boldsymbol{\kappa}$ by

$$\operatorname{ord}(\boldsymbol{\kappa}) := \min\{\omega(n) : \kappa_n \neq 0\}.$$

If $r \geq 0$ we define

$$\partial^r(\boldsymbol{\kappa}) := \max\{j : \kappa_n \in p^j H^1_{\mathcal{F}(n)}(\mathbf{Q}, A_{\nu(n)}) \otimes G_n \text{ for all } n \in \mathcal{N} \text{ with } \omega(n) = r\}.$$

In particular $\partial^0(\boldsymbol{\kappa}) = \max\{j : \kappa_1 \in p^j H^1_{\mathcal{F}}(\mathbf{Q}, \mathcal{A})\}$, and $\partial^r(\boldsymbol{\kappa}) = \infty$ if $r < \operatorname{ord}(\boldsymbol{\kappa})$. For $i > \operatorname{ord}(\boldsymbol{\kappa})$ define the sequence of *elementary divisors*

$$e_i(\boldsymbol{\kappa}) := \partial^{i-1}(\boldsymbol{\kappa}) - \partial^i(\boldsymbol{\kappa}).$$

Theorem 3.2.6. *Suppose $\chi(\mathcal{A}) = 1$ and $\boldsymbol{\kappa} \in \mathbf{KS}(\mathcal{A})$ is nonzero. Let $r = \operatorname{ord}(\boldsymbol{\kappa})$. Then* $\operatorname{corank}_{\mathbf{Z}_p}(H^1_{\mathcal{F}^*}(\mathbf{Q}, \mathcal{A}^*)) = r$, *$\partial^i(\boldsymbol{\kappa}) \neq \infty$ if $i \geq r$, the sequence $e_i(\boldsymbol{\kappa})$ is nonincreasing, nonnegative, equal to zero for large i, and*

$$H^1_{\mathcal{F}^*}(\mathbf{Q}, \mathcal{A}^*) \cong (\mathbf{Q}_p/\mathbf{Z}_p)^r \times \bigoplus_{i>r} \mathbf{Z}/p^{e_i(\boldsymbol{\kappa})}\mathbf{Z}.$$

Exercise 3.2.7. Deduce Theorem 3.2.6 from Propositions 2.9.8 and 2.4.5 and Exercise 3.2.2.

Corollary 3.2.8. *The value of* $\operatorname{ord}(\boldsymbol{\kappa})$ *and the sequence of $e_i(\boldsymbol{\kappa})$ are independent of (the nonzero) $\boldsymbol{\kappa}$.*

Corollary 3.2.9. *Suppose $\chi(\mathcal{A}) = 1$. Then:*

(1) $\operatorname{length}(H^1_{\mathcal{F}^*}(\mathbf{Q}, \mathcal{A}^*)) \leq \partial^0(\boldsymbol{\kappa})$, *with equality if $\boldsymbol{\kappa}$ is primitive.*
(2) *If $\kappa_1 \neq 0$, then $H^1_{\mathcal{F}^*}(\mathbf{Q}, \mathcal{A}^*)$ is finite.*
(3) *If $\boldsymbol{\kappa} \neq 0$ but $\kappa_1 = 0$, then $H^1_{\mathcal{F}^*}(\mathbf{Q}, \mathcal{A}^*)$ is infinite.*

3.3. Example: units and ideal class groups.

Fix a character $\rho : G_{\mathbf{Q}} \to \mathbf{Z}_p^\times$ of finite order. Assume that $p > 2$, so that the order of ρ is prime to p. Let L be the cyclic extension of $\mathbf{Q}$ which is the fixed field of ρ, and $\Delta = \operatorname{Gal}(L/\mathbf{Q})$. We will also view ρ as a (primitive) Dirichlet character in the usual way, so for every prime ℓ,

$$\rho(\ell) = 0 \iff \rho \text{ is ramified at } \ell \iff L/\mathbf{Q} \text{ is ramified at } \ell,$$
$$\rho(\ell) = 1 \iff \rho(G_{\mathbf{Q}_\ell}) = 1 \iff \ell \text{ splits completely in } L/\mathbf{Q}.$$

If B is a $\mathbf{Z}_p[\Delta]$-module, we define

$$B^\rho := \{b \in B : \delta b = \rho(\delta) b \text{ for every } \delta \in \Delta\}.$$

Let

$$\mathcal{A} := \mathbf{Z}_p(1) \otimes \rho^{-1} = \varprojlim \boldsymbol{\mu}_{p^m} \otimes \rho^{-1},$$

a free, rank-one $\mathbf{Z}_p$-module with $G_{\mathbf{Q}}$ acting via the product of ρ^{-1} and the cyclotomic character. (We write "$\otimes\rho^{-1}$" as an abbreviation for tensoring with a free, rank-one module on which $G_{\mathbf{Q}}$ acts as ρ^{-1}.) The dual representation

$$\mathcal{A}^* = \operatorname{Hom}(\mathcal{A}, \boldsymbol{\mu}_{p^\infty}) \cong \mathbf{Q}_p/\mathbf{Z}_p \otimes \rho,$$

i.e., $\mathcal{A}^*$ is isomorphic as a group to $\mathbf{Q}_p/\mathbf{Z}_p$, with $G_{\mathbf{Q}}$ acting via ρ.

For every $m \geq 1$, let $A_m = \mathcal{A}/p^m\mathcal{A} = \boldsymbol{\mu}_{p^m} \otimes \rho^{-1}$, so $A_m^* = \mathcal{A}^*[p^m] = \mathbf{Z}/p^m\mathbf{Z} \otimes \rho$.

Exercise 3.3.1. Describe $H^0(\mathbf{Q}_\ell, \mathcal{A}^*)$ for every prime ℓ, and show in particular that it is a divisible $\mathbf{Z}_p$-module.

Exercise 3.3.2. Suppose $\ell \neq p$ and ρ is ramified at ℓ. Show that $H^1(\mathbf{Q}_\ell, \mathcal{A}) = 0$, $H^1(\mathbf{Q}_\ell, \mathcal{A}^*) = 0$, and for every m, $H^1(\mathbf{Q}_\ell, A_m) = 0$ and $H^1(\mathbf{Q}_\ell, A_m^*) = 0$. (Hint: show that $H^1_{\mathrm{u}}(\mathbf{Q}_\ell, A_m) = 0$ and $H^1_{\mathrm{u}}(\mathbf{Q}_\ell, A_m^*) = 0$, and use Proposition 1.9.1.)

We have an inflation-restriction exact sequence

$$0 \to H^1(L/\mathbf{Q}, (A_m)^{G_L}) \to H^1(\mathbf{Q}, A_m) \to H^1(L, A_m)^\Delta \to H^2(L/\mathbf{Q}, (A_m)^{G_L})$$

and similarly with A_m replaced by A_m^*. Since $[L : \mathbf{Q}]$ is prime to p, Exercise 1.4.3 shows that $H^i(L/\mathbf{Q}, A_m) = H^i(L/\mathbf{Q}, A_m^*) = 0$ for $i = 1, 2$, so

$$H^1(\mathbf{Q}, A_m) \cong H^1(L, A_m)^\Delta \cong (H^1(L, \boldsymbol{\mu}_{p^m}) \otimes \rho^{-1})^\Delta,$$
$$H^1(\mathbf{Q}, A_m^*) \cong H^1(L, A_m^*)^\Delta \cong (H^1(L, \mathbf{Z}/p^m\mathbf{Z}) \otimes \rho)^\Delta$$

Since (using Theorem 1.6.3)

$$H^1(L, \boldsymbol{\mu}_{p^m}) = L^\times/(L^\times)^{p^m}, \quad H^1(L, \mathbf{Z}/p^m\mathbf{Z}) = \mathrm{Hom}(G_L, \mathbf{Z}/p^m\mathbf{Z})$$

we conclude that

$$(3.2) \qquad H^1(\mathbf{Q}, A_m) \cong (L^\times/(L^\times)^{p^m})^\rho, \quad H^1(\mathbf{Q}, A_m^*) \cong \mathrm{Hom}(G_L, \mathbf{Z}/p^m\mathbf{Z})^{\rho^{-1}}$$

the isomorphisms depending on a choice of generators of the free rank-one $\mathbf{Z}_p$-modules ρ and ρ^{-1}.

Exercise 3.3.3. Show that (3.2) has the following generalization. If F is a number field and $L \cap F = \mathbf{Q}$, then

$$H^1(F, A_m) \cong (FL^\times/(FL^\times)^{p^m})^\rho, \quad H^1(F, A_m^*) \cong \mathrm{Hom}(G_{FL}, \mathbf{Z}/p^m\mathbf{Z})^{\rho^{-1}}.$$

For every prime ℓ, let $L_\ell := L \otimes_{\mathbf{Q}} \mathbf{Q}_\ell$, the sum of the completions of L at primes above ℓ, let $\mathcal{O}_{L,\ell} := \mathcal{O}_L \otimes_{\mathbf{Z}} \mathbf{Z}_\ell$, and if λ is a prime of L above ℓ, let $I_{L_\lambda} \subset G_{L_\lambda}$ be the inertia group.

Exercise 3.3.4. Show that for every prime ℓ,

(1) $H^1(\mathbf{Q}_\ell, A_m) \cong (\bigoplus_{\lambda|\ell} (L_\lambda^\times/(L_\lambda^\times))^{p^m})^\rho = (L_\ell^\times/(L_\ell^\times)^{p^m})^\rho$,

(2) $H^1(\mathbf{Q}_\ell, A_m^*) \cong (\bigoplus_{\lambda|\ell} \mathrm{Hom}(G_{L_\lambda}, \mathbf{Z}/p^m\mathbf{Z}))^{\rho^{-1}}$.

Exercise 3.3.5. Show that since $p \neq 2$, $H^1(\mathbf{R}, A_m) = H^1(\mathbf{R}, A_m^*) = 0$ for every m, so we can safely ignore the infinite place in our definition of Selmer group and Selmer structure.

Let $\mathcal{F}$ be the canonical Selmer structure $\mathcal{F}_{\mathrm{can}}$ defined in Example 3.1.5. Let $\mathbb{1}$ denote the trivial character of $G_{\mathbf{Q}}$, and $\epsilon_{\mathrm{cycl}} : G_{\mathbf{Q}} \to \mathrm{Aut}(\boldsymbol{\mu}_p) \cong \mathbf{F}_p^\times \hookrightarrow \mathbf{Z}_p^\times$ the mod-p cyclotomic character.

Exercise 3.3.6. Show that if $\rho \neq \mathbb{1}$ and $\rho \neq \epsilon_{\mathrm{cycl}}$, then $\mathcal{A}$ and $\mathcal{F}$ satisfy assumptions (H.1) through (H.5) of §3.1.

Exercise 3.3.7. Use Proposition 3.1.9 to show that with the canonical Selmer structure (and with $\rho \neq \mathbb{1}, \epsilon_{\mathrm{cycl}}$), $\chi(\mathcal{A})$ is given by the following table:

$$\chi(\mathcal{A}) = \left\{ \begin{array}{c|c|c} & \rho \text{ even} & \rho \text{ odd} \\ \hline \rho(p) \neq 1 & 1 & 0 \\ \hline \rho(p) = 1 & 2 & 1 \end{array} \right.$$

(recall that ρ is even if and only if $\rho(-1) = 1$).

Proposition 3.3.8. *For every $m \geq 1$, with the identifications of Exercise 3.3.4 we have*

(1) $H^1_{\mathcal{F}}(\mathbf{Q}_p, A_m) = (L_p^\times/(L_p^\times)^{p^m})^\rho$ *and* $H^1_{\mathcal{F}^*}(\mathbf{Q}_p, A_m^*) = 0$,
(2) *If* $\ell \neq p$, *then*

$$H^1_{\mathcal{F}}(\mathbf{Q}_\ell, A_m) = (\mathcal{O}_{L,\ell}^\times/(\mathcal{O}_{L,\ell}^\times)^{p^m})^\rho,$$
$$H^1_{\mathcal{F}^*}(\mathbf{Q}_\ell, A_m^*) = (\bigoplus_{\lambda|\ell}\mathrm{Hom}(G_{L_\lambda}/I_{L_\lambda}, \mathbf{Z}/p^m\mathbf{Z}))^{\rho^{-1}}$$

Proof. By Exercises 3.1.8 and 3.3.1, $H^1_{\mathcal{F}}(\mathbf{Q}_p, A_m) = H^1(\mathbf{Q}_p, A_m)$, so the first part of (1) follows from Exercise 3.3.4(1).

For (2) we consider three cases.

Case 1: $\ell \neq p$ and $\rho(\ell) = 1$. In this case ℓ splits completely in L/K, and $A_m \cong \boldsymbol{\mu}_{p^m}$ as a $G_{\mathbf{Q}_\ell}$-module, so (using Exercise 1.6.8)

$$H^1_{\mathrm{u}}(\mathbf{Q}_\ell, A_m) = H^1_{\mathrm{u}}(\mathbf{Q}_\ell, \boldsymbol{\mu}_{p^m}) = \mathbf{Z}_\ell^\times/(\mathbf{Z}_\ell^\times)^{p^m} = (\mathcal{O}_{L,\ell}^\times/(\mathcal{O}_{L,\ell}^\times)^{p^m})^\rho.$$

Since ρ is unramified at ℓ, Exercise 3.1.6(2) shows that $H^1_{\mathcal{F}}(\mathbf{Q}_\ell, A_m) = H^1_{\mathrm{u}}(\mathbf{Q}_\ell, A_m)$. This proves the first part of (2) in this case.

Case 2: $\ell \neq p$, $\mathcal{A}$ is unramified at ℓ, and $\rho(\ell) \neq 1$. In this case Exercise 3.1.6(2) shows that $H^1_{\mathcal{F}}(\mathbf{Q}_\ell, A_m) = H^1_{\mathrm{u}}(\mathbf{Q}_\ell, A_m)$. On the other hand, the inflation-restriction exact sequence gives

$$0 \longrightarrow H^1_{\mathrm{u}}(\mathbf{Q}_\ell, A_m) \longrightarrow H^1(\mathbf{Q}_\ell, A_m) \longrightarrow H^1(\mathbf{Q}_\ell^{\mathrm{ur}}, A_m)^{G_{\mathbf{Q}_\ell}}. \tag{3.3}$$

We have

$$H^1(\mathbf{Q}_\ell^{\mathrm{ur}}, A_m) = H^1(\mathbf{Q}_\ell^{\mathrm{ur}}, \boldsymbol{\mu}_{p^m} \otimes \rho^{-1})$$
$$= H^1(\mathbf{Q}_\ell^{\mathrm{ur}}, \boldsymbol{\mu}_{p^m}) \otimes \rho^{-1} = ((\mathbf{Q}_\ell^{\mathrm{ur}})^\times \otimes \mathbf{Z}/p^m\mathbf{Z}) \otimes \rho^{-1}$$

and, writing $\mathbf{Z}_\ell^{\mathrm{ur}}$ for the ring of integers of $\mathbf{Q}_\ell^{\mathrm{ur}}$,

$$0 \longrightarrow (\mathbf{Z}_\ell^{\mathrm{ur}})^\times \longrightarrow (\mathbf{Q}_\ell^{\mathrm{ur}})^\times \xrightarrow{\mathrm{ord}_\ell} \mathbf{Z} \longrightarrow 0.$$

The unit group $(\mathbf{Z}_\ell^{\mathrm{ur}})^\times$ is p-divisible, so $(\mathbf{Q}_\ell^{\mathrm{ur}})^\times \otimes \mathbf{Z}/p^m\mathbf{Z} \cong \mathbf{Z}/p^m\mathbf{Z}$ as $G_{\mathbf{Q}_\ell}$-modules, and so

$$H^1(\mathbf{Q}_\ell^{\mathrm{ur}}, A_m)^{G_{\mathbf{Q}_\ell}} = (\mathbf{Z}/p^m\mathbf{Z} \otimes \rho)^{G_{\mathbf{Q}_\ell}} = (\mathbf{Z}/p^m\mathbf{Z})^{\rho^{-1}} = 0.$$

Thus (3.3) and Exercise 3.3.4 yield

$$H^1_{\mathcal{F}}(\mathbf{Q}_\ell, A_m) = H^1_{\mathrm{u}}(\mathbf{Q}_\ell, A_m) = H^1(\mathbf{Q}_\ell, A_m) = (L_\ell^\times/(L_\ell^\times)^{p^m})^\rho.$$

In the exact sequence

$$0 \longrightarrow \mathcal{O}_{L,\ell}^\times \longrightarrow L_\ell^\times \xrightarrow{\oplus \mathrm{ord}_\lambda} \bigoplus_{\lambda|\ell} \mathbf{Z}\lambda \longrightarrow 0,$$

the decomposition group Δ_ℓ of ℓ in Δ acts trivially on $\oplus_\lambda \mathbf{Z}\lambda$, but $\rho(\Delta_\ell) \neq 1$, so $(\oplus_\lambda(\mathbf{Z}/p^m\mathbf{Z})\lambda)^\rho = 0$. Therefore $(L_\ell^\times/(L_\ell^\times)^{p^m})^\rho = (\mathcal{O}_{L,\ell}^\times/(\mathcal{O}_{L,\ell}^\times)^{p^m})^\rho$, and this proves (2) for A_m in this case.

Case 3: $\ell \neq p$ and $\mathcal{A}$ is ramified at ℓ. Exercise 3.3.2 shows that $H^1_{\mathcal{F}}(\mathbf{Q}_\ell, A_m) = 0$. On the other hand, the kernel of the reduction map

$$\mathcal{O}_{L,\ell}^\times \twoheadrightarrow \bigoplus_{\lambda|\ell} (\mathcal{O}_L/\lambda)^\times$$

is a pro-ℓ-group, so tensoring with $\mathbf{Z}/p^m\mathbf{Z}$ gives an isomorphism

$$\mathcal{O}_{L,\ell}^\times/(\mathcal{O}_{L,\ell}^\times)^{p^m} \xrightarrow{\sim} \bigoplus_{\lambda|\ell}(\mathcal{O}_L/\lambda)^\times/((\mathcal{O}_L/\lambda)^\times)^{p^m}.$$

Since the inertia group I_ℓ acts trivially on the right-hand side, but $\rho(I_\ell) \neq 1$, we conclude that $(\mathcal{O}_{L,\ell}^\times/(\mathcal{O}_{L,\ell}^\times)^{p^m})^\rho = 0$, which proves assertion (2) for A_m.

The proof of the proposition for A_m^* is left as an exercise. □

Let $\mathcal{C}_L$ denote the p-part of the ideal class group of L, and

$$\mathcal{C}_L' := \mathrm{Pic}(\mathcal{O}_L[1/p])[p^\infty],$$

the quotient of $\mathcal{C}_L$ by the classes of primes above p.

Exercise 3.3.9. Suppose $\rho(p) \neq 1$. Show that $H^1_\mathcal{F}(\mathbf{Q}_p, A_m) = (\mathcal{O}_{L,p}^\times/(\mathcal{O}_{L,p}^\times)^{p^m})^\rho$, $(\mathcal{O}_L[1/p]^\times/(\mathcal{O}_L[1/p]^\times)^{p^m})^\rho = (\mathcal{O}_L^\times/(\mathcal{O}_L^\times)^{p^m})^\rho$, and $(\mathcal{C}_L')^\rho = \mathcal{C}_L^\rho$.

Proposition 3.3.10. *For every m there is a natural exact sequence*

$$0 \longrightarrow (\mathcal{O}_L[1/p]^\times/(\mathcal{O}_L[1/p]^\times)^{p^m}))^\rho \longrightarrow H^1_\mathcal{F}(\mathbf{Q}, A_m) \longrightarrow \mathcal{C}_L'[p^m]^\rho \longrightarrow 0$$

and an isomorphism

$$H^1_{\mathcal{F}^*}(\mathbf{Q}, A_m^*) \cong \mathrm{Hom}((\mathcal{C}_L')^\rho, \mathbf{Z}/p^m\mathbf{Z}).$$

If $\rho(p) \neq 1$ we can replace $\mathcal{O}_L[1/p]^\times/(\mathcal{O}_L[1/p]^\times)^{p^m}$ by $\mathcal{O}_L^\times/(\mathcal{O}_L^\times)^{p^m}$ and $\mathcal{C}_L'$ by $\mathcal{C}_L$.

Proof. Under the identification (3.2), using Proposition 3.3.8 we have

$$H^1_\mathcal{F}(\mathbf{Q}, A_m) = \{x \in (L^\times/(L^\times)^{p^m})^\rho : \mathrm{ord}_\lambda(x) \equiv 0 \pmod{p^m} \text{ for every } \lambda \nmid p\}.$$

If $x \in L^\times$ represents a class in $c \in H^1_\mathcal{F}(\mathbf{Q}, A_m)$, then the ideal $x\mathcal{O}_L[1/p] = \mathfrak{a}^{p^m}$ for some fractional ideal $\mathfrak{a}$ of $\mathcal{O}_L[1/p]$. The map that sends c to the class of $\mathfrak{a}$ is a well-defined surjection from $H^1_\mathcal{F}(\mathbf{Q}, A_m)$ to $\mathcal{C}_L'[p^m]$, and its kernel is $\mathcal{O}_L[1/p]^\times/(\mathcal{O}_L[1/p]^\times)^{p^m}$. This gives the desired exact sequence.

Under the identification (3.2), Proposition 3.3.8 shows that $H^1_{\mathcal{F}^*}(\mathbf{Q}, A_m^*)$ consists of everywhere unramified homomorphisms in $\mathrm{Hom}(G_L, \mathbf{Z}/p^m\mathbf{Z})^{\rho^{-1}}$ that are trivial on the decomposition group at p, i.e.,

$$H^1_{\mathcal{F}^*}(\mathbf{Q}, A_m^*) = \mathrm{Hom}(\mathcal{C}_L', \mathbf{Z}/p^m\mathbf{Z})^{\rho^{-1}}.$$

The isomorphism of the proposition follows from this (for the final assertion use Exercise 3.3.9). □

Exercise 3.3.11. Use Proposition 3.3.10 to give a direct proof (i.e., without using Proposition 3.1.9 as in Exercise 3.3.7) that if $\rho \neq \mathbb{1}, \epsilon_{\mathrm{cyl}}$, then $\chi(\mathcal{A})$ is given by the table in Exercise 3.3.7.

Corollary 3.3.12. *Suppose ρ is even, $\rho(p) \neq 1$, and $\boldsymbol{\kappa} \in \mathbf{KS}(\mathcal{A})$ is nonzero. Then $\kappa_1 \neq 0$ and*

$$|\mathcal{C}_L^\rho| \leq [(\mathcal{O}_L^\times \otimes \mathbf{Z}_p)^\rho : \mathbf{Z}_p\kappa_1],$$

with equality if and only if $\boldsymbol{\kappa}$ is primitive.

PROOF. We will deduce this from Corollary 3.2.9. By Proposition 3.3.10 we have $H^1(\mathbf{Q}, \mathcal{A}) = (\mathcal{O}_L^\times \otimes \mathbf{Z}_p)^\rho$. The proof of the Dirichlet Unit Theorem shows that this is a free $\mathbf{Z}_p$-module of rank one, so

$$\partial^0(\boldsymbol{\kappa}) = [(\mathcal{O}_L^\times \otimes \mathbf{Z}_p)^\rho : \mathbf{Z}_p \kappa_1].$$

Now the corollary follows directly from Corollary 3.2.9(1) and Proposition 3.3.10. □

Proposition 3.3.13. *Suppose $\ell \notin \Sigma$ and $1 \le m \le \nu(\ell)$. Then*

(1) *$\rho(\ell) = 1$ and ℓ splits completely in $\mathbf{Q}(\boldsymbol{\mu}_{p^m})/\mathbf{Q}$,*
(2) *the isomorphism $H^1(\mathbf{Q}_\ell, A_m) \cong H^1(\mathbf{Q}_\ell, \boldsymbol{\mu}_{p^m}) \cong \mathbf{Q}_\ell^\times/(\mathbf{Q}_\ell^\times)^{p^m}$ identifies $H^1_{\mathrm{u}}(\mathbf{Q}_\ell, A_m)$ with $\mathbf{Z}_\ell^\times/(\mathbf{Z}_\ell^\times)^{p^m}$ and $H^1_{\mathrm{t}}(\mathbf{Q}_\ell, A_m)$ with the subgroup generated by ℓ,*
(3) *there is a commutative diagram (recall that $G_\ell := \mathrm{Gal}(\mathbf{Q}(\boldsymbol{\mu}_\ell)/\mathbf{Q})$)*

$$\begin{array}{ccc} H^1_{\mathrm{u}}(\mathbf{Q}_\ell, A_m) & \xrightarrow{\ \phi_\ell^{\mathrm{ut}}\ } & H^1_{\mathrm{t}}(\mathbf{Q}_\ell, A_m) \otimes G_\ell \\ \downarrow \cong & & \downarrow \cong \\ \mathbf{Z}_\ell^\times/(\mathbf{Z}_\ell^\times)^{p^m} & \xrightarrow{\ \ell \otimes [\,\cdot\,, \mathbf{Q}_\ell(\boldsymbol{\mu}_\ell)/\mathbf{Q}_\ell]\ } & \langle \ell \rangle \otimes G_\ell \end{array}$$

where the vertical isomorphisms are from (2) and $[\,\cdot\,, \mathbf{Q}_\ell(\boldsymbol{\mu}_\ell)/\mathbf{Q}_\ell]$ is the local Artin map.

PROOF. Since $m \le \nu(\ell)$, we have $\ell \equiv 1 \pmod{p^m}$ and $(\mathrm{Fr}_\ell - 1)A_m = 0$, so $\ell\rho^{-1}(\ell) \equiv 1 \pmod{p^m}$. Hence $\rho(\ell) \equiv 1 \pmod{p^m}$, and so $\rho(\ell) = 1$. This proves (1).

Since $\rho(\ell) = 1$,

$$H^1(\mathbf{Q}_\ell, A_m) = H^1(\mathbf{Q}_\ell, \boldsymbol{\mu}_{p^m}) = \mathbf{Q}_\ell^\times/(\mathbf{Q}_\ell^\times)^{p^m},$$

and the description of $H^1_{\mathrm{u}}(\mathbf{Q}_\ell, A_m)$ in (2) is similar to Proposition 3.3.8. By definition,

$$\begin{aligned} H^1_{\mathrm{t}}(\mathbf{Q}_\ell, A_m) &= \ker\big[H^1(\mathbf{Q}_\ell, A_m) \to H^1(\mathbf{Q}_\ell(\boldsymbol{\mu}_\ell), A_m)\big] \\ &= \ker\big[\mathbf{Q}_\ell^\times/(\mathbf{Q}_\ell^\times)^{p^m} \to \mathbf{Q}_\ell(\boldsymbol{\mu}_\ell)^\times/(\mathbf{Q}_\ell(\boldsymbol{\mu}_\ell)^\times)^{p^m}\big] \end{aligned}$$

Let ζ_ℓ denote a primitive ℓ-th root of unity in $\mathbf{Q}_\ell(\boldsymbol{\mu}_\ell)$. We have

$$\ell = \mathbf{N}_{\mathbf{Q}_\ell(\boldsymbol{\mu}_\ell)/\mathbf{Q}_\ell}(1 - \zeta_\ell) = \prod_{i=1}^{\ell-1}(1 - \zeta_\ell^i) = (1 - \zeta_\ell)^{\ell-1} \prod_{i=1}^{\ell-1} \frac{1 - \zeta_\ell^i}{1 - \zeta_\ell}$$

Since $\prod_{i=1}^{\ell-1} \frac{1-\zeta_\ell^i}{1-\zeta_\ell} \equiv \prod_{i=1}^{\ell-1} i \equiv -1 \pmod{\zeta_\ell - 1}$, we also see that $\prod_{i=1}^{\ell-1} \frac{1-\zeta_\ell^i}{1-\zeta_\ell} \in (\mathbf{Q}_\ell(\boldsymbol{\mu}_\ell)^\times)^{p^m}$. Hence $\ell \in (\mathbf{Q}_\ell(\boldsymbol{\mu}_\ell)^\times)^{p^m}$, so $\ell \in H^1_{\mathrm{t}}(\mathbf{Q}_\ell, A_m)$. On the other hand, since $G_{\mathbf{Q}_\ell}$ acts trivially on A_m,

$$H^1_{\mathrm{t}}(\mathbf{Q}_\ell, A_m) = H^1(\mathbf{Q}_\ell(\boldsymbol{\mu}_\ell)/\mathbf{Q}_\ell, A_m) = \mathrm{Hom}(\mathrm{Gal}(\mathbf{Q}_\ell(\boldsymbol{\mu}_\ell)/\mathbf{Q}_\ell), A_m)$$

is cyclic of order p^m, and (2) follows.

The proof of (3) is left as an exercise. □

3.4. Example: Tate modules of elliptic curves.

Now suppose that E is an elliptic curve defined over $\mathbf{Q}$, and

$$\mathcal{A} = T_p(E) := \varprojlim E[p^m]$$

is the p-adic Tate module of E. Then the Weil pairing shows that $\mathcal{A}^* = E[p^\infty]$, and for every m, $A_m = A_m^* = E[p^m]$. We will also assume that

(3.4)
- $p \geq 5$,
- the p-adic representation $G_\mathbf{Q} \to \mathrm{Aut}(E[p^\infty]) \cong \mathrm{GL}_2(\mathbf{Z}_p)$ is surjective

and we let $\mathcal{F}$ be the Selmer structure $\mathcal{F}_{\mathrm{can}}$ on $T_p(E)$.

Exercise 3.4.1. Show that under the assumptions of (3.4), $T_p(E)$ and $\mathcal{F}$ satisfy hypotheses (H.1) through (H.5) of §3.1.

Exercise 3.4.2. Show using Proposition 3.1.9 that $\chi(T_p(E)) = 1$.

Proposition 3.4.3.
(1) *If $\ell \neq p$, then $H^1(\mathbf{Q}_\ell, T_p(E))$ and $H^1(\mathbf{Q}_\ell, E[p^\infty])$ are finite.*
(2) *For every ℓ, $H^1_\mathcal{F}(\mathbf{Q}_\ell, T_p(E)) = H^1(\mathbf{Q}_\ell, T_p(E))$ and $H^1_{\mathcal{F}^*}(\mathbf{Q}_\ell, E[p^\infty]) = 0$.*

PROOF. Suppose first that $\ell \neq p$, and $m > 0$. Writing I_ℓ for the inertia group in $G_{\mathbf{Q}_\ell}$, we have (see (1.2) and Proposition 1.4.13(1))

$$H^1_\mathrm{u}(\mathbf{Q}_\ell, E[p^m]) = H^1(\mathbf{Q}_\ell^{\mathrm{ur}}/\mathbf{Q}_\ell, E[p^m]^{I_\ell}) \cong E[p^m]^{I_\ell}/(\mathrm{Fr}_\ell - 1)(E[p^m]^{I_\ell}).$$

It follows from the exact sequence

$$0 \to E(\mathbf{Q}_\ell)[p^m] \to E[p^m]^{I_\ell} \xrightarrow{\mathrm{Fr}_\ell - 1} E[p^m]^{I_\ell} \to E[p^m]^{I_\ell}/(\mathrm{Fr}_\ell - 1)(E[p^m]^{I_\ell}) \to 0$$

that $E[p^m]^{I_\ell}/(\mathrm{Fr}_\ell - 1)(E[p^m]^{I_\ell})$ and $E(\mathbf{Q}_\ell)[p^m]$ have the same order, so

$$|H^1_\mathrm{u}(\mathbf{Q}_\ell, E[p^m])| = |E(\mathbf{Q}_\ell)[p^m]|.$$

Since $E[p^m]^* = E[p^m]$, Proposition 1.9.1 shows that $H^1_\mathrm{u}(\mathbf{Q}_\ell, E[p^m])$ is its own orthogonal complement in $H^1(\mathbf{Q}_\ell, E[p^m]$, so $|H^1(\mathbf{Q}_\ell, E[p^m])| = |E(\mathbf{Q}_\ell)[p^m]|^2$. Since $|E(\mathbf{Q}_\ell)[p^m]|$ is bounded independently of m, we conclude that both $H^1(\mathbf{Q}_\ell, T_p(E))$ and $H^1(\mathbf{Q}_\ell, E[p^\infty])$ are finite. This proves (1), and (2) follows from the definition of $\mathcal{F}_{\mathrm{can}}$ (for $H^1_\mathcal{F}(\mathbf{Q}_\ell, T_p(E))$) and Proposition 3.1.7(2) (for $H^1_{\mathcal{F}^*}(\mathbf{Q}_\ell, E[p^\infty])$).

If $\ell = p$, then (2) is the definition of $\mathcal{F}_{\mathrm{can}}$ (for $H^1_\mathcal{F}(\mathbf{Q}_\ell, T_p(E))$) and Proposition 3.1.7(1) (for $H^1_{\mathcal{F}^*}(\mathbf{Q}_\ell, E[p^\infty])$). □

Exercise 3.4.4. Show that if $E(\mathbf{Q}_\ell)[p] = 0$, then

$$H^1_{\mathcal{F}^*}(\mathbf{Q}_\ell, E[p^m]) = 0, \quad H^1_\mathcal{F}(\mathbf{Q}_\ell, E[p^m]) = H^1(\mathbf{Q}_\ell, E[p^m]).$$

(Hint: use Exercise 3.1.8.)

Recall (Definition 1.6.13) that for every ℓ and m, $H^1_\mathrm{f}(\mathbf{Q}_\ell, E[p^m])$ is the image of the Kummer map $E(\mathbf{Q}_\ell)/p^m E(\mathbf{Q}_\ell) \hookrightarrow H^1(\mathbf{Q}_\ell, E[p^m])$, and $\mathrm{Sel}_{p^m}(E/\mathbf{Q})$ is the corresponding Selmer group.

Proposition 3.4.5. *Suppose $\ell \neq p$ and $m > 0$. Then*

$$H^1_\mathcal{F}(\mathbf{Q}_\ell, E[p^m]) = H^1_{\mathcal{F}^*}(\mathbf{Q}_\ell, E[p^m]) = H^1_\mathrm{f}(\mathbf{Q}_\ell, E[p^m]).$$

PROOF. By Exercise 3.1.8 and Proposition 3.4.3, there is a natural isomorphism

$$E(\mathbf{Q}_\ell)[p^\infty]/p^m E(\mathbf{Q}_\ell)[p^\infty] \xrightarrow{\sim} H^1_{\mathcal{F}^*}(\mathbf{Q}_\ell, E[p^m]). \tag{3.5}$$

By [**Si**, Theorem IV.6.4 and Proposition VII.2.2], there is a (noncanonical) decomposition

$$E(\mathbf{Q}_\ell) \cong \mathbf{Z}_\ell \times E(\mathbf{Q}_\ell)_{\mathrm{tors}}$$

so there is a canonical isomorphism

$$E(\mathbf{Q}_\ell)/p^m E(\mathbf{Q}_\ell) \xrightarrow{\sim} E(\mathbf{Q}_\ell)_{\mathrm{tors}}/p^m E(\mathbf{Q}_\ell)_{\mathrm{tors}} = E(\mathbf{Q}_\ell)[p^\infty]/p^m E(\mathbf{Q}_\ell)[p^\infty].$$

Combined with (3.5) (and checking that the obvious diagram commutes) this proves that $H^1_{\mathcal{F}^*}(\mathbf{Q}_\ell, E[p^m]) = H^1_{\mathrm{f}}(\mathbf{Q}_\ell, E[p^m])$. By Example 1.8.4, $H^1_{\mathrm{f}}(\mathbf{Q}_\ell, E[p^m])$ is its own orthogonal complement in $H^1(\mathbf{Q}_\ell, E[p^m])$ under the Tate pairing, so we have $H^1_{\mathcal{F}}(\mathbf{Q}_\ell, E[p^m]) = H^1_{\mathrm{f}}(\mathbf{Q}_\ell, E[p^m])$ as well. □

Corollary 3.4.6. *There are exact sequences*

$$0 \longrightarrow \mathrm{Sel}_{p^m}(E/\mathbf{Q}) \longrightarrow H^1_{\mathcal{F}}(\mathbf{Q}, E[p^m]) \longrightarrow H^1_{\mathcal{F}}(\mathbf{Q}_p, E[p^m])/H^1_{\mathrm{f}}(\mathbf{Q}_p, E[p^m])$$

$$0 \longrightarrow H^1_{\mathcal{F}^*}(\mathbf{Q}, E[p^m]) \longrightarrow \mathrm{Sel}_{p^m}(E/\mathbf{Q}) \longrightarrow H^1_{\mathrm{f}}(\mathbf{Q}_p, E[p^m])/H^1_{\mathcal{F}^*}(\mathbf{Q}_p, E[p^m]).$$

Exercise 3.4.7. Show directly from the definition that $\chi(T_p(E)) = 1$. Hint: use Corollary 3.4.6, Theorem 1.10.2, Exercise 3.1.8, and the fact that $E(\mathbf{Q}_p) \cong E(\mathbf{Q}_p)_{\mathrm{tors}} \times \mathbf{Z}_p$.

LECTURE 4

Euler systems

The preceding lectures have shown that Kolyvagin systems are useful for describing the structure of Selmer groups, and even that Kolyvagin systems exist for certain p-adic representations. It still remains to construct Kolyvagin systems and to relate them to special values of L-functions (something that has been done only in very special cases). In this lecture we introduce Euler systems, and describe a general version of Kolyvagin's machinery [**Ko**] for using Euler systems to construct Kolyvagin systems.

4.1. Definition of an Euler system.

Fix a rational prime p and a p-adic representation $\mathcal{A}$ (a free $\mathbf{Z}_p$-module of finite rank with a continuous action of $G_{\mathbf{Q}}$, ramified at only finitely many primes) as in the previous lecture. Fix a finite set Σ of places containing ∞, p, and all primes ℓ where $\mathcal{A}$ is ramified.

If $\ell \notin \Sigma$, define

$$P_\ell(x) := \det(1 - \mathrm{Fr}_\ell x | \mathcal{A}) \in \mathbf{Z}_p[x].$$

Let

$$\mathcal{N} := \{np^k : n \text{ is a squarefree product of primes } \ell \notin \Sigma,\ k \geq 0\}.$$

An *Euler system* for $\mathcal{A}$ is a collection

$$\boldsymbol{\xi} = \{\xi_n \in H^1(\mathbf{Q}(\boldsymbol{\mu}_n), \mathcal{A}) : n \in \mathcal{N}\}$$

such that for $n\ell \in \mathcal{N}$,

$$\mathrm{Cor}_{n\ell/n}\xi_{n\ell} = \begin{cases} P_\ell(\mathrm{Fr}_\ell^{-1})\xi_n & \text{if } \ell \neq p, \\ \xi_n & \text{if } \ell = p \end{cases} \tag{4.1}$$

where

$$\mathrm{Cor}_{n\ell/n} = \mathrm{Cor}_{\mathbf{Q}(\boldsymbol{\mu}_{n\ell})/\mathbf{Q}(\boldsymbol{\mu}_n)} : H^1(\mathbf{Q}(\boldsymbol{\mu}_{n\ell}), \mathcal{A}) \to H^1(\mathbf{Q}(\boldsymbol{\mu}_n), \mathcal{A})$$

is the corestriction map. Let $\mathbf{ES}(\mathcal{A})$ denote the $\mathbf{Z}_p[G_{\mathbf{Q}}]$-module of Euler systems for $\mathcal{A}$.

4.2. Example: the cyclotomic unit Euler system

Fix a nontrivial character $\rho : G_{\mathbf{Q}} \twoheadrightarrow \mathrm{Gal}(\mathbf{Q}(\boldsymbol{\mu}_p)/\mathbf{Q}) \to \mathbf{Z}_p^\times$, and let $\mathcal{A} = \mathbf{Z}_p(1) \otimes \rho^{-1}$ as in §3.3. Then for every $\ell \neq p$, Fr_ℓ acts on $\mathcal{A}$ as $\ell\rho^{-1}(\ell)$, so

$$P_\ell(x) = 1 - \ell\rho^{-1}(\ell)x.$$

Let $\Sigma = \{\infty, p\}$.

If $p \mid n$ then $\rho(G_{\mathbf{Q}(\boldsymbol{\mu}_n)}) = 1$, so there are isomorphisms

$$H^1(\mathbf{Q}(\boldsymbol{\mu}_n), \mathcal{A}) \cong H^1(\mathbf{Q}(\boldsymbol{\mu}_n), \mathbf{Z}_p(1)) \otimes \rho^{-1} \cong (\mathbf{Q}(\boldsymbol{\mu}_n)^\times \otimes \mathbf{Z}_p) \otimes \rho^{-1} \tag{4.2}$$

and commutative diagrams for every ℓ

$$\begin{array}{ccc} (\mathbf{Q}(\boldsymbol{\mu}_{n\ell})^\times \otimes \mathbf{Z}_p) \otimes \rho^{-1} & \xrightarrow{\sim} & H^1(\mathbf{Q}(\boldsymbol{\mu}_{n\ell}), \mathcal{A}) \\ \Big\downarrow{\scriptstyle \mathbf{N}_{\mathbf{Q}(\boldsymbol{\mu}_{n\ell})/\mathbf{Q}(\boldsymbol{\mu}_n)}\otimes 1} & & \Big\downarrow{\scriptstyle \mathrm{Cor}_{n\ell/n}} \\ (\mathbf{Q}(\boldsymbol{\mu}_n)^\times \otimes \mathbf{Z}_p) \otimes \rho^{-1} & \xrightarrow{\sim} & H^1(\mathbf{Q}(\boldsymbol{\mu}_n), \mathcal{A}). \end{array} \tag{4.3}$$

If $p \nmid n$, then Exercise 3.3.3 gives a commutative diagram

$$\begin{array}{ccc} (\mathbf{Q}(\boldsymbol{\mu}_{np})^\times \otimes \mathbf{Z}_p) \otimes \rho^{-1} & \xrightarrow{\sim} & H^1(\mathbf{Q}(\boldsymbol{\mu}_{np}), \mathcal{A}) \\ \Big\downarrow{\scriptstyle \sum_{\delta\in\Delta}\rho^{-1}(\delta)\delta\otimes 1} & & \Big\downarrow{\scriptstyle \mathrm{Cor}_{np/n}} \\ ((\mathbf{Q}(\boldsymbol{\mu}_{np})^\times \otimes \mathbf{Z}_p) \otimes \rho^{-1})^\Delta & \xrightarrow{\sim} & H^1(\mathbf{Q}(\boldsymbol{\mu}_n), \mathcal{A}). \end{array} \tag{4.4}$$

For every $n \geq 1$ fix a primitive n-th root of unity ζ_n such that for every m, $\zeta_{mn}^m = \zeta_n$. If $n \in \mathcal{N}$ and $p \mid n$, we identify $\Delta := \mathrm{Gal}(\mathbf{Q}(\boldsymbol{\mu}_p)/\mathbf{Q})$ with a subgroup of $\mathrm{Gal}(\mathbf{Q}(\boldsymbol{\mu}_n)/\mathbf{Q})$ as follows. Write $n = p^k n'$ with $k \geq 1$ and $p \nmid n'$, and identify Δ with the (unique) subgroup of order $p-1$ in $\mathrm{Gal}(\mathbf{Q}(\boldsymbol{\mu}_n)/\mathbf{Q}(\boldsymbol{\mu}_{n'})) \cong \mathrm{Gal}(\mathbf{Q}(\boldsymbol{\mu}_{p^k})/\mathbf{Q}) \cong (\mathbf{Z}/p^k\mathbf{Z})^\times$.

If $n \in \mathcal{N}$, define

$$\alpha_n := \begin{cases} \prod_{\delta\in\Delta} (\zeta_n^\delta - 1)^{\rho^{-1}(\delta)/(p-1)} \in (\mathbf{Q}(\boldsymbol{\mu}_n)^\times \otimes \mathbf{Z}_p)^\rho & \text{if } p \mid n, \\ \prod_{\delta\in\Delta} (\zeta_{np}^\delta - 1)^{\rho^{-1}(\delta)} \in (\mathbf{Q}(\boldsymbol{\mu}_{np})^\times \otimes \mathbf{Z}_p)^\rho & \text{if } p \nmid n, \end{cases}$$

$$\beta_n := \prod_{d \mid n, p \nmid d} \alpha_{n/d}^{\prod_{\ell\mid d}(-\mathrm{Fr}_\ell^{-1})}.$$

Note that $\alpha_n^\delta = \alpha_n^{\rho(\delta)}$ and $\beta_n^\delta = \beta_n^{\rho(\delta)}$ for every $\delta \in \Delta$.

Lemma 4.2.1. *Suppose $n\ell \in \mathcal{N}$. Then*

$$\mathbf{N}_{\mathbf{Q}(\boldsymbol{\mu}_{n\ell})/\mathbf{Q}(\boldsymbol{\mu}_n)} \beta_{n\ell} = \begin{cases} \beta_n^{1-\ell\mathrm{Fr}_\ell^{-1}} & \text{if } \ell \neq p, \\ \beta_n & \text{if } \ell = p \text{ and } p \mid n. \end{cases}$$

If $p \nmid n$, then $\prod_{\delta\in\Delta} \beta_{np}^{\rho^{-1}(\delta)\delta} = \beta_n$.

Sketch of proof. First suppose $\ell \neq p$, and note that $\eta_0 := \zeta_n^{\mathrm{Fr}_\ell^{-1}}/\zeta_{n\ell} \in \boldsymbol{\mu}_\ell$ (see this by raising both sides to the ℓ-th power). Then

$$\mathbf{N}_{\mathbf{Q}(\boldsymbol{\mu}_{n\ell})/\mathbf{Q}(\boldsymbol{\mu}_n)}(\zeta_{n\ell} - 1) = \prod_{\eta\in\boldsymbol{\mu}_\ell, \eta\neq\eta_0} (\eta\zeta_{n\ell} - 1) = \frac{\zeta_n - 1}{\zeta_n^{\mathrm{Fr}_\ell^{-1}} - 1} = (\zeta_n - 1)^{1-\mathrm{Fr}_\ell^{-1}}$$

$$\Longrightarrow \quad \mathbf{N}_{\mathbf{Q}(\boldsymbol{\mu}_{n\ell})/\mathbf{Q}(\boldsymbol{\mu}_n)} \alpha_{n\ell} = \alpha_n^{1-\mathrm{Fr}_\ell^{-1}}$$

$$\Longrightarrow \quad \mathbf{N}_{\mathbf{Q}(\boldsymbol{\mu}_{n\ell})/\mathbf{Q}(\boldsymbol{\mu}_n)} \beta_{n\ell} = \beta_n^{1-\ell\mathrm{Fr}_\ell^{-1}}.$$

If $\ell = p$ and $p \mid n$, a similar argument shows that $\mathbf{N}_{\mathbf{Q}(\boldsymbol{\mu}_{np})/\mathbf{Q}(\boldsymbol{\mu}_n)}(\zeta_{np} - 1) = \zeta_n - 1$. If $\ell = p$ and $p \nmid n$, the desired equality follows from the definition of α_n. □

Definition 4.2.2. Fix a generator u_ρ of the free, rank-one $\mathbf{Z}_p$-module "ρ^{-1}". For every $n \in \mathcal{N}'$ define $\xi_n^{\mathrm{cycl}} \in H^1(\mathbf{Q}(\boldsymbol{\mu}_n), \mathcal{A})$ to be the image of $\beta_n \otimes u_\rho$ under the isomorphism (4.2). Let

$$\boldsymbol{\xi}^{\mathrm{cycl}} := \{\xi_n^{\mathrm{cycl}} : n \in \mathcal{N}\}.$$

Proposition 4.2.3. *The collection $\boldsymbol{\xi}^{\mathrm{cycl}}$ is an Euler system for $\mathcal{A}$.*

PROOF. This follows from Lemma 4.2.1 and the commutative diagrams (4.3) and (4.4). □

Exercise 4.2.4. If $\boldsymbol{\xi}^{\mathrm{cycl}}$ is the cyclotomic unit Euler system defined above, show that

$$\xi_1^{\mathrm{cycl}} = \beta_1 = \alpha_1 = \prod_{\delta\in\Delta}(\zeta_p^\delta - 1)^{\rho^{-1}(\delta)} \in (\mathbf{Q}(\boldsymbol{\mu}_p)^\times \otimes \mathbf{Z}_p)^\rho.$$

What can you say about ξ_1^{cycl} when ρ is trivial? When ρ is odd? When ρ is even?

Exercise 4.2.5. What would change in this section if we wanted to consider Euler systems $\boldsymbol{\xi}^{\mathrm{cycl}}$ for $\mathbf{Z}_p(1)\otimes\rho^{-1}$ with an arbitrary character $\rho : G_{\mathbf{Q}} \to \mathbf{Z}_p^\times$?

What would we have to do to consider characters ρ of order not dividing $p-1$?

4.3. Euler systems and Kolyvagin systems.

Theorem 4.3.1. *Suppose*

1. *for every $\ell \notin \Sigma$, $P_\ell(x)$ has no roots in $\boldsymbol{\mu}_{p^\infty}$,*
2. *$H^0(\mathbf{Q}_p, \mathcal{A}^*)$ is a divisible $\mathbf{Z}_p$-module.*

Then there is a canonical map $\mathbf{ES}(\mathcal{A}) \to \mathbf{KS}(\mathcal{A}, \mathcal{F}_{\mathrm{can}})$ such that if $\boldsymbol{\xi}$ maps to $\boldsymbol{\kappa}$, then $\kappa_1 = \xi_1 \in H^1_{\mathcal{F}_{\mathrm{can}}}(\mathbf{Q}, \mathcal{A})$, where $\mathcal{F}_{\mathrm{can}}$ is the canonical Selmer structure of Example 3.1.5.

For the proof of Theorem 4.3.1, see [**MR1**, Theorem 3.2.4] and [**R2**, Chapter 4]. Here we will describe the construction of the map of Theorem 4.3.1, and give some details for the cyclotomic unit Euler system of §4.2 where the construction is particularly explicit.

Let $\mathcal{N}' = \{n \in \mathcal{N} : p \nmid n\}$. For every $n \in \mathcal{N}'$, let $\Gamma_n = \mathrm{Gal}(\mathbf{Q}(\boldsymbol{\mu}_n)/\mathbf{Q})$. If $n = n_1 n_2$, then there is a canonical isomorphism $\Gamma_n \cong \Gamma_{n_1} \times \Gamma_{n_2}$, so we can view Γ_{n_1} as either a subgroup or a quotient of Γ_n. With this identification we have

$$\Gamma_n = \prod_{\ell | n} \Gamma_\ell.$$

For every $\ell \notin \Sigma$, fix a generator σ_ℓ of Γ_ℓ. Define Kolyvagin's derivative operators

$$D_\ell := \sum_{i=1}^{\ell-2} i\sigma_\ell^i \in \mathbf{Z}[\Gamma_\ell]$$

and for $n \in \mathcal{N}'$,

$$D_n := \prod_{\ell | n} D_\ell \in \mathbf{Z}[\Gamma_n]$$

where we view $\mathbf{Z}[\Gamma_\ell] \subset \mathbf{Z}[\Gamma_n]$ as above. Also define the norm elements

$$N_n := \sum_{\gamma\in\Gamma_n} \gamma \in \mathbf{Z}[\Gamma_n]$$

for $n \in \mathcal{N}'$.

Lemma 4.3.2. *For every $\ell \notin \Sigma$, we have $(\sigma_\ell - 1)D_\ell = (\ell - 1) - N_\ell$.*

PROOF. Exercise. □

As in §3.2, if $\ell \notin \Sigma$, let

$$\nu(\ell) := \max\{m : \ell \equiv 1 \pmod{p^m} \\ \text{and } A_m/(\mathrm{Fr}_\ell - 1)A_m \text{ is free of rank 1 over } \mathbf{Z}/p^m\mathbf{Z}\}.$$

and if $n \in \mathcal{N}'$ define $\nu(n) := \min\{\nu(\ell) : \ell \mid n\}$. By convention we let $\nu(1) = \infty$, and $A_\infty := \mathcal{A}$.

Exercise 4.3.3. Show that $p^{\nu(\ell)}$ divides $P_\ell(1)$ for every $\ell \notin \Sigma$. Deduce that the reduction of $P_\ell(x)$ modulo $p^{\nu(\ell)}$ lies in $(x-1)(\mathbf{Z}/p^{\nu(\ell)}\mathbf{Z})[x]$.

Proposition 4.3.4. *If $\boldsymbol{\xi} \in \mathbf{ES}(\mathcal{A})$ and $n \in \mathcal{N}'$, then the image of $D_n\xi_n$ under the map $H^1(\mathbf{Q}(\boldsymbol{\mu}_n), \mathcal{A}) \to H^1(\mathbf{Q}(\boldsymbol{\mu}_n), A_{\nu(n)})$ lies in $H^1(\mathbf{Q}(\boldsymbol{\mu}_n), A_{\nu(n)})^{\Gamma_n}$.*

Proof. We need to show that $(\gamma - 1)D_n\xi_n$ maps to zero in $H^1(\mathbf{Q}(\boldsymbol{\mu}_n), A_{\nu(n)})$ for every $\gamma \in \Gamma_n$. Since Γ_n is generated by $\{\sigma_\ell : \ell \mid n\}$, it is enough to check this for $\gamma = \sigma_\ell$. We will show, by induction on the number of primes dividing n, that $(\sigma_\ell - 1)D_n\xi_n \in p^{\nu(n)}H^1(\mathbf{Q}(\boldsymbol{\mu}_n), \mathcal{A})$. This will prove the lemma.

Using Lemma 4.3.2 we have

$$(\sigma_\ell - 1)D_n\xi_n = ((\ell - 1) - N_\ell)D_{n/\ell}\xi_n = (\ell - 1)D_{n/\ell}\xi_n - D_{n/\ell}N_\ell\xi_n.$$

The composition

$$H^1(\mathbf{Q}(\boldsymbol{\mu}_n), \mathcal{A}) \xrightarrow{\mathrm{Cor}_{\mathbf{Q}(\boldsymbol{\mu}_n)/\mathbf{Q}(\boldsymbol{\mu}_{n/\ell})}} H^1(\mathbf{Q}(\boldsymbol{\mu}_{n/\ell}), \mathcal{A}) \xrightarrow{\mathrm{Res}} H^1(\mathbf{Q}(\boldsymbol{\mu}_n), \mathcal{A})$$

is multiplication by N_ℓ, so by definition of an Euler system we conclude

$$(\sigma_\ell - 1)D_n\xi_n = (\ell - 1)D_{n/\ell}\xi_n - \mathrm{Res}(P_\ell(\mathrm{Fr}_\ell^{-1})D_{n/\ell}\xi_{n/\ell}).$$

We have $p^{\nu(\ell)} \mid \ell - 1$, so $p^{\nu(n)} \mid \ell - 1$. By Exercise 4.3.3, we have

$$P_\ell(x) = (x - 1)g(x) + p^{\nu(\ell)}h(x)$$

with $h(x), g(x) \in \mathbf{Z}[x]$, so we conclude by induction that

$$P_\ell(\mathrm{Fr}_\ell^{-1})D_{n/\ell}\xi_{n/\ell} \in (p^{\nu(n/\ell)}, p^{\nu(\ell)})H^1(\mathbf{Q}(\boldsymbol{\mu}_{n/\ell}), \mathcal{A}) = p^{\nu(n)}H^1(\mathbf{Q}(\boldsymbol{\mu}_{n/\ell}), \mathcal{A}).$$

This completes the proof. □

Proposition 4.3.5. *Suppose $\boldsymbol{\xi} \in \mathbf{ES}(\mathcal{A})$ and $n \in \mathcal{N}'$. The image of $D_n\xi_n$ in $H^1(\mathbf{Q}(\boldsymbol{\mu}_n), A_{\nu(n)})$ has a canonical inverse image in $H^1(\mathbf{Q}, A_{\nu(n)})$ under the restriction map*

$$H^1(\mathbf{Q}, A_{\nu(n)}) \longrightarrow H^1(\mathbf{Q}(\boldsymbol{\mu}_n), A_{\nu(n)})^{\Gamma_n}.$$

Sketch of proof. The inflation-restriction exact sequence of Theorem 1.4.5 shows that the kernel and cokernel of the restriction map of the proposition are $H^1(\mathbf{Q}(\boldsymbol{\mu}_n)/\mathbf{Q}, A_{\nu(n)}^{G_{\mathbf{Q}(\boldsymbol{\mu}_n)}})$ and $H^2(\mathbf{Q}(\boldsymbol{\mu}_n)/\mathbf{Q}, A_{\nu(n)}^{G_{\mathbf{Q}(\boldsymbol{\mu}_n)}})$, respectively. If $A_{\nu(n)}^{G_{\mathbf{Q}(\boldsymbol{\mu}_n)}} = 0$, then it follows that the the restriction map is an isomorphism, and the proposition follows directly from Proposition 4.3.4.

For the proof in the general case, see [**R2**, §4.4] □

Exercise 4.3.6. Show that if $\mathcal{A} = \mathbf{Z}_p(1) \otimes \rho^{-1}$ as in §4.2, and ρ is not the mod p cyclotomic character, then $A_m^{G_{\mathbf{Q}(\boldsymbol{\mu}_n)}} = 0$ for every $n \in \mathcal{N}'$ and every m.

Example 4.3.7. Let $\boldsymbol{\xi}^{\rm cycl}$ be the cyclotomic unit Euler system of §4.2. We will describe the construction of Proposition 4.3.5 explicitly in this case.

Keep the notation of §4.2; in particular for every $n \in \mathcal{N}'$ we identify

$$\xi_n^{\rm cycl} = \beta_n \in (\mathbf{Q}(\boldsymbol{\mu}_{np})^\times \otimes \mathbf{Z}_p)^\rho.$$

Fix $n \in \mathcal{N}'$ and let $m = \nu(n)$. The proof of Proposition 4.3.4 shows that for every $\sigma \in \Gamma_n$,

$$\beta_n^{(\sigma-1)D_n} \in (\mathbf{Q}(\boldsymbol{\mu}_{np})^\times \otimes \mathbf{Z}_p)^{p^m}$$

but even more, that argument gives a *canonical* p^m-th root

$$c_n(\sigma) \in (\mathbf{Q}(\boldsymbol{\mu}_{np})^\times \otimes \mathbf{Z}_p)^\rho$$

of $\beta_n^{(\sigma-1)D_n}$. The map $\sigma \mapsto c_n(\sigma)$ is a 1-cocycle in $Z^1(\Gamma_n, (\mathbf{Q}(\boldsymbol{\mu}_{np})^\times \otimes \mathbf{Z}_p)^\rho)$. Since $H^1(\Gamma_n, (\mathbf{Q}(\boldsymbol{\mu}_{np})^\times \otimes \mathbf{Z}_p)^\rho) = 0$ (see Proposition 1.6.1) it follows that there is a $\gamma_n \in (\mathbf{Q}(\boldsymbol{\mu}_{np})^\times \otimes \mathbf{Z}_p)^\rho$ and $x_n \in (\mathbf{Q}(\boldsymbol{\mu}_p)^\times \otimes \mathbf{Z}_p)^\rho$ such that

$$\beta_n^{D_n} = \gamma_n^{p^m} x_n.$$

Then γ_n is well-defined modulo $(\mathbf{Q}(\boldsymbol{\mu}_p)^\times \otimes \mathbf{Z}_p)^\rho$, so x_n is well-defined element of $(\mathbf{Q}(\boldsymbol{\mu}_p)^\times/(\mathbf{Q}(\boldsymbol{\mu}_p)^\times)^{p^m})^\rho = H^1(\mathbf{Q}, A_m)$ that restricts to $D_n\xi_n^{\rm cycl} \in H^1(\mathbf{Q}(\boldsymbol{\mu}_n), A_m)$.

Recall (Definition 2.1.3) that if $n \in \mathcal{N}'$, then $G_n := \otimes_{\ell \mid n}\Gamma_\ell$.

Fix an Euler system $\boldsymbol{\xi} \in \mathbf{ES}(\mathcal{A})$. For every $n \in \mathcal{N}'$, let $x_n \in H^1(\mathbf{Q}, A_{\nu(n)})$ be the canonical inverse image of $D_n\xi_n \in H^1(\mathbf{Q}(\boldsymbol{\mu}_n), A_{\nu(n)})$ given by Proposition 4.3.5, and define

$$\xi'_n := x_n \otimes (\otimes_{\ell \mid n}\sigma_\ell) \in H^1(\mathbf{Q}(\boldsymbol{\mu}_n), A_{\nu(n)}) \otimes G_n.$$

Exercise 4.3.8. Show that, although $D_n\xi_n$ depends on the choice of the generators σ_ℓ of Γ_ℓ, ξ'_n does not.

Definition 4.3.9. Suppose $n \in \mathcal{N}'$ and $\ell \mid n$. By Proposition 1.9.5(3) we have a decomposition

$$H^1(\mathbf{Q}_\ell, A_{\nu(n)}) \cong H^1_{\rm u}(\mathbf{Q}_\ell, A_{\nu(n)}) \oplus H^1_{\rm t}(\mathbf{Q}_\ell, A_{\nu(n)})$$

and if $x \in H^1(\mathbf{Q}, A_{\nu(n)})$ we let $x_{\ell,\rm u}$ and $x_{\ell,\rm t}$ denote the projections of $\mathrm{Res}_\ell(x) \in H^1(\mathbf{Q}_\ell, A_{\nu(n)})$ into $H^1_{\rm u}(\mathbf{Q}_\ell, A_{\nu(n)}) \otimes G_n$ and $H^1_{\rm t}(\mathbf{Q}_\ell, A_{\nu(n)}) \otimes G_n$, respectively.

The natural next step is to try to show that the collection

$$\boldsymbol{\xi}' = \{\xi'_n : n \in \mathcal{N}'\}$$

is a Kolyvagin system for $\mathcal{A}$ as defined in §3.2. Unfortunately this is not true in general, but a slight modification of $\boldsymbol{\xi}'$ will be a Kolyvagin system. To check that $\boldsymbol{\xi}'$ is a Kolyvagin system we would need to verify that for every $n \in \mathcal{N}'$:

- $\mathrm{Res}_\ell(\xi'_n) \in H^1_{\mathcal{F}_{\rm can}}(\mathbf{Q}_\ell, A_{\nu(n)}) \otimes G_n$ if $\ell \nmid n$,
- $\mathrm{Res}_\ell(\xi'_n) \in H^1_{\rm t}(\mathbf{Q}_\ell, A_{\nu(n)}) \otimes G_n$ if $\ell \mid n$,
- $\psi_\ell^{\rm ut} \circ \mathrm{Res}_\ell(\xi'_{n/\ell}) = (\xi'_n)_{\ell,\rm t} \in H^1_{\rm t}(\mathbf{Q}_\ell, A_{\nu(n)}) \otimes G_n$ if $\ell \mid n$.

The following theorem deals with the first and third of these conditions, and Theorem 4.3.12 below investigates the second.

Theorem 4.3.10. *Suppose $\boldsymbol{\xi} \in \mathbf{ES}(\mathcal{A})$ and $H^0(\mathbf{Q}_p, \mathcal{A}^*)$ is a divisible $\mathbf{Z}_p$-module, and $n \in \mathcal{N}'$.*

(1) *If $\ell \nmid n$, then $\mathrm{Res}_\ell(\xi'_n) \in H^1_{\mathcal{F}_{\rm can}}(\mathbf{Q}_\ell, A_{\nu(n)}) \otimes G_n$.*

(2) *If* $\ell \mid n$, *then* $\phi_\ell^{\mathrm{ut}} \circ \mathrm{Res}_\ell(\xi'_{n/\ell}) = (\xi'_n)_{\ell,\mathrm{t}}$ *in* $H^1_{\mathrm{t}}(\mathbf{Q}_\ell, A_{\nu(n)}) \otimes G_n$.

Proof. For the proof of Proposition 4.3.10, see [**R2**, Theorem 4.5.1] for (1) when $\ell \neq p$, and [**R2**, Theorem 4.5.4] for (2). Note that since $H^0(\mathbf{Q}_p, \mathcal{A}^*)$ is divisible, Exercise 3.1.8 shows that $H^1_{\mathcal{F}_{\mathrm{can}}}(\mathbf{Q}_p, A_{\nu(n)}) = H^1(\mathbf{Q}_p, A_{\nu(n)})$, so (1) is trivially satisfied when $\ell = p$.

See also [**R1**, Proposition 2.4] for the proof in the case of the cyclotomic unit Euler system. □

Note that if $\ell \mid n$ and $x \in H^1(\mathbf{Q}, A_{\nu(n)})$, then

$$\mathrm{Res}_\ell(x) \in H^1_{\mathrm{t}}(\mathbf{Q}_\ell, A_{\nu(n)}) \iff x_{\ell,\mathrm{u}} = 0.$$

The $(\xi'_n)_{\ell,\mathrm{u}}$ are not zero in general, but they are described by Theorem 4.3.12 below.

Definition 4.3.11. If $\ell \notin \Sigma$, let I_ℓ be the augmentation ideal of $(\mathbf{Z}/p^{\nu(\ell)}\mathbf{Z})[G_\ell]$. Then there is a canonical isomorphism

$$\rho_\ell : I_\ell/I_\ell^2 \longrightarrow G_\ell \otimes (\mathbf{Z}/p^{\nu(\ell)}\mathbf{Z})$$

that sends $g-1$ to g for every $g \in G_\ell$.

If $n \in \mathcal{N}'$ let $\mathfrak{S}(n)$ denote the set of permutations of the primes dividing n, and if $\ell \mid n$ let $\mathfrak{S}^1_\ell(n) \subset \mathfrak{S}(n)$ be the subset

$$\{\pi \in \mathfrak{S}(n) : \text{the } \ell \text{ not fixed by } \pi \text{ form a single } \pi\text{-orbit, and } \pi(\ell) \neq \ell\}.$$

If $\pi \in \mathfrak{S}(n)$ let $d_\pi := \prod_{\pi(q)=q} q$.

Recall that $\omega(d)$ is the number of prime factors of a positive integer d.

Theorem 4.3.12. *Suppose* $\boldsymbol{\xi} \in \mathbf{ES}(\mathcal{A})$ *and for every* $\ell \notin \Sigma$, $P_\ell(x)$ *has no roots in* $\boldsymbol{\mu}_{p^\infty}$. *If* $n \in \mathcal{N}'$ *and* $\ell \mid n$ *then*

$$(\xi'_n)_{\ell,\mathrm{u}} = \sum_{\pi \in \mathfrak{S}^1_\ell(n)} (-1)^{\omega(n/d_\pi)} \mathrm{Res}_\ell(\xi'_{d_\pi}) \bigotimes_{q \mid (n/d_\pi)} \rho_q(P_q(\mathrm{Fr}^{-1}_{\pi(q)}|_{\mathbf{Q}(\boldsymbol{\mu}_q)})).$$

Proof. For the proof, see [**MR1**, Theorem A.4]. For a proof in the special case of the cyclotomic unit Euler system, see [**MR2**, Appendix A]. □

Corollary 4.3.13. *Suppose* $\boldsymbol{\xi} \in \mathbf{KS}(\mathcal{A})$ *and*

(1) *for every* $\ell \notin \Sigma$, $P_\ell(x)$ *has no roots in* $\boldsymbol{\mu}_{p^\infty}$,

(2) $H^0(\mathbf{Q}_p, \mathcal{A}^*)$ *is a divisible* $\mathbf{Z}_p$*-module.*

For $n \in \mathcal{N}'$ *define*

$$\kappa_n := \sum_{\pi \in \mathfrak{S}(n)} \mathrm{sign}(\pi) \xi'_{d_\pi} \bigotimes_{\ell \mid (n/d_\pi)} \rho_\ell(P_\ell(\mathrm{Fr}^{-1}_{\pi(\ell)}|_{\mathbf{Q}(\boldsymbol{\mu}_\ell)})) \in H^1(\mathbf{Q}, A_{\nu(n)}) \otimes G_n.$$

Then the collection $\boldsymbol{\kappa} := \{\kappa_n : n \in \mathcal{N}'\}$ *is a Kolyvagin system for* $(\mathcal{A}, \mathcal{F}_{\mathrm{can}})$, *and* $\kappa_1 = \xi_1$.

Proof. By Theorem 4.3.10, it is enough to prove that $(\kappa_n)_{\ell,\mathrm{u}} = 0$ whenever $n \in \mathcal{N}'$ and $\ell \mid n$. This follows from Theorem 4.3.12; for the details see the proof of Theorem 3.2.4 on pages 80–81 of [**MR1**]. □

The map $\mathbf{ES}(\mathcal{A}) \to \mathbf{KS}(\mathcal{A}, \mathcal{F}_{\mathrm{can}})$ of Theorem 4.3.1 is given by $\boldsymbol{\xi} \mapsto \boldsymbol{\kappa}$ as described above.

LECTURE 5

Applications

5.1. Cyclotomic units and ideal class groups.

Fix a rational prime $p > 2$, and let $F = \mathbf{Q}(\boldsymbol{\mu}_p)^+$, the real subfield of $\mathbf{Q}(\boldsymbol{\mu}_p)$. Let $\mathcal{C}_F$ denote the p-part of the ideal class group of F. Let $\mathcal{E}_F \subset \mathcal{O}_F^\times$ denote the group of cyclotomic units of F, the intersection of $\mathcal{O}_F^\times$ with the group generated by

$$\{\zeta, \zeta - 1 : \zeta \in \boldsymbol{\mu}_p, \zeta \neq 1\}.$$

The following theorem was first proved (by a different method) by Mazur and Wiles [**MW**].

Theorem 5.1.1. *Let $\rho : \mathrm{Gal}(F/\mathbf{Q}) \to \mathbf{Z}_p^\times$ be a nontrivial (necessarily even) character. Then*

$$|\mathcal{C}_F^\rho| = |((\mathcal{O}_F^\times/\mathcal{E}_F) \otimes \mathbf{Z}_p)^\rho|.$$

Proof. As in §4.2, let $\boldsymbol{\xi}^{\mathrm{cycl}} \in \mathbf{ES}(\mathbf{Z}_p(1) \otimes \rho^{-1})$ be the Euler system constructed there. Let $\boldsymbol{\kappa}^{\mathrm{cycl}} \in \mathbf{KS}(\mathbf{Z}_p(1) \otimes \rho^{-1})$ be the Kolyvagin system constructed from $\boldsymbol{\xi}^{\mathrm{cycl}}$ in Theorem 4.3.1. Then we have

$$\kappa_1^{\mathrm{cycl}} = \xi_1^{\mathrm{cycl}} = \prod_{\delta \in \mathrm{Gal}(\mathbf{Q}(F)/\mathbf{Q})} (\zeta_p^\delta - 1)^{\rho^{-1}(\delta)} \in (\mathcal{E}_F \otimes \mathbf{Z}_p)^\rho$$

$$\subset (F^\times \otimes \mathbf{Z}_p)^\rho = H^1(\mathbf{Q}, \mathbf{Z}_p(1) \otimes \rho^{-1}).$$

In fact, κ_1^{cycl} is a $\mathbf{Z}_p$-generator of $(\mathcal{E}_F \otimes \mathbf{Z}_p)^\rho$, so by Corollary 3.3.12,

$$|\mathcal{C}_F^\rho| \leq [(\mathcal{O}_F^\times \otimes \mathbf{Z}_p)^\rho : \mathbf{Z}_p \kappa_1] = |((\mathcal{O}_F^\times/\mathcal{E}_F) \otimes \mathbf{Z}_p)^\rho|. \tag{5.1}$$

Note that (5.1) also holds when $\rho = 1$, since in that case both sides are equal to 1. Taking the product over all ρ, we have

$$|\mathcal{C}_F| = \prod_\rho |\mathcal{C}_F^\rho| \leq \prod_\rho |((\mathcal{O}_F^\times/\mathcal{E}_F) \otimes \mathbf{Z}_p)^\rho| = |(\mathcal{O}_F^\times/\mathcal{E}_F) \otimes \mathbf{Z}_p|.$$

The analytic class number formula (see for example [**L**, Theorem 5.1 of Chapter 3]) says that $|\mathcal{C}_F| = |(\mathcal{O}_F^\times/\mathcal{E}_F) \otimes \mathbf{Z}_p|$, so we must have equality in (5.1) for every ρ. □

Remark 5.1.2. Note that using Corollary 3.3.12, the proof of Theorem 5.1.1 also shows that the cyclotomic unit Kolyvagin system $\boldsymbol{\kappa}^{\mathrm{cycl}}$ is primitive.

5.2. Elliptic curves.

Return now to the example of §3.4. Namely, E is an elliptic curve defined over $\mathbf{Q}$, $p \geq 5$ is a rational prime, and the p-adic representation $G_\mathbf{Q} \to \mathrm{Aut}(E[p^\infty]) \cong \mathrm{GL}_2(\mathbf{Z}_p)$ is surjective.

In this setting, an Euler system for $T_p(E)$ was constructed by Kato in [**Ka**]. We will not touch on his construction here, but we will discuss briefly the essential properties of Kato's Euler system, and some of its consequences. For more details, see [**Ka**], [**R2**, §3.5], and [**Sc**].

For simplicity, we also assume from now on that

$$E \text{ has good reduction at } p \text{ and } p \nmid |E(\mathbf{F}_p)|. \tag{5.2}$$

Recall that for every $m > 0$, $H^1_{\mathrm{f}}(\mathbf{Q}_p, E[p^m])$ is the image of $E(\mathbf{Q}_p)/p^m E(\mathbf{Q}_p)$ in $H^1(\mathbf{Q}_p, E[p^m])$ under the (injective) Kummer map as in Definition 1.6.13. We define

$$H^1_{\mathrm{f}}(\mathbf{Q}_p, T_p(E)) := \varprojlim H^1_{\mathrm{f}}(\mathbf{Q}_p, E[p^m]) \subset H^1(\mathbf{Q}_p, T_p(E))$$

so $H^1_{\mathrm{f}}(\mathbf{Q}_p, T_p(E)) \cong E(\mathbf{Q}_p) \otimes \mathbf{Z}_p$. It follows from (5.2) that $E(\mathbf{Q}_p)$ has no p-torsion, so $H^1_{\mathrm{f}}(\mathbf{Q}_p, T_p(E))$ is a free, rank-one $\mathbf{Z}_p$-module.

Fix a minimal Weierstrass model of E. Let ω_E and Ω_E be the corresponding holomorphic differential on E, and real period. Corresponding to ω_E there is a *dual exponential map* $\exp_{\omega_E}$ and an exact sequence

$$0 \longrightarrow H^1_{\mathrm{f}}(\mathbf{Q}_p, T_p(E)) \longrightarrow H^1(\mathbf{Q}_p, T_p(E)) \xrightarrow{\exp_{\omega_E}} p^{-1}\mathbf{Z}_p \longrightarrow 0. \tag{5.3}$$

Let $L(E, s)$ denote the Hasse-Weil L-function of E, and

$$L_p(E, s) = (1 - a_p p^{-s} + p^{1-2s}) L(E, s),$$

the L-function with the Euler factor at p removed (here $a_p := p + 1 - |E(\mathbf{F}_p)|$).

We make no attempt here to prove the following theorem of Kato.

Theorem 5.2.1 (Kato [**Ka**]). *There is a nonzero Euler system* $\boldsymbol{\xi}^{\mathrm{Kato}} \in \mathbf{ES}(T_p(E))$ *such that*

$$\exp_{\omega_E}(\mathrm{Res}_p(\xi_1^{\mathrm{Kato}})) = \frac{L_p(E,1)}{\Omega_E}.$$

Theorem 5.2.2 (Kato [**Ka**]). *With assumptions as above, we have:*

(1) *If* $L(E,1) \neq 0$ *then* $E(\mathbf{Q})$ *is finite and* $|Ш(E/\mathbf{Q})[p^\infty]|$ *divides* $L(E,1)/\Omega_E$.
(2) *If* $L(E,1) = 0$ *and* $\xi_1^{\mathrm{Kato}} \neq 0$, *then either* $E(\mathbf{Q})$ *or* $Ш(E/\mathbf{Q})[p^\infty]$ *is infinite.*

Proof. Let $\mathcal{F}$ denote the canonical Selmer structure on $T_p(E)$, and let $\boldsymbol{\kappa}^{\mathrm{Kato}} \in \mathbf{KS}(T_p(E), \mathcal{F})$ be the Kolyvagin system attached to $\boldsymbol{\xi}^{\mathrm{Kato}}$ by Theorem 4.3.1. (Note that the hypotheses of Theorem 4.3.1 hold, thanks to (5.2).)

Suppose first that $L(E,1) \neq 0$. Define integers $-1 \leq b \leq c$ by

$$\begin{array}{ccc} p^b\mathbf{Z}_p & = & \exp_{\omega_E}(\mathrm{Res}_p(H^1_{\mathcal{F}}(\mathbf{Q}, T_p(E)))) \\ \cup & & \cup \\ p^c\mathbf{Z}_p & = & \exp_{\omega_E}(\mathbf{Z}_p \mathrm{Res}_p(\xi_1^{\mathrm{Kato}})) \end{array}$$

so $c = \mathrm{ord}_p(L_p(E,1)/\Omega_E)$. Then (see Definition 3.2.5) $\partial^0(\boldsymbol{\kappa}^{\mathrm{Kato}}) \leq c - b$, so by Corollary 3.2.9(1)

$$|H^1_{\mathcal{F}^*}(\mathbf{Q}, E[p^\infty])| \text{ divides } p^{c-b}. \tag{5.4}$$

We need to compare $H^1_{\mathcal{F}^*}(\mathbf{Q}, E[p^\infty])$ with the classical Selmer group

$$\mathrm{Sel}_{p^\infty}(E/\mathbf{Q}) := H^1_{\mathrm{f}}(\mathbf{Q}, E[p^\infty])$$

defined using the Selmer structure of local cohomology groups $H^1_f(\mathbf{Q}_\ell, E[p^\infty])$ (see Definition 1.6.13). By Proposition 3.4.5, $H^1_{\mathcal{F}}(\mathbf{Q}_\ell, E[p^\infty]) = H^1_f(\mathbf{Q}_\ell, E[p^\infty])$ for every $\ell \neq p$. Passing to the limit in Corollary 3.4.6 gives

$$0 \longrightarrow H^1_f(\mathbf{Q}, T_p(E)) \longrightarrow H^1_{\mathcal{F}}(\mathbf{Q}, T_p(E)) \longrightarrow H^1(\mathbf{Q}_p, T_p(E))/H^1_f(\mathbf{Q}_p, T_p(E))$$

$$0 \longrightarrow H^1_{\mathcal{F}^*}(\mathbf{Q}, E[p^\infty]) \longrightarrow \mathrm{Sel}_{p^\infty}(E/\mathbf{Q}) \longrightarrow H^1_f(\mathbf{Q}_p, E[p^\infty]).$$

By global duality (Theorem 1.10.2), together with the fact that $H^1_f(\mathbf{Q}_\ell, E[p^m])$ is its own orthogonal complement under the Tate pairing for every ℓ (Example 1.8.4), the images of the two right-hand maps are orthogonal complements under the limit of the local Tate pairings. By definition of b and (5.3), the cokernel of the upper right-hand map has order p^{b+1}. Hence we conclude from the bottom sequence that

$$[\mathrm{Sel}_{p^\infty}(E/\mathbf{Q}) : H^1_{\mathcal{F}^*}(\mathbf{Q}, E[p^\infty])] = p^{b+1},$$

and combining this with (5.4) we see that

$$|\mathrm{Sel}_{p^\infty}(E/\mathbf{Q})| \text{ divides } p^{c-b} \cdot p^{b+1} = p^{c+1}.$$

Further

$$\begin{aligned} c+1 = \mathrm{ord}_p(pL_p(E,1)/\Omega_E) &= \mathrm{ord}_p((p - a_p + 1)L(E,1)/\Omega_E) \\ &= \mathrm{ord}_p(|E(\mathbf{F}_p)|L(E,1)/\Omega_E) = \mathrm{ord}_p(L(E,1)/\Omega_E) \end{aligned}$$

since we assumed that $p \nmid |E(\mathbf{F}_p)|$. Now (1) follows from the exact sequence

$$0 \longrightarrow E(\mathbf{Q}) \otimes \mathbf{Q}_p/\mathbf{Z}_p \longrightarrow \mathrm{Sel}_{p^\infty}(E/\mathbf{Q}) \longrightarrow Ш(E/\mathbf{Q})[p^\infty] \longrightarrow 0,$$

the direct limit of the exact sequences (1.3).

Suppose now that $L(E,1) = 0$ and $\xi_1^{\mathrm{Kato}} \neq 0$. By (5.3) we have $\mathrm{Res}_p(\xi_1^{\mathrm{Kato}}) \in H^1_f(\mathbf{Q}_p, T_p(E))$, so

$$\xi_1^{\mathrm{Kato}} \in H^1_f(\mathbf{Q}, T_p(E))$$

and in particular $H^1_f(\mathbf{Q}, T_p(E)) \neq 0$. But the inverse limit of the exact sequences (1.3) give

$$0 \longrightarrow E(\mathbf{Q}) \otimes \mathbf{Z}_p \longrightarrow H^1_f(\mathbf{Q}, T_p(E)) \longrightarrow \varprojlim Ш(E/\mathbf{Q})[p^m] \longrightarrow 0.$$

The inverse limit on the right is with respect to the maps "multiplication by p", so this inverse limit is either zero or infinite. Similarly, since we assumed that the p-adic representation $G_{\mathbf{Q}} \to \mathrm{Aut}(E[p^\infty])$ is surjective, $E(\mathbf{Q})$ has no p-torsion so $E(\mathbf{Q}) \otimes \mathbf{Z}_p$ is either zero or infinite. Hence if $\xi_1^{\mathrm{Kato}} \neq 0$ then one of those two groups is infinite, and this completes the proof of (2). □

Exercise 5.2.3. Using Corollary 3.2.9(1) and the proof of Theorem 5.2.2, show that if $L(E,1) \neq 0$ then

$$\mathrm{length}(Ш(E/\mathbf{Q})[p^\infty]) = \mathrm{ord}_p(L(E,1)/\Omega_E) \iff \boldsymbol{\kappa}^{\mathrm{Kato}} \text{ is primitive.}$$

Note that according to the Birch and Swinnerton-Dyer conjecture, it is *not* always the case that $\mathrm{length}(Ш(E/\mathbf{Q})[p^\infty]) = \mathrm{ord}_p(L(E,1)/\Omega_E)$, and therefore it is *not* always the case that $\boldsymbol{\kappa}^{\mathrm{Kato}}$ is primitive.

5.3. General speculation.

In this section we sketch a general framework for the example of the previous section.

Suppose $\mathcal{A}$ is a p-adic representation of $G_{\mathbf{Q}}$ as in Lecture 3: a free $\mathbf{Z}_p$-module of finite rank with a continuous action of $G_{\mathbf{Q}}$, ramified at only finitely many primes. Let $\mathcal{F}$ denote the canonical Selmer structure on $\mathcal{A}$, and suppose $\mathcal{G}$ is another Selmer structure on $\mathcal{A}$, with the property that $H^1_{\mathcal{G}}(\mathbf{Q}_\ell, \mathcal{A}) = H^1_{\mathcal{F}}(\mathbf{Q}_\ell, \mathcal{A})$ for all ℓ different from p. Then $H^1_{\mathcal{G}}(\mathbf{Q}, \mathcal{A}) \subset H^1_{\mathcal{F}}(\mathbf{Q}, \mathcal{A})$, and we have a diagram

$$\begin{array}{ccccccc} 0 & \longrightarrow & H^1_{\mathcal{G}}(\mathbf{Q}, \mathcal{A}) & \longrightarrow & H^1_{\mathcal{F}}(\mathbf{Q}, \mathcal{A}) & \xrightarrow{\mathrm{Res}_p} & H^1(\mathbf{Q}_p, \mathcal{A})/H^1_{\mathcal{G}}(\mathbf{Q}_p, \mathcal{A}) \\ 0 & \longrightarrow & H^1_{\mathcal{F}^*}(\mathbf{Q}, \mathcal{A}^*) & \longrightarrow & H^1_{\mathcal{G}^*}(\mathbf{Q}, \mathcal{A}^*) & \xrightarrow{\mathrm{Res}_p} & H^1_{\mathcal{G}^*}(\mathbf{Q}_p, \mathcal{A}^*). \end{array}$$

Passing to the limit in Theorem 1.10.2, we see that

$$[H^1_{\mathcal{G}^*}(\mathbf{Q}, \mathcal{A}^*) : H^1_{\mathcal{F}^*}(\mathbf{Q}, \mathcal{A}^*)] = [H^1(\mathbf{Q}_p, \mathcal{A}) : H^1_{\mathcal{G}}(\mathbf{Q}_p, \mathcal{A}) + \mathrm{Res}_p(H^1_{\mathcal{F}}(\mathbf{Q}, \mathcal{A}))].$$

Suppose in addition there is an isomorphism

$$\mathfrak{L} : H^1(\mathbf{Q}_p, \mathcal{A})/H^1_{\mathcal{G}}(\mathbf{Q}_p, \mathcal{A}) \xrightarrow{\sim} \mathbf{Z}_p.$$

Then we conclude that

$$[H^1_{\mathcal{G}^*}(\mathbf{Q}, \mathcal{A}^*) : H^1_{\mathcal{F}^*}(\mathbf{Q}, \mathcal{A}^*)] = [\mathbf{Z}_p : \mathfrak{L}(\mathrm{Res}_p(H^1_{\mathcal{F}}(\mathbf{Q}, \mathcal{A})))]. \tag{5.5}$$

Theorem 5.3.1. *Suppose that $\mathcal{A}$, $\mathcal{G}$, and $\mathfrak{L}$ are as above, that $\mathcal{A}$ satisfies hypotheses (H.1-4) of §3.1 and the hypotheses of Theorem 4.3.1, and that the core rank $\chi(\mathcal{A}) = 1$. If $\boldsymbol{\xi} \in \mathbf{ES}(\mathcal{A})$ is an Euler system for $\mathcal{A}$, then*

$$\mathrm{length}_{\mathbf{Z}_p}(H^1_{\mathcal{G}^*}(\mathbf{Q}, \mathcal{A}^*)) \le \mathrm{ord}_p(\mathfrak{L}(\mathrm{Res}_p(\xi_1))).$$

Proof. Let $\boldsymbol{\kappa} \in \mathbf{KS}(\mathcal{A}, \mathcal{F})$ be the Kolyvagin system corresponding to $\boldsymbol{\xi}$ under Theorem 4.3.1. By Corollary 3.2.9, using that $\kappa_1 = \xi_1$,

$$|H^1_{\mathcal{F}^*}(\mathbf{Q}, \mathcal{A}^*)| \le p^{\partial^0(\boldsymbol{\kappa})} \le [\mathfrak{L}(\mathrm{Res}_p(H^1_{\mathcal{F}}(\mathbf{Q}, \mathcal{A}))) : \mathfrak{L}(\mathrm{Res}_p(\xi_1))\mathbf{Z}_p].$$

Combining this with (5.5), we see that

$$|H^1_{\mathcal{G}^*}(\mathbf{Q}, \mathcal{A}^*)| \le [\mathbf{Z}_p : \mathfrak{L}(\mathrm{Res}_p(\xi_1))\mathbf{Z}_p],$$

which proves the theorem. □

Example 5.3.2. Suppose E is an elliptic curve as in §5.2. Let $\mathcal{A} = T_p(E)$, let $H^1_{\mathcal{G}}(\mathbf{Q}_\ell, T_p(E)) = H^1_{\mathrm{f}}(\mathbf{Q}_\ell, T_p(E)$ for every ℓ, let $\mathfrak{L} = p\exp_{\omega_E}$, and let $\boldsymbol{\xi} = \boldsymbol{\xi}^{\mathrm{Kato}}$. Then $H^1_{\mathcal{G}}(\mathbf{Q}, \mathcal{A}^*) = \mathrm{Sel}_{p^\infty}(E/\mathbf{Q})$, $\mathfrak{L}(\mathrm{Res}_p(\xi_1)) = |E(\mathbf{F}_p)|L(E,1)/\Omega_E$, and Theorem 5.3.1 is part of Theorem 5.2.2(1).

Example 5.3.3. Suppose ρ is a nontrivial character of $\mathrm{Gal}(\mathbf{Q}(\boldsymbol{\mu}_p)^+/\mathbf{Q})$, $\mathcal{A} = \mathbf{Z}_p(1) \otimes \rho^{-1}$, and $\boldsymbol{\xi}^{\mathrm{cycl}} \in \mathbf{ES}(\mathbf{Z}_p(1) \otimes \rho^{-1})$ is the cyclotomic unit Euler system of §4.2. Let $\mathcal{F}$ be the canonical Selmer structure on $\mathbf{Z}_p(1) \otimes \rho^{-1}$, and let $\mathcal{G}$ be the Selmer structure defined by

$$H^1_{\mathcal{G}}(\mathbf{Q}_\ell, \mathbf{Z}_p(1) \otimes \rho^{-1}) := \begin{cases} H^1_{\mathcal{F}}(\mathbf{Q}_\ell, \mathbf{Z}_p(1) \otimes \rho^{-1}) & \text{if } \ell \ne p, \\ 0 & \text{if } \ell = p. \end{cases}$$

Let M is the maximal abelian p-extension of $\mathbf{Q}(\boldsymbol{\mu}_p)^+$ unramified outside of p (with no local restriction at p), and $\mathcal{X} := \mathrm{Gal}(M/\mathbf{Q}(\boldsymbol{\mu}_p)^+)$. Then $H^1_{\mathcal{G}^*}(\mathbf{Q}_p, \mathbf{Q}_p/\mathbf{Z}_p \otimes \rho) = H^1(\mathbf{Q}_p, \mathbf{Q}_p/\mathbf{Z}_p \otimes \rho)$, so as in Proposition 3.3.10 we have

$$H^1_{\mathcal{G}^*}(\mathbf{Q}, \mathbf{Q}_p/\mathbf{Z}_p \otimes \rho) = \mathrm{Hom}(\mathcal{X}^\rho, \mathbf{Q}_p/\mathbf{Z}_p).$$

Let $F := \mathbf{Q}(\boldsymbol{\mu}_p)^+$. By Exercise 3.3.9, $H^1_{\mathcal{F}}(\mathbf{Q}_p, \mathbf{Z}_p(1) \otimes \rho^{-1}) \cong (\mathcal{O}_{F,p}^\times \otimes \mathbf{Z}_p)^\rho$. Let $\log_p : \mathcal{O}_{F,p}^\times \to F_p$ denote the p-adic logarithm map, let $\tau(\rho^{-1}) : \sum_{a=1}^{p-1} \rho^{-1}(a)\zeta_p^a \in F_p$ be the Gauss sum, and define

$$\mathfrak{L} := \tfrac{1}{\tau(\rho^{-1})} \log_p : H^1_{\mathcal{F}}(\mathbf{Q}_p, \mathbf{Z}_p(1) \otimes \rho^{-1}) \xrightarrow{\sim} (\mathcal{O}_{F,p}^\times \otimes \mathbf{Z}_p)^\rho \longrightarrow F_p.$$

Exercise 5.3.4. Show that $\mathfrak{L}$ is an isomorphism from $H^1_{\mathcal{F}}(\mathbf{Q}_p, \mathbf{Z}_p(1) \otimes \rho^{-1})$ onto $\mathbf{Z}_p$.

By Exercise 4.2.4 we have

$$\mathfrak{L}(\mathrm{Res}_p(\xi_1^{\mathrm{cycl}})) = -L_p(1, \rho)$$

where $L_p(s, \rho)$ is the Kubota-Leopoldt p-adic L-function attached to ρ (see for example [**L**, Theorem 4.3.6]). Therefore we can apply Theorem 5.3.1 to conclude that

$$\mathrm{length}(\mathcal{X}^\rho) \le \mathrm{ord}_p(L_p(1, \rho)). \tag{5.6}$$

Exercise 5.3.5. Use the fact that $\boldsymbol{\kappa}^{\mathrm{cycl}}$ is primitive (Remark 5.1.2) to show that equality holds in (5.6).

Remark 5.3.6. Returning to the case of a more general p-adic representation $\mathcal{A}$, it is clear what we would like to do in general: find an Euler system $\boldsymbol{\xi} \in \mathbf{ES}(\mathcal{A})$, and a map

$$\mathfrak{L} : H^1(\mathbf{Q}_p, \mathcal{A}) \twoheadrightarrow \mathbf{Z}_p$$

such that $\mathfrak{L}(\mathrm{Res}_p(\xi_1))$ is (related to) a special value of a (p-adic) L-function attached to $\mathcal{A}$. Then (under suitable assumptions) Theorem 5.3.1 will bound the length of the Selmer group $H^1_{\mathcal{G}^*}(\mathbf{Q}, \mathcal{A}^*)$ in terms of that L-value, where the Selmer structure $\mathcal{G}$ is defined by $H^1_{\mathcal{G}}(\mathbf{Q}_p, \mathcal{A}) := \ker(\mathfrak{L})$.

Although very little general progress has been made in this direction, one can formulate a more precise conjecture. See for example [**PR**], or [**R2**, Chapter 8].

Bibliography

[AT] M. Artin, J. Tate, Class field theory. New York: Benjamin (1967).

[AW] M. Atiyah, C.T.C. Wall, Cohomology of groups. In: Algebraic Number Theory, J. W. S. Cassels and A. Fröhlich, eds., London: Academic Press (1967) 94–115.

[Ca] J.W.S. Cassels, Arithmetic on curves of genus 1. VIII. On conjectures of Birch and Swinnerton-Dyer. *J. Reine Angew. Math.* **217** (1965) 180–199.

[Gr] K. Gruenberg, Profinite groups. In: Algebraic Number Theory, J. W. S. Cassels and A. Fröhlich, eds., London: Academic Press (1967) 116–127.

[Ka] K. Kato, p-adic Hodge theory and values of zeta functions of modular forms. *Astérisque* **295** (2004) 117–290.

[Ko] V. A. Kolyvagin, Euler systems. In: The Grothendieck Festschrift (Vol. II), P. Cartier et al., eds., *Prog. in Math* **87**, Boston: Birkhäuser (1990) 435–483.

[L] S. Lang, Cyclotomic fields I and II. *Graduate Texts in Math.* **121**, New York: Springer-Verlag (1990).

[MR1] B. Mazur, K. Rubin, Kolyvagin systems. *Memoirs of the Amer. Math. Soc.* **799** (2004).

[MR2] B. Mazur, K. Rubin, Introduction to Kolyvagin systems. In: Starks Conjectures: Recent Work and New Directions, *Contemp. Math.* **358**, Providence: Amer. Math. Soc. (2004) 207-221.

[MW] B. Mazur, A. Wiles, Class fields of abelian extensions of **Q**, *Invent. math.* **76** (1984) 179–330.

[Mi] J.S. Milne, Arithmetic duality theorems, *Perspectives in Math.* **1**, Orlando: Academic Press (1986).

[PR] B. Perrin-Riou, Fonctions L p-adiques des représentations p-adiques. *Astérisque* **229** (1995).

[R1] K. Rubin, The main conjecture. Appendix to: Cyclotomic fields I and II, S. Lang, *Graduate Texts in Math.* **121**, New York: Springer-Verlag (1990) 397–419.

[R2] K. Rubin, Euler systems. *Annals of Math. Studies* **147**, Princeton: Princeton University Press (2000).

[Sc] A. Scholl, An introduction to Kato's Euler systems. In: Galois representations in arithmetic algebraic geometry, A. J. Scholl and R. L. Taylor, eds., *London Math. Soc. Lect. Notes* **254** Cambridge: Cambridge Univ. Press (1998) 379–460.

[Se1] J-P. Serre, Cohomologie Galoisienne, Fifth edition. *Lecture Notes in Math.* **5**, Berlin: Springer-Verlag (1994).

[Se2] J-P. Serre, Corps Locaux, 2nd edition. Paris: Hermann (1968).

[Si] J.H. Silverman, The Arithmetic of Elliptic Curves, Graduate Texts in Mathematics **106**, New York: Springer-Verlag (1986).

[Ta1] J. Tate, WC-groups over $\mathfrak{p}$-adic fields. *Séminaire Bourbaki* **4**, Exp. 156, Paris: Soc. Math. France (1995) 265–277.

[Ta2] J. Tate, Duality theorems in Galois cohomology over number fields. In: *Proc. Intern. Cong. Math.*, Stockholm (1962) 234–241.

[Ta3] J. Tate, Galois cohomology. In: Arithmetic algebraic geometry (Park City, UT, 1999), *IAS/Park City Math. Ser.* **9**, Providence: Amer. Math. Soc. (2001) 465–479.

[Wi] A. Wiles, Modular elliptic curves and Fermat's Last Theorem. *Annals of Math.* **141** (1995) 443–551.

Titles in This Series

18 **Cristian Popescu, Karl Rubin, and Alice Silverberg, Editors,** Arithmetic of L-functions, 2011

17 **Jeffery McNeal and Mircea Mustaţă, Editors,** Analytic and Algebraic Geometry, 2010

16 **Scott Sheffield and Thomas Spencer, Editors,** Statistical Mechanics, 2009

15 **Tomasz S. Mrowka and Peter S. Ozsváth, Editors,** Low Dimensional Topology, 2009

14 **Mark A. Lewis, Mark A. J. Chaplain, James P. Keener, and Philip K. Maini, Editors,** Mathematical Biology, 2009

13 **Ezra Miller, Victor Reiner, and Bernd Sturmfels, Editors,** Geometric Combinatorics, 2007

12 **Peter Sarnak and Freydoon Shahidi, Editors,** Automorphic Forms and Applications, 2007

11 **Daniel S. Freed, David R. Morrison, and Isadore Singer, Editors,** Quantum Field Theory, Supersymmetry, and Enumerative Geometry, 2006

10 **Steven Rudich and Avi Wigderson, Editors,** Computation Complexity Theory, 2004

9 **Brian Conrad and Karl Rubin, Editors,** Arithmetic Algebraic Geometry, 2001

8 **Jeffrey Adams and David Vogan, Editors,** Representation Theory of Lie Groups, 2000

7 **Yakov Eliashberg and Lisa Traynor, Editors,** Symplectic Geometry and Topology, 1999

6 **Elton P. Hsu and S. R. S. Varadhan, Editors,** Probability Theory and Applications, 1999

5 **Luis Caffarelli and Weinan E, Editors,** Hyperbolic Equations and Frequency Interactions, 1999

4 **Robert Friedman and John W. Morgan, Editors,** Gauge Theory and the Topology of Four-Manifolds, 1998

3 **János Kollár, Editor,** Complex Algebraic Geometry, 1997

2 **Robert Hardt and Michael Wolf, Editors,** Nonlinear Partial Differential Equations in Differential Geometry, 1996

1 **Daniel S. Freed and Karen K. Uhlenbeck, Editors,** Geometry and Quantum Field Theory, 1995